Lecture Notes in Physics

New Series m: Monographs

The Editorial Policy for Monographs

The series Lecture Notes in Physics reports new developments in physical research and teaching - quickly, informally, and at a high level. The type of material considered for publication in the New Series m includes monographs presenting original research or new angles in a classical field. The timeliness of a manuscript is more important than its form, which may be preliminary or tentative. Manuscripts should be reasonably self-contained. They will often present not only results of the author(s) but also related work by other people and will provide sufficient motivation, examples, and applications.
The manuscripts or a detailed description thereof should be submitted either to one of the series editors or to the managing editor. The proposal is then carefully refereed. A final decision concerning publication can often only be made on the basis of the complete manuscript, but otherwise the editors will try to make a preliminary decision as definite as they can on the basis of the available information.
Manuscripts should be no less than 100 and preferably no more than 400 pages in length. Final manuscripts should preferably be in English, or possibly in French or German. They should include a table of contents and an informative introduction accessible also to readers not particularly familiar with the topic treated. Authors are free to use the material in other publications. However, if extensive use is made elsewhere, the publisher should be informed. Authors receive jointly 50 complimentary copies of their book. They are entitled to purchase further copies of their book at a reduced rate. As a rule no reprints of individual contributions can be supplied. No royalty is paid on Lecture Notes in Physics volumes. Commitment to publish is made by letter of interest rather than by signing a formal contract. Springer-Verlag secures the copyright for each volume.

The Production Process

The books are hardbound, and quality paper appropriate to the needs of the author(s) is used. Publication time is about ten weeks. More than twenty years of experience guarantee authors the best possible service. To reach the goal of rapid publication at a low price the technique of photographic reproduction from a camera-ready manuscript was chosen. This process shifts the main responsibility for the technical quality considerably from the publisher to the author. We therefore urge all authors to observe very carefully our guidelines for the preparation of camera-ready manuscripts, which we will supply on request. This applies especially to the quality of figures and halftones submitted for publication. Figures should be submitted as originals or glossy prints, as very often Xerox copies are not suitable for reproduction. In addition, it might be useful to look at some of the volumes already published or, especially if some atypical text is planned, to write to the Physics Editorial Department of Springer-Verlag direct. This avoids mistakes and time-consuming correspondence during the production period.
As a special service, we offer free of charge LaTeX and TeX macro packages to format the text according to Springer-Verlag's quality requirements. We strongly recommend authors to make use of this offer, as the result will be a book of considerably improved technical quality. The typescript will be reduced in size (75% of the original). Therefore, for example, any writing within figures should not be smaller than 2.5 mm.
Manuscripts not meeting the technical standard of the series will have to be returned for improvement.
For further information please contact Springer-Verlag, Physics Editorial Department II, Tiergartenstrasse 17, D-69121 Heidelberg, FRG.

Alexandre Ern Vincent Giovangigli

Multicomponent Transport Algorithms

Springer-Verlag
Berlin Heidelberg GmbH

Authors

Alexandre Ern
Centre d'Enseignement et de Recherche en
Modélisation Informatique et Calcul Scientifique
ENPC, La Courtine
F-93167 Noisy-Le-Grand Cedex, France
and
Mechanical Engineering Department
Yale University, P. O. Box 208286
New Haven, CT 06520-8286, USA

Vincent Giovangigli
Centre de Mathématiques Appliquées, CNRS
Ecole Polytechnique
F-91128 Palaiseau Cedex, France

Please address correspondence to the authors at CMAP, Ecole Polytechnique.

ISBN 978-3-662-14510-4 ISBN 978-3-540-48650-3 (eBook)
DOI 10.1007/978-3-540-48650-3

CIP data applied for.

Originally published by Springer-Verlag Berlin Heidelberg New York in 1994
Softcover reprint of the hardcover 1st edition 1994

Typesetting: Camera-ready by the authors
SPIN: 10080280 55/3140-543210 - Printed on acid-free paper

Preface

With the advent of sophisticated computer technology and the development of efficient computational algorithms, numerical modeling of complex multicomponent laminar reacting flows has emerged as an increasingly popular and firmly established area of scientific research. Progress in this area aims at obtaining better resolved and more accurate solutions of specific technological problems in less computer time. Therefore, it strongly relies upon the ability of evaluating fundamental parameters appearing in the physical models.

Transport properties constitute a typical example of the above characterization. Evaluating transport coefficients of dilute polyatomic gas mixtures is often critical in many engineering applications, including chemical reactors, hypersonic flows, combustion phenomena, and chemical vapor deposition. Using the kinetic theory of dilute polyatomic gas mixtures as a starting point, this book offers a systematic development of a mathematical and numerical theory for the evaluation of transport properties in dilute polyatomic gas mixtures. The present investigation is not specifically about the kinetic theory of gases, for which there are plenty of excellent and thoroughly documented textbooks; it is rather geared toward the development of new, efficient, and general algorithms with which to evaluate transport properties of dilute polyatomic gas mixtures at a reasonable computational cost.

In this book we compute the coefficients of the transport linear systems, i.e., the linear systems to be solved in order to obtain the transport coefficients, in their natural constrained singular symmetric form. New transport linear systems, corresponding to lower computational costs for practical applications, are also introduced. From a theoretical viewpoint, we extract the structural properties of the transport linear systems directly from the Boltzmann equation and use them systematically in order to derive a mathematical framework for iterative algorithms. As a result, we express all the transport coefficients as convergent series, and truncating these series then yields analytical approximate expressions for all the transport coefficients.

We hope that the present algorithms will be of extensive interest in theoretical calculations and numerical modeling for fluid mechanics, combustion, crystal growth, and other engineering applications. The material covered in this book is intended for people who are not familiar with multicomponent transport property evaluation as well as for experienced readers interested in the relevant areas of modern research. Readers need only be familiar with introductory linear algebra concepts and some basic ideas of the kinetic theory of dilute gases. The more elaborate sections of the book should still be readable at the same level since we restate the more sophisticated results as needed.

The logical ordering of the chapters closely follows the development of the present theory, from kinetic theory concepts, through the derivation of the transport linear system structure to convergence theorems for various iterative algorithms, followed by several numerical examples. Readers only interested in practical applications may consider a different reading path, which only includes the introductory parts of all chapters, Sections 2.1 and 2.2, the notation and the summaries of results presented in Section 2.3, the derivation of the transport linear systems described in Sections 2.4–2.8, the convergence theorems stated in Chapter 5, and the numerical results of Chapter 6. All the transport linear systems are treated explicitly, which contributes significantly to the length of the manuscript, but provides a complete source of reference for any transport coefficient that may be needed in a given application. A more detailed reading of the book includes the derivation of the transport linear system properties directly from the kinetic theory as detailed in Section 2.3, the mathematical framework of Chapter 4, the proofs of the theorems in Chapters 4 and 5, and the singular limit of zero concentrations treated in Chapter 3 and also considered in the subsequent chapters.

The authors owe their sincere thanks to many of their colleagues for several stimulating discussions. They are also grateful to Professor I. Kuščer, University of Ljubljana, for constructive remarks about chemically reacting flows, and to Professor M. D. Smooke and Doctor M. A. Tanoff, Yale University, and Professor V. Ern, University of Strasbourg, for numerous comments. The assistance of Mr. M. Multan from the Prêt Inter-Bibliothèques, Ecole Polytechnique, is also gratefully acknowledged.

Finally, Professor W. Beiglböck and his assistants, Ms. S. Landgraf and Ms. B. Reichel-Mayer, deserve special thanks for an excellent job in editing the book.

Paris, May 1994

Alexandre Ern
Vincent Giovangigli

Contents

1 Introduction

1.1 Detailed Modeling of Dilute Polyatomic Gas Mixtures

The equations governing multicomponent gaseous laminar flows are derived from the kinetic theory of dilute gases and contain the terms for transport fluxes, that is, the pressure tensor, the species diffusion velocities, and the heat flux vector. More specifically, the momentum, species mass, and energy conservation equations for these flows can be written in the form

$$\partial_t(\rho v) + \nabla\cdot(\rho v \otimes v) = -\nabla\cdot\Pi + \sum_{i\in S} \rho Y_i b_i, \tag{1.1.1}$$

$$\partial_t(\rho Y_i) + \nabla\cdot(\rho v Y_i) = -\nabla\cdot(\rho Y_i V_i), \qquad i \in S, \tag{1.1.2}$$

$$\partial_t(\tfrac{1}{2}\rho v\cdot v + \rho u) + \nabla\cdot\big(\rho v(\tfrac{1}{2}v\cdot v + u)\big) = -\nabla\cdot(q + \Pi v) + \sum_{i\in S} \rho Y_i (v + V_i)\cdot b_i, \tag{1.1.3}$$

where ∂_t is the time derivative operator, ∇ the space derivative operator, "$\cdot$" the scalar product in three dimensions, ρ the density, v the mass averaged flow velocity, $v\otimes v$ the velocity tensor of rank two, Π the pressure tensor and $\nabla\cdot\Pi$ its divergence, Y_i the mass fraction of the ith species, b_i the external force per unit mass on the ith species, V_i the diffusion velocity of the ith species, $S = [1, n]$ the set of species indices, n the number of species, u the internal energy per unit mass of the mixture, and q the heat flux vector. In order to solve these governing equations, the transport fluxes Π, V_i, for $i \in S$, and q must be determined first. The expressions rigorously obtained from the kinetic theory of dilute polyatomic gas mixtures can be written, to the first approximation in the Enskog-Chapman expansion, as

$$\Pi = \bar{p}I - (\kappa - \tfrac{2}{3}\eta)(\nabla\cdot v)I - \eta(\nabla v + (\nabla v)^t), \tag{1.1.4}$$

$$V_i = -\sum_{j\in S} D_{ij} d_j - \theta_i \nabla \log \overline{T}, \qquad i \in S, \tag{1.1.5}$$

$$q = \sum_{i\in S} \rho h_i Y_i V_i - \lambda' \nabla \overline{T} - \bar{p} \sum_{i\in S} \theta_i d_i, \tag{1.1.6}$$

where $\bar{p}$ is the thermodynamic pressure, I the identity matrix, κ the volume viscosity—also termed bulk viscosity—, η the shear viscosity, $D = (D_{ij})_{i,j\in\mathcal{S}}$ the diffusion matrix, d_i the diffusion driving force of the ith species, $\theta = (\theta_i)_{i\in\mathcal{S}}$ the thermal diffusion vector, $\overline{T}$ the absolute temperature, h_i the enthalpy per unit mass of the ith species, and λ' the partial thermal conductivity. The vectors d_i incorporate the effects of various state variable gradients and external forces and are given by

$$d_i = \nabla X_i + (X_i - Y_i)\frac{\nabla\bar{p}}{\bar{p}} + \frac{\rho}{\bar{p}}\sum_{j\in\mathcal{S}} Y_i Y_j (b_j - b_i), \qquad i \in \mathcal{S}, \tag{1.1.7}$$

where X_i denotes the mole fraction of the ith species. Alternate expressions for V_i, $i \in \mathcal{S}$, and q are

$$V_i = -\sum_{j\in\mathcal{S}} D_{ij}(d_j + \chi_j \nabla \log \overline{T}), \qquad i \in \mathcal{S}, \tag{1.1.8}$$

$$q = \sum_{i\in\mathcal{S}} \rho h_i Y_i V_i - \lambda\nabla\overline{T} + \bar{p}\sum_{i\in\mathcal{S}} \chi_i V_i, \tag{1.1.9}$$

where $\chi = (\chi_i)_{i\in\mathcal{S}}$ are the thermal diffusion ratios and λ is the thermal conductivity. For reactive mixtures in which characteristic chemistry times are larger than the corresponding mean free times of the molecules, the transport fluxes due to macroscopic variable gradients are also given by (1.1.4)–(1.1.9).

It follows from these expressions that detailed modeling of a polyatomic gas mixture requires the evaluation of its transport properties, that is, the volume viscosity κ, the shear viscosity η, the diffusion matrix D, and either the thermal diffusion vector θ and the partial thermal conductivity λ', or, alternatively, the thermal diffusion ratios χ and the thermal conductivity λ, if the expressions (1.1.8)(1.1.9) are used instead of (1.1.5)(1.1.6). These coefficients, in turn, are functions of the state variables $\bar{p}$, $\overline{T}$, and $Y_1, \ldots, Y_n$. However, the transport coefficients are not explicitly given by the kinetic theory. Their evaluation, indeed, requires solving linear systems derived from orthogonal polynomial expansions of the species perturbed distribution functions [WT62]. Numerical evaluations of multicomponent transport properties using Gaussian elimination have been considered, in particular, by Dixon-Lewis [Di68], Coffee and Heimerl [CHe81], Kee, Dixon-Lewis, Warnatz, Coltrin, and Miller [KDWCM86], Lebedev [Le89], and Warnatz [Wr82]. Nevertheless, solving these linear systems by direct methods may become computationally expensive since their size can be relatively large and since transport properties have to be evaluated at each computational cell in space and time. As a consequence, the use of iterative techniques can be an interesting and appealing alternative. Moreover, iterative methods provide a rigorous way to define approximate transport coefficients by truncating the resulting convergent series.

Iterative schemes have been implicitly considered, in particular, by Hirschfelder, Curtiss, and Bird [HCB54] and Brokaw [Br58] [Br64] in deriving approximate formulas for the shear viscosity and the thermal conductivity of monatomic gas mixtures. These two transport coefficients are such that the associated linear systems are naturally nonsingular, so that iterative algorithms present no technical difficulties. For diffusion velocities, which involve the solution of a constrained singular system, iterative schemes have been introduced later by Jones and Boris [JB81], who have shown numerically the convergence of their algorithms. In order to select the proper diffusion velocities, a corrector term needs to be added after convergence [JB81] [OB81]. This corrector term was written explicitly by Oran and Boris [OB81] and corresponds to applying a projector matrix as shown in [Gi91]. The convergence of the Jones-Oran-Boris algorithm has been proven rigorously by Giovangigli [Gi91] who also established that the corresponding iteration matrix has a spectral radius unity. Additional algorithms have been introduced for the multicomponent diffusion matrices in [Gi91] for which the iteration matrix has a spectral radius strictly lower than unity. These algorithms are obtained from the theory of iterative methods for singular systems and require the application of a projection matrix at each step. In particular, it was shown in [Gi91] that the first-order multicomponent diffusion matrix can be written as a symmetric convergent series for which each partial sum is symmetric, conserves mass, and yields a positive entropy production on the hyperplane of zero sum driving forces.

The principal objective of this book is now a systematic development of a mathematical and numerical theory of iterative algorithms for solving the transport linear systems, i.e., the linear systems associated with the evaluation of all the transport properties. The transport linear systems are first evaluated in the framework of the isotropic semi-classical theory of dilute polyatomic gas mixtures [Wa58] [WT62]. The structure of the linear systems is deduced from the Boltzmann equation and, in particular, from the properties of the kinetic integral bracket operator. We also consider the singular case of vanishing mass fractions and introduce appropriately rescaled transport linear systems. Several mathematical properties of the systems and of the transport coefficients are then established. Standard iterative and conjugate gradient methods are shown to be convergent by using the theory of iterative algorithms for singular systems and, in particular, for symmetric positive semi-definite matrices. As a result, we express the transport coefficients of dilute polyatomic gas mixtures, i.e., the shear viscosity, the volume viscosity, the diffusion matrix, the partial thermal conductivity, the thermal conductivity, the thermal diffusion vector, and the thermal diffusion ratios as convergent series. Truncating these series then provides a rigorous way to define accurate analytic approximations for all the transport coefficients. Matrices obtained with inexact collision integrals resulting from practical approximations [MM62] [MPM65] are shown to possess the mathematical structure deduced from the kinetic theory, also. Finally, numerical tests are performed for various gas mixtures arising in combustion applications, and the optimization of transport property evaluation in multicomponent flow calculations is also discussed.

1.2 Outline of Subsequent Chapters

In Chapter 2 the transport linear systems are derived, and their mathematical structure is deduced from the kinetic theory.

Section 2.1 summarizes some of the aspects of the isotropic semi-classical theory for nonreactive dilute polyatomic gas mixtures as given by Waldmann and Trübenbacher [WT62]. Since reacting flows are important for practical applications, we also discuss, in this section, the extension of the Waldmann and Trübenbacher theory to the case of reactive mixtures in which the characteristic chemistry times are larger than the corresponding mean free times of the molecules. In this situation, the transport coefficients can be evaluated as if there were no chemical reactions. Section 2.2 then describes the transport linear systems using the formalism of Waldmann and Trübenbacher [WT62] who obtained these systems in their natural constrained singular symmetric form. On the contrary, Monchick, Yun, and Mason [MYM63] have eliminated the singularities by explicitly using the linear constraints and zeroing the diagonal coefficients of the system matrices, following a procedure introduced by Curtiss and Hirschfelder [CH49], thereby destroying the natural symmetries of the model. In the same section we proceed to calculate the bracket integrals associated with the variational procedure since [WT62] presented only a formal theory. In Section 2.3 the general mathematical structure of the transport linear systems is derived directly from the Boltzmann equation. All the properties that will be needed in the following chapters to establish the well posedness of the systems and the convergence of various iterative algorithms are obtained. To the authors' knowledge, this is the first time that such an analysis is made. These results are applied specifically to the shear viscosity in Section 2.4, the volume viscosity in Section 2.5, the diffusion matrix in Section 2.6, the partial thermal conductivity and the thermal diffusion vector in Section 2.7, and the thermal conductivity and the thermal diffusion ratios in Section 2.8. In these sections we also present various simplified formulations obtained by using smaller approximation functional spaces for the perturbed species distribution functions. For completeness, in Section 2.9 we compare our transport linear systems to the ones of [MYM63] after elimination of the singularities, as done in [MYM63]. The agreement between both calculations is complete except for two misprints identified in the paper of Monchick, Yun, and Mason. Finally, in Section 2.10, we review practical approximations [MM62] [MPM65] for various collision integrals appearing in the linear system coefficients.

In Chapter 3 we investigate the case of vanishing mass fractions and we introduce rescaled versions of the transport linear systems obtained for positive mass fractions in Chapter 2. These rescaled systems will be used in Chapter 4 in order to obtain the smoothness of the transport coefficients when some mass fractions become arbitrarily small, except for diffusion matrices for which some components diverge [Gi91]. In the case of zero mass fractions, the fluxes $F_k = Y_k V_k$, $k \in \mathcal{S}$, are indeed the proper quantities to be evaluated [Gi91] instead of the diffusion velocities V_k, $k \in \mathcal{S}$. For this purpose, the flux diffusion coefficients $\widetilde{D}_{kl} = Y_k D_{kl}$ are introduced in this chapter and are related to rescaled systems. The rescaled systems will also be used in Chapter 5 in order to investigate the numerical stability of various iterative algorithms for vanishing mass fractions.

The rescaled systems are introduced in Section 3.1 and their properties are again directly deduced from the kinetic theory. In particular, the corresponding matrices allow elimination of the artificial singularities arising with zero mass fractions, but are no longer symmetric. The main tools for investigating these rescaled systems are a block-partitioning between positive and zero mass fraction species and the use of the results obtained in Chapter 2 for positive concentrations. The results are applied specifically to the shear viscosity in Section 3.2, the volume viscosity in Section 3.3, the flux diffusion matrix in Section 3.4, the partial thermal conductivity and the thermal diffusion vector in Section 3.5, and the thermal conductivity and the thermal diffusion ratios in Section 3.6. Finally, rescaled versions of the approximate systems, given in Section 2.10, are presented in Section 3.7.

In Chapter 4 we establish various mathematical results for the transport linear systems and for the transport coefficients. The structure properties of these systems are extracted from the kinetic theory investigations of Chapters 2 and 3. In Chapters 4 and 5, however, these properties are recast into a set of assumptions written in a mathematical framework. The purpose of this mathematical framework is first to investigate rigorously the well posedness of the transport linear systems and the singular limit of vanishing mass fractions. In addition, when collision integrals are not exact but estimated, it is then sufficient to verify that the approximate system matrix satisfies the corresponding mathematical assumptions in order for the theoretical results derived in Chapters 4 and 5 to apply. This will be verified systematically in the case of the practical approximations presented in Section 2.10 [MM62] [MPM65]. The mathematical framework will be fundamental in Chapter 5 also, in order to prove the convergence of various iterative algorithms.

In Section 4.1 we present some mathematical preliminaries and, for completeness, we restate several results on generalized inverses and constrained singular systems. These results are applied specifically to the shear viscosity in Section 4.2, the volume viscosity in Section 4.3, the diffusion matrix and the flux diffusion matrix in Section 4.4, the partial thermal conductivity and the thermal diffusion vector in Section 4.5, and the thermal conductivity and the thermal diffusion ratios in Section 4.6. In each case, we first consider the transport linear systems for positive mass fractions—including the simplified formulations—and establish their well posedness as well as various properties of the corresponding transport coefficients. We then consider the case of nonnegative mass fractions and establish the well posedness of the rescaled transport linear systems. We also investigate the smoothness of the transport coefficients as functions of the state variables, including the singular limit of vanishing mass fractions. In particular, provided diffusion matrices are replaced by flux diffusion matrices [Gi91], all the transport coefficients are shown to be smooth rational functions of the mass fractions.

In Chapter 5 we investigate, within a mathematical framework, the convergence of iterative schemes for the solution of the transport linear systems. As a result, all the transport coefficients are expressed as convergent series, and truncating these series then yields new, explicit, and rigorously derived expressions for the transport coefficients. As in Chapter 4, all the convergence theorems are valid for any general system matrix that satisfies a given set of structure properties extracted from the kinetic theory investigations of Chapters 2 and 3. In particular, we systematically verify that all the structure properties still hold when collision integrals are estimated using the practical approximations presented in Section 2.10 [MM62] [MPM65].

In Section 5.1 we restate, for completeness, several mathematical results on convergent iterative methods for constrained singular systems and, in particular, for symmetric positive semi-definite matrices. These results are applied specifically to the shear viscosity in Section 5.2, the volume viscosity in Section 5.3, the diffusion matrix and the flux diffusion matrix in Section 5.4, the partial thermal conductivity and the thermal diffusion vector in Section 5.5, and the thermal conductivity and the thermal diffusion ratios in Section 5.6. For each transport linear system we consider standard iterative and generalized conjugate gradient methods and also discuss iterative algorithms obtained by using Schur complements. Furthermore, we investigate the numerical behavior of these algorithms in the case of vanishing mass fractions. To the authors' knowledge, all the convergence theorems applied to the transport linear systems are new with the exception of standard iterative methods for the first-order diffusion matrix [Gi91].

In Chapter 6 we present numerical experiments illustrating the convergence of the iterative algorithms derived in Chapter 5. We also discuss the accuracy of the transport coefficients associated with simplified formulations of the transport linear systems as well as the accuracy of several empirical mixture-averaged formulas for the transport coefficients. Some of these empirical formulas correspond to approximations frequently used in numerical simulations, but new empirical mixture-averaged formulas are also introduced. The numerical experiments are performed for typical gas mixtures associated with hydrogen and methane combustion applications, and we consider mixtures in both positive and nonnegative mass fractions states.

In Section 6.1 we discuss the computational cost of transport property evaluation in multicomonent flow calculations. We compare iterative methods versus direct inversions for the transport linear systems associated with a given gas mixture. We also consider the use of empirical mixture-averaged formulas as opposed to analytic expressions that are rigorously derived from the kinetic theory. We then examine efficient implementation of transport property evaluation in reacting flow numerical simulations. The numerical experiments are presented for the shear viscosity in Section 6.2, the volume viscosity in Section 6.3, the diffusion matrix and the flux diffusion matrix in Section 6.4, the partial thermal conductivity and the thermal diffusion vector in Section 6.5, and the thermal conductivity and the thermal diffusion ratios in Section 6.6. For all the transport linear systems, including the simplified formulations, we present the convergence results observed for standard iterative and conjugate gradient methods. High convergence rates are achieved in all cases.

Concluding remarks are given in Chapter 7. The main advantages of iterative methods applied to transport linear systems are summarized and several possible extensions of the present theory are mentioned.

2 Transport Linear Systems

In this chapter the transport linear systems, i.e., the linear systems that are to be solved to evaluate the transport coefficients, are obtained and their mathematical structure is deduced from the kinetic theory.

In Sections 2.1 and 2.2 we summarize some of the results of Waldmann [Wa58] and Waldmann and Trübenbacher [WT62] for dilute polyatomic gas mixtures. An Enskog-Chapman procedure is performed which leads to linearized Boltzmann equations. Using a variational procedure with polynomial expansions, these equations then give rise to the transport linear systems. The general form of the transport linear systems is given and a calculation of the bracket integrals associated with the variational procedure is performed. This calculation was needed since Waldmann and Trübenbacher only derived a formal theory. The resulting expressions of the bracket integrals have been compared formally with the results of Köhler and 't Hooft [MBKK91] obtained for linear molecules in a fully quantum mechanical framework, and the agreement has been found to be complete. We also present the extension of the Waldmann-Trübenbacher theory to reacting flows in which the mean free times of the molecules are an order of magnitude smaller than the corresponding characteristic chemistry times.

In Section 2.3 we next derive the general mathematical structure of the transport linear systems directly from the properties of the Boltzmann equation and of the variational framework. All the properties that will be needed in the following chapters to establish the well posedness of the systems and the convergence of various iterative algorithms are obtained. It is the first time, to the authors' knowledge, that such an analysis is made.

The results of Sections 2.1–2.3 are then applied specifically to the shear viscosity in Section 2.4, the volume viscosity in Section 2.5, the diffusion matrix in Section 2.6, the partial thermal conductivity and the thermal diffusion vector in Section 2.7, and the thermal conductivity and the thermal diffusion ratios in Section 2.8. In these sections, simplified formulations associated with the use of smaller variational approximation spaces for the perturbed distribution functions are also presented. These formulations extend the ideas proposed by Thijsse et al. [TTCKB79] and Van den Oord and Korving [VK88] for the thermal conductivity of a polyatomic gas and by McCourt et al. [MBKK90] for the partial thermal conductivity of binary mixtures. To the authors' knowledge, the systematic extension of these ideas to all the transport coefficients of dilute polyatomic gas mixtures is new.

In Section 2.9 the expressions of the linear systems are compared with those of Monchick, Yun, and Mason [MYM63] by systematically eliminating the singularities, making use of the linear constraints, and zeroing the diagonal coefficients of the system matrices. The agreement between both approaches is complete except for two misprints identified in the paper of Monchick, Yun, and Mason. Note that according to [MYM63], Offerhaus also evaluated these linear system coefficients, but, to the authors' knowledge, these calculations were never published.

Finally, in order to compute explicitly the transport coefficients, various collision integrals appearing in the linear system coefficients have to be evaluated, and practical approximations [MM62] [MPM65] are presented in Section 2.10. The validity of these approximations is sometimes questioned [WV92], but we point out that the general theory presented in this book applies to many other systems, using for instance different approximations.

2.1 Transport Coefficients of Dilute Polyatomic Gas Mixtures

2.1.1 Kinetic Theories

The kinetic theory of dilute gases with molecules having internal degrees of freedom was first developed by Wang Chang and Uhlenbeck and De Boer [WU51] [WUD64] [WU70] in a semi-classical framework, in which the translational motion is treated classically and the internal motion quantum mechanically. In this treatment, one assumes a symmetry condition on the quantum cross-sections, which is only valid if the molecular states are nondegenerate and which is analogous to the classical assumption of the existence of inverse collisions. Waldmann [Wa58] and then Mason and Monchick [MM62] have further shown that although the treatment required the assumption of nondegeneracy and detailed balance, its results were still valid for all molecules, provided the quantum mechanical cross-sections were replaced by degeneracy averaged quantum mechanical cross-sections. The required symmetry property of the cross-sections is then obtained from the invariance of the Hamiltonian under the combined operation of space inversion and time reversal rather than by requiring the assumption of detailed balance [Wa58] [MS64].

A fully quantum mechanical treatment was given by McCourt and Snider [MS64] [MS65] and Snider and Sanctuary [SS71] using the quantum mechanical Boltzmann equation derived by Waldmann [Wa57] and Snider [Sn60]. This fully quantum mechanical transport theory was able to describe the Kagan-Affassanaev [KA61] polarizations associated with the Senftleben-Beenaker effects [MBKK90] [MBKK91], i.e., the effect of magnetic fields on transport properties, but the corresponding macroscopic equations are more complicated than the Navier-Stokes equations since they involve now a macroscopic angular momentum conservation equation [MS64] [FK72]. Since these polarization effects are only important in the presence of applied magnetic or electric fields which will not be considered in this book, we will use throughout the so-called isotropic or Pidduck approximation [MVW88] valid in the absence of polarization effects. In such situation, the quantum mechanical theory yields the same formal results as the semi-classical approach [MS64] [MVW88] which will be used in this book. It should also be noted that a fully classical theory was later developed by Kagan and Maksimov [KM66] [MBKK90] and Curtiss [Cu81]. The corresponding Boltzmann equation is the classical limit of the quantum mechanical Waldmann-Snider equation [MBKK90]. This formulation may lead, however, to infinite integral cross-sections [MBKK90] and will not be considered in this book.

The extension of the semi-classical theory to dilute polyatomic gas mixtures was given by Waldmann and Trübenbacher [WT62] and by Monchick, Yun, and Mason [MYM63]. Although the transport coefficients obtained from both treatments are identical, there is an important difference, however, in the final structure of the linear systems that need to be solved in order to obtain the transport properties. Indeed, Monchick, Yun, and Mason have systematically eliminated the singularities arising in the naturally singular and symmetric linear systems, obtained from the variational approximation procedure, by explicitly using the linear constraints and zeroing the diagonal coefficients of the system matrices, following a procedure introduced by Curtiss and Hirschfelder [CH49] [HCB54]. Although such formulation of the linear systems may be used for direct numerical inversions, we will obtain in this book symmetric positive definite forms of the linear systems which can be inverted at half the computational cost. Moreover, the original constrained singular symmetric forms obtained in [WT62] are preferable for iterative techniques and the symmetric systems also have simpler analytic expressions so that they are better suited for analytic approximations of the transport coefficients. Furthermore, Waldmann and Trübenbacher [WT62] have used symmetric diffusion coefficients, which are formally compatible with Onsager reciprocal relations. Note that symmetric diffusion coefficients have also been considered by Waldmann [Wa58], Chapman and Cowling [CC70], Ferziger and Kapper [FK72], and Curtiss [Cu68], at variance with Monchick, Yun, and Mason [MYM63], Curtiss and Hirschfelder [CH49], and Hirschfelder, Curtiss, and Bird [HCB54], who have artificially destroyed this symmetry [Va67]. In this book, we will therefore use the elegant formalism of [WT62] which fully respects the natural symmetries appearing in the model.

2.1.2 The Waldmann and Trübenbacher Isotropic Semi-Classical Theory

In the theory of Waldmann and Trübenbacher [WT62], the translational motion is treated classically, the internal motion quantum mechanically, there are no polarization effects, i.e., no macroscopic angular momentum and no external magnetic and electric fields, so that the distribution functions are isotropic, and there are no chemical reactions [Wa58] [WT62]. The corresponding Boltzmann equation, which was first introduced by Waldmann, preaverages the cross-sections over all the magnetic quantum numbers and can be derived from the Waldmann-Snider quantum mechanical Boltzmann equation [Wa58] [KHCBV93]. This form of the Boltzmann equation is also equivalent to the Wang Chang and Uhlenbeck type equation considered in [MYM63].

The Boltzmann Equation. For a mixture of n species indexed by $\mathcal{S} = [1,n]$, where $n \geq 1$ unless explicitly stated, we denote by f_i the distribution function of the ith species, $i \in \mathcal{S}$, and the dependence on (r,t) of the distribution functions $f_i = f_i(r,t,c_i,\mathrm{I})$, $i \in \mathcal{S}$, is made implicit, in order to avoid notational complexities. The Boltzmann equation in [WT62] then reads

$$\mathfrak{D}_i(f_i) = \sum_{j\in\mathcal{S}} \mathfrak{B}_{ij}(f_i,f_j), \qquad i \in \mathcal{S}, \tag{2.1.1}$$

where $\mathfrak{D}_i$ is the usual differential operator

$$\mathfrak{D}_i(f_i) = \partial_t(f_i) + (c_i\cdot\nabla)f_i + (b_i\cdot\nabla_{c_i})f_i, \tag{2.1.2}$$

and $\mathfrak{B}_{ij}$ the Boltzmann binary collision operator for the species pair (i,j)

$$\mathfrak{B}_{ij}(f_i,f_j) =$$

$$\sum_{\substack{I' \in \mathcal{E}_i \\ J,J' \in \mathcal{E}_j}} \iiint \Big(f_i(c_i', I') f_j(\tilde{c}_j', J') \frac{a_{iI} a_{jJ}}{a_{iI'} a_{jJ'}} - f_i(c_i, I) f_j(\tilde{c}_j, J)\Big) \bar{\sigma}_{ij}^{IJI'J'} g \sin\chi d\chi d\varphi d\tilde{c}_j. \quad (2.1.3)$$

In these expressions, ∂_t denotes the time derivative operator, ∇ the spatial derivative operator, ∇_{c_i} the derivative operator with respect to the velocity c_i, "$\cdot$" the three-dimensional scalar product, c_i and $\tilde{c}_j$ the velocities of the incoming particles—of species i and j respectively—before collision, and c_i' and $\tilde{c}_j'$ the particle velocities after collision. The extra superscript $\tilde{}$ is used to distinguish one of the collision partners from the other in the case where i and j are the same. In (2.1.3), $\mathcal{E}_i$ also denotes the set of quantum energy shells of the ith species, I and J the indices for the quantum energy shells of the ith and j^{th} species before collision, I' and J' the corresponding numbers after collision, a_{iI} the degeneracy of the Ith quantum energy shell of the ith species, g the modulus of the relative velocity $c_i - \tilde{c}_j$ of the colliding particles, χ and φ the polar and azimuthal angles which describe the orientation of $c_i' - \tilde{c}_j'$ with respect to $c_i - \tilde{c}_j$, and $\bar{\sigma}_{ij}^{IJI'J'}$ the collision cross-section which is a function $\bar{\sigma}_{ij}^{IJI'J'}(g, \chi, \varphi)$ of g, χ, and φ. The modulus g' of the relative velocity after collision $c_i' - \tilde{c}_j'$ satisfies

$$\frac{m_{ij}}{2}(g^2 - g'^2) = E_{iI'} + E_{jJ'} - E_{iI} - E_{jJ}, \quad (2.1.4)$$

where E_{iI} is the energy of the Ith energy shell of the ith species, $m_{ij} = m_i m_j/(m_i + m_j)$ the reduced mass, and m_i the mass of the ith species. The symmetry property of the collision cross-section can be written $a_{iI} a_{jJ} g^2 \, \bar{\sigma}_{ij}^{IJI'J'} = a_{iI'} a_{iJ'} g'^2 \, \bar{\sigma}_{ij}^{I'J'IJ}$, and since $dc_i d\tilde{c}_j / g = dc_i' d\tilde{c}_j' / g'$, we have the following reciprocity relation for the collision cross-section $\bar{\sigma}_{ij}^{IJI'J'}$ [Wa58] [WT62]

$$a_{iI} a_{jJ} g \, \bar{\sigma}_{ij}^{IJI'J'} dc_i d\tilde{c}_j = a_{iI'} a_{iJ'} g' \, \bar{\sigma}_{ij}^{I'J'IJ} dc_i' d\tilde{c}_j'. \quad (2.1.5)$$

The Enskog-Chapman Procedure. The solution of the Boltzmann equation is carried out by adopting the Enskog-Chapman procedure [CC70] [FK72] [HCB54] [MBBK90] [WT62]. The equations (2.1.1) are rewritten in the form

$$\mathfrak{D}_i(f_i) = \frac{1}{\mathfrak{e}} \sum_{j \in \mathcal{S}} \mathfrak{B}_{ij}(f_i, f_j), \qquad i \in \mathcal{S}, \quad (2.1.6)$$

where $\mathfrak{e}$ is a formal expansion parameter and the distribution functions are expanded in the form

$$f_i = f_i^0 (1 + \mathfrak{e}\phi_i + \mathcal{O}(\mathfrak{e}^2)), \qquad i \in \mathcal{S}. \quad (2.1.7)$$

Substitution of the expansion (2.1.7) into the Boltzmann equation (2.1.6) yields at the zeroth order that

$$\sum_{j\in\mathcal{S}} \mathfrak{B}_{ij}(f_i^0, f_j^0) = 0, \qquad i \in \mathcal{S}, \tag{2.1.8}$$

so that f_i^0 is a locally Maxwellian distribution [CC70] [FK72] [HCB54] [MBBK90] [WT62]. This distribution is given by

$$f_i^0 = \frac{n_i a_{iI}}{Q_i}\Big(\frac{m_i}{2\pi k_B \overline{T}}\Big)^{3/2} \exp\Big(-w_i{\cdot}w_i - \epsilon_{iI}\Big), \tag{2.1.9}$$

with

$$w_i = \Big(\frac{m_i}{2k_B\overline{T}}\Big)^{1/2}(c_i - v), \qquad \epsilon_{iI} = \frac{E_{iI}}{k_B\overline{T}}, \qquad Q_i = \sum_{I\in\mathcal{E}_i} a_{iI}\exp(-\epsilon_{iI}), \tag{2.1.10}$$

where k_B the Boltzmann constant, n_i the number density of the ith species, v the mass averaged flow velocity, and $\overline{T}$ the macroscopic mixture temperature [WT62] [FK72]. The arbitrary parameters $(n_i)_{i\in\mathcal{S}}$, v, and $\overline{T}$, appearing in f^0, are thus constrained to be the local macroscopic number densities, velocity vector, and temperature. The zero-order approximation f^0 is, therefore, a locally Maxwellian distribution corresponding to the local macroscopic properties. These constraints yield the relations

$$n_i = \sum_{I\in\mathcal{E}_i}\int f_i dc_i = \sum_{I\in\mathcal{E}_i}\int f_i^0 dc_i, \qquad i \in \mathcal{S}, \tag{2.1.11}$$

$$\rho v = \sum_{\substack{i\in\mathcal{S}\\ I\in\mathcal{E}_i}}\int m_i c_i f_i dc_i = \sum_{\substack{i\in\mathcal{S}\\ I\in\mathcal{E}_i}}\int m_i c_i f_i^0 dc_i, \tag{2.1.12}$$

and

$$\tfrac{1}{2}\rho v{\cdot}v + \rho u = \sum_{\substack{i\in\mathcal{S}\\ I\in\mathcal{E}_i}}\int (\tfrac{1}{2}m_i c_i{\cdot}c_i + E_{iI}) f_i dc_i = \sum_{\substack{i\in\mathcal{S}\\ I\in\mathcal{E}_i}}\int (\tfrac{1}{2}m_i c_i{\cdot}c_i + E_{iI}) f_i^0 dc_i. \tag{2.1.13}$$

The constraints (2.1.11)–(2.1.13) are now rewritten in terms of collisional invariants [WT62].

2.1.3 Collisional Invariants

Notation. We first introduce some notation that will be used throughout this book. For a family of functions ξ_i, $i \in \mathcal{S}$, where ξ_i depends on c_i and I, and is either a scalar, a three-dimensional vector or a three by three matrix, we introduce the compact notation $\xi = (\xi_i)_{i\in\mathcal{S}}$. The perturbed distribution function ϕ, defined by $\phi = (\phi_i)_{i\in\mathcal{S}}$, is an example of a scalar family. The vector and matrix families will be needed later when expanding ϕ in terms of macroscopic variable gradients. The components ξ_i of ξ are therefore tensors of rank zero, one or two with respect to the physical three-dimensional space $\mathbb{R}^3$. We denote by a this tensorial rank and by $\mathcal{T}_\nu$, $\nu \in [1,\tau]$, where $\tau = 3^a$, the canonical basis of the space of tensors of rank a over the three-dimensional space. The components ξ_i, $i \in \mathcal{S}$, of ξ have components with respect to the basis $\mathcal{T}_\nu$, $\nu \in [1,\tau]$, which are denoted by $\xi_{\nu i}$, $\nu \in [1,\tau]$, so that

$$\xi_i = \sum_{\nu\in[1,\tau]} \xi_{\nu i}\, \mathcal{T}_\nu. \tag{2.1.14}$$

We will correspondingly denote by ξ_ν, $\nu \in [1,\tau]$, the scalar families whose components are $\xi_{\nu i}$, $i \in \mathcal{S}$, so that $\xi_\nu = (\xi_{\nu i})_{i\in\mathcal{S}}$. Various properties of tensors of rank zero, one and two over the space $\mathbb{R}^3$ are restated in Appendix A.

In addition, let $\xi = (\xi_i)_{i\in\mathcal{S}}$ and $\zeta = (\zeta_i)_{i\in\mathcal{S}}$ be any scalar, three-dimensional vector or three by three matrix families. We denote by $\xi_i \odot \zeta_i$ the maximum contracted product between the tensors ξ_i and ζ_i. More specifically, we have $\xi_i\odot\zeta_i = \xi_i\zeta_i$ if either ξ_i or ζ_i is a scalar, $\xi_i\odot\zeta_i = \xi_i{\cdot}\zeta_i$ if both ξ_i and ζ_i are three-dimensional vectors, $\xi_i\odot\zeta_i = \xi_i{:}\zeta_i$ if both ξ_i and ζ_i are three by three matrices, $\xi_i\odot\zeta_i = \xi_i\zeta_i$ if ξ_i is a matrix and ζ_i a vector, and $\xi_i\odot\zeta_i = \zeta_i^t\xi_i$ if ξ_i is a vector and ζ_i a matrix, where ζ_i^t denotes the transpose of ζ_i. We point out that the notation "$\cdot$", "$:$", and "$\odot$" is restricted, in this book, to tensors of rank zero, one or two with respect to the three-dimensional physical space $\mathbb{R}^3$.

Let $\xi = (\xi_i)_{i\in\mathcal{S}}$ and $\zeta = (\zeta_i)_{i\in\mathcal{S}}$ be then any scalar, three-dimensional vector or three by three matrix families. We define the bilinear form $\langle\!\langle \xi, \zeta \rangle\!\rangle$ by setting

$$\langle\!\langle \xi, \zeta \rangle\!\rangle = \sum_{\substack{i\in\mathcal{S}\\ \mathrm{I}\in\mathcal{E}_i}} \int \xi_i\odot\zeta_i f_i^0\, dc_i, \tag{2.1.15}$$

and we have $\langle\!\langle \xi, \zeta \rangle\!\rangle = \langle\!\langle \zeta, \xi \rangle\!\rangle$ when ξ and ζ are of the same tensorial type or when ξ or ζ is a scalar or when ξ and ζ are a symmetric matrix and a vector, respectively. This bilinear form is also positive definite.

Collisional Invariants. The collisional invariants are now the families $\xi = (\xi_i)_{i\in\mathcal{S}}$ such that for any pair of indices $i, j \in \mathcal{S}$ and any collision between molecules of types i and j, we have the identity

$$\xi_i(c_i, \mathrm{I}) + \xi_j(\tilde{c}_j, \mathrm{J}) - \xi_i(c_i', \mathrm{I}') - \xi_j(\tilde{c}_j', \mathrm{J}') = 0, \tag{2.1.16}$$

keeping the notation of Section 2.1.2. We will write $\xi \in \mathcal{I}_{\mathcal{S}}^a$ in order to indicate that ξ is a collisional invariant family with respect to the mixture composed of the species indexed by $\mathcal{S}$ and that its components ξ_i are tensors of rank a with respect to $\mathbb{R}^3$. We also introduce the $n+4$ independent scalar collisional invariants ψ^l, $l \in [1, n+4]$, given by

$$\begin{cases} \psi^k = (\delta_{ki})_{i\in\mathcal{S}}, & k \in \mathcal{S}, \\ \psi^{n+\nu} = (m_i c_{\nu i})_{i\in\mathcal{S}}, & \nu = 1,2,3, \\ \psi^{n+4} = (\tfrac{1}{2} m_i c_i \cdot c_i + E_{iI})_{i\in\mathcal{S}}, \end{cases} \tag{2.1.17}$$

where δ_{ij} denotes the Kronecker symbol and $c_{\nu i}$, $\nu = 1,2,3$, the three components of the particle velocity c_i. Any scalar collisional invariant is a linear combination of the ψ^l, $l \in [1, n+4]$, [WT62]. For convenience, we also introduce the new collisional invariants $\widehat{\psi}_l$, $l \in [1, n+4]$, defined by

$$\begin{cases} \widehat{\psi}^k = (\delta_{ki})_{i\in\mathcal{S}}, & k \in \mathcal{S}, \\ \widehat{\psi}^{n+\nu} = \big(m_i(c_{\nu i} - v_\nu)\big)_{i\in\mathcal{S}}, & \nu = 1,2,3, \\ \widehat{\psi}^{n+4} = \big(\tfrac{3}{2} - w_i \cdot w_i + \bar{\epsilon}_i - \epsilon_{iI}\big)_{i\in\mathcal{S}}, \end{cases} \tag{2.1.18}$$

where $\bar{\epsilon}_i$ is the averaged reduced internal energy of the ith species given by $\bar{\epsilon}_i = \sum_{I\in\mathcal{E}_i} a_{iI}\epsilon_{iI} \exp(-\epsilon_{iI})/Q_i$, and these collisional invariants are easily shown to be independent linear combinations of the ψ^l, $l \in [1, n+4]$, and orthogonal with respect to the bilinear form $\langle\!\langle \; , \; \rangle\!\rangle$ [WT62]. As a consequence [WT62], any scalar collisional invariant is a linear combination of the $\widehat{\psi}^l$, $l \in [1, n+4]$, and therefore any tensorial collisional invariant is a linear combination of the $\mathcal{T}_\nu \widehat{\psi}^l$, $(l, \nu) \in [1, n+4] \times [1, \tau]$, so that

$$\mathcal{I}_{\mathcal{S}}^a = \operatorname{span}\{\, \mathcal{T}_\nu \widehat{\psi}^l, \; (l, \nu) \in [1, n+4] \times [1, \tau] \,\}. \tag{2.1.19}$$

Constraints for the Perturbed Distribution Functions. Denoting by $\mathbb{I} = (1)_{k\in\mathcal{S}}$ the family whose components are ones, the constraints (2.1.11)–(2.1.13) can now be rewritten in the compact form

$$\begin{cases} n_k = \langle\!\langle \mathbb{I} + \phi, \psi^k \rangle\!\rangle = \langle\!\langle \mathbb{I}, \psi^k \rangle\!\rangle, & k \in \mathcal{S}, \\ \rho v_\nu = \langle\!\langle \mathbb{I} + \phi, \psi^{n+\nu} \rangle\!\rangle = \langle\!\langle \mathbb{I}, \psi^{n+\nu} \rangle\!\rangle, & \nu = 1,2,3, \\ \tfrac{1}{2}\rho v \cdot v + \rho u = \langle\!\langle \mathbb{I} + \phi, \psi^{n+4} \rangle\!\rangle = \langle\!\langle \mathbb{I}, \psi^{n+4} \rangle\!\rangle. \end{cases} \tag{2.1.20}$$

The family ϕ is thus subjected to the $n+4$ scalar constraints $\langle\langle \phi, \psi^l \rangle\rangle = 0$, $l \in [1, n+4]$, which are rewritten for convenience in the form $\langle\langle \phi, \widehat{\psi}^l \rangle\rangle = 0$, $l \in [1, n+4]$.

2.1.4 Integral Equations for the Perturbed Distribution Functions

Linearized Boltzmann Equation. Substitution of the expansion (2.1.7) into the equations (2.1.6) yields at the first order that

$$\sum_{j \in \mathcal{S}} \Big(\mathfrak{B}_{ij}(f_i^0 \phi_i, f_j^0) + \mathfrak{B}_{ij}(f_i^0, f_j^0 \phi_j) \Big) = \mathfrak{D}_i(f_i^0), \qquad i \in \mathcal{S}. \tag{2.1.21}$$

Denoting by $\mathfrak{I}$ the linearized Boltzmann collision operator defined by $\mathfrak{I}(\xi) = \big(\mathfrak{I}_i(\xi)\big)_{i \in \mathcal{S}}$ with $\xi = (\xi_i)_{i \in \mathcal{S}}$, and

$$\mathfrak{I}_i(\xi) = \sum_{\substack{j \in \mathcal{S} \\ \mathrm{I}' \in \mathcal{E}_i \\ \mathrm{J}, \mathrm{J}' \in \mathcal{E}_j}} \iiint f_j^0 \big(\xi_i(c_i, \mathrm{I}) + \xi_j(\tilde{c}_j, \mathrm{J}) - \xi_i(c_i', \mathrm{I}') - \xi_j(\tilde{c}_j', \mathrm{J}')\big) \tilde{\sigma}_{ij}^{\mathrm{IJI'J'}} g \sin\chi d\chi d\varphi d\tilde{c}_j, \tag{2.1.22}$$

and denoting by Ψ_i the function $\Psi_i = -\mathfrak{D}_i(\log f_i^0)$, the relations (2.1.21) can be written in the form

$$\mathfrak{I}_i(\phi) = \Psi_i, \qquad i \in \mathcal{S}. \tag{2.1.23}$$

An explicit evaluation then yields that

$$\Psi_i = -\Psi_i^{\eta} : \nabla v - \Psi_i^{\kappa} (1/3) \nabla \cdot v - \sum_{l \in \mathcal{S}} \Psi_i^{D_l} \cdot (\nabla \overline{p}_l - \rho Y_l b_l) - \Psi_i^{\lambda'} \cdot \nabla (1/k_{\mathrm{B}} \overline{T}), \tag{2.1.24}$$

where

$$\begin{cases} \Psi_i^{\eta} = 2(w_i \otimes w_i - \frac{1}{3} w_i \cdot w_i I), \\ \Psi_i^{\kappa} = (2c^{\mathrm{int}}/c_{\mathrm{v}})(w_i \cdot w_i - \frac{3}{2}) + (2c_{\mathrm{v}}^{\mathrm{tr}}/c_{\mathrm{v}})(\overline{\epsilon}_i - \epsilon_{i\mathrm{I}}), \\ \Psi_i^{D_l} = (1/\overline{p}_i)(\delta_{il} - Y_i)(c_i - v), \qquad l \in \mathcal{S}, \\ \Psi_i^{\lambda'} = k_{\mathrm{B}} \overline{T} \big(\frac{5}{2} - w_i \cdot w_i + \overline{\epsilon}_i - \epsilon_{i\mathrm{I}} \big)(c_i - v), \end{cases} \tag{2.1.25}$$

and where c^{int} is the mean internal specific heat per molecule, $c_{\mathrm{v}}^{\mathrm{tr}} = \frac{3}{2} k_{\mathrm{B}}$ the translational constant volume specific heat per molecule, $c_{\mathrm{v}} = c_{\mathrm{v}}^{\mathrm{tr}} + c^{\mathrm{int}}$ the mean constant volume specific heat per molecule, $\overline{p}_i = X_i \overline{p}$ the partial pressure of the ith species, and X_i the mole fraction of the ith species. The mean internal specific heat per molecule c^{int} is given by $c^{\mathrm{int}} = \sum_{i \in \mathcal{S}} X_i c_i^{\mathrm{int}}$ where c_i^{int} is the internal heat capacity of the molecules of the ith species.

Finally, introducing the families $\Psi = (\Psi_i)_{i \in S}$ and $\phi = (\phi_i)_{i \in S}$, the integral equations (2.1.23) read [WT62]

$$\Im(\phi) = \Psi, \tag{2.1.26}$$

and ϕ is subjected to the $n+4$ scalar constraints (2.1.20) written in the form

$$\langle\!\langle \phi, \widehat{\psi}^k \rangle\!\rangle = 0, \qquad k \in [1, n+4]. \tag{2.1.27}$$

Properties of the Integral Operator. The integral operator $\Im$ has important symmetry and positivity properties. Between two scalar, vector or matrix families ξ and ζ, let us define the integral bracket operator $[\xi, \zeta]$ by setting

$$[\xi, \zeta] = \langle\!\langle \xi, \Im(\zeta) \rangle\!\rangle, \tag{2.1.28}$$

which can be written in the form

$$4[\xi, \zeta] = \sum_{\substack{i,j \in \mathcal{S} \\ I,I' \in \mathcal{E}_i \\ J,J' \in \mathcal{E}_j}} \iiiint f_i^0 f_j^0 (\xi_i + \tilde{\xi}_j - \xi_i' - \tilde{\xi}_j') \odot (\zeta_i + \tilde{\zeta}_j - \zeta_i' - \tilde{\zeta}_j') \bar{\sigma}_{ij}^{IJI'J'} g \sin\chi d\chi d\varphi dc_i d\tilde{c}_j, \tag{2.1.29}$$

using the obvious notation $\tilde{\xi}_j = \xi_j(\tilde{c}_j, J)$, $\xi_i' = \xi_i(c_i', I')$, and $\tilde{\xi}_j' = \xi_j(\tilde{c}_j', J')$. The bilinear form $[\xi, \zeta]$ is then symmetric, positive semi-definite, and its kernel is composed of the collisional invariants and hence spanned by the $\mathcal{T}_\nu \widehat{\psi}^l$, $(l, \nu) \in [1, n+4] \times [1, \tau]$. We can thus write

$$\left\{ \begin{aligned} & [\xi, \zeta] = \langle\!\langle \xi, \Im(\zeta) \rangle\!\rangle = \langle\!\langle \Im(\xi), \zeta \rangle\!\rangle = [\zeta, \xi], \\ & [\xi, \xi] \geq 0, \\ & [\xi, \xi] = 0 \iff \xi = \sum_{\substack{l \in [1, n+4] \\ \nu \in [1, \tau]}} u_{l\nu} \mathcal{T}_\nu \widehat{\psi}^l, \end{aligned} \right. \tag{2.1.30}$$

where $\mathcal{T}_\nu$, $\nu \in [1, \tau]$, have the same tensorial rank as ξ and the scalar coefficients $u_{l\nu}$ only depend on time and spatial location [WT62].

The bracket bilinear form [,] is by construction positive definite on the functional subspace associated with (2.1.27), and the right-hand side Ψ is orthogonal to its kernel since it can be shown [WT62] that $\langle\!\langle \Psi, \widehat{\psi}^k \rangle\!\rangle = 0$, $k \in [1, n+4]$, so that the integral equation system (2.1.26)(2.1.27) is generally well-posed [FK72] [Ce88].

2.1.5 Transport Coefficients

Transport Fluxes. Multiplying the Boltzmann equation (2.1.1) by the collisional invariants ψ^l, $l \in [1, n+4]$, integrating with respect to the particle velocities, and summing over the quantum energy states yields as usual the macroscopic governing equations (1.1.1)–(1.1.3) and the expressions of the transport fluxes Π, V_i, $i \in \mathcal{S}$, and q, in terms of the perturbed distribution function ϕ [CC70] [FK72] [HCB54] [WT62]. After some algebra, these transport fluxes are easily shown to be given by

$$\Pi = \overline{p} I + k_B \overline{T} \langle\langle \Psi^\eta, \phi \rangle\rangle + \tfrac{1}{3} k_B \overline{T} \langle\langle \Psi^\kappa, \phi \rangle\rangle \, I, \tag{2.1.31}$$

$$V_i = k_B \overline{T} \langle\langle \Psi^{D_i}, \phi \rangle\rangle, \qquad i \in \mathcal{S}, \tag{2.1.32}$$

$$q = \sum_{i \in \mathcal{S}} \rho h_i Y_i V_i - \langle\langle \Psi^{\lambda'}, \phi \rangle\rangle. \tag{2.1.33}$$

In order to express these fluxes in terms of linearly independent macroscopic variable gradients, the perturbed distribution function $\phi = (\phi_i)_{i \in \mathcal{S}}$ is further expanded in the form

$$\phi_i = -\phi_i^\eta : \nabla v - \phi_i^\kappa (1/3) \nabla \cdot v - \sum_{l \in \mathcal{S}} \phi_i^{D_l} \cdot (\nabla \overline{p}_l - \rho Y_l b_l) - \phi_i^{\lambda'} \cdot \nabla (1/k_B \overline{T}), \tag{2.1.34}$$

where ϕ_i^η is a traceless symmetric matrix function, ϕ_i^κ a scalar function, and $\phi_i^{D_l}$ and $\phi_i^{\lambda'}$ are vector functions.

Introducing the families $\Psi^\eta = (\Psi_i^\eta)_{i \in \mathcal{S}}$, $\Psi^\kappa = (\Psi_i^\kappa)_{i \in \mathcal{S}}$, $\Psi^{D_l} = (\Psi_i^{D_l})_{i \in \mathcal{S}}$, $l \in \mathcal{S}$, and $\Psi^{\lambda'} = (\Psi_i^{\lambda'})_{i \in \mathcal{S}}$, and similarly the families $\phi^\eta = (\phi_i^\eta)_{i \in \mathcal{S}}$, $\phi^\kappa = (\phi_i^\kappa)_{i \in \mathcal{S}}$, $\phi^{D_l} = (\phi_i^{D_l})_{i \in \mathcal{S}}$, $l \in \mathcal{S}$, and $\phi^{\lambda'} = (\phi_i^{\lambda'})_{i \in \mathcal{S}}$, and substituting the expansions (2.1.24) and (2.1.34) into the linearized Boltzmann equation (2.1.26) and the constraints (2.1.27) gives rise to equations for each of the expansion coefficients ϕ^η, ϕ^κ, ϕ^{D_l}, $l \in \mathcal{S}$, and $\phi^{\lambda'}$ [WT62]. All of these subsystems are of the form

$$\mathfrak{I}(\phi^\mu) = \Psi^\mu, \tag{2.1.35}$$

where μ stands for η, κ, D_l, $l \in \mathcal{S}$, or λ', and ϕ^μ is subjected to the constraints

$$\langle\langle \phi^\mu, \widehat{\psi}^k \rangle\rangle = 0, \qquad k \in [1, n+4]. \tag{2.1.36}$$

The subsystems (2.1.35) are now of tensorial type since ϕ_i^μ and Ψ_i^μ are three by three traceless symmetric matrix functions for $\mu = \eta$, scalar functions for $\mu = \kappa$, and three-dimensional vector functions for $\mu = D_l$, $l \in \mathcal{S}$, and $\mu = \lambda'$. Correspondingly, the relations (2.1.36) are tensorial constraints which apply to each tensorial component of ϕ^{μ}.

Transport Coefficients. Substitution of the expansion (2.1.34) into the relations (2.1.31)–(2.1.33) then yields the relations (1.1.4)–(1.1.6) and the following expressions of the transport coefficients in terms of the perturbed distribution functions ϕ^{η}, ϕ^{κ}, ϕ^{D_l}, $l \in \mathcal{S}$, and $\phi^{\lambda'}$ [WT62]

$$\begin{cases} \eta = \dfrac{k_{\mathrm{B}}\overline{T}}{10}[\phi^{\eta}, \phi^{\eta}], \\ \kappa = \dfrac{k_{\mathrm{B}}\overline{T}}{9}[\phi^{\kappa}, \phi^{\kappa}], \\ D_{kl} = \dfrac{\overline{p}k_{\mathrm{B}}\overline{T}}{3}[\phi^{D_k}, \phi^{D_l}], \qquad k,l \in \mathcal{S}, \\ \lambda' = \dfrac{1}{3k_{\mathrm{B}}\overline{T}^2}[\phi^{\lambda'}, \phi^{\lambda'}], \\ \theta_k = -\dfrac{1}{3}[\phi^{D_k}, \phi^{\lambda'}], \qquad k \in \mathcal{S}. \end{cases} \tag{2.1.37}$$

The evaluation of the transport coefficients therefore reduces to the calculation of various bracket products $[\phi^{\mu}, \phi^{\mu'}]$, where μ stands for η, κ, D_l, $l \in \mathcal{S}$, or λ', and where ϕ^{μ} satisfies the system (2.1.35)(2.1.36).

In addition, the symmetry of the bracket operator (2.1.30) and the relations (2.1.37) imply the symmetry of the diffusion coefficients $D_{kl} = D_{lk}$, $k,l \in \mathcal{S}$. An explicit evaluation also yields the relation $\sum_{k\in\mathcal{S}} Y_k \Psi^{D_k} = 0$ which implies, by linearity, that $\sum_{k\in\mathcal{S}} Y_k \phi^{D_k} = 0$. Therefore, from (2.1.37), we obtain the mass constraints $\sum_{k\in\mathcal{S}} Y_k D_{kl} = 0$, $l \in \mathcal{S}$, and $\sum_{k\in\mathcal{S}} Y_k \theta_k = 0$.

Unconstrained Diffusion Driving Forces. Substitution of the expansion (2.1.34) into the relations (2.1.31)–(2.1.33) actually yields the transport fluxes V_i, $i \in \mathcal{S}$, and q in terms of the unconstrained diffusion driving forces $\widehat{d}_j$, $j \in \mathcal{S}$, [WT62] given by

$$\widehat{d}_j = \frac{\nabla \overline{p}_j - \rho Y_j b_j}{\overline{p}} = \nabla X_j + X_j \frac{\nabla \overline{p}}{\overline{p}} - \frac{\rho}{\overline{p}} Y_j b_j, \qquad j \in \mathcal{S}.$$

On the other hand, the diffusion driving forces d_j, $j \in \mathcal{S}$, defined by (1.1.7), satisfy the constraint $\sum_{j\in\mathcal{S}} d_j = 0$, and are given by

$$d_j = \widehat{d}_j - Y_j \sum_{k\in\mathcal{S}} \widehat{d}_k, \qquad j \in \mathcal{S}.$$

By using the mass constraints $\sum_{j\in S} D_{ij}Y_j = 0$, $i \in S$, and $\sum_{i\in S} \theta_j Y_j = 0$, it is then straightforward to check that the relations (1.1.5), (1.1.6), and (1.1.8) are also valid. Note that unconstrained diffusion driving forces can also be important from a computational or a mathematical point of view.

The Thermal Conductivity and the Thermal Diffusion Ratios. Defining then the thermal diffusion ratios $\chi = (\chi_k)_{k\in S}$ by the relations [Wa47] [WT62]

$$\begin{cases} D\chi = \theta, \\ \sum_{k\in S} \chi_k = 0, \end{cases} \tag{2.1.38}$$

and the thermal conductivity λ by [WT62]

$$\lambda = \lambda' - (\bar{p}/\overline{T}) \sum_{k\in S} \theta_k \chi_k, \tag{2.1.39}$$

we easily obtain the relations (1.1.8) and (1.1.9).

It is also possible to express the thermal conductivity λ and the thermal diffusion ratios $\chi = (\chi_k)_{k\in S}$ in terms of bracket products. Indeed, let ϕ^λ and Ψ^λ be given by

$$\begin{cases} \phi^\lambda = \phi^{\lambda'} + \bar{p}k_B\overline{T} \sum_{l\in S} \chi_l \phi^{D_l}, \\ \Psi^\lambda = \Psi^{\lambda'} + \bar{p}k_B\overline{T} \sum_{l\in S} \chi_l \Psi^{D_l}. \end{cases} \tag{2.1.40}$$

By linearity, ϕ^λ then satisfies (2.1.35)(2.1.36) for $\mu = \lambda$. Furthermore, after some algebra, one may establish from (2.1.37)–(2.1.40) that

$$\begin{cases} \lambda = \dfrac{1}{3k_B\overline{T}^2}[\phi^\lambda, \phi^\lambda], \\ \chi_k = \dfrac{m_k}{3\bar{p}k_B\overline{T}}[\mathfrak{V}^k, \phi^\lambda], \qquad k \in S, \end{cases} \tag{2.1.41}$$

where $\mathfrak{V}^k$ is defined by $\mathfrak{V}^k = \big((c_k - v)\delta_{ki}\big)_{i\in S}$. This expression for λ is classical but the one for χ is, to the authors' knowledge, new. These characterizations of λ and χ will be useful in Section 2.8 in order to evaluate directly λ and χ without the intermediate calculations of D, λ', and θ. The equivalence of definitions (2.1.38)(2.1.39) with (2.1.40)(2.1.41) will also be discussed in Section 2.8.

Remark. The coefficient λ' is not accessible to direct experimental measurement since, in a mixture of gases, a temperature gradient induces thermal diffusion and thus concentration gradients. Both coefficients λ' and λ are still important in practice since any of the expressions (1.1.6) or (1.1.9) can be used to evaluate the heat flux vector. Following [FK72], the coefficient λ' is termed the partial thermal conductivity of the mixture, although λ' is larger than λ.

Remark. In the special case $n = 1$ and $\mathcal{S} = \{1\}$, we have $\Psi^{D_1} = 0$ from (2.1.25) and consequently $\phi^{D_1} = 0$ from (2.1.35)(2.1.36), so that $D = (D_{11}) = 0$ and $\theta = (\theta_1) = 0$ from (2.1.37). As a consequence, we obtain that $\chi = (\chi_1) = 0$ from the constraint on χ in (2.1.38), and the relation (2.1.39) then yields that $\lambda = \lambda'$ in this case.

Entropy Production. The nonnegativity of the bracket operator yields the nonnegativity of the entropy production due to particle collisions, σ_{col}, since one can show that $\sigma_{\mathrm{col}} = k_B[\phi, \phi]$. An explicit evaluation in terms of the macroscopic variable gradients also yields the expression

$$\begin{aligned}\sigma_{\mathrm{col}} = (\overline{p}/\overline{T}) \sum_{k,l\in\mathcal{S}} D_{kl}(d_k + \chi_k \nabla \log \overline{T})\cdot(d_l + \chi_l \nabla \log \overline{T}) \\ + \lambda(\nabla \log \overline{T})\cdot(\nabla \log \overline{T}) + (\kappa/\overline{T})(\nabla\cdot v)^2 \\ + (\eta/2\overline{T})\big(\nabla v + (\nabla v)^t - \tfrac{2}{3}(\nabla\cdot v)I\big):\big(\nabla v + (\nabla v)^t - \tfrac{2}{3}(\nabla\cdot v)I\big). \qquad (2.1.42)\end{aligned}$$

The positivity of σ_{col} therefore requires that κ, η, and λ are positive, whereas the matrix D has to be positive definite on the hyperplane of zero sum driving forces [Wa58] [FK72] [Gi91].

2.1.6 The Case of Reactive Mixtures

The Waldmann and Trübenbacher theory summarized in the previous sections only considers the case of nonreactive mixtures [WT62]. Because reacting flows are important in practical applications, we discuss, in this section, the extension of the Waldmann and Trübenbacher theory to the case of reactive mixtures in which all the characteristic chemical times are larger than the corresponding mean free times of the molecules. In this case, the transport coefficients can be evaluated as if there were no chemical reactions. We also discuss the evaluation of the chemistry source terms.

Chemical Reactions. We assume that there are chemical reactions among the species of the mixture, in the form

$$\sum_{i\in\mathcal{S}} \sum_{\mathrm{I}\in\mathcal{E}_i} \mathfrak{n}'_{\mathfrak{r}i\mathrm{I}} \mathfrak{M}_{i\mathrm{I}} \rightleftharpoons \sum_{i\in\mathcal{S}} \sum_{\mathrm{I}\in\mathcal{E}_i} \mathfrak{n}''_{\mathfrak{r}i\mathrm{I}} \mathfrak{M}_{i\mathrm{I}}, \qquad \mathfrak{r} \in \mathfrak{R}, \qquad (2.1.43)$$

where $\mathfrak{M}_{iI}$ is the symbol of species i in energy state I, and where $\mathfrak{R}$ denotes the set of chemical reaction indices. The integer coefficients $\mathfrak{n}'_{\mathfrak{r}iI}$ and $\mathfrak{n}''_{\mathfrak{r}iI}$, $i \in \mathcal{S}$, $I \in \mathcal{E}_i$, are the stoichiometric coefficients of reaction $\mathfrak{r}$. Note that any reaction taking place in the mixture is reversible and that the number of reactants and products may vary between reactions. In particular, even though triple nonreactive collisions are neglected in the collision operator (2.1.3), it is important in the applications to consider triple reactive collisions since recombinations cannot often proceed otherwise [LH60] [Ku91]. In order to avoid notational complexities associated with the formal description of arbitrary chemical reactions such as (2.1.43), we will often consider the typical example of the bimolecular reaction

$$\mathfrak{M}_{iI} + \mathfrak{M}_{jJ} \rightleftharpoons \mathfrak{M}_{kK} + \mathfrak{M}_{lL}, \tag{2.1.44}$$

where the species indices i, j, k, l are assumed to be distinct.

The Boltzmann Equation. In order to investigate the transport properties of the reactive mixture, the Boltzmann equation (2.1.1) must be generalized first [PX49] [PM50] [Ta51] [HCB54] [Pr59] [LH60] [RM61] [Ku91]. Indeed, each chemical reaction taking place in the mixture yields a source term in the species Boltzmann equation of the corresponding reactants and products. Denoting by $\mathfrak{C}_i(f)$ the source term for the ith species, the new Boltzmann equation describing the reactive mixture takes the form

$$\mathfrak{D}_i(f_i) = \sum_{j\in\mathcal{S}} \mathfrak{B}_{ij}(f_i, f_j) + \mathfrak{C}_i(f), \qquad i \in \mathcal{S}. \tag{2.1.45}$$

The source term $\mathfrak{C}_i(f)$ is a function of c_i and I obtained by summing the contribution of all the chemical reactions

$$\mathfrak{C}_i(f)(c_i, I) = \sum_{\mathfrak{r}\in\mathfrak{R}} \mathfrak{P}_{\mathfrak{r}iI}, \tag{2.1.46}$$

where $\mathfrak{P}_{\mathfrak{r}iI}$ denotes the production term for species i in energy state I due to reaction $\mathfrak{r}$. For instance, reaction (2.1.44) yields the source term

$$\mathfrak{P}_{\mathfrak{r}iI} = \iiint \Big(f'_k f'_l \frac{a_{iI}a_{jJ}}{a_{kK}a_{lL}} \frac{m_i^3 m_j^3}{m_k^3 m_l^3} - f_i f_j\Big) g_{ij} \mathfrak{a}_{ijkl}^{IJKL} \sin\chi d\chi d\varphi d\tilde{c}_j, \tag{2.1.47}$$

where $\mathfrak{a}_{ijkl}^{IJKL}$ denotes the chemical cross-section and g_{ij} the modulus of the relative velocity $c_i - \tilde{c}_j$ before collision. Note that, following Waldmann, the chemical cross-sections have been preaveraged over the magnetic quantum numbers. From quantum mechanics, these cross-sections satisfy the reciprocity relations $a_{iI}a_{jJ}m_i^2 m_j^2 g_{ij}^2\, \mathfrak{a}_{ijkl}^{IJKL} =$

$a_{kK} a_{lL} m_k^2 m_l^2 g_{kl}^2 \mathfrak{a}_{klij}^{KLIJ}$ [Wa58] and from the conservation of mass, momentum, and energy in reactive collisions [Wa58] we have $m_i m_j dc_i d\tilde{c}_j / g_{ij} = m_k m_l dc'_k d\tilde{c}'_l / g_{kl}$, so that the following reciprocity relation holds for the collision cross-section $\mathfrak{a}_{ijkl}^{IJKL}$

$$a_{iI} a_{jJ} m_i^3 m_j^3 g_{ij}\, \mathfrak{a}_{ijkl}^{IJKL}\, dc_i d\tilde{c}_j = a_{kK} a_{lL} m_k^3 m_l^3 g_{kl}\, \mathfrak{a}_{klij}^{KLIJ}\, dc'_k d\tilde{c}'_l. \tag{2.1.48}$$

The relation between $\mathfrak{a}_{ijkl}^{IJKL}$ and $\mathfrak{a}_{klij}^{KLIJ}$ can also be obtained by using micro-reversibility at equilibrium [PM50] and the equilibrium constants derived from statistical thermodynamics [Fo36] [An89]. Similar relations to (2.1.47) and (2.1.48) can be written for three body recombination reactions and we refer to Ludwig and Heil [LH60], Kuščer [Ku91], and Alexeev et al. [ACG94] for more details.

The Tempered Reaction Regime. It is then possible to distinguish a number of different regimes for reactive mixtures. For extremely fast reactions, where the characteristic chemistry times and the mean free times—the times of free flight—are of the same order of magnitude, there is a kinetic equilibrium regime, first considered by Ludwig and Heil [LH60], in which the transport coefficients can be defined unambiguously. There is then an intermediate regime for which there are no unambiguous definition of transport coefficients. Finally, in this book, following Curtiss [HCB54], Prigogine and Xhrouet [PX49], Prigogine and Mathieu [PM50], Takayanagi [Ta51], Present [Pr59], Ross and Mazur [RM61], and Shizgal and Karplus [SK70] [SK71a] [SK71b], we make the fundamental assumption that the chemistry characteristic times are larger by an order of magnitude than the corresponding mean free times, and this regime is also considered in Ludwig and Heil [LH60]. This regime will be referred to as the "tempered reaction regime".

We will thus write formally that

$$\mathfrak{D}_i(f_i) = \frac{1}{\mathfrak{e}} \sum_{j \in \mathcal{S}} \mathfrak{B}_{ij}(f_i, f_j) + \mathfrak{C}_i(f), \qquad i \in \mathcal{S}, \tag{2.1.49}$$

where $\mathfrak{e}$ is a formal expansion parameter and the distribution functions are again expanded in the form

$$f_i = f_i^0 \big(1 + \mathfrak{e}\phi_i + \mathcal{O}(\mathfrak{e}^2)\big), \qquad i \in \mathcal{S}. \tag{2.1.50}$$

As a consequence, at the zeroth order, we recover the equations (2.1.8), so that the f_i^0, $i \in \mathcal{S}$, are the locally Maxwellian distributions given by (2.1.9) and (2.1.10). Therefore, in the tempered reaction regime, a frozen local equilibrium is achieved at the zeroth order.

Linearized Boltzmann Equation. At the first order, we next obtain that

$$\sum_{j \in \mathcal{S}} \Big(\mathfrak{B}_{ij}(f_i^0 \phi_i, f_j^0) + \mathfrak{B}_{ij}(f_i^0, f_j^0 \phi_j) \Big) = \mathfrak{D}_i(f_i^0) - \mathfrak{C}_i(f^0), \qquad i \in \mathcal{S}. \tag{2.1.51}$$

In these equations, the source terms $\mathfrak{C}_i(f^0)$ are thus evaluated with Maxwellian distributions

$$\mathfrak{C}_i(f^0)(c_i, I) = \sum_{\mathfrak{r}\in\mathfrak{R}} \mathfrak{P}^0_{\mathfrak{r}iI}, \tag{2.1.52}$$

where, for instance,

$$\mathfrak{P}^0_{\mathfrak{r}iI} = \iiint \Big(f_k^{0\prime} f_l^{0\prime} \frac{a_{iI}a_{jJ}}{a_{kK}a_{lL}} \frac{m_i^3 m_j^3}{m_k^3 m_l^3} - f_i^0 f_j^0\Big) g_{ij} \mathfrak{a}_{ijkl}^{IJKL} \sin\chi d\chi d\varphi d\tilde{c}_j, \tag{2.1.53}$$

for the bimolecular reaction (2.1.44). The equations (2.1.51) are rewritten for convenience in the form $\mathfrak{I}_i(\phi) = \Psi_i$, where now $\Psi_i = -\mathfrak{D}_i(\log f_i^0) + \mathfrak{C}_i(f^0)/f_i^0$. A straightforward calculation [LH60] then yields that

$$\mathfrak{I}_i(\phi) = \Psi_i^{\mathrm{f}} + \Psi_i^{\mathrm{r}}, \qquad i \in \mathcal{S}, \tag{2.1.54}$$

where Ψ_i^{f} is the frozen—or nonreactive—right member already given in (2.1.24) and (2.1.25) and Ψ_i^{r} is solely due to chemical reactions and reads

$$\Psi_i^{\mathrm{r}} = \frac{\mathfrak{C}_i(f^0)}{f_i^0} - \frac{\mathfrak{c}_i^0}{n_i} - \frac{\sum_{j\in\mathcal{S}}(\frac{3}{2} + \bar{\epsilon}_j + \mathfrak{u}_j)\mathfrak{c}_j^0}{n c_{\mathrm{v}}/k_{\mathrm{B}}}\big(\tfrac{3}{2} - w_i{\cdot}w_i + \bar{\epsilon}_i - \epsilon_{iI}\big), \qquad i \in \mathcal{S}, \tag{2.1.55}$$

where $\mathfrak{c}_i^0$ is the Maxwellian total production rate of the ith species

$$\mathfrak{c}_i^0 = \sum_{I\in\mathcal{E}_i} \int \mathfrak{C}_i(f^0) dc_i, \tag{2.1.56}$$

$\mathfrak{U}_i$ the energy of formation of the ith species at the reference temperature, and $\mathfrak{u}_i = \mathfrak{U}_i/k_{\mathrm{B}}\overline{T}$ the corresponding reduced energy of formation. Notice that the $n+4$ scalar collisional invariants ψ^l, $l \in [1, n+4]$, are now given by

$$\begin{cases} \psi^k = (\delta_{ki})_{i\in\mathcal{S}}, & k \in \mathcal{S}, \\ \psi^{n+\nu} = (m_i c_{\nu i})_{i\in\mathcal{S}}, & \nu = 1,2,3, \\ \psi^{n+4} = (\frac{1}{2} m_i c_i{\cdot}c_i + E_{iI} + \mathfrak{U}_i)_{i\in\mathcal{S}}, \end{cases} \tag{2.1.57}$$

whereas the collisional invariants $\widehat{\psi}_l$, $l \in [1, n+4]$, are unchanged. By using the conservation of mass, momentum, and energy in chemical reactions, it is also straightforward to check that $\langle\!\langle \Psi^{\mathrm{r}}, \widehat{\psi}^l \rangle\!\rangle = 0$, $l \in [1, n+4]$, and we have already seen that $\langle\!\langle \Psi^{\mathrm{f}}, \widehat{\psi}^l \rangle\!\rangle = 0$, $l \in [1, n+4]$. Therefore, the right member Ψ is still such that $\langle\!\langle \Psi, \widehat{\psi}^l \rangle\!\rangle = 0$, $l \in [1, n+4]$.

Global Reactions. In order to expand the right member Ψ^{r} in terms of macroscopic quantities which play the same role as the macroscopic variable gradients for Ψ^{f}, we now introduce global chemical reactions. For a given reaction (2.1.43) we consider an associated global reaction

$$\sum_{i\in\mathcal{S}}\Big(\sum_{I\in\mathcal{E}_i}\mathfrak{n}'_{\mathfrak{r}iI}\Big)\mathfrak{S}_i \rightleftharpoons \sum_{i\in\mathcal{S}}\Big(\sum_{I\in\mathcal{E}_i}\mathfrak{n}''_{\mathfrak{r}iI}\Big)\mathfrak{S}_i, \tag{2.1.58}$$

where $\mathfrak{S}_i$ is the symbol of the ith species. The set of global reactions is now denoted by

$$\sum_{i\in\mathcal{S}}\mathfrak{m}'_{\mathfrak{g}i}\mathfrak{S}_i \rightleftharpoons \sum_{i\in\mathcal{S}}\mathfrak{m}''_{\mathfrak{g}i}\mathfrak{S}_i, \qquad \mathfrak{g}\in\mathfrak{G}, \tag{2.1.59}$$

where $\mathfrak{m}'_{\mathfrak{g}i}$ and $\mathfrak{m}''_{\mathfrak{g}i}$, $i\in\mathcal{S}$, are the stoichiometric coefficients of global reaction $\mathfrak{g}$ and where $\mathfrak{G}$ is the set of global reaction indices. For each global reaction, it is then convenient to introduce the set of indices of all the corresponding elementary reactions

$$\mathfrak{R}_\mathfrak{g} = \{\ \mathfrak{r}\in\mathfrak{R},\ \mathfrak{m}'_{\mathfrak{g}i} = \sum_{I\in\mathcal{E}_i}\mathfrak{n}'_{\mathfrak{r}iI},\ \mathfrak{m}''_{\mathfrak{g}i} = \sum_{I\in\mathcal{E}_i}\mathfrak{n}''_{\mathfrak{r}iI},\ i\in\mathcal{S}\ \}. \tag{2.1.60}$$

We also introduce the equilibrium constant of global reaction $\mathfrak{g}$ [Fo36] [An89]

$$\mathfrak{K}_\mathfrak{g} = \prod_{i\in\mathcal{S}}\Big[\Big(\frac{2\pi m_i k_B \overline{T}}{h_P{}^2}\Big)^{3/2} Q_i\Big]^{\mathfrak{m}_{\mathfrak{g}i}} \exp\Big(-\sum_{i\in\mathcal{S}}\mathfrak{m}_{\mathfrak{g}i}\mathfrak{u}_i\Big), \tag{2.1.61}$$

where h_P is the Planck constant and $\mathfrak{m}_{\mathfrak{g}i}$, $i\in\mathcal{S}$, the algebraic stoichiometric coefficient of global reaction $\mathfrak{g}$ given by

$$\mathfrak{m}_{\mathfrak{g}i} = \mathfrak{m}''_{\mathfrak{g}i} - \mathfrak{m}'_{\mathfrak{g}i}, \qquad i\in\mathcal{S}. \tag{2.1.62}$$

We also introduce the quantities

$$\mathfrak{A}_\mathfrak{g} = \frac{1}{\mathfrak{K}_\mathfrak{g}}\prod_{i\in\mathcal{S}} n_i^{\mathfrak{m}''_{\mathfrak{g}i}} - \prod_{i\in\mathcal{S}} n_i^{\mathfrak{m}'_{\mathfrak{g}i}}, \qquad \mathfrak{g}\in\mathfrak{G}, \tag{2.1.63}$$

which are related to the affinity of reaction $\mathfrak{g}$. These quantities will act as driving forces, as the macroscopic variable gradients for Ψ^{f} [RM61].

Expansion of Ψ^{r}. By using the conservation of mass, momentum, and energy in chemically reactive collisions, one can show that the quantity $\mathfrak{A}_\mathfrak{g}$ can be factorized in the Maxwellian production terms of the reactions $\mathfrak{r}\in\mathfrak{R}$ such that $\mathfrak{r}\in\mathfrak{R}_\mathfrak{g}$. Therefore, we have

$$\mathfrak{P}^0_{\mathfrak{r}iI} = \mathfrak{A}_\mathfrak{g}\Omega_{\mathfrak{r}iI}, \qquad i\in\mathcal{S}, \quad I\in\mathcal{E}_i, \quad \mathfrak{r}\in\mathfrak{R}_\mathfrak{g}. \tag{2.1.64}$$

For the bimolecular reaction (2.1.44) we have for instance

$$\Omega_{\mathfrak{r}iI} = \iiint \tilde{f}_i^0 \tilde{f}_j^0 g_{ij}\mathfrak{a}_{ijkl}^{IJKL}\sin\chi\, d\chi\, d\varphi\, d\tilde{c}_j, \tag{2.1.65}$$

where $\tilde{f}_i^0$ is the normalized Maxwellian distribution such that $n_i \tilde{f}_i^0 = f_i^0$.

We can then define the family $\mathfrak{D}_\mathfrak{g} = (\mathfrak{D}_{\mathfrak{g}i})_{i\in\mathcal{S}}$ by

$$\mathfrak{D}_{\mathfrak{g}i}(c_i, I) = \sum_{\mathfrak{r}\in\mathfrak{R}_\mathfrak{g}} \Omega_{\mathfrak{r}iI}, \tag{2.1.66}$$

in such a way that

$$\mathfrak{C}_i(f^0) = \sum_{\mathfrak{g}\in\mathfrak{G}} \mathfrak{A}_\mathfrak{g} \mathfrak{D}_{\mathfrak{g}i}. \tag{2.1.67}$$

By integration, we also obtain the quantity

$$\mathfrak{o}_{\mathfrak{g}i} = \sum_{I\in\mathcal{E}_i} \int \mathfrak{D}_{\mathfrak{g}i} dc_i, \tag{2.1.68}$$

such that $\mathfrak{c}_i^0 = \sum_{\mathfrak{g}\in\mathfrak{G}} \mathfrak{A}_\mathfrak{g} \mathfrak{o}_{\mathfrak{g}i}$. We further define the family $\Psi^\mathfrak{g} = (\Psi_i^\mathfrak{g})_{i\in\mathcal{S}}$ by

$$\Psi_i^\mathfrak{g} = \frac{\mathfrak{D}_{\mathfrak{g}i}}{f_i^0} - \frac{\mathfrak{o}_{\mathfrak{g}i}}{n_i} - \frac{\sum_{j\in\mathcal{S}}(\frac{3}{2} + \bar{\epsilon}_j + \mathfrak{u}_j)\mathfrak{o}_{\mathfrak{g}j}}{nc_\mathrm{v}/k_\mathrm{B}} \left(\tfrac{3}{2} - w_i{\cdot}w_i + \bar{\epsilon}_i - \epsilon_{iI}\right), \qquad i \in \mathcal{S}, \tag{2.1.69}$$

and we finally obtain that

$$\Psi^\mathfrak{r} = \sum_{\mathfrak{g}\in\mathfrak{G}} \mathfrak{A}_\mathfrak{g} \Psi^\mathfrak{g}. \tag{2.1.70}$$

This expansion of $\Psi^\mathfrak{r}$ plays a similar role to the expansion (2.1.24) for $\Psi^\mathfrak{f}$, with the quantities $\mathfrak{A}_\mathfrak{g}$ acting as driving forces. Moreover, from the conservation of mass, momentum, and energy in a reactive collision, it is shown that $\langle\!\langle \Psi^\mathfrak{g}, \widehat{\psi}^l \rangle\!\rangle = 0$, for $l \in [1, n{+}4]$ and $\mathfrak{g} \in \mathfrak{G}$.

Decomposition of ϕ and Expansion of $\phi^\mathfrak{r}$. By linearity, we now deduce from (2.1.54) that the perturbation ϕ can be decomposed into

$$\phi_i = \phi_i^\mathfrak{f} + \phi_i^\mathfrak{r}, \qquad i \in \mathcal{S}, \tag{2.1.71}$$

where $\phi_i^\mathfrak{f}$ is given by the expansion (2.1.34) whereas $\phi^\mathfrak{r} = (\phi_i^\mathfrak{r})_{i\in\mathcal{S}}$ is the perturbation due to chemical reactions and satisfies the equation $\mathfrak{I}(\phi^\mathfrak{r}) = \Psi^\mathfrak{r}$ and the constraints $\langle\!\langle \phi^\mathfrak{r}, \widehat{\psi}^l \rangle\!\rangle = 0$, $l \in [1, n{+}4]$. Furthermore, introducing, for $\mathfrak{g} \in \mathfrak{G}$, the solution $\phi^\mathfrak{g}$ of the scalar integral equation

$$\mathfrak{I}(\phi^\mathfrak{g}) = \Psi^\mathfrak{g}, \tag{2.1.72}$$

subjected to the constraints

$$\langle\!\langle \phi^\mathfrak{g}, \widehat{\psi}^k \rangle\!\rangle = 0, \qquad k \in [1, n{+}4], \tag{2.1.73}$$

we deduce from (2.1.70) that ϕ^{r} can be expanded in the form

$$\phi^{\mathrm{r}} = \sum_{\mathfrak{g}\in\mathfrak{G}} \mathfrak{A}_{\mathfrak{g}} \phi^{\mathfrak{g}}. \tag{2.1.74}$$

Governing Equations and Transport Fluxes. Multiplying the Boltzmann equation (2.1.45) by the collisional invariants (2.1.57), integrating with respect to the particle velocities, and summing over the quantum energy states yields the reactive macroscopic governing equations and the expressions of the transport fluxes Π, V_i, $i \in \mathcal{S}$, and q, in terms of the perturbed distribution function ϕ [LH60]. After some algebra, the macroscopic equations expressing the conservation of momentum and total energy are found to be unchanged and are thus given by (1.1.1) and (1.1.3). On the other hand, the mass conservation equation for the ith species now reads

$$\partial_t(\rho Y_i) + \nabla\cdot(\rho v Y_i) = -\nabla\cdot(\rho Y_i V_i) + \varpi_i, \qquad i \in \mathcal{S}, \tag{2.1.75}$$

where ϖ_i is the macroscopic mass production rate for the ith species. Finally, the relations expressing the fluxes Π, V_i, $i \in \mathcal{S}$, and q, in terms of the perturbed distribution function ϕ, are also found to be unchanged and are thus given by (2.1.31)–(2.1.33).

Substitution of the decomposition $\phi = \phi^{\mathrm{f}} + \phi^{\mathrm{r}}$ into the relations (2.1.31)–(2.1.33) then yields the relations

$$\Pi = \Pi^{\mathrm{f}} + \Pi^{\mathrm{r}}, \tag{2.1.76}$$

$$V_i = V_i^{\mathrm{f}}, \qquad i \in \mathcal{S}, \tag{2.1.77}$$

$$q = q^{\mathrm{f}}, \tag{2.1.78}$$

where the superscripts ^{f} and ^{r} refer to the fluxes induced by ϕ^{f} and ϕ^{r}, respectively, since by spatial isotropy the scalar family ϕ^{r} is such that $\langle\!\langle \Psi^{\eta}, \phi^{\mathrm{r}} \rangle\!\rangle = 0$, $\langle\!\langle \Psi^{D_l}, \phi^{\mathrm{r}} \rangle\!\rangle = 0$, $l \in \mathcal{S}$, and $\langle\!\langle \Psi^{\lambda'}, \phi^{\mathrm{r}} \rangle\!\rangle = 0$. As a consequence, the only modification to the transport fluxes appears in the pressure tensor Π through the new term $\Pi^{\mathrm{r}} = \frac{1}{3} k_{\mathrm{B}} \overline{T} \langle\!\langle \Psi^{\kappa}, \phi^{\mathrm{r}} \rangle\!\rangle\, I$, which can further be expressed as

$$\Pi^{\mathrm{r}} = \Big(\sum_{\mathfrak{g}\in\mathfrak{G}} \tfrac{1}{3} k_{\mathrm{B}} \overline{T} \langle\!\langle \Psi^{\kappa}, \phi^{\mathfrak{g}} \rangle\!\rangle \mathfrak{A}_{\mathfrak{g}} \Big)\, I. \tag{2.1.79}$$

Therefore, the transport fluxes due to macroscopic variable gradients are unchanged by the chemical reactions, since the nonreactive term ϕ^{f} is unchanged and since no macroscopic variable gradients appear in the expansion of Π^{r}. The transport coefficients are thus identical to the transport coefficients obtained as if there were no chemical reactions.

Macroscopic Species Source Terms. The macroscopic mass production rate appearing in the ith species governing equation (2.1.75) is also shown to be given by

$$\varpi_i = m_i\big(\mathfrak{c}_i^0 + \mathfrak{w}_i\big), \tag{2.1.80}$$

where $\mathfrak{c}_i^0$ is estimated by using Maxwellian distributions whereas the perturbed term $\mathfrak{w}_i$ reads

$$\mathfrak{w}_i = -\sum_{\mathfrak{g}\in\mathfrak{G}} \langle\!\langle \mathcal{H}^{\mathfrak{g}i}, \phi \rangle\!\rangle, \tag{2.1.81}$$

where $\mathcal{H}^{\mathfrak{g}i}$, $\mathfrak{g}\in\mathfrak{G}$, are functionals arising from the linearization of the chemistry source term $\mathfrak{C}_i(f)$. Substitution of the decomposition $\phi = \phi^{\mathrm{f}} + \phi^{\mathrm{r}}$ into (2.1.81) then yields

$$\mathfrak{w}_i = \mathfrak{w}_i^{\mathrm{f}} + \mathfrak{w}_i^{\mathrm{r}}, \tag{2.1.82}$$

and from (2.1.34) and (2.1.74) we further deduce that

$$\mathfrak{w}_i^{\mathrm{f}} = \Big(\sum_{\mathfrak{g}\in\mathfrak{G}} \tfrac{1}{3}\langle\!\langle \mathcal{H}^{\mathfrak{g}i}, \phi^{\kappa} \rangle\!\rangle\Big)\nabla\!\cdot\! v, \tag{2.1.83}$$

and

$$\mathfrak{w}_i^{\mathrm{r}} = -\sum_{\mathfrak{g}\in\mathfrak{G}}\sum_{\mathfrak{g}'\in\mathfrak{G}} \langle\!\langle \mathcal{H}^{\mathfrak{g}i}, \phi^{\mathfrak{g}'} \rangle\!\rangle\, \mathfrak{A}^{\mathfrak{g}'}. \tag{2.1.84}$$

From these relations, it is then shown that $\mathfrak{w}_i^{\mathrm{f}}$ is a polynomial in n_i, $i\in\mathcal{S}$, of degree lower than three multiplied by $\nabla\!\cdot\! v$, whereas $\mathfrak{w}_i^{\mathrm{f}}$ is a polynomial in n_i, $i\in\mathcal{S}$, of degree lower than six.

Simplifying Assumptions. To the authors' knowledge, the perturbed chemistry source terms $\mathfrak{w}_i^{\mathrm{r}}$, $i\in\mathcal{S}$, were first mentioned by Prigogine and Xhrouet [PX49]. These terms have been estimated in a number of simplified situations by Prigogine and Xhrouet [PX49], Prigogine and Mathieu [PM50], Takayanagi [Ta51], Present [Pr59], Shizgal and Karplus [SK70] [SK71a] [SK71b], and are generally believed to be small [HCB54]. To the authors' knowledge, the cross terms $\mathfrak{w}_i^{\mathrm{f}}$, $i\in\mathcal{S}$, and Π^{r} are first mentioned in Ludwig and Heil [LH60]. These terms are associated with the coupling of internal energy distribution and chemical reactions. Although complete estimates of Π^{r}, $\mathfrak{w}_i^{\mathrm{f}}$, $i\in\mathcal{S}$, and $\mathfrak{w}_i^{\mathrm{r}}$, $i\in\mathcal{S}$, are still missing for realistic reactive mixtures, these terms are usually neglected.

With these simplifications, the mass production rates ϖ_i, $i \in \mathcal{S}$, appearing in the species conservation equations (2.1.75) simply become $\varpi_i = m_i \mathfrak{c}_i^0$, $i \in \mathcal{S}$. From stoichiometry, it is also shown that

$$\mathfrak{o}_{\mathfrak{g}i} = -\mathfrak{m}_{\mathfrak{g}i}\mathfrak{F}_{\mathfrak{g}}, \tag{2.1.85}$$

where $\mathfrak{F}_{\mathfrak{g}}$ is the forward molecular rate constant of reaction $\mathfrak{g}$, so that the source term $\varpi_i = m_i \sum_{\mathfrak{g}\in\mathfrak{G}} \mathfrak{o}_{\mathfrak{g}i}\mathfrak{A}_{\mathfrak{g}}$ finally reads

$$\varpi_i = m_i \sum_{\mathfrak{g}\in\mathfrak{G}} \mathfrak{m}_{\mathfrak{g}i}\Big(\mathfrak{F}_{\mathfrak{g}} \prod_{i\in\mathcal{S}} n_i^{\mathfrak{m}'_{\mathfrak{g}i}} - \frac{\mathfrak{F}_{\mathfrak{g}}}{\mathfrak{K}_{\mathfrak{g}}} \prod_{i\in\mathcal{S}} n_i^{\mathfrak{m}''_{\mathfrak{g}i}}\Big), \tag{2.1.86}$$

and is, therefore, compatible with the usual phenomenological rates obtained by the law of mass action [RM61].

In addition, the total mass conservation in reactive collisions, associated with the mass collisional invariant $(m_i)_{i\in\mathcal{S}} = \sum_{k\in\mathcal{S}} m_k \widehat{\psi}^k$, yields the relation

$$\sum_{i\in\mathcal{S}} \varpi_i = 0, \tag{2.1.87}$$

and similar relations can be written for conservation of the elements.

Applicability of the Tempered Reaction Regime. In order to investigate the validity of the tempered reaction regime, it is necessary to compare the characteristic chemistry times with the mean free times in the mixture. This corresponds to estimate the order of magnitude of the scattering and chemistry contributions to the Boltzmann equation.

For a given species pair (i, j), we first have to evaluate the mean free time between successive collisions of a molecule of the ith species with a molecule of the jth species. This time can be evaluated as

$$t_{ij}^{\text{coll}} = \Big(\frac{N_{ij}^{\text{coll}}}{n_i}\Big)^{-1}, \tag{2.1.88}$$

where N_{ij}^{coll} is the number of collisions between molecules of the ith species and the jth species per unit volume and unit time. This number can be estimated by using a rigid sphere model

$$N_{ij}^{\text{coll}} = \pi\sigma_{ij}^2 \sum_{I\in\mathcal{E}_i} \sum_{J\in\mathcal{E}_j} \iint f_i^0 f_j^0 g_{ij} dc_i d\tilde{c}_j = \sigma_{ij}^2 n_i n_j \Big(\frac{8\pi k_{\text{B}} \overline{T}}{m_{ij}}\Big)^{1/2}, \tag{2.1.89}$$

where σ_{ij} the collision diameter of the species pair (i,j) [FK72]. On the other hand, for any reaction $\mathfrak{g}$ in the form $\mathfrak{S}_i + \mathfrak{S}_j \rightleftharpoons$ Products, the reaction characteristic time is given by

$$t_{\mathfrak{g}i}^{\mathrm{chem}} = \left(\frac{N_{\mathfrak{g}i}^{\mathrm{chem}}}{n_i}\right)^{-1}, \tag{2.1.90}$$

where $N_{\mathfrak{g}i}^{\mathrm{chem}}$ is the number of reactive collisions between molecules of the ith species and the jth species per unit volume and unit time due to reaction $\mathfrak{g}$

$$N_{\mathfrak{g}i}^{\mathrm{chem}} = \mathfrak{F}_{\mathfrak{g}} n_i n_j, \tag{2.1.91}$$

and $\mathfrak{F}_{\mathfrak{g}}$ the molecular forward rate constant of chemical reaction $\mathfrak{g}$. Similar estimates are easily written for three body recombination reactions. The tempered reaction regime is thus a reasonable approximation whenever the inequalities

$$t_{ij}^{\mathrm{coll}}/t_{\mathfrak{g}i}^{\mathrm{chem}} \;=\; N_{\mathfrak{g}i}^{\mathrm{chem}}/N_{ij}^{\mathrm{coll}} \;\ll\; 1, \tag{2.1.92}$$

are satisfied for each reaction $\mathfrak{g}$ involving the species pair (i,j) and each species pair (i,j) of the mixture. Notice that only elastic collisions are taken into account in these estimates. A typical example will be given in Appendix E for combustion applications.

An additional requirement for the tempered reaction regime to be valid is that the characteristic chemistry times are larger than the characteristic relaxation times of the internal degrees of freedom. In other words, energy transfer in inelastic colisions must be sufficiently rapid to more than counterbalance the reactive loss from higher energy states [RM61] [Ze84]. This usually holds for rotational degrees of freedom which are equilibrated within a few collisions. Vibrational relaxation, however, is comparatively slow and typically requires several thousands of collisions, so that an equilibrium distribution of vibrational states may not be maintained [Ze84]. These internal desequilibrium may then modify the chemical production rates as discussed for instance by Zwolinsky and Eyring [ZE47], Montroll and Shuler [MS58], and Zellner [Ze84]. In these situations, it becomes necessary to consider each molecule in a given internal state as a distinct "species" of the mixture and to take into account internal energy desequilibrium at the zeroth order. The conservation equations and transport properties of such mixtures will not be addressed in this book.

Chemical Equilibrium. In the tempered reaction regime, the macroscopic variables $\overline{p}$, $\overline{T}$, and Y_i, $i \in S$, are independent and each corresponding macroscopic gradient must be taken into account in the expressions for the transport fluxes (1.1.4)–(1.1.9).

However, a situation often considered in engineering applications is that of chemical equilibrium for which these gradients are no longer independent. A first regime taking into account chemical equilibrium is the fast equilibrium regime in which the chemistry characteristic times are of the same order of magnitude than the mean free times. A theoretical model for this regime was established by Ludwig and Heil [LH60], but will not be addressed in this book. Another regime, which may sometimes be of interest, arises when the chemical characteristic times are larger than the mean free times but smaller than the flow time so that chemical equilibrium is again achieved.

In this case, the required assumptions for the tempered reaction regime are met, but the mass fractions Y_i, $i \in S$, are no longer independent. Assuming for simplicity that the concentrations of the elements are known, the mass fractions Y_i, $i \in S$, are then functions of the pressure $\bar{p}$ and temperature $\overline{T}$, given by the equilibrium values $Y_i^{\mathrm{eq}}(\bar{p},\overline{T})$. Consequently, the mole fractions X_i, $i \in S$, are given by the equilibrium values

$$X_i = X_i^{\mathrm{eq}}(\bar{p},\overline{T}), \qquad i \in S,$$

so that the following relations hold

$$\nabla X_i = \frac{\partial X_i^{\mathrm{eq}}}{\partial \bar{p}} \nabla \bar{p} + \frac{\partial X_i^{\mathrm{eq}}}{\partial \overline{T}} \nabla \overline{T}, \qquad i \in S.$$

Since the heat flux vector q involves terms in ∇X_i, these terms will add new contributions involving $\nabla \bar{p}$ and $\nabla \overline{T}$. It is customary to regroup all the terms in $\nabla \bar{p}$ and $\nabla \overline{T}$ by introducing new transport coefficients [An89]. For instance, if we regroup all the $\nabla \overline{T}$ terms in the heat flux vector q given by (1.1.9), we obtain a contribution $-\lambda^{\mathrm{eq}} \nabla \overline{T}$, where we have introduced

$$\lambda^{\mathrm{eq}} = \lambda + \sum_{i \in S} (\rho h_i Y_i^{\mathrm{eq}} + \bar{p}\chi_i) \Big(\sum_{j \in S} D_{ij} \frac{\partial X_j^{\mathrm{eq}}}{\partial \overline{T}} + \frac{\theta_i}{\overline{T}} \Big).$$

If the chemical equilibrium values X_i^{eq} are evaluated explicitly, by using a chemical reaction network for instance, the equilibrium thermal conductivity λ^{eq} then depends explicitly on the chemical reactions, at variance with the nonequilibrium thermal conductivity λ. Therefore, this type of equilibrium transport coefficients should not be confused with nonequilibrium coefficients.

2.2 Derivation of the Transport Linear Systems

2.2.1 Variational Procedure

The Functional Space $\mathcal{A}^\mu$. The linear integral equations (2.1.35)(2.1.36) are solved approximately by using a variational procedure [WT62]. More specifically, a finite dimensional functional space $\mathcal{A}^\mu$ is first selected

$$\mathcal{A}^\mu = \text{span}\{ \xi^{rk},\ (r,k) \in \mathcal{B}^\mu \}, \tag{2.2.1}$$

where ξ^{rk}, $(r,k) \in \mathcal{B}^\mu$, are basis functions and where $\mathcal{B}^\mu$ is a set of basis function indices. The set of basis function indices $\mathcal{B}^\mu$ is a subset of the product set $\mathcal{F} \times \mathcal{S}$

$$\mathcal{B}^\mu \subset \mathcal{F} \times \mathcal{S}, \tag{2.2.2}$$

where $\mathcal{F}$ denotes a set of function type indices and $\mathcal{S}$ the set of species indices $\mathcal{S} = [1,n]$, where $n \geq 1$ unless explicitly stated. The set $\mathcal{B}^\mu$ differs from $\mathcal{F} \times \mathcal{S}$ since some types of functions do not appear for certain species. For instance, functions in the reduced internal energy must not be considered for the monatomic species of the mixture. It will be convenient to define the subsets

$$\begin{cases} \mathcal{S}_r = \{ k \in \mathcal{S},\ \ (r,k) \in \mathcal{B}^\mu \}, & r \in \mathcal{F}, \\ \mathcal{F}_k = \{ r \in \mathcal{F},\ \ (r,k) \in \mathcal{B}^\mu \}, & k \in \mathcal{S}, \end{cases} \tag{2.2.3}$$

in such a way that

$$\mathcal{B}^\mu = \bigcup_{r \in \mathcal{F}} \{ r \} \times \mathcal{S}_r = \bigcup_{k \in \mathcal{S}} \mathcal{F}_k \times \{ k \}. \tag{2.2.4}$$

The subset $\mathcal{S}_r$ is simply the set of species indices for which the rth function type is selected, and $\mathcal{F}_k$ is the set of function indices that are considered for the kth species. Denoting by ω_r the number of elements of $\mathcal{S}_r$ and letting $\omega = \sum_{r \in \mathcal{F}} \omega_r$, the set $\mathcal{B}^\mu$ has ω elements, and the functional space $\mathcal{A}^\mu$ has dimension ω.

The basis functions ξ^{rk}, $(r,k) \in \mathcal{B}^\mu$, have the same tensorial rank as ϕ^μ and Ψ^μ with respect to the three-dimensional physical space. We denote by a_μ the tensorial rank of ϕ^μ, Ψ^μ, and of the basis functions ξ^{rk}, so that the components ξ_i^{rk}, $i \in \mathcal{S}$, are tensors of rank a_μ with respect to $\mathbb{R}^3$, and we denote by $\tau_\mu = 3^{a_\mu}$ the dimension of the space of tensors of rank a_μ over $\mathbb{R}^3$. In particular, we have $a_\mu = 0$, $\tau_\mu = 1$, and $\mu = \kappa$ in the scalar case, $a_\mu = 1$, $\tau_\mu = 3$, and $\mu = D_l$, $l \in \mathcal{S}$, $\mu = \lambda'$, or $\mu = \lambda$, in the vector case, and $a_\mu = 2$, $\tau_\mu = 9$, and $\mu = \eta$ in the three by three traceless symmetric matrix case. As stated previously, $\mathcal{T}_\nu$, $\nu \in [1, \tau_\mu]$, denotes the canonical basis of the space of tensors of order a_μ over the three-dimensional space $\mathbb{R}^3$. The components ξ_i^{rk}, $i \in S$, of the functions ξ^{rk}, $(r,k) \in \mathcal{B}^\mu$, have components with respect to the basis $\mathcal{T}_\nu$, $\nu \in [1,\tau_\mu]$, which are denoted by $\xi_{\nu i}^{rk}$, $\nu \in [1,\tau_\mu]$, so that

$$\xi_i^{rk} = \sum_{\nu\in[1,\tau_\mu]} \xi_{\nu i}^{rk}\, \mathcal{T}_\nu. \tag{2.2.5}$$

Similarly, we decompose ϕ^μ into $\phi_i^\mu = \sum_{\nu\in[1,\tau_\mu]} \phi_{\nu i}^\mu \mathcal{T}_\nu$ and we correspondingly denote by ϕ_ν and ξ_ν^{rk} the scalar families whose components are $\phi_{\nu i}^\mu$ and $\xi_{\nu i}^{rk}$, respectively, so that $\phi_\nu^\mu = (\phi_{\nu i}^\mu)_{i\in S}$ and $\xi_\nu^{rk} = (\xi_{\nu i}^{rk})_{i\in S}$.

Variational Procedure. The distribution function ϕ^μ is next expanded in the form

$$\phi^\mu = \sum_{(r,k)\in\mathcal{B}^\mu} \alpha_k^{r\mu} \xi^{rk}, \tag{2.2.6}$$

where the $\alpha_k^{r\mu}$ are scalars. In the notation $\alpha_k^{r\mu}$, the superscript μ refers to the coefficient μ, the superscript r to the type of function that is considered, and the subscript k to the species.

The variational procedure applied to the integral equation (2.1.35) then yields the system

$$[\xi^{rk}, \phi^\mu] = \langle\!\langle \xi^{rk}, \Psi^\mu \rangle\!\rangle, \qquad (r,k) \in \mathcal{B}^\mu, \tag{2.2.7}$$

which must be solved under the constraints (2.1.36) [WT62]. The relations (2.2.7) yield a linear system of size ω in the form

$$\sum_{(s,l)\in\mathcal{B}^\mu} G_{kl}^{rs} \alpha_l^{s\mu} = \beta_k^{r\mu}, \qquad (r,k) \in \mathcal{B}^\mu, \tag{2.2.8}$$

where the unknowns are the ω coefficients $\alpha_k^{r\mu}$, $(r,k) \in \mathcal{B}^\mu$, and where

$$\begin{cases} G_{kl}^{rs} = [\xi^{rk}, \xi^{sl}], \\ \beta_k^{r\mu} = \langle\!\langle \xi^{rk}, \Psi^\mu \rangle\!\rangle. \end{cases} \tag{2.2.9}$$

The $n+4$ tensorial constraints (2.1.36) also yield the $(n+4)\tau_\mu$ scalar constraints

$$\langle\!\langle \phi^\mu, \mathcal{T}_\nu \widehat{\psi}^l \rangle\!\rangle = 0, \qquad (l,\nu) \in [1,n+4]\times[1,\tau_\mu], \tag{2.2.10}$$

since

$$\langle\!\langle \phi^\mu, \widehat{\psi}^l \rangle\!\rangle = \sum_{\nu\in[1,\tau_\mu]} \langle\!\langle \phi_\nu^\mu, \widehat{\psi}^l \rangle\!\rangle\, \mathcal{T}_\nu = \sum_{\nu\in[1,\tau_\mu]} \langle\!\langle \phi^\mu, \mathcal{T}_\nu \widehat{\psi}^l \rangle\!\rangle\, \mathcal{T}_\nu. \tag{2.2.11}$$

These constraints are rewritten in terms of α^μ in the form

$$\sum_{(r,k)\in\mathcal{B}^\mu} \mathcal{G}_k^{rl\nu}\alpha_k^{r\mu} = 0, \qquad (l,\nu) \in [1,n+4]\times[1,\tau_\mu], \tag{2.2.12}$$

where

$$\mathcal{G}_k^{rl\nu} = \langle\!\langle \xi^{rk}, \mathcal{T}_\nu \widehat{\psi}^l \rangle\!\rangle, \qquad (l,\nu) \in [1,n+4]\times[1,\tau_\mu], \quad (r,k) \in \mathcal{B}^\mu. \tag{2.2.13}$$

Thanks to the orthogonality properties of the basis functions ξ^{rk}, $(r,k) \in \mathcal{B}^\mu$, and of the tensorial collisional invariants $\mathcal{T}_\nu \widehat{\psi}^l$, $(l,\nu) \in [1,n+4]\times[1,\tau_\mu]$, most of these relations will be found to be trivial, i.e., will yield zero constraint coefficients $\mathcal{G}_k^{rl\nu} = 0$, $(r,k) \in \mathcal{B}^\mu$, as detailed in the next sections.

Finally, solving the system (2.2.8)(2.2.12) will yield the products $[\phi^\mu, \phi^\mu]$ from the relation

$$[\phi^\mu, \phi^\mu] = \langle\!\langle \Psi^\mu, \phi^\mu \rangle\!\rangle = \sum_{(r,k)\in\mathcal{B}^\mu} \alpha_k^{r\mu}\beta_k^{r\mu}, \tag{2.2.14}$$

from which the transport coefficients (2.1.37) are easily evaluated.

In the remaining part of Section 2.2, we present the basis functions and discuss the practical evaluation of the system coefficients in terms of collision integrals. The structure and mathematical properties of the system (2.2.8)(2.2.12) will be analyzed in Section 2.3.

2.2.2 Basis Functions

The basis functions ξ^{rk}, $(r,k) \in \mathcal{B}^\mu$, are generally chosen as simple linear combinations of the functions ϕ^{a0cdk} defined by

$$\phi^{a0cdk}(c_k, K) = \Big(S^c_{a+\frac{1}{2}}(w_k \cdot w_k)\, W_k^d(\epsilon_{kK})\, \overline{\otimes^a w_k}\, \delta_{ki} \Big)_{i\in\mathcal{S}}, \tag{2.2.15}$$

where a, c, and d are integers, $S^c_{a+1/2}$ is the Laguerre and Sonine polynomial of order c with parameter $a + 1/2$, W_k^d the Wang Chang and Uhlenbeck polynomial of order d for the kth species, and $\overline{\otimes^a w_k}$ a tensor of rank a with respect to the three-dimensional space given by $\overline{\otimes^0 w_k} = 1$, $\overline{\otimes^1 w_k} = w_k$, and $\overline{\otimes^2 w_k} = w_k \otimes w_k - \frac{1}{3} w_k \cdot w_k I$ [WT62]. In the notation ϕ^{abcdk}, the first index a thus refers to the tensorial rank with respect to $\mathbb{R}^3$, the second index $b = 0$ to the absence of polarization effects [MBKK90], the third index c to the Laguerre and Sonine polynomial, the fourth index d to the Wang Chang and Uhlenbeck polynomial, and the last index k to the species. Keeping in mind that ξ^{rk} is generally a simple linear combination of the functions ϕ^{a0cdk}, the index r in ξ^{rk} is usually a multi-index. Note that the functions ϕ^{a0cdk} are only defined provided that $d < \text{card}(\mathcal{E}_k)$ where $\text{card}(\mathcal{E}_k)$ is the number of elements of $\mathcal{E}_k$, that is to say, the number of different energy levels of the molecules of the kth species. It is convenient, however, to extend this definition by setting that $\phi^{a0cdk} = 0$ whenever $d \geq \text{card}(\mathcal{E}_k)$. The notation in (2.2.15) is similar to the one in [MBKK90] [MBKK91], but the basis functions (2.2.15) are not normalized since it would lead to artificial notational complexities and introduce concentration dependent functions. These functions have important orthogonality properties since we have the relations [WT62] [MBKK90] [MBKK91]

$$\langle\!\langle \phi^{a0cdk}, \phi^{a'0c'd'l} \rangle\!\rangle = \langle\!\langle \phi^{a0cdk}, \phi^{a0cdk} \rangle\!\rangle \delta_{aa'}\delta_{cc'}\delta_{dd'}\delta_{kl}, \tag{2.2.16}$$

for $a, a', c, c', d, d' \geq 0$ and $k, l \in \mathcal{S}$. Various properties of the Laguerre and Sonine polynomials, the Wang Chang and Uhlenbeck polynomials, and the functions ϕ^{a0cdk} are summarized in Appendix B.

In particular, the following scalar basis functions will be used for the scalar integral equation in ϕ^{κ}

$$\begin{cases} \phi^{0010k} = \left((\frac{3}{2} - w_k{\cdot}w_k)\delta_{ki} \right)_{i\in\mathcal{S}}, & k \in \mathcal{S}, \\ \phi^{0001k} = \left((\bar{\epsilon}_k - \epsilon_{kK})\delta_{ki} \right)_{i\in\mathcal{S}}, & k \in \mathcal{P}, \end{cases} \tag{2.2.17}$$

where $\mathcal{P}$ denotes the set of species which have at least two different energy levels

$$\mathcal{P} = \{\ k \in \mathcal{S},\ \text{card}(\mathcal{E}_k) \geq 2\ \}, \tag{2.2.18}$$

so that $\bar{\epsilon}_k - \epsilon_{kK}$ is not identically zero when K takes its values in $\mathcal{E}_k$. This set is generally taken to be the set of polyatomic species, and we denote by p the number of elements of the set $\mathcal{P}$. The following vector basis functions will also be used for the vector integral equations in ϕ^{D_l}, $l \in \mathcal{S}$, $\phi^{\lambda'}$, and ϕ^{λ}

$$\begin{cases} \phi^{1000k} = (w_k\delta_{ki})_{i\in\mathcal{S}}, & k \in \mathcal{S}, \\ \phi^{1010k} = \left((\frac{5}{2} - w_k{\cdot}w_k)w_k\delta_{ki} \right)_{i\in\mathcal{S}}, & k \in \mathcal{S}, \\ \phi^{1001k} = \left((\bar{\epsilon}_k - \epsilon_{kK})w_k\delta_{ki} \right)_{i\in\mathcal{S}}, & k \in \mathcal{P}, \end{cases} \tag{2.2.19}$$

and finally the following traceless symmetric matrix basis functions will be used for the traceless symmetric matrix integral equation in ϕ^{η}

$$\phi^{2000k} = \left((w_k{\otimes}w_k - \tfrac{1}{3}w_k{\cdot}w_k I)\delta_{ki} \right)_{i\in\mathcal{S}}, \qquad k \in \mathcal{S}. \tag{2.2.20}$$

Since the basis functions ϕ^{a0cdk} involve the polynomials W_k^d in the internal energy ϵ_{kK} of the molecules, all the internal modes of the molecules, e.g., rotation and vibration, are forced to have the same internal temperature [Mo64]. However, it is possible to consider polynomials in the energies of the various internal modes, which leads to larger variational approximation spaces for the species perturbed distribution functions [Ah72] [Mo64]. This would only complicate the transport linear systems, but the general theory that is presented would equally apply. In addition, considering polynomials in the internal energy of the molecules only may sometimes lead to a faster convergence of the orthogonal polynomial expansions of the perturbed distribution functions. In particular, it has been observed experimentally by Van den Oord et al. [VDBK88] that, for iodine at room temperature, in the presence of a temperature gradient, the species distribution function $\phi^{\lambda'}$ is linear in the total internal energy, but not in the different internal energy modes, i.e., rotation and vibration, for which the quadratic terms become significant in the orthogonal polynomial expansions [VDBK88].

Finally, we point out that an interesting extension of this work would be to take into account the full species vibrational desequilibrium as needed for reentry problems [An89] [Bu88] [Mo64]. In this regime, a strong vibrational desequilibrium appears already at the zeroth order governing equations [Bu88].

2.2.3 Evaluation of the System Coefficients

In order to evaluate the integral bracket products $[\xi^{rk}, \xi^{sl}]$ which are needed to form the transport linear systems (2.2.8), it is convenient to introduce the partial brackets of Waldmann [Wa58] [WT62]. For any species pair (i, j) and for any functions ξ_{ij} and ζ_{ij} of c_i, $\tilde{c}_j$, I, J, c_i', $\tilde{c}_j'$, I', and J', we first define the partial collision product $\{\xi_{ij}, \zeta_{ij}\}_{ij}$ by setting

$$n_i n_j \{\xi_{ij}, \zeta_{ij}\}_{ij} = \sum_{\substack{I,I' \in \mathcal{E}_i \\ J,J' \in \mathcal{E}_j}} \iiiint f_i^0 f_j^0 \, \xi_{ij} \odot \zeta_{ij} \, \bar{\sigma}_{ij}^{IJI'J'} g \sin\chi d\chi d\varphi dc_i d\tilde{c}_j. \tag{2.2.21}$$

For two families $\xi = (\xi_i)_{i \in S}$ and $\zeta = (\zeta_i)_{i \in S}$, where ξ_i and ζ_i only depend on c_i and I, the partial brackets $[\xi, \zeta]_{ij}'$ and $[\xi, \zeta]_{ij}''$ introduced by Waldmann [Wa58] [WT62] are then given by

$$\begin{cases} [\xi, \zeta]_{ij}' = \{\xi_i, \zeta_i - \zeta_i'\}_{ij}, \\ [\xi, \zeta]_{ij}'' = \{\xi_i, \tilde{\zeta}_j - \tilde{\zeta}_j'\}_{ij}, \end{cases} \tag{2.2.22}$$

where, as usual, $\zeta_i' = \zeta_i(c_i', I')$, $\tilde{\zeta}_j = \zeta_j(\tilde{c}_j, J)$, and $\tilde{\zeta}_j' = \zeta_j(\tilde{c}_j', J')$. From these relations, one may check that

$$[\xi, \zeta] = \sum_{i,j\in S} n_i n_j \big([\xi,\zeta]_{ij}' + [\xi,\zeta]_{ij}''\big), \tag{2.2.23}$$

and

$$\begin{cases} [\xi,\zeta]_{ij}' = [\zeta,\xi]_{ij}' = \frac{1}{2}\{\xi_i - \xi_i', \zeta_i - \zeta_i'\}_{ij}, \\ [\xi,\zeta]_{ij}'' = [\zeta,\xi]_{ji}'' = \frac{1}{2}\{\xi_i - \xi_i', \tilde{\zeta}_j - \tilde{\zeta}_j'\}_{ij}. \end{cases} \tag{2.2.24}$$

An important consequence of (2.2.23) is that if the basis functions are "localized with respect to the species", i.e., satisfy the property

$$\xi_i^{rk} = 0, \quad \text{for} \ \ i \neq k, \tag{2.2.25}$$

we then have

$$\begin{cases} [\xi^{rk}, \xi^{sk}] = \sum_{l\in S} n_k n_l [\xi^{rk}, \xi^{sk}]_{kl}' + n_k^2 [\xi^{rk}, \xi^{sk}]_{kk}'', \qquad (r,k),(s,k) \in \mathcal{B}^\mu, \\ [\xi^{rk}, \xi^{sl}] = n_k n_l [\xi^{rk}, \xi^{sl}]_{kl}'', \qquad (r,k),(s,l) \in \mathcal{B}^\mu, \quad k \neq l, \end{cases} \tag{2.2.26}$$

so that only the partial brackets (2.2.22) need to be evaluated. These relations also show that the coefficients $G_{kl}^{rs} = [\xi^{rk}, \xi^{sl}]$ are quadratic functions of the number densities $(n_i)_{i\in S}$ since the partial brackets $[\xi^{rk}, \xi^{sk}]_{kl}'$ and $[\xi^{rk}, \xi^{sl}]_{kl}''$ are independent of $(n_i)_{i\in S}$ and only depend on the temperature. Note that the partial brackets $[\xi^{rk}, \xi^{sk}]_{kl}'$ are defined for $(r,k),(s,k) \in \mathcal{B}^\mu$ and $l \in S$, whereas the partial brackets $[\xi^{rk}, \xi^{sl}]_{kl}''$ are defined for $(r,k),(s,l) \in \mathcal{B}^\mu$.

Since Waldmann and Trübenbacher only derived a formal theory, the calculation of the partial bracket products has been performed for the basis functions (2.2.17)–(2.2.20). The corresponding expressions are given in Appendix C in terms of collision integrals. These expressions have been compared formally with the results of Köhler and 't Hooft [MBKK91] obtained for linear molecules in a fully quantum mechanical framework, and the agreement has been found to be complete.

Finally, the right member coefficients $\beta_k^{r\mu} = \langle\langle \xi^{rk}, \Psi^\mu \rangle\rangle$ of the linear systems (2.2.8) are easily evaluated by expressing the functions Ψ^μ in terms of the basis functions (2.2.17)-(2.2.20) and by using the associated orthogonality properties (2.2.16) [WT62] and the corresponding scalar products $\langle\langle \phi^{a0cdk}, \phi^{a0cdk} \rangle\rangle$ given in Appendix B. The linear constraint coefficients $\mathcal{G}_k^{rl\nu} = \langle\langle \xi^{rk}, T_\nu \widehat{\psi}^l \rangle\rangle = \langle\langle \xi_\nu^{rk}, \widehat{\psi}^l \rangle\rangle$ are also easily calculated since the collisional invariants $\widehat{\psi}^l$, $l \in [1, n{+}4]$, are linear combinations of the basis functions ϕ^{00rk}, $r = 00, 10, 01$, and of the components ϕ_ν^{1000k}, $\nu = 1,2,3$, of ϕ^{1000k}. The explicit formulas for the transport linear systems (2.2.8)(2.2.12) corresponding to each transport coefficient are given in Sections 2.4–2.8.

2.2.4 Collision Integrals

The partial bracket products given in Appendix C are expressed in terms of collision integrals. These collision integrals are defined from the averaging operator $[\![\ ]\!]$ given by

$$[\![\alpha]\!]_{ij} = \Big(\frac{k_{\rm B}\overline{T}}{2\pi m_{ij}}\Big)^{1/2} \sum_{\substack{I,I'\in\mathcal{E}_i \\ J,J'\in\mathcal{E}_j}} \frac{a_{iI}a_{jJ}}{Q_iQ_j} \iiint \alpha\gamma^3 \exp(-\gamma^2-\epsilon_{iI}-\epsilon_{jJ})\bar{\sigma}_{ij}^{IJI'J'} \sin\chi d\chi d\varphi d\gamma, \quad (2.2.27)$$

where $\gamma = g(m_{ij}/2k_{\rm B}\overline{T})^{1/2}$ is integrated over $(0,+\infty)$, χ over $(0,\pi)$, φ over $(0,2\pi)$, and where α stands for any function of γ, $\gamma' = g'(m_{ij}/2k_{\rm B}\overline{T})^{1/2}$, χ, φ, ϵ_{iI}, $\epsilon_{iI'}$, ϵ_{jJ}, and $\epsilon_{jJ'}$, keeping the notation of Section 2.1.2. Notice also that collision integrals only depend on the temperature and that the averaging operator is equivalent to the one introduced in [MYM63]. In particular, the following collision integrals will be used in the next sections

$$\begin{cases} \Omega_{ij}^{(1,1)} = [\![\gamma^2 - \gamma\gamma'\cos\chi]\!]_{ij}, \\ \Omega_{ij}^{(2,2)} = [\![\gamma^4 - \gamma^2\gamma'^2\cos^2\chi - \frac{1}{6}(\Delta\epsilon_{ij})^2]\!]_{ij}, \\ \Omega_{ij}^{(1,2)} = [\![\gamma^4 - \gamma^3\gamma'\cos\chi]\!]_{ij}, \\ \Omega_{ij}^{(1,3)} = [\![\gamma^6 - \gamma^3\gamma'^3\cos\chi]\!]_{ij}, \end{cases} \quad (2.2.28)$$

and generalize the usual monatomic integrals. In these relations, we have used the notation

$$\begin{cases} \Delta\epsilon_{ij} = \Delta\epsilon_i + \widetilde{\Delta}\epsilon_j, \\ \Delta\epsilon_i = \epsilon_{iI'} - \epsilon_{iI}, \\ \widetilde{\Delta}\epsilon_j = \epsilon_{jJ'} - \epsilon_{jJ}, \end{cases} \quad (2.2.29)$$

where the extra superscript $\tilde{}$ is again used to distinguish one of the collision partner from the other in the case where i and j are the same, so that for $i=j$ we have

$$\begin{cases} \widetilde{\Delta}\epsilon_i = \epsilon_{iJ'} - \epsilon_{iJ}, \\ \Delta\epsilon_{ii} = \Delta\epsilon_i + \widetilde{\Delta}\epsilon_i. \end{cases}$$

In the next sections, we will also need the usual quantities

$$\begin{cases} \bar{A}_{ij} = \dfrac{1}{2}\dfrac{\Omega_{ij}^{(2,2)}}{\Omega_{ij}^{(1,1)}}, \\[2ex] \bar{B}_{ij} = \dfrac{1}{3}\dfrac{5\Omega_{ij}^{(1,2)} - \Omega_{ij}^{(1,3)}}{\Omega_{ij}^{(1,1)}}, \\[2ex] \bar{C}_{ij} = \dfrac{1}{3}\dfrac{\Omega_{ij}^{(1,2)}}{\Omega_{ij}^{(1,1)}}, \end{cases} \tag{2.2.30}$$

and the binary diffusion coefficient $\mathcal{D}_{ij}$ of the species pair (i,j) which is given by

$$\mathcal{D}_{ij} = \frac{3k_B\overline{T}}{16nm_{ij}}\frac{1}{\Omega_{ij}^{(1,1)}}. \tag{2.2.31}$$

We will also use the following diffusion coefficients for internal energy [MPM65]

$$\mathcal{D}_{i\ \mathrm{int},j} = \frac{3k_B\overline{T}}{16nm_{ij}}\frac{1}{\Omega_{i\ \mathrm{int},j}^{(1,1)}}, \tag{2.2.32}$$

with

$$\begin{cases} \dfrac{c_i^{\mathrm{int}}}{k_B}\Omega_{i\ \mathrm{int},i}^{(1,1)} = \left[\epsilon_{iI}^0\left((\epsilon_{iI}^0 - \epsilon_{iJ}^0)\gamma^2 - (\epsilon_{iI'}^0 - \epsilon_{iJ'}^0)\gamma\gamma'\cos\chi\right)\right]_{ii}, & i \in \mathcal{P}, \\[2ex] \dfrac{c_i^{\mathrm{int}}}{k_B}\Omega_{i\ \mathrm{int},j}^{(1,1)} = \left[\epsilon_{iI}^0(\epsilon_{iI}^0\gamma^2 - \epsilon_{iI'}^0\gamma\gamma'\cos\chi)\right]_{ij}, & i \in \mathcal{P},\quad j \in \mathcal{S},\quad i \neq j, \end{cases} \tag{2.2.33}$$

where c_i^{int} denotes the internal heat capacity of the molecules of the ith species

$$c_i^{\mathrm{int}} = k_B \sum_{I\in\mathcal{E}_i} a_{iI}(\epsilon_{iI} - \bar{\epsilon}_i)^2 \exp(-\epsilon_{iI})/Q_i, \tag{2.2.34}$$

so that $c^{\mathrm{int}} = \sum_{i\in\mathcal{S}} X_i c_i^{\mathrm{int}}$ [WT62], and where

$$\epsilon_{iI}^0 = \epsilon_{iI} - \bar{\epsilon}_i, \tag{2.2.35}$$

is a shifted reduced internal energy.

2.3 Mathematical Structure of the Transport Linear Systems

The purpose of this section is to obtain the mathematical structure of the transport linear systems (2.2.8)(2.2.12) directly from the properties of the Boltzmann equation (2.1.1), the integral bracket operator (2.1.29), the functional space $\mathcal{A}^\mu$, the basis functions ξ^{rk}, $(r,k) \in \mathcal{B}^\mu$, and the tensorial collisional invariants $\mathcal{T}_\nu\widehat{\psi}^l$, $(l,\nu) \in [1,n+4]\times[1,\tau_\mu]$.

These results are of fundamental importance for the iterative theory of the transport linear systems. In addition, the notation that is introduced will be used throughout the book. Therefore, readers only interested in practical applications should at least read the notation and the summary of results of Sections 2.3.5 and 2.3.6 before skipping to Sections 2.4–2.10.

2.3.1 Notation and Block-Structure of the Transport Linear Systems

General Notation. For a vector $x \in \mathbb{R}^\nu$, we denote by $x = (x_i)_{i\in[1,\nu]}$ its components and by $\mathbb{R}x$ the subspace $\mathrm{span}(x) = \{\, tx;\ t \in \mathbb{R}\,\}$. The canonical basis of $\mathbb{R}^\nu$ is also denoted by e^i, $i \in [1,\nu]$, so that $x = \sum_{i\in[1,\nu]} x_i e^i$. For $x,y \in \mathbb{R}^\nu$, $\langle x,y\rangle$ denotes the scalar product $\langle x,y\rangle = \sum_{i\in[1,\nu]} x_i y_i$. Recall that the notation "$\cdot$" is restricted to three-dimensional scalar products between physical variables in this book. For $x \in \mathbb{R}^\nu$, $x \neq 0$, we denote by $x^\perp$ the subspace $x^\perp = \{\, y \in \mathbb{R}^\nu;\ \langle x,y\rangle = 0\,\}$. If S_1 and S_2 are two complementary subspaces of $\mathbb{R}^\nu$, i.e., $S_1 \oplus S_2 = \mathbb{R}^\nu$, we denote by P_{S_1,S_2} the oblique projector onto the subspace S_1 along the subspace S_2. Finally, if each component of a vector $x \in \mathbb{R}^\nu$ is nonnegative (positive) we shall write $x \geq 0$ ($x > 0$).

We denote by $\mathbb{R}^{\nu_1,\nu_2}$ the set of matrices with ν_1 rows and ν_2 columns. For $A \in \mathbb{R}^{\nu_1,\nu_2}$, we write $A = (A_{ij})_{i\in[1,\nu_1],j\in[1,\nu_2]}$ the coefficients of the matrix A and A^t the transpose of A. The nullspace—or kernel—and the range of A are denoted by $N(A)$ and $R(A)$, respectively, and the dimension of the range of A is referred to as the rank of A and is denoted by $\mathrm{rank}(A)$. For $u,v \in \mathbb{R}^\nu$, $u\otimes v$ denotes the tensor product matrix $u\otimes v = (u_i v_j)_{i,j\in[1,\nu]} \in \mathbb{R}^{\nu,\nu}$. The identity matrix is denoted by I, $\mathrm{diag}(\lambda_1,\ldots,\lambda_\nu)$ is the diagonal matrix with diagonal elements $\lambda_1,\ldots,\lambda_\nu$. For a vector $x \in \mathbb{R}^\nu$, we use the notation $\mathrm{diag}\big((x_i)_{i\in[1,\nu]}\big) = \mathrm{diag}(x_1,\ldots,x_\nu)$, and for a matrix $A \in \mathbb{R}^{\nu_1,\nu_2}$ not necessarily being square, we define $\mathrm{diag}(A) \in \mathbb{R}^{\nu_1,\nu_2}$ by $\big(\mathrm{diag}(A)\big)_{ij} = A_{ij}\delta_{ij}$, $(i,j) \in [1,\nu_1]\times[1,\nu_2]$, where δ_{ij} is the Kronecker symbol.

Block-Decomposition of Vectors and Matrices. Let $\mathcal{A}^\mu = \mathrm{span}\{\ \xi^{rk},\ (r,k) \in \mathcal{B}^\mu\ \}$ be the finite dimensional functional space used in the variational procedure, where

ξ^{rk}, $(r,k) \in \mathcal{B}^\mu$, are the basis functions, and $\mathcal{B}^\mu$ is the set of basis function indices. As stated previously, $\mathcal{B}^\mu$ is a subset of $\mathcal{F} \times \mathcal{S}$ where $\mathcal{F}$ denotes the set of function type indices and $\mathcal{S} = [1,n]$ the set of species indices, with $n \geq 1$ unless explicitly stated. We have already defined the subsets $\mathcal{S}_r = \{\, k \in \mathcal{S},\ (r,k) \in \mathcal{B}^\mu \,\}$, for $r \in \mathcal{F}$, and $\mathcal{F}_k = \{\, r \in \mathcal{F},\ (r,k) \in \mathcal{B}^\mu \,\}$, for $k \in \mathcal{S}$, such that (2.2.4) holds. We also restate that ω_r denotes the number of elements of $\mathcal{S}_r$ and $\omega = \sum_{r \in \mathcal{F}} \omega_r$ so that the set $\mathcal{B}^\mu$ has ω elements, and the functional space $\mathcal{A}^\mu$ has dimension ω.

The components with respect to the basis ξ^{rk}, $(r,k) \in \mathcal{B}^\mu$, of the functions $\Xi = \sum_{(r,k)\in\mathcal{B}^\mu} x_k^r \xi^{rk}$ of $\mathcal{A}^\mu$ now form a vector of $\mathbb{R}^\omega$ denoted by $x = (x_k^r)_{(r,k)\in\mathcal{B}^\mu}$. Ordering the set $\mathcal{B}^\mu$ with the lexicographical order, the components of any vector $x \in \mathbb{R}^\omega$ are correspondingly denoted by $x = (x_k^r)_{(r,k)\in\mathcal{B}^\mu}$, thereby identifying $\mathbb{R}^\omega$ and $\mathbb{R}^{\mathcal{B}^\mu}$. The set $\mathcal{B}^\mu$ can then be used as a natural indexing set. In particular, the canonical basis of $\mathbb{R}^\omega$ is denoted by e^{rk}, $(r,k) \in \mathcal{B}^\mu$, so that

$$x = \sum_{(r,k)\in\mathcal{B}^\mu} x_k^r e^{rk}.$$

For $x, y \in \mathbb{R}^\omega$, the scalar product $\langle x, y\rangle$ is also given by $\langle x, y\rangle = \sum_{(r,k)\in\mathcal{B}^\mu} x_k^r y_k^r$. Since $\mathcal{B}^\mu$ is a subset of $\mathcal{F} \times \mathcal{S}$, we can further introduce the block-decomposition of a vector $x \in \mathbb{R}^\omega$, $x = (x_k^r)_{(r,k)\in\mathcal{B}^\mu}$, by defining the vectors $x^r = (x_k^r)_{k\in\mathcal{S}_r}$, $r \in \mathcal{F}$, so that $x^r \in \mathbb{R}^{\omega_r}$, $r \in \mathcal{F}$, and $x = (x^r)_{r\in\mathcal{F}}$.

For $A \in \mathbb{R}^{\omega,\omega}$, we write $A = (A_{kl}^{rs})_{(r,k),(s,l)\in\mathcal{B}^\mu}$ the coefficients of the matrix A. If $x \in \mathbb{R}^\omega$, $x = (x_k^r)_{(r,k)\in\mathcal{B}^\mu}$, then $\mathrm{diag}\big((x_k^r)_{(r,k)\in\mathcal{B}^\mu}\big)$ is the diagonal matrix of $\mathbb{R}^{\omega,\omega}$ whose diagonal elements are x_k^r, $(r,k) \in \mathcal{B}^\mu$, ordered as $\mathcal{B}^\mu$. Similarly, the matrix $y \otimes z$ is now given by $y \otimes z = (y_k^r z_l^s)_{(r,k),(s,l)\in\mathcal{B}^\mu}$. Furthermore, a matrix $A = (A_{kl}^{rs})_{(r,k),(s,l)\in\mathcal{B}^\mu}$ in $\mathbb{R}^{\omega,\omega}$ can be partitioned into the blocks $A^{rs} = (A_{kl}^{rs})_{k\in\mathcal{S}_r,\, l\in\mathcal{S}_s}$ of size $\omega_r * \omega_s$, $r, s \in \mathcal{F}$, so that $A = (A^{rs})_{r,s\in\mathcal{F}}$. We then denote by $db(A) \in \mathbb{R}^{\omega,\omega}$ the matrix formed by the diagonals of all the rectangular blocks A^{rs}, $r, s \in \mathcal{F}$, of A, so that $db(A)_{kl}^{rs} = A_{kl}^{rs}\delta_{kl}$, $(r,k),(s,l) \in \mathcal{B}^\mu$. Notice that in the special case $n = 1$ and $\mathcal{S} = \{1\}$, we have $db(A) = A$ for any matrix $A \in \mathbb{R}^{\omega,\omega}$.

Block-Decomposition of the Transport Linear Systems. Making use of the new formalism, the relations (2.2.8) can now be written in the compact form

$$G\alpha^\mu = \beta^\mu, \tag{2.3.1}$$

where we have defined

$$\begin{cases} \alpha^\mu = (\alpha_k^{r\mu})_{(r,k)\in\mathcal{B}^\mu} \in \mathbb{R}^\omega, \\ \beta^\mu = (\beta_k^{r\mu})_{(r,k)\in\mathcal{B}^\mu} \in \mathbb{R}^\omega, \end{cases} \tag{2.3.2}$$

and

$$G = (G_{kl}^{rs})_{(r,k),(s,l)\in\mathcal{B}^\mu} \in \mathbb{R}^{\omega,\omega}. \tag{2.3.3}$$

Similarly, we introduce the constraint vectors $\mathcal{G}^{l\nu} \in \mathbb{R}^\omega$, $(l,\nu) \in [1,n+4]\times[1,\tau_\mu]$, defined by

$$\mathcal{G}^{l\nu} = \left(\mathcal{G}_k^{rl\nu}\right)_{(r,k)\in\mathcal{B}^\mu}, \tag{2.3.4}$$

where $\mathcal{G}_k^{rl\nu} = \langle\!\langle \xi^{rk}, \mathcal{T}_\nu \widehat{\psi}^l \rangle\!\rangle$, so that the linear constraints (2.2.12) can be written as

$$\langle \mathcal{G}^{l\nu}, \alpha^\mu \rangle = 0, \qquad (l,\nu) \in [1,n+4]\times[1,\tau_\mu]. \tag{2.3.5}$$

Note that most of the constraint vectors $\mathcal{G}^{l\nu}$ are usually zero thanks to the properties of the collisional invariants and of the basis functions, as detailed in the next sections. The products $[\phi^\mu,\phi^\mu]$ can now be written in terms of $\langle \alpha^\mu, \beta^\mu \rangle$ since

$$[\phi^\mu,\phi^\mu] = \langle\!\langle \Psi^\mu, \phi^\mu \rangle\!\rangle = \langle \alpha^\mu, \beta^\mu \rangle. \tag{2.3.6}$$

Introducing then, for $r \in \mathcal{F}$, the vectors

$$\begin{cases} \alpha^{r\mu} = (\alpha_k^{r\mu})_{k\in\mathcal{S}_r}, \\ \beta^{r\mu} = (\beta_k^{r\mu})_{k\in\mathcal{S}_r}, \\ \mathcal{G}^{rl\nu} = (\mathcal{G}_k^{rl\nu})_{k\in\mathcal{S}_r}, \qquad (l,\nu) \in [1,n+4]\times[1,\tau_\mu], \end{cases} \tag{2.3.7}$$

so that $\alpha^\mu = (\alpha^{r\mu})_{r\in\mathcal{F}}$, $\beta^\mu = (\beta^{r\mu})_{r\in\mathcal{F}}$, and $\mathcal{G}^{l\nu} = (\mathcal{G}^{rl\nu})_{r\in\mathcal{F}}$, $(l,\nu) \in [1,n+4]\times[1,\tau_\mu]$, and the blocks

$$G^{rs} = (G_{kl}^{rs})_{k\in\mathcal{S}_r,\, l\in\mathcal{S}_s}, \tag{2.3.8}$$

of size $\omega_r{*}\omega_s$, $r,s \in \mathcal{F}$, we can also write the system (2.3.1)(2.3.5) in the block-form

$$\sum_{s\in\mathcal{F}} G^{rs}\alpha^{s\mu} = \beta^{r\mu}, \qquad r \in \mathcal{F}, \tag{2.3.9}$$

with the block-constraints

$$\sum_{r\in\mathcal{F}} \langle \mathcal{G}^{rl\nu}, \alpha^{r\mu} \rangle = 0, \qquad (l,\nu) \in [1,n+4]\times[1,\tau_\mu], \tag{2.3.10}$$

and the products $[\phi^\mu,\phi^\mu]$ can now be written in the block-form

$$[\phi^\mu,\phi^\mu] = \langle \alpha^\mu, \beta^\mu \rangle = \sum_{r\in\mathcal{F}} \langle \alpha^{r\mu}, \beta^{r\mu} \rangle. \tag{2.3.11}$$

This block structure will be often used in this book.

2.3.2 Properties of G and β^μ

Symmetry. The symmetry of the matrix $G \in \mathbb{R}^{\omega,\omega}$ is first deduced from the symmetry of the bracket operator (2.1.30) and from (2.2.9) since

$$G_{kl}^{rs} = [\xi^{rk}, \xi^{sl}] = [\xi^{sl}, \xi^{rk}] = G_{lk}^{sr}, \tag{2.3.12}$$

so that $G = G^t$.

Positive Definiteness. We now show that G is positive semi-definite on $\mathbb{R}^\omega$ and positive definite on the "constrained" subspace C of $\mathbb{R}^\omega$ defined by

$$C = \Big(\text{span}\{\ \mathcal{G}^{l\nu},\ (l,\nu) \in [1,n{+}4]{\times}[1,\tau_\mu]\ \}\Big)^\perp. \tag{2.3.13}$$

Notice that $x \in C$ if and only if x is orthogonal to the constraints vectors $\mathcal{G}^{l\nu}$, $(l,\nu) \in [1,n{+}4]{\times}[1,\tau_\mu]$, so that (2.3.5) is equivalent to $\alpha^\mu \in C$. Let now $x \in \mathbb{R}^\omega$, $x = (x_k^r)_{(r,k)\in\mathcal{B}^\mu}$, let Ξ be its associated function $\Xi = \sum_{(r,k)\in\mathcal{B}^\mu} x_k^r \xi^{rk}$, and consider the scalar product $\langle Gx, x\rangle$. From (2.2.9) we have

$$\langle Gx, x\rangle = \sum_{\substack{(r,k)\in\mathcal{B}^\mu\\(s,l)\in\mathcal{B}^\mu}} G_{kl}^{rs} x_k^r x_l^s = \sum_{\substack{(r,k)\in\mathcal{B}^\mu\\(s,l)\in\mathcal{B}^\mu}} [\xi^{rk}, \xi^{sl}] x_k^r x_l^s = [\Xi, \Xi], \tag{2.3.14}$$

which is nonnegative so that G is positive semi-definite. Moreover we have $\langle Gx, x\rangle = 0$ if and only if Ξ is a collisional invariant, that is, $\Xi \in \mathcal{I}_S^{a_\mu}$. Keeping in mind that G is positive semi-definite, we then have $Gx = 0$ if and only if $\langle Gx, x\rangle = 0$, so that the nullspace of G is spanned by the components in the basis ξ^{rk}, $(r,k) \in \mathcal{B}^\mu$, of the collisional invariants that are in $\mathcal{A}^\mu$

$$N(G) = \{\ x \in \mathbb{R}^\omega,\ \sum_{(r,k)\in\mathcal{B}^\mu} x_k^r \xi^{rk} \in \mathcal{I}_S^{a_\mu} \cap \mathcal{A}^\mu\ \}. \tag{2.3.15}$$

As stated previously, $\mathcal{I}_S^{a_\mu}$ is the space of collisional invariants of the mixture that are of tensorial rank a_μ over $\mathbb{R}^3$

$$\mathcal{I}_S^{a_\mu} = \text{span}\{\ \mathcal{T}_\nu \widehat{\psi}^l,\ (l,\nu) \in [1,n{+}4]{\times}[1,\tau_\mu]\ \}. \tag{2.3.16}$$

As a consequence, for $x \in N(G)$ and $\Xi = \sum_{(r,k)\in\mathcal{B}^\mu} x_k^r \xi^{rk}$, there exist scalars $u_{l\nu}$, $(l,\nu) \in [1,n{+}4]{\times}[1,\tau_\mu]$, such that

$$\Xi = \sum_{\substack{l\in[1,n+4]\\ \nu\in[1,\tau_\mu]}} u_{l\nu} \mathcal{T}_\nu \widehat{\psi}^l,$$

which implies that

$$\langle\!\langle \Xi, \Xi \rangle\!\rangle = \sum_{\substack{l\in[1,n+4] \\ \nu\in[1,\tau_\mu]}} u_{l\nu} \langle x, \mathcal{G}^{l\nu} \rangle,$$

since $\langle\!\langle \Xi, \mathcal{T}_\nu \widehat{\psi}^l \rangle\!\rangle = \langle x, \mathcal{G}^{l\nu} \rangle$. Therefore $x \in N(G) \cap \mathcal{C}$ implies that $\Xi = 0$ and $x = 0$ so that

$$N(G) \cap \mathcal{C} = \{0\}, \tag{2.3.17}$$

and hence we obtain that G is positive definite on $\mathcal{C}$.

Right Member. We now claim that β^μ is in the range of G. One may first easily establish [WT62] from (2.1.18)(2.1.25) that

$$\langle\!\langle \Psi^\mu, \widehat{\psi}^l \rangle\!\rangle = 0, \qquad l \in [1, n+4], \tag{2.3.18}$$

which is easily shown to be equivalent to

$$\langle\!\langle \Psi^\mu, \mathcal{T}_\nu \widehat{\psi}^l \rangle\!\rangle = 0, \qquad (l,\nu) \in [1, n+4] \times [1, \tau_\mu].$$

For $x \in N(G)$, we deduce from (2.3.15) that

$$\sum_{(r,k)\in\mathcal{B}^\mu} x_k^r \xi^{rk} = \sum_{\substack{l\in[1,n+4] \\ \nu\in[1,\tau_\mu]}} u_{l\nu} \mathcal{T}_\nu \widehat{\psi}^l,$$

for scalars $u_{l\nu}$, $(l,\nu) \in [1, n+4] \times [1, \tau_\mu]$, so that

$$\begin{aligned} \langle x, \beta^\mu \rangle &= \sum_{(r,k)\in\mathcal{B}^\mu} x_k^r \, \langle\!\langle \Psi^\mu, \xi^{rk} \rangle\!\rangle = \langle\!\langle \Psi^\mu, \sum_{(r,k)\in\mathcal{B}^\mu} x_k^r \xi^{rk} \rangle\!\rangle \\ &= \langle\!\langle \Psi^\mu, \sum_{\substack{l\in[1,n+4] \\ \nu\in[1,\tau]}} u_{l\nu} \mathcal{T}_\nu \widehat{\psi}^l \rangle\!\rangle = \sum_{\substack{l\in[1,n+4] \\ \nu\in[1,\tau]}} u_{l\nu} \langle\!\langle \Psi^\mu, \mathcal{T}_\nu \widehat{\psi}^l \rangle\!\rangle = 0. \end{aligned}$$

Hence we have $\beta^\mu \in N(G)^\perp$ so that $\beta^\mu \in R(G)$ since G is symmetric.

2.3.3 Properties of the Constrained Subspace $\mathcal{C}$

Well Posedness. We will establish in Chapter 4 that the linear system (2.3.1)(2.3.5) is well posed, i.e., admits a unique solution α^μ for any $\beta^\mu \in R(G)$, if and only if the kernel of G and the constrained subspace $\mathcal{C} = \big(\mathrm{span}\{\, \mathcal{G}^{l\nu},\ (l,\nu) \in [1, n+4] \times [1, \tau_\mu] \,\}\big)^\perp$ are complementary spaces

$$N(G) \oplus \mathcal{C} = \mathbb{R}^\omega. \tag{2.3.19}$$

As a consequence, the property (2.3.19) will be termed "the well posedness condition".

The Perpendicularity Property. We now establish that the well posedness condition (2.3.19) holds if and only if the space $\mathcal{I}_S^{a_\mu}$ is perpendicular to $\mathcal{A}^\mu$

$$\mathcal{I}_S^{a_\mu} = \mathcal{I}_S^{a_\mu} \cap \mathcal{A}^\mu \quad \oplus \quad \mathcal{I}_S^{a_\mu} \cap \mathcal{A}^{\mu\perp}, \tag{2.3.20}$$

where $\mathcal{I}_S^{a_\mu} \cap \mathcal{A}^{\mu\perp}$ denotes the elements of $\mathcal{I}_S^{a_\mu}$ that are orthogonal to $\mathcal{A}^\mu$ with respect to the bilinear form $\langle\!\langle\, , \,\rangle\!\rangle$

$$\mathcal{I}_S^{a_\mu} \cap \mathcal{A}^{\mu\perp} = \{\, \xi \in \mathcal{I}_S^{a_\mu}, \;\; \forall \zeta \in \mathcal{A}^\mu \;\; \langle\!\langle \xi, \zeta \rangle\!\rangle = 0 \,\}. \tag{2.3.21}$$

The property (2.3.20) will be termed "the perpendicularity property" in the following sections. This property will be satisfied for all the particular cases that will be considered.

We first establish that the well posedness condition (2.3.19) is a consequence of the perpendicularity property (2.3.20). For $x = (x_k^r)_{(r,k)\in\mathcal{B}^\mu} \in \mathbb{R}^\omega$ and $\Xi = \sum_{(r,k)\in\mathcal{B}^\mu} x_k^r \xi^{rk}$, we first note that

$$\begin{aligned} x \in C &\iff \forall (l,\nu) \in [1, n{+}4]\times[1, \tau_\mu] \;\; \langle \mathcal{G}^{l\nu}, x \rangle = 0, \\ &\iff \forall (l,\nu) \in [1, n{+}4]\times[1, \tau_\mu] \;\; \langle\!\langle \mathcal{T}_\nu \widehat{\psi}^l, \Xi \rangle\!\rangle = 0, \\ &\iff \forall \zeta \in \mathcal{I}_S^{a_\mu} \;\; \langle\!\langle \zeta, \Xi \rangle\!\rangle = 0, \end{aligned}$$

so that (2.3.20) implies that

$$x \in C \iff \forall \zeta \in \mathcal{I}_S^{a_\mu} \cap \mathcal{A}^\mu, \quad \langle\!\langle \zeta, \Xi \rangle\!\rangle = 0. \tag{2.3.22}$$

Therefore, C has the same dimension as the subspace of $\mathcal{A}^\mu$ which is orthogonal to $\mathcal{I}_S^{a_\mu} \cap \mathcal{A}^\mu$ with respect to the positive definite quadratic form $\langle\!\langle\, , \,\rangle\!\rangle$, so that

$$\dim(C) = \omega - \dim(\mathcal{I}_S^{a_\mu} \cap \mathcal{A}^\mu).$$

On the other hand, from (2.3.15), we already know that

$$\dim(N(G)) = \dim(\mathcal{I}_S^{a_\mu} \cap \mathcal{A}^\mu),$$

so that $\dim\big(N(G)\big) + \dim\big(C\big) = \omega$. Finally, we have $N(G) \cap C = \{0\}$ from (2.3.17) so that $N(G) \oplus C = \mathbb{R}^\omega$.

Conversely, we now establish that the perpendicularity property (2.3.20) is a consequence of the well posedness condition (2.3.19). Let $\zeta \in \mathcal{I}_S^{a_\mu}$ and consider the solution x of the linear system

$$\sum_{(r,k)\in\mathcal{B}^\mu} x_k^r \langle\!\langle \xi^{rk}, \xi^{sl} \rangle\!\rangle = \langle\!\langle \zeta, \xi^{sl} \rangle\!\rangle, \qquad (s,l) \in \mathcal{B}^\mu.$$

This solution x is well defined since the quadratic form $\langle\!\langle\ ,\ \rangle\!\rangle$ is positive definite so that the matrix in $\mathbb{R}^{\omega,\omega}$ with components $\langle\!\langle \xi^{rk}, \xi^{sl} \rangle\!\rangle$, $(r,k),(s,l) \in \mathcal{B}^\mu$, is nonsingular. On the other hand, letting now $\Xi = \sum_{(r,k)\in\mathcal{B}^\mu} x_k^r \xi^{rk}$ and $\zeta = \Xi + \mathcal{U}$, we have $\Xi \in \mathcal{A}^\mu$ and $\langle\!\langle \mathcal{U}, \xi^{rk} \rangle\!\rangle = 0$, $(r,k) \in \mathcal{B}^\mu$, by construction. From (2.3.19) we may now decompose $x \in \mathbb{R}^\omega$ into $x = y + z$ where $y \in N(G)$ and $z \in \mathcal{C}$. Thanks to (2.3.15), $y \in N(G)$ implies that $\sum_{(r,k)\in\mathcal{B}^\mu} y_k^r \xi^{rk} \in \mathcal{I}_S^{a_\mu} \cap \mathcal{A}^\mu$. As a consequence, we have

$$\Big\langle\!\!\Big\langle \sum_{(r,k)\in\mathcal{B}^\mu} z_k^r \xi^{rk}, \sum_{(r,k)\in\mathcal{B}^\mu} z_k^r \xi^{rk} \Big\rangle\!\!\Big\rangle = \Big\langle\!\!\Big\langle \zeta - \sum_{(r,k)\in\mathcal{B}^\mu} y_k^r \xi^{rk} - \mathcal{U}, \sum_{(r,k)\in\mathcal{B}^\mu} z_k^r \xi^{rk} \Big\rangle\!\!\Big\rangle = 0,$$

since $\zeta - \sum_{(r,k)\in\mathcal{B}^\mu} y_k^r \xi^{rk} \in \mathcal{I}_S^{a_\mu}$ and $z \in \mathcal{C}$, and since $\langle\!\langle \mathcal{U}, \xi^{rk} \rangle\!\rangle = 0$, $(r,k) \in \mathcal{B}^\mu$, by definition of $\mathcal{U}$. This shows that $z = 0$ and hence that $x = y \in N(G)$ so that $\Xi \in \mathcal{I}_S^{a_\mu} \cap \mathcal{A}^\mu$ and hence $\mathcal{U} = \zeta - \Xi \in \mathcal{I}_S^{a_\mu}$ so that $\mathcal{U} \in \mathcal{I}_S^{a_\mu} \cap \mathcal{A}^{\mu\perp}$. We have thus shown that any $\zeta \in \mathcal{I}_S^{a_\mu}$ can we written in the form $\zeta = \Xi + \mathcal{U}$ where $\Xi \in \mathcal{I}_S^{a_\mu} \cap \mathcal{A}^\mu$ and $\mathcal{U} \in \mathcal{I}_S^{a_\mu} \cap \mathcal{A}^{\mu\perp}$, and since such a decomposition is obviously unique, we have established that (2.3.20) holds.

Remark. Notice that under the perpendicularity property (2.3.20), the $(n+4)\tau_\mu$ constraints $\langle \mathcal{G}^{l\nu}, x \rangle = 0$, $(l,\nu) \in [1,n+4]\times[1,\tau_\mu]$, can be simplified by using any set of functions spanning the subspace $\mathcal{I}_S^{a_\mu} \cap \mathcal{A}^\mu$ since from (2.3.22) we have

$$x \in \mathcal{C} \iff \forall \zeta \in \mathcal{I}_S^{a_\mu} \cap \mathcal{A}^\mu, \quad \sum_{(r,k)\in\mathcal{B}^\mu} x_k^r \langle\!\langle \xi^{rk}, \zeta \rangle\!\rangle = 0. \tag{2.3.23}$$

Summary. We have established that whenever the perpendicularity property (2.3.20) holds, then

$$\begin{cases} G \ \text{ is symmetric positive semi-definite,} \\[2ex] N(G) = \{\, x \in \mathbb{R}^\omega, \ \displaystyle\sum_{(r,k)\in\mathcal{B}^\mu} x_k^r \xi^{rk} \in \mathcal{I}_S^{a_\mu} \cap \mathcal{A}^\mu \,\}, \\[2ex] G \ \text{ is positive definite on } \mathcal{C}, \\[2ex] N(G) \oplus \mathcal{C} = \mathbb{R}^\omega, \\[2ex] \beta^\mu \in R(G). \end{cases} \tag{2.3.24}$$

Notice that the results (2.3.24) have been derived from the properties of the spaces $\mathcal{A}^\mu$ and $\mathcal{I}_S^{a_\mu}$, and of the bilinear forms $[\ ,\ ]$ and $\langle\!\langle\ ,\ \rangle\!\rangle$, and do not depend on the particular choice of the basis functions ξ^{rk}, $(r,k) \in \mathcal{B}^\mu$.

2.3.4 Properties of $db(G)$ and $2db(G) - G$

The Matrix $db(G)$. In order to establish the convergence of various iterative algorithms in Chapter 5, we will need to know whether the matrices $db(G)$ and $2db(G) - G$ are positive definite on $\mathbb{R}^\omega$. As stated previously, the matrix $db(G)$ is composed of the diagonal of all the rectangular blocks G^{rs}, $r, s \in \mathcal{F}$, of G, so that

$$db(G)_{kl}^{rs} = G_{kl}^{rs}\delta_{kl}, \qquad (r,k),(s,l) \in \mathcal{B}^\mu.$$

Under sufficient conditions, we establish in this section that $2db(G) - G$ is positive definite when $n \geq 3$ and we identify its nullspace when $n = 2$ and $n = 1$. Similarly, we establish that $db(G)$ is positive definite when $n \geq 2$ and we identify its nullspace when $n = 1$.

The Species Localization Property. First note that we are now interested in some specific coefficients of the matrix G, that is to say, in the submatrix $db(G)$ and not in the full matrix G as in Section 2.3.3. As a consequence, the properties of $db(G)$ and $2db(G) - G$ will depend on the particular choice of the basis functions ξ^{rk}, $(r,k) \in \mathcal{B}^\mu$. Indeed, we now assume that the basis functions are "localized with respect to the species"

$$\xi_i^{rk} = 0 \qquad \text{for} \qquad i \neq k, \tag{2.3.25}$$

so that $\xi^{rk} = (\xi_k^{rk}\delta_{ki})_{i\in S}$. Under the assumption (2.3.25), we first establish that the matrix $2db(G) - G$ is positive semi-definite and we characterize its kernel.

Positive Semi-Definiteness of $2db(G) - G$. Recall that under assumption (2.3.25) we have established in Section 2.2.3 the fundamental relations

$$\begin{cases} G_{kk}^{rs} = \sum\limits_{l\in S} n_k n_l G_{kl}^{\prime rs} + n_k^2 G_{kk}^{\prime\prime rs}, & (r,k),(s,k) \in \mathcal{B}^\mu, \\ G_{kl}^{rs} = n_k n_l G_{kl}^{\prime\prime rs}, \quad (r,k),(s,l) \in \mathcal{B}^\mu, \quad k \neq l, \end{cases} \tag{2.3.26}$$

with

$$\begin{cases} G_{kl}^{\prime rs} = [\xi^{rk}, \xi^{sk}]_{kl}^{\prime}, & (r,k),(s,k) \in \mathcal{B}^\mu, \quad l \in S, \\ G_{kl}^{\prime\prime rs} = [\xi^{rk}, \xi^{sl}]_{kl}^{\prime\prime}, & (r,k),(s,l) \in \mathcal{B}^\mu. \end{cases} \tag{2.3.27}$$

As a consequence, for $x = (x_k^r)_{(r,k)\in\mathcal{B}^\mu}$, $\Xi = \sum_{(r,k)\in\mathcal{B}^\mu} x_k^r \xi^{rk}$, and $\Xi = (\Xi_k)_{k\in\mathcal{S}}$, the identity

$$\langle (2db(G) - G)x, x\rangle = \sum_{\substack{(r,k)\in\mathcal{B}^\mu\\(s,k)\in\mathcal{B}^\mu}} G_{kk}^{rs} x_k^r x_k^s - \sum_{\substack{(r,k)\in\mathcal{B}^\mu\\(s,l)\in\mathcal{B}^\mu\\k\neq l}} G_{kl}^{rs} x_k^r x_l^s$$

$$= \sum_{k\in\mathcal{S}} \sum_{\substack{r\in\mathcal{F}_k\\s\in\mathcal{F}_k}} G_{kk}^{rs} x_k^r x_k^s - \sum_{\substack{k,l\in\mathcal{S}\\k\neq l}} \sum_{\substack{r\in\mathcal{F}_k\\s\in\mathcal{F}_l}} G_{kl}^{rs} x_k^r x_l^s,$$

is easily transformed into

$$\langle (2db(G) - G)x, x\rangle = \sum_{k\in\mathcal{S}} \sum_{\substack{r\in\mathcal{F}_k\\s\in\mathcal{F}_k}} \Big(\sum_{l\in\mathcal{S}} n_k n_l [\xi^{rk}, \xi^{sk}]'_{kl} + n_k^2 [\xi^{rk}, \xi^{sk}]''_{kk}\Big) x_k^r x_k^s$$

$$- \sum_{\substack{k,l\in\mathcal{S}\\k\neq l}} \sum_{\substack{r\in\mathcal{F}_k\\s\in\mathcal{F}_l}} n_k n_l [\xi^{rk}, \xi^{sl}]''_{kl} x_k^r x_l^s.$$

The right member of this identity can then be rewritten in the form

$$\sum_{\substack{k,l\in\mathcal{S}\\k\neq l}} n_k n_l \bigg(\Big[\sum_{r\in\mathcal{F}_k} x_k^r \xi^{rk}, \sum_{s\in\mathcal{F}_k} x_k^s \xi^{sk}\Big]'_{kl} - \Big[\sum_{r\in\mathcal{F}_k} x_k^r \xi^{rk}, \sum_{s\in\mathcal{F}_l} x_l^s \xi^{sl}\Big]''_{kl}\bigg)$$

$$+ \sum_{k\in\mathcal{S}} n_k^2 \bigg(\Big[\sum_{r\in\mathcal{F}_k} x_k^r \xi^{rk}, \sum_{s\in\mathcal{F}_k} x_k^s \xi^{sk}\Big]'_{kk} + \Big[\sum_{r\in\mathcal{F}_k} x_k^r \xi^{rk}, \sum_{s\in\mathcal{F}_k} x_k^s \xi^{sk}\Big]''_{kk}\bigg),$$

so that

$$\langle (2db(G) - G)x, x\rangle = \sum_{\substack{k,l\in\mathcal{S}\\k\neq l}} n_k n_l \Big([\Xi, \Xi]'_{kl} - [\Xi, \Xi]''_{kl}\Big) + \sum_{k\in\mathcal{S}} n_k^2 \Big([\Xi, \Xi]'_{kk} + [\Xi, \Xi]''_{kk}\Big),$$

keeping in mind that $\sum_{r\in\mathcal{F}_k} x_k^r \xi^{rk}$ and Ξ have the same k^{th} component Ξ_k. Using the identities (2.2.24) then yields

$$[\Xi, \Xi]'_{kl} - [\Xi, \Xi]''_{kl} = \tfrac{1}{2}\{\Xi_k - \Xi'_k, \Xi_k - \Xi'_k - \widetilde{\Xi}_l + \widetilde{\Xi}'_l\}_{kl},$$

and

$$[\Xi, \Xi]'_{kk} + [\Xi, \Xi]''_{kk} = \tfrac{1}{2}\{\Xi_k - \Xi'_k, \Xi_k - \Xi'_k + \widetilde{\Xi}_k - \widetilde{\Xi}'_k\}_{kk}.$$

Making use of inverse collisions and symmetrizing, we finally obtain

$$\langle (2db(G) - G)x, x\rangle = \sum_{\substack{k,l\in\mathcal{S}\\k\neq l}} \tfrac{1}{4} n_k n_l \big\{\Xi_k - \Xi'_k - \widetilde{\Xi}_l + \widetilde{\Xi}'_l,\, \Xi_k - \Xi'_k - \widetilde{\Xi}_l + \widetilde{\Xi}'_l\big\}_{kl}$$

$$+ \sum_{k\in\mathcal{S}} \tfrac{1}{4} n_k^2 \big\{\Xi_k - \Xi'_k + \widetilde{\Xi}_k - \widetilde{\Xi}'_k,\, \Xi_k - \Xi'_k + \widetilde{\Xi}_k - \widetilde{\Xi}'_k\big\}_{kk}.$$

(2.3.28)

This is a sum of nonnegative terms, so that the symmetric matrix $2db(G)-G$ is positive semi-definite on $\mathbb{R}^\omega$ for any $n \geq 1$. As a comparison, it is interesting to obtain from (2.3.14), (2.1.29), and (2.2.21) that

$$\langle Gx, x\rangle = \sum_{k,l\in S} \tfrac{1}{4} n_k n_l \{\Xi_k - \Xi_k' + \widetilde{\Xi}_l - \widetilde{\Xi}_l', \Xi_k - \Xi_k' + \widetilde{\Xi}_l - \widetilde{\Xi}_l'\}_{kl}.$$

We then observe that the terms in the simple sum indexed by $k \in S$ in (2.3.28) are equal to the corresponding term in the expression of $\langle Gx, x\rangle$ whereas the terms in the double sum indexed by $k, l \in S$, $k \neq l$, differ in the sign of the contributions from the collision partner l.

Characterization of $N(2db(G)-G)$. By using the expression (2.3.28) we can now characterize the nullspace $N(2db(G)-G)$. Since for $n=1$ we have $db(G)=G$ and $2db(G)-G=G$, we only have to consider the particular case $n \geq 2$. In this situation, keeping in mind that the number densities are positive, we deduce from (2.3.28) that $x \in N(2db(G)-G)$ if and only if for all species pair (k,l), with $k \neq l$, the subfamily $(\Xi_k, -\Xi_l)$ is a collisional invariant of the binary submixture (k,l), which means that the collision invariance property (2.1.16) holds for $i, j \in \{k,l\}$ only. Denoting by $\mathcal{I}_{kl}^{a_\mu}$ the space of collisional invariants of the submixture (k,l) which are tensorial of order a_μ over $\mathbb{R}^3$, we have thus established that for $n \geq 2$

$$N(2db(G)-G) = \{\, x \in \mathbb{R}^\omega,\ \forall k,l \in S,\ k \neq l,\ (\Xi_k, -\Xi_l) \in \mathcal{I}_{kl}^{a_\mu} \,\}. \tag{2.3.29}$$

As a comparison it is interesting to note that for $n \geq 2$ we also have

$$N(G) = \{\, x \in \mathbb{R}^\omega,\ \forall k,l \in S,\ k \neq l,\ (\Xi_k, \Xi_l) \in \mathcal{I}_{kl}^{a_\mu} \,\}.$$

The space $\mathcal{I}_{kl}^{a_\mu}$ is spanned by the functions $\mathcal{T}_\nu \widehat{\psi}_{kl}^m$, $(m,\nu) \in (\{k,l\} \cup [n+1, n+4]) \times [1, \tau_\mu]$,

$$\mathcal{I}_{kl}^{a_\mu} = \mathrm{span}\{\, \mathcal{T}_\nu \widehat{\psi}_{kl}^m,\ (m,\nu) \in (\{k,l\} \cup [n+1, n+4]) \times [1, \tau_\mu] \,\}, \tag{2.3.30}$$

where $\widehat{\psi}_{kl}^m$, $m \in \{k,l\} \cup [n+1, n+4]$, are the scalar collisional invariants of the submixture (k,l). These collisional invariants are related to the full mixture collisional invariants $\widehat{\psi}^m$, $m \in \{k,l\} \cup [n+1, n+4]$, since we have

$$\widehat{\psi}_{kl}^m = (\widehat{\psi}_k^m, \widehat{\psi}_l^m), \qquad m \in \{k,l\} \cup [n+1, n+4], \tag{2.3.31}$$

by definition of collisional invariants and from the properties (2.1.18)(2.1.19) applied to the binary mixture (k,l).

The Species Orthogonality Property and the Nullspace $N(2db(G)-G)$. We now identify the nullspace $N(2db(G)-G)$ by using the characterization (2.3.29). This is done by assuming that the perpendicularity property (2.3.20) holds, that the localization property (2.3.25) holds, and that the species collisional invariants $\mathcal{T}_\nu\widehat{\psi}^l$, $l\in\mathcal{S}$ and $\nu\in[1,\tau_\mu]$, are in $\mathcal{I}_{\mathcal{S}}^{a_\mu}\cap\mathcal{A}^{\mu\perp}$, that is,

$$\mathcal{T}_\nu\widehat{\psi}^l\in\mathcal{I}_{\mathcal{S}}^{a_\mu}\cap\mathcal{A}^{\mu\perp},\qquad (l,\nu)\in\mathcal{S}\times[1,\tau_\mu]. \tag{2.3.32}$$

This property will be termed "the species orthogonality property" in this book.

Assume first that $n\geq 3$, let $x=(x_k^r)_{(r,k)\in\mathcal{B}^\mu}\in\mathbb{R}^\omega$ and $\Xi=\sum_{(r,k)\in\mathcal{B}^\mu}x_k^r\xi^{rk}$, and assume that $\langle(2db(G)-G)x,x\rangle=0$. From (2.3.29), we know that for any given species pair (k,l) with $k\neq l$, there exist scalars $u_{j\nu}$, $(j,\nu)\in(\{k,l\}\cup[n+1,n+4])\times[1,\tau_\mu]$, such that

$$(\Xi_k,-\Xi_l)=\sum_{\substack{j\in\{k,l\}\\ \nu\in[1,\tau_\mu]}}u_{j\nu}\mathcal{T}_\nu\widehat{\psi}^j_{kl}+\sum_{\substack{j\in[n+1,n+4]\\ \nu\in[1,\tau_\mu]}}u_{j\nu}\mathcal{T}_\nu\widehat{\psi}^j_{kl}.$$

Introducing the corresponding full mixture collisional invariant

$$\mathcal{U}=\sum_{\substack{j\in\{k,l\}\\ \nu\in[1,\tau_\mu]}}u_{j\nu}\mathcal{T}_\nu\widehat{\psi}^j+\sum_{\substack{j\in[n+1,n+4]\\ \nu\in[1,\tau_\mu]}}u_{j\nu}\mathcal{T}_\nu\widehat{\psi}^j,$$

and identifying the kth and l^{th} components yields that $\Xi_k=\mathcal{U}_k$ and $\Xi_l=-\mathcal{U}_l$. From $\Xi_k=\mathcal{U}_k$ and the species localization property (2.3.25), we obtain that $\langle\!\langle\mathcal{U},\mathcal{T}_\nu\widehat{\psi}^k\rangle\!\rangle=\langle\!\langle\Xi,\mathcal{T}_\nu\widehat{\psi}^k\rangle\!\rangle$ and therefore that $\langle\!\langle\mathcal{U},\mathcal{T}_\nu\widehat{\psi}^k\rangle\!\rangle=0$ thanks to $\Xi\in\mathcal{A}^\mu$ and the species orthogonality property. Since $\langle\!\langle\mathcal{U},\mathcal{T}_\nu\widehat{\psi}^k\rangle\!\rangle=u_{k\nu}\langle\!\langle\mathcal{T}_\nu\widehat{\psi}^k,\mathcal{T}_\nu\widehat{\psi}^k\rangle\!\rangle=u_{k\nu}n_k$, we deduce that $u_{k\nu}=0$, for $\nu\in[1,\tau_\mu]$. Similarly, one can show that $u_{l\nu}=0$, for $\nu\in[1,\tau_\mu]$, so that we have

$$\mathcal{U}=\sum_{\substack{j\in[n+1,n+4]\\ \nu\in[1,\tau_\mu]}}u_{j\nu}\mathcal{T}_\nu\widehat{\psi}^j.$$

Since $n\geq 3$ we can consider k, l, and m such that $k\neq l$, $l\neq m$, and $m\neq k$. From the preceding analysis, we can first write for the species pair (k,l) that $\Xi_k=\mathcal{U}_k$ and $\Xi_l=-\mathcal{U}_l$ where $\mathcal{U}$ is a collisional invariant of the full mixture. Similarly, for the species pair (l,m), we can then write that $\Xi_l=\mathcal{V}_l$ and $\Xi_m=-\mathcal{V}_m$ where $\mathcal{V}$ is another collisional invariant of the full mixture in the form

$$\mathcal{V}=\sum_{\substack{j\in[n+1,n+4]\\ \nu\in[1,\tau_\mu]}}v_{j\nu}\mathcal{T}_\nu\widehat{\psi}^j.$$

However, since the scalar functions $\mathcal{T}_\nu \widehat{\psi}_l^j$, $(j,\nu) \in [n{+}1, n{+}4]\times[1,\tau_\mu]$, are linearly independent, we deduce from $\mathcal{U}_l = -\Xi_l = -\mathcal{V}_l$ that $u_{j\nu} = -v_{j\nu}$, $(j,\nu) \in [n{+}1, n{+}4]\times[1,\tau_\mu]$, so that we indeed have $\mathcal{U} = -\mathcal{V}$. By repeating the argument for the species pair (m,k), we finally obtain that

$$\begin{cases} \Xi_k = \mathcal{U}_k, & \Xi_l = -\mathcal{U}_l, \\ \Xi_l = -\mathcal{U}_l, & \Xi_m = \mathcal{U}_m, \\ \Xi_m = \mathcal{U}_m, & \Xi_k = -\mathcal{U}_k, \end{cases}$$

so that $\Xi_k = \mathcal{U}_k = -\mathcal{U}_k = 0$. This now implies that $\Xi_k = 0$ for any $k \in \mathcal{S}$, so that $x = 0$ and $2db(G) - G$ is positive definite if $n \geq 3$.

On the other hand, in the special case $n = 2$, we may write $\mathcal{S} = \{1,2\}$ and we directly deduce from (2.3.29) that we have $x \in N\big(2db(G) - G\big)$ if and only if $x^* \in N(G)$, where

$$x^* = \sum_{r\in\mathcal{F}_1} x_1^r e^{r1} - \sum_{r\in\mathcal{F}_2} x_2^r e^{r2},$$

for $x = \sum_{r\in\mathcal{F}_1} x_1^r e^{r1} + \sum_{r\in\mathcal{F}_2} x_2^r e^{r2}$. Finally, in the special case $n = 1$, we have $2db(G) - G = G$ so that $N\big(2db(G) - G\big) = N(G)$.

Properties of the Matrix $db(G)$. We now want to investigate the properties of the matrix $db(G)$. In order to identify the nullspace of the matrix $db(G)$, we will need to know that when the perpendicularity property (2.2.20), the species orthogonality property (2.3.32), and the species localization property (2.3.25) hold, there are no nonzero elements $\Xi = (\Xi_k)_{k\in\mathcal{S}}$ in $\mathcal{I}_\mathcal{S}^{a_\mu} \cap \mathcal{A}^\mu$ with a zero component, say $\Xi_k = 0$. Indeed, let $\Xi \in \mathcal{I}_\mathcal{S}^{a_\mu} \cap \mathcal{A}^\mu$ given by

$$\Xi = \sum_{\substack{j\in[1,n+4] \\ \nu\in[1,\tau_\mu]}} u_{j\nu} \mathcal{T}_\nu \widehat{\psi}^j,$$

where the $u_{j\nu}$, $(j,\nu) \in [1, n{+}4]\times[1,\tau_\mu]$, are scalars. From the species orthogonality property (2.3.32), we first deduce that $\langle\!\langle \Xi, \mathcal{T}_\nu \widehat{\psi}^j \rangle\!\rangle = 0$. As a consequence, we have $\langle\!\langle \Xi, \mathcal{T}_\nu \widehat{\psi}^j \rangle\!\rangle = u_{j\nu} \langle\!\langle \mathcal{T}_\nu \widehat{\psi}^j, \mathcal{T}_\nu \widehat{\psi}^j \rangle\!\rangle = u_{j\nu} n_j$, so that $u_{j\nu} = 0$, for $j \in \mathcal{S}$ and $\nu \in [1,\tau_\mu]$. This shows that

$$\Xi = \sum_{\substack{j\in[n+1,n+4] \\ \nu\in[1,\tau_\mu]}} u_{j\nu} \mathcal{T}_\nu \widehat{\psi}^j,$$

so that if the component Ξ_k is zero, we then have $u_{j\nu} = 0$ for $(j,\nu) \in [n{+}1, n{+}4]\times[1,\tau_\mu]$, since the scalar functions $\mathcal{T}_\nu \widehat{\psi}_k^j$, $(j,\nu) \in [n{+}1, n{+}4]\times[1,\tau_\mu]$, are linearly independent. Therefore, Ξ must also be zero, and we have established that

$$\forall Z \in \mathcal{I}_\mathcal{S}^{a_\mu} \cap \mathcal{A}^\mu, \qquad Z \neq 0 \implies \forall k \in \mathcal{S},\ \ Z_k \neq 0. \tag{2.3.33}$$

Using this result, we now investigate the positive definiteness of the matrix $db(G)$. We first note that

$$db(G) = \tfrac{1}{2}\big(2db(G) - G\big) + \tfrac{1}{2}G,$$

and that both $2db(G) - G$ and G are positive semi-definite. This shows that $db(G)$ is positive semi-definite for any $n \geq 1$ and that

$$N\big(db(G)\big) = N\big(2db(G) - G\big) \cap N(G).$$

Since $2db(G)-G$ is positive definite for $n \geq 3$, we thus obtain that $db(G)$ is also positive definite in this case. In the case $n = 2$, we claim that the matrix $db(G)$ is also positive definite since then $N\big(2db(G) - G\big) \cap N(G) = \{0\}$. Let indeed x be in $N\big(2db(G) - G\big) \cap N(G)$ and denote $S = \{1, 2\}$. Since $x \in N\big(2db(G) - G\big)$, we have already seen that $x^* \in N(G)$ where we have defined $x^* = \sum_{r\in\mathcal{F}_1} x_1^r e^{r1} - \sum_{r\in\mathcal{F}_2} x_2^r e^{r2}$ for $x = \sum_{r\in\mathcal{F}_1} x_1^r e^{r1} + \sum_{r\in\mathcal{F}_2} x_2^r e^{r2}$. Letting $\Xi = \sum_{(r,k)\in\mathcal{B}^\mu} x_k^r \xi^{rk}$ and $\Xi^* = \sum_{(r,k)\in\mathcal{B}^\mu} x_k^{*r} \xi^{rk}$, we then deduce that $\Xi \in \mathcal{I}_S^{a_\mu} \cap \mathcal{A}^\mu$ and $\Xi^* \in \mathcal{I}_S^{a_\mu} \cap \mathcal{A}^\mu$ since $x \in N(G)$ and $x^* \in N(G)$. This implies that $\Xi + \Xi^*$ and $\Xi - \Xi^*$ are in $\mathcal{I}_S^{a_\mu} \cap \mathcal{A}^\mu$ which implies from (2.3.33) that $\Xi + \Xi^*$ and $\Xi - \Xi^*$ are zero since both have a zero component. Hence $\Xi = 0$ and $x = 0$ so that $db(G)$ is positive definite for $n = 2$. Finally, in the case $n = 1$ we have $db(G) = G$, so that, in this situation, $N\big(db(G)\big) = N(G)$.

Summary. We have shown that whenever the perpendicularity property (2.3.20), the species orthogonality property (2.3.32), and the species localization property (2.3.25) hold, then the matrix $2db(G) - G$ satisfies

$$\begin{cases} 2db(G) - G \ \text{ is symmetric positive semi-definite for } n \geq 1, \\[2ex] 2db(G) - G \ \text{ is positive definite for } n \geq 3, \\[2ex] N\big(2db(G) - G\big) = \{\ x^*,\ x \in N(G)\ \} \text{ for } n = 2, \\[2ex] N\big(2db(G) - G\big) = N(G) \text{ for } n = 1, \end{cases} \tag{2.3.34}$$

and the matrix $db(G)$ satisfies

$$\begin{cases} db(G) \ \text{ is symmetric positive semi-definite for } n \geq 1, \\[2ex] db(G) \ \text{ is positive definite for } n \geq 2, \\[2ex] N\big(db(G)\big) = N(G) \text{ for } n = 1, \end{cases} \tag{2.3.35}$$

where for $n = 2$ and $\mathcal{S} = \{1, 2\}$, we have defined $x^* = \sum_{r\in\mathcal{F}_1} x_1^r e^{r1} - \sum_{r\in\mathcal{F}_2} x_2^r e^{r2}$ for $x = \sum_{r\in\mathcal{F}_1} x_1^r e^{r1} + \sum_{r\in\mathcal{F}_2} x_2^r e^{r2}$.

Remark. The structure of the matrix $2db(G) - G$ reveals that the general case for mixtures is $n \geq 3$ and that binary mixtures are a degenerate case inadequate for a general theory.

2.3.5 The Special Case $\dim(\mathcal{I}_{\mathcal{S}}^{a_\mu} \cap \mathcal{A}^\mu) = 0$

We examine in this section the particular case where $\dim(\mathcal{I}_{\mathcal{S}}^{a_\mu} \cap \mathcal{A}^\mu) = 0$. More specifically, we assume that the perpendicularity property (2.3.20), the species orthogonality property (2.3.32), and the species localization property (2.3.25) are satisfied, and we also assume that

$$\mathcal{I}_{\mathcal{S}}^{a_\mu} \cap \mathcal{A}^\mu = \{0\}. \tag{2.3.36}$$

In the next section we will examine the case where $\dim(\mathcal{I}_{\mathcal{S}}^{a_\mu} \cap \mathcal{A}^\mu) = 1$ and both situations will actually cover all the particular cases that will be investigated in the following sections. Notice that (2.3.20)(2.3.32)(2.3.36) are also equivalent to $\mathcal{I}_{\mathcal{S}}^{a_\mu} = \mathcal{I}_{\mathcal{S}}^{a_\mu} \cap \mathcal{A}^{\mu\perp}$.

From the assumptions we first deduce that the results (2.3.24), (2.3.34), and (2.3.35) established in Sections 2.3.3 and 2.3.4 apply. As a consequence we first obtain from (2.3.24) and (2.3.36) that $\dim(N(G)) = \dim(\mathcal{I}_{\mathcal{S}}^{a_\mu} \cap \mathcal{A}^\mu) = 0$ so that $N(G) = \{0\}$ and the matrix G is symmetric positive definite for any $n \geq 1$. In addition, we also obtain from (2.3.24) that $C = \mathbb{R}^\omega$ and therefore all the constraint vectors are zero vectors from the definition (2.3.13) of the constrained space C. Furthermore, we deduce from (2.3.34) and (2.3.35) that the matrices $2db(G) - G$ and $db(G)$ are symmetric positive definite for any $n \geq 1$.

Summary. We have shown that whenever the perpendicularity property (2.3.20), the species orthogonality property (2.3.32), the species localization property (2.3.25), and the equality $\dim(\mathcal{I}_{\mathcal{S}}^{a_\mu} \cap \mathcal{A}^\mu) = 0$ are satisfied, then

$$\begin{cases} G \ \text{ is positive definite for } n \geq 1, \\ \\ 2db(G) - G \ \text{ is positive definite for } n \geq 1, \\ \\ db(G) \ \text{ is positive definite for } n \geq 1. \end{cases} \tag{2.3.37}$$

2.3.6 The Special Case $\dim(\mathcal{I}_{\mathcal{S}}^{a_\mu} \cap \mathcal{A}^\mu) = 1$

We examine in this section the particular case where $\dim(\mathcal{I}_{\mathcal{S}}^{a_\mu} \cap \mathcal{A}^\mu) = 1$. More specifically, we assume that the perpendicularity property (2.3.20), the species orthogonality property (2.3.32), and the species localization property (2.3.25) are satisfied, and we also assume that

$$\mathcal{I}_{\mathcal{S}}^{a_\mu} \cap \mathcal{A}^\mu = \mathbb{R}Z, \tag{2.3.38}$$

where Z is a nonzero collisional invariant. Note also that (2.3.20)(2.3.38) imply that $\mathcal{I}_{\mathcal{S}}^{a_\mu} = \mathbb{R}Z \;\oplus\; \mathcal{I}_{\mathcal{S}}^{a_\mu} \cap \mathcal{A}^{\mu\perp}$. We will denote by $\mathcal{Z} = (\mathcal{Z}_k^r)_{(r,k)\in\mathcal{B}^\mu}$ the components of the collisional invariant Z with respect to the basis functions ξ^{rk}

$$Z = \sum_{(r,k)\in\mathcal{B}^\mu} \mathcal{Z}_k^r \xi^{rk}. \tag{2.3.39}$$

From the assumptions we first deduce that the results (2.3.24), (2.3.34), and (2.3.35) established in Sections 2.3.3 and 2.3.4 apply. As a consequence, combining (2.3.24) and (2.3.38) first yields that $\dim(N(G)) = \dim(\mathcal{I}_{\mathcal{S}}^{a_\mu} \cap \mathcal{A}^\mu) = 1$ and from (2.3.39) we have $N(G) = \mathbb{R}\mathcal{Z}$. From (2.3.23) we also obtain that

$$x \in \mathcal{C} \iff \sum_{(r,k)\in\mathcal{B}^\mu} x_k^r \,\langle\!\langle \xi^{rk}, Z \rangle\!\rangle = 0,$$

since $\mathcal{I}_{\mathcal{S}}^{a_\mu} \cap \mathcal{A}^\mu = \mathbb{R}Z$, so that

$$\mathcal{C} = \mathcal{G}^\perp,$$

where we have defined the vector $\mathcal{G}$ by

$$\mathcal{G} = (\mathcal{G}_k^r)_{(r,k)\in\mathcal{B}^\mu} = \big(\langle\!\langle \xi^{rk}, Z \rangle\!\rangle\big)_{(r,k)\in\mathcal{B}^\mu}. \tag{2.3.40}$$

All the constraints vectors $\mathcal{G}^{l\nu}$, $(l,\nu) \in [1, n{+}4]\times[1,\tau_\mu]$, can thus be replaced by the single vector $\mathcal{G}$. Note that from (2.3.24) we have $\mathbb{R}\mathcal{Z} \oplus \mathcal{G}^\perp = \mathbb{R}^\omega$ so that we must have $\langle \mathcal{Z}, \mathcal{G} \rangle \neq 0$. Indeed, a direct evaluation yields that

$$\langle \mathcal{Z}, \mathcal{G} \rangle = \langle\!\langle Z, Z \rangle\!\rangle,$$

so that $\langle \mathcal{Z}, \mathcal{G} \rangle > 0$ since $Z \neq 0$. We also obtain from (2.3.33) that none of the subvectors $(\mathcal{Z}_k^r)_{r\in\mathcal{F}_k}$, $k \in \mathcal{S}$, can be zero and from (2.3.24) we also obtain that G is positive definite on $\mathcal{G}^\perp = \mathcal{C}$ and that $\langle \mathcal{Z}, \beta^\mu \rangle = 0$ since $\beta^\mu \in R(G) = (N(G))^\perp$.

From (2.3.34) we next obtain that $2db(G) - G$ is positive definite for $n \geq 3$, whereas for $n = 2$ and $\mathcal{S} = \{1,2\}$, we have $N(2db(G) - G) = \mathbb{R}\mathcal{Z}^*$ with $\mathcal{Z}_1^{r*} = \mathcal{Z}_1^r$, $r \in \mathcal{F}_1$,

$\mathcal{Z}_2^{r*} = -\mathcal{Z}_2^r$, $r \in \mathcal{F}_2$, and, finally, for $n = 1$ and $\mathcal{S} = \{1\}$, we have $N\big(2db(G)-G\big) = \mathbb{R}\mathcal{Z}$. From (2.3.35) we also obtain that the matrix $db(G)$ is positive definite for $n \geq 2$, whereas for $n = 1$ and $\mathcal{S} = \{1\}$, we have $N\big(db(G)\big) = \mathbb{R}\mathcal{Z}$.

Summary. We have shown that whenever the perpendicularity property (2.3.20), the species orthogonality property (2.3.32), the species localization property (2.3.25), and the equality $\dim\big(\mathcal{I}_S^{a_\mu} \cap \mathcal{A}^\mu\big) = 1$ are satisfied, then the matrix G satisfies

$$\begin{cases} G \;\text{ is symmetric positive semi-definite for } n \geq 1, \\ N(G) = \mathbb{R}\mathcal{Z} \;\text{ for } n \geq 1, \\ G \;\text{ is positive definite on } \mathcal{C} = \mathcal{G}^\perp \text{ for } n \geq 1, \\ \langle \mathcal{Z}, \mathcal{G} \rangle \neq 0 \text{ for } n \geq 1, \\ \langle \mathcal{Z}, \beta^\mu \rangle = 0 \text{ for } n \geq 1, \end{cases} \tag{2.3.41}$$

the matrix $2db(G) - G$ satisfies

$$\begin{cases} 2db(G) - G \;\text{ is symmetric positive semi-definite for } n \geq 1, \\ 2db(G) - G \;\text{ is positive definite for } n \geq 3, \\ N\big(2db(G) - G\big) = \mathbb{R}\mathcal{Z}^* \text{ for } n = 2, \\ N\big(2db(G) - G\big) = \mathbb{R}\mathcal{Z} \text{ for } n = 1, \end{cases} \tag{2.3.42}$$

and the matrix $db(G)$ satisfies

$$\begin{cases} db(G) \;\text{ is symmetric positive semi-definite for } n \geq 1, \\ db(G) \;\text{ is positive definite for } n \geq 2, \\ N\big(db(G)\big) = \mathbb{R}\mathcal{Z} \text{ for } n = 1. \end{cases} \tag{2.3.43}$$

where for $n = 2$ and $\mathcal{S} = \{1, 2\}$, we have defined $\mathcal{Z}^* = \sum_{r\in\mathcal{F}_1} \mathcal{Z}_1^r e^{r1} - \sum_{r\in\mathcal{F}_2} \mathcal{Z}_2^r e^{r2}$.

Remark. In practical applications, the basis functions ξ^{rk}, $(r, k) \in \mathcal{B}^\mu$, are usually orthogonal with respect to the quadratic form $\langle\!\langle\, , \rangle\!\rangle$. In this situation, there is a simpler relation between $\mathcal{Z} \in N(G)$ and the constraint vector $\mathcal{G}$ since we then obtain that $\mathcal{G}_k^r = \langle\!\langle \xi^{rk}, Z \rangle\!\rangle = \mathcal{Z}_k^r \langle\!\langle \xi^{rk}, \xi^{rk} \rangle\!\rangle$. More generally, the components of any vectors spanning $N(G)$ and $\mathcal{C}^\perp$ only differ by the mixture dependent multiplicative factor $\langle\!\langle \xi^{rk}, \xi^{rk} \rangle\!\rangle$ up to a multiplicative constant.

2.3.7 Functional Spaces

In this section we establish that the perpendicularity property (2.3.20) and the species orthogonality property (2.3.32), introduced in the previous sections, hold for very general functional spaces. We make use of the functions ϕ^{a0cdk} introduced in (2.2.15) and of their orthogonality properties (2.2.16).

The Scalar Case. We first investigate the scalar case, where $a_\mu = 0$, $\tau_\mu = 1$, and $\mu = \kappa$, and we consider a finite dimensional space $\mathcal{A}^\mu$ such that

$$\mathcal{A}^\mu \subset \operatorname{span}\{\, \phi^{00cdk},\ (cd,k) \in \mathcal{B}^\mu \,\}, \tag{2.3.44}$$

where $\mathcal{B}^\mu \subset \mathbb{N}^2 \times \mathcal{S}$ and where the indices $(cd,k) \in \mathcal{B}^\mu$ are such that $cd \neq 00$ and $d < \operatorname{card}(\mathcal{E}_k)$. Keeping in mind that in the scalar case we have $\mathcal{T}_1 = (1)$ and using the orthogonality properties (2.2.16) of the functions ϕ^{a0cdk}, one may then easily establish that $\mathcal{T}_\nu \widehat{\psi}^l \in \mathcal{I}_S^0 \cap \mathcal{A}^{\mu\perp}$ for $(l,\nu) \in [1,n+3] \times \{1\}$. In addition, if $\{10\} \times \mathcal{S} \subset \mathcal{B}^\mu$ and $\{01\} \times \mathcal{P} \subset \mathcal{B}^\mu$, we have $\mathcal{T}_1 \widehat{\psi}^{n+4} \in \operatorname{span}\{\, \phi^{00cdk},\ (cd,k) \in \mathcal{B}^\mu \,\}$ since

$$\widehat{\psi}^{n+4} = \sum_{k \in \mathcal{S}} \phi^{0010k} + \sum_{k \in \mathcal{P}} \phi^{0001k}. \tag{2.3.45}$$

Assuming that $\widehat{\psi}^{n+4} \in \mathcal{A}^\mu$, we obtain the decomposition

$$\mathcal{I}_S^0 = \mathbb{R}\widehat{\psi}^{n+4} \ \oplus\ \mathcal{I}_S^0 \cap \mathcal{A}^{\mu\perp}, \tag{2.3.46}$$

where $\mathbb{R}\widehat{\psi}^{n+4} = \mathcal{I}_S^0 \cap \mathcal{A}^\mu$ and where the space $\mathcal{I}_S^0 \cap \mathcal{A}^{\mu\perp}$ is spanned by the collisional invariants $\mathcal{T}_\nu \widehat{\psi}^l$, $(l,\nu) \in [1,n+3] \times \{1\}$. The perpendicularity property (2.3.20), the species orthogonality property (2.3.32), and the relation (2.3.38) are thus satisfied. As a consequence, if the basis functions of $\mathcal{A}^\mu$ are also localized with respect to the species, i.e., satisfy (2.3.25), we conclude that the results (2.3.41)–(2.3.43) established in Section 2.3.6 apply.

Another interesting case arises when

$$\mathcal{A}^\mu \subset \operatorname{span}\{\, \widehat{\phi}^{00cdk},\ (cd,k) \in \mathcal{B}^\mu \,\}, \tag{2.3.47}$$

where $\mathcal{B}^\mu \subset \mathbb{N}^2 \times \mathcal{S}$, the indices $(cd,k) \in \mathcal{B}^\mu$ are such that $cd \neq 00$ and $d < \operatorname{card}(\mathcal{E}_k)$, and

$$\widehat{\phi}^{00cdk} = \phi^{00cdk} - \frac{\langle\!\langle \phi^{00cdk}, \widehat{\psi}^{n+4} \rangle\!\rangle}{\langle\!\langle \widehat{\psi}^{n+4}, \widehat{\psi}^{n+4} \rangle\!\rangle} \widehat{\psi}^{n+4},$$

in which case $\widehat{\psi}^{n+4} \in \mathcal{I}_S^0 \cap \mathcal{A}^{\mu\perp}$ and

$$\mathcal{I}_S^0 = \mathcal{I}_S^0 \cap \mathcal{A}^{\mu\perp}, \tag{2.3.48}$$

so that (2.3.20) holds. In this situation, the results (2.3.24) of Section 2.3.3 apply, but the results on $2db(G) - G$ do not apply directly, since the basis functions of $\mathcal{A}^\mu$ are not necessarily localized, i.e., do not satisfy the species localization property (2.3.25). In this case, the properties of $2db(G) - G$ will be obtained from matrices corresponding to the general case (2.3.44) as will be shown in Section 2.3.8.

The Vector Case. In the vector case, we have $a_\mu = 1$, $\tau_\mu = 3$, and either $\mu = D_l$, $l \in \mathcal{S}$, $\mu = \lambda'$, or $\mu = \lambda$. By isotropy, the solutions of (2.1.35)(2.1.36) are necessarily of the form $\phi_i^\mu = \varphi_i^\mu w_i$, $i \in \mathcal{S}$, where φ_i^μ is a scalar function [WT62] [MYM63]. We will thus generally consider functional spaces such that

$$\mathcal{A}^\mu \subset \operatorname{span}\{\ \phi^{10cdk},\ (cd, k) \in \mathcal{B}^\mu\ \}, \tag{2.3.49}$$

where $\mathcal{B}^\mu \subset \mathbb{N}^2 \times \mathcal{S}$ and where the indices $(cd, k) \in \mathcal{B}^\mu$ are such that $d < \operatorname{card}(\mathcal{E}_k)$. Keeping in mind that, in the vector case, we now have $\mathcal{T}_1 = (1,0,0)$, $\mathcal{T}_2 = (0,1,0)$, and $\mathcal{T}_3 = (0,0,1)$, and by using the orthogonality properties (2.2.16) of the functions ϕ^{a0cdk}, one may then easily establish that we have $\mathcal{T}_\nu \widehat{\psi}^l \in \mathcal{I}_S^1 \cap \mathcal{A}^{\mu\perp}$ for $(l, \nu) \in ([1,n] \cup \{n+4\}) \times [1,3]$ and $(l, \nu) = (n+\nu', \nu)$, with $\nu, \nu' \in [1,3]$ and $\nu \neq \nu'$. Moreover, we also have $\mathcal{T}_1 \widehat{\psi}^{n+1} - \mathcal{T}_2 \widehat{\psi}^{n+2} \in \mathcal{I}_S^1 \cap \mathcal{A}^{\mu\perp}$ and $\mathcal{T}_2 \widehat{\psi}^{n+2} - \mathcal{T}_3 \widehat{\psi}^{n+3} \in \mathcal{I}_S^1 \cap \mathcal{A}^{\mu\perp}$, by isotropy. In addition, if $\{00\} \times \mathcal{S} \subset \mathcal{B}^\mu$, then the linear combination $\mathcal{T}_1 \widehat{\psi}^{n+1} + \mathcal{T}_2 \widehat{\psi}^{n+2} + \mathcal{T}_3 \widehat{\psi}^{n+3}$ is in the functional space $\operatorname{span}\{\ \phi^{10cdk},\ (cd, k) \in \mathcal{B}^\mu\ \}$ since

$$\mathcal{T}_1 \widehat{\psi}^{n+1} + \mathcal{T}_2 \widehat{\psi}^{n+2} + \mathcal{T}_3 \widehat{\psi}^{n+3} = \sum_{k \in \mathcal{S}} \sqrt{2k_{\mathrm{B}} \overline{T} m_k}\ \phi^{1000k}. \tag{2.3.50}$$

Assuming then that $\mathcal{T}_1 \widehat{\psi}^{n+1} + \mathcal{T}_2 \widehat{\psi}^{n+2} + \mathcal{T}_3 \widehat{\psi}^{n+3} \in \mathcal{A}^\mu$, we obtain the decomposition

$$\mathcal{I}_S^1 = \mathbb{R}(\mathcal{T}_1 \widehat{\psi}^{n+1} + \mathcal{T}_2 \widehat{\psi}^{n+2} + \mathcal{T}_3 \widehat{\psi}^{n+3}) \ \oplus\ \mathcal{I}_S^1 \cap \mathcal{A}^{\mu\perp}, \tag{2.3.51}$$

where $\mathbb{R}(\mathcal{T}_1 \widehat{\psi}^{n+1} + \mathcal{T}_2 \widehat{\psi}^{n+2} + \mathcal{T}_3 \widehat{\psi}^{n+3}) = \mathcal{I}_S^1 \cap \mathcal{A}^\mu$ and where $\mathcal{I}_S^1 \cap \mathcal{A}^{\mu\perp}$ is spanned by the tensorial collisional invariants $\mathcal{T}_\nu \widehat{\psi}^l$ for $(l, \nu) \in ([1,n] \cup \{n+4\}) \times [1,3]$ and $(l, \nu) = (n+\nu', \nu)$ with $\nu, \nu' \in [1,3]$, $\nu \neq \nu'$, and by the collisional invariants $\mathcal{T}_1 \widehat{\psi}^{n+1} - \mathcal{T}_2 \widehat{\psi}^{n+2}$ and $\mathcal{T}_2 \widehat{\psi}^{n+2} - \mathcal{T}_3 \widehat{\psi}^{n+3}$. The perpendicularity property (2.3.20), the species orthogonality property (2.3.32), and the relation (2.3.38) are thus satisfied. As a consequence, if the the basis functions of $\mathcal{A}^\mu$ are also localized with respect to the species, i.e., satisfy (2.3.25), we conclude that the results (2.3.41)–(2.3.43) established in Section 2.3.6 apply.

Another interesting case, which will be used for $\mu = \lambda$, is when $\mathcal{B}^\mu \subset \mathbb{N}^2 \times \mathcal{S}$, with the indices $(cd, k) \in \mathcal{B}^\mu$ such that $d < \mathrm{card}(\mathcal{E}_k)$, but with $\mathcal{B}^\mu \cap (\{00\} \times \mathcal{S}) = \emptyset$, in which case we have

$$\mathcal{I}^1_{\mathcal{S}} = \mathcal{I}^1_{\mathcal{S}} \cap \mathcal{A}^{\mu\perp}. \tag{2.3.52}$$

The perpendicularity property (2.3.20), the species orthogonality property (2.3.32), and the relation (2.3.36) are then satisfied. As a consequence, if the the basis functions of $\mathcal{A}^\mu$ are also localized with respect to the species, i.e., satisfy (2.3.25), we conclude that the results (2.3.37) established in Section 2.3.5 apply.

The Matrix Case. Finally, in the matrix case, we have $a_\mu = 2$, $\tau_\mu = 9$, and $\mu = \eta$. By isotropy, the solutions of (2.1.35)(2.1.36) are necessarily of the form $\phi^\mu_i = \varphi^\mu_i (w_i \otimes w_i - \frac{1}{3} w_i \cdot w_i I)$, $i \in \mathcal{S}$, where φ^μ_i is a scalar function [WT62] [MYM63]. We will thus generally consider functional spaces such that

$$\mathcal{A}^\mu \subset \ \mathrm{span}\{\ \phi^{20cdk},\ (cd, k) \in \mathcal{B}^\mu\ \}, \tag{2.3.53}$$

where $\mathcal{B}^\mu \subset \mathbb{N}^2 \times \mathcal{S}$ and where the indices $(cd, k) \in \mathcal{B}^\mu$ are such that $d < \mathrm{card}(\mathcal{E}_k)$. From the orthogonality properties of the functions ϕ^{a0cdk}, one may then easily establish that $\mathcal{T}_\nu \widehat{\psi}^l \in \mathcal{I}^2_{\mathcal{S}} \cap \mathcal{A}^{\mu\perp}$ for any $(l, \nu) \in [1, n+4] \times [1, 9]$, so that

$$\mathcal{I}^2_{\mathcal{S}} = \mathcal{I}^2_{\mathcal{S}} \cap \mathcal{A}^{\mu\perp}, \tag{2.3.54}$$

and $\mathcal{I}^2_{\mathcal{S}} \cap \mathcal{A}^\mu = \{0\}$. The perpendicularity property (2.3.20), the species orthogonality property (2.3.32), and the relation (2.3.36) are thus satisfied. As a consequence, if the the basis functions of $\mathcal{A}^\mu$ are also localized with respect to the species, i.e., satisfy (2.3.25), we conclude that the results (2.3.37) established in Section 2.3.5 apply.

Remark. In practice, only a few function type indices $r \in \mathcal{F}$ are retained in the polynomial expansions of the species perturbed distribution functions. As a consequence, the dimension of $\mathcal{A}^\mu \subset \mathcal{F} \times \mathcal{S}$ is usually of the order of a small multiple of n. We observe, however, that the results established in this section apply for an arbitrary number of function type indices.

As a conclusion, we have shown in this section that the particular cases described in Sections 2.3.5 and 2.3.6 cover most of the cases encountered in practical applications. The exceptional case (2.3.47) is treated as a particular case in the next section.

2.3.8 Reduced Transport Linear Systems

In this section we consider simplified formulations of the transport linear systems associated with the use of smaller variational approximation spaces for the perturbed distribution functions ϕ^μ. Using functional spaces of lower dimension will indeed reduce the size of the transport linear systems and hence simplify the transport algorithms. These investigations generalize the ideas proposed by Thijsse et al. [TTCKB79] and Van den Oord and Korving [VK88] for the thermal conductivity of a polyatomic gas and by McCourt et al. [MBKK90] for the partial thermal conductivity of binary mixtures.

Evaluation of Reduced Systems. We consider a subspace $\mathcal{A}^\mu_{[\mathrm{red}]}$ of $\mathcal{A}^\mu$ spanned by the vectors ζ^{rk}, for $(r,k) \in \mathcal{B}^\mu_{[\mathrm{red}]}$, where $\mathcal{B}^\mu_{[\mathrm{red}]}$ is a reduced set of basis function indices. We write

$$\zeta^{rk} = \sum_{(s,l)\in\mathcal{B}^\mu} \mathcal{R}^{rs}_{kl}\xi^{sl}, \qquad (r,k) \in \mathcal{B}^\mu_{[\mathrm{red}]}, \tag{2.3.55}$$

where $\mathcal{R} = (\mathcal{R}^{rs}_{kl})_{(r,k)\in\mathcal{B}^\mu_{[\mathrm{red}]},\,(s,l)\in\mathcal{B}^\mu}$ is the transformation matrix. We denote by $\omega_{[\mathrm{red}]}$ the number of elements of $\mathcal{B}^\mu_{[\mathrm{red}]}$ so that we have $\omega_{[\mathrm{red}]} \le \omega$. In order to compare the transport coefficient $\mu_{[\mathrm{red}]}$—obtained with the reduced space $\mathcal{A}^\mu_{[\mathrm{red}]}$—with μ, we need to solve the corresponding reduced constrained linear system in $\alpha^\mu_{[\mathrm{red}]}$, where ϕ^μ is now. approximated as $\sum_{(r,k)\in\mathcal{B}^\mu_{[\mathrm{red}]}} \alpha^{r\mu}_{[\mathrm{red}]k}\,\zeta^{rk}$. This system reads

$$\begin{cases} G_{[\mathrm{red}]}\alpha^\mu_{[\mathrm{red}]} = \beta^\mu_{[\mathrm{red}]}, \\ \langle \alpha^\mu_{[\mathrm{red}]}, \mathcal{G}^{l\nu}_{[\mathrm{red}]}\rangle = 0, \qquad (l,\nu) \in [1,n{+}4]\times[1,\tau_\mu], \end{cases} \tag{2.3.56}$$

with

$$\begin{cases} G_{[\mathrm{red}]} = \big([\![\zeta^{rk},\zeta^{sl}]\!]\big)_{(r,k),(s,l)\in\mathcal{B}^\mu_{[\mathrm{red}]}}, \\ \beta^\mu_{[\mathrm{red}]} = \big(\langle\!\langle \zeta^{rk},\Psi^\mu\rangle\!\rangle\big)_{(r,k)\in\mathcal{B}^\mu_{[\mathrm{red}]}}, \\ \mathcal{G}^{l\nu}_{[\mathrm{red}]} = \big(\langle\!\langle \zeta^{rk},\mathcal{T}_\nu\widehat{\psi}^l\rangle\!\rangle\big)_{(r,k)\in\mathcal{B}^\mu_{[\mathrm{red}]}}, \qquad (l,\nu) \in [1,n{+}4]\times[1,\tau_\mu], \end{cases} \tag{2.3.57}$$

and its properties are directly obtained from the general results derived in Sections 2.3.1 to 2.3.6 provided that the corresponding assumptions are satisfied, in particular the species localization property. The associated simplified transport coefficients are subsequently calculated from the scalar products $\langle \alpha^\mu_{[\mathrm{red}]}, \beta^\mu_{[\mathrm{red}]}\rangle$.

However, one may also easily check from (2.3.55)–(2.3.57) that the reduced system is simply given by

$$\begin{cases} G_{[\mathrm{red}]} = \mathcal{R}G\mathcal{R}^t, \\ \beta^\mu_{[\mathrm{red}]} = \mathcal{R}\beta^\mu, \\ \mathcal{G}^{l\nu}_{[\mathrm{red}]} = \mathcal{R}\mathcal{G}^{l\nu}, \qquad (l,\nu) \in [1,n{+}4]\times[1,\tau_\mu], \end{cases} \tag{2.3.58}$$

so that the reduced transport linear systems are easily evaluated from the larger ones.

Non Localized Basis Functions. We now analyze a more general situation which includes, in particular, the spaces considered in (2.3.47). Indeed, we consider a subspace $\mathcal{A}^{\mu}_{[\mathrm{red}]}$ of $\mathcal{A}^{\mu}$ spanned by the functions $\widehat{\zeta}^{rk}$, for $(r,k) \in \mathcal{B}^{\mu}_{[\mathrm{red}]}$, where $\mathcal{B}^{\mu}_{[\mathrm{red}]}$ is a set of basis function indices, that are written in the form

$$\widehat{\zeta}^{rk} = \zeta^{rk} + \Sigma^{rk}, \tag{2.3.59}$$

where $\Sigma^{rk} \in \mathcal{I}_S^{a_\mu}$ is a collisional invariant of $\mathcal{A}^{\mu}$, and where the functions $\zeta^{rk} \in \mathcal{A}^{\mu}$, $(r,k) \in \mathcal{B}^{\mu}_{[\mathrm{red}]}$, form a basis and satisfy the species localization property (2.3.25). We also assume that the basis functions ξ^{rk}, $(r,k) \in \mathcal{B}^{\mu}$, of $\mathcal{A}^{\mu}$ satisfy the species localization property (2.3.25). Note that we do not assume that the functions $\widehat{\zeta}^{rk}$, $(r,k) \in \mathcal{B}^{\mu}_{[\mathrm{red}]}$, satisfy this property, and this is exactly the situation considered in (2.3.47). We can then write

$$\widehat{\zeta}^{rk} = \sum_{(s,l)\in\mathcal{B}^{\mu}} \widehat{\mathcal{R}}^{rs}_{kl}\xi^{sl}, \qquad (r,k) \in \mathcal{B}^{\mu}_{[\mathrm{red}]}, \tag{2.3.60}$$

where $\widehat{\mathcal{R}} = (\widehat{\mathcal{R}}^{rs}_{kl})_{(r,k)\in\mathcal{B}^{\mu}_{[\mathrm{red}]},\,(s,l)\in\mathcal{B}^{\mu}}$ is the transformation matrix, and

$$\zeta^{rk} = \sum_{(s,l)\in\mathcal{B}^{\mu}} \mathcal{R}^{rs}_{kl}\xi^{sl}, \qquad (r,k) \in \mathcal{B}^{\mu}_{[\mathrm{red}]}, \tag{2.3.61}$$

where $\mathcal{R} = (\mathcal{R}^{rs}_{kl})_{(r,k)\in\mathcal{B}^{\mu}_{[\mathrm{red}]},\,(s,l)\in\mathcal{B}^{\mu}}$ is the transformation matrix. With ϕ^{μ} approximated as $\sum_{(r,k)\in\mathcal{B}^{\mu}_{[\mathrm{red}]}} \alpha^{r\mu}_{[\mathrm{red}]k}\widehat{\zeta}^{rk}$, the relations (2.3.57) then yield that

$$\begin{cases} G_{[\mathrm{red}]} = \big([\widehat{\zeta}^{rk},\widehat{\zeta}^{sl}]\big)_{(r,k),(s,l)\in\mathcal{B}^{\mu}_{[\mathrm{red}]}}, \\ \beta^{\mu}_{[\mathrm{red}]} = \big(\langle\!\langle\widehat{\zeta}^{rk},\Psi^{\mu}\rangle\!\rangle\big)_{(r,k)\in\mathcal{B}^{\mu}_{[\mathrm{red}]}}, \end{cases} \tag{2.3.62}$$

which imply that $G_{[\mathrm{red}]} = \widehat{\mathcal{R}}G\widehat{\mathcal{R}}^t$ and $\beta^{\mu}_{[\mathrm{red}]} = \widehat{\mathcal{R}}\beta^{\mu}$. However, since the collisional invariants Σ^{rk}, $(r,k) \in \mathcal{B}^{\mu}_{[\mathrm{red}]}$, are in the kernel of the bilinear bracket operator $[\ ,]$ and since they are also orthogonal to Ψ^{μ} with respect to the bilinear form $\langle\!\langle\ ,\ \rangle\!\rangle$ from (2.3.18), we have $[\widehat{\zeta}^{rk},\widehat{\zeta}^{sl}] = [\zeta^{rk},\zeta^{sl}]$ and $[\widehat{\zeta}^{rk},\Psi^{\mu}] = [\zeta^{rk},\Psi^{\mu}]$, so that

$$\begin{cases} G_{[\mathrm{red}]} = \big([\zeta^{rk},\zeta^{sl}]\big)_{(r,k),(s,l)\in\mathcal{B}^{\mu}_{[\mathrm{red}]}}, \\ \beta^{\mu}_{[\mathrm{red}]} = \big(\langle\!\langle\zeta^{rk},\Psi^{\mu}\rangle\!\rangle\big)_{(r,k)\in\mathcal{B}^{\mu}_{[\mathrm{red}]}}. \end{cases} \tag{2.3.63}$$

Therefore, we have the simpler relations

$$\begin{cases} G_{[\mathrm{red}]} = \mathcal{R}G\mathcal{R}^t, \\ \beta^{\mu}_{[\mathrm{red}]} = \mathcal{R}\beta^{\mu}. \end{cases} \tag{2.3.64}$$

In addition, since the basis functions ζ^{rk}, $(r,k) \in \mathcal{B}^{\mu}_{[\mathrm{red}]}$, and ξ^{rk}, $(r,k) \in \mathcal{B}^{\mu}$, satisfy the species localization property (2.3.25), one may easily check that the matrix $\mathcal{R}$ has the property

$$\mathcal{R}^{rs}_{kl} = \mathcal{R}^{rs}_{kk}\,\delta_{kl}, \qquad (r,k) \in \mathcal{B}^{\mu}_{[\mathrm{red}]}, \quad (s,l) \in \mathcal{B}^{\mu}, \tag{2.3.65}$$

so that $\mathcal{R}$ leaves invariant the species indices. As a consequence, we first deduce from (2.3.65) that the coefficients of the reduced systems still satisfy the fundamental relations (2.3.26)(2.3.27). In addition, by using $G_{[\mathrm{red}]} = \mathcal{R}G\mathcal{R}^t$, a straightforward calculation now yields the important relations

$$\begin{cases} 2db(G_{[\mathrm{red}]}) - G_{[\mathrm{red}]} = \mathcal{R}\big(2db(G) - G\big)\mathcal{R}^t, \\ db(G_{[\mathrm{red}]}) = \mathcal{R}db(G)\mathcal{R}^t. \end{cases} \tag{2.3.66}$$

The relations (2.3.64)(2.3.66) can then be used to investigate the properties of $G_{[\mathrm{red}]}$, $2db(G_{[\mathrm{red}]}) - G_{[\mathrm{red}]}$, and $db(G_{[\mathrm{red}]})$. In particular, for any $x_{[\mathrm{red}]} \in \mathbb{R}^{\omega_{[\mathrm{red}]}}$, letting $x = \mathcal{R}^t x_{[\mathrm{red}]}$, we have the relations

$$\langle x_{[\mathrm{red}]}, G_{[\mathrm{red}]}x_{[\mathrm{red}]}\rangle = \langle x, Gx\rangle,$$

and

$$\langle x_{[\mathrm{red}]}, \big(2db(G_{[\mathrm{red}]}) - G_{[\mathrm{red}]}\big)x_{[\mathrm{red}]}\rangle = \langle x, \big(2db(G) - G\big)x\rangle.$$

As a consequence, we have $x_{[\mathrm{red}]} \in N(G_{[\mathrm{red}]})$ if and only if $\mathcal{R}^t x_{[\mathrm{red}]} \in N(G)$, and $x_{[\mathrm{red}]} \in N\big(2db(G_{[\mathrm{red}]}) - G_{[\mathrm{red}]}\big)$ if and only if $\mathcal{R}^t x_{[\mathrm{red}]} \in N\big(2db(G) - G\big)$. Assuming that $N(G) \cap R(\mathcal{R}^t) = \{0\}$ yields in particular that $N(G_{[\mathrm{red}]}) = \{0\}$ since $\mathcal{R}^t x_{[\mathrm{red}]} \in N(G)$ implies that $\mathcal{R}^t x_{[\mathrm{red}]} = 0$ so that $x_{[\mathrm{red}]} = 0$ because $\mathcal{R}^t$ is into thanks to the fact that the ζ^{rk}, $(r,k) \in \mathcal{B}^{\mu}_{[\mathrm{red}]}$, form a basis. Therefore, we have shown that

$$N(G) \cap R(\mathcal{R}^t) = \{0\} \quad \Longrightarrow \quad N(G_{[\mathrm{red}]}) = \{0\}. \tag{2.3.67}$$

Similarly, we have

$$N\big(2db(G) - G\big) \cap R(\mathcal{R}^t) = \{0\} \quad \Longrightarrow \quad N\big(2db(G_{[\mathrm{red}]}) - G_{[\mathrm{red}]}\big) = \{0\}. \tag{2.3.68}$$

In this situation $G_{[\mathrm{red}]}$, $2db(G_{[\mathrm{red}]}) - G_{[\mathrm{red}]}$, and $db(G_{[\mathrm{red}]})$ are then symmetric positive definite for any $n \geq 1$. The results (2.3.64)–(2.3.68) will be very useful in order to obtain the properties of the matrices corresponding to the functional spaces introduced in (2.3.47).

Remark. Note that we should distinguish the exact solution ϕ^μ of the integral equation (2.1.35)(2.1.36) from its approximation in the functional space $\mathcal{A}^\mu$ given by (2.2.6), which should indeed be denoted by $\phi^\mu_{[\mathcal{A}^\mu]}$, and also from its approximation in the reduced functional space $\mathcal{A}^\mu_{[\mathrm{red}]}$, which should correspondingly be denoted by $\phi^\mu_{[\mathcal{A}^\mu_{[\mathrm{red}]}]}$. Similarly, the corresponding transport coefficients should be written μ, $\mu_{[\mathcal{A}^\mu]}$, and $\mu_{[\mathcal{A}^\mu_{[\mathrm{red}]}]}$, respectively. In this book, however, these notational complexities will be avoided. The coefficients $\mu_{[\mathcal{A}^\mu]}$ corresponding to the largest functional approximation space $\mathcal{A}^\mu$ used for ϕ^μ will simply be denoted by μ and the coefficients $\mu_{[\mathcal{A}^\mu_{[\mathrm{red}]}]}$ by $\mu_{[\mathrm{red}]}$, e.g., by $\mu_{[\mathrm{x}]}$, where x stands for a simple symbol associated with the reduced space $\mathcal{A}^\mu_{[\mathrm{red}]}$.

2.4 The Shear Viscosity

2.4.1 The System $H\alpha^\eta = \beta^\eta$

The matrix integral equation corresponding to ϕ^η is

$$\Im(\phi^\eta) = \Psi^\eta, \tag{2.4.1}$$

where Ψ^η can be written

$$\Psi^\eta = 2\sum_{k\in\mathcal{S}} \phi^{2000k}, \tag{2.4.2}$$

from (2.1.25)(2.2.20). By isotropy, the functional space $\mathcal{A}^\eta$ to be considered in the first place is the space spanned by ϕ^{2000k}, $k \in \mathcal{S}$, [WT62] [MYM63]

$$\mathcal{A}^\eta = \mathrm{span}\{\ \phi^{2000k},\ k \in \mathcal{S}\ \}, \tag{2.4.3}$$

and the corresponding indexing set $\mathcal{B}^\eta$ is

$$\mathcal{B}^\eta = \{00\}\times\mathcal{S}. \tag{2.4.4}$$

For convenience, ϕ^η is taken in the form

$$\phi^\eta = (2/\bar{p})\sum_{k\in\mathcal{S}} \alpha_k^{00\eta}\phi^{2000k}. \tag{2.4.5}$$

The matrix associated with the variational procedure is denoted by H and is rescaled such that $H_{kl}^{rs} = (2/5\bar{n}\bar{p})\left[\phi^{20rk}, \phi^{20sl}\right]$, $(r,k),(s,l) \in \mathcal{B}^\eta$, where $\bar{n}$ denotes the mixture number density given by $\bar{n} = \sum_{k\in\mathcal{S}} n_k$. We also rescale the right member $\beta_k^{r\eta} = (1/5\bar{n})\langle\!\langle \phi^{20rk}, \Psi^\eta \rangle\!\rangle$, $(r,k) \in \mathcal{B}^\eta$. We then have $H \in \mathbb{R}^{n,n}$, $\beta^\eta \in \mathbb{R}^n$, and the linear system for $\alpha^\eta \in \mathbb{R}^n$ reads

$$H\alpha^\eta = \beta^\eta. \tag{2.4.6}$$

An explicit calculation yields that

$$H_{kk}^{0000} = \sum_{\substack{l\in\mathcal{S}\\ l\neq k}} \frac{16}{5k_{\mathrm{B}}\overline{T}} X_k X_l \left[\frac{10}{3}\frac{m_k m_l}{(m_k+m_l)^2}\Omega_{kl}^{(1,1)} + \frac{m_l^2}{(m_k+m_l)^2}\Omega_{kl}^{(2,2)}\right] + \frac{8}{5k_{\mathrm{B}}\overline{T}} X_k^2 \Omega_{kk}^{(2,2)}, \qquad k\in\mathcal{S}, \tag{2.4.7}$$

$$H_{kl}^{0000} = \frac{16}{5k_{\mathrm{B}}\overline{T}} X_k X_l \frac{m_k m_l}{(m_k+m_l)^2}\left[-\frac{10}{3}\Omega_{kl}^{(1,1)} + \Omega_{kl}^{(2,2)}\right], \quad k,l\in\mathcal{S}, \quad k\neq l, \tag{2.4.8}$$

and that

$$\beta_k^{00\eta} = X_k, \qquad k\in\mathcal{S}. \tag{2.4.9}$$

All the constraint vectors are found to be zero vectors so that the constraints (2.3.5) are automatically satisfied. The shear viscosity is finally given by

$$\eta = \langle \alpha^\eta, \beta^\eta \rangle = \langle \alpha^{00\eta}, \beta^{00\eta} \rangle = \sum_{k\in\mathcal{S}} X_k \alpha_k^{00\eta}. \tag{2.4.10}$$

2.4.2 Mathematical Structure of the System $H\alpha^\eta = \beta^\eta$

Properties of $\mathcal{A}^\eta$. The approximation space $\mathcal{A}^\eta$ introduced in (2.4.3) satisfies the perpendicularity property (2.3.20), the species orthogonality property (2.3.32), and the relation (2.3.36) from (2.3.54). In addition, the basis functions of $\mathcal{A}^\eta$ satisfy the species localization property (2.3.25). As a consequence, the results (2.3.37) established in Section 2.3.5 apply.

Properties of H. We now deduce from (2.3.37) that the matrices H, $2db(H)-H$, and $db(H)$ are symmetric positive definite for any $n \geq 1$. Note also that in our particular case, only one mode has been selected in the basis functions, i.e., the mode $\{00\}$, so that the matrix $db(H)$ is simply the diagonal of H.

2.4.3 Alternative Formulations

Higher order expansions could easily be generated for ϕ^η by considering additional basis functions, e.g., ϕ^{2010k}, $k \in \mathcal{S}$, and ϕ^{2001k}, $k \in \mathcal{P}$, as suggested in [MMW83]. These additional terms are believed to influence η only weakly [MM63] and will not be considered in this work. On the other hand, the one-dimensional functional space spanned by Ψ^η could have been considered and corresponds to the first iterate of the conjugate gradient algorithm applied to the system (2.4.6). This algorithm and its preconditioned version will be presented in Section 5.1.4.

2.5 The Volume Viscosity

First note that if there are no polyatomic species then $\kappa = 0$ and the classical result of monatomic gas mixtures [FK72] is recovered. Indeed, in this situation, we have $c_k^{\text{int}} = 0$, $k \in \mathcal{S}$, so that $c^{\text{int}} = \sum_{k\in\mathcal{S}} X_k c_k^{\text{int}} = 0$, and $\bar{\epsilon}_k - \epsilon_k = 0$, $k \in \mathcal{S}$, so that $\Psi_k^\kappa = 0$, $k \in \mathcal{S}$, from (2.1.25). The unique solution to the integral equation in ϕ^κ is then $\phi^\kappa = 0$ and thus $\kappa = 0$ from (2.1.37) [WT62]. In this section we may thus assume that there is at least one polyatomic species so that $p \geq 1$.

2.5.1 The System $K\alpha^\kappa = \beta^\kappa$

The scalar integral equation corresponding to ϕ^κ is

$$\mathfrak{I}(\phi^\kappa) = \Psi^\kappa, \tag{2.5.1}$$

where Ψ^κ can be written

$$\Psi^\kappa = -(2c^{\text{int}}/c_{\text{v}}) \sum_{k\in\mathcal{S}} \phi^{0010k} + (2c_{\text{v}}^{\text{tr}}/c_{\text{v}}) \sum_{k\in\mathcal{P}} \phi^{0001k}, \tag{2.5.2}$$

from (2.1.25)(2.2.17). A first approximation space for ϕ^κ is the functional space $\mathcal{A}^\kappa$ spanned by ϕ^{0010k}, $k \in \mathcal{S}$, and ϕ^{0001k}, $k \in \mathcal{P}$,

$$\mathcal{A}^\kappa = \text{span}\{\ \phi^{0010k},\ k \in \mathcal{S},\quad \phi^{0001k},\ k \in \mathcal{P}\ \}, \tag{2.5.3}$$

and the corresponding indexing set $\mathcal{B}^\kappa$ is

$$\mathcal{B}^\kappa = \{10\}\times\mathcal{S}\ \cup\ \{01\}\times\mathcal{P}. \tag{2.5.4}$$

For convenience, ϕ^κ is taken in the form

$$\phi^\kappa = -\frac{3}{\overline{p}}\Big(\sum_{k\in\mathcal{S}} \alpha_k^{10\kappa}\phi^{0010k} + \sum_{k\in\mathcal{P}} \alpha_k^{01\kappa}\phi^{0001k}\Big). \tag{2.5.5}$$

The matrix associated with the variational procedure will be denoted by K and is rescaled such that $K_{kl}^{rs} = (1/\overline{n}\overline{p})[\phi^{00rk}, \phi^{00sl}]$, $(r,k),(s,l) \in \mathcal{B}^\kappa$, where $\overline{n}$ denotes the mixture number density. The right member is also rescaled such that $\beta_k^{r\kappa} = -(1/3\overline{n})\langle\!\langle \phi^{00rk}, \Psi^\kappa \rangle\!\rangle$, $(r,k) \in \mathcal{B}^\kappa$. We then have $K \in \mathbb{R}^{n+p,n+p}$, $\beta^\kappa \in \mathbb{R}^{n+p}$, and the system in $\alpha^\kappa \in \mathbb{R}^{n+p}$ finally reads

$$K\alpha^\kappa = \beta^\kappa. \tag{2.5.6}$$

An explicit calculation yields that

$$\begin{aligned} K_{kk}^{1010} = & \sum_{\substack{l\in\mathcal{S}\\ l\neq k}} \frac{4}{k_\mathrm{B}\overline{T}} X_k X_l \Big[4\frac{m_k m_l}{(m_k+m_l)^2}\Omega_{kl}^{(1,1)} + \frac{m_l^2}{(m_k+m_l)^2}\big[(\Delta\epsilon_{kl})^2\big]_{kl} \Big] \\ & + \frac{2}{k_\mathrm{B}\overline{T}} X_k^2 \big[(\Delta\epsilon_{kk})^2\big]_{kk}, \qquad k\in\mathcal{S}, \end{aligned} \tag{2.5.7}$$

$$\begin{aligned} K_{kl}^{1010} = & \frac{4}{k_\mathrm{B}\overline{T}} X_k X_l \frac{m_k m_l}{(m_k+m_l)^2}\Big[-4\Omega_{kl}^{(1,1)} + \big[(\Delta\epsilon_{kl})^2\big]_{kl}\Big], \\ & k,l\in\mathcal{S}, \quad k\neq l, \end{aligned} \tag{2.5.8}$$

$$\begin{aligned} K_{kk}^{1001} = & -\sum_{\substack{l\in\mathcal{S}\\ l\neq k}} \frac{4}{k_\mathrm{B}\overline{T}} X_k X_l \frac{m_l}{(m_k+m_l)}\big[\Delta\epsilon_k \Delta\epsilon_{kl}\big]_{kl} \\ & - \frac{2}{k_\mathrm{B}\overline{T}} X_k^2 \big[(\Delta\epsilon_{kk})^2\big]_{kk}, \qquad k\in\mathcal{P}, \end{aligned} \tag{2.5.9}$$

$$K_{kl}^{1001} = -\frac{4}{k_\mathrm{B}\overline{T}} X_k X_l \frac{m_l}{m_k+m_l}[\![\widetilde{\Delta}\epsilon_l \Delta\epsilon_{kl}]\!]_{kl}, \qquad k\in\mathcal{S}, \quad l\in\mathcal{P}, \quad k\neq l, \tag{2.5.10}$$

$$K_{kk}^{0101} = \sum_{\substack{l\in\mathcal{S}\\ l\neq k}} \frac{4}{k_\mathrm{B}\overline{T}} X_k X_l \big[(\Delta\epsilon_k)^2\big]_{kl} + \frac{2}{k_\mathrm{B}\overline{T}} X_k^2 \big[(\Delta\epsilon_{kk})^2\big]_{kk}, \qquad k\in\mathcal{P}, \tag{2.5.11}$$

$$K_{kl}^{0101} = \frac{4}{k_\mathrm{B}\overline{T}} X_k X_l \big[\Delta\epsilon_k \widetilde{\Delta}\epsilon_l\big]_{kl}, \qquad k,l\in\mathcal{P}, \quad k\neq l, \tag{2.5.12}$$

and that

$$\beta_k^{10\kappa} = \frac{c^{\mathrm{int}}}{c_v} X_k, \qquad k\in\mathcal{S}, \tag{2.5.13}$$

$$\beta_k^{01\kappa} = -\frac{c_k^{\mathrm{int}}}{c_v} X_k, \qquad k\in\mathcal{P}. \tag{2.5.14}$$

The constraint vectors $\mathcal{G}^{l\nu}$, $(l,\nu) \in [1,n+4]\times\{1\}$, associated with the variational space $\mathcal{A}^\kappa$ are then zero vectors for $1 \le l \le n+3$, whereas $\mathcal{G}^{(n+4)1}$, corresponding to $\mathcal{T}_1\widehat{\psi}^{n+4}$, is proportional to $\mathcal{K} \in \mathbb{R}^{n+p}$, where $\mathcal{K}_k^{10} = X_k c_{\rm V}^{\rm tr}$, $k \in \mathcal{S}$, and $\mathcal{K}_k^{01} = X_k c_k^{\rm int}$, $k \in \mathcal{P}$. We restate that $c_{\rm V}^{\rm tr} = \frac{3}{2}k_{\rm B}$ and $c_k^{\rm int}$ are the translational constant volume specific heat per molecule and the internal specific heat per molecule of the $k^{\rm th}$ species, respectively. This yields the linear relation

$$\langle \mathcal{K}, \alpha^\kappa \rangle = \sum_{k\in\mathcal{S}} X_k c_{\rm V}^{\rm tr} \alpha_k^{10\kappa} + \sum_{k\in\mathcal{P}} X_k c_k^{\rm int} \alpha_k^{01\kappa} = 0. \tag{2.5.15}$$

The volume viscosity is finally given by

$$\kappa = \langle \alpha^\kappa, \beta^\kappa \rangle = \langle \alpha^{10\kappa}, \beta^{10\kappa} \rangle + \langle \alpha^{01\kappa}, \beta^{01\kappa} \rangle, \tag{2.5.16}$$

which can be simplified into the relations

$$\kappa = \sum_{k\in\mathcal{S}} X_k \alpha_k^{10\kappa} = -\sum_{k\in\mathcal{P}} X_k (c_k^{\rm int}/c_{\rm V}^{\rm tr}) \alpha_k^{01\kappa}, \tag{2.5.17}$$

by explicitly using the constraint (2.5.15). Note also that under the particular approximation (2.5.5), we recover that $\kappa = 0$ when there are no polyatomic species since then $c_k^{\rm int} = 0$, $k \in \mathcal{S}$, so that $c^{\rm int} = \sum_{k\in\mathcal{S}} X_k c_k^{\rm int} = 0$ and thus $\beta^\kappa = 0$ from (2.5.13) (2.5.14) and $\kappa = 0$ from (2.5.16).

2.5.2 Mathematical Structure of the System $K\alpha^\kappa = \beta^\kappa$

Properties of $\mathcal{A}^\kappa$. The approximation space $\mathcal{A}^\kappa$ introduced in (2.5.3) satisfies the perpendicularity property (2.3.20), the species orthogonality property (2.3.32), and the relation (2.3.38) from (2.3.46). In addition, the basis functions satisfy the species localization property (2.3.25). As a consequence, the results (2.3.41)–(2.3.43) established in Section 2.3.6 apply.

Properties of K. The constrained space $\mathcal{C}$ is given by $\mathcal{C} = \mathcal{K}^\perp$, where $\mathcal{K}$ has already been evaluated in Section 2.5.1. The matrix K is therefore symmetric positive semi-definite and positive definite on $\mathcal{K}^\perp$ for any $n \ge 1$. We also deduce from

$$\mathcal{I}_S^0 \cap \mathcal{A}^\kappa = \mathbb{R}\Big(\sum_{k\in\mathcal{S}} \phi^{0010k} + \sum_{k\in\mathcal{P}} \phi^{0001k}\Big),$$

that the nullspace of K is $N(K) = \mathbb{R}\mathcal{V}$, where $\mathcal{V} \in \mathbb{R}^{n+p}$ is the vector whose components are unity, i.e., $\mathcal{V}_k^{10} = 1$, $k \in \mathcal{S}$, and $\mathcal{V}_k^{01} = 1$, $k \in \mathcal{P}$. We also have the important property that $\langle \mathcal{K}, \mathcal{V} \rangle \neq 0$.

Furthermore, the matrix $2db(K) - K$ is symmetric positive semi-definite for any $n \geq 1$. In addition, the matrix $2db(K) - K$ is positive definite if $n \geq 3$ whereas in the special case $n = 2$ and $\mathcal{S} = \{1, 2\}$, the nullspace of $2db(K)-K$ is $N\big(2db(K)-K\big) = \mathbb{R}\mathcal{V}^*$, where $\mathcal{V}^* \in \mathbb{R}^{n+p}$ is the vector whose components are given by $\mathcal{V}_1^{r*} = 1$, $r \in \mathcal{F}_1$, and $\mathcal{V}_2^{r*} = -1$, $r \in \mathcal{F}_2$. In other words, we have $\mathcal{V}^* = (1, -1, 1, -1)$ if $p = 2$, and either $\mathcal{V}^* = (1, -1, 1)$ or $\mathcal{V}^* = (1, -1, -1)$ if $p = 1$, depending on which molecule is polyatomic, the case $p = 0$ being excluded by assumption. Finally, in the special case $n = p = 1$, the nullspace of $2db(K) - K$ is $N\big(2db(K) - K\big) = \mathbb{R}\mathcal{V}$, so that $N\big(2db(K) - K\big) = \mathbb{R}(1, 1)$.

Finally, the matrix $db(K)$ is symmetric positive semi-definite for any $n \geq 1$. In addition, the matrix $db(K)$ is positive definite for $n \geq 2$, and in the special case $n = 1$, the nullspace of $db(K)$ is $N\big(db(K)\big) = \mathbb{R}\mathcal{V}$, so that $N\big(db(K)\big) = \mathbb{R}(1, 1)$.

2.5.3 The System $K_{[01]}\alpha^{\kappa}_{[01]} = \beta^{\kappa}_{[01]}$

A first simplification for evaluating the volume viscosity is suggested from the relation $\kappa = -\sum_{k\in\mathcal{P}} X_k (c_k^{\mathrm{int}}/c_{\mathrm{v}}^{\mathrm{tr}})\alpha_k^{01\kappa}$. This relation shows that any approximation of $\alpha^{01\kappa}$ will yield a formula for κ and thus suggests the use of the basis functions ϕ^{0001k}, $k \in \mathcal{P}$. However, because of the energy constraint $\langle\!\langle \phi^{\kappa}, \widehat{\psi}^{n+4} \rangle\!\rangle = 0$ that must be satisfied, we have to use instead the projected basis functions $\widehat{\phi}^{0001k} = \phi^{0001k} - a_k \widehat{\psi}^{n+4}$ where a_k is evaluated in such a way that $\langle\!\langle \widehat{\phi}^{0001k}, \widehat{\psi}^{n+4} \rangle\!\rangle = 0$. The terms in $\widehat{\psi}^{n+4}$ guarantee the well posedness of the linear system and that the energy constraint is satisfied for the approximate perturbed distribution function ϕ^{κ}. The functional space to be considered is thus the space spanned by $\widehat{\phi}^{0001k}$, $k \in \mathcal{P}$,

$$\mathcal{A}^{\kappa}_{[01]} = \mathrm{span}\{\ \widehat{\phi}^{0001k},\ k \in \mathcal{P}\ \}, \tag{2.5.18}$$

and the corresponding indexing set $\mathcal{B}^{\kappa}_{[01]}$ is

$$\mathcal{B}^{\kappa}_{[01]} = \{01\} \times \mathcal{P}. \tag{2.5.19}$$

The associated linear system is obtained from the formalism (2.3.59)–(2.3.68) which directly applies to our case. The transformation matrix $\widehat{\mathcal{R}}$ expressing the new basis functions $\widehat{\phi}^{0001k}$, $k \in \mathcal{P}$, as linear combinations of the old basis functions ϕ^{0010k}, $k \in \mathcal{S}$, and ϕ^{0001k}, $k \in \mathcal{P}$, has its coefficients $\widehat{\mathcal{R}}_{kl}^{rs}$, $(r,k) \in \mathcal{B}^{\kappa}_{[01]}$, $(s,l) \in \mathcal{B}^{\kappa}$, given by

$$\begin{cases} \widehat{\mathcal{R}}_{kl}^{0110} = -\, X_k c_k^{\mathrm{int}}/c_{\mathrm{v}}, & k \in \mathcal{P},\quad l \in \mathcal{S}, \\ \widehat{\mathcal{R}}_{kl}^{0101} = \delta_{kl} - X_k c_k^{\mathrm{int}}/c_{\mathrm{v}}, & k \in \mathcal{P},\quad l \in \mathcal{P}. \end{cases} \tag{2.5.20}$$

From (2.3.64) we also know that the simplified transformation matrix $\mathcal{R}$ given by

$$\begin{cases} \mathcal{R}_{kl}^{0110} = 0, & k \in \mathcal{P}, \quad l \in \mathcal{S}, \\ \mathcal{R}_{kl}^{0101} = \delta_{kl}, & k \in \mathcal{P}, \quad l \in \mathcal{P}, \end{cases} \tag{2.5.21}$$

can be used instead of $\widehat{\mathcal{R}}$ in order to evaluate the reduced transport linear system. This reduced transport linear system will be denoted by

$$K_{[01]}\alpha_{[01]}^{\kappa} = \beta_{[01]}^{\kappa}, \tag{2.5.22}$$

and from $K_{[01]} = \mathcal{R}K\mathcal{R}^t$ and $\beta_{[01]}^{\kappa} = \mathcal{R}\beta^{\kappa}$ we obtain that

$$K_{[01]} = K^{0101}, \tag{2.5.23}$$

and

$$\beta_{[01]}^{\kappa} = \beta^{01\kappa}. \tag{2.5.24}$$

The associated volume viscosity is then given by

$$\kappa_{[01]} = \langle \alpha_{[01]}^{\kappa}, \beta_{[01]}^{\kappa} \rangle = \langle \alpha_{[01]}^{01\kappa}, \beta_{[01]}^{01\kappa} \rangle, \tag{2.5.25}$$

thus involving the solution of a linear system of size p instead of size $n+p$.

Remark. The corresponding volume viscosity $\kappa_{[01]}$ will be shown to be accurate in Section 6.3.2. This reveals the importance of the components of ϕ^{κ} along the ϕ^{0001k}, $k \in \mathcal{P}$, whereas a global energy constraint using $\widehat{\psi}^{n+4}$ can accurately replace the components along the ϕ^{0010k}, $k \in \mathcal{S}$.

2.5.4 Mathematical Structure of the System $K_{[01]}\alpha_{[01]}^{\kappa} = \beta_{[01]}^{\kappa}$

Properties of $\mathcal{A}_{[01]}^{\kappa}$. The approximation space $\mathcal{A}_{[01]}^{\kappa}$ introduced in (2.5.18) satisfies the assumptions associated with (2.3.59). Furthermore, we have

$$R(\mathcal{R}^t) = \operatorname{span}\{\, e^{01k},\ k \in \mathcal{P} \,\} = \{\, x \in \mathbb{R}^{n+p},\ x_k^{10} = 0,\ k \in \mathcal{S} \,\},$$

so that $R(\mathcal{R}^t) \cap N(K) = \{0\}$ and $R(\mathcal{R}^t) \cap N\big(2db(K) - K\big) = \{0\}$ for any $n \geq 1$ since the vectors $\mathcal{V}$ and $\mathcal{V}^*$ always have nonzero components along the vectors e_k^{10}, $k \in \mathcal{S}$.

Properties of $K_{[01]}$. We now deduce from (2.3.67)(2.3.68) that the matrices $K_{[01]}$ and $2db(K_{[01]}) - K_{[01]}$ are symmetric positive definite for any $n \geq 1$, and therefore that the matrix $db(K_{[01]})$ is also symmetric positive definite for any $n \geq 1$.

2.5.5 The System $K_{[10]}\alpha^{\kappa}_{[10]} = \beta^{\kappa}_{[10]}$

A second simplification, analogous to the previous one, is suggested from the relation $\kappa = \sum_{k\in\mathcal{S}} X_k \alpha_k^{10\kappa}$ deduced from (2.5.17). This relation indicates that any approximation of $\alpha^{10\kappa}$ will yield a formula for κ and thus suggests the use of the basis functions ϕ^{0010k}, $k \in \mathcal{S}$. Due to the energy constraint $\langle\langle \phi^{\kappa}, \widehat{\psi}^{n+4} \rangle\rangle = 0$ that must be satisfied, we have to use the projected basis functions $\widehat{\phi}^{0010k} = \phi^{0010k} - a_k\widehat{\psi}^{n+4}$ where a_k is evaluated in such a way that $\langle\langle \widehat{\phi}^{0010k}, \widehat{\psi}^{n+4} \rangle\rangle = 0$. The functional space to be considered is thus the space spanned by $\widehat{\phi}^{0010k}$, $k \in \mathcal{S}$,

$$\mathcal{A}^{\kappa}_{[10]} = \mathrm{span}\{\ \widehat{\phi}^{0010k},\ k \in \mathcal{S}\ \}, \tag{2.5.26}$$

and the corresponding indexing set $\mathcal{B}^{\kappa}_{[10]}$ is

$$\mathcal{B}^{\kappa}_{[10]} = \{10\}\times\mathcal{S}. \tag{2.5.27}$$

The associated linear system is obtained from the formalism (2.3.59)–(2.3.68) which directly applies to our case. The transformation matrix $\widehat{\mathcal{R}}$ expressing the new basis functions $\widehat{\phi}^{0010k}$, $k \in \mathcal{S}$, as linear combinations of the old basis functions ϕ^{0010k}, $k \in \mathcal{S}$, and ϕ^{0001k}, $k \in \mathcal{P}$, has its coefficients $\widehat{\mathcal{R}}^{rs}_{kl}$, $(r,k) \in \mathcal{B}^{\kappa}_{[10]}$, $(s,l) \in \mathcal{B}^{\kappa}$, given by

$$\begin{cases} \widehat{\mathcal{R}}^{1010}_{kl} = \delta_{kl} - X_k c_v^{\mathrm{tr}}/c_v, & k \in \mathcal{S},\quad l \in \mathcal{S}, \\ \widehat{\mathcal{R}}^{1001}_{kl} = \ - X_k c_v^{\mathrm{tr}}/c_v, & k \in \mathcal{S},\quad l \in \mathcal{P}. \end{cases} \tag{2.5.28}$$

From (2.3.64) we also know that the simplified transformation matrix $\mathcal{R}$ given by

$$\begin{cases} \mathcal{R}^{1010}_{kl} = \delta_{kl}, & k \in \mathcal{S},\quad l \in \mathcal{S}, \\ \mathcal{R}^{1001}_{kl} = 0, & k \in \mathcal{S},\quad l \in \mathcal{P}, \end{cases} \tag{2.5.29}$$

can be used instead of $\widehat{\mathcal{R}}$ in order to evaluate the reduced transport linear system. This reduced transport linear system will be denoted by

$$K_{[10]}\alpha^{\kappa}_{[10]} = \beta^{\kappa}_{[10]}, \tag{2.5.30}$$

and from $K_{[10]} = \mathcal{R}K\mathcal{R}^t$ and $\beta^{\kappa}_{[10]} = \mathcal{R}\beta^{\kappa}$ we obtain that

$$K_{[10]} = K^{1010}, \tag{2.5.31}$$

and

$$\beta^{\kappa}_{[10]} = \beta^{10\kappa}. \tag{2.5.32}$$

The associated volume viscosity is then given by

$$\kappa_{[10]} = \langle \alpha^{\kappa}_{[10]}, \beta^{\kappa}_{[10]} \rangle = \langle \alpha^{10\kappa}_{[10]}, \beta^{10\kappa}_{[10]} \rangle, \tag{2.5.33}$$

thus involving the solution of a linear system of size n instead of size $n+p$.

Remark. We will see in Section 6.3.2 that, although the volume viscosity $\kappa_{[01]}$ is accurate, the volume viscosity $\kappa_{[10]}$ is not.

2.5.6 Mathematical Structure of the System $K_{[10]}\alpha^{\kappa}_{[10]} = \beta^{\kappa}_{[10]}$

Properties of $\mathcal{A}^{\kappa}_{[10]}$. The approximation space $\mathcal{A}^{\kappa}_{[10]}$ satisfies the assumptions associated with (2.3.59). In addition we have

$$R(\mathcal{R}^t) = \operatorname{span}\{\, e^{10k},\ k \in \mathcal{S} \,\} = \{\, x \in \mathbb{R}^{n+p},\ x^{01}_k = 0,\ k \in \mathcal{P} \,\},$$

so that $R(\mathcal{R}^t) \cap N(K) = \{0\}$ and $R(\mathcal{R}^t) \cap N\big(2db(K) - K\big) = \{0\}$ for any $n \geq 1$ since the vectors $\mathcal{V}$ and $\mathcal{V}^*$ always have nonzero components along the vectors e^{01}_k, $k \in \mathcal{P}$, keeping in mind that $p \geq 1$ in this section.

Properties of $K_{[10]}$. We now deduce from (2.3.67)(2.3.68) that the matrices $K_{[10]}$ and $2db(K_{[10]}) - K_{[10]}$ are symmetric positive definite for any $n \geq 1$, and therefore that the matrix $db(K_{[10]})$ is also symmetric positive definite for any $n \geq 1$.

2.5.7 The System $K_{[\mathrm{d}]}\alpha^{\kappa}_{[\mathrm{d}]} = \beta^{\kappa}_{[\mathrm{d}]}$

A third simplification can be obtained by rewriting the right member Ψ^{κ} in the form $\Psi^{\kappa} = -(2c^{\mathrm{int}}/c_{\mathrm{v}}) \sum_{k\in\mathcal{S}} \phi^{00\mathrm{d}k}$, where we have defined

$$\phi^{00\mathrm{d}k} = \phi^{0010k} - (c^{\mathrm{tr}}_{\mathrm{v}}/c^{\mathrm{int}})\delta_{\mathcal{P}}(k)\phi^{0001k}, \qquad k \in \mathcal{S}, \tag{2.5.34}$$

and

$$\delta_{\mathcal{P}}(k) = \begin{cases} 1 & \text{if } k \in \mathcal{P}, \\ 0 & \text{otherwise.} \end{cases} \tag{2.5.35}$$

The superscript d indicates that a weighted difference between ϕ^{0010k} and ϕ^{0001k} is used to form $\phi^{00\mathrm{d}k}$. The expression $\Psi^{\kappa} = -(2c^{\mathrm{int}}/c_{\mathrm{v}}) \sum_{k\in\mathcal{S}} \phi^{00\mathrm{d}k}$ now suggests the use of the approximation space spanned by $\phi^{00\mathrm{d}k}$, $k \in \mathcal{S}$. This method of simplification extends the one of Thijsse et al. [TTCKB79] and Van den Oord and Korving [VK88] for the thermal conductivity of pure mixtures and McCourt et al. [MBKK90] for the partial thermal conductivity of binary mixtures. Because of the constraint $\langle\!\langle \phi^\kappa, \widehat{\psi}^{n+4} \rangle\!\rangle = 0$ that must be satisfied, it is necessary to use the projected basis functions $\widehat{\phi}^{00\mathrm{d}k} = \phi^{00\mathrm{d}k} - a_k \widehat{\psi}^{n+4}$ where a_k is evaluated in such a way that $\langle\!\langle \widehat{\phi}^{00\mathrm{d}k}, \widehat{\psi}^{n+4} \rangle\!\rangle = 0$. The functional space to be considered is thus the space spanned by $\widehat{\phi}^{00\mathrm{d}k}$, $k \in \mathcal{S}$,

$$\mathcal{A}^\kappa_{[\mathrm{d}]} = \mathrm{span}\{\ \widehat{\phi}^{00\mathrm{d}k},\ k \in \mathcal{S}\ \}, \tag{2.5.36}$$

and the corresponding indexing set $\mathcal{B}^\kappa_{[\mathrm{d}]}$ is

$$\mathcal{B}^\kappa_{[\mathrm{d}]} = \{\mathrm{d}\} \times \mathcal{S}. \tag{2.5.37}$$

The associated linear system is obtained from the formalism (2.3.59)–(2.3.68) which directly applies to our case. The transformation matrix $\widehat{\mathcal{R}}$ expressing the new basis functions $\widehat{\phi}^{00\mathrm{d}k}$, $k \in \mathcal{S}$, as linear combinations of the old basis functions ϕ^{0010k}, $k \in \mathcal{S}$, and ϕ^{0001k}, $k \in \mathcal{P}$, has its coefficients $\widehat{\mathcal{R}}^{rs}_{kl}$, $(r,k) \in \mathcal{B}^\kappa_{[\mathrm{d}]}$, $(s,l) \in \mathcal{B}^\kappa$, given by

$$\begin{cases} \widehat{\mathcal{R}}^{\mathrm{d}10}_{kl} = \delta_{kl} - (c_{\mathrm{v}}^{\mathrm{tr}}/c_{\mathrm{v}}) X_k (1 - c_k^{\mathrm{int}}/c^{\mathrm{int}}), & k,l \in \mathcal{S}, \\ \widehat{\mathcal{R}}^{\mathrm{d}01}_{kl} = -(c_{\mathrm{v}}^{\mathrm{tr}}/c^{\mathrm{int}})\delta_{kl} - (c_{\mathrm{v}}^{\mathrm{tr}}/c_{\mathrm{v}}) X_k (1 - c_k^{\mathrm{int}}/c^{\mathrm{int}}), & k \in \mathcal{S}, \quad l \in \mathcal{P}. \end{cases} \tag{2.5.38}$$

From (2.3.64) we also know that the simplified transformation matrix $\mathcal{R}$ given by

$$\begin{cases} \mathcal{R}^{\mathrm{d}10}_{kl} = \delta_{kl}, & k,l \in \mathcal{S}, \\ \mathcal{R}^{\mathrm{d}01}_{kl} = -(c_{\mathrm{v}}^{\mathrm{tr}}/c^{\mathrm{int}})\delta_{kl}, & k \in \mathcal{S}, \quad l \in \mathcal{P}, \end{cases} \tag{2.5.39}$$

can be used instead of $\widehat{\mathcal{R}}$ in order to evaluate the reduced transport linear system. This reduced transport linear system will be denoted by

$$K_{[\mathrm{d}]} \alpha^\kappa_{[\mathrm{d}]} = \beta^\kappa_{[\mathrm{d}]}, \tag{2.5.40}$$

and from $K_{[\mathrm{d}]} = \mathcal{R} K \mathcal{R}^t$ and $\beta^\kappa_{[\mathrm{d}]} = \mathcal{R}\beta^\kappa$ we obtain that

$$\begin{aligned} K^{\mathrm{dd}}_{[\mathrm{d}]kl} = K^{1010}_{kl} &- \frac{c_{\mathrm{v}}^{\mathrm{tr}}}{c^{\mathrm{int}}} \delta_{\mathcal{P}}(k) K^{0110}_{kl} - \frac{c_{\mathrm{v}}^{\mathrm{tr}}}{c^{\mathrm{int}}} \delta_{\mathcal{P}}(l) K^{1001}_{kl} \\ &+ \left(\frac{c_{\mathrm{v}}^{\mathrm{tr}}}{c^{\mathrm{int}}}\right)^2 \delta_{\mathcal{P}}(k) \delta_{\mathcal{P}}(l) K^{0101}_{kl}, \qquad k,l \in \mathcal{S}, \end{aligned} \tag{2.5.41}$$

and

$$\beta^{\mathrm{d}\kappa}_{[\mathrm{d}]k} = \beta^{10\kappa}_k - \frac{c_{\mathrm{v}}^{\mathrm{tr}}}{c^{\mathrm{int}}} \delta_{\mathcal{P}}(k) \beta^{01\kappa}_k, \qquad k \in \mathcal{S}. \tag{2.5.42}$$

The associated volume viscosity is then given by

$$\kappa_{[\mathrm{d}]} = \langle \alpha^\kappa_{[\mathrm{d}]}, \beta^\kappa_{[\mathrm{d}]} \rangle, = \langle \alpha^{\mathrm{d}\kappa}_{[\mathrm{d}]}, \beta^{\mathrm{d}\kappa}_{[\mathrm{d}]} \rangle, \tag{2.5.43}$$

thus involving the solution of a linear system of size n instead of size $n+p$.

Remark. We will see in Section 6.3.2 that the volume viscosity $\kappa_{[\mathrm{d}]}$ is not accurate.

2.5.8 Mathematical Structure of the System $K_{[\mathrm{d}]}\alpha^{\kappa}_{[\mathrm{d}]} = \beta^{\kappa}_{[\mathrm{d}]}$

Properties of $\mathcal{A}^{\kappa}_{[\mathrm{d}]}$. The approximation space $\mathcal{A}^{\kappa}_{[\mathrm{d}]}$ introduced in (2.5.36) satisfies the assumptions associated with (2.3.59). In addition we have

$$R(\mathcal{R}^t) = \mathrm{span}\{\ e^{10k} - (c_{\mathrm{v}}^{\mathrm{tr}}/c^{\mathrm{int}})\delta_{\mathcal{P}}(k)e^{01k},\ k \in \mathcal{S}\ \},$$

so that

$$R(\mathcal{R}^t) = \{\ x \in \mathbb{R}^{n+p},\ x_k^{10} = -x_k^{01}(c^{\mathrm{int}}/c_{\mathrm{v}}^{\mathrm{tr}}),\ k \in \mathcal{P}\ \},$$

and thus $R(\mathcal{R}^t) \cap N(K) = \{0\}$ and $R(\mathcal{R}^t) \cap N\big(2db(K) - K\big) = \{0\}$ for any $n \geq 1$. Indeed, the vectors $\mathcal{V}$ and $\mathcal{V}^*$ always have nonzero components along the vectors e^{10k} and e^{01k}, $k \in \mathcal{P}$, which have identical signs, keeping in mind that $p \geq 1$ in this section.

Properties of $K_{[\mathrm{d}]}$. We then deduce from (2.3.67)(2.3.68) that the matrices $K_{[\mathrm{d}]}$ and $2db(K_{[\mathrm{d}]}) - K_{[\mathrm{d}]}$ are symmetric positive definite for any $n \geq 1$, and therefore that the matrix $db(K_{[\mathrm{d}]})$ is also symmetric positive definite for any $n \geq 1$.

2.5.9 Alternative Formulations

We note that higher order approximations for κ could easily be generated by considering additional basis functions, as, for instance, the functions ϕ^{0020k}, $k \in \mathcal{S}$, ϕ^{0011k}, $k \in \mathcal{P}$, and ϕ^{0002k}, $k \in \{\, k \in \mathcal{S},\ \mathrm{card}(\mathcal{E}_k) \geq 3\,\}$. The influence of these additional basis functions on the volume viscosity is not known and will not be considered in this book. Finally, we note that the approximation obtained by using the two-dimensional space spanned by Ψ^{κ} and $\widehat{\psi}^{n+4}$ is equivalent to the first iterate of the conjugate gradient algorithm for singular systems applied to the system (2.5.6). This algorithm and its preconditioned version will be presented in Section 5.1.4.

2.6 The Diffusion Matrix

2.6.1 The System $L\alpha^{D_l} = \beta^{D_l}$

The vector integral equation corresponding to ϕ^{D_l}, $l \in \mathcal{S}$, is

$$\Im(\phi^{D_l}) = \Psi^{D_l}, \qquad l \in \mathcal{S}, \tag{2.6.1}$$

and Ψ^{D_l} can be written

$$\Psi^{D_l} = \sum_{k\in\mathcal{S}} \frac{\sqrt{2}}{\sqrt{m_k k_B T}} \frac{1}{n_k} (\delta_{lk} - Y_k)\phi^{1000k}, \qquad l \in \mathcal{S}, \tag{2.6.2}$$

from (2.1.25)(2.2.19). The simplest approximation for ϕ^{D_l}, $l \in \mathcal{S}$, is obtained by using the variational approximation space spanned by ϕ^{1000k}, $k \in \mathcal{S}$. The corresponding linear system will be discussed in Sections 2.6.5–2.6.6 and has been investigated in [Gi91]. In this section we focus on the second-order diffusion matrix obtained by using the functional space $\mathcal{A}^D$ spanned by ϕ^{1000k}, $k \in \mathcal{S}$, ϕ^{1010k}, $k \in \mathcal{S}$, and ϕ^{1001k}, $k \in \mathcal{P}$,

$$\mathcal{A}^D = \mathrm{span}\{\ \phi^{1000k},\ k \in \mathcal{S}, \quad \phi^{1010k},\ k \in \mathcal{S}, \quad \phi^{1001k},\ k \in \mathcal{P}\ \}, \tag{2.6.3}$$

and the corresponding indexing set $\mathcal{B}^D$ is

$$\mathcal{B}^D = \{00, 10\}\times\mathcal{S} \ \cup\ \{01\}\times\mathcal{P}. \tag{2.6.4}$$

For convenience, ϕ^{D_l} will be taken in the form

$$\phi^{D_l} = \frac{\sqrt{2}}{\overline{p}\sqrt{k_B T}}\Big(\sum_{k\in\mathcal{S}} \sqrt{m_k}\big(\alpha_k^{00D_l}\phi^{1000k} + \alpha_k^{10D_l}\phi^{1010k}\big) + \sum_{k\in\mathcal{P}} \sqrt{m_k}\alpha_k^{01D_l}\phi^{1001k}\Big). \tag{2.6.5}$$

The matrix associated with the variational procedure is denoted by L and is rescaled such that $L_{kl}^{rs} = (2\sqrt{m_k m_l}/3\overline{p})[\phi^{10rk}, \phi^{10sl}]$, $(r,k),(s,l) \in \mathcal{B}^D$, and the right member is rescaled such that $\beta_k^{rD_l} = (\sqrt{2k_B T m_k}/3)\langle\!\langle \phi^{10rk}, \Psi^{D_l}\rangle\!\rangle$, $(r,k) \in \mathcal{B}^D$. We then have $L \in \mathbb{R}^{2n+p,2n+p}$, $\beta^{D_l} \in \mathbb{R}^{2n+p}$, $l \in \mathcal{S}$, and the system in $\alpha^{D_l} \in \mathbb{R}^{2n+p}$, $l \in \mathcal{S}$, finally reads

$$L\alpha^{D_l} = \beta^{D_l}, \qquad l \in \mathcal{S}. \tag{2.6.6}$$

An explicit calculation yields that

$$L_{kk}^{0000} = \sum_{\substack{l\in\mathcal{S}\\ l\neq k}} \frac{X_k X_l}{\mathcal{D}_{kl}}, \qquad k \in \mathcal{S}, \tag{2.6.7}$$

$$L_{kl}^{0000} = -\frac{X_k X_l}{\mathcal{D}_{kl}}, \qquad k,l \in \mathcal{S}, \quad k \neq l, \tag{2.6.8}$$

$$L_{kk}^{0010} = -\sum_{\substack{l\in\mathcal{S}\\ l\neq k}} \frac{X_k X_l}{2\mathcal{D}_{kl}} \frac{m_l}{m_k+m_l}(6\bar{C}_{kl}-5), \qquad k \in \mathcal{S}, \tag{2.6.9}$$

$$L_{kl}^{0010} = \frac{X_k X_l}{2\mathcal{D}_{kl}} \frac{m_k}{m_k+m_l}(6\bar{C}_{kl}-5), \qquad k,l \in \mathcal{S}, \quad k \neq l, \tag{2.6.10}$$

$$L_{kk}^{0001} = -\sum_{\substack{l\in\mathcal{S}\\ l\neq k}} X_k X_l \frac{\left[\epsilon_{kK}^0(\gamma^2 - \gamma\gamma'\cos\chi)\right]_{kl}}{\Omega_{kl}^{(1,1)}\mathcal{D}_{kl}}, \qquad k \in \mathcal{P}, \tag{2.6.11}$$

$$L_{kl}^{0001} = X_k X_l \frac{\left[\epsilon_{lL}^0(\gamma^2 - \gamma\gamma'\cos\chi)\right]_{kl}}{\Omega_{kl}^{(1,1)}\mathcal{D}_{kl}}, \qquad k \in \mathcal{S}, \quad l \in \mathcal{P}, \quad k \neq l, \tag{2.6.12}$$

$$L_{kk}^{1010} = \sum_{\substack{l\in\mathcal{S}\\ l\neq k}} \frac{X_k X_l}{\mathcal{D}_{kl}} \frac{m_k m_l}{(m_k+m_l)^2}\Big[\frac{15}{2}\frac{m_k}{m_l} + \frac{25}{4}\frac{m_l}{m_k} - 3\frac{m_l}{m_k}\bar{B}_{kl} + 4\bar{A}_{kl}$$

$$+ \frac{25}{12}\frac{\left[(\Delta\epsilon_{kl})^2\right]_{kl}}{\Omega_{kl}^{(1,1)}}\Big] + \frac{X_k^2}{2\mathcal{D}_{kk}}\Big[4\bar{A}_{kk} + \frac{25}{12}\frac{\left[(\Delta\epsilon_{kk})^2\right]_{kk}}{\Omega_{kk}^{(1,1)}}\Big], \qquad k \in \mathcal{S}, \tag{2.6.13}$$

$$L_{kl}^{1010} = -\frac{X_k X_l}{\mathcal{D}_{kl}} \frac{m_k m_l}{(m_k+m_l)^2}\Big[\frac{55}{4} - 3\bar{B}_{kl} - 4\bar{A}_{kl}$$

$$- \frac{25}{12}\frac{\left[(\Delta\epsilon_{kl})^2\right]_{kl}}{\Omega_{kl}^{(1,1)}}\Big], \qquad k,l \in \mathcal{S}, \quad k \neq l, \tag{2.6.14}$$

$$L_{kk}^{1001} = -\sum_{\substack{l\in\mathcal{S}\\ l\neq k}} \frac{X_k X_l}{\mathcal{D}_{kl}} \frac{m_k}{m_k+m_l}\Big[\frac{5}{4}\frac{\left[\Delta\epsilon_k\Delta\epsilon_{kl}\right]_{kl}}{\Omega_{kl}^{(1,1)}} + \frac{5}{2}\frac{m_l}{m_k}\frac{\left[\epsilon_{kK}^0(\gamma^2 - \gamma\gamma'\cos\chi)\right]_{kl}}{\Omega_{kl}^{(1,1)}}$$

$$- \frac{m_l}{m_k}\frac{\left[\epsilon_{kK}^0(\gamma^4 - \gamma\gamma'^3\cos\chi)\right]_{kl}}{\Omega_{kl}^{(1,1)}}\Big] - \frac{5}{8}\frac{X_k^2}{\mathcal{D}_{kk}}\frac{\left[(\Delta\epsilon_{kk})^2\right]_{kk}}{\Omega_{kk}^{(1,1)}}, \qquad k \in \mathcal{P}, \tag{2.6.15}$$

$$L_{kl}^{1001} = \frac{X_k X_l}{\mathcal{D}_{kl}} \frac{m_l}{m_k+m_l}\Big[-\frac{5}{4}\frac{\left[\widetilde{\Delta}\epsilon_l\Delta\epsilon_{kl}\right]_{kl}}{\Omega_{kl}^{(1,1)}} + \frac{5}{2}\frac{\left[\epsilon_{lL}^0(\gamma^2 - \gamma\gamma'\cos\chi)\right]_{kl}}{\Omega_{kl}^{(1,1)}}$$

$$- \frac{\left[\epsilon_{lL}^0(\gamma^4 - \gamma\gamma'^3\cos\chi)\right]_{kl}}{\Omega_{kl}^{(1,1)}}\Big], \qquad k \in \mathcal{S}, \quad l \in \mathcal{P}, \quad k \neq l, \tag{2.6.16}$$

$$L_{kk}^{0101} = \sum_{\substack{l\in\mathcal{S}\\ l\neq k}} X_k X_l \Big[\frac{c_k^{\text{int}}}{k_B \mathcal{D}_{k\,\text{int},l}} + \frac{3}{4}\frac{m_k}{m_l}\frac{\left[(\Delta\epsilon_k)^2\right]_{kl}}{\Omega_{kl}^{(1,1)}\mathcal{D}_{kl}}\Big]$$

$$+ X_k^2 \Big[\frac{c_k^{\rm int}}{k_{\rm B} \mathcal{D}_{k\,{\rm int},k}} + \frac{3}{8} \frac{\big[(\Delta\epsilon_{kk})^2\big]_{kk}}{\Omega_{kk}^{(1,1)} \mathcal{D}_{kk}} \Big], \qquad k \in \mathcal{P}, \tag{2.6.17}$$

$$L_{kl}^{0101} = - X_k X_l \Big[\frac{\big[\epsilon_{kK}^0 (\epsilon_{lL}^0 \gamma^2 - \epsilon_{lL'}^0 \gamma\gamma' \cos\chi)\big]_{kl}}{\Omega_{kl}^{(1,1)} \mathcal{D}_{kl}}$$

$$- \frac{3}{4} \frac{\big[\Delta\epsilon_k \widetilde{\Delta}\epsilon_l\big]_{kl}}{\Omega_{kl}^{(1,1)} \mathcal{D}_{kl}} \Big], \qquad k, l \in \mathcal{P}, \quad k \neq l, \tag{2.6.18}$$

and that

$$\beta_k^{00D_l} = \delta_{kl} - Y_k, \qquad k, l \in \mathcal{S}, \tag{2.6.19}$$

$$\beta_k^{10D_l} = 0, \qquad k, l \in \mathcal{S}, \tag{2.6.20}$$

$$\beta_k^{01D_l} = 0, \qquad k \in \mathcal{P}, \quad l \in \mathcal{S}. \tag{2.6.21}$$

The constraint vectors $\mathcal{G}^{l\nu}$, $(l,\nu) \in [1, n+4] \times [1,3]$, associated with the variational space $\mathcal{A}^D$ are then zero vectors for $(l,\nu) \in \big(\mathcal{S} \cup \{n+4\}\big) \times [1,3]$ and for $(l,\nu) = (n+\nu', \nu)$, with $\nu, \nu' \in [1,3]$ and $\nu \neq \nu'$. Furthermore, the constraint vectors $\mathcal{G}^{l\nu}$, for $(l,\nu) = (n+\nu, \nu)$ and $\nu \in [1,3]$, are all proportional to $\mathcal{H} \in \mathbb{R}^{2n+p}$ given by $\mathcal{H}_k^{00} = \sqrt{m_k}\, n_k$, $k \in \mathcal{S}$, $\mathcal{H}_k^{10} = 0$, $k \in \mathcal{S}$, and $\mathcal{H}_k^{01} = 0$, $k \in \mathcal{P}$. Taking into account the rescaling of ϕ^{D_l}, $l \in \mathcal{S}$, we thus obtain the linear constraint

$$\langle \mathcal{L}, \alpha^{D_l} \rangle = \langle \mathcal{L}^{00}, \alpha^{00D_l} \rangle = \sum_{k \in \mathcal{S}} Y_k \alpha_k^{00D_l} = 0, \qquad l \in \mathcal{S}, \tag{2.6.22}$$

where the vector $\mathcal{L} \in \mathbb{R}^{2n+p}$ is given by $\mathcal{L}_k^{00} = Y_k$, $k \in \mathcal{S}$, $\mathcal{L}_k^{10} = 0$, $k \in \mathcal{S}$, and $\mathcal{L}_k^{01} = 0$, $k \in \mathcal{P}$. Finally, the diffusion coefficients are given by

$$D_{kl} = \langle \alpha^{D_l}, \beta^{D_k} \rangle = \langle \alpha^{D_k}, \beta^{D_l} \rangle = \langle \alpha^{00D_l}, \beta^{00D_k} \rangle = \langle \alpha^{00D_k}, \beta^{00D_l} \rangle, \tag{2.6.23}$$

which can be simplified into

$$D_{kl} = \alpha_k^{00D_l} = \alpha_l^{00D_k}, \tag{2.6.24}$$

by explicitly using the constraint (2.6.22).

2.6.2 Mathematical Structure of the System $L\alpha^{D_l} = \beta^{D_l}$

Properties of $\mathcal{A}^D$. The approximation space $\mathcal{A}^D$ introduced in (2.6.3) satisfies the perpendicularity property (2.3.20), the species orthogonality property (2.3.32), and the relation (2.3.38) from (2.3.51). In addition, the basis functions of $\mathcal{A}^D$ satisfy the species localization property (2.3.25). As a consequence, the results (2.3.41)–(2.3.43) established in Section 2.3.6 apply.

Properties of L. The constrained space $\mathcal{C}$ is given by $\mathcal{C} = \mathcal{L}^\perp$, where $\mathcal{L}$ has already been evaluated in Section 2.6.1. The matrix L is therefore symmetric positive semi-definite and positive definite on $\mathcal{L}^\perp$ for any $n \geq 1$. Taking into account the rescaling of L, we also deduce from

$$\mathcal{I}_S^1 \cap \mathcal{A}^D = \mathbb{R}\Big(\sum_{k\in S} \sqrt{m_k}\phi^{1000k}\Big),$$

that the nullspace of L is $N(L) = \mathbb{R}\mathcal{U}$ where the vector $\mathcal{U} \in \mathbb{R}^{2n+p}$ has components $\mathcal{U}_k^{00} = 1$, $k \in S$, $\mathcal{U}_k^{10} = 0$, $k \in S$, and $\mathcal{U}_k^{01} = 0$, $k \in \mathcal{P}$. In addition, we also have the important property that $\langle \mathcal{L},\mathcal{U}\rangle \neq 0$.

Furthermore, the matrix $2db(L)-L$ symmetric positive semi-definite for any $n \geq 1$. In addition, the matrix $2db(L) - L$ is positive definite if $n \geq 3$ whereas in the special case $n = 2$ and $S = \{1, 2\}$, the nullspace of $2db(L) - L$ is $N\big(2db(L) - L\big) = \mathbb{R}\mathcal{U}^*$, where $\mathcal{U}^* \in \mathbb{R}^{2n+p}$ is the vector whose components are given by $\mathcal{U}_1^{00*} = 1$, $\mathcal{U}_1^{r*} = 0$ if $r \neq 00$, $\mathcal{U}_2^{00*} = -1$, and $\mathcal{U}_2^{r*} = 0$ if $r \neq 00$. In other words, we have $\mathcal{U}^* = (1,-1,0,0,0,0)$ if $p = 2$, $\mathcal{U}^* = (1,-1,0,0,0)$ if $p = 1$, and $\mathcal{U}^* = (1,-1,0,0)$ if $p = 0$. Finally, in the special case $n = 1$, the nullspace of $2db(L) - L$ is $N\big(2db(L) - L\big) = \mathbb{R}\mathcal{U}$, so that $N\big(2db(L) - L\big) = \mathbb{R}(1,0,0)$ if $p = 1$ and $N\big(2db(L) - L\big) = \mathbb{R}(1,0)$ if $p = 0$.

On the other hand, the matrix $db(L)$ is symmetric positive semi-definite for any $n \geq 1$. In addition, the matrix $db(L)$ is positive definite for $n \geq 2$, and in the special case $n = 1$, the nullspace of $db(L)$ is $N\big(db(L)\big) = \mathbb{R}\mathcal{U}$, so that $N\big(db(L)\big) = \mathbb{R}(1,0,0)$ if $p = 1$ and $N\big(db(L)\big) = \mathbb{R}(1,0)$ if $p = 0$.

2.6.3 The System $L_{[\mathrm{e}]}\alpha_{[\mathrm{e}]}^{D_l} = \beta_{[\mathrm{e}]}^{D_l}$

In this section we consider an approximation space for ϕ^{D_l}, $l \in S$, which arises naturally for $\phi^{\lambda'}$ as detailed in Section 2.7.2. More specifically, we introduce the basis functions $\phi^{10\mathrm{e}k}$, $k \in S$, defined by

$$\phi^{10\mathrm{e}k} = \phi^{1010k} + \delta_{\mathcal{P}}(k)\phi^{1001k}, \qquad k \in S, \tag{2.6.25}$$

where $\delta_{\mathcal{P}}(k) = 1$ if $k \in \mathcal{P}$ and $\delta_{\mathcal{P}}(k) = 0$ otherwise. The superscript e is used here because $\phi^{10\mathrm{e}k}$ is associated with the energy of the molecules, i.e., the sum of the kinetic and internal energy. We then consider the approximation space $\mathcal{A}^D_{[e]}$ spanned by ϕ^{1000k}, $k \in \mathcal{S}$, and ϕ^{10ek}, $k \in \mathcal{S}$,

$$\mathcal{A}^D_{[e]} = \mathrm{span}\{ \ \phi^{1000k},\ k \in \mathcal{S}, \quad \phi^{10ek},\ k \in \mathcal{S} \ \}, \tag{2.6.26}$$

with the corresponding indexing set $\mathcal{B}^D_{[e]}$

$$\mathcal{B}^D_{[e]} = \{00, e\} \times \mathcal{S}. \tag{2.6.27}$$

The associated linear system is obtained from the formalism (2.3.55)–(2.3.58) which directly applies to our case. The transformation matrix $\mathcal{R}$ expressing the new basis functions ϕ^{1000k}, $k \in \mathcal{S}$, and ϕ^{10ek}, $k \in \mathcal{S}$, as linear combinations of the old basis functions ϕ^{1000k}, $k \in \mathcal{S}$, ϕ^{1010k}, $k \in \mathcal{S}$, and ϕ^{1001k}, $k \in \mathcal{P}$, has its coefficients $\mathcal{R}^{rs}_{kl}$, $(r,k) \in \mathcal{B}^D_{[e]}$, $(s,l) \in \mathcal{B}^D$, given by

$$\begin{cases} \mathcal{R}^{0000}_{kl} = \delta_{kl}, & k,l \in \mathcal{S}, \\ \mathcal{R}^{0010}_{kl} = 0, & k,l \in \mathcal{S}, \\ \mathcal{R}^{0001}_{kl} = 0, & k \in \mathcal{S}, \quad l \in \mathcal{P}, \\ \mathcal{R}^{e00}_{kl} = 0, & k,l \in \mathcal{S}, \\ \mathcal{R}^{e10}_{kl} = \delta_{kl}, & k,l \in \mathcal{S}, \\ \mathcal{R}^{e01}_{kl} = \delta_{kl}, & k \in \mathcal{S}, \quad l \in \mathcal{P}. \end{cases} \tag{2.6.28}$$

The transport linear system corresponding to the diffusion matrix $D_{[e]}$ will be denoted by

$$L_{[e]}\alpha^{D_l}_{[e]} = \beta^{D_l}_{[e]}, \qquad l \in \mathcal{S}, \tag{2.6.29}$$

and from $L_{[e]} = \mathcal{R}L\mathcal{R}^t$ and $\beta^{D_l}_{[e]} = \mathcal{R}\beta^{D_l}$, we obtain that

$$L^{0000}_{[e]kl} = L^{0000}_{kl}, \qquad k,l \in \mathcal{S}, \tag{2.6.30}$$

$$L^{00e}_{[e]kl} = L^{0010}_{kl} + \delta_{\mathcal{P}}(l)L^{0001}_{kl}, \qquad k,l \in \mathcal{S}, \tag{2.6.31}$$

$$L^{ee}_{[e]kl} = L^{1010}_{kl} + \delta_{\mathcal{P}}(k)L^{0110}_{kl} + \delta_{\mathcal{P}}(l)L^{1001}_{kl} + \delta_{\mathcal{P}}(k)\delta_{\mathcal{P}}(l)L^{0101}_{kl}, \qquad k,l \in \mathcal{S}, \tag{2.6.32}$$

from which the matrix $L_{[e]}$ is easily evaluated, and that

$$\beta^{00D_l}_{[e]k} = \beta^{00D_l}_{k} = \delta_{kl} - Y_k, \qquad k,l \in \mathcal{S}, \tag{2.6.33}$$

$$\beta^{eD_l}_{[e]k} = \beta^{10D_l}_{k} + \delta_{\mathcal{P}}(k)\beta^{01D_l}_{k} = 0, \qquad k,l \in \mathcal{S}. \tag{2.6.34}$$

The system (2.6.29) must also be completed by the constraint

$$\langle \mathcal{L}_{[e]}, \alpha_{[e]}^{D_l} \rangle = \langle \mathcal{L}_{[e]}^{00}, \alpha_{[e]}^{00D_l} \rangle = \sum_{k \in S} Y_k \alpha_{[e]k}^{00D_l} = 0, \tag{2.6.35}$$

where we have introduced the vector $\mathcal{L}_{[e]} \in \mathbb{R}^{2n}$ defined by $\mathcal{L}_{[e]k}^{00} = Y_k$, $k \in \mathcal{S}$, and $\mathcal{L}_{[e]k}^{e} = 0$, $k \in \mathcal{S}$. The corresponding diffusion coefficients are then given by

$$D_{[e]kl} = \langle \alpha_{[e]}^{D_l}, \beta_{[e]}^{D_k} \rangle = \langle \alpha_{[e]}^{D_k}, \beta_{[e]}^{D_l} \rangle = \langle \alpha_{[e]}^{00D_l}, \beta_{[e]}^{00D_k} \rangle = \langle \alpha_{[e]}^{00D_k}, \beta_{[e]}^{00D_l} \rangle, \tag{2.6.36}$$

which can be simplified into

$$D_{[e]kl} = \alpha_{[e]k}^{00D_l} = \alpha_{[e]l}^{00D_k}, \tag{2.6.37}$$

by explicitly using the constraint (2.6.35). The diffusion matrix $D_{[e]}$ will be shown to be accurate in Section 6.4.2.

2.6.4 Mathematical Structure of the System $L_{[e]}\alpha_{[e]}^{D_l} = \beta_{[e]}^{D_l}$

Properties of $\mathcal{A}_{[e]}^{D}$. The approximation space $\mathcal{A}_{[e]}^{D}$ introduced in (2.6.26) satisfies the perpendicularity property (2.3.20), the species orthogonality property (2.3.32), and the relation (2.3.38) from (2.3.51). In addition, the basis functions of $\mathcal{A}_{[e]}^{D}$ satisfy the species localization property (2.3.25). As a consequence, the results (2.3.41)–(2.3.43) established in Section 2.3.6 apply.

Properties of $L_{[e]}$. The constrained space C is given by $C = \mathcal{L}_{[e]}^{\perp}$, where $\mathcal{L}_{[e]}$ has already been evaluated in Section 2.6.3. As a consequence, for $n \geq 1$, the matrix $L_{[e]}$ is symmetric positive semi-definite and positive definite on $\mathcal{L}_{[e]}^{\perp}$. In addition, taking into account the rescaling of $L_{[e]}$, we deduce from

$$\mathcal{I}_{\mathcal{S}}^{1} \cap \mathcal{A}_{[e]}^{D} = \mathbb{R}\Big(\sum_{k \in \mathcal{S}} \sqrt{m_k}\,\phi^{1000k}\Big),$$

that the nullspace of $L_{[e]}$ is $N(L_{[e]}) = \mathbb{R}\,\mathcal{U}_{[e]}$, where we have defined the vector $\mathcal{U}_{[e]} \in \mathbb{R}^{2n}$ by $\mathcal{U}_{[e]k}^{00} = 1$, $k \in \mathcal{S}$, and $\mathcal{U}_{[e]k}^{e} = 0$, $k \in \mathcal{S}$. We also have the important property that $\langle \mathcal{L}_{[e]}, \mathcal{U}_{[e]} \rangle \neq 0$.

Furthermore, the matrix $2db(L_{[e]}) - L_{[e]}$ is symmetric positive semi-definite for any $n \geq 1$. In addition, the matrix $2db(L_{[e]}) - L_{[e]}$ is positive definite if $n \geq 3$, and if $n = 2$ and $\mathcal{S} = \{1, 2\}$, then the nullspace of $2db(L_{[e]}) - L_{[e]}$ is $N\big(2db(L_{[e]}) - L_{[e]}\big) = \mathbb{R}\mathcal{U}_{[e]}^{*}$, where $\mathcal{U}_{[e]}^{*} \in \mathbb{R}^{2n}$ is the vector whose components are given by $\mathcal{U}_{[e]1}^{00*} = 1$, $\mathcal{U}_{[e]1}^{e*} = 0$,

$\mathcal{U}_{[e]2}^{00*} = -1$, and $\mathcal{U}_{[e]2}^{e*} = 0$, so that $\mathcal{U}_{[e]}^{*} = (1, -1, 0, 0)$. Finally, for $n = 1$, the nullspace of $2db(L_{[e]}) - L_{[e]}$ is $N\big(2db(L_{[e]}) - L_{[e]}\big) = \mathbb{R}\mathcal{U}_{[e]} = \mathbb{R}(1, 0)$.

On the other hand, the matrix $db(L_{[e]})$ is symmetric, positive semi-definite for any $n \geq 1$. In addition, the matrix $db(L_{[e]})$ is positive definite for $n \geq 2$, and for $n = 1$, the nullspace of $db(L_{[e]})$ is $N\big(db(L_{[e]})\big) = \mathbb{R}\mathcal{U}_{[e]} = \mathbb{R}(1, 0)$.

2.6.5 The System $L_{[00]}\alpha_{[00]}^{D_l} = \beta_{[00]}^{D_l}$

In this section we consider the approximation space $\mathcal{A}_{[00]}^{D}$ spanned by ϕ^{1000k}, $k \in \mathcal{S}$,

$$\mathcal{A}_{[00]}^{D} = \text{span}\{\ \phi^{1000k},\ k \in \mathcal{S}\ \}, \tag{2.6.38}$$

with the corresponding indexing set $\mathcal{B}_{[00]}^{D}$

$$\mathcal{B}_{[00]}^{D} = \{00\} \times \mathcal{S}. \tag{2.6.39}$$

The subscript 00 is associated with this approximation, since it is derived by only considering the basis functions ϕ^{1000k}, $k \in \mathcal{S}$. The corresponding transport linear system is easily obtained from the formalism (2.3.55)–(2.3.58) which directly applies to our case. The transformation matrix $\mathcal{R}$ expressing the new basis functions ϕ^{1000k}, $k \in \mathcal{S}$, as linear combinations of the old basis functions ϕ^{1000k}, $k \in \mathcal{S}$, ϕ^{1010k}, $k \in \mathcal{S}$, and ϕ^{1001k}, $k \in \mathcal{P}$, has its coefficients $\mathcal{R}_{kl}^{rs}$, $(r, k) \in \mathcal{B}_{[00]}^{D}$, $(s, l) \in \mathcal{B}^{D}$, given by

$$\begin{cases} \mathcal{R}_{kl}^{0000} = \delta_{kl}, & k, l \in \mathcal{S}, \\ \mathcal{R}_{kl}^{0010} = 0, & k, l \in \mathcal{S}, \\ \mathcal{R}_{kl}^{0001} = 0, & k \in \mathcal{S}, \quad l \in \mathcal{P}. \end{cases} \tag{2.6.40}$$

The transport linear system corresponding to the diffusion matrix $D_{[00]}$ will be denoted by

$$L_{[00]}\alpha_{[00]}^{D_l} = \beta_{[00]}^{D_l}, \qquad l \in \mathcal{S}, \tag{2.6.41}$$

and from $L_{[00]} = \mathcal{R}L\mathcal{R}^t$ and $\beta_{[00]}^{D_l} = \mathcal{R}\beta^{D_l}$, we obtain that

$$L_{[00]} = L^{0000}, \tag{2.6.42}$$

and

$$\beta_{[00]k}^{00D_l} = \beta_{k}^{00D_l} = \delta_{kl} - Y_k, \qquad k, l \in \mathcal{S}. \tag{2.6.43}$$

The system (2.6.41) must also be completed by the constraint

$$\langle Y, \alpha_{[00]}^{D_l} \rangle = \langle Y, \alpha_{[00]}^{00D_l} \rangle = \sum_{k \in \mathcal{S}} Y_k \alpha_{[00]k}^{00D_l} = 0, \tag{2.6.44}$$

where the vector $Y \in \mathbb{R}^n$ is given by $Y = (Y_k)_{k\in S}$. Finally, the corresponding diffusion coefficients are given by

$$D_{[00]kl} = \langle \alpha^{D_l}_{[00]}, \beta^{D_k}_{[00]} \rangle = \langle \alpha^{D_k}_{[00]}, \beta^{D_l}_{[00]} \rangle = \langle \alpha^{00D_l}_{[00]}, \beta^{00D_k}_{[00]} \rangle = \langle \alpha^{00D_k}_{[00]}, \beta^{00D_l}_{[00]} \rangle, \tag{2.6.45}$$

which can be simplified into

$$D_{[00]kl} = \alpha^{00D_l}_{[00]k} = \alpha^{00D_k}_{[00]l}, \tag{2.6.46}$$

by explicitly using the constraint (2.6.45).

Note that from (2.6.43) we can write $\big(\beta^{D_l}_{[00]k}\big)_{k,l\in S} = I - Y\otimes U$, so that the relations (2.6.41)–(2.6.46) imply that

$$L_{[00]}D_{[00]} = I - Y\otimes U, \tag{2.6.47}$$

and

$$D_{[00]}Y = 0, \tag{2.6.48}$$

where $U = (1)_{k\in S}$ and transposing (2.6.47) also yields that $D_{[00]}L_{[00]} = I - U\otimes Y$ [Gi90] [Gi91]. From (2.6.47) and the mass constraint $\sum_{j\in S} d_j = 0$, we will deduce in Section 4.4.7 that the first-order diffusion velocities defined by $V_{[00]i} = \sum_{j\in S} D_{[00]ij} d_j$, $i \in S$, satisfy the well-known Stefan-Maxwell-Boltzmann equations. Generalized forms of the Stefan-Maxwell-Boltzmann equations will also be described in Section 4.4.7 by using the matrices L and $L_{[e]}$.

2.6.6 Mathematical Structure of the System $L_{[00]}\alpha^{D_l}_{[00]} = \beta^{D_l}_{[00]}$

Properties of $\mathcal{A}^D_{[00]}$. The approximation space $\mathcal{A}^D_{[00]}$ introduced in (2.6.38) satisfies the perpendicularity property (2.3.20), the species orthogonality property (2.3.32), and the relation (2.3.38) from (2.3.51). In addition, the basis functions of $\mathcal{A}^D_{[00]}$ satisfy the species localization property (2.3.25). As a consequence, the results (2.3.41)–(2.3.43) established in Section 2.3.6 apply.

Properties of $L_{[00]}$. The constrained space C is given by $C = Y^\perp$, where Y has already been evaluated in Section 2.6.5. As a consequence, for any $n \geq 1$, the matrix $L_{[00]}$ is symmetric positive semi-definite, and positive definite on $Y^\perp$. Taking into account the rescaling of $L_{[00]}$, we also deduce from

$$\mathcal{I}^1_S \cap \mathcal{A}^D_{[00]} = \mathbb{R}\Big(\sum_{k\in S} \sqrt{m_k}\,\phi^{1000k}\Big),$$

that the nullspace of $L_{[00]}$ is $N(L_{[00]}) = \mathbb{R}U$, where the vector $U \in \mathbb{R}^n$ has components $U_k = 1$, $k \in \mathcal{S}$. We also have the important property that $\langle Y, U \rangle \neq 0$.

Furthermore, the matrix $2db(L_{[00]}) - L_{[00]}$ is symmetric positive semi-definite for any $n \geq 1$. In addition, the matrix $2db(L_{[00]}) - L_{[00]}$ is positive definite if $n \geq 3$, and if $n = 2$ and $\mathcal{S} = \{1, 2\}$, the nullspace of $2db(L_{[00]}) - L_{[00]}$ is $N\big(2db(L_{[00]}) - L_{[00]}\big) = \mathbb{R}U^*$, where $U^* \in \mathbb{R}^n$ is the vector whose components are given by $U_1^{00*} = 1$ and $U_2^{00*} = -1$, so that $U^* = (1, -1)$. Finally, for $n = 1$, we have $L_{[00]} = \{0\}$ and $N\big(2db(L_{[00]}) - L_{[00]}\big) = \mathbb{R}U = \mathbb{R}(1)$.

On the other hand, the matrix $db(L_{[00]})$ is symmetric positive semi-definite for any $n \geq 1$. In addition, the matrix $db(L_{[00]})$ is positive definite for $n \geq 2$, and for $n = 1$, we have $N\big(db(L_{[00]})\big) = \mathbb{R}U = \mathbb{R}(1)$.

2.6.7 Alternative Formulations

To the authors' knowledge, higher order approximations have not been used for polyatomic gas mixtures. In addition, we note that a monatomic approximation is obtained by using the functional space $\mathcal{A}^D_{[\mathrm{mon}]}$ with

$$\mathcal{A}^D_{[\mathrm{mon}]} = \mathrm{span}\{\ \phi^{1000k},\ k \in \mathcal{S},\quad \phi^{1010k},\ k \in \mathcal{S}\ \}.$$

The corresponding linear systems $L_{[\mathrm{mon}]}\alpha^{D_l}_{[\mathrm{mon}]} = \beta^{D_l}_{[\mathrm{mon}]}$, $l \in \mathcal{S}$, and the diffusion matrix $D_{[\mathrm{mon}]}$ are easily obtained, but are omitted for brevity. Finally, we note that the approximate coefficients obtained by using the two-dimensional space spanned by Ψ^{D_l} and the collisional invariant $\sum_{k\in\mathcal{S}} \sqrt{m_k}\,\phi^{1000k}$ are equivalently obtained by applying the conjugate gradient algorithm to the constrained singular system (2.6.6). This algorithm and its preconditioned version will be presented in Section 5.1.4.

2.7 The Partial Thermal Conductivity and the Thermal Diffusion Vector

2.7.1 The System $L\alpha^{\lambda'} = \beta^{\lambda'}$

The vector integral equation corresponding to $\phi^{\lambda'}$ is

$$\mathfrak{I}(\phi^{\lambda'}) = \Psi^{\lambda'}, \tag{2.7.1}$$

where $\Psi^{\lambda'}$ can be written as

$$\Psi^{\lambda'} = \sum_{k\in\mathcal{S}} \sqrt{2(k_B\overline{T})^3/m_k}\,\phi^{1010k} + \sum_{k\in\mathcal{P}} \sqrt{2(k_B\overline{T})^3/m_k}\,\phi^{1001k}, \tag{2.7.2}$$

from (2.1.25)(2.2.19). A first natural approximation space for $\phi^{\lambda'}$ is the space $\mathcal{A}^{\lambda'} = \mathcal{A}^D$ spanned by ϕ^{1000k}, $k \in \mathcal{S}$, ϕ^{1010k}, $k \in \mathcal{S}$, and ϕ^{1001k}, $k \in \mathcal{P}$,

$$\mathcal{A}^{\lambda'} = \mathrm{span}\{\ \phi^{1000k},\ k \in \mathcal{S}, \quad \phi^{1010k},\ k \in \mathcal{S}, \quad \phi^{1001k},\ k \in \mathcal{P}\ \}, \tag{2.7.3}$$

already introduced in (2.6.3) so that the corresponding indexing set is $\mathcal{B}^{\lambda'} = \mathcal{B}^D$, i.e.,

$$\mathcal{B}^{\lambda'} = \{00, 10\}\times\mathcal{S}\ \cup\ \{01\}\times\mathcal{P}. \tag{2.7.4}$$

For convenience, $\phi^{\lambda'}$ is taken in the form

$$\phi^{\lambda'} = \sqrt{2k_{\mathrm{B}}\overline{T}}\Big(\sum_{k\in\mathcal{S}} \sqrt{m_k}\big(\alpha_k^{00\lambda'}\phi^{1000k} + \alpha_k^{10\lambda'}\phi^{1010k}\big) + \sum_{k\in\mathcal{P}} \sqrt{m_k}\alpha_k^{01\lambda'}\phi^{1001k}\Big). \tag{2.7.5}$$

The matrix associated with the variational procedure is denoted by L and is rescaled such that $L_{kl}^{rs} = (2\sqrt{m_k m_l}/3\bar{p})[\phi^{10rk}, \phi^{10sl}]$, $(r,k),(s,l) \in \mathcal{B}^{\lambda'}$, and similarly, the right member is rescaled such that $\beta_k^{r\lambda'} = \big(\sqrt{2m_k}/3\bar{p}\sqrt{k_{\mathrm{B}}\overline{T}}\big)\langle\!\langle\phi^{10rk}, \Psi^{\lambda'}\rangle\!\rangle$, $(r,k) \in \mathcal{B}^{\lambda'}$. We have $L \in \mathbb{R}^{2n+p,2n+p}$, $\beta^{\lambda'} \in \mathbb{R}^{2n+p}$, and the final system in $\alpha^{\lambda'} \in \mathbb{R}^{2n+p}$ reads

$$L\alpha^{\lambda'} = \beta^{\lambda'}. \tag{2.7.6}$$

The matrix L has already been given in Section 2.6.1, and an explicit calculation yields

$$\beta_k^{00\lambda'} = 0, \qquad k \in \mathcal{S}, \tag{2.7.7}$$

$$\beta_k^{10\lambda'} = \frac{c_{\mathrm{p}}^{\mathrm{tr}}}{k_{\mathrm{B}}} X_k, \qquad k \in \mathcal{S}, \tag{2.7.8}$$

$$\beta_k^{01\lambda'} = \frac{c_k^{\mathrm{int}}}{k_{\mathrm{B}}} X_k, \qquad k \in \mathcal{P}, \tag{2.7.9}$$

where $c_{\mathrm{p}}^{\mathrm{tr}} = \frac{5}{2}k_{\mathrm{B}}$ is the constant pressure specific heat capacity per molecule. The constraints on the vector $\alpha^{\lambda'}$ also reduce to the only relation

$$\langle \mathcal{L}, \alpha^{\lambda'}\rangle = \langle \mathcal{L}^{00}, \alpha^{00\lambda'}\rangle = \sum_{k\in\mathcal{S}} Y_k \alpha_k^{00\lambda'} = 0, \tag{2.7.10}$$

recalling that $\mathcal{L} \in \mathbb{R}^{2n+p}$ is given by $\mathcal{L}_k^{00} = Y_k$, $k \in \mathcal{S}$, $\mathcal{L}_k^{10} = 0$, $k \in \mathcal{S}$, and $\mathcal{L}_k^{01} = 0$, $k \in \mathcal{P}$. The properties of the matrix L have already been investigated in Section 2.6.2.

Finally, the partial thermal conductivity λ' is given by

$$\begin{aligned}\lambda' &= \frac{\bar{p}}{\overline{T}}\langle\alpha^{\lambda'}, \beta^{\lambda'}\rangle = \frac{\bar{p}}{\overline{T}}\langle\alpha^{10\lambda'}, \beta^{10\lambda'}\rangle + \frac{\bar{p}}{\overline{T}}\langle\alpha^{01\lambda'}, \beta^{01\lambda'}\rangle \\ &= \frac{\bar{p}}{k_{\mathrm{B}}\overline{T}}\Big(\sum_{k\in\mathcal{S}} X_k c_{\mathrm{p}}^{\mathrm{tr}} \alpha_k^{10\lambda'} + \sum_{k\in\mathcal{P}} X_k c_k^{\mathrm{int}} \alpha_k^{01\lambda'}\Big),\end{aligned} \tag{2.7.11}$$

and the thermal diffusion vector θ by

$$\theta_k = -\langle \alpha^{\lambda'}, \beta^{D_k} \rangle = -\langle \alpha^{00\lambda'}, \beta^{00D_k} \rangle, \qquad k \in \mathcal{S}. \tag{2.7.12}$$

This formula can be simplified into

$$\theta_k = -\alpha_k^{00\lambda'}, \qquad k \in \mathcal{S}, \tag{2.7.13}$$

by explicitly using the constraint (2.7.10), and we also have the alternative formulation

$$\theta_k = -\langle \alpha^{D_k}, \beta^{\lambda'} \rangle, \qquad k \in \mathcal{S}, \tag{2.7.14}$$

which requires more linear systems to be solved. Note that the formulas (2.7.12) and (2.7.14) yield the same result since the same variational space $\mathcal{A}^{\lambda'} = \mathcal{A}^D$ is used for $\phi^{\lambda'}$ and ϕ^{D_l}, $l \in \mathcal{S}$.

2.7.2 The System $L_{[e]}\alpha^{\lambda'}_{[e]} = \beta^{\lambda'}_{[e]}$

A simplified formulation is derived by rewriting the right member $\Psi^{\lambda'}$ in the form

$$\Psi^{\lambda'} = \sum_{k\in\mathcal{S}} \sqrt{2(k_B\overline{T})^3/m_k}\ \phi^{10ek}, \tag{2.7.15}$$

where $\phi^{10ek} = \phi^{1010k} + \delta_{\mathcal{P}}(k)\phi^{1001k}$, $k \in \mathcal{S}$, has been introduced in Section 2.6.3. The superscript e is used here because ϕ^{10ek} is associated with the energy of the molecules. The expression (2.7.15) then suggests the use of the approximation space $\mathcal{A}^{\lambda'}_{[e]} = \mathcal{A}^{D}_{[e]}$ spanned by ϕ^{1000k}, $k \in \mathcal{S}$, and ϕ^{10ek}, $k \in \mathcal{S}$,

$$\mathcal{A}^{\lambda'}_{[e]} = \mathrm{span}\{\ \phi^{1000k},\ k \in \mathcal{S},\quad \phi^{10ek},\ k \in \mathcal{S}\ \}, \tag{2.7.16}$$

already introduced in (2.6.26) so that the corresponding indexing set is $\mathcal{B}^{\lambda'}_{[e]} = \mathcal{B}^{D}_{[e]}$, i.e.,

$$\mathcal{B}^{\lambda'}_{[e]} = \{00, \mathrm{e}\} \times \mathcal{S}. \tag{2.7.17}$$

This procedure extends the ideas of Thijsse et al. [TTCKB79] and Van den Oord and Korving [VK88] for the thermal conductivity of pure mixtures and McCourt et al. [MBKK90] for the partial thermal conductivity of binary mixtures. Note that we have kept the basis functions ϕ^{1000k}, $k \in \mathcal{S}$, in the expansion of $\phi^{\lambda'}$ since their suppression would yield that $\theta = 0$ from (2.7.13) and therefore that $\lambda = \lambda'$. The associated transport linear system is then easily obtained from the formalism (2.3.55)–(2.3.58) which directly applies to our case. We have already evaluated in (2.6.28) the coefficients $\mathcal{R}_{kl}^{rs}$, $(r,k) \in \mathcal{B}_{[e]}^{\lambda'}$, $(s,l) \in \mathcal{B}^{\lambda'}$, of the transformation matrix $\mathcal{R}$ expressing the new basis functions ϕ^{1000k}, $k \in \mathcal{S}$, and ϕ^{10ek}, $k \in \mathcal{S}$, as linear combinations of the old basis functions ϕ^{1000k}, $k \in \mathcal{S}$, ϕ^{1010k}, $k \in \mathcal{S}$, and ϕ^{1001k}, $k \in \mathcal{P}$.

The corresponding transport linear system will be denoted

$$L_{[e]}\alpha_{[e]}^{\lambda'} = \beta_{[e]}^{\lambda'}, \tag{2.7.18}$$

and from (2.3.58) we have $L_{[e]} = \mathcal{R}L\mathcal{R}^t$ and $\beta_{[e]}^{\lambda'} = \mathcal{R}\beta^{\lambda'}$. The matrix $L_{[e]}$ has already been given in (2.6.30)–(2.6.32) whereas $\beta_{[e]}^{\lambda'}$ is given by

$$\beta_{[e]k}^{00\lambda'} = \beta_k^{00\lambda'} = 0, \qquad k \in \mathcal{S}, \tag{2.7.19}$$

$$\beta_{[e]k}^{e\lambda'} = \beta_k^{10\lambda'} + \delta_{\mathcal{P}}(k)\beta_k^{01\lambda'} = \frac{c_{\mathrm{p}}^{\mathrm{tr}} + c_k^{\mathrm{int}}}{k_{\mathrm{B}}} X_k, \qquad k \in \mathcal{S}. \tag{2.7.20}$$

The system (2.7.18) must also be completed by the constraint

$$\langle \mathcal{L}_{[e]}, \alpha_{[e]}^{\lambda'} \rangle = \langle \mathcal{L}_{[e]}^{00}, \alpha_{[e]}^{00\lambda'} \rangle = \sum_{k\in\mathcal{S}} Y_k \alpha_{[e]k}^{00\lambda'} = 0, \tag{2.7.21}$$

recalling that $\mathcal{L}_{[e]} \in \mathbb{R}^{2n}$ is given by $\mathcal{L}_{[e]k}^{00} = Y_k$, $k \in \mathcal{S}$, and $\mathcal{L}_{[e]k}^{e} = 0$, $k \in \mathcal{S}$. The properties of the matrix $L_{[e]}$ have already been investigated in Section 2.6.4.

Finally, the partial thermal conductivity $\lambda'_{[e]}$ is given by

$$\begin{aligned} \lambda'_{[e]} &= \frac{\bar{p}}{\overline{\overline{T}}} \langle \alpha_{[e]}^{\lambda'}, \beta_{[e]}^{\lambda'} \rangle = \frac{\bar{p}}{\overline{\overline{T}}} \langle \alpha_{[e]}^{e\lambda'}, \beta_{[e]}^{e\lambda'} \rangle \\ &= \frac{\bar{p}}{k_{\mathrm{B}}\overline{\overline{T}}} \sum_{k\in\mathcal{S}} X_k (c_{\mathrm{p}}^{\mathrm{tr}} + c_k^{\mathrm{int}}) \alpha_{[e]k}^{e\lambda'}, \end{aligned} \tag{2.7.22}$$

and the thermal diffusion vector $\theta_{[e]}$ by

$$\theta_{[e]k} = -\langle \alpha_{[e]}^{\lambda'}, \beta_{[e]}^{D_k} \rangle = -\langle \alpha_{[e]}^{00\lambda'}, \beta_{[e]}^{00D_k} \rangle, \qquad k \in \mathcal{S}. \tag{2.7.23}$$

This formula can be simplified into

$$\theta_{[e]k} = -\alpha_{[e]k}^{00\lambda'}, \qquad k \in \mathcal{S}, \tag{2.7.24}$$

by explicitly using the constraint (2.7.21), and we also have the alternative formulation

$$\theta_{[e]k} = -\langle \alpha_{[e]}^{D_k}, \beta_{[e]}^{\lambda'} \rangle, \qquad k \in \mathcal{S}, \tag{2.7.25}$$

which requires more linear systems to be solved. Note that the formulas (2.7.23) and (2.7.25) yield the same result since the same variational space $\mathcal{A}^{\lambda'}_{[e]} = \mathcal{A}^{D}_{[e]}$ is used for $\phi^{\lambda'}_{[e]}$ and $\phi^{D_l}_{[e]}$, $l \in \mathcal{S}$.

The partial thermal coefficient $\lambda'_{[e]}$ and the thermal diffusion vector $\theta_{[e]}$ will be shown to be accurate in Section 6.5.2. This shows that the complete energy modes ϕ^{10ek}, $k \in \mathcal{S}$, are sufficient to accurately model heat conduction and thermal diffusion.

2.7.3 Alternative Formulations

Higher order formulations will not be investigated in this book. Notice also that a monatomic approximation is obtained by using the functional space $\mathcal{A}^{\lambda'}_{[\mathrm{mon}]}$ with

$$\mathcal{A}^{\lambda'}_{[\mathrm{mon}]} = \mathrm{span}\{\ \phi^{1000k},\ k \in \mathcal{S},\quad \phi^{1010k},\ k \in \mathcal{S}\ \}.$$

The corresponding linear system $L_{[\mathrm{mon}]}\alpha^{\lambda'}_{[\mathrm{mon}]} = \beta^{\lambda'}_{[\mathrm{mon}]}$, the partial thermal conductivity $\lambda'_{[\mathrm{mon}]}$, and the thermal diffusion vector $\theta_{[\mathrm{mon}]}$ are easily obtained, but are omitted for brevity. On the other hand, approximate coefficients obtained by using the two-dimensional space spanned by $\Psi^{\lambda'}$ and the collisional invariant $\sum_{k\in\mathcal{S}} \sqrt{m_k}\,\phi^{1000k}$ are equivalently obtained by applying the conjugate gradient algorithm to the constrained singular system (2.7.6). This algorithm and its preconditioned version will be presented in Section 5.1.4.

2.8 The Thermal Conductivity and the Thermal Diffusion Ratios

2.8.1 The System $\Lambda\alpha^{\lambda} = \beta^{\lambda}$

We have seen in Section 2.1 that the thermal conductivity λ and the thermal diffusion ratios χ are defined in terms of the partial thermal conductivity λ' and the thermal diffusion vector θ [WT62] from (2.1.38)(2.1.39). Indeed, the thermal diffusion ratios are given by the solution of the constrained singular system

$$\begin{cases} D\chi = \theta, \\ \langle U, \chi\rangle = 0, \end{cases} \tag{2.8.1}$$

where D is the diffusion matrix and θ the thermal diffusion vector. In Section 4.6.1 we will show that since $\theta \in Y^{\perp}$ and $\langle U, Y\rangle \neq 0$, the system (2.8.1) is well posed and thus properly defines the thermal diffusion ratios χ. The thermal conductivity λ is then given in terms of the partial thermal conductivity, the thermal diffusion vector, and the thermal diffusion ratios by the expression

$$\lambda = \lambda' - (\bar{p}/\overline{T}) \sum_{k \in \mathcal{S}} \theta_k \chi_k. \tag{2.8.2}$$

These definitions, however, may not be convenient in practice. In particular, one may want to evaluate directly the thermal conductivity λ and the thermal diffusion ratios χ without the intermediate calculations of λ' and θ. This can be done by using the auxiliary functions ϕ^λ and Ψ^λ introduced in (2.1.40) and the relations (2.1.41), although the thermal diffusion ratios appearing in Ψ^λ are unknown.

Indeed, we now introduce a variational framework associated with the functions ϕ^λ which is, to the authors' knowledge, new. We consider the scalar integral equation corresponding to ϕ^λ

$$\mathcal{I}(\phi^\lambda) = \Psi^\lambda, \tag{2.8.3}$$

where Ψ^λ is given by

$$\Psi^\lambda = \Psi^{\lambda'} + \bar{p} k_{\mathrm{B}} \overline{T} \sum_{l \in \mathcal{S}} \chi_l \Psi^{D_l}. \tag{2.8.4}$$

From the expressions of $\Psi^{\lambda'}$ and Ψ^{D_l}, $l \in \mathcal{S}$, a first natural approximation space for ϕ^λ would be the space spanned by ϕ^{1000k}, $k \in \mathcal{S}$, ϕ^{1010k}, $k \in \mathcal{S}$, and ϕ^{1001k}, $k \in \mathcal{P}$. However, the function ϕ^λ has necessarily to be orthogonal to the basis functions ϕ^{1000k}, $k \in \mathcal{S}$. Indeed, making use of (2.1.37), (2.1.38), and (2.1.40), we first have the relations $[\phi^{D_l}, \phi^\lambda] = 0$, $l \in \mathcal{S}$, which imply that $\langle\!\langle \Psi^{D_l}, \phi^\lambda \rangle\!\rangle = 0$, $l \in \mathcal{S}$. On the other hand, it is straightforward to check that ϕ^{1000l} is a linear combination of Ψ^{D_l} and the vector collisional invariant $\sum_{k \in \mathcal{S}} \sqrt{m_k} \phi^{1000k}$. Since ϕ^λ must also satisfy the constraint $\langle\!\langle \sum_{k \in \mathcal{S}} \sqrt{m_k} \phi^{1000k}, \phi^\lambda \rangle\!\rangle = 0$, we conclude that we must have $\langle\!\langle \phi^{1000l}, \phi^\lambda \rangle\!\rangle = 0$, $l \in \mathcal{S}$. The first natural variational approximation space for ϕ^λ is therefore the space $\mathcal{A}^\lambda$ spanned by ϕ^{1010k}, $k \in \mathcal{S}$, and ϕ^{1001k}, $k \in \mathcal{P}$,

$$\mathcal{A}^\lambda = \mathrm{span}\{\ \phi^{1010k},\ k \in \mathcal{S}, \quad \phi^{1001k},\ k \in \mathcal{P}\ \}, \tag{2.8.5}$$

and the corresponding indexing set $\mathcal{B}^\lambda$ is

$$\mathcal{B}^\lambda = \{10\} \times \mathcal{S} \ \cup\ \{01\} \times \mathcal{P}. \tag{2.8.6}$$

For convenience, ϕ^λ is taken in the form

$$\phi^\lambda = \sqrt{2 k_{\mathrm{B}} \overline{T}} \Big(\sum_{k \in \mathcal{S}} \sqrt{m_k} \alpha_k^{10\lambda} \phi^{1010k} + \sum_{k \in \mathcal{P}} \sqrt{m_k} \alpha_k^{01\lambda} \phi^{1001k} \Big). \tag{2.8.7}$$

The matrix associated with the variational procedure will be denoted by Λ and is rescaled such that $\Lambda_{kl}^{rs} = (2\sqrt{m_k m_l}/3\bar{p})[\phi^{10rk}, \phi^{10sl}]$, $(r,k),(s,l) \in \mathcal{B}^\lambda$, and the right member is rescaled such that $\beta_k^{r\lambda} = (\sqrt{2m_k}/3\bar{p}\sqrt{k_\mathrm{B}\bar{T}})\langle\langle\phi^{10rk}, \Psi^\lambda\rangle\rangle$, $(r,k) \in \mathcal{B}^\lambda$.

It is then fundamental to observe that all the functions Ψ^{D_l}, $l \in \mathcal{S}$, are orthogonal to the variational approximation space so that we have $\langle\langle\phi^{10rk}, \Psi^\lambda\rangle\rangle = \langle\langle\phi^{10rk}, \Psi^{\lambda'}\rangle\rangle$, $(r,k) \in \mathcal{B}^\lambda$. As a consequence, the thermal diffusion ratios are eliminated and the right member can be evaluated explicitly. Notice again that this elimination has been made possible by constraining ϕ^λ to be orthogonal to the ϕ^{1000l}, $l \in \mathcal{S}$. We finally have $\Lambda \in \mathbb{R}^{n+p,n+p}$, $\beta^\lambda \in \mathbb{R}^{n+p}$, and the system in $\alpha^\lambda \in \mathbb{R}^{n+p}$ reads

$$\Lambda\alpha^\lambda = \beta^\lambda. \tag{2.8.8}$$

An explicit calculation simply yields that Λ is a submatrix of L

$$\Lambda^{1010} = L^{1010}, \tag{2.8.9}$$

$$\Lambda^{1001} = L^{1001}, \tag{2.8.10}$$

$$\Lambda^{0101} = L^{0101}, \tag{2.8.11}$$

and that β^λ is a subvector of $\beta^{\lambda'}$

$$\beta^{10\lambda} = \beta^{10\lambda'}, \tag{2.8.12}$$

$$\beta^{01\lambda} = \beta^{01\lambda'}. \tag{2.8.13}$$

By using the relations (2.1.41), we then obtain that the thermal conductivity is given by

$$\begin{aligned}\lambda &= \frac{\bar{p}}{\bar{T}}\langle\alpha^\lambda, \beta^\lambda\rangle = \frac{\bar{p}}{\bar{T}}\langle\alpha^{10\lambda}, \beta^{10\lambda}\rangle + \frac{\bar{p}}{\bar{T}}\langle\alpha^{01\lambda}, \beta^{01\lambda}\rangle \\ &= \frac{\bar{p}}{k_\mathrm{B}\bar{T}}\Big(\sum_{k\in\mathcal{S}} X_k c_\mathrm{p}^\mathrm{tr}\alpha_k^{10\lambda} + \sum_{k\in\mathcal{P}} X_k c_k^\mathrm{int}\alpha_k^{01\lambda}\Big),\end{aligned} \tag{2.8.14}$$

and that the thermal diffusion ratios χ are expressed in terms of the blocks of the matrix L and the vector α^λ with the relation

$$\chi = [L^{0010}, L^{0001}]\alpha^\lambda = L^{0010}\alpha^{10\lambda} + L^{0001}\alpha^{01\lambda}. \tag{2.8.15}$$

For the particular choice (2.6.3), (2.7.3), and (2.8.5) of the variational approximation spaces, the equivalence of definitions (2.8.14)(2.8.15) with definitions (2.8.1)(2.8.2) will be also established in Section 4.6.1.

Remark. Introducing the functional space

$$\mathfrak{A}^{\lambda'} = \{ \, \xi \in \mathcal{A}^{\lambda'}, \ \langle\!\langle \xi, \sum_{k\in\mathcal{S}} \sqrt{m_k}\phi^{1000k} \rangle\!\rangle = 0 \, \},$$

and the functional $\mathfrak{f}$ defined by

$$\mathfrak{f}(\xi) = \tfrac{1}{2}\langle\!\langle \xi, \Im(\xi) \rangle\!\rangle - \langle\!\langle \Psi^{\lambda'}, \xi \rangle\!\rangle, \qquad \xi \in \mathcal{A}^{\lambda'},$$

we note that $\phi^{\lambda'}$ minimizes $\mathfrak{f}$ on $\mathfrak{A}^{\lambda'}$ whereas ϕ^{λ} minimizes $\mathfrak{f}$ on $\mathcal{A}^{\lambda}$. Furthermore, it is straightforward to check that

$$\mathcal{A}^{\lambda} = \{ \, \xi \in \mathfrak{A}^{\lambda'}, \ \langle\!\langle \Psi^{D_l}, \xi \rangle\!\rangle = 0, \ 1 \le l \le n-1 \, \},$$

keeping in mind that $\sum_{l\in\mathcal{S}} Y_l \Psi^{D_l} = 0$ and that ϕ^{1000l}, $l \in \mathcal{S}$, is a linear combination of Ψ^{D_l} and of the collisional invariant $\sum_{k\in\mathcal{S}} \sqrt{m_k}\phi^{1000k}$. As a consequence, there exist Lagrange multipliers $\mathfrak{y}_1, \dots, \mathfrak{y}_{n-1}$ such that

$$\Im(\phi^{\lambda}) - \Psi^{\lambda'} = \sum_{1\le l\le n-1} \mathfrak{y}_l \Psi^{D_l}.$$

Letting now $\mathfrak{y}_n = 0$ and $(\bar{p}k_{\mathrm{B}}\overline{T})\, \mathfrak{v}_l = (\mathfrak{y}_l - Y_l \sum_{k\in\mathcal{S}} \mathfrak{y}_k)$, $l \in \mathcal{S}$, we deduce that

$$\Im(\phi^{\lambda}) = \Psi^{\lambda'} + \bar{p}k_{\mathrm{B}}\overline{T} \sum_{l\in\mathcal{S}} \mathfrak{v}_l \Psi^{D_l},$$

where $\sum_{l\in\mathcal{S}} \mathfrak{v}_l = 0$ since $\sum_{l\in\mathcal{S}} Y_l \Psi^{D_l} = 0$. By linearity we obtain that

$$\phi^{\lambda} = \phi^{\lambda'} + \bar{p}k_{\mathrm{B}}\overline{T} \sum_{l\in\mathcal{S}} \mathfrak{v}_l \phi^{D_l},$$

and from (2.1.41) we then deduce that $\mathfrak{v}_l = \chi_l$, $l \in \mathcal{S}$. As a consequence, the thermal diffusion ratios appear as Lagrange multipliers associated with the linearly dependent constraint functions Ψ^{D_l}, $l \in \mathcal{S}$, and constrained by the relation $\sum_{l\in\mathcal{S}} \chi_l = 0$.

2.8.2 Mathematical Structure of the System $\Lambda\alpha^{\lambda} = \beta^{\lambda}$

Properties of $\mathcal{A}^{\lambda}$. The approximation space $\mathcal{A}^{\lambda}$ introduced in (2.8.5) satisfies the perpendicularity property (2.3.20), the species orthogonality property (2.3.32), and the relation (2.3.36) from (2.3.52). In addition, the basis functions satisfy the species localization property (2.3.25). As a consequence, the results (2.3.37) established in Section 2.3.5 apply.

Properties of Λ. This shows that the matrices Λ and $2db(\Lambda) - \Lambda$ are symmetric positive definite for $n \geq 1$, and so is $db(\Lambda)$.

Remark. The properties of the matrix Λ could also be derived from the properties of the matrix L.

2.8.3 The System $\Lambda_{[e]}\alpha^{\lambda}_{[e]} = \beta^{\lambda}_{[e]}$

A simplified formulation for λ and χ can easily be derived from the one considered in Section 2.7.2 for λ' and θ. This formulation is associated with the use of the basis functions ϕ^{10ek}, $k \in \mathcal{S}$. We thus consider the approximation space $\mathcal{A}^{\lambda}_{[e]}$ spanned by ϕ^{10ek}, $k \in \mathcal{S}$,

$$\mathcal{A}^{\lambda}_{[e]} = \text{span}\{\ \phi^{10ek},\ k \in \mathcal{S}\ \}, \tag{2.8.16}$$

with the corresponding indexing set $\mathcal{B}^{\lambda}_{[e]}$

$$\mathcal{B}^{\lambda}_{[e]} = \{e\}\times\mathcal{S}. \tag{2.8.17}$$

The transformation matrix $\mathcal{R}$ expressing the new basis functions ϕ^{10ek}, $k \in \mathcal{S}$, in terms of the old basis functions ϕ^{1010k}, $k \in \mathcal{S}$, and ϕ^{1001k}, $k \in \mathcal{P}$, has its coefficients $\mathcal{R}^{rs}_{kl}$, $(r,k) \in \mathcal{B}^{\lambda}_{[e]}$, $(s,l) \in \mathcal{B}^{\lambda}$, given by

$$\begin{cases} \mathcal{R}^{e10}_{kl} = \delta_{kl}, & k,l \in \mathcal{S} \\ \mathcal{R}^{e01}_{kl} = \delta_{kl}, & k \in \mathcal{S}, \quad l \in \mathcal{P}. \end{cases} \tag{2.8.18}$$

The corresponding transport linear system will be denoted by

$$\Lambda_{[e]}\alpha^{\lambda}_{[e]} = \beta^{\lambda}_{[e]}, \tag{2.8.19}$$

and from (2.3.58) we have $\Lambda_{[e]} = \mathcal{R}\Lambda\mathcal{R}^t$ and $\beta^{\lambda}_{[e]} = \mathcal{R}\beta^{\lambda}$. As a consequence, the matrix $\Lambda_{[e]}$ is simply a submatrix of $L_{[e]}$

$$\Lambda^{ee}_{[e]} = L^{ee}_{[e]}, \tag{2.8.20}$$

and the right member $\beta^{\lambda}_{[e]}$ is simply a subvector of $\beta^{\lambda'}_{[e]}$

$$\beta^{e\lambda}_{[e]} = \beta^{e\lambda'}_{[e]}. \tag{2.8.21}$$

Finally, the thermal conductivity $\lambda_{[e]}$ is given by

$$\begin{aligned} \lambda_{[e]} &= \frac{\bar{p}}{\bar{T}}\langle \alpha^{\lambda}_{[e]}, \beta^{\lambda}_{[e]} \rangle = \frac{\bar{p}}{\bar{T}}\langle \alpha^{e\lambda}_{[e]}, \beta^{e\lambda}_{[e]} \rangle \\ &= \frac{\bar{p}}{k_B\bar{T}} \sum_{k\in\mathcal{S}} X_k(c^{tr}_p + c^{int}_k)\alpha^{e\lambda}_{[e]k}, \end{aligned} \tag{2.8.22}$$

and the thermal diffusion ratios $\chi_{[e]}$ are expressed in terms of a block of the matrix $L_{[e]}$ and the vector $\alpha^{\lambda}_{[e]}$ with the relation

$$\chi_{[e]} = [L^{00e}_{[e]}]\alpha^{\lambda}_{[e]} = L^{00e}_{[e]}\alpha^{e\lambda}_{[e]}. \tag{2.8.23}$$

Remark. We will also establish in Section 4.6.2 that the thermal diffusion ratios $\chi_{[e]}$ are the unique solution of the linear system

$$\begin{cases} D_{[e]}\chi_{[e]} = \theta_{[e]}, \\ \langle U, \chi_{[e]}\rangle = 0, \end{cases} \tag{2.8.24}$$

and that we have

$$\lambda_{[e]} = \lambda'_{[e]} - (\bar{p}/\overline{T}) \sum_{k\in\mathcal{S}} \theta_{[e]k}\chi_{[e]k}. \tag{2.8.25}$$

Remark. Following Section 2.8.1, the thermal diffusion ratios $\chi_{[e]}$ can also be interpreted as Lagrange multipliers.

2.8.4 Mathematical Structure of the System $\Lambda_{[e]}\alpha^{\lambda}_{[e]} = \beta^{\lambda}_{[e]}$

Properties of $\mathcal{A}^{\lambda}_{[e]}$. The approximation space $\mathcal{A}^{\lambda}_{[e]}$ introduced in (2.8.16) satisfies the perpendicularity property (2.3.20), the species orthogonality property (2.3.32), and the relation (2.3.36) from (2.3.52). In addition, the basis functions satisfy the species localization property (2.3.25). As a consequence, the results (2.3.37) established in Section 2.3.5 apply.

Properties of $\Lambda_{[e]}$. We have thus established that $\Lambda_{[e]}$ and $2db(\Lambda_{[e]}) - \Lambda_{[e]}$ are symmetric positive definite for $n \geq 1$, and so is $db(\Lambda_{[e]})$.

2.8.5 Alternative Formulations

Higher order formulations will not be investigated in this book. Notice also that a monatomic approximation is obtained by using the functional space $\mathcal{A}^{\lambda}_{[\mathrm{mon}]}$ with

$$\mathcal{A}^{\lambda}_{[\mathrm{mon}]} = \mathrm{span}\{\ \phi^{1010k},\ k \in \mathcal{S}\ \}.$$

The corresponding linear system $\Lambda_{[\mathrm{mon}]}\alpha^{\lambda}_{[\mathrm{mon}]} = \beta^{\lambda}_{[\mathrm{mon}]}$, the thermal conductivity $\lambda_{[\mathrm{mon}]}$, and the thermal diffusion ratios $\chi_{[\mathrm{mon}]}$ are easily obtained, but are omitted for brevity. On the other hand, approximate coefficients obtained by using the one-dimensional space spanned by $\Psi^{\lambda'}$ are equivalently obtained by applying the conjugate gradient algorithm to the system (2.8.8). This algorithm and its preconditioned version will be presented in Section 5.1.4.

2.9 Comparison with the Systems of Monchick, Yun, and Mason

In this section we examine the link between the transport coefficients as they result from the variational approach described in Sections 2.1-2.8 and the transport coefficients of the linear systems considered by Monchick, Yun, and Mason in [MYM63]. This comparison is motivated, since the structure of the linear systems obtained by Monchick, Yun, and Mason is different from that of Waldmann and Trübenbacher [WT62]. Indeed, Monchick, Yun, and Mason have eliminated the singularities by explicitly using the linear constraints and zeroing the diagonal coefficients of the system matrices, following a procedure introduced by Curtiss and Hirschfelder [CH49] [HCB54], thereby destroying the natural symmetries of the model. In this section we show that the same value is obtained for all the transport coefficients, provided that two misprints are corrected in the expressions for the matrix K and L given in [MYM63].

Finally, all the quantities introduced in [MYM63] will be denoted with the superscript ^{MYM} in this section.

2.9.1 The Boltzmann Equation

The starting point of the Monchick, Yun, and Mason approach is the Boltzmann equation of Wang Chang and Uhlenbeck generalized to mixtures [MYM63]. Keeping the notation of Section 2.1 and letting f_i^{MYM} be the corresponding distribution function of the ith species, this equation can be written in the form

$$\begin{aligned}
&\partial_t(f_i^{\text{MYM}}) + (c_i.\nabla)f_i^{\text{MYM}} + (b_i.\nabla_{c_i})f_i^{\text{MYM}} = \\
&\sum_{\substack{j\in\mathcal{S}\\ I'\in\mathcal{E}_i\\ J,J'\in\mathcal{E}_j}} \sum_{\substack{\mathcal{I}'\in\mathcal{M}_{iI'}\\ \mathcal{J}\in\mathcal{M}_{jJ}\\ \mathcal{J}'\in\mathcal{M}_{jJ'}}} \iiint \Big(f_i^{\text{MYM}}(c_i', I', \mathcal{I}') f_j^{\text{MYM}}(\tilde{c}_j', J', \mathcal{J}') g' \mathfrak{I}_{ij}\left[\begin{smallmatrix} I & J & \mathcal{I} & \mathcal{J} \\ I' & J' & \mathcal{I}' & \mathcal{J}' \end{smallmatrix}\right] \\
&\qquad - f_i^{\text{MYM}}(c_i, I, \mathcal{I}) f_j^{\text{MYM}}(\tilde{c}_j, J, \mathcal{J}) g \mathfrak{I}_{ij}\left[\begin{smallmatrix} I' & J' & \mathcal{I}' & \mathcal{J}' \\ I & J & \mathcal{I} & \mathcal{J} \end{smallmatrix}\right]\Big) \sin\chi d\chi d\varphi d\tilde{c}_j,
\end{aligned} \tag{2.9.1}$$

where $\mathcal{M}_{iI}$ denotes the set of magnetic quantum numbers associated with the Ith quantum energy shell of the ith species, $\mathcal{I}$ and $\mathcal{J}$ the magnetic quantum numbers of the

i^{th} and j^{th} species before collision, $\mathcal{I}'$ and $\mathcal{J}'$ the corresponding numbers after collision, $a_{iI} = \text{card}(\mathcal{M}_{iI})$ the degeneracy of the Ith quantum energy shell of the ith species, and $\mathfrak{I}_{ij}\left[\begin{smallmatrix} I' & J' & \mathcal{I}' & \mathcal{J}' \\ I & J & \mathcal{I} & \mathcal{J} \end{smallmatrix}\right]$ the collision cross-section which is a function $\mathfrak{I}_{ij}\left[\begin{smallmatrix} I' & J' & \mathcal{I}' & \mathcal{J}' \\ I & J & \mathcal{I} & \mathcal{J} \end{smallmatrix}\right](g, \chi, \varphi)$ of g, χ, and φ. Notice that the two indices I and $\mathcal{I}$ describing the quantum state of the ith species are regrouped into a single index in [MYM63]. In the Wang Chang and Uhlenbeck theory, the collision cross-sections also satisfy the detailed balance property

$$g\mathfrak{I}_{ij}\left[\begin{smallmatrix} I' & J' & \mathcal{I}' & \mathcal{J}' \\ I & J & \mathcal{I} & \mathcal{J} \end{smallmatrix}\right] = g'\mathfrak{I}_{ij}\left[\begin{smallmatrix} I & J & \mathcal{I} & \mathcal{J} \\ I' & J' & \mathcal{I}' & \mathcal{J}' \end{smallmatrix}\right].$$

The link between the Waldmann and Wang Chang and Uhlenbeck formulations is that, in the absence of macroscopic angular momentum and external magnetic and electric fields, there are no polarization phenomena, and the mixture is isotropic in such a way that f_i^{MYM} does not depend on $\mathcal{I}$, so that $f_i^{\text{MYM}}(c_i, I, \mathcal{I}) = f_i^{\text{MYM}}(c_i, I)$ [Wa58]. We may thus introduce the degeneracy averaged distribution function

$$f_i(c_i, I) = \sum_{\mathcal{I} \in \mathcal{M}_{iI}} f_i^{\text{MYM}}(c_i, I, \mathcal{I}) = a_{iI} f_i^{\text{MYM}}(c_i, I), \tag{2.9.2}$$

and the averaged cross-sections over all magnetic quantum numbers

$$\bar{\sigma}_{ij}^{IJI'J'} = \frac{1}{a_{iI}a_{jJ}} \sum_{\substack{\mathcal{I} \in \mathcal{M}_{iI} \\ \mathcal{J} \in \mathcal{M}_{jJ}}} \sum_{\substack{\mathcal{I}' \in \mathcal{M}_{iI'} \\ \mathcal{J}' \in \mathcal{M}_{jJ'}}} \mathfrak{I}_{ij}\left[\begin{smallmatrix} I' & J' & \mathcal{I}' & \mathcal{J}' \\ I & J & \mathcal{I} & \mathcal{J} \end{smallmatrix}\right], \tag{2.9.3}$$

so that by summing the a_{iI} equations (2.9.1) for each $\mathcal{I} \in \mathcal{M}_{iI}$ we reobtain the Boltzmann equation (2.1.1)–(2.1.3) introduced by Waldmann. Notice that more general symmetry relations than the detailed balance property can also be introduced, but still lead to the reciprocity relation (2.1.5) which is directly obtained from quantum investigations [Wa58] [WT62] [MYM63]. For isotropic mixtures, both forms of the Boltzmann equation are therefore equivalent [Wa58] [MM63] [MYM63].

2.9.2 Linear Algebra Preliminaries

We note that in [MYM63], the transport coefficients are expressed as ratios of determinants, e.g.,

$$\mu = -\begin{vmatrix} A & b \\ c^t & 0 \end{vmatrix} \times |A|^{-1},$$

where A is a nonsingular matrix, b and c are vectors, c^t is the transpose of c, and where $|M|$ denotes the determinant of any matrix M. By expanding the larger determinant with respect to the last row and by using Cramer's rule, this expression is readily seen to be equivalent to

$$\mu = \langle x, c \rangle,$$

where x is the unique solution of the linear system

$$Ax = b,$$

so that the Monchick, Yun, and Mason formulas can easily be transformed into scalar products involving solutions of linear systems. We will freely use these equivalent formulations in the following part of Section 2.9 without further notice.

2.9.3 The Shear Viscosity

For the shear viscosity, a direct examination reveals that the matrix H and the right-hand side β^η given in (2.4.7)–(2.4.9) exactly coincide with the ones given in (20)(21) of [MYM63] and that the formulas used to evaluate η are identical.

2.9.4 The Volume Viscosity

From (34) of [MYM63], κ^{MYM} is given by

$$\kappa^{\text{MYM}} = (k_{\text{B}}\overline{T}/c_{\text{v}}) \sum_{k\in\mathcal{S}} X_k \alpha_k^{10\kappa\text{MYM}},$$

where $\alpha^{\kappa\text{MYM}}$ satisfies the linear system $K^{\text{MYM}}\alpha^{\kappa\text{MYM}} = \beta^{\kappa\text{MYM}}$. The matrix K^{MYM} and the vector β^{MYM} are given from (34) and (37-40) of [MYM63]. Our calculations have shown that one misprint is present in (37) of [MYM63] in the term $-2x_q x_{q'} \langle \Delta\epsilon_{qq'}^2 \rangle_{qq'}$ which should be replaced by $-2x_q x_{q'} \langle \Delta\epsilon_{qq}^2 \rangle_{qq}$. Assuming that this correction has been made, one can then verify that the matrix K^{MYM} is given by

$$K^{\text{MYM}} = (k_{\text{B}}\overline{T}) \mathcal{Q} K \mathcal{Q} + \Re \otimes (1/k_{\text{B}}) \mathcal{Q}\mathcal{K}, \tag{2.9.4}$$

where $\mathcal{Q}$ is the diagonal matrix $\mathcal{Q} = \text{diag}\big((x_k^r)_{(r,k)\in\mathcal{B}^\kappa}\big)$ with $x_k^{10} = 1$, $k \in \mathcal{S}$, and $x_k^{01} = -1$, $k \in \mathcal{P}$, and where $\Re$ is the vector given by

$$\begin{aligned}
\Re_k^{10} = & -\frac{8}{3} \sum_{\substack{l\in\mathcal{S}\\ l\neq k}} X_l \Big(4\frac{m_k m_l}{(m_k+m_l)^2} \Omega_{kl}^{(1,1)} + \frac{m_l^2}{(m_k+m_l)^2} \big[(\Delta\epsilon_{kl})^2\big]_{kl} \Big) \\
& - \frac{4}{3} X_k \big[\Delta\epsilon_{kk}^2\big]_{kk}, \qquad k \in \mathcal{S},
\end{aligned}$$

$$\Re_k^{01} = \frac{4k_{\text{B}}}{c_k^{\text{int}}} \Big(\sum_{\substack{l\in\mathcal{S}\\ l\neq k}} X_l \big[(\Delta\epsilon_k)^2\big]_{kl} + X_k \big[\Delta\epsilon_k \Delta\epsilon_{kk}\big]_{kk} \Big), \qquad k \in \mathcal{P}.$$

In addition, one can also verify that the right member is given by

$$\beta^{\kappa\text{MYM}} = c_v \mathcal{Q}\beta^{\kappa}. \tag{2.9.5}$$

Multiplying now the relation $K^{\text{MYM}}\alpha^{\kappa\text{MYM}} = \beta^{\kappa\text{MYM}}$ on the left by $(\mathcal{Q}\mathcal{V})^t$ and noting that $(\mathcal{Q}\mathcal{V})^t\,\mathcal{Q}K\mathcal{Q} = 0$ since $\mathcal{Q}^t\mathcal{Q} = I$ and $\mathcal{V}^tK = 0$, and that $(\mathcal{Q}\mathcal{V})^t\,\mathcal{Q}\beta^{\kappa} = 0$ since $\mathcal{V}^t\beta^{\kappa} = 0$, we get that $\langle \mathcal{Q}\mathcal{K}, \alpha^{\kappa\text{MYM}}\rangle = 0$ since $(\mathcal{Q}\mathcal{V})^t\Re = \langle \mathcal{Q}\mathcal{V}, \Re\rangle \neq 0$, as may easily be checked. This, in turn, shows that $(k_B\overline{T})\mathcal{Q}K\mathcal{Q}\alpha^{\kappa\text{MYM}} = \beta^{\kappa\text{MYM}}$. Defining now $x = (k_B\overline{T}/c_v)\mathcal{Q}\alpha^{\kappa\text{MYM}}$, we deduce, after some algebra, that x satisfies $Kx = \beta^{\kappa}$ and $\langle \mathcal{K}, x\rangle = 0$, so that $x = \alpha^{\kappa}$ and, therefore, $\alpha^{\kappa} = (k_B\overline{T}/c_v)\mathcal{Q}\alpha^{\kappa\text{MYM}}$. Finally, since the volume viscosity in [MYM63] is evaluated as $\kappa^{\text{MYM}} = (k_B\overline{T}/c_v)\sum_{k\in\mathcal{S}} X_k\alpha_k^{10\kappa\text{MYM}}$, we obtain $\kappa^{\text{MYM}} = \sum_{k\in\mathcal{S}} X_k\alpha_k^{10\kappa} = \kappa$ and both definitions coincide.

2.9.5 The Partial Thermal Conductivity

From (75)(76) of [MYM63] it is readily seen that λ'^{MYM} is given by

$$\lambda'^{\text{MYM}} = -4\Big(\sum_{k\in\mathcal{S}} X_k\alpha_k^{10\lambda'\text{MYM}} + \sum_{k\in\mathcal{P}} X_k\alpha_k^{01\lambda'\text{MYM}}\Big),$$

where the vector $\alpha^{\lambda'\text{MYM}}$ satisfies the linear system $L^{\text{MYM}}\alpha^{\lambda'\text{MYM}} = \beta^{\lambda'\text{MYM}}$. The matrix L^{MYM} and the vector $\beta^{\lambda'\text{MYM}}$ are given from (75)(76) and (82-90) of [MYM63]. Our calculations have shown that one misprint is present in (85) of [MYM63] in the term $\langle(\epsilon_{qi} - \bar{\epsilon}_q)(\gamma^2 - \gamma\gamma'\cos\chi)\rangle_{qq''}$ which should be replaced by $\langle(\epsilon_{q'i} - \bar{\epsilon}_{q'})(\gamma^2 - \gamma\gamma'\cos\chi)\rangle_{qq''}$. Note that this term is usually assumed to vanish [MPM65], so that this misprint has no consequences on the various numerical calculations that can be found in the literature. Assuming that this correction has been made, one can then easily check that the matrix L^{MYM} can be expressed as

$$L^{\text{MYM}} = -(16\overline{T}/25\bar{p})\mathcal{Q}L\mathcal{Q} + \Re\otimes\mathcal{L}, \tag{2.9.6}$$

where $\mathcal{Q}$ is the diagonal matrix $\mathcal{Q} = \text{diag}\big((x_k^r)_{(r,k)\in\mathcal{B}^{\lambda'}}\big)$ with $x_k^{00} = 1$, $k\in\mathcal{S}$, $x_k^{10} = 1$, $k\in\mathcal{S}$, and $x_k^{01} = (5k_B/2c_k^{\text{int}})$, $k\in\mathcal{P}$, and where $\Re$ is the vector given by

$$\Re_k^{00} = \frac{16\overline{T}}{25\bar{p}}\frac{X_k}{Y_k}\sum_{\substack{l\in\mathcal{S}\\ l\neq k}}\frac{X_l}{\mathcal{D}_{kl}}, \qquad k\in\mathcal{S},$$

$$\Re_k^{10} = 0, \quad k\in\mathcal{S},$$

$$\Re_k^{01} = 0, \quad k\in\mathcal{P}.$$

In addition, one can also verify that the right member is given by

$$\beta^{\lambda'\text{MYM}} = (2/5)\mathcal{Q}\beta^{\lambda'}. \tag{2.9.7}$$

Multiplying now the equation $L^{\text{MYM}}\alpha^{\lambda'\text{MYM}} = \beta^{\lambda'\text{MYM}}$ on the left by $\mathcal{U}^t$ and noting that $\mathcal{U}^t\mathcal{Q}L\mathcal{Q} = 0$ since $\mathcal{U}^t\mathcal{Q} = \mathcal{U}^t$ and $\mathcal{U}^tL = 0$, and that $\mathcal{U}^t\mathcal{Q}\beta^{\lambda'} = 0$ since $\mathcal{U}^t\beta^{\lambda'} = 0$, we get that $\langle\mathcal{L}, \alpha^{\lambda'\text{MYM}}\rangle = 0$ since $\mathcal{U}^t\mathfrak{R} = \langle\mathcal{U}, \mathfrak{R}\rangle \neq 0$, as may easily be checked. This, in turn, shows that we have $-(16\overline{T}/25\overline{p})\mathcal{Q}L\mathcal{Q}\alpha^{\lambda'\text{MYM}} = \beta^{\lambda'\text{MYM}}$. Defining now $x = -(8\overline{T}/5\overline{p})\mathcal{Q}\alpha^{\lambda'\text{MYM}}$, we deduce, after some algebra, that x satisfies $Lx = \beta^{\lambda'}$ and $\langle\mathcal{L}, x\rangle = 0$, so that $x = \alpha^{\lambda'}$ and therefore $\alpha^{\lambda'} = -(8\overline{T}/5\overline{p})\mathcal{Q}\alpha^{\lambda'\text{MYM}}$. From this relation, we finally obtain that

$$\lambda'^{\text{MYM}} = \frac{5\overline{p}}{2\overline{T}}\Big(\sum_{k\in\mathcal{S}} X_k\alpha_k^{10\lambda'} + \sum_{k\in\mathcal{P}} \frac{2c_k^{\text{int}}}{5k_{\text{B}}} X_k\alpha_k^{01\lambda'}\Big) = \frac{\overline{p}}{\overline{T}}\langle\alpha^{\lambda'}, \beta^{\lambda'}\rangle = \lambda',$$

so that both definitions are equivalent.

2.9.6 The Thermal Conductivity

From (93)(94) of [MYM63] it is readily seen that λ^{MYM} is given by

$$\lambda^{\text{MYM}} = -4\Big(\sum_{k\in\mathcal{S}} X_k\alpha_k^{10\lambda\text{MYM}} + \sum_{k\in\mathcal{P}} X_k\alpha_k^{01\lambda\text{MYM}}\Big),$$

where the vector $\alpha^{\lambda\text{MYM}}$ satisfies the linear system $\Lambda^{\text{MYM}}\alpha^{\lambda\text{MYM}} = \beta^{\lambda\text{MYM}}$. The matrix Λ^{MYM} and the vector $\beta^{\lambda\text{MYM}}$ are given from (93)(94) and (87-90) of [MYM63], and one may easily check that the matrix Λ^{MYM} reads

$$\Lambda^{\text{MYM}} = -\frac{16\overline{T}}{25\overline{p}}\mathcal{Q}\Lambda\mathcal{Q}, \tag{2.9.8}$$

where $\mathcal{Q}$ is the diagonal matrix $\mathcal{Q} = \text{diag}\big((x_k^r)_{(r,k)\in\mathcal{B}^\lambda}\big)$ with $x_k^{10} = 1$, $k \in \mathcal{S}$, and $x_k^{01} = (5k_{\text{B}}/2c_k^{\text{int}})$, $k \in \mathcal{P}$, and that the right member is given by

$$\beta^{\lambda\text{MYM}} = (2/5)\mathcal{Q}\beta^{\lambda}. \tag{2.9.9}$$

Defining now $x = -(8\overline{T}/5\overline{p})\mathcal{Q}\alpha^{\lambda\text{MYM}}$, we deduce, after some algebra, that x satisfies $\Lambda x = \beta^\lambda$, so that $x = \alpha^\lambda$ and therefore $\alpha^\lambda = -(8\overline{T}/5\overline{p})\mathcal{Q}\alpha^{\lambda\text{MYM}}$. From these relations, we finally obtain that

$$\lambda^{\text{MYM}} = \frac{5\overline{p}}{2\overline{T}}\Big(\sum_{k\in\mathcal{S}} X_k\alpha_k^{10\lambda} + \sum_{k\in\mathcal{P}} \frac{2c_k^{\text{int}}}{5k_{\text{B}}} X_k\alpha_k^{01\lambda}\Big) = \frac{\overline{p}}{\overline{T}}\langle\alpha^{\lambda}, \beta^{\lambda}\rangle = \lambda,$$

so that the thermal conductivity λ^{MYM} coincides with λ.

2.9.7 The Definition of Mass Fluxes

Before comparing the species diffusion coefficients and the thermal diffusion coefficients, we have to take into account different mass flux definitions. In the work of Monchick, Yun, and Mason, following Curtiss and Hirschfelder [CH49], the mass fluxes are defined by

$$\rho Y_k V_k = (\bar{n}^2 m_k/\rho) \sum_{l\in\mathcal{S}} m_l D_{kl}^{\text{MYM}} d_l - \theta_k^{\text{MYM}} \nabla \log \overline{T}, \qquad k \in \mathcal{S}. \tag{2.9.10}$$

On the contrary, in this book, following Chapman and Cowling [CC70], Curtiss [Cu68], Ferziger and Kapper [FK72], Waldmann [Wa58], and Waldmann and Trübenbacher [WT62], we have used the definition

$$\rho Y_k V_k = -\rho Y_k \sum_{l\in\mathcal{S}} D_{kl} d_l - \rho Y_k \theta_k \nabla \log \overline{T}, \qquad k \in \mathcal{S}. \tag{2.9.11}$$

In particular, the species diffusion coefficients used by Monchick, Yun, and Mason are not symmetric and, therefore, are not formally compatible with Onsager reciprocal relations [Va67].

Identifying now both definitions, we first deduce that

$$\theta_k^{\text{MYM}} = \rho Y_k \theta_k, \qquad k \in \mathcal{S}. \tag{2.9.12}$$

In addition, keeping in mind that the diffusion driving forces $(d_k)_{k\in\mathcal{S}}$ are constrained by the linear relation $\sum_{k\in\mathcal{S}} d_k = 0$, we next deduce that if A is the matrix whose coefficients are $A_{kl} = -\bar{n}^2 m_l D_{kl}^{\text{MYM}}/\rho n_k$, the matrix $D - A$ is then zero on $U^\perp$, where $U = (1)_{k\in\mathcal{S}}$. Therefore, there exists a vector $C \in \mathbb{R}^n$ with $D = A + C \otimes U$ which is easily identified since, by definition [MYM63], we have $D_{kk}^{\text{MYM}} = 0$, $k \in \mathcal{S}$, so that $C_k = D_{kk}$. This finally yields that

$$D_{kl}^{\text{MYM}} = -(\rho n_k/\bar{n}^2 m_l)(D_{kl} - D_{kk}), \qquad k, l \in \mathcal{S}. \tag{2.9.13}$$

The formulas (2.9.12)(2.9.13) relate the species diffusion coefficients and thermal diffusion coefficients and have to be recovered from the explicit expressions given in [MYM63].

2.9.8 The Thermal Diffusion Vector

From (101) of [MYM63] it is readily seen that θ_k^{MYM} is given by

$$\theta_k^{\text{MYM}} = \frac{8}{5}\frac{m_k}{k_B} X_k \alpha_k^{00\lambda'\text{MYM}},$$

where the vector $\alpha^{\lambda'\text{MYM}}$ satisfies the linear system $L^{\text{MYM}}\alpha^{\lambda'\text{MYM}} = \beta^{\lambda'\text{MYM}}$ which has been discussed previously in connection with the partial thermal conductivity. In particular, we have already seen that $\alpha^{\lambda'} = -(8\overline{T}/5\overline{p})\mathcal{Q}\alpha^{\lambda'\text{MYM}}$. From this relation, we deduce that

$$\theta_k^{\text{MYM}} = \frac{8}{5}\frac{m_k}{k_B} X_k \alpha_k^{00\lambda',\text{MYM}} = -\overline{n} m_k X_k \alpha_k^{00\lambda'} = \rho Y_k \theta_k,$$

and we recover the relations (2.9.12).

2.9.9 The Diffusion Matrix

Although the thermal diffusion coefficients considered in [MYM63] involve a three-term expansion of the species perturbed distribution functions, the species diffusion coefficients that are introduced in [MYM63] are only associated with one term expansions. These coefficients have been discussed in Section 2.6.5 and have also been investigated in [Gi91]. As a consequence, we will recover in this section the relations (2.9.13) relating the matrix D^{MYM} and the first-order diffusion matrix $D_{[00]}$.

From (64)(65) of [MYM63], we have

$$D_{kl}^{\text{MYM}} = \frac{\rho}{\overline{n} m_l}(A_{kl}^{-1} - A_{kk}^{-1}),$$

where the matrix A can be written as [MYM63]

$$A_{kl} = \Big(\frac{X_k}{\mathcal{D}_{kl}} + \frac{m_l}{m_k}\sum_{\substack{i\in\mathcal{S}\\ i\neq k}} \frac{X_i}{\mathcal{D}_{ki}}\Big)(1-\delta_{kl}), \qquad k,l \in \mathcal{S}.$$

From these relations, we first deduce that A is such that

$$A\mathcal{X} = -(L^{0000} - \mathcal{H}\otimes Y),$$

where L^{0000} is given in (2.6.7)(2.6.8), $\mathcal{X} = \text{diag}(X_1, \ldots, X_n)$ and the vector $\mathcal{H}$ has components $\mathcal{H}_k = L_{kk}^{0000}/Y_k$. We now claim that the matrix $L^{0000} - \mathcal{H}\otimes Y$ is invertible. Indeed, assuming that $x \in \mathbb{R}^n$ is such that $(L^{0000} - \mathcal{H}\otimes Y)x = 0$, we get, by multiplying on the left by U^t, that $\langle Y, x\rangle = 0$ since $U^t L^{0000} = 0$ and $\langle \mathcal{H}, U\rangle \neq 0$, as may easily be checked. Therefore, we also have $L^{0000}x = (L^{0000} - \mathcal{H}\otimes Y)x = 0$ so that $x \in \mathbb{R}U =$

$N(L^{0000})$. This shows that $x \in \mathbb{R}U \cap Y^{\perp} = \{0\}$ so that $x = 0$ and $L^{0000} - \mathcal{H}\otimes Y$ is nonsingular. Therefore, we can introduce the vector Z such that $(L^{0000} - \mathcal{H}\otimes Y)Z = -Y$ and we have $(L^{0000} - \mathcal{H}\otimes Y)(D_{[00]} - Z\otimes U) = L^{0000}D_{[00]} + Y\otimes U$ since $\mathcal{H}\otimes Y\, D_{[00]} = \mathcal{H}\otimes(D^t_{[00]}Y) = \mathcal{H}\otimes(D_{[00]}Y) = 0$ because $D_{[00]}Y = 0$. However, from (2.6.47) we know that $L^{0000}D_{[00]} = I - Y\otimes U$, so that we finally obtain $(L^{0000} - \mathcal{H}\otimes Y)(D_{[00]} - Z\otimes U) = I$. This shows that $(L^{0000} - \mathcal{H}\otimes Y)^{-1} = (D_{[00]} - Z\otimes U)$ so that $A^{-1} = -\mathcal{X}^{-1}(D_{[00]} - Z\otimes U)$. As a consequence, we obtain that

$$D^{\text{MYM}}_{kl} = -\frac{\rho n_k}{\bar{n}^2 m_l}(D_{[00]kl} - D_{[00]kk}),$$

which are exactly the relations (2.9.13) specialized to first-order coefficients.

2.10 The Mason and Monchick Approximations

In order to calculate numerically the coefficients of the matrices H, K, and L, various collision integrals have to be evaluated. Although, in principle, these collision integrals could be evaluated computationally, this has seldom been done for polyatomic species pairs over wide temperature ranges. Usually, these collision integrals are either related to some known quantity, or evaluated from experimental measurements, or even estimated. In this book, we have used the approximations of Mason and Monchick [MM62] and Monchick, Pereira, and Mason [MPM65] for our numerical applications. The quality of these approximations is sometimes questioned [WV92], but we point out that the general theory presented in this book applies to many other systems, using for instance different approximations.

2.10.1 Approximate Collision Integrals

In the framework of the Mason and Monchick [MM62] and Monchick, Pereira, and Mason [MPM65] approximations, we first neglect complex collisions, i.e., collisions in which there are more than one quantum jump. In addition, we assume that the reduced internal energies ϵ_{iI} can be split into different internal energy modes as for instance rotation and vibration, so that we have $\epsilon_{iI} = \sum_{\mathfrak{m}\in\mathfrak{C}} \epsilon^{\mathfrak{m}}_{iI}$, where the modes are indexed by $\mathfrak{m} \in \mathfrak{C}$ and where $\epsilon^{\mathfrak{m}}_{iI}$ is the reduced internal energy in mode $\mathfrak{m}$ for the particle of species i in state I. In this situation, one can state [MM62] [MPM65] that

$$\begin{cases} \left[(\Delta\epsilon_{ii})^2\right]_{ii} = \sum_{\mathfrak{m}\in\mathfrak{C}} \left[(\Delta\epsilon_{ii}^{\mathfrak{m}})^2\right]_{ii} = \frac{\overline{T}}{2\overline{p}} \sum_{\mathfrak{m}\in\mathfrak{C}} \frac{c_i^{\mathfrak{m}}}{\tau_{ii}^{\mathfrak{m}}}, \\ \left[\Delta\epsilon_i\Delta\epsilon_{ij}\right]_{ij} = \left[(\Delta\epsilon_i)^2\right]_{ij} = \sum_{\mathfrak{m}\in\mathfrak{C}} \left[(\Delta\epsilon_i^{\mathfrak{m}})^2\right]_{ij} = \frac{\overline{T}}{4\overline{p}} \sum_{\mathfrak{m}\in\mathfrak{C}} \frac{c_i^{\mathfrak{m}}}{\tau_{ij}^{\mathfrak{m}}}, \end{cases} \tag{2.10.1}$$

where $c_i^{\mathfrak{m}}$ is the specific internal capacity of species i associated with the internal energy mode $\mathfrak{m}$ and where $\tau_{ij}^{\mathfrak{m}}$ is the relaxation time of the energy mode $\mathfrak{m}$ for species i colliding with species j. These relaxation times are expressed in terms of collision numbers $\xi_{ij}^{\mathfrak{m}}$ by

$$\xi_{ij}^{\mathfrak{m}} = \frac{4\overline{p}}{\pi\eta_{ij}}\tau_{ij}^{\mathfrak{m}}, \tag{2.10.2}$$

where $\eta_{ij} = (5/8)k_{\mathrm{B}}\overline{T}/\Omega_{ij}^{(2,2)}$. Usually, all the collision numbers $\xi_{ij}^{\mathfrak{m}}$, $j \in \mathcal{S}$, are approximated by a single number $\xi_i^{\mathfrak{m}}$, independent of j. Note that the formulas (2.10.1) yield a global relaxation time τ_{ij} which is a weighted harmonic mean between the different internal energy mode relaxation times

$$\frac{c_i^{\mathrm{int}}}{\tau_{ij}} = \sum_{\mathfrak{m}\in\mathfrak{C}} \frac{c_i^{\mathfrak{m}}}{\tau_{ij}^{\mathfrak{m}}},$$

where $c_i^{\mathrm{int}} = \sum_{\mathfrak{m}\in\mathfrak{C}} c_i^{\mathfrak{m}}$, and this is because the modes are constrained to have the same temperature, since only polynomials in ϵ_{iI} are considered in the functional space basis. When polynomials in $\epsilon_{iI}^{\mathfrak{m}}$, $\mathfrak{m} \in \mathfrak{C}$, are considered, weighted arithmetic means are obtained instead of harmonic means, together with considerably more collision integrals, as detailed, for instance, in [Mo64] for pure gases.

We only consider the rotational and vibrational modes, and assuming that the collision numbers ξ_{ij}^{rot} and ξ_{ij}^{vib} are independent of $j \in \mathcal{S}$, i.e., $\xi_{ij}^{\mathrm{rot}} = \xi_i^{\mathrm{rot}}$, and $\xi_{ij}^{\mathrm{vib}} = \xi_i^{\mathrm{vib}}$, for $i, j \in \mathcal{S}$, we define the overall internal energy collision number ξ_i^{int} by

$$\frac{c_i^{\mathrm{int}}}{\xi_i^{\mathrm{int}}} = \frac{c_i^{\mathrm{rot}}}{\xi_i^{\mathrm{rot}}} + \frac{c_i^{\mathrm{vib}}}{\xi_i^{\mathrm{vib}}}. \tag{2.10.3}$$

Usually $c_i^{\mathrm{rot}} \gg c_i^{\mathrm{vib}}$ and $\xi_i^{\mathrm{rot}} \ll \xi_i^{\mathrm{vib}}$ so that the vibrational contribution is small. The temperature dependence of ξ_i^{rot} and ξ_i^{vib} is generally estimated from [Pa59]. In addition, in order to avoid notational complexity, we define $\xi_i^{\mathrm{int}} = 1$ for $i \notin \mathcal{P}$, recalling that $c_i^{\mathrm{int}} = 0$ in this situation. The formulas (2.10.1) then become

$$\begin{cases} \left[(\Delta\epsilon_{ii})^2\right]_{ii} = \frac{2\overline{T}}{\pi\eta_i}\frac{c_i^{\mathrm{int}}}{\xi_i^{\mathrm{int}}}, \\ \left[\Delta\epsilon_i\Delta\epsilon_{ij}\right]_{ij} = \left[(\Delta\epsilon_i)^2\right]_{ij} = \frac{\overline{T}}{\pi\eta_{ij}}\frac{c_i^{\mathrm{int}}}{\xi_i^{\mathrm{int}}}. \end{cases} \tag{2.10.4}$$

Following [MYM63], we also have the estimates

$$\begin{cases} \left[\epsilon^0_{iI}\gamma^r(\gamma^s - \gamma'^s \cos\chi)\right]_{ij} = 0, & i,j \in \mathcal{S}, \quad r,s \geq 0, \\ \left[\epsilon^0_{iI}\gamma^r(\epsilon^0_{jJ}\gamma^s - \epsilon^0_{jJ'}\gamma'^s \cos\chi)\right]_{ij} = 0, & i,j \in \mathcal{S}, \quad i \neq j, \qquad r,s \geq 0, \end{cases} \tag{2.10.5}$$

and the diffusion coefficients of internal energy are approximated by the binary diffusion coefficients

$$\mathcal{D}_{i\ \mathrm{int},j} = \mathcal{D}_{ij}, \tag{2.10.6}$$

except for polar molecules for which resonant collisions are important. In this situation, the coefficients $\mathcal{D}_{i\ \mathrm{int},i}$ are approximated as $\mathcal{D}_{i\ \mathrm{int},i} = \mathcal{D}_{ii}/(1+\delta_i')$ where δ_i' is a function of the temperature [Di68]. Furthermore, in order to avoid notational complexity, we also define $\mathcal{D}_{i\ \mathrm{int},j} = 1$ for $i \notin \mathcal{P}$ and $j \in \mathcal{S}$.

Finally, the collision integrals $\Omega_{ij}^{(r,s)}$, the collision numbers ξ_i^{rot}, and the thermodynamic properties of the mixture components are assumed to be given functions of the temperature.

Remark. It is important to realize that by using approximate collision integrals, the coefficients of the linear systems are no longer "noble" constants from theoretical physics, but merely "ordinary" numerical parameters. As a consequence, it will be necessary to verify that the structure properties of the exact transport linear systems obtained in this chapter are still valid for the approximate systems.

2.10.2 Expressions for the System Matrices

We give here the expressions of the matrices H, K, L, and $L_{[e]}$ under the approximations [MM62] [MPM65] since they are often used in practice and will be used in the following chapters. Note that the matrices $K_{[01]}$, $L_{[00]}$, Λ, and $\Lambda_{[e]}$ are then easily obtained as blocks of K, L, and $L_{[e]}$.

Under the preceding approximations, the matrix H is given by

$$H_{kk}^{0000} = \sum_{\substack{l\in\mathcal{S}\\ l\neq k}} 2\frac{X_k X_l}{\eta_{kl}}\Big[\frac{5}{3\bar{A}_{kl}}\frac{m_k m_l}{(m_k+m_l)^2} + \frac{m_l^2}{(m_k+m_l)^2}\Big] + \frac{X_k^2}{\eta_k}, \qquad k \in \mathcal{S}, \tag{2.10.7}$$

$$H_{kl}^{0000} = 2\frac{X_k X_l}{\eta_{kl}}\frac{m_k m_l}{(m_k+m_l)^2}\Big[-\frac{5}{3\bar{A}_{kl}} + 1\Big], \qquad k,l \in \mathcal{S}, \qquad k \neq l, \tag{2.10.8}$$

the matrix K is given by

$$K_{kk}^{1010} = \sum_{\substack{l\in\mathcal{S}\\ l\neq k}} 5\frac{X_k X_l}{\bar{A}_{kl}\eta_{kl}}\frac{m_k m_l}{(m_k+m_l)^2} + \frac{4}{k_B\pi}\frac{X_k X_l}{\eta_{kl}}\frac{m_l^2}{(m_k+m_l)^2}\Big(\frac{c_k^{\mathrm{int}}}{\xi_k^{\mathrm{int}}} + \frac{c_l^{\mathrm{int}}}{\xi_l^{\mathrm{int}}}\Big)$$

$$+\frac{4}{k_{\rm B}\pi}\frac{X_k^2}{\eta_k}\frac{c_k^{\rm int}}{\xi_k^{\rm int}}, \qquad k\in\mathcal{S}, \tag{2.10.9}$$

$$K_{kl}^{1010} = -5\frac{X_kX_l}{\bar{A}_{kl}\eta_{kl}}\frac{m_km_l}{(m_k+m_l)^2} + \frac{4}{k_{\rm B}\pi}\frac{X_kX_l}{\eta_{kl}}\frac{m_km_l}{(m_k+m_l)^2}\Big(\frac{c_k^{\rm int}}{\xi_k^{\rm int}}+\frac{c_l^{\rm int}}{\xi_l^{\rm int}}\Big), \qquad k,l\in\mathcal{S}, \quad k\neq l, \tag{2.10.10}$$

$$K_{kk}^{1001} = \sum_{\substack{l\in\mathcal{S}\\ l\neq k}} -\frac{4}{k_{\rm B}\pi}\frac{X_kX_l}{\eta_{kl}}\frac{m_l}{(m_k+m_l)}\frac{c_k^{\rm int}}{\xi_k^{\rm int}} - \frac{4}{k_{\rm B}\pi}\frac{X_k^2}{\eta_k}\frac{c_k^{\rm int}}{\xi_k^{\rm int}}, \qquad k\in\mathcal{P}, \tag{2.10.11}$$

$$K_{kl}^{1001} = -\frac{4}{k_{\rm B}\pi}\frac{X_kX_l}{\eta_{kl}}\frac{m_l}{(m_k+m_l)}\frac{c_l^{\rm int}}{\xi_l^{\rm int}}, \qquad k\in\mathcal{S}, \quad l\in\mathcal{P}, \quad k\neq l, \tag{2.10.12}$$

$$K_{kk}^{0101} = \sum_{\substack{l\in\mathcal{S}\\ l\neq k}} \frac{4}{k_{\rm B}\pi}\frac{X_kX_l}{\eta_{kl}}\frac{c_k^{\rm int}}{\xi_k^{\rm int}} + \frac{4}{k_{\rm B}\pi}\frac{X_k^2}{\eta_k}\frac{c_k^{\rm int}}{\xi_k^{\rm int}}, \qquad k\in\mathcal{P}, \tag{2.10.13}$$

$$K_{kl}^{0101} = 0, \qquad k,l\in\mathcal{P}, \quad k\neq l, \tag{2.10.14}$$

the matrix L is given by

$$L_{kk}^{0000} = \sum_{\substack{l\in\mathcal{S}\\ l\neq k}} \frac{X_kX_l}{\mathcal{D}_{kl}}, \qquad k\in\mathcal{S}, \tag{2.10.15}$$

$$L_{kl}^{0000} = -\frac{X_kX_l}{\mathcal{D}_{kl}}, \qquad k,l\in\mathcal{S}, \quad k\neq l, \tag{2.10.16}$$

$$L_{kk}^{0010} = -\sum_{\substack{l\in\mathcal{S}\\ l\neq k}} \frac{X_kX_l}{2\mathcal{D}_{kl}}\frac{m_l}{m_k+m_l}(6\bar{C}_{kl}-5), \qquad k\in\mathcal{S}, \tag{2.10.17}$$

$$L_{kl}^{0010} = \frac{X_kX_l}{2\mathcal{D}_{kl}}\frac{m_k}{m_k+m_l}(6\bar{C}_{kl}-5), \qquad k,l\in\mathcal{S}, \quad k\neq l, \tag{2.10.18}$$

$$L_{kl}^{0001} = 0, \qquad k\in\mathcal{S}, \quad l\in\mathcal{P}, \tag{2.10.19}$$

$$L_{kk}^{1010} = \sum_{\substack{l\in\mathcal{S}\\ l\neq k}} \frac{X_kX_l}{\mathcal{D}_{kl}}\frac{m_km_l}{(m_k+m_l)^2}\Big[\frac{15}{2}\frac{m_k}{m_l}+\frac{25}{4}\frac{m_l}{m_k}-3\frac{m_l}{m_k}\bar{B}_{kl}+4\bar{A}_{kl} + \frac{20}{3}\frac{\bar{A}_{kl}}{k_{\rm B}\pi}\Big(\frac{c_k^{\rm int}}{\xi_k^{\rm int}}+\frac{c_l^{\rm int}}{\xi_l^{\rm int}}\Big)\Big] + \frac{X_k^2}{\mathcal{D}_{kk}}\Big[2\bar{A}_{kk}+\frac{20}{3}\frac{\bar{A}_{kk}}{k_{\rm B}\pi}\frac{c_k^{\rm int}}{\xi_k^{\rm int}}\Big], \qquad k\in\mathcal{S}, \tag{2.10.20}$$

$$L_{kl}^{1010} = -\frac{X_kX_l}{\mathcal{D}_{kl}}\frac{m_km_l}{(m_k+m_l)^2}\Big[\frac{55}{4}-3\bar{B}_{kl}-4\bar{A}_{kl}$$

$$- \frac{20}{3} \frac{\bar{A}_{kl}}{k_B \pi} \Big(\frac{c_k^{\text{int}}}{\xi_k^{\text{int}}} + \frac{c_l^{\text{int}}}{\xi_l^{\text{int}}} \Big) \Big], \qquad k, l \in \mathcal{S}, \quad k \neq l, \tag{2.10.21}$$

$$L_{kk}^{1001} = - \sum_{\substack{l \in \mathcal{S} \\ l \neq k}} 4 \frac{\bar{A}_{kl}}{k_B \pi} \frac{X_k X_l}{\mathcal{D}_{kl}} \frac{m_k}{m_k + m_l} \frac{c_k^{\text{int}}}{\xi_k^{\text{int}}} - 4 \frac{\bar{A}_{kk}}{k_B \pi} \frac{X_k^2}{\mathcal{D}_{kk}} \frac{c_k^{\text{int}}}{\xi_k^{\text{int}}}, \qquad k \in \mathcal{P}, \tag{2.10.22}$$

$$L_{kl}^{1001} = - 4 \frac{\bar{A}_{kl}}{k_B \pi} \frac{X_k X_l}{\mathcal{D}_{kl}} \frac{m_l}{m_k + m_l} \frac{c_l^{\text{int}}}{\xi_l^{\text{int}}}, \qquad k \in \mathcal{S}, \quad l \in \mathcal{P}, \quad k \neq l, \tag{2.10.23}$$

$$\begin{aligned} L_{kk}^{0101} = {} & \sum_{\substack{l \in \mathcal{S} \\ l \neq k}} X_k X_l \frac{c_k^{\text{int}}}{k_B \mathcal{D}_{k\,\text{int},l}} + X_k^2 \frac{c_k^{\text{int}}}{k_B \mathcal{D}_{k\,\text{int},k}} \\ & + \sum_{\substack{l \in \mathcal{S} \\ l \neq k}} \frac{12}{5} \frac{\bar{A}_{kl}}{k_B \pi} \frac{X_k X_l}{\mathcal{D}_{kl}} \frac{m_k}{m_l} \frac{c_k^{\text{int}}}{\xi_k^{\text{int}}} + \frac{12}{5} \frac{\bar{A}_{kk}}{k_B \pi} \frac{X_k^2}{\mathcal{D}_{kk}} \frac{c_k^{\text{int}}}{\xi_k^{\text{int}}}, \qquad k \in \mathcal{P}, \end{aligned} \tag{2.10.24}$$

$$L_{kl}^{0101} = 0, \qquad k, l \in \mathcal{P}, \quad k \neq l. \tag{2.10.25}$$

and, finally, the matrix $L_{[e]}$ is given by

$$L_{[e]kk}^{0000} = \sum_{\substack{l \in \mathcal{S} \\ l \neq k}} \frac{X_k X_l}{\mathcal{D}_{kl}}, \qquad k \in \mathcal{S}, \tag{2.10.26}$$

$$L_{[e]kl}^{0000} = - \frac{X_k X_l}{\mathcal{D}_{kl}}, \qquad k, l \in \mathcal{S}, \quad k \neq l, \tag{2.10.27}$$

$$L_{[e]kk}^{00e} = - \sum_{\substack{l \in \mathcal{S} \\ l \neq k}} \frac{X_k X_l}{2\mathcal{D}_{kl}} \frac{m_l}{m_k + m_l} (6\bar{C}_{kl} - 5), \qquad k \in \mathcal{S}, \tag{2.10.28}$$

$$L_{[e]kl}^{00e} = \frac{X_k X_l}{2\mathcal{D}_{kl}} \frac{m_k}{m_k + m_l} (6\bar{C}_{kl} - 5), \qquad k, l \in \mathcal{S}, \quad k \neq l, \tag{2.10.29}$$

$$\begin{aligned} L_{[e]kk}^{ee} = {} & \sum_{\substack{l \in \mathcal{S} \\ l \neq k}} \frac{X_k X_l}{\mathcal{D}_{kl}} \frac{m_k m_l}{(m_k + m_l)^2} \Big[\frac{15}{2} \frac{m_k}{m_l} + \frac{25}{4} \frac{m_l}{m_k} - 3 \frac{m_l}{m_k} \bar{B}_{kl} + 4 \bar{A}_{kl} \\ & + \frac{4}{15} \frac{(3m_k - 2m_l)^2}{m_l^2} \frac{\bar{A}_{kl}}{k_B \pi} \frac{c_k^{\text{int}}}{\xi_k^{\text{int}}} + \frac{20}{3} \frac{\bar{A}_{kl}}{k_B \pi} \frac{c_l^{\text{int}}}{\xi_l^{\text{int}}} + \frac{(m_k + m_l)^2}{m_k m_l} \frac{c_k^{\text{int}} \mathcal{D}_{kl}}{k_B \mathcal{D}_{k\,\text{int},l}} \Big] \\ & + \frac{X_k^2}{\mathcal{D}_{kk}} \Big[2 \bar{A}_{kk} + \frac{16}{15} \frac{\bar{A}_{kk}}{k_B \pi} \frac{c_k^{\text{int}}}{\xi_k^{\text{int}}} + \frac{c_k^{\text{int}} \mathcal{D}_{kk}}{k_B \mathcal{D}_{k\,\text{int},k}} \Big], \qquad k \in \mathcal{S}, \end{aligned} \tag{2.10.30}$$

$$\begin{aligned} L_{[e]kl}^{ee} = {} & - \frac{X_k X_l}{\mathcal{D}_{kl}} \frac{m_k m_l}{(m_k + m_l)^2} \Big[\frac{55}{4} - 3\bar{B}_{kl} - 4\bar{A}_{kl} + \frac{4}{3} \frac{\bar{A}_{kl}}{k_B \pi} \frac{3m_k - 2m_l}{m_l} \frac{c_k^{\text{int}}}{\xi_k^{\text{int}}} \\ & + \frac{4}{3} \frac{\bar{A}_{kl}}{k_B \pi} \frac{3m_l - 2m_k}{m_k} \frac{c_l^{\text{int}}}{\xi_l^{\text{int}}} \Big], \qquad k, l \in \mathcal{S}, \quad k \neq l. \end{aligned} \tag{2.10.31}$$

3 Rescaled Transport Linear Systems

Although zero mass fractions may be considered unrealizable from a thermochemistry point of view, they are convenient to use in various models. From a computational point of view, it is also important to understand the mathematical and numerical behavior of the transport coefficients when some mass fractions become arbitrarily small. The results derived in Chapter 2, however, were obtained under the assumption of positive mass fractions, i.e., positive number densities, or, equivalently, positive mole fractions. In particular, lines and columns of zeros are obtained in the various transport linear systems described in the previous chapter when mass fractions are allowed to vanish, which leads to various singularities.

In order to overcome these problems, rescaled versions of the previous transport linear systems have to be investigated. These modified versions of the transport linear systems allow elimination of the artificial singularities arising with vanishing mass fractions, but are no longer symmetric. The main tools for investigating these rescaled systems are a block-partitioning between positive and zero mass fraction species and the use of the results obtained in Chapter 2 for positive concentrations. Using the modified transport linear systems, we will establish, in Chapter 4, that the transport coefficients are smooth rational functions of the mass fractions Y into the whole physical domain $\{ Y \in \mathbb{R}^n, Y \geq 0, \langle Y, U \rangle = 1 \}$, except for diffusion matrices, for which some components diverge [Gi91]. The fluxes $F_k = Y_k V_k$, $k \in S$, are then the proper quantities to be evaluated [Gi91] instead of the diffusion velocities V_k, $k \in S$. For this purpose, the flux diffusion coefficients $\widetilde{D}_{kl} = Y_k D_{kl}$ are introduced in this chapter and related to rescaled systems. In Chapter 5 we will also use the rescaled systems to investigate the numerical behavior of various iterative algorithms for arbitrarily small mass fractions.

In Section 3.1 we introduce the rescaled transport linear systems and investigate their mathematical structure. We then apply these results specifically to the shear viscosity in Section 3.2, the volume viscosity in Section 3.3, the flux diffusion matrix in Section 3.4, the partial thermal conductivity and the thermal diffusion vector in Section 3.5, and the thermal conductivity and the thermal diffusion ratios in Section 3.6.

3.1 Transport Linear Systems for Nonnegative Mass Fractions

3.1.1 Derivation of the Rescaled Systems $\widetilde{G}\alpha^\mu = \widetilde{\beta}^\mu$

In this chapter we consider a mixture of n species indexed by $\mathcal{S} = [1, n]$ and we assume that the number densities are nonnegative, $n_i \geq 0$, $i \in \mathcal{S}$, but not all zero, so that $\bar{n} = \sum_{i\in\mathcal{S}} n_i > 0$. Equivalently, the vector $Y = (Y_i)_{i\in\mathcal{S}}$ is such that $Y \geq 0$ and $Y \neq 0$. We also consider an approximation variational space $\mathcal{A}^\mu$ spanned by the basis functions ξ^{rk}, $(r,k) \in \mathcal{B}^\mu$, where $\mathcal{B}^\mu$ is the indexing set, and we want to investigate the corresponding transport linear system in this situation. The main assumptions introduced in Chapter 2 are supposed to hold in this section, unless explicitly stated. More specifically, keeping the notation of Chapter 2, we assume that the perpendicularity condition (2.3.20) holds

$$\mathcal{I}_\mathcal{S}^{a_\mu} = \mathcal{I}_\mathcal{S}^{a_\mu} \cap \mathcal{A}^\mu \;\oplus\; \mathcal{I}_\mathcal{S}^{a_\mu} \cap \mathcal{A}^{\mu\perp}, \tag{3.1.1}$$

that the species orthogonality property (2.3.32) holds

$$\mathcal{T}_\nu \widehat{\psi}^k \in \mathcal{I}_\mathcal{S}^{a_\mu} \cap \mathcal{A}^{\mu\perp}, \qquad (k,\nu) \in \mathcal{S}\times[1,\tau_\mu], \tag{3.1.2}$$

and we assume that the basis functions ξ^{rk}, $(r,k) \in \mathcal{B}^\mu$, satisfy the species localization property (2.3.25)

$$\xi_i^{rk} = 0, \quad \text{for} \quad i \neq k. \tag{3.1.3}$$

As stated previously, $\mathcal{I}_\mathcal{S}^{a_\mu}$ is the space of collisional invariants $\mathcal{I}_\mathcal{S}^{a_\mu}$ that are of tensorial rank a_μ over $\mathbb{R}^3$

$$\mathcal{I}_\mathcal{S}^{a_\mu} = \text{span}\{\ \mathcal{T}_\nu \widehat{\psi}^l,\ (l,\nu) \in [1, n{+}4]\times[1,\tau_\mu]\ \},$$

where $\tau_\mu = 3^{a_\mu}$ and $\mathcal{I}_\mathcal{S}^{a_\mu} \cap \mathcal{A}^{\mu\perp}$ denotes the elements of $\mathcal{I}_\mathcal{S}^{a_\mu}$ that are orthogonal to $\mathcal{A}$ with respect to the bilinear form $\langle\!\langle\, , \rangle\!\rangle$

$$\mathcal{I}_\mathcal{S}^{a_\mu} \cap \mathcal{A}^{\mu\perp} = \{\ \xi \in \mathcal{I}_\mathcal{S}^{a_\mu},\ \ \forall \zeta \in \mathcal{A}^\mu\ \ \langle\!\langle \xi, \zeta \rangle\!\rangle = 0\ \}.$$

We have shown in Section 2.3.7 that the properties (3.1.1)–(3.1.3) are valid for most functional spaces in practical applications.

Under these assumptions, the variational procedure introduced in Section 2.2—for solving the integral equation (2.1.25) with the constraints (2.1.26)—still yields a system similar to (2.3.1)(2.3.5)

$$\begin{cases} G\alpha^\mu = \beta^\mu, \\ \langle \mathcal{G}^{l\nu}, \alpha^\mu \rangle = 0, \qquad (l,\nu) \in [1, n{+}4]\times[1,\tau_\mu], \end{cases} \tag{3.1.4}$$

whose coefficients are given by

$$\begin{cases} G_{kl}^{rs} = \langle\!\langle \xi^{rk}, \Im(\xi^{sl}) \rangle\!\rangle = [\xi^{rk}, \xi^{sl}], \\ \beta_k^{r\mu} = \langle\!\langle \xi^{rk}, \Psi^\mu \rangle\!\rangle, \\ \mathcal{G}_k^{rl\nu} = \langle\!\langle \xi^{rk}, \mathcal{T}_\nu \widehat{\psi}^l \rangle\!\rangle = \langle\!\langle \xi_\nu^{rk}, \widehat{\psi}^l \rangle\!\rangle, \end{cases} \tag{3.1.5}$$

but which only possesses some of the properties established in Chapter 2. In particular, the block structure described in (2.3.9)–(2.3.11) is valid, the symmetry property (2.3.12) holds, and from the identities (2.3.14) and (2.3.28), the matrices G and $2db(G) - G$ are positive semi-definite.

We further introduce the rescaled Maxwellian distributions $\widetilde{f}^0 = (\widetilde{f}_i^0)_{i\in\mathcal{S}}$ given by

$$\widetilde{f}_i^0 = \frac{a_{i\mathrm{I}}}{Q_i}\Big(\frac{m_i}{2\pi k_{\mathrm{B}} T}\Big)^{3/2} \exp\Big(-w_i{\cdot}w_i - \epsilon_{i\mathrm{I}}\Big), \tag{3.1.6}$$

in such a way that $n_i \widetilde{f}_i^0 = f_i^0$, $i \in \mathcal{S}$. The associated scalar product $\langle\!\langle\!\langle , \rangle\!\rangle\!\rangle$, similar to $\langle\!\langle , \rangle\!\rangle$, is also defined by

$$\langle\!\langle\!\langle \xi, \zeta \rangle\!\rangle\!\rangle = \sum_{\substack{i\in\mathcal{S} \\ \mathrm{I}\in\mathcal{E}_i}} \int \xi_i {\odot} \zeta_i \widetilde{f}_i^0 \, dc_i, \tag{3.1.7}$$

where the families $\xi = (\xi_i)_{i\in\mathcal{S}}$ and $\zeta = (\zeta_i)_{i\in\mathcal{S}}$ are such that ξ_i and ζ_i only depend on c_i and I. The rescaled matrix $\widetilde{G} \in \mathbb{R}^{\omega,\omega}$ and the rescaled right member $\widetilde{\beta}^\mu \in \mathbb{R}^\omega$ are then defined by

$$\begin{cases} \widetilde{G}_{kl}^{rs} = \langle\!\langle\!\langle \xi^{rk}, \Im(\xi^{sl}) \rangle\!\rangle\!\rangle, \qquad (r,k),(s,l) \in \mathcal{B}^\mu, \\ \widetilde{\beta}_k^{r\mu} = \langle\!\langle\!\langle \xi^{rk}, \Psi^\mu \rangle\!\rangle\!\rangle, \qquad (r,k) \in \mathcal{B}^\mu. \end{cases} \tag{3.1.8}$$

Note that the rescaled right member $\widetilde{\beta}^\mu$ is defined as long as Ψ^μ is defined. More specifically, the functions Ψ^μ and the vectors $\widetilde{\beta}^\mu$ are defined for μ equal to η, κ, D_l, $l \in \mathcal{S}$, and λ' when all the mass fractions are positive, but are only defined for μ equal to η, κ and λ', when there are zero mass fractions. This results from the fact that for $\mu = D_l$, the function Ψ^{D_l} has a factor n_l in the denominator of its lth component $\Psi_l^{D_l}$, so that Ψ^{D_l} blows up when $n_l = 0$. The case of diffusion coefficients for zero mass fractions will thus be considered separately.

From the relations (3.1.8) and since $\xi_i^{rk} = 0$ for $i \neq k$, we also obtain that

$$n_k \widetilde{G}_{kl}^{rs} = n_k \langle\!\langle\!\langle \xi^{rk}, \Im(\xi^{sl}) \rangle\!\rangle\!\rangle = \langle\!\langle \xi^{rk}, \Im(\xi^{sl}) \rangle\!\rangle = G_{kl}^{rs},$$

and that

$$n_k \widetilde{\beta}_k^{r\mu} = n_k \langle\!\langle\!\langle \xi^{rk}, \Psi^{\mu} \rangle\!\rangle\!\rangle = \langle\!\langle \xi^{rk}, \Psi^{\mu} \rangle\!\rangle = \beta_k^{r\mu}.$$

As a consequence, denoting by $\mathcal{N}$ the diagonal matrix of order ω whose $(r,k)^{\text{th}}$ diagonal element is n_k

$$\mathcal{N} = \operatorname{diag}\big((n_k)_{(r,k)\in\mathcal{B}^{\mu}}\big), \tag{3.1.9}$$

we have the relations

$$\begin{cases} \mathcal{N}\widetilde{G} = G, \\ \mathcal{N}\widetilde{\beta}^{\mu} = \beta^{\mu}. \end{cases} \tag{3.1.10}$$

When all the mass fractions are positive, the matrix $\widetilde{G}$ and the right member $\widetilde{\beta}^{\mu}$ are easily evaluated, since the matrix $\mathcal{N}$ is then invertible so that $\widetilde{G} = \mathcal{N}^{-1} G$ and $\widetilde{\beta}^{\mu} = \mathcal{N}^{-1}\beta^{\mu}$. Only definitions (3.1.8) are valid, however, when some mass fractions are zero, since $\mathcal{N}$ is then singular. Note also that the matrix $\widetilde{G}$ is no more symmetric. Using now the partial brackets (2.2.22), the definitions (3.1.5), and the definitions (3.1.8), one may easily establish that the matrix G still satisfies the fundamental relations (2.2.26)

$$\begin{cases} G_{kk}^{rs} = \sum_{l\in S} n_k n_l G_{kl}^{\prime rs} + n_k^2 G_{kk}^{\prime\prime rs}, & (r,k),(s,k) \in \mathcal{B}^{\mu}, \\ G_{kl}^{rs} = n_k n_l G_{kl}^{\prime\prime rs}, \quad (r,k),(s,l) \in \mathcal{B}^{\mu}, \quad k \neq l, \end{cases} \tag{3.1.11}$$

and that the matrix $\widetilde{G}$ satisfies the relations

$$\begin{cases} \widetilde{G}_{kk}^{rs} = \sum_{l\in S} n_l G_{kl}^{\prime rs} + n_k G_{kk}^{\prime\prime rs}, & (r,k),(s,k) \in \mathcal{B}^{\mu}, \\ \widetilde{G}_{kl}^{rs} = n_l G_{kl}^{\prime\prime rs}, \quad (r,k),(s,l) \in \mathcal{B}^{\mu}, \quad k \neq l, \end{cases} \tag{3.1.12}$$

where

$$\begin{cases} G_{kl}^{\prime rs} = [\xi^{rk}, \xi^{sk}]_{kl}^{\prime}, & (r,k),(s,k) \in \mathcal{B}^{\mu}, \quad l \in S, \\ G_{kl}^{\prime\prime rs} = [\xi^{rk}, \xi^{sl}]_{kl}^{\prime\prime}, & (r,k),(s,l) \in \mathcal{B}^{\mu}. \end{cases} \tag{3.1.13}$$

As a consequence, the coefficients of $\widetilde{G}$ are linear expressions in the number densities $(n_k)_{k\in\mathcal{S}}$ whereas the coefficients of the matrix G are quadratics in the $(n_k)_{k\in\mathcal{S}}$.

The modified version of the general transport linear system (3.1.1) is then simply the rescaled system

$$\widetilde{G}\alpha^\mu = \widetilde{\beta}^\mu, \tag{3.1.14}$$

under the same constraints

$$\langle \mathcal{G}^{l\nu}, \alpha^\mu \rangle = 0, \qquad (l,\nu) \in [1, n{+}4]\times[1, \tau_\mu], \tag{3.1.15}$$

and we have the equivalence result that when all the mass fractions are positive, i.e., when the matrix $\mathcal{N}$ is invertible, then α^μ satisfies (3.1.4) if and only if α^μ satisfies (3.1.14) (3.1.15). When some mass fractions are zero, however, only the system (3.1.14)(3.1.15) has a nonsingular behavior whereas (3.1.4) is singular because of the singularity of $\mathcal{N}$.

Notice also that we could have considered different rescaled versions of the transport linear systems. In particular, we could have used the symmetric matrices $\widehat{G} = \mathcal{N}^{-1/2} G \mathcal{N}^{-1/2}$ instead of $\widetilde{G} = \mathcal{N}^{-1} G$. These matrices will be briefly considered in Section 3.1.8, but their coefficients are not differentiable functions of the mass fractions. On the other hand, the system matrices $G\mathcal{N}^{-1}$ are of limited interest since they do not provide the quantities α^μ but rather $\mathcal{N}\alpha^\mu$. The matrices $G\mathcal{N}^{-1}$ may be interesting, however, for evaluating the flux diffusion coefficients $\widetilde{D}_{kl}$ and will be briefly discussed in Section 3.4.8.

3.1.2 Partitioning of the Transport Linear Systems

Notation. We denote by n^+ and p^+ the number of species and polyatomic species with positive mass fraction, respectively, and similarly by n^- and p^- the number of species and polyatomic species with zero mass fraction, respectively, so that we have $n = n^+ + n^-$ and $p = p^+ + p^-$, and we assume that $n^+ \geq 1$ and $n^- \geq 0$. We define the sets $\mathcal{S}^+ = \{\ k \in \mathcal{S},\ Y_k > 0\ \}$ and $\mathcal{S}^- = \{\ k \in \mathcal{S},\ Y_k = 0\ \}$, and we denote by Υ the permutation matrix associated with the reordering of $\mathcal{S}$ into $(\mathcal{S}^+, \mathcal{S}^-)$. The permutation matrix Υ can be used to decompose any family indexed by $\mathcal{S}$ into its $\mathcal{S}^+$ and $\mathcal{S}^-$ subfamilies. In particular, for any $\Xi \in \mathcal{A}^\mu$, $\Xi = (\Xi_i)_{i\in\mathcal{S}}$, we may consider the subfamilies of functions $\Xi^+ = (\Xi_i)_{i\in\mathcal{S}^+}$ and $\Xi^- = (\Xi_i)_{i\in\mathcal{S}^-}$, and we then have $\Xi = \Upsilon(\Xi^+, \Xi^-)$ and $(\Xi^+, \Xi^-) = \Upsilon^t\,\Xi$, making use of $\Upsilon^t = \Upsilon^{-1}$. Similarly, we define the sets $\mathcal{P}^+ = \{ k \in \mathcal{P},\ Y_k > 0 \}$ and $\mathcal{P}^- = \{ k \in \mathcal{P},\ Y_k = 0 \}$. Finally, the submixture indexed by $\mathcal{S}^+$ and composed of the n^+ species with positive mass fraction will be termed "the $\mathcal{S}^+$ mixture" in this book.

We also introduce the partitioning $\mathcal{B}^\mu = \mathcal{B}^{\mu+} \cup \mathcal{B}^{\mu-}$ where

$$\begin{cases} \mathcal{B}^{\mu+} = \{ (r,k) \in \mathcal{B}^\mu,\ k \in \mathcal{S}^+ \}, \\ \mathcal{B}^{\mu-} = \{ (r,k) \in \mathcal{B}^\mu,\ k \in \mathcal{S}^- \}, \end{cases} \tag{3.1.16}$$

and we denote by Γ^μ the permutation matrix associated with the reordering of $\mathcal{B}^\mu$ into $(\mathcal{B}^{\mu+}, \mathcal{B}^{\mu-})$. We denote by ω^+ the number of elements of $\mathcal{B}^{\mu+}$ and ω^- the number of elements of $\mathcal{B}^{\mu-}$, so that $\omega = \omega^+ + \omega^-$. Using this partitioning, we may associate with each vector $x \in \mathbb{R}^\omega$ the vectors $x^+ \in \mathbb{R}^{\omega^+}$ and $x^- \in \mathbb{R}^{\omega^-}$ defined by $x^+ = (x_k^r)_{(r,k)\in\mathcal{B}^{\mu+}}$ and $x^- = (x_k^r)_{(r,k)\in\mathcal{B}^{\mu-}}$ in such a way that

$$x = \Gamma^\mu(x^+, x^-), \qquad x \in \mathbb{R}^\omega. \tag{3.1.17}$$

In addition, if $\Xi \in \mathcal{A}^\mu$, $\Xi = \sum_{(r,k)\in\mathcal{B}^\mu} x_k^r \xi^{rk}$, and $(\Xi^+, \Xi^-) = \Upsilon^t \Xi$, we have by linearity that $\Xi^+ = \sum_{(r,k)\in\mathcal{B}^{\mu+}} x_k^r \xi^{rk+}$ and $\Xi^- = \sum_{(r,k)\in\mathcal{B}^{\mu-}} x_k^r \xi^{rk-}$, since we have from (3.1.3) $(\xi^{rk+}, 0) = \Upsilon^t \xi^{rk}$ for $(r,k) \in \mathcal{B}^{\mu+}$, and $(0, \xi^{rk-}) = \Upsilon^t \xi^{rk}$ for $(r,k) \in \mathcal{B}^{\mu-}$, so that the components x, x^+, and x^- of Ξ, Ξ^+, and Ξ^-, respectively, satisfy $x = \Gamma^\mu(x^+, x^-)$.

Correspondingly, we decompose each matrix $A \in \mathbb{R}^{\omega,\omega}$ into the blocks $A^{++} \in \mathbb{R}^{\omega^+,\omega^+}$, $A^{+-} \in \mathbb{R}^{\omega^+,\omega^-}$, $A^{-+} \in \mathbb{R}^{\omega^-,\omega^+}$, and $A^{--} \in \mathbb{R}^{\omega^-,\omega^-}$ in such a way that $y = Ax$ if and only if $y^+ = A^{++}x^+ + A^{+-}x^-$ and $y^- = A^{-+}x^+ + A^{--}x^-$. This block-decomposition is obtained from the matrix A and the permutation matrix Γ^μ by

$$(\Gamma^\mu)^t A \Gamma^\mu = \begin{bmatrix} A^{++} & A^{+-} \\ A^{-+} & A^{--} \end{bmatrix}, \tag{3.1.18}$$

and it is straightforward to check that for any matrix $A \in \mathbb{R}^{\omega,\omega}$, we have

$$(\Gamma^\mu)^t\, db(A)\, \Gamma^\mu = \begin{bmatrix} db(A^{++}) & 0 \\ 0 & db(A^{--}) \end{bmatrix}, \tag{3.1.19}$$

since the blocks A^{+-} and A^{-+} involve coefficients A_{kl}^{rs} such that $k \neq l$.

Partitioning of G and $\widetilde{G}$. We now investigate the partitioning of G and $\widetilde{G}$. Let thus G be the matrix of the transport linear system (3.1.4) and $\widetilde{G}$ the corresponding rescaled matrix (3.1.8). Then, after the reordering induced by Γ^μ, the matrices G and $\widetilde{G}$ admit the block-decompositions

$$(\Gamma^\mu)^t G \Gamma^\mu = \begin{bmatrix} G^{++} & 0 \\ 0 & 0 \end{bmatrix}, \tag{3.1.20}$$

and

$$(\Gamma^\mu)^t \widetilde{G} \Gamma^\mu = \begin{bmatrix} \widetilde{G}^{++} & 0 \\ \widetilde{G}^{-+} & \widetilde{G}^{--} \end{bmatrix}. \tag{3.1.21}$$

The block-partitioning of G and $\widetilde{G}$ is indeed a direct consequence of the relations (3.1.11)(3.1.12). Furthermore, we also deduce that

$$\begin{cases} (\widetilde{G}^{++})_{kk}^{rs} = \sum_{l\in\mathcal{S}^+} n_l[\xi^{rk},\xi^{sk}]'_{kl} + n_k[\xi^{rk},\xi^{sk}]''_{kk}, & (r,k),(s,k)\in\mathcal{B}^{\mu+}, \\ (\widetilde{G}^{++})_{kl}^{rs} = n_l[\xi^{rk},\xi^{sl}]''_{kl}, \quad (r,k),(s,l)\in\mathcal{B}^{\mu+}, \quad k\neq l, \\ (\widetilde{G}^{+-})_{kl}^{rs} = 0, \quad (r,k)\in\mathcal{B}^{\mu+}, \quad (s,l)\in\mathcal{B}^{\mu-}, \\ (\widetilde{G}^{-+})_{kl}^{rs} = n_l[\xi^{rk},\xi^{sl}]''_{kl}, \quad (r,k)\in\mathcal{B}^{\mu-}, \quad (s,l)\in\mathcal{B}^{\mu+}, \\ (\widetilde{G}^{--})_{kk}^{rs} = \sum_{l\in\mathcal{S}^+} n_l[\xi^{rk},\xi^{sk}]'_{kl}, \quad (r,k),(s,k)\in\mathcal{B}^{\mu-}, \\ (\widetilde{G}^{--})_{kl}^{rs} = 0, \quad (r,k),(s,l)\in\mathcal{B}^{\mu-}, \quad k\neq l. \end{cases} \tag{3.1.22}$$

Note that only collisions between molecule pairs of type $(\mathcal{S}^+,\mathcal{S}^+)$ and $(\mathcal{S}^+,\mathcal{S}^-)$ are involved in (3.1.22), but no collisions between molecule pairs of type $(\mathcal{S}^-,\mathcal{S}^-)$.

We now deduce from (3.1.22) that the matrix $\widetilde{G}^{--}$ is symmetric and block-diagonal. We further show that $\widetilde{G}^{--}$ is positive definite on $\mathbb{R}^{\omega^-}$. Let indeed $x^-\in\mathbb{R}^{\omega^-}$, $x=\Gamma^\mu(0,x^-)\in\mathbb{R}^\omega$, and consider the scalar product $\langle x^-,\widetilde{G}^{--}x^-\rangle$. From the relations (3.1.22) we have

$$\begin{aligned} \langle x^-,\widetilde{G}^{--}x^-\rangle &= \sum_{\substack{(r,k)\in\mathcal{B}^{\mu-}\\(s,k)\in\mathcal{B}^{\mu-}}} \sum_{l\in\mathcal{S}^+} n_l[\xi^{rk},\xi^{sk}]'_{kl}\, x_k^r x_k^s, \\ &= \sum_{\substack{k\in\mathcal{S}^-\\ l\in\mathcal{S}^+}} n_l\Big[\sum_{r\in\mathcal{F}_k} x_k^r\xi^{rk}, \sum_{s\in\mathcal{F}_k} x_k^s\xi^{sk}\Big]'_{kl}, \\ &= \sum_{\substack{k\in\mathcal{S}^-\\ l\in\mathcal{S}^+}} n_l[\Xi,\Xi]'_{kl} = \sum_{\substack{k\in\mathcal{S}^-\\ l\in\mathcal{S}^+}} \tfrac{1}{2}n_l\{\Xi_k-\Xi'_k,\, \Xi_k-\Xi'_k\}_{kl}, \end{aligned}$$

keeping in mind that $\sum_{r\in\mathcal{F}_k} x_k^r\xi^{rk}$ and $\Xi=(\Xi_k)_{k\in\mathcal{S}}=\sum_{(r,k)\in\mathcal{B}^{\mu-}} x_k^r\xi^{rk}$ have the same k^{th} component. On the other hand, we have

$$2[\Xi,\Xi]'_{kl} = \sum_{\substack{K,K'\in\mathcal{E}_k\\ L,L'\in\mathcal{E}_l}} \iiiint \widetilde{f}_k^0\widetilde{f}_l^0(\Xi_k-\Xi'_k)\odot(\Xi_k-\Xi'_k)\bar{\sigma}_{kl}^{KLK'L'}\, g\sin\chi d\chi d\varphi dc_k d\widetilde{c}_l,$$

so that $\langle x^-, \widetilde{G}^{--}x^-\rangle$ is a sum of nonnegative terms, and therefore, $\widetilde{G}^{--}$ is positive semi-definite. Moreover, if this sum is zero, we deduce that for any $l \in \mathcal{S}^+$ and any $k \in \mathcal{S}^-$, the relation $\Xi_k(c_k, \kappa) = \Xi_k(c_k', \kappa')$ holds, where c_k' and κ' are the velocity and the quantum energy shell, respectively, of the molecule of the kth species, after any admissible collision with a molecule of the lth species, i.e., a collision such that $\bar{\sigma}_{kl}^{\kappa L \kappa' L'} \neq 0$. It is always possible, however, to find admissible collisions between molecules of the kth and l^{th} species such that c_k' is arbitrary and κ is invariant, i.e., $\kappa' = \kappa$, which first shows that Ξ_k does not depend on c_k, so that $\Xi_k = \Xi_k(\kappa) = \Xi_k(\kappa')$. In addition, it is always possible to find admissible collisions such that κ' is arbitrary, which now shows that Ξ_k is actually a constant. Denoting by y_k, $k \in \mathcal{S}^-$, this constant tensor and decomposing y_k into $y_k = \sum_{\nu\in[1,\tau_\mu]} y_{k\nu}\mathcal{T}_\nu$, we deduce that $\Xi = \sum_{(k,\nu)\in\mathcal{S}^-\times[1,\tau_\mu]} y_{k\nu}\mathcal{T}_\nu\widehat{\psi}^k$. Since the species collisional invariants $\mathcal{T}_\nu\widehat{\psi}^k$, $(k,\nu) \in \mathcal{S}^-\times[1,\tau_\mu]$ are all in $\mathcal{I}_{\mathcal{S}}^{a_\mu} \cap \mathcal{A}^{\mu\perp}$ from (3.1.2), we now obtain that $\langle\!\langle \Xi, \mathcal{T}_\nu\widehat{\psi}^k\rangle\!\rangle = y_{k\nu}\langle\!\langle \mathcal{T}_\nu\widehat{\psi}^k, \mathcal{T}_\nu\widehat{\psi}^k\rangle\!\rangle = 0$ so that $y_{k\nu} = 0$, $(k,\nu) \in \mathcal{S}^-\times[1,\tau_\mu]$, and $\Xi = 0$, which yields that $x^- = 0$ so that $\widetilde{G}^{--}$ is positive definite on $\mathbb{R}^{\omega^-}$.

The diagonal matrix $\mathcal{N}$ has the block-decomposition

$$(\Gamma^\mu)^t\mathcal{N}\Gamma^\mu = \begin{bmatrix} \mathcal{N}^{++} & 0 \\ 0 & 0 \end{bmatrix}, \tag{3.1.23}$$

where $\mathcal{N}^{++} = \operatorname{diag}\big((n_k)_{(r,k)\in\mathcal{B}^{\mu+}}\big)$ is invertible by definition of $\mathcal{B}^{\mu+}$ and $\mathcal{S}^+$. The decomposition (3.1.23) then implies from (3.1.20) and (3.1.21) that $\mathcal{N}^{++}\widetilde{G}^{++} = G^{++}$, so that

$$\widetilde{G}^{++} = \big(\mathcal{N}^{++}\big)^{-1}G^{++}. \tag{3.1.24}$$

Moreover, from (3.1.11)–(3.1.13) the matrices G^{++} and $\widetilde{G}^{++}$ exactly correspond to the ones that would have been obtained by considering the $\mathcal{S}^+$ mixture. Furthermore, since the matrix $\mathcal{N}$ is diagonal, we also have $\mathcal{N}db(\widetilde{G}) = db(G)$ which implies that $\mathcal{N}^{++}db(\widetilde{G}^{++}) = db(G^{++})$ so that

$$db(\widetilde{G}^{++}) = \big(\mathcal{N}^{++}\big)^{-1}db(G^{++}). \tag{3.1.25}$$

In addition, a straightforward calculation shows that $db(G^{++})$ commutes with the matrix $\mathcal{N}^{++}$ and therefore with the matrices $\big(\mathcal{N}^{++}\big)^{-1}$ and $\big(\mathcal{N}^{++}\big)^{-1/2}$. This yields, in particular, that

$$db(\widetilde{G}^{++}) = \big(\mathcal{N}^{++}\big)^{-1/2}db(G^{++})\big(\mathcal{N}^{++}\big)^{-1/2}, \tag{3.1.26}$$

so that $db(\widetilde{G}^{++})$ inherits the symmetry property of $db(G^{++})$. Notice also that from (3.1.19) we have

$$(\Gamma^\mu)^t \, db(\widetilde{G})\Gamma^\mu = \begin{bmatrix} db(\widetilde{G}^{++}) & 0 \\ 0 & db(\widetilde{G}^{--}) \end{bmatrix}, \qquad (3.1.27)$$

which will be useful in the following sections. We have thus established that whenever the perpendicularity property (3.1.1), the species orthogonality property (3.1.2), and the species localization property (3.1.3) hold, then the matrices G, $\widetilde{G}$, G^{++}, and $\widetilde{G}^{++}$ satisfy

$$\begin{cases} G \text{ and } \widetilde{G} \text{ admit the partitioning (3.1.20)(3.1.21)}, \\ G^{++} \text{ and } \widetilde{G}^{++} \text{ correspond to the } \mathcal{S}^+ \text{ mixture}, \\ \widetilde{G}^{++} = \left(\mathcal{N}^{++}\right)^{-1} G^{++}, \\ \widetilde{G}^{--} \text{ is block-diagonal and symmetric positive definite}, \\ db(\widetilde{G}^{++}) = \left(\mathcal{N}^{++}\right)^{-1/2} db(G^{++}) \left(\mathcal{N}^{++}\right)^{-1/2}. \end{cases} \qquad (3.1.28)$$

Partitioning of the Linear Systems. From the decomposition (3.1.23) and the relation $\beta^\mu = \mathcal{N}\widetilde{\beta}^\mu$, we also obtain that $\mathcal{N}^{++}\widetilde{\beta}^{\mu+} = \beta^{\mu+}$, so that

$$\widetilde{\beta}^{\mu+} = \left(\mathcal{N}^{++}\right)^{-1} \beta^{\mu+}, \qquad (3.1.29)$$

and we also deduce that

$$\beta^{\mu-} = 0. \qquad (3.1.30)$$

Similarly, introducing

$$\widetilde{\mathcal{G}}_k^{rl\nu} = (\widetilde{\mathcal{G}}_k^{rl\nu})_{(r,k)\in\mathcal{B}^\mu} = \left(\langle\!\langle\!\langle \xi^{rk}, \mathcal{T}_\nu \widehat{\psi}^l \rangle\!\rangle\!\rangle\right)_{(r,k)\in\mathcal{B}^\mu}, \qquad (3.1.31)$$

yields that

$$\mathcal{N}\widetilde{\mathcal{G}}^{l\nu} = \mathcal{G}^{l\nu}, \qquad (l,\nu) \in [1, n+4]\times[1, \tau_\mu], \qquad (3.1.32)$$

so that

$$\mathcal{G}^{l\nu-} = 0. \qquad (3.1.33)$$

As a consequence, it is now straightforward to obtain that

$$G\alpha^\mu = \beta^\mu \iff G^{++}\alpha^{\mu+} = \beta^{\mu+},$$

and that

$$\begin{aligned}
&\langle \mathcal{G}^{l\nu}, \alpha^{\mu} \rangle = 0, \quad (l,\nu) \in [1, n+4] \times [1, \tau_\mu], \\
&\qquad \Longleftrightarrow \quad \langle \mathcal{G}^{l\nu+}, \alpha^{\mu+} \rangle = 0, \quad (l,\nu) \in [1, n+4] \times [1, \tau_\mu], \\
&\qquad \Longleftrightarrow \quad \langle \mathcal{G}^{l\nu+}, \alpha^{\mu+} \rangle = 0, \quad (l,\nu) \in \big(\mathcal{S}^+ \cup [n+1, n+4]\big) \times [1, \tau_\mu],
\end{aligned}$$

since $\mathcal{G}^{l\nu+} = 0$ for $l \in \mathcal{S}^-$, so that the system (3.1.4) is equivalent to the subsystem

$$\begin{cases} G^{++}\alpha^{\mu+} = \beta^{\mu+}, \\ \langle \mathcal{G}^{l\nu+}, \alpha^{\mu+} \rangle = 0, \qquad (l,\nu) \in \big(\mathcal{S}^+ \cup [n+1, n+4]\big) \times [1, \tau_\mu]. \end{cases} \tag{3.1.34}$$

Similarly, the system (3.1.14)(3.1.15) is easily shown to be equivalent to the two uncoupled subsystems

$$\begin{cases} \widetilde{G}^{++}\alpha^{\mu+} = \widetilde{\beta}^{\mu+}, \\ \langle \mathcal{G}^{l\nu+}, \alpha^{\mu+} \rangle = 0, \qquad (l,\nu) \in \big(\mathcal{S}^+ \cup [n+1, n+4]\big) \times [1, \tau_\mu]. \end{cases} \tag{3.1.35}$$

and

$$\widetilde{G}^{--}\alpha^{\mu-} = \widetilde{\beta}^{\mu-} - \widetilde{G}^{-+}\alpha^{\mu+}. \tag{3.1.36}$$

Notice that from (3.1.24) and (3.1.29), the systems (3.1.34) and (3.1.35) are equivalent, whereas (3.1.36) is nonsingular since $\widetilde{G}^{--}$ is invertible from (3.1.28).

3.1.3 Properties of the Subsystem $G^{++}\alpha^{\mu+} = \beta^{\mu+}$

In this section we investigate the properties of the subsystem (3.1.34) corresponding to positive mass fractions. We establish that this subsystem corresponds to a variational formulation having the perpendicularity property (2.3.20), the species orthogonality property (2.3.32), and the species localization property (2.3.25). As a consequence, we will obtain that the results of Sections 2.3.2, 2.3.3, and 2.3.4 apply to this subsystem.

The Variational Structure. We introduce the families ξ^{rk+}, $(r,k) \in \mathcal{B}^{\mu+}$, defined by

$$\xi^{rk+} = (\xi_i^{rk})_{i \in \mathcal{S}^+}, \qquad (r,k) \in \mathcal{B}^{\mu+}, \tag{3.1.37}$$

which satisfy $(\xi^{rk+}, 0) = \Upsilon^t \xi^{rk}$, $(r,k) \in \mathcal{B}^{\mu+}$, from the species localization property. The ξ^{rk+}, $(r,k) \in \mathcal{B}^{\mu+}$, satisfy the species localization property from (3.1.3) and are linearly independent from the relations $(\xi^{rk+}, 0) = \Upsilon^t \xi^{rk}$, $(r,k) \in \mathcal{B}^{\mu+}$, and since the

ξ^{rk}, $(r,k) \in \mathcal{B}^\mu$, are linearly independent. Furthermore, we denote by $\mathcal{A}^{\mu+}$ the space spanned by the functions ξ^{rk+}, $(r,k) \in \mathcal{B}^{\mu+}$,

$$\mathcal{A}^{\mu+} = \mathrm{span}\{ \ \xi^{rk+}, \ (r,k) \in \mathcal{B}^{\mu+} \ \}. \tag{3.1.38}$$

Similarly, we introduce the collisional invariants $\widehat{\psi}^{l+}$, $l \in \mathcal{S}^+ \cup [n+1, n+4]$, of the $\mathcal{S}^+$ mixture

$$\widehat{\psi}^{l+} = (\widehat{\psi}^l_i)_{i \in \mathcal{S}^+}, \qquad l \in \mathcal{S}^+ \cup [n+1, n+4], \tag{3.1.39}$$

which also satisfy $(\widehat{\psi}^{l+}, \widehat{\psi}^{l-}) = \Upsilon^t \widehat{\psi}^l$, where $\widehat{\psi}^l$, $l \in \mathcal{S}^+ \cup [n+1, n+4]$, is the full mixture collisional invariant. This results from the definition of the collisional invariants and from the properties (2.1.12)(2.1.13) applied to the $\mathcal{S}^+$ mixture. The corresponding space of collisional invariants which are of tensorial rank a_μ over $\mathbb{R}^3$ is denoted by $\mathcal{I}^{a_\mu}_{\mathcal{S}^+}$ and is spanned by the $\mathcal{T}_\nu \widehat{\psi}^{l+}$, $(l,\nu) \in \big(\mathcal{S}^+ \cup [n+1, n+4]\big) \times [1, \tau_\mu]$,

$$\mathcal{I}^{a_\mu}_{\mathcal{S}^+} = \mathrm{span}\{ \ \mathcal{T}_\nu \widehat{\psi}^{l+}, \ (l,\nu) \in \big(\mathcal{S}^+ \cup [n+1, n+4]\big) \times [1, \tau_\mu] \ \}. \tag{3.1.40}$$

Furthermore, we introduce $\Psi^{\mu+}$

$$\Psi^{\mu+} = (\Psi^\mu_i)_{i \in \mathcal{S}^+}, \tag{3.1.41}$$

such that $(\Psi^{\mu+}, \Psi^{\mu-}) = \Upsilon^t \Psi^\mu$, and $\phi^{\mu+}$

$$\phi^{\mu+} = (\phi^\mu_i)_{i \in \mathcal{S}^+}, \tag{3.1.42}$$

which satisfies $(\phi^{\mu+}, \phi^{\mu-}) = \Upsilon^t \phi^\mu$. Finally, we still denote by $\langle\!\langle \ , \ \rangle\!\rangle$ and $[\ , \ ]$ the usual bilinear forms acting on subfamilies of functions indexed by $\mathcal{S}^+$.

We now claim that the variational procedure for the $\mathcal{S}^+$ mixture with variational approximation space $\mathcal{A}^{\mu+}$ and with $\phi^{\mu+}$ taken in the form

$$\phi^{\mu+} = \sum_{(r,k) \in \mathcal{B}^{\mu+}} \alpha^{r\mu+}_k \xi^{rk+}, \tag{3.1.43}$$

exactly yields the system (3.1.34). Indeed, this variational procedure yields the constrained linear system

$$\begin{cases} \displaystyle\sum_{(s,l) \in \mathcal{B}^{\mu+}} [\xi^{rk+}, \xi^{sl+}] \alpha^{s\mu+}_l = \langle\!\langle \xi^{rk+}, \Psi^{\mu+} \rangle\!\rangle, \qquad (r,k) \in \mathcal{B}^{\mu+}, \\ \displaystyle\sum_{(s,l) \in \mathcal{B}^{\mu+}} \langle\!\langle \xi^{rk+}, \mathcal{T}_\nu \widehat{\psi}^{l+} \rangle\!\rangle \alpha^{r\mu+}_k = 0, \qquad (l,\nu) \in \big(\mathcal{S}^+ \cup [n+1, n+4]\big) \times [1, \tau_\mu]. \end{cases}$$

By using the species localization property (3.1.3), we can then express the system coefficients as $[\xi^{rk+},\xi^{sl+}] = [\xi^{rk},\xi^{sl}]$ for $(r,k),(s,l) \in \mathcal{B}^{\mu+}$. Similarly, the right members can be written $\langle\!\langle \xi^{rk+},\Psi^{\mu+}\rangle\!\rangle = \langle\!\langle \xi^{rk},\Psi^{\mu}\rangle\!\rangle$ for $(r,k) \in \mathcal{B}^{\mu+}$. In addition, the constraint coefficients can also be expressed as $\langle\!\langle \xi^{rk+},\mathcal{T}_\nu\widehat{\psi}^{l+}\rangle\!\rangle = \langle\!\langle \xi^{rk},\mathcal{T}_\nu\widehat{\psi}^{l}\rangle\!\rangle$ for $(r,k) \in \mathcal{B}^{\mu+}$ and $(l,\nu) \in \big(\mathcal{S}^+ \cup [n{+}1,n{+}4]\big) \times [1,\tau_\mu]$. This shows that the above system coincides with (3.1.34). Finally, we also deduce from (2.3.18) that the compatibility conditions $\langle\!\langle \Psi^{\mu+},\widehat{\psi}^{l+}\rangle\!\rangle = 0$, $l \in \mathcal{S}^+ \cup [n{+}1,n{+}4]$, are satisfied since it is straightforward to obtain that $\langle\!\langle \Psi^{\mu+},\widehat{\psi}^{l+}\rangle\!\rangle = \langle\!\langle \Psi^{\mu},\widehat{\psi}^{l}\rangle\!\rangle$.

Properties of $\mathcal{A}^{\mu+}$. Since the basis functions (3.1.37) satisfy the species localization property (2.3.25), it only remains to show that the perpendicularity property (2.3.20) and the species orthogonality property (2.3.32) hold for the variational procedure with the $\mathcal{S}^+$ mixture. More specifically, we have to show that

$$\mathcal{I}_{\mathcal{S}^+}^{a_\mu} = \mathcal{I}_{\mathcal{S}^+}^{a_\mu} \cap \mathcal{A}^{\mu+} \;\oplus\; \mathcal{I}_{\mathcal{S}^+}^{a_\mu} \cap \big(\mathcal{A}^{\mu+}\big)^{\perp}, \tag{3.1.44}$$

and that

$$\mathcal{T}_\nu\widehat{\psi}^{l+} \in \mathcal{I}_{\mathcal{S}^+}^{a_\mu} \cap \big(\mathcal{A}^{\mu+}\big)^{\perp}, \qquad (l,\nu) \in \mathcal{S}^+ \times [1,\tau_\mu]. \tag{3.1.45}$$

In order to establish these properties, we consider $\zeta^+ \in \mathcal{I}_{\mathcal{S}^+}^{a_\mu}$

$$\zeta^+ = \sum_{\substack{l\in\mathcal{S}^+\cup[n+1,n+4]\\ \nu\in[1,\tau_\mu]}} y_{l\nu}\mathcal{T}_\nu\widehat{\psi}^{l+},$$

and we introduce the corresponding full mixture collisional invariant

$$\zeta = \sum_{\substack{l\in\mathcal{S}^+\cup[n+1,n+4]\\ \nu\in[1,\tau_\mu]}} y_{l\nu}\mathcal{T}_\nu\widehat{\psi}^{l}.$$

From (3.1.1) we can write that $\zeta = \Xi + \mathcal{U}$ where $\Xi \in \mathcal{I}_{\mathcal{S}}^{a_\mu} \cap \mathcal{A}^\mu$ and $\mathcal{U} \in \mathcal{I}_{\mathcal{S}}^{a_\mu} \cap \mathcal{A}^{\mu\perp}$. Projecting on the $\mathcal{S}^+$ components first yields that

$$\zeta^+ = \Xi^+ + \mathcal{U}^+,$$

where $(\Xi^+,\Xi^-) = \Upsilon^t\Xi$ and $(\mathcal{U}^+,\mathcal{U}^-) = \Upsilon^t\mathcal{U}$. Since $\Xi \in \mathcal{I}_{\mathcal{S}}^{a_\mu} \cap \mathcal{A}^\mu$, there exist scalars x_k^r, $(r,k) \in \mathcal{B}^\mu$, and $u_{l\nu}$, $(l,\nu) \in [1,n{+}4] \times [1,\tau_\mu]$, such that

$$\Xi = \sum_{(r,k)\in\mathcal{B}^\mu} x_k^r\xi^{rk} = \sum_{\substack{l\in[1,n+4]\\ \nu\in[1,\tau_\mu]}} u_{l\nu}\mathcal{T}_\nu\widehat{\psi}^{l},$$

and projecting on the $\mathcal{S}^+$ components yields that

$$\Xi^+ = \sum_{(r,k)\in\mathcal{B}^{\mu+}} x_k^r \xi^{rk+} = \sum_{\substack{l\in\mathcal{S}^+\cup[n+1,n+4]\\ \nu\in[1,\tau_\mu]}} u_{l\nu} \mathcal{T}_\nu \widehat{\psi}^{l+},$$

since $\xi^{rk+} = 0$ and $\widehat{\psi}^{k+} = 0$ if $k \in \mathcal{S}^-$. These relations show that $\Xi^+ \in \mathcal{I}_{\mathcal{S}^+}^{a_\mu} \cap \mathcal{A}^{\mu+}$. Furthermore, consider now $\mathcal{U}^+$ and any given $\xi^+ \in \mathcal{A}^{\mu+}$. Letting $\xi^+ = \sum_{(r,k)\in\mathcal{B}^{\mu+}} x_k^{r+} \xi^{rk+}$ and $\xi = \Upsilon(\xi^+, 0)$, we obtain that $\xi = \sum_{(r,k)\in\mathcal{B}^{\mu+}} x_k^{r+} \xi^{rk}$ and $\xi \in \mathcal{A}^\mu$. From the species localization property (3.1.3), we now have $\langle\!\langle \xi^+, \mathcal{U}^+ \rangle\!\rangle = \langle\!\langle \xi, \mathcal{U} \rangle\!\rangle$ and since $\mathcal{U} \in \mathcal{I}_{\mathcal{S}}^{a_\mu} \cap \mathcal{A}^{\mu\perp}$ and $\xi \in \mathcal{A}^\mu$ we also have $\langle\!\langle \xi, \mathcal{U} \rangle\!\rangle = 0$. This shows that $\mathcal{U}^+ \in \mathcal{I}_{\mathcal{S}^+}^{a_\mu} \cap (\mathcal{A}^{\mu+})^\perp$, and therefore, we have shown that any $\zeta^+ \in \mathcal{I}_{\mathcal{S}^+}^{a_\mu}$ can be written in the form $\zeta^+ = \Xi^+ + \mathcal{U}^+$ where $\Xi^+ \in \mathcal{I}_{\mathcal{S}^+}^{a_\mu} \cap \mathcal{A}^{\mu+}$ and $\mathcal{U}^+ \in \mathcal{I}_{\mathcal{S}^+}^{a_\mu} \cap (\mathcal{A}^{\mu+})^\perp$. Since the intersection of $\mathcal{I}_{\mathcal{S}^+}^{a_\mu} \cap \mathcal{A}^{\mu+}$ and $\mathcal{I}_{\mathcal{S}^+}^{a_\mu} \cap (\mathcal{A}^{\mu+})^\perp$ obviously reduces to zero, we have established that (3.1.44) holds.

Let now $\mathcal{T}_\nu \widehat{\psi}^{l+}$ for $(l,\nu) \in \mathcal{S}^+ \times [1,\tau_\mu]$ and consider any $\xi^+ \in \mathcal{A}^{\mu+}$ and $\xi = \Upsilon(\xi^+, 0)$. Since $\xi^+ = \sum_{(r,k)\in\mathcal{B}^{\mu+}} x_k^{r+} \xi^{rk+}$, we deduce by linearity that $\xi = \sum_{(r,k)\in\mathcal{B}^{\mu+}} x_k^{r+} \xi^{rk}$ so that $\xi \in \mathcal{A}^\mu$. From the species localization property (3.1.3), we have $\langle\!\langle \xi^+, \mathcal{T}_\nu \widehat{\psi}^{l+} \rangle\!\rangle = \langle\!\langle \xi, \mathcal{T}_\nu \widehat{\psi}^{l} \rangle\!\rangle$ and we also have $\langle\!\langle \xi, \mathcal{T}_\nu \widehat{\psi}^{l} \rangle\!\rangle = 0$ since $\mathcal{T}_\nu \widehat{\psi}^{l} \in \mathcal{I}_{\mathcal{S}}^{a_\mu} \cap \mathcal{A}^{\mu\perp}$ from (3.1.2). This shows that $\mathcal{T}_\nu \widehat{\psi}^{l+} \in \mathcal{I}_{\mathcal{S}^+}^{a_\mu} \cap (\mathcal{A}^{\mu+})^\perp$ and (3.1.45) is established.

Properties of the Subsystem $G^{++}\alpha^{\mu+} = \beta^{\mu+}$. All the results of Sections 2.3.2, 2.3.3, and 2.3.4 therefore apply to the subsystem (3.1.34).

In particular, for any $n^+ \geq 1$, G^{++} is symmetric positive semi-definite, and $x^+ \in N(G^{++})$ if and only if $\sum_{(r,k)\in\mathcal{B}^{\mu+}} x_k^{r+} \xi^{rk+}$ is in the space $\mathcal{I}_{\mathcal{S}^+}^{a_\mu} \cap \mathcal{A}^{\mu+}$. The matrix G^{++} is also positive definite on the constrained space C^+ where

$$C^+ = \Big(\mathrm{span}\{\ \mathcal{G}^{l\nu+},\ (l,\nu) \in \big(\mathcal{S}^+\cup[n+1,n+4]\big)\times[1,\tau_\mu]\ \}\Big)^\perp,$$

and

$$N(G^{++}) \oplus C^+ = \mathbb{R}^{\omega^+}.$$

In addition, we also have $\beta^{\mu+} \in R(G^{++})$. Notice that the constraints $\langle \alpha^{\mu+}, \mathcal{G}^{l\nu+} \rangle = 0$, $(l,\nu) \in \big(\mathcal{S}^+\cup[n+1,n+4]\big) \times [1,\tau_\mu]$, can also be simplified by using any set of functions spanning the subspace $\mathcal{I}_{\mathcal{S}^+}^{a_\mu} \cap \mathcal{A}^{\mu+}$ from (2.3.23) and (3.1.44). These properties and the relations (3.1.24) and (3.1.29) also imply that $N(\widetilde{G}^{++}) = N(G^{++})$, $R(\widetilde{G}^{++}) = (\mathcal{N}^{++})^{-1} R(G^{++})$, and $\widetilde{\beta}^{\mu+} \in R(\widetilde{G}^{++})$.

Furthermore, the matrix $2db(G^{++}) - G^{++}$ is positive semi-definite for any $n^+ \geq 1$ and this matrix is positive definite if $n^+ \geq 3$. If $n^+ = 2$ and $\mathcal{S}^+ = \{1,2\}$, then $N\big(2db(G^{++}) - G^{++}\big) = \{x^{*+},\ x^+ \in N(G^{++})\}$, where for $x^+ \in \mathbb{R}^{\omega^+}$ we have defined

x^{*+} by $x_1^{r*+} = x_1^{r+}$, $r \in \mathcal{F}_1$, and $x_2^{r*+} = -x_2^{r+}$, $r \in \mathcal{F}_2$. Finally, if $n^+ = 1$, we have $N\big(2db(G^{++}) - G^{++}\big) = N(G^{++})$.

In addition, the matrix $db(G^{++})$ is symmetric positive semi-definite for any $n^+ \geq 1$ and positive definite if $n^+ \geq 2$. In the special case $n^+ = 1$, we also have $N\big(db(G^{++})\big) = N(G^{++})$.

Summary. We have shown that whenever the perpendicularity property (3.1.1), the species orthogonality property (3.1.2), and the species localization property (3.1.3) hold, then the matrix G^{++} satisfies

$$\begin{cases} G^{++} \ \text{ is symmetric positive semi-definite for } n^+ \geq 1, \\ N(G^{++}) = \{\, x^+ \in \mathbb{R}^{\omega^+}, \ \displaystyle\sum_{(r,k)\in\mathcal{B}^{\mu+}} x_k^{r+}\xi^{rk+} \in \mathcal{I}_{S^+}^{a_\mu} \cap \mathcal{A}^{\mu+} \,\} \text{ for } n^+ \geq 1, \\ G^{++} \ \text{ is positive definite on } C^+ \text{ for } n^+ \geq 1, \\ N(G^{++}) \oplus C^+ = \mathbb{R}^{\omega^+} \text{ for } n^+ \geq 1, \\ \beta^{\mu+} \in R(G^{++}) \text{ for } n^+ \geq 1, \end{cases} \tag{3.1.46}$$

the matrix $2db(G^{++}) - G^{++}$ satisfies

$$\begin{cases} 2db(G^{++}) - G^{++} \ \text{ is symmetric positive semi-definite for } n^+ \geq 1, \\ 2db(G^{++}) - G^{++} \ \text{ is positive definite for } n^+ \geq 3, \\ N\big(2db(G^{++}) - G^{++}\big) = \{\, x^{*+},\ x^+ \in N(G^{++}) \,\} \text{ for } n^+ = 2, \\ N\big(2db(G^{++}) - G^{++}\big) = N(G^{++}) \text{ for } n^+ = 1, \end{cases} \tag{3.1.47}$$

and the matrix $db(G^{++})$ satisfies

$$\begin{cases} db(G^{++}) \ \text{ is symmetric positive semi-definite for } n^+ \geq 1, \\ db(G^{++}) \ \text{ is positive definite for } n^+ \geq 2, \\ N\big(db(G^{++})\big) = N(G^{++}) \text{ for } n^+ = 1, \end{cases} \tag{3.1.48}$$

where for $n^+ = 2$, $S^+ = \{1,2\}$, and $x^+ \in \mathbb{R}^{\omega^+}$, we have defined x^{*+} by $x_1^{r*+} = x_1^{r+}$, $r \in \mathcal{F}_1$, and $x_2^{r*+} = -x_2^{r+}$, $r \in \mathcal{F}_2$.

3.1.4 Properties of the System $\widetilde{G}\alpha^\mu = \widetilde{\beta}^\mu$

In this section we use the partitioning of the transport linear systems introduced in Section 3.1.2 and the properties of the subsystem $G^{++}\alpha^{\mu+} = \beta^{\mu+}$ established in Section 3.1.3 in order to obtain the properties of the system $\widetilde{G}\alpha^\mu = \widetilde{\beta}^\mu$.

The Range $R(\widetilde{G})$. In order to identify the range $R(\widetilde{G})$, we first characterize the nullspace $N(\widetilde{G}^t)$ of $\widetilde{G}^t$. We claim that $\widetilde{G}^t x = 0$ if and only if x admits the partitioning $x = \Gamma^\mu(x^+, 0)$ where $x^+ \in \mathcal{N}^{++}(N(G^{++}))$, so that

$$N(\widetilde{G}^t) = \{ \, x \in \mathbb{R}^\omega, \; x = \Gamma^\mu(x^+,0), \; x^+ \in \mathcal{N}^{++}(N(G^{++})) \, \}. \tag{3.1.49}$$

Indeed, from the block-partitioning of $\widetilde{G}$, we first obtain a block-partitioning for $\widetilde{G}^t$. This partitioning of $\widetilde{G}^t$ yields that $\widetilde{G}^t x = 0$ if and only if $(\widetilde{G}^t)^{++}x^+ + (\widetilde{G}^t)^{+-}x^- = 0$ and $(\widetilde{G}^t)^{--}x^- = 0$, where $x = \Gamma^\mu(x^+, x^-)$, $x^+ \in \mathbb{R}^{\omega^+}$ and $x^- \in \mathbb{R}^{\omega^-}$. Since $(\widetilde{G}^t)^{--} = \widetilde{G}^{--}$ and since this matrix is nonsingular according to (3.1.28), we thus get that $x^- = 0$. This implies now that $(\widetilde{G}^t)^{++}x^+ = 0$, and since from (3.1.28) $(\widetilde{G}^t)^{++}\mathcal{N}^{++} = G^{++}$ and $\mathcal{N}^{++}$ is invertible, we get that $x^+ \in \mathcal{N}^{++}(N(G^{++}))$ and (3.1.49) is proven. Note that a direct investigation of $R(\widetilde{G})$ can also be done and yields that

$$R(\widetilde{G}) = \{ \, y \in \mathbb{R}^\omega, \; y = \Gamma^\mu(y^+, y^-), \; y^+ \in (\mathcal{N}^{++})^{-1}(R(G^{++})) \, \}, \tag{3.1.50}$$

which is precisely the orthogonal of $\{ \, x \in \mathbb{R}^\omega, x = \Gamma^\mu(x^+,0), \; x^+ \in \mathcal{N}^{++}(N(G^{++})) \, \}$. Indeed, from the block-partitioning of $\widetilde{G}$, we first obtain that $y = \widetilde{G}x$ if and only if $y^+ = \widetilde{G}^{++}x^+$ and $y^- = \widetilde{G}^{-+}x^+ + \widetilde{G}^{--}x^-$, where $y = \Gamma^\mu(y^+, y^-)$ and $x = \Gamma^\mu(x^+, x^-)$. Since $\widetilde{G}^{--}$ is nonsingular according to (3.1.28), we can always find a vector x^- such that $y^- = \widetilde{G}^{-+}x^+ + \widetilde{G}^{--}x^-$. This implies that $y \in R(\widetilde{G})$ if and only if $y^+ \in R(\widetilde{G}^{++})$ and since $\mathcal{N}^{++}\widetilde{G}^{++} = G^{++}$ and $\mathcal{N}^{++}$ is invertible, we obtain (3.1.50).

Since $\beta^{\mu+} \in R(G^{++})$ from (3.1.46), we deduce from (3.1.24) and (3.1.29) that $\widetilde{\beta}^{\mu+} \in R(\widetilde{G}^{++})$ and hence that $\widetilde{\beta}^\mu \in R(\widetilde{G})$ from (3.1.50).

The Nullspace $N(\widetilde{G})$. In order to identify the nullspace of $\widetilde{G}$, we first establish the inclusion

$$\{ \, x \in \mathbb{R}^\omega, \, \Xi = \sum_{(r,k)\in\mathcal{B}^\mu} x_k^r \xi^{rk} \in \mathcal{I}_S^{a_\mu} \cap \mathcal{A}^\mu \, \} \subset N(\widetilde{G}). \tag{3.1.51}$$

For $x \in \mathbb{R}^\omega$, $\Xi = \sum_{(r,k)\in\mathcal{B}^\mu} x_k^r \xi^{rk}$, and $(r,k) \in \mathcal{B}^\mu$, we obtain from (3.1.12) and (3.1.13) that

$$(\widetilde{G}x)_k^r = \sum_{s\in\mathcal{F}_k} \Big(\sum_{l\in\mathcal{S}} n_l [\xi^{rk}, \xi^{sk}]'_{kl} + n_k [\xi^{rk}, \xi^{sk}]''_{kk} \Big) x_k^s + \sum_{\substack{(s,l)\in\mathcal{B}^\mu \\ l\neq k}} n_l [\xi^{rk}, \xi^{sl}]''_{kl} x_l^s.$$

The right member of this equality can easily be rewritten in the form

$$\sum_{l\in\mathcal{S}} n_l\Big[\xi^{rk},\sum_{s\in\mathcal{F}_k} x_k^s\xi^{sk}\Big]'_{kl} + n_k\Big[\xi^{rk},\sum_{s\in\mathcal{F}_k} x_k^s\xi^{sk}\Big]''_{kk} + \sum_{\substack{l\in\mathcal{S}\\ l\neq k}} n_l\Big[\xi^{rk},\sum_{s\in\mathcal{F}_l} x_l^s\xi^{sl}\Big]''_{kl},$$

so that we have

$$(\widetilde{G}x)_k^r = \sum_{l\in\mathcal{S}} n_l\Big([\xi^{rk},\Xi]'_{kl} + [\xi^{rk},\Xi]''_{kl}\Big) = \sum_{l\in\mathcal{S}} n_l\Big\{\xi_k^{rk},\Xi_k - \Xi_k' + \widetilde{\Xi}_l - \widetilde{\Xi}_l'\Big\}_{kl},$$

where we have used the symmetry properties of the partial brackets $[\,,\,]'$ and $[\,,\,]''$, and the fact that Ξ and $\sum_{r\in\mathcal{F}_k} x_k^r\xi^{rk}$ have the same $k^{\rm th}$ component Ξ_k. This relation now shows that $\widetilde{G}x = 0$ when Ξ is a collisional invariant, i.e., when $\Xi \in \mathcal{I}_{\mathcal{S}}^{a_\mu}$, so that (3.1.51) is proven.

On the other hand, keeping in mind that $\dim\big(N(\widetilde{G})\big) = \dim\big(N(\widetilde{G}^t)\big)$, we deduce from (3.1.49) that $\dim\big(N(\widetilde{G})\big) = \dim\big(N(G^{++})\big)$, and we also obtain from (3.1.46) that $\dim\big(N(G^{++})\big) = \dim(\mathcal{I}_{\mathcal{S}+}^{a_\mu} \cap \mathcal{A}^{\mu+})$. Assuming temporarily that

$$\dim(\mathcal{I}_{\mathcal{S}}^{a_\mu} \cap \mathcal{A}^{\mu}) = \dim(\mathcal{I}_{\mathcal{S}+}^{a_\mu} \cap \mathcal{A}^{\mu+}), \tag{3.1.52}$$

we thus obtain that $\dim\big(N(G^{++})\big) = \dim(\mathcal{I}_{\mathcal{S}}^{a_\mu} \cap \mathcal{A}^{\mu})$ so that the inclusion (3.1.51) is actually an equality.

The Dimension of $\mathcal{I}_{\mathcal{S}+}^{a_\mu} \cap \mathcal{A}^{\mu+}$. We now establish the important property (3.1.52) which has already been used and which will be useful in the next sections. We denote $m = \dim(\mathcal{I}_{\mathcal{S}}^{a_\mu} \cap \mathcal{A}^{\mu})$ and we consider a basis $\Xi^1,\ldots,\Xi^m$ of $\mathcal{I}_{\mathcal{S}}^{a_\mu} \cap \mathcal{A}^{\mu}$. We first note that each Ξ^i, $i = 1,\ldots,m$, is a linear combination of the ξ^{rk}, $(r,k) \in \mathcal{B}^\mu$, and also of the $\mathcal{T}_\nu\widehat{\psi}^l$, $(l,\nu) \in [1,n+4]\times[1,\tau_\mu]$, so that projecting on the $\mathcal{S}^+$ components yields that each Ξ^{i+} is a linear combination of the ξ^{rk+}, $(r,k) \in \mathcal{B}^{\mu+}$, and also of the $\mathcal{T}_\nu\widehat{\psi}^{l+}$, $(l,\nu) \in \big(\mathcal{S}^+\cup[n+1,n+4]\big)\times[1,\tau_\mu]$. As a consequence, we deduce that $\Xi^{1+},\ldots,\Xi^{m+}$ are in $\mathcal{I}_{\mathcal{S}+}^{a_\mu} \cap \mathcal{A}^{\mu+}$. We now claim that $\Xi^{1+},\ldots,\Xi^{m+}$ form a basis of $\mathcal{I}_{\mathcal{S}+}^{a_\mu} \cap \mathcal{A}^{\mu+}$.

Indeed, let first $\Xi^+ \in \mathcal{I}_{\mathcal{S}+}^{a_\mu} \cap \mathcal{A}^{\mu+}$, which by definition can be written in the form

$$\Xi^+ = \sum_{(r,k)\in\mathcal{B}^{\mu+}} x_k^{r+}\xi^{rk+} = \sum_{\substack{l\in\mathcal{S}^+\cup[n+1,n+4]\\ \nu\in[1,\tau_\mu]}} u_{l\nu}\mathcal{T}_\nu\widehat{\psi}^{l+},$$

where x_k^{r+}, $(r,k) \in \mathcal{B}^{\mu+}$, and $u_{l\nu}$, $(l,\nu) \in \big(\mathcal{S}^+\cup[n+1,n+4]\big)\times[1,\tau_\mu]$, are scalars. By taking the scalar product with $\mathcal{T}_\nu\widehat{\psi}^{l+}$ and using the species orthogonality property

(3.1.45), we then get $\langle\!\langle \Xi^+, \mathcal{T}_\nu\widehat{\psi}^{l+}\rangle\!\rangle = u_{l\nu}\langle\!\langle \mathcal{T}_\nu\widehat{\psi}^{l+}, \mathcal{T}_\nu\widehat{\psi}^{l+}\rangle\!\rangle = u_{l\nu}n_l = 0$ so that $u_{l\nu} = 0$, $(l,\nu) \in \mathcal{S}^+ \times [1,\tau_\mu]$, and hence

$$\Xi^+ = \sum_{\substack{l\in[n+1,n+4]\\ \nu\in[1,\tau_\mu]}} u_{l\nu}\mathcal{T}_\nu\widehat{\psi}^{l+}.$$

We then consider the full mixture collisional invariant

$$\Xi = \sum_{\substack{l\in[n+1,n+4]\\ \nu\in[1,\tau_\mu]}} u_{l\nu}\mathcal{T}_\nu\widehat{\psi}^{l},$$

and we claim that $\Xi \in \mathcal{I}_{\mathcal{S}}^{a_\mu} \cap \mathcal{A}^\mu$. We may indeed write $\Xi = \mathfrak{X} + \mathcal{U}$ where $\mathfrak{X} \in \mathcal{I}_{\mathcal{S}}^{a_\mu} \cap \mathcal{A}^\mu$ and $\mathcal{U} \in \mathcal{I}_{\mathcal{S}}^{a_\mu} \cap \mathcal{A}^{\mu\perp}$. Since $\mathfrak{X} \in \mathcal{I}_{\mathcal{S}}^{a_\mu} \cap \mathcal{A}^\mu$, by using the species orthogonality property (3.1.2), there exist scalars y_k^r, $(r,k) \in \mathcal{B}^\mu$, and $v_{l\nu}$, $(l,\nu) \in [n+1,n+4]\times[1,\tau_\mu]$, such that

$$\mathfrak{X} = \sum_{(r,k)\in\mathcal{B}^\mu} y_k^r\xi^{rk} = \sum_{\substack{l\in[n+1,n+4]\\ \nu\in[1,\tau_\mu]}} v_{l\nu}\mathcal{T}_\nu\widehat{\psi}^{l}.$$

We also have $\langle\!\langle \Xi, \xi^{rk}\rangle\!\rangle = \langle\!\langle \mathfrak{X}, \xi^{rk}\rangle\!\rangle$ for any $(r,k) \in \mathcal{B}^{\mu+}$ since $\mathcal{U} \in \mathcal{I}_{\mathcal{S}}^{a_\mu} \cap \mathcal{A}^{\mu\perp}$ so that from the species localization property, we obtain that $\langle\!\langle \Xi^+, \xi^{rk+}\rangle\!\rangle = \langle\!\langle \mathfrak{X}^+, \xi^{rk+}\rangle\!\rangle$ so that $\langle\!\langle \Xi^+ - \mathfrak{X}^+, \xi^{rk+}\rangle\!\rangle = 0$, for any $(r,k) \in \mathcal{B}^{\mu+}$. Since Ξ^+ and $\mathfrak{X}^+$ are both in $\mathcal{I}_{\mathcal{S}^+}^{a_\mu} \cap \mathcal{A}^{\mu+}$, $\Xi^+ - \mathfrak{X}^+$ is a linear combination of the ξ^{rk+}, $(r,k) \in \mathcal{B}^{\mu+}$, and thus $\langle\!\langle \Xi^+ - \mathfrak{X}^+, \Xi^+ - \mathfrak{X}^+\rangle\!\rangle = 0$ so that $\Xi^+ = \mathfrak{X}^+$. Since we have

$$\Xi - \mathfrak{X} = \sum_{\substack{l\in[n+1,n+4]\\ \nu\in[1,\tau_\mu]}} (u_{l\nu} - v_{l\nu})\mathcal{T}_\nu\widehat{\psi}^{l},$$

and since $\Xi^+ - \mathfrak{X}^+ = 0$, we deduce that

$$\sum_{\substack{l\in[n+1,n+4]\\ \nu\in[1,\tau_\mu]}} (u_{l\nu} - v_{l\nu})\mathcal{T}_\nu\widehat{\psi}^{l+} = 0,$$

which implies that $u_{l\nu} = v_{l\nu}$, $(l,\nu) \in [n+1,n+4]\times[1,\tau_\mu]$, since the $\mathcal{T}_\nu\widehat{\psi}^{l+}$, $(l,\nu) \in [n+1,n+4]\times[1,\tau_\mu]$, are linearly independent. As a consequence, we have $\Xi = \mathfrak{X}$ and thus $\Xi \in \mathcal{I}_{\mathcal{S}}^{a_\mu} \cap \mathcal{A}^\mu$, and hence Ξ is a linear combination of the $\Xi^1,\dots,\Xi^m$, i.e., $\Xi = \sum_{i\in[1,m]}\gamma_i\Xi^i$. Projecting on the $\mathcal{S}^+$ components now yields that $\Xi^+ = \sum_{i\in[1,m]}\gamma_i\Xi^{i+}$, so that $\mathcal{I}_{\mathcal{S}^+}^{a_\mu} \cap \mathcal{A}^{\mu+}$ is spanned by $\Xi^{1+},\dots,\Xi^{m+}$.

In addition, if we have a linear relation $\sum_{i\in[1,m]}\gamma_i\Xi^{i+} = 0$ between $\Xi^{1+},\dots,\Xi^{m+}$, then from the property (2.3.33), we must have $\sum_{i\in[1,m]}\gamma_i\Xi^i = 0$ so that by assumption

$\gamma_i = 0$, $i \in [1,m]$, since $\Xi^1,\dots,\Xi^m$ form a basis of $\mathcal{I}_S^{a_\mu} \cap \mathcal{A}^\mu$. We have therefore established that under the assumptions (3.1.1)–(3.1.3)

$$\begin{aligned} \Xi^1,\dots,\Xi^m \text{ form a basis of } \mathcal{I}_S^{a_\mu} \cap \mathcal{A}^\mu \quad &\Longrightarrow \\ \Xi^{1+},\dots,\Xi^{m+} \text{ form a basis of } \mathcal{I}_{S+}^{a_\mu} \cap \mathcal{A}^{\mu+} \end{aligned} \tag{3.1.53}$$

which trivially implies (3.1.52).

The Nullspace $N\big(db(\widetilde{G})\big)$. Finally, we establish that $db(\widetilde{G})x = 0$ if and only if x admits the partitioning $x = \Gamma^\mu(x^+,0)$ where $x^+ \in N\big(db(G^{++})\big)$, so that

$$N\big(db(\widetilde{G})\big) = \{\ x \in \mathbb{R}^\omega, x = \Gamma^\mu(x^+,0),\ x^+ \in N\big(db(G^{++})\big)\ \}. \tag{3.1.54}$$

Indeed, the block-partitioning of $db(\widetilde{G})$ first yields that $db(\widetilde{G})x = 0$ if and only if $db(\widetilde{G}^{++})x^+ = 0$ and $db(\widetilde{G}^{--})x^- = 0$, where $x = \Gamma^\mu(x^+,x^-)$, $x^+ \in \mathbb{R}^{\omega^+}$ and $x^- \in \mathbb{R}^{\omega^-}$. Since $db(\widetilde{G}^{--}) = \widetilde{G}^{--}$ and since this matrix is nonsingular according to (3.1.28), we thus get that $x^- = 0$. In addition $db(\widetilde{G}^{++})x^+ = 0$ if and only if $db(G^{++})x^+ = 0$ from (3.1.25), and hence (3.1.54) is proven.

Summary. We have established that whenever the perpendicularity property (3.1.1), the species orthogonality property (3.1.2), and the species localization property (3.1.3) hold, then the matrix $\widetilde{G}$ satisfies

$$\begin{cases} N(\widetilde{G}) = \{\ x \in \mathbb{R}^\omega,\ \Xi = \displaystyle\sum_{(r,k)\in\mathcal{B}^\mu} x_k^r \xi^{rk} \in \mathcal{I}_S^{a_\mu} \cap \mathcal{A}^\mu\ \}, \\ R(\widetilde{G}) = \{\ x \in \mathbb{R}^\omega, x = \Gamma^\mu(x^+,0),\ x^+ \in \mathcal{N}^{++}\big(N(G^{++})\big)\ \}^\perp, \\ N\big(db(\widetilde{G})\big) = \{\ x \in \mathbb{R}^\omega, x = \Gamma^\mu(x^+,0),\ x^+ \in N\big(db(G^{++})\big)\ \}, \\ \widetilde{\beta}^\mu \in R(\widetilde{G}). \end{cases} \tag{3.1.55}$$

3.1.5 The Special Case $\dim\big(\mathcal{I}_S^{a_\mu} \cap \mathcal{A}^\mu\big) = 0$

We consider in this section the special case where $\dim\big(\mathcal{I}_S^{a_\mu} \cap \mathcal{A}^\mu\big) = 0$ as in Section 2.3.5. More specifically, we assume that the perpendicularity property (3.1.1), the species orthogonality property (3.1.2), and the species localization property (3.1.3) are satisfied, and we also assume that

$$\mathcal{I}_S^{a_\mu} \cap \mathcal{A}^\mu = \{0\}. \tag{3.1.56}$$

In order to derive the properties of G and $\widetilde{G}$ in this special case, we simply show that the results established in Section 2.3.5 directly apply to G^{++}. All we have to show is that

$$\mathcal{I}_{S+}^{a_\mu} \cap \mathcal{A}^{\mu+} = \{0\}, \tag{3.1.57}$$

but this is a direct consequence of (3.1.52). Therefore, the results of Section 2.3.5 are valid for the subsystem (3.1.34) and the results of Sections 3.1.2 and 3.1.4 apply to the rescaled system (3.1.14)(3.1.15).

Properties of G. The matrix G admits the block-decomposition (3.1.20). In addition, the matrices G^{++}, $db(G^{++})$, and $2db(G^{++}) - G^{++}$ are symmetric positive definite for any $n^+ \geq 1$.

Properties of $\widetilde{G}$. The matrix $\widetilde{G}$ admits the block-decomposition (3.1.21). From (3.1.28) the matrix $\widetilde{G}^{++}$ is also nonsingular for any $n^+ \geq 1$. In addition, the matrix $\widetilde{G}^{--}$ is block-diagonal, i.e., $\widetilde{G}^{--} = db(\widetilde{G}^{--})$, and is symmetric positive definite for any $n^+ \geq 1$. As a consequence, the matrix $\widetilde{G}$ is also nonsingular for any $n^+ \geq 1$.

The matrix $db(\widetilde{G})$ admits the block-decomposition (3.1.27). From (3.1.28) the matrix $db(\widetilde{G}^{++})$ is symmetric positive definite for any $n^+ \geq 1$, and so is the matrix $db(\widetilde{G}^{--})$. As a consequence, the matrix $db(\widetilde{G})$ is symmetric positive definite for any $n^+ \geq 1$.

Summary. We have shown that whenever the perpendicularity property (3.1.1), the species orthogonality property (3.1.2), the species localization property (3.1.3), and the equality $\dim\big(\mathcal{I}_S^{a_\mu} \cap \mathcal{A}^\mu\big) = 0$ are satisfied, then the matrix G^{++} satisfies

$$\begin{cases} G^{++} \ \text{ is symmetric positive definite for } n^+ \geq 1, \\ 2db(G^{++}) - G^{++} \ \text{ is symmetric positive definite for } n^+ \geq 1, \\ db(G^{++}) \ \text{ is symmetric positive definite for } n^+ \geq 1. \end{cases} \tag{3.1.58}$$

and the matrix $\widetilde{G}$ satisfies

$$\begin{cases} \widetilde{G} \ \text{ is nonsingular for } n^+ \geq 1, \\ db(\widetilde{G}) \ \text{ is symmetric positive definite for } n^+ \geq 1. \end{cases} \tag{3.1.59}$$

3.1.6 The Special Case $\dim\big(\mathcal{I}_S^{a_\mu} \cap \mathcal{A}^\mu\big) = 1$

We consider in this section the special case where $\dim\big(\mathcal{I}_S^{a_\mu} \cap \mathcal{A}^\mu\big) = 1$ as in Section 2.3.6. More specifically, we assume that the perpendicularity property (3.1.1), the species orthogonality property (3.1.2), and the species localization property (3.1.3) are satisfied, and we also assume that

$$\mathcal{I}_{\mathcal{S}}^{a_\mu} \cap \mathcal{A}^\mu = \mathbb{R}Z, \tag{3.1.60}$$

where Z is a nonzero collisional invariant. We denote by $\mathcal{Z} = (\mathcal{Z}_k^r)_{(r,k)\in\mathcal{B}^\mu}$ the components of the collisional invariant Z with respect to the basis functions ξ^{rk}

$$Z = \sum_{(r,k)\in\mathcal{B}^\mu} \mathcal{Z}_k^r \xi^{rk}. \tag{3.1.61}$$

In order to derive the properties of $\widetilde{G}$ in this special case, we simply show that the results established in Section 2.3.6 directly apply to G^{++}. All we have to show is that

$$\mathcal{I}_{\mathcal{S}+}^{a_\mu} \cap \mathcal{A}^{\mu+} = \mathbb{R}Z^+, \tag{3.1.62}$$

where $Z^+ = (Z_i)_{i\in\mathcal{S}+}$, but this is a direct consequence of (3.1.53). Therefore, the results of Section 2.3.6 are valid for the subsystem (3.1.34) and the results of Sections 3.1.2 and 3.1.4 apply to the rescaled system (3.1.14)(3.1.15).

Properties of G. The matrix G admits the block-decomposition (3.1.20). In addition, G^{++} is symmetric positive semi-definite and $N(G^{++}) = \mathbb{R}\mathcal{Z}^+$, with $\mathcal{Z}^+ = (\mathcal{Z}_k^r)_{(r,k)\in\mathcal{B}^{\mu+}}$. The matrix G^{++} is positive definite on the constrained space $\mathcal{C}^+ = (\mathcal{G}^+)^\perp$ where $\mathcal{G}^+ = (\mathcal{G}_k^r)_{(r,k)\in\mathcal{B}^{\mu+}}$ and $\mathcal{G}_k^r = \langle\!\langle \xi^{rk}, Z \rangle\!\rangle$. Moreover, we have $\langle \mathcal{G}^+, \mathcal{Z}^+ \rangle \neq 0$ from (2.3.41) and a direct evaluation indeed yields that $\langle \mathcal{G}^+, \mathcal{Z}^+ \rangle = \langle\!\langle Z^+, Z^+ \rangle\!\rangle$. We also have $\beta^{\mu+} \in R(G^{++})$, that is, $\langle \beta^{\mu+}, \mathcal{Z}^+ \rangle = 0$ from (2.3.41).

The matrix $2db(G^{++}) - G^{++}$ is also symmetric positive semi-definite for any $n^+ \geq 1$ from (2.3.42). In addition, the matrix $2db(G^{++}) - G^{++}$ is positive definite if $n^+ \geq 3$, whereas its nullspace is $N\big(2db(G^{++}) - G^{++}\big) = \mathbb{R}\mathcal{Z}^{*+}$ for $n^+ = 2$, $\mathcal{S}^+ = \{1,2\}$, where $\mathcal{Z}^{*+}$ is defined by $\mathcal{Z}_1^{r*+} = \mathcal{Z}_1^r$, $r \in \mathcal{F}_1$, and $\mathcal{Z}_2^{r*+} = -\mathcal{Z}_2^r$, $r \in \mathcal{F}_2$, from (2.3.42), and its nullspace is $N\big(2db(G^{++}) - G^{++}\big) = \mathbb{R}\mathcal{Z}^+$ for $n^+ = 1$.

Finally, the matrix $db(G^{++})$ is also symmetric positive semi-definite for any $n^+ \geq 1$ from (2.3.43). In addition, the matrix $db(G^{++})$ is positive definite for any $n^+ \geq 2$ whereas for $n^+ = 1$, the nullspace of $db(G^{++})$ is $N\big(db(G^{++})\big) = \mathbb{R}\mathcal{Z}^+$.

Properties of $\widetilde{G}$. The matrix $\widetilde{G}$ admits the block-decomposition (3.1.21). From (3.1.28) we deduce that $\widetilde{G}^{++}$ is such that $N(\widetilde{G}^{++}) = N(G^{++}) = \mathbb{R}\mathcal{Z}^+$ and $R(\widetilde{G}^{++}) = (\mathcal{N}^{++}\mathcal{Z}^+)^\perp$. In addition, the matrix $\widetilde{G}^{--}$ is block-diagonal, i.e., $\widetilde{G}^{--} = db(\widetilde{G}^{--})$, and is symmetric positive definite for any $n^+ \geq 1$.

We next obtain from (3.1.49) that $N(\widetilde{G}^t) = \Gamma^\mu(\mathcal{N}^{++}(\mathbb{R}\mathcal{Z}^+),0)$ which implies that $N(\widetilde{G}^t) = \mathbb{R}\mathcal{N}(\Gamma^\mu(\mathcal{Z}^+,0)) = \mathbb{R}\mathcal{N}\mathcal{Z}$ since $\mathcal{N}\mathcal{Z} = \mathcal{N}(\Gamma^\mu(\mathcal{Z}^+,0))$ because $\mathcal{N}^{--} = 0$. This implies that $R(\widetilde{G}) = (\widetilde{\mathcal{Z}})^\perp$ where we have defined

$$\widetilde{\mathcal{Z}} = \mathcal{N}\mathcal{Z}. \tag{3.1.63}$$

Finally, from (3.1.55) we also get that $N(\widetilde{G}) = \mathbb{R}\mathcal{Z}$.

The matrix $db(\widetilde{G})$ admits the block-decomposition (3.1.27). From (3.1.28), the matrix $db(\widetilde{G}^{++})$ is also symmetric, positive semi-definite, positive definite if $n^+ \geq 2$ and its nullspace is given by $N(db(\widetilde{G}^{++})) = \mathbb{R}\mathcal{Z}^+$ if $n^+ = 1$ from (3.1.25). In addition, the matrix $db(\widetilde{G}^{--})$ is symmetric positive definite for any $n^+ \geq 1$. As a consequence, the matrix $db(\widetilde{G})$ is symmetric positive definite for $n^+ \geq 2$, whereas $db(\widetilde{G})$ is symmetric positive semi-definite for $n^+ = 1$ with nullspace $N(db(\widetilde{G})) = \mathbb{R}\Gamma^\mu(\mathcal{Z}^+,0)$ from (3.1.54).

Summary. We have shown that whenever the perpendicularity property (3.1.1), the species orthogonality property (3.1.2), the species localization property (3.1.3), and the equality $\dim(\mathcal{I}_S^{a_\mu} \cap \mathcal{A}^\mu) = 1$ are satisfied, then the matrix G^{++} satisfies

$$\begin{cases} G^{++} \ \text{ is symmetric positive semi-definite for } n^+ \geq 1, \\ N(G^{++}) = \mathbb{R}\mathcal{Z}^+ \ \text{ for } \ n^+ \geq 1, \\ G^{++} \ \text{ is positive definite on } \mathcal{C}^+ = (\mathcal{G}^+)^\perp \ \text{ for } \ n^+ \geq 1, \\ \langle \mathcal{G}^+, \mathcal{Z}^+ \rangle \neq 0 \ \text{ for } \ n^+ \geq 1, \\ \beta^{\mu+} \in R(G^{++}) \ \text{ for } \ n^+ \geq 1, \end{cases} \tag{3.1.64}$$

the matrix $2db(G^{++}) - G^{++}$ satisfies

$$\begin{cases} 2db(G^{++}) - G^{++} \ \text{ is symmetric positive semi-definite for } n^+ \geq 1, \\ 2db(G^{++}) - G^{++} \ \text{ is positive definite for } n^+ \geq 3, \\ N\big(2db(G^{++}) - G^{++}\big) = \mathbb{R}\mathcal{Z}^{*+} \text{ for } n^+ = 2, \\ N\big(2db(G^{++}) - G^{++}\big) = \mathbb{R}\mathcal{Z}^{+} \text{ for } n^+ = 1, \end{cases} \tag{3.1.65}$$

and the matrix $db(G^{++})$ satisfies

$$\begin{cases} db(G^{++}) \ \text{ is symmetric positive semi-definite for } n^+ \geq 1, \\ db(G^{++}) \ \text{ is positive definite for } n^+ \geq 2, \\ N\big(db(G^{++})\big) = \mathbb{R}\mathcal{Z}^+ \text{ for } n^+ = 1, \end{cases} \tag{3.1.66}$$

where for $n^+ = 2$ and $\mathcal{S}^+ = \{1,2\}$, we have defined $\mathcal{Z}^{*+}$ by $\mathcal{Z}_1^{r*+} = \mathcal{Z}_1^{r+}$, $r \in \mathcal{F}_1$, and $\mathcal{Z}_2^{r*+} = -\mathcal{Z}_2^{r+}$, $r \in \mathcal{F}_2$. In addition, the matrix $\widetilde{G}$ satisfies

$$\begin{cases} N(\widetilde{G}) = \mathbb{R}\mathcal{Z} \quad \text{for } n^+ \geq 1, \\ R(\widetilde{G}) = (\widetilde{\mathcal{Z}})^\perp \quad \text{for } n^+ \geq 1, \\ db(\widetilde{G}) \text{ is positive definite for } n^+ \geq 2, \\ N\big(db(\widetilde{G})\big) = \mathbb{R}\Gamma^\mu(\mathcal{Z}^+, 0) \quad \text{for } n^+ = 1. \end{cases} \tag{3.1.67}$$

3.1.7 Rescaled Reduced Transport Linear Systems

In this section we consider the rescaled transport linear systems associated with the simplified formulations described in Section 2.3.8.

Evaluation of Rescaled Reduced Systems. We first consider the simplified transport linear system $G_{[\mathrm{red}]}\alpha^\mu_{[\mathrm{red}]} = \beta^\mu_{[\mathrm{red}]}$ associated with a subspace $\mathcal{A}^\mu_{[\mathrm{red}]}$ of $\mathcal{A}^\mu$ spanned by the vectors ζ^{rk}, $(r,k) \in \mathcal{B}^\mu_{[\mathrm{red}]}$, as described in (2.3.55)–(2.3.58). The basis functions ζ^{rk}, $(r,k) \in \mathcal{B}^\mu_{[\mathrm{red}]}$, are written in the form

$$\zeta^{rk} = \sum_{(s,l)\in\mathcal{B}^\mu} \mathcal{R}^{rs}_{kl}\xi^{sl}, \qquad (r,k) \in \mathcal{B}^\mu_{[\mathrm{red}]}, \tag{3.1.68}$$

and $\mathcal{R} = (\mathcal{R}^{rs}_{kl})_{(r,k)\in\mathcal{B}^\mu_{[\mathrm{red}]},\,(s,l)\in\mathcal{B}^\mu}$ is the corresponding transformation matrix. We assume that the functional space $\mathcal{A}^\mu$, the basis functions ξ^{rk}, $(r,k) \in \mathcal{B}^\mu$, the reduced space $\mathcal{A}^\mu_{[\mathrm{red}]}$, and the basis functions ζ^{rk}, $(r,k) \in \mathcal{B}^\mu_{[\mathrm{red}]}$, satisfy the properties (3.1.1)–(3.1.3). The rescaled linear systems $\widetilde{G}\alpha^\mu = \widetilde{\beta}^\mu$ and $\widetilde{G}_{[\mathrm{red}]}\alpha^\mu_{[\mathrm{red}]} = \widetilde{\beta}^\mu_{[\mathrm{red}]}$ are thus well defined from Section 3.1.1. We have in particular

$$\begin{cases} \widetilde{G}^{rs}_{[\mathrm{red}]kl} = \langle\!\langle\!\langle \zeta^{rk}, \Im(\zeta^{sl}) \rangle\!\rangle\!\rangle, \qquad (r,k),(s,l) \in \mathcal{B}^\mu_{[\mathrm{red}]}, \\ \widetilde{\beta}^{r\mu}_{[\mathrm{red}]k} = \langle\!\langle\!\langle \zeta^{rk}, \Psi^\mu \rangle\!\rangle\!\rangle, \qquad (r,k) \in \mathcal{B}^\mu_{[\mathrm{red}]}, \end{cases} \tag{3.1.69}$$

and the properties of the corresponding reduced linear system are directly obtained from the general results derived in Sections 3.1.1 to 3.1.6, provided that the corresponding assumptions are satisfied. The associated simplified transport coefficients are subsequently calculated from the scalar products $\langle \alpha^\mu_{[\mathrm{red}]}, \beta^\mu_{[\mathrm{red}]} \rangle$.

However, one may also easily check that from these expressions, from (3.1.8), and from the linear relations (3.1.68), the rescaled reduced system is simply given by

$$\begin{cases} \widetilde{G}_{[\mathrm{red}]} = \mathcal{R}\widetilde{G}\mathcal{R}^t, \\ \widetilde{\beta}^\mu_{[\mathrm{red}]} = \mathcal{R}\widetilde{\beta}^\mu, \end{cases} \tag{3.1.70}$$

so that the rescaled reduced transport linear systems are easily evaluated from the larger ones.

Non Localized Basis Functions. We now analyze the more general situation (2.3.59) which includes, in particular, the spaces considered in (2.3.47). We thus consider a subspace $\mathcal{A}^{\mu}_{[\mathrm{red}]}$ of $\mathcal{A}^{\mu}$ spanned by the functions $\widehat{\zeta}^{rk}$, $(r,k) \in \mathcal{B}^{\mu}_{[\mathrm{red}]}$, which are written in the form

$$\widehat{\zeta}^{rk} = \zeta^{rk} + \Sigma^{rk}, \tag{3.1.71}$$

where $\Sigma^{rk} \in \mathcal{I}_S^{a_\mu}$ is a collisional invariant of $\mathcal{A}^{\mu}$, and where the vectors $\zeta^{rk} \in \mathcal{A}^{\mu}$, $(r,k) \in \mathcal{B}^{\mu}_{[\mathrm{red}]}$, form a basis and satisfy the species localization property (3.1.3). We also assume that the functional space $\mathcal{A}^{\mu}$ and the basis functions ξ^{rk}, $(r,k) \in \mathcal{B}^{\mu}$, satisfy the properties (3.1.1)–(3.1.3), so that, in particular, the rescaled linear system $\widetilde{G}\alpha^{\mu} = \widetilde{\beta}^{\mu}$ is well defined. On the other hand, since we do not assume that the functions $\widehat{\zeta}^{rk} \in \mathcal{A}^{\mu}$, $(r,k) \in \mathcal{B}^{\mu}_{[\mathrm{red}]}$, satisfy the species localization property, the theory described in Sections 3.1.1 to 3.1.6 does not apply to the reduced system described in (2.3.59)–(2.3.68). In particular, we note that the matrix $\widetilde{G}_{[\mathrm{red}]}$ is not defined.

In order to still define a rescaled reduced system, we first restate that if the auxiliary basis functions $\zeta^{rk} \in \mathcal{A}^{\mu}$, $(r,k) \in \mathcal{B}^{\mu}_{[\mathrm{red}]}$, are written in the form

$$\zeta^{rk} = \sum_{(s,l)\in\mathcal{B}^{\mu}} \mathcal{R}^{rs}_{kl}\xi^{sl}, \qquad (r,k) \in \mathcal{B}^{\mu}_{[\mathrm{red}]}, \tag{3.1.72}$$

and if $\mathcal{R} = (\mathcal{R}^{rs}_{kl})_{(r,k)\in\mathcal{B}^{\mu}_{[\mathrm{red}]},\,(s,l)\in\mathcal{B}^{\mu}}$ is the corresponding transformation matrix, the reduced transport linear system can then be evaluated from

$$\begin{cases} G_{[\mathrm{red}]} = \mathcal{R} G \mathcal{R}^t, \\ \beta^{\mu}_{[\mathrm{red}]} = \mathcal{R}\beta^{\mu}. \end{cases} \tag{3.1.73}$$

In addition, since the basis functions ζ^{rk}, $(r,k) \in \mathcal{B}^{\mu}_{[\mathrm{red}]}$, and ξ^{rk}, $(r,k) \in \mathcal{B}^{\mu}$, satisfy the species localization property (3.1.3), we already know that the matrix $\mathcal{R}$ has the fundamental property that

$$\mathcal{R}^{rs}_{kl} = \mathcal{R}^{rs}_{kk}\,\delta_{kl}, \qquad (r,k) \in \mathcal{B}^{\mu}_{[\mathrm{red}]}, \quad (s,l) \in \mathcal{B}^{\mu}, \tag{3.1.74}$$

so that $\mathcal{R}$ leaves invariant the species indices. Denoting then

$$\mathcal{N}_{[\mathrm{red}]} = \mathrm{diag}\big((n_k)_{(r,k)\in\mathcal{B}^{\mu}_{[\mathrm{red}]}}\big), \tag{3.1.75}$$

it is straightforward to obtain from (3.1.74)(3.1.75) that

$$\mathcal{N}_{[\mathrm{red}]}\mathcal{R} = \mathcal{R}\mathcal{N}. \tag{3.1.76}$$

The relations (3.1.73)–(3.1.76) now yield that

$$G_{[\mathrm{red}]} = \mathcal{R}G\mathcal{R}^t = \mathcal{R}\mathcal{N}\widetilde{G}\mathcal{R}^t = \mathcal{N}_{[\mathrm{red}]}\mathcal{R}\widetilde{G}\mathcal{R}^t,$$

and that

$$\beta^\mu_{[\mathrm{red}]} = \mathcal{R}\beta^\mu = \mathcal{R}\mathcal{N}\widetilde{\beta}^\mu = \mathcal{N}_{[\mathrm{red}]}\mathcal{R}\widetilde{\beta}^\mu,$$

so that we may define $\widetilde{G}_{[\mathrm{red}]}$ and $\widetilde{\beta}^\mu_{[\mathrm{red}]}$ from the expressions

$$\begin{cases} \widetilde{G}_{[\mathrm{red}]} = \mathcal{R}\widetilde{G}\mathcal{R}^t, \\ \widetilde{\beta}^\mu_{[\mathrm{red}]} = \mathcal{R}\widetilde{\beta}^\mu. \end{cases} \tag{3.1.77}$$

From the above analysis, we then deduce that

$$\begin{cases} G_{[\mathrm{red}]} = \mathcal{N}_{[\mathrm{red}]}\widetilde{G}_{[\mathrm{red}]}, \\ \beta^\mu_{[\mathrm{red}]} = \mathcal{N}_{[\mathrm{red}]}\widetilde{\beta}^\mu_{[\mathrm{red}]}. \end{cases} \tag{3.1.78}$$

The relations (3.1.77) also imply that

$$db(\widetilde{G}_{[\mathrm{red}]}) = \mathcal{R}db(\widetilde{G})\mathcal{R}^t, \tag{3.1.79}$$

since $\mathcal{R}$ satisfies (3.1.74). From (3.1.74) we also deduce that the rescaled reduced systems still satisfy the relations (3.1.12).

In order to derive the properties of the rescaled matrix $\widetilde{G}_{[\mathrm{red}]}$, we will now use the relations (3.1.77), the relation (3.1.79), and the results obtained in Sections 3.1.1 to 3.1.6 for the system $\widetilde{G}\alpha^\mu = \widetilde{\beta}^\mu$. We first introduce the partitioning $\mathcal{B}^\mu_{[\mathrm{red}]} = \mathcal{B}^{\mu+}_{[\mathrm{red}]} \cup \mathcal{B}^{\mu-}_{[\mathrm{red}]}$ where

$$\begin{cases} \mathcal{B}^{\mu+}_{[\mathrm{red}]} = \{\ (r,k) \in \mathcal{B}^\mu_{[\mathrm{red}]},\ k \in \mathcal{S}^+\ \}, \\ \mathcal{B}^{\mu-}_{[\mathrm{red}]} = \{\ (r,k) \in \mathcal{B}^\mu_{[\mathrm{red}]},\ k \in \mathcal{S}^-\ \}, \end{cases} \tag{3.1.80}$$

and we denote by $\Gamma^\mu_{[\mathrm{red}]}$ the permutation matrix associated with the reordering of $\mathcal{B}^\mu_{[\mathrm{red}]}$ into $(\mathcal{B}^{\mu+}_{[\mathrm{red}]}, \mathcal{B}^{\mu-}_{[\mathrm{red}]})$. We denote by $\omega^+_{[\mathrm{red}]}$ the number of elements of $\mathcal{B}^{\mu+}_{[\mathrm{red}]}$ and $\omega^-_{[\mathrm{red}]}$ the number of elements of $\mathcal{B}^{\mu-}_{[\mathrm{red}]}$, so that $\omega_{[\mathrm{red}]} = \omega^+_{[\mathrm{red}]} + \omega^-_{[\mathrm{red}]}$. It is then easy to check that we have the block-decomposition

$$(\Gamma^\mu_{[\mathrm{red}]})^t\mathcal{R}\Gamma^\mu = \begin{bmatrix} \mathcal{R}^{++} & 0 \\ 0 & \mathcal{R}^{--} \end{bmatrix}, \tag{3.1.81}$$

where $\mathcal{R}^{++}$ and $\mathcal{R}^{--}$ are the transformation matrices such that

$$\begin{cases} \zeta^{rk} = \displaystyle\sum_{(s,l)\in\mathcal{B}^{\mu+}} \mathcal{R}^{rs++}_{kl}\xi^{sl}, & (r,k) \in \mathcal{B}^{\mu+}_{[\mathrm{red}]}, \\ \zeta^{rk} = \displaystyle\sum_{(s,l)\in\mathcal{B}^{\mu-}} \mathcal{R}^{rs--}_{kl}\xi^{sl}, & (r,k) \in \mathcal{B}^{\mu-}_{[\mathrm{red}]}. \end{cases} \tag{3.1.82}$$

Notice that there are no blocks $\mathcal{R}^{+-}$ and $\mathcal{R}^{-+}$ in the decomposition (3.1.81) since the matrix $\mathcal{R}$ satisfies (3.1.74), or, equivalently, since the basis functions ζ^{rk}, $(r,k) \in \mathcal{B}^{\mu}_{[\mathrm{red}]}$, and ξ^{rk}, $(r,k) \in \mathcal{B}^{\mu}$, satisfy the species localization property (3.1.3). For the same reason, the partial transformation matrices $\mathcal{R}^{++}$ and $\mathcal{R}^{--}$ satisfy

$$\begin{cases} \mathcal{R}^{rs++}_{kl} = \mathcal{R}^{rs++}_{kk}\delta_{kl}, & (r,k) \in \mathcal{B}^{\mu+}_{[\mathrm{red}]}, \quad (s,l) \in \mathcal{B}^{\mu+}, \\ \mathcal{R}^{rs--}_{kl} = \mathcal{R}^{rs--}_{kk}\delta_{kl}, & (r,k) \in \mathcal{B}^{\mu-}_{[\mathrm{red}]}, \quad (s,l) \in \mathcal{B}^{\mu-}. \end{cases} \tag{3.1.83}$$

As a consequence, we deduce from the block-decompositions of G, $\widetilde{G}$, and $db(\widetilde{G})$ that we have the following block-decompositions for $G_{[\mathrm{red}]}$, $\widetilde{G}_{[\mathrm{red}]}$, and $db(\widetilde{G}_{[\mathrm{red}]})$

$$(\Gamma^{\mu}_{[\mathrm{red}]})^t G_{[\mathrm{red}]} \Gamma^{\mu}_{[\mathrm{red}]} = \begin{bmatrix} G^{++}_{[\mathrm{red}]} & 0 \\ 0 & 0 \end{bmatrix}, \tag{3.1.84}$$

$$(\Gamma^{\mu}_{[\mathrm{red}]})^t \widetilde{G}_{[\mathrm{red}]} \Gamma^{\mu}_{[\mathrm{red}]} = \begin{bmatrix} \widetilde{G}^{++}_{[\mathrm{red}]} & 0 \\ \widetilde{G}^{-+}_{[\mathrm{red}]} & \widetilde{G}^{--}_{[\mathrm{red}]} \end{bmatrix}, \tag{3.1.85}$$

and

$$(\Gamma^{\mu}_{[\mathrm{red}]})^t \, db(\widetilde{G}_{[\mathrm{red}]}) \Gamma^{\mu}_{[\mathrm{red}]} = \begin{bmatrix} db(\widetilde{G}^{++}_{[\mathrm{red}]}) & 0 \\ 0 & db(\widetilde{G}^{--}_{[\mathrm{red}]}) \end{bmatrix}, \tag{3.1.86}$$

where

$$\begin{cases} G^{++}_{[\mathrm{red}]} = \mathcal{R}^{++} G^{++} (\mathcal{R}^{++})^t, \\ \widetilde{G}^{++}_{[\mathrm{red}]} = \mathcal{R}^{++} \widetilde{G}^{++} (\mathcal{R}^{++})^t, \\ \widetilde{G}^{-+}_{[\mathrm{red}]} = \mathcal{R}^{--} \widetilde{G}^{-+} (\mathcal{R}^{++})^t, \\ \widetilde{G}^{--}_{[\mathrm{red}]} = \mathcal{R}^{--} \widetilde{G}^{--} (\mathcal{R}^{--})^t, \end{cases} \tag{3.1.87}$$

and

$$\begin{cases} db(\widetilde{G}^{++}_{[\mathrm{red}]}) = \mathcal{R}^{++} db(\widetilde{G}^{++}) (\mathcal{R}^{++})^t, \\ db(\widetilde{G}^{--}_{[\mathrm{red}]}) = \mathcal{R}^{--} db(\widetilde{G}^{--}) (\mathcal{R}^{--})^t, \end{cases} \tag{3.1.88}$$

so that we also have

$$2db(G^{++}_{[\mathrm{red}]}) - G^{++}_{[\mathrm{red}]} = \mathcal{R}^{++}\big(2db(G^{++}) - G^{++}\big)(\mathcal{R}^{++})^t. \tag{3.1.89}$$

In addition, it is straightforward to check that

$$\begin{cases} G^{++}_{[\mathrm{red}]} = \mathcal{N}^{++}_{[\mathrm{red}]} \widetilde{G}^{++}_{[\mathrm{red}]}, \\ db(G^{++}_{[\mathrm{red}]}) = \mathcal{N}^{++}_{[\mathrm{red}]} db(\widetilde{G}^{++}_{[\mathrm{red}]}), \\ db(G^{++}_{[\mathrm{red}]}) = (\mathcal{N}^{++}_{[\mathrm{red}]})^{1/2} db(\widetilde{G}^{++}_{[\mathrm{red}]}) (\mathcal{N}^{++}_{[\mathrm{red}]})^{1/2}, \end{cases}$$

so that we have

$$\left\{\begin{aligned} \widetilde{G}^{++}_{[\mathrm{red}]} &= (\mathcal{N}^{++}_{[\mathrm{red}]})^{-1} G^{++}_{[\mathrm{red}]}, \\ db(\widetilde{G}^{++}_{[\mathrm{red}]}) &= (\mathcal{N}^{++}_{[\mathrm{red}]})^{-1} db(G^{++}_{[\mathrm{red}]}), \\ db(\widetilde{G}^{++}_{[\mathrm{red}]}) &= (\mathcal{N}^{++}_{[\mathrm{red}]})^{-1/2} db(G^{++}_{[\mathrm{red}]}) (\mathcal{N}^{++}_{[\mathrm{red}]})^{-1/2}. \end{aligned}\right. \tag{3.1.90}$$

As a consequence, we can now obtain the properties of the various reduced blocks from the properties of the corresponding blocks of $\widetilde{G}$. From (3.1.83)(3.1.87) we deduce in particular that $\widetilde{G}^{--}_{[\mathrm{red}]}$ is block-diagonal and symmetric positive definite on $\mathbb{R}^{\omega^-_{[\mathrm{red}]}}$. Similarly, from (3.1.87), we deduce that we have $x^+_{[\mathrm{red}]} \in N(G^{++}_{[\mathrm{red}]})$ if and only if $(\mathcal{R}^{++})^t x^+_{[\mathrm{red}]} \in N(G^{++})$. Assuming that $N(G^{++}) \cap R((\mathcal{R}^{++})^t) = \{0\}$ yields in particular that $N(G^{++}_{[\mathrm{red}]}) = \{0\}$ since $(\mathcal{R}^{++})^t x_{[\mathrm{red}]} = 0$ implies $x^+_{[\mathrm{red}]} = 0$ thanks to the fact that the ζ^{rk}, $(r,k) \in \mathcal{B}^{\mu+}_{[\mathrm{red}]}$, form a basis, so that

$$N(G^{++}) \cap R((\mathcal{R}^{++})^t) = \{0\} \quad \Longrightarrow \quad N(G^{++}_{[\mathrm{red}]}) = \{0\}. \tag{3.1.91}$$

Similarly, we have

$$N\big(2db(G^{++}) - G^{++}\big) \cap R((\mathcal{R}^{++})^t) = \{0\} \quad \Longrightarrow \quad N\big(2db(G^{++}_{[\mathrm{red}]}) - G^{++}_{[\mathrm{red}]}\big) = \{0\}. \tag{3.1.92}$$

In this situation, $G^{++}_{[\mathrm{red}]}$, $2db(G^{++}_{[\mathrm{red}]}) - G^{++}_{[\mathrm{red}]}$, and $db(G^{++}_{[\mathrm{red}]})$ are symmetric positive definite for any $n^+ \geq 1$. In addition, $\widetilde{G}^{++}_{[\mathrm{red}]}$ is then nonsingular and $db(\widetilde{G}^{++}_{[\mathrm{red}]})$ is symmetric positive definite. Finally, the matrix $\widetilde{G}_{[\mathrm{red}]}$ is nonsingular and the matrix $db(\widetilde{G}_{[\mathrm{red}]})$ is symmetric positive definite for any $n^+ \geq 1$. The properties (3.1.71)–(3.1.92) will be very useful in order to obtain the properties of the rescaled systems corresponding to the functional spaces described in (2.3.47).

Remark. The reduced rescaled systems have been defined from (3.1.77), and this is the only straightforward definition. In particular, we point out that they cannot be defined directly from the expressions $\langle\!\langle\!\langle \widehat{\zeta}^{rk}, \Im(\widehat{\zeta}^{sl}) \rangle\!\rangle\!\rangle$, $(r,k),(s,l) \in \mathcal{B}^{\mu}_{[\mathrm{red}]}$, since, in general, we have $\langle\!\langle\!\langle \widehat{\zeta}^{rk}, \Im(\widehat{\zeta}^{sl}) \rangle\!\rangle\!\rangle \neq \langle\!\langle\!\langle \zeta^{rk}, \Im(\zeta^{sl}) \rangle\!\rangle\!\rangle$, $(r,k),(s,l) \in \mathcal{B}^{\mu}_{[\mathrm{red}]}$, as opposed to $\langle\!\langle \widehat{\zeta}^{rk}, \Im(\widehat{\zeta}^{sl}) \rangle\!\rangle = \langle\!\langle \zeta^{rk}, \Im(\zeta^{sl}) \rangle\!\rangle$, $(r,k),(s,l) \in \mathcal{B}^{\mu}_{[\mathrm{red}]}$.

3.1.8 The Symmetric Rescaled System $\widehat{G}\widehat{\alpha}^{\mu} = \widehat{\beta}^{\mu}$

In this section we introduce a new rescaled system denoted by $\widehat{G}\widehat{\alpha}^{\mu} = \widehat{\beta}^{\mu}$ and briefly summarize its main properties. This system will be useful in Chapter 5 in order to derive stabilized versions, for vanishing mass fractions, of preconditioned conjugate gradient algorithms.

The modified matrix $\widehat{G} \in \mathbb{R}^{\omega,\omega}$, the modified right member $\widehat{\beta}^{\mu} \in \mathbb{R}^{\omega}$, and the modified constraint vectors $\widehat{\mathcal{G}}^{l\nu} \in \mathbb{R}^{\omega}$, $(l,\nu) \in [1,n+4]\times[1,\tau_{\mu}]$, associated with the transport linear system (3.1.4), are simply defined by

$$\begin{cases} \widehat{G}_{kk}^{rs} = \sum_{l\in\mathcal{S}} n_l[\xi^{rk},\xi^{sk}]'_{kl} + n_k[\xi^{rk},\xi^{sk}]''_{kk}, & (r,k),(s,k) \in \mathcal{B}^{\mu}, \\ \widehat{G}_{kl}^{rs} = \sqrt{n_k n_l}\,[\xi^{rk},\xi^{sl}]''_{kl}, & (r,k),(s,l) \in \mathcal{B}^{\mu}, \quad k \neq l, \\ \widehat{\beta}_k^{r\mu} = \sqrt{n_k}\,\widetilde{\beta}_k^{r\mu}, & (r,k) \in \mathcal{B}^{\mu}, \\ \widehat{\mathcal{G}}_k^{rl\nu} = \sqrt{n_k}\,\widetilde{\mathcal{G}}_k^{rl\nu}, & (r,k) \in \mathcal{B}^{\mu}, \quad (l,\nu) \in [1,n+4]\times[1,\tau_{\mu}], \end{cases} \tag{3.1.93}$$

where we have already defined $\widetilde{G}$ and $\widetilde{\beta}^{\mu}$ in (3.1.8) and $\widetilde{\mathcal{G}}^{l\nu}$ in (3.1.31). Note that the rescaled right member $\widehat{\beta}^{\mu}$ is defined as long as $\widetilde{\beta}^{\mu}$ is defined, that is, for μ equal to η, κ, D_l, $l \in \mathcal{S}$, and λ' when all the mass fractions are positive, but only for μ equal to η, κ and λ', when there are zero mass fractions. The case of diffusion coefficients for zero mass fractions is again to be considered separately.

From these relations and by using (3.1.10) and (3.1.32), we obtain that

$$\sqrt{n_k n_l}\,\widehat{G}_{kl}^{rs} = G_{kl}^{rs},$$

$$\sqrt{n_k}\,\widehat{\beta}_k^{r\mu} = n_k\,\widetilde{\beta}_k^{r\mu} = \beta_k^{r\mu},$$

and

$$\sqrt{n_k}\,\widehat{\mathcal{G}}_k^{rl\nu} = n_k\,\widetilde{\mathcal{G}}_k^{rl\nu} = \mathcal{G}_k^{rl\nu},$$

so that we have the relations

$$\begin{cases} \mathcal{N}^{1/2}\,\widehat{G}\,\mathcal{N}^{1/2} = G, \\ \mathcal{N}^{1/2}\,\widehat{\beta}^{\mu} = \beta^{\mu}, \\ \mathcal{N}^{1/2}\,\widehat{\mathcal{G}}^{l\nu} = \mathcal{G}^{l\nu}, & (l,\nu) \in [1,n+4]\times[1,\tau_{\mu}]. \end{cases} \tag{3.1.94}$$

When all the mass fractions are positive, the matrix $\widehat{G}$, the right members $\widehat{\beta}^{\mu}$, and the constraints vectors $\widehat{\mathcal{G}}^{l\nu}$, $(l,\nu) \in [1,n+4]\times[1,\tau_{\mu}]$, are easily evaluated from (3.1.94), since the matrix $\mathcal{N}$ is then invertible and therefore, $\widehat{G} = \mathcal{N}^{-1/2}\,G\,\mathcal{N}^{-1/2}$, $\widehat{\beta}^{\mu} = \mathcal{N}^{-1/2}\beta^{\mu}$, and $\widehat{\mathcal{G}}^{l\nu} = \mathcal{N}^{-1/2}\mathcal{G}^{l\nu}$. Only definitions (3.1.94) are valid, however, when some mass fractions are zero, since $\mathcal{N}^{1/2}$ is then singular. Note also that the matrix $\widehat{G}$ is symmetric, but not smooth with respect to the number densities.

The symmetric rescaled version of the general transport linear system (3.1.4) is then the rescaled system

$$\widehat{G}\widehat{\alpha}^{\mu} = \widehat{\beta}^{\mu}, \tag{3.1.95}$$

under the rescaled constraints

$$\langle \widehat{\mathcal{G}}^{l\nu}, \widehat{\alpha}^{\mu} \rangle = 0, \qquad (l,\nu) \in [1, n{+}4] \times [1, \tau_{\mu}], \tag{3.1.96}$$

and we have the equivalence result that when the mass fractions are positive, i.e., when the matrix $\mathcal{N}$ is invertible, then α^{μ} satisfies (3.1.4) if and only if $\widehat{\alpha}^{\mu} = \mathcal{N}^{1/2}\alpha^{\mu}$ satisfies (3.1.95)(3.1.96).

From (3.1.93), the matrix $\widehat{G}$, after the reordering induced by Γ^{μ}, admits the block-decomposition

$$(\Gamma^{\mu})^t \widehat{G} \Gamma^{\mu} = \begin{bmatrix} \widehat{G}^{++} & 0 \\ 0 & \widehat{G}^{--} \end{bmatrix}, \tag{3.1.97}$$

and $\widehat{G}^{++}$ and $\widehat{G}^{--}$ are symmetric. Moreover, the matrix $\widehat{G}^{--}$ is block-diagonal, symmetric positive definite on $\mathbb{R}^{\omega^-}$ since $\widehat{G}^{--} = db(\widehat{G}^{--}) = \widetilde{G}^{--}$ from (3.1.93) and since $\widetilde{G}^{--}$ is positive definite from (3.1.28). We also have $db(\widehat{G}^{++}) = db(\widetilde{G}^{++})$ from (3.1.93) so that

$$db(\widehat{G}) = db(\widetilde{G}). \tag{3.1.98}$$

From (3.1.97) we also deduce that

$$\left\{ \begin{aligned} \widehat{G}^{++} &= \big(\mathcal{N}^{++}\big)^{-1/2} G^{++} \big(\mathcal{N}^{++}\big)^{-1/2}, \\ \widehat{G}^{++} &= \big(\mathcal{N}^{++}\big)^{1/2} \widetilde{G}^{++} \big(\mathcal{N}^{++}\big)^{-1/2}, \\ db(\widehat{G}^{++}) &= \big(\mathcal{N}^{++}\big)^{-1/2} db(G^{++}) \big(\mathcal{N}^{++}\big)^{-1/2}, \end{aligned} \right. \tag{3.1.99}$$

and from (3.1.93) we also obtain that $\widehat{\beta}^{\mu+} = \big(\mathcal{N}^{++}\big)^{-1/2}\beta^{\mu+}$. As a consequence, the results obtained in Sections 3.1.1 to 3.1.7 are easily rewritten for $\widehat{G}$. This will be done when needed in Chapter 4 and Chapter 5.

3.1.9 Number Density Independent Characterizations

In this section we investigate how some properties of the matrices G, $2db(G) - G$, and $\widetilde{G}$ can be expressed independently of the number densities by using the relations

(3.1.11)–(3.1.13). We restate that the coefficients G'^{rs}_{kl}, $(r,k),(s,k) \in \mathcal{B}^\mu$, $l \in \mathcal{S}$, and G''^{rs}_{kl}, $(r,k),(s,l) \in \mathcal{B}^\mu$, are independent of the number densities $(n_k)_{k\in\mathcal{S}}$, and that they satisfy the symmetry relations $G'^{rs}_{kl} = G'^{sr}_{kl}$, $(r,k),(s,k) \in \mathcal{B}^\mu$, $l \in \mathcal{S}$, and $G''^{rs}_{kl} = G''^{sr}_{lk}$, $(r,k),(s,l) \in \mathcal{B}^\mu$.

For $x \in \mathbb{R}^\omega$, a direct calculation yields that

$$\langle Gx,x\rangle = \sum_{k,l\in\mathcal{S}} \tfrac{1}{2} n_k n_l \Big(\sum_{\substack{r\in\mathcal{F}_k\\ s\in\mathcal{F}_k}} G'^{rs}_{kl} x^r_k x^s_k + 2 \sum_{\substack{r\in\mathcal{F}_k\\ s\in\mathcal{F}_l}} G''^{rs}_{kl} x^r_k x^s_l + \sum_{\substack{r\in\mathcal{F}_l\\ s\in\mathcal{F}_l}} G'^{rs}_{lk} x^r_l x^s_l \Big),$$

and if we define, for any pair $k,l \in \mathcal{S}$, the quadratic form

$$\mathcal{Q}_{kl} = \sum_{\substack{r\in\mathcal{F}_k\\ s\in\mathcal{F}_k}} G'^{rs}_{kl} x^r_k x^s_k + 2 \sum_{\substack{r\in\mathcal{F}_k\\ s\in\mathcal{F}_l}} G''^{rs}_{kl} x^r_k x^s_l + \sum_{\substack{r\in\mathcal{F}_l\\ s\in\mathcal{F}_l}} G'^{rs}_{lk} x^r_l x^s_l,$$

we then have

$$\langle Gx,x\rangle = \sum_{k,l\in\mathcal{S}} \tfrac{1}{2} n_k n_l \mathcal{Q}_{kl}.$$

Since the quadratic forms $\mathcal{Q}_{kl}$, $k,l \in \mathcal{S}$, are independent of the number densities, and since the number densities may be chosen positive, but arbitrarily small, we deduce that G is positive semi-definite for any positive number densities if and only if all the quadratic forms $\mathcal{Q}_{kl}$ are positive semi-definite. The global property of G therefore depends on the quadratics $\mathcal{Q}_{kl}$, $k,l \in \mathcal{S}$, which are independent of $(n_k)_{k\in\mathcal{S}}$.

Similarly, for $x \in \mathbb{R}^\omega$, a direct calculation yields that

$$\Big\langle \big(2db(G) - G\big)x, x \Big\rangle =$$

$$\sum_{\substack{k,l\in\mathcal{S}\\ k\neq l}} \tfrac{1}{2} n_k n_l \Big(\sum_{\substack{r\in\mathcal{F}_k\\ s\in\mathcal{F}_k}} G'^{rs}_{kl} x^r_k x^s_k - 2 \sum_{\substack{r\in\mathcal{F}_k\\ s\in\mathcal{F}_l}} G''^{rs}_{kl} x^r_k x^s_l + \sum_{\substack{r\in\mathcal{F}_l\\ s\in\mathcal{F}_l}} G'^{rs}_{lk} x^r_l x^s_l \Big)$$

$$+ \sum_{k\in\mathcal{S}} n_k^2 \Big(\sum_{\substack{r\in\mathcal{F}_k\\ s\in\mathcal{F}_k}} G'^{rs}_{kk} x^r_k x^s_k + \sum_{\substack{r\in\mathcal{F}_k\\ s\in\mathcal{F}_k}} G''^{rs}_{kk} x^r_k x^s_k \Big),$$

so that we have

$$\Big\langle \big(2db(G) - G\big)x, x \Big\rangle = \sum_{\substack{k,l\in\mathcal{S}\\ k\neq l}} \tfrac{1}{2} n_k n_l \widetilde{\mathcal{Q}}_{kl} + \sum_{k\in\mathcal{S}} \tfrac{1}{2} n_k^2 \mathcal{Q}_{kk},$$

where we have defined

$$\widetilde{\mathcal{Q}}_{kl} = \sum_{\substack{r\in\mathcal{F}_k\\ s\in\mathcal{F}_k}} G'^{rs}_{kl} x^r_k x^s_k - 2 \sum_{\substack{r\in\mathcal{F}_k\\ s\in\mathcal{F}_l}} G''^{rs}_{kl} x^r_k x^s_l + \sum_{\substack{r\in\mathcal{F}_l\\ s\in\mathcal{F}_l}} G'^{rs}_{lk} x^r_l x^s_l,$$

and where we have used the property

$$\sum_{\substack{r\in\mathcal{F}_k\\ s\in\mathcal{F}_k}} G'^{rs}_{kk} x^r_k x^s_k + \sum_{\substack{r\in\mathcal{F}_k\\ s\in\mathcal{F}_k}} G''^{rs}_{kk} x^r_k x^s_k = \tfrac{1}{2}\mathcal{Q}_{kk}.$$

This shows that the matrix $2db(G)-G$ is positive semi-definite for any positive number densities if and only if all the quadratic forms $\widetilde{\mathcal{Q}}_{kl}$, $k,l\in\mathcal{S}$, $k\neq l$, and $\mathcal{Q}_{kk}$, $k\in\mathcal{S}$, are positive semi-definite. However, since we have

$$\widetilde{\mathcal{Q}}_{kl}\big((x^r_k)_{r\in\mathcal{F}_k},(x^s_l)_{s\in\mathcal{F}_l}\big) = \mathcal{Q}_{kl}\big((x^r_k)_{r\in\mathcal{F}_k},(-x^s_l)_{s\in\mathcal{F}_l}\big),$$

with obvious notation, we deduce that the matrix $2db(G)-G$ is positive semi-definite for any positive number densities if and only if all the quadratic forms $\mathcal{Q}_{kl}$, $k,l\in\mathcal{S}$, are positive semi-definite. On the other hand, we also have that $2db(G)-G$ is positive definite for any positive number densities if and only if the quadratic form

$$\sum_{\substack{k,l\in\mathcal{S}\\ k\neq l}} \tfrac{1}{2}\widetilde{\mathcal{Q}}_{kl} + \sum_{k\in\mathcal{S}} \tfrac{1}{2}\mathcal{Q}_{kk},$$

which is independent of $(n_k)_{k\in\mathcal{S}}$, is positive definite.

Finally, a direct calculation shows that for nonnegative mass fractions we have the relation

$$\langle x^-, \widetilde{G}^{--}x^-\rangle = \sum_{\substack{k\in\mathcal{S}^-\\ l\in\mathcal{S}^+}} n_l\Big(\sum_{\substack{r\in\mathcal{F}_k\\ s\in\mathcal{F}_k}} G'^{rs}_{kl} x^r_k x^s_k\Big),$$

so that

$$\langle x^-, \widetilde{G}^{--}x^-\rangle = \sum_{\substack{k\in\mathcal{S}^-\\ l\in\mathcal{S}^+}} n_l \mathcal{H}_{kl},$$

where we have introduced, for $k,l\in\mathcal{S}$, the quadratic forms

$$\mathcal{H}_{kl} = \sum_{\substack{r\in\mathcal{F}_k\\ s\in\mathcal{F}_k}} G'^{rs}_{kl} x^r_k x^s_k.$$

We thus deduce that the matrices $\widetilde{G}^{--}$ are positive definite if and only if all the quadratic forms $\mathcal{H}_{kl}$, which are independent of $(n_k)_{k\in\mathcal{S}}$, are positive definite.

3.2 The Shear Viscosity

3.2.1 The Rescaled System $\widetilde{H}\alpha^\eta = \widetilde{\beta}^\eta$

The rescaled system associated with $H\alpha^\eta = \beta^\eta$ is denoted by $\widetilde{H}\alpha^\eta = \widetilde{\beta}^\eta$ and is defined, for convenience, by $\widetilde{H}_{kl}^{rs} = (2/5\bar{p})\langle\!\langle\!\langle \phi^{20rk}, \Im(\phi^{20sl})\rangle\!\rangle\!\rangle$, $(r,k),(s,l) \in \mathcal{B}^\eta$, and $\widetilde{\beta}_k^{r\eta} = (1/5)\langle\!\langle\!\langle \phi^{20rk}, \Psi^\eta\rangle\!\rangle\!\rangle$, $(r,k) \in \mathcal{B}^\eta$. Denoting by $\mathcal{X}^\eta$ the diagonal matrix $\mathcal{X}^\eta = \mathrm{diag}\big((X_k)_{(r,k)\in\mathcal{B}^\eta}\big)$

$$\mathcal{X}^\eta = \mathrm{diag}(X_1, \ldots, X_n), \tag{3.2.1}$$

we then have

$$\mathcal{X}^\eta \widetilde{H} = H, \tag{3.2.2}$$

and

$$\mathcal{X}^\eta \widetilde{\beta}^\eta = \beta^\eta. \tag{3.2.3}$$

A direct calculation yields that

$$\widetilde{H}_{kk}^{0000} = \sum_{\substack{l\in\mathcal{S}\\ l\neq k}} \frac{16}{5k_{\mathrm{B}}\overline{T}} X_l \Big[\frac{10}{3}\frac{m_k m_l}{(m_k+m_l)^2}\Omega_{kl}^{(1,1)} + \frac{m_l^2}{(m_k+m_l)^2}\Omega_{kl}^{(2,2)}\Big]$$
$$+ \frac{8}{5k_{\mathrm{B}}\overline{T}} X_k \Omega_{kk}^{(2,2)}, \qquad k\in\mathcal{S}, \tag{3.2.4}$$

$$\widetilde{H}_{kl}^{0000} = \frac{16}{5k_{\mathrm{B}}\overline{T}} X_l \frac{m_k m_l}{(m_k+m_l)^2}\Big[-\frac{10}{3}\Omega_{kl}^{(1,1)} + \Omega_{kl}^{(2,2)}\Big], \qquad k,l\in\mathcal{S}, \quad k\neq l, \tag{3.2.5}$$

and

$$\widetilde{\beta}_k^{00\eta} = 1, \quad k\in\mathcal{S}. \tag{3.2.6}$$

Finally, denoting by α^η the solution of

$$\widetilde{H}\alpha^\eta = \widetilde{\beta}^\eta, \tag{3.2.7}$$

the shear viscosity η is defined by

$$\eta = \langle \alpha^\eta, \beta^\eta\rangle = \langle \alpha^{00\eta}, \beta^{00\eta}\rangle = \sum_{k\in\mathcal{S}} X_k \alpha_k^{00\eta}. \tag{3.2.8}$$

When the mass fractions are positive, $\mathcal{X}^\eta$ is invertible, and we have $\widetilde{H}\alpha^\eta = \widetilde{\beta}^\eta$ if and only if $H\alpha^\eta = \beta^\eta$, since $\widetilde{H} = (\mathcal{X}^\eta)^{-1}H$ and $\widetilde{\beta}^\eta = (\mathcal{X}^\eta)^{-1}\beta^\eta$. As a consequence, the definition (2.4.10) of the shear viscosity is equivalent to the definition (3.2.8).

3.2.2 Mathematical Structure of the Rescaled System $\widetilde{H}\alpha^\eta = \widetilde{\beta}^\eta$

Properties of $\mathcal{A}^\eta$. The functional space $\mathcal{A}^\eta$ introduced in Section 2.4 to approximate ϕ^η satisfies the perpendicularity property (3.1.1), the species orthogonality property (3.1.2), and the relation (3.1.56) from (2.3.54). In addition, the basis functions satisfy the species localization property (3.1.3). As a consequence, the results (3.1.58)(3.1.59) established in Section 3.1.5 apply.

Properties of H. Denoting by Γ^η the permutation matrix associated with the reordering of $\mathcal{B}^\eta$ into $(\mathcal{B}^{\eta+}, \mathcal{B}^{\eta-})$, the matrix H admits the block-decomposition

$$(\Gamma^\eta)^t H \Gamma^\eta = \begin{bmatrix} H^{++} & 0 \\ 0 & 0 \end{bmatrix}.$$

Furthermore, the matrices H^{++}, $db(H^{++})$, and $2db(H^{++}) - H^{++}$ are symmetric positive definite for any $n^+ \geq 1$.

Properties of $\widetilde{H}$. The matrix $\widetilde{H}$ admits the block-decomposition

$$(\Gamma^\eta)^t \widetilde{H} \Gamma^\eta = \begin{bmatrix} \widetilde{H}^{++} & 0 \\ \widetilde{H}^{-+} & \widetilde{H}^{--} \end{bmatrix}.$$

The matrix $\widetilde{H}^{++}$ is nonsingular and the matrix $\widetilde{H}^{--}$ is block-diagonal, i.e., $\widetilde{H}^{--} = db(\widetilde{H}^{--})$, and is symmetric positive definite, so that $\widetilde{H}$ is nonsingular for any $n^+ \geq 1$.

The matrix $db(\widetilde{H})$ admits the block-decomposition

$$(\Gamma^\eta)^t\, db(\widetilde{H}) \Gamma^\eta = \begin{bmatrix} db(\widetilde{H}^{++}) & 0 \\ 0 & db(\widetilde{H}^{--}) \end{bmatrix}.$$

The matrix $db(\widetilde{H}^{++})$ is symmetric positive definite for any $n^+ \geq 1$, and so is the matrix $db(\widetilde{H}^{--})$. As a consequence, the matrix $db(\widetilde{H})$ is symmetric positive definite for any $n^+ \geq 1$.

3.3 The Volume Viscosity

In this section we investigate the rescaled systems $\widetilde{K}\alpha^\kappa = \widetilde{\beta}^\kappa$ and $\widetilde{K}_{[01]}\alpha^\kappa_{[01]} = \widetilde{\beta}^\kappa_{[01]}$ associated with the evaluation of the volume viscosities κ and $\kappa_{[01]}$. The rescaled systems $\widetilde{K}_{[10]}\alpha^\kappa_{[10]} = \widetilde{\beta}^\kappa_{[10]}$ and $\widetilde{K}_{[d]}\alpha^\kappa_{[d]} = \widetilde{\beta}^\kappa_{[d]}$, associated with the volume viscosities

$\kappa_{[10]}$ and $\kappa_{[\mathrm{d}]}$, respectively, are not considered, since these volume viscosities are not accurate.

3.3.1 The Rescaled System $\widetilde{K}\alpha^\kappa = \widetilde{\beta}^\kappa$

The rescaled system associated with $K\alpha^\kappa = \beta^\kappa$ is denoted by $\widetilde{K}\alpha^\kappa = \widetilde{\beta}^\kappa$ and is defined, for convenience, by $\widetilde{K}_{kl}^{rs} = (1/\bar{p})\langle\!\langle\!\langle \phi^{00rk}, \Im(\phi^{00sl})\rangle\!\rangle\!\rangle$, $(r,k),(s,l) \in \mathcal{B}^\kappa$, and $\widetilde{\beta}_k^\kappa = (-1/3)\langle\!\langle\!\langle \phi^{00rk}, \Psi^\kappa\rangle\!\rangle\!\rangle$, $(r,k) \in \mathcal{B}^\kappa$. If $\mathcal{X}^\kappa$ denotes the diagonal matrix $\mathcal{X}^\kappa = \mathrm{diag}\big((X_k)_{(r,k)\in\mathcal{B}^\kappa}\big)$

$$\mathcal{X}^\kappa = \mathrm{diag}(X_1, \dots, X_n, X_{i_1}, \dots, X_{i_p}), \tag{3.3.1}$$

where $\{i_1, \dots, i_p\} = \mathcal{P}$ are the polyatomic species, we have

$$\mathcal{X}^\kappa \widetilde{K} = K, \tag{3.3.2}$$

and

$$\mathcal{X}^\kappa \widetilde{\beta}^\kappa = \beta^\kappa. \tag{3.3.3}$$

A direct calculation then yields that

$$\widetilde{K}_{kk}^{1010} = \sum_{\substack{l\in\mathcal{S}\\ l\neq k}} \frac{4}{k_{\mathrm{B}}\overline{T}} X_l \Big[4\frac{m_k m_l}{(m_k+m_l)^2}\Omega_{kl}^{(1,1)} + \frac{m_l^2}{(m_k+m_l)^2}\big[(\Delta\epsilon_{kl})^2\big]_{kl}\Big] + \frac{2}{k_{\mathrm{B}}\overline{T}} X_k \big[(\Delta\epsilon_{kk})^2\big]_{kk}, \qquad k\in\mathcal{S}, \tag{3.3.4}$$

$$\widetilde{K}_{kl}^{1010} = \frac{4}{k_{\mathrm{B}}\overline{T}} X_l \frac{m_k m_l}{(m_k+m_l)^2}\Big[-4\Omega_{kl}^{(1,1)} + \big[(\Delta\epsilon_{kl})^2\big]_{kl}\Big], \quad k,l\in\mathcal{S}, \quad k\neq l, \tag{3.3.5}$$

$$\widetilde{K}_{kk}^{1001} = -\sum_{\substack{l\in\mathcal{S}\\ l\neq k}} \frac{4}{k_{\mathrm{B}}\overline{T}} X_l \frac{m_l}{(m_k+m_l)}\big[\Delta\epsilon_k \Delta\epsilon_{kl}\big]_{kl} - \frac{2}{k_{\mathrm{B}}\overline{T}} X_k \big[(\Delta\epsilon_{kk})^2\big]_{kk}, \qquad k\in\mathcal{P}, \tag{3.3.6}$$

$$\widetilde{K}_{kl}^{1001} = -\frac{4}{k_{\mathrm{B}}\overline{T}} X_l \frac{m_l}{m_k+m_l}[\![\widetilde{\Delta}\epsilon_l \Delta\epsilon_{kl}]\!]_{kl}, \qquad k\in\mathcal{S}, \quad l\in\mathcal{P}, \quad k\neq l, \tag{3.3.7}$$

$$\widetilde{K}_{kk}^{0110} = -\sum_{\substack{l\in\mathcal{S}\\ l\neq k}} \frac{4}{k_{\mathrm{B}}\overline{T}} X_l \frac{m_l}{(m_k+m_l)}\big[\Delta\epsilon_k \Delta\epsilon_{kl}\big]_{kl}$$

$$-\frac{2}{k_{\mathrm{B}}T}X_k\left[(\Delta\epsilon_{kk})^2\right]_{kk}, \qquad k\in\mathcal{P}, \tag{3.3.8}$$

$$\widetilde{K}^{0110}_{kl} = -\frac{4}{k_{\mathrm{B}}T}X_l\frac{m_k}{m_k+m_l}[\![\Delta\epsilon_k\Delta\epsilon_{kl}]\!]_{kl}, \qquad k\in\mathcal{P},\quad l\in\mathcal{S},\quad k\neq l, \tag{3.3.9}$$

$$\widetilde{K}^{0101}_{kk} = \sum_{\substack{l\in\mathcal{S}\\ l\neq k}}\frac{4}{k_{\mathrm{B}}T}X_l\left[(\Delta\epsilon_k)^2\right]_{kl} + \frac{2}{k_{\mathrm{B}}T}X_k\left[(\Delta\epsilon_{kk})^2\right]_{kk}, \qquad k\in\mathcal{P}, \tag{3.3.10}$$

$$\widetilde{K}^{0101}_{kl} = \frac{4}{k_{\mathrm{B}}T}X_l\left[\Delta\epsilon_k\widetilde{\Delta}\epsilon_l\right]_{kl}, \qquad k,l\in\mathcal{P},\quad k\neq l, \tag{3.3.11}$$

and that

$$\widetilde{\beta}^{10\kappa}_k = \frac{c^{\mathrm{int}}}{c_v}, \qquad k\in\mathcal{S}, \tag{3.3.12}$$

$$\widetilde{\beta}^{01\kappa}_k = -\frac{c^{\mathrm{int}}_k}{c_v}, \qquad k\in\mathcal{P}. \tag{3.3.13}$$

Finally, denoting by α^κ the solution of the system

$$\begin{cases} \widetilde{K}\alpha^\kappa = \widetilde{\beta}^\kappa, \\ \langle\mathcal{K},\alpha^\kappa\rangle = 0, \end{cases} \tag{3.3.14}$$

where the constraint vector $\mathcal{K}$ has already been introduced in Section 2.5.1, the volume viscosity κ is defined by

$$\kappa = \langle\alpha^\kappa,\beta^\kappa\rangle = \langle\alpha^{10\kappa},\beta^{10\kappa}\rangle + \langle\alpha^{01\kappa},\beta^{01\kappa}\rangle, \tag{3.3.15}$$

which can be simplified into the relations

$$\kappa = \sum_{k\in\mathcal{S}}X_k\alpha^{10\kappa}_k = -\sum_{k\in\mathcal{P}}X_k(c^{\mathrm{int}}_k/c^{\mathrm{tr}}_v)\alpha^{01\kappa}_k, \tag{3.3.16}$$

by explicitly using the constraint in (3.3.14). When the mass fractions are positive, $\mathcal{X}^\kappa$ is invertible and we have the relations $\widetilde{K} = (\mathcal{X}^\kappa)^{-1}K$ and $\widetilde{\beta}^\kappa = (\mathcal{X}^\kappa)^{-1}\beta^\kappa$ so that $\widetilde{K}\alpha^\kappa = \widetilde{\beta}^\kappa$ if and only if $K\alpha^\kappa = \beta^\kappa$. As a consequence, the definition (2.5.16) of the volume viscosity is equivalent to the definition (3.3.15).

3.3.2 Mathematical Structure of the Rescaled System $\widetilde{K}\alpha^\kappa = \widetilde{\beta}^\kappa$

Properties of $\mathcal{A}^\kappa$. The functional space $\mathcal{A}^\kappa$ introduced in Section 2.5 to approximate ϕ^κ satisfies the perpendicularity property (3.1.1), the species orthogonality property

(3.1.2), and the relation (3.1.60) from (2.3.46). In addition, the basis functions satisfy the species localization property (3.1.3). As a consequence, the results (3.1.64)–(3.1.67) established in Section 3.1.6 apply.

Properties of K. Denoting by Γ^κ the permutation matrix associated with the reordering of $\mathcal{B}^\kappa$ into $(\mathcal{B}^{\kappa +}, \mathcal{B}^{\kappa -})$, the matrix K admits the block-decomposition

$$(\Gamma^\kappa)^t K \Gamma^\kappa = \begin{bmatrix} K^{++} & 0 \\ 0 & 0 \end{bmatrix}.$$

The matrix K^{++} is symmetric positive semi-definite, and its nullspace is given by $N(K^{++}) = \mathbb{R}\mathcal{V}^+$, where $\mathcal{V}_k^{10} = 1$, $k \in \mathcal{S}$, and $\mathcal{V}_k^{01} = 1$, $k \in \mathcal{P}$. The matrix K^{++} is positive definite on the constrained space $(\mathcal{K}^+)^\perp$ where $\mathcal{K}$ is given by $\mathcal{K}_k^{10} = X_k c_k^{\mathrm{tr}}$, $k \in \mathcal{S}$, $\mathcal{K}_k^{01} = X_k c_k^{\mathrm{int}}$, $k \in \mathcal{P}$. Moreover we have $\langle \mathcal{K}^+, \mathcal{V}^+ \rangle \neq 0$. We also have $\beta^{\kappa +} \in R(K^{++})$, that is, $\langle \beta^{\kappa +}, \mathcal{V}^+ \rangle = 0$.

The matrix $db(K^{++})$ is positive definite for any $n^+ \geq 2$, whereas for $n^+ = 1$ we have $N(db(K^{++})) = \mathbb{R}\mathcal{V}^+$. Finally, the matrix $2db(K^{++}) - K^{++}$ is positive definite if $n^+ \geq 3$, whereas its nullspace is $N(2db(K^{++}) - K^{++}) = \mathbb{R}\mathcal{V}^{*+}$ for $n^+ = 2$, $\mathcal{S}^+ = \{1, 2\}$, where $\mathcal{V}^{*+}$ is defined by $\mathcal{V}_1^{r*+} = 1$, $r \in \mathcal{F}_1$, and $\mathcal{V}_2^{r*+} = -1$, $r \in \mathcal{F}_2$, and its nullspace is $N(2db(K^{++}) - K^{++}) = \mathbb{R}\mathcal{V}^+$ if $n^+ = 1$.

Properties of $\widetilde{K}$. The matrix $\widetilde{K}$ admits the block-decomposition

$$(\Gamma^\kappa)^t \widetilde{K} \Gamma^\kappa = \begin{bmatrix} \widetilde{K}^{++} & 0 \\ \widetilde{K}^{-+} & \widetilde{K}^{--} \end{bmatrix}.$$

The matrix $\widetilde{K}^{++}$ satisfies $N(\widetilde{K}^{++}) = \mathbb{R}\mathcal{V}^+$ and $R(\widetilde{K}^{++}) = (\widetilde{\mathcal{V}}^+)^\perp$, where we have introduced the notation

$$\widetilde{\mathcal{V}} = \mathcal{X}^\kappa \mathcal{V},$$

so that $\widetilde{\mathcal{V}}_k^{10} = X_k$, $k \in \mathcal{S}$, and $\widetilde{\mathcal{V}}_k^{01} = X_k$, $k \in \mathcal{P}$. In addition, we also have $\widetilde{\beta}^{\kappa +} \in R(\widetilde{K}^{++})$, that is, $\langle \widetilde{\beta}^{\kappa +}, \widetilde{\mathcal{V}}^+ \rangle = 0$. Furthermore, the matrix $\widetilde{K}^{--}$ is block-diagonal, i.e., $\widetilde{K}^{--} = db(\widetilde{K}^{--})$, and is symmetric positive definite for any $n^+ \geq 1$. In addition, we have $R(\widetilde{K}) = \widetilde{\mathcal{V}}^\perp$, $N(\widetilde{K}) = \mathbb{R}\mathcal{V}$, and $\widetilde{\beta}^\kappa \in R(\widetilde{K})$.

The matrix $db(\widetilde{K})$ admits the block-decomposition

$$(\Gamma^\kappa)^t \, db(\widetilde{K}) \Gamma^\kappa = \begin{bmatrix} db(\widetilde{K}^{++}) & 0 \\ 0 & db(\widetilde{K}^{--}) \end{bmatrix}.$$

The matrix $db(\widetilde{K}^{++})$ is symmetric, positive semi-definite, positive definite if $n^+ \geq 2$, and its nullspace is given by $N(db(\widetilde{K}^{++})) = \mathbb{R}\mathcal{V}^+$ if $n^+ = 1$. Furthermore, the matrix

$db(\widetilde{K}^{--})$ is symmetric positive definite for any $n^+ \geq 1$. As a consequence, the matrix $db(\widetilde{K})$ is symmetric positive definite for $n^+ \geq 2$, whereas it is symmetric positive semi-definite for $n^+ = 1$ with nullspace $N\big(db(\widetilde{K})\big) = \mathbb{R}\Gamma^\kappa(\mathcal{V}^+, 0)$.

3.3.3 The Rescaled System $\widetilde{K}_{[01]}\alpha^\kappa_{[01]} = \widetilde{\beta}^\kappa_{[01]}$

The rescaled transport linear system associated with $K_{[01]}\alpha^\kappa_{[01]} = \beta^\kappa_{[01]}$ is denoted by $\widetilde{K}_{[01]}\alpha^\kappa_{[01]} = \widetilde{\beta}^\kappa_{[01]}$ and is defined by using the formalism (3.1.71)–(3.1.92). More specifically, we have $\widetilde{K}_{[01]} = \mathcal{R}\widetilde{K}\mathcal{R}^t$ and $\widetilde{\beta}^\kappa_{[01]} = \mathcal{R}\widetilde{\beta}^\kappa$, where $\mathcal{R}$ is the transformation matrix given by (2.5.21), i.e., $\mathcal{R}^{0110}_{kl} = 0$, $k \in \mathcal{P}$, $l \in \mathcal{S}$, and $\mathcal{R}^{0101}_{kl} = \delta_{kl}$, $k, l \in \mathcal{P}$. If $\mathcal{X}^\kappa_{[01]}$ denotes the matrix $\mathcal{X}^\kappa_{[01]} = \operatorname{diag}\big((X_k)_{k\in\mathcal{B}^\kappa_{[01]}}\big)$

$$\mathcal{X}^\kappa_{[01]} = \operatorname{diag}(X_{i_1}, \dots, X_{i_p}), \tag{3.3.17}$$

where $\{i_1, \dots, i_p\} = \mathcal{P}$ are the polyatomic species, we then have

$$\mathcal{X}^\kappa_{[01]}\widetilde{K}_{[01]} = K_{[01]}, \tag{3.3.18}$$

$$\mathcal{X}^\kappa_{[01]}\widetilde{\beta}^\kappa_{[01]} = \beta^\kappa_{[01]}. \tag{3.3.19}$$

An explicit evaluation also yields that

$$\widetilde{K}_{[01]} = \widetilde{K}^{0101}, \tag{3.3.20}$$

$$\widetilde{\beta}^\kappa_{[01]} = \widetilde{\beta}^{01\kappa}. \tag{3.3.21}$$

Finally, denoting by $\alpha^\kappa_{[01]}$ the solution of the system

$$\widetilde{K}_{[01]}\alpha^\kappa_{[01]} = \widetilde{\beta}^\kappa_{[01]}, \tag{3.3.22}$$

the volume viscosity $\kappa_{[01]}$ is defined by

$$\kappa_{[01]} = \langle \alpha^\kappa_{[01]}, \beta^\kappa_{[01]} \rangle = \langle \alpha^{01\kappa}_{[01]}, \beta^{01\kappa}_{[01]} \rangle. \tag{3.3.23}$$

When the mass fractions are positive, $\mathcal{X}^\kappa_{[01]}$ is invertible, and we have the relations $\widetilde{K}_{[01]} = (\mathcal{X}^\kappa_{[01]})^{-1}K_{[01]}$ and $\widetilde{\beta}^\kappa_{[01]} = (\mathcal{X}^\kappa_{[01]})^{-1}\beta^\kappa_{[01]}$ so that $\widetilde{K}_{[01]}\alpha^\kappa_{[01]} = \widetilde{\beta}^\kappa_{[01]}$ if and only if $K_{[01]}\alpha^\kappa_{[01]} = \beta^\kappa_{[01]}$. As a consequence, the definition (3.3.23) of the volume viscosity is equivalent to the definition (2.5.25).

3.3.4 Mathematical Structure of the Rescaled System $\widetilde{K}_{[01]}\alpha^{\kappa}_{[01]} = \widetilde{\beta}^{\kappa}_{[01]}$

Properties of $\mathcal{A}^{\kappa}_{[01]}$. The approximation space $\mathcal{A}^{\kappa}_{[01]}$ satisfies the assumptions introduced in (3.1.71) from (2.3.47). In addition, the partial transformation matrix $\mathcal{R}^{++}$ defined from the relations (3.1.83) is given by $\mathcal{R}^{0110++}_{kl} = 0$, $k \in \mathcal{P}^+$, $l \in \mathcal{S}^+$, and $\mathcal{R}^{0101++}_{kl} = \delta_{kl}$, $k, l \in \mathcal{P}^+$. As a consequence, we have

$$R\big((\mathcal{R}^{++})^t\big) = \mathrm{span}\{\ e^{01k+},\ k \in \mathcal{P}^+\ \} = \{\ x^+ \in \mathbb{R}^{n^+ + p^+},\ x^{10+}_k = 0,\ k \in \mathcal{S}^+\ \},$$

so that $R\big((\mathcal{R}^{++})^t\big) \cap N(K^{++}) = \{0\}$ and $R\big((\mathcal{R}^{++})^t\big) \cap N\big(2db(K^{++}) - K^{++}\big) = \{0\}$ for any $n^+ \geq 1$, since the vectors $\mathcal{V}^+$ and $\mathcal{V}^{*+}$ always have nonzero components along the vectors e^{10k+}, $k \in \mathcal{S}^+$. The properties of the rescaled system can therefore be obtained by using the formalism (3.1.71)–(3.1.92).

Properties of $K_{[01]}$. Denoting by $\Gamma^{\kappa}_{[01]}$ the permutation matrix corresponding to the reordering of $\mathcal{B}^{\kappa}_{[01]}$ into $\big(\mathcal{B}^{\kappa+}_{[01]}, \mathcal{B}^{\kappa-}_{[01]}\big)$, we obtain from (3.1.84) the block-decomposition

$$(\Gamma^{\kappa}_{[01]})^t\, K_{[01]}\, \Gamma^{\kappa}_{[01]} = \begin{bmatrix} K^{++}_{[01]} & 0 \\ 0 & 0 \end{bmatrix}.$$

From (3.1.91)(3.1.92) we obtain that the matrices $K^{++}_{[01]}$ and $2db(K^{++}_{[01]}) - K^{++}_{[01]}$ are symmetric positive definite for any $n^+ \geq 1$, and therefore, that the matrix $db(K^{++}_{[01]})$ is also positive definite for any $n^+ \geq 1$.

Properties of $\widetilde{K}_{[01]}$. From (3.1.85), the matrix $\widetilde{K}_{[01]}$ admits the block-decomposition

$$(\Gamma^{\kappa}_{[01]})^t\, \widetilde{K}_{[01]}\, \Gamma^{\kappa}_{[01]} = \begin{bmatrix} \widetilde{K}^{++}_{[01]} & 0 \\ \widetilde{K}^{-+}_{[01]} & \widetilde{K}^{--}_{[01]} \end{bmatrix}.$$

In addition, from (3.1.78), the matrix $\widetilde{K}^{++}_{[01]}$ is nonsingular, and the matrix $\widetilde{K}^{--}_{[01]}$ is block-diagonal, i.e., $\widetilde{K}^{--}_{[01]} = db(\widetilde{K}^{--}_{[01]})$, and is symmetric positive definite for any $n^+ \geq 1$. As a consequence, $\widetilde{K}_{[01]}$ is nonsingular for any $n^+ \geq 1$.

Furthermore, from (3.1.86) we have the block-decomposition

$$(\Gamma^{\kappa}_{[01]})^t\, db(\widetilde{K}_{[01]})\, \Gamma^{\kappa}_{[01]} = \begin{bmatrix} db(\widetilde{K}^{++}_{[01]}) & 0 \\ 0 & db(\widetilde{K}^{--}_{[01]}) \end{bmatrix}.$$

The matrix $db(\widetilde{K}^{++}_{[01]})$ is symmetric positive definite, and so is $db(\widetilde{K}^{--}_{[01]})$. As a consequence, $db(\widetilde{K}_{[01]})$ is symmetric positive definite for any $n^+ \geq 1$.

3.4 The Flux Diffusion Matrix

3.4.1 The Mass Fluxes

In the case of vanishing mass fractions, the diffusion matrix D and the diffusion velocities V are no longer defined [Gi91]. Nevertheless, the quantities that are needed to formulate the multicomponent flow equations are the fluxes $F_k = Y_k V_k$, $k \in \mathcal{S}$. We express these fluxes in the form

$$F_k = -\sum_{l\in\mathcal{S}} \widetilde{D}_{kl} d_l - Y_k \theta_k \nabla \log \overline{T}, \qquad k \in \mathcal{S}, \tag{3.4.1}$$

so that the flux diffusion coefficients $\widetilde{D}_{kl}$, $k,l \in \mathcal{S}$, are then the proper quantities to be evaluated. These coefficients are defined such that

$$\sum_{k\in\mathcal{S}} \widetilde{D}_{kl} = 0, \qquad l \in \mathcal{S}, \tag{3.4.2}$$

that is, such that $\widetilde{D}^t U = 0$. In the case of positive mass fractions, we also deduce from the definition properties of the diffusion matrix (1.1.2), (2.6.22), and (2.6.24) that

$$\widetilde{D}_{kl} = Y_k D_{kl}, \qquad k,l \in \mathcal{S}. \tag{3.4.3}$$

In this section, the flux transport coefficients $\widetilde{D}_{kl}$, $k,l \in \mathcal{S}$, will be directly related to new quantities $\widetilde{\alpha}^{D_l}$ and will extend the definition of the coefficients $Y_k D_{kl}$, $k,l \in \mathcal{S}$, to the whole mass fraction physical domain $\{\, Y \in \mathbb{R}^n,\ Y \geq 0,\ \langle Y, U\rangle = 1 \,\}$.

3.4.2 The Rescaled System $\widetilde{L}\widetilde{\alpha}^{D_l} = \widetilde{\beta}^{D_l}$

The rescaled matrix associated with L is denoted by $\widetilde{L}$ and is defined, for convenience, by $\widetilde{L}_{kl}^{rs} = (2\overline{n}\sqrt{m_k m_l}/3\overline{p})\langle\!\langle \phi^{10rk}, \Im(\phi^{10sl})\rangle\!\rangle$, $(r,k),(s,l) \in \mathcal{B}^D$. Keeping in mind that the vectors β^{D_l}, $l \in \mathcal{S}$, cannot be rescaled by the matrix $\mathcal{X}^D$, we introduce new quantities $\widetilde{\beta}^{D_l} \in \mathbb{R}^{2n+p}$ defined by

$$\widetilde{\beta}_k^{00D_l} = \frac{m_k}{m}\delta_{kl} - \frac{m_k}{m} Y_l, \qquad k,l \in \mathcal{S}, \tag{3.4.4}$$

$$\widetilde{\beta}_k^{10D_l} = 0, \qquad k,l \in \mathcal{S}, \tag{3.4.5}$$

$$\widetilde{\beta}_k^{01D_l} = 0, \qquad k \in \mathcal{P}, \quad l \in \mathcal{S}. \tag{3.4.6}$$

Denoting by $\mathcal{X}^D \in \mathbb{R}^{2n+p,2n+p}$ the diagonal matrix $\mathcal{X}^D = \mathrm{diag}\big((X_k)_{(r,k)\in\mathcal{B}^D}\big)$

$$\mathcal{X}^D = \mathrm{diag}(X_1,\ldots,X_n,X_1,\ldots,X_n,X_{i_1},\ldots,X_{i_p}), \tag{3.4.7}$$

where $\{i_1,\ldots,i_p\} = \mathcal{P}$ are the polyatomic species, we then have

$$\mathcal{X}^D \widetilde{L} = L, \tag{3.4.8}$$

and

$$\mathcal{X}^D \widetilde{\beta}^{D_l} = Y_l \beta^{D_l}, \qquad l \in \mathcal{S}. \tag{3.4.9}$$

Notice that the vectors β^{D_l}, $l \in \mathcal{S}$, can always be defined from

$$\beta_k^{rD_l} = (\sqrt{2k_{\text{B}}\overline{T}m_k}/3)\langle\!\langle\!\langle \xi^{10rk}, \widetilde{\Psi}^{D_l} \rangle\!\rangle\!\rangle,$$

where $\widetilde{\Psi}^{D_l}$ has its components given by $\widetilde{\Psi}_i^{D_l} = (1/\overline{p})(\delta_{il} - Y_i)(c_i - v)$, $i \in \mathcal{S}$, since for positive mass fractions we have

$$\langle\!\langle \xi^{10rk}, \Psi^{D_l} \rangle\!\rangle = n_k \,\langle\!\langle\!\langle \xi^{10rk}, \Psi^{D_l} \rangle\!\rangle\!\rangle = \langle\!\langle\!\langle \xi^{10rk}, \widetilde{\Psi}^{D_l} \rangle\!\rangle\!\rangle.$$

In Section 2.6 we have already evaluated that

$$\beta_k^{00D_l} = \delta_{kl} - Y_k, \qquad k, l \in \mathcal{S}, \tag{3.4.10}$$

$$\beta_k^{10D_l} = 0, \qquad k, l \in \mathcal{S}, \tag{3.4.11}$$

$$\beta_k^{01D_l} = 0, \qquad k \in \mathcal{P}, \quad l \in \mathcal{S}. \tag{3.4.12}$$

A direct calculation yields that

$$\widetilde{L}_{kk}^{0000} = \sum_{\substack{l \in \mathcal{S} \\ l \neq k}} \frac{X_l}{\mathcal{D}_{kl}}, \qquad k \in \mathcal{S}, \tag{3.4.13}$$

$$\widetilde{L}_{kl}^{0000} = -\frac{X_l}{\mathcal{D}_{kl}}, \qquad k, l \in \mathcal{S}, \quad k \neq l, \tag{3.4.14}$$

$$\widetilde{L}_{kk}^{0010} = -\sum_{\substack{l \in \mathcal{S} \\ l \neq k}} \frac{X_l}{2\mathcal{D}_{kl}} \frac{m_l}{m_k + m_l}(6\bar{C}_{kl} - 5), \qquad k \in \mathcal{S}, \tag{3.4.15}$$

$$\widetilde{L}_{kl}^{0010} = \frac{X_l}{2\mathcal{D}_{kl}} \frac{m_k}{m_k + m_l}(6\bar{C}_{kl} - 5), \qquad k, l \in \mathcal{S}, \quad k \neq l, \tag{3.4.16}$$

$$\widetilde{L}_{kk}^{0001} = -\sum_{\substack{l \in \mathcal{S} \\ l \neq k}} X_l \frac{\big[(\epsilon_{kK}^0(\gamma^2 - \gamma\gamma' \cos\chi)\big]_{kl}}{\Omega_{kl}^{(1,1)} \mathcal{D}_{kl}}, \qquad k \in \mathcal{P}, \tag{3.4.17}$$

$$\widetilde{L}_{kl}^{0001} = X_l \frac{\big[(\epsilon_{lL}^0(\gamma^2 - \gamma\gamma' \cos\chi)\big]_{kl}}{\Omega_{kl}^{(1,1)} \mathcal{D}_{kl}}, \qquad k \in \mathcal{S}, \quad l \in \mathcal{P}, \quad k \neq l, \tag{3.4.18}$$

$$\widetilde{L}_{kk}^{1000} = -\sum_{\substack{l\in\mathcal{S}\\ l\neq k}} \frac{X_l}{2\mathcal{D}_{kl}} \frac{m_l}{m_k+m_l}(6\bar{C}_{kl}-5), \qquad k\in\mathcal{S}, \tag{3.4.19}$$

$$\widetilde{L}_{kl}^{1000} = \frac{X_l}{2\mathcal{D}_{kl}} \frac{m_l}{m_k+m_l}(6\bar{C}_{kl}-5), \qquad k,l\in\mathcal{S}, \quad k\neq l, \tag{3.4.20}$$

$$\widetilde{L}_{kk}^{1010} = \sum_{\substack{l\in\mathcal{S}\\ l\neq k}} \frac{X_l}{\mathcal{D}_{kl}} \frac{m_k m_l}{(m_k+m_l)^2}\Big[\frac{15}{2}\frac{m_k}{m_l} + \frac{25}{4}\frac{m_l}{m_k} - 3\frac{m_l}{m_k}\bar{B}_{kl} + 4\bar{A}_{kl} + \frac{25}{12}\frac{\big[(\Delta\epsilon_{kl})^2\big]_{kl}}{\Omega_{kl}^{(1,1)}}\Big]$$

$$+ \frac{X_k}{2\mathcal{D}_{kk}}\Big[4\bar{A}_{kk} + \frac{25}{12}\frac{\big[(\Delta\epsilon_{kk})^2\big]_{kk}}{\Omega_{kk}^{(1,1)}}\Big], \qquad k\in\mathcal{S}, \tag{3.4.21}$$

$$\widetilde{L}_{kl}^{1010} = -\frac{X_l}{\mathcal{D}_{kl}} \frac{m_k m_l}{(m_k+m_l)^2}\Big[\frac{55}{4} - 3\bar{B}_{kl} - 4\bar{A}_{kl}$$

$$- \frac{25}{12}\frac{\big[(\Delta\epsilon_{kl})^2\big]_{kl}}{\Omega_{kl}^{(1,1)}}\Big] \qquad k,l\in\mathcal{S}, \quad k\neq l, \tag{3.4.22}$$

$$\widetilde{L}_{kk}^{1001} = -\sum_{\substack{l\in\mathcal{S}\\ l\neq k}} \frac{X_l}{\mathcal{D}_{kl}} \frac{m_k}{m_k+m_l}\Big[\frac{5}{4}\frac{\big[(\Delta\epsilon_k\Delta\epsilon_{kl}\big]_{kl}}{\Omega_{kl}^{(1,1)}} + \frac{5}{2}\frac{m_l}{m_k}\frac{\big[(\epsilon_{kK}^0(\gamma^2-\gamma\gamma'\cos\chi)\big]_{kl}}{\Omega_{kl}^{(1,1)}}$$

$$- \frac{m_l}{m_k}\frac{\big[(\epsilon_{kK}^0(\gamma^4-\gamma\gamma'^3\cos\chi)\big]_{kl}}{\Omega_{kl}^{(1,1)}}\Big] - \frac{5}{8}\frac{X_k}{\mathcal{D}_{kk}}\frac{\big[(\Delta\epsilon_{kk})^2\big]_{kk}}{\Omega_{kk}^{(1,1)}}, \qquad k\in\mathcal{P}, \tag{3.4.23}$$

$$\widetilde{L}_{kl}^{1001} = \frac{X_l}{\mathcal{D}_{kl}} \frac{m_l}{m_k+m_l}\Big[-\frac{5}{4}\frac{\big[(\widetilde{\Delta}\epsilon_l\Delta\epsilon_{kl}\big]_{kl}}{\Omega_{kl}^{(1,1)}} + \frac{5}{2}\frac{\big[(\epsilon_{lL}^0(\gamma^2-\gamma\gamma'\cos\chi)\big]_{kl}}{\Omega_{kl}^{(1,1)}}$$

$$- \frac{\big[(\epsilon_{lL}^0(\gamma^4-\gamma\gamma'^3\cos\chi)\big]_{kl}}{\Omega_{kl}^{(1,1)}}\Big], \qquad k\in\mathcal{S}, \quad l\in\mathcal{P}, \quad k\neq l, \tag{3.4.24}$$

$$\widetilde{L}_{kk}^{0100} = -\sum_{\substack{l\in\mathcal{S}\\ l\neq k}} \frac{X_l}{\mathcal{D}_{kl}} \frac{\big[(\epsilon_{kK}^0(\gamma^2-\gamma\gamma'\cos\chi)\big]_{kl}}{\Omega_{kl}^{(1,1)}}, \qquad k\in\mathcal{P}, \tag{3.4.25}$$

$$\widetilde{L}_{kl}^{0100} = \frac{X_l}{\mathcal{D}_{kl}} \frac{\big[(\epsilon_{kK}^0(\gamma^2-\gamma\gamma'\cos\chi)\big]_{kl}}{\Omega_{kl}^{(1,1)}}, \qquad k\in\mathcal{P}, \quad l\in\mathcal{S}, \quad k\neq l, \tag{3.4.26}$$

$$\widetilde{L}_{kk}^{0110} = -\sum_{\substack{l\in\mathcal{S}\\ l\neq k}} \frac{X_l}{\mathcal{D}_{kl}} \frac{m_k}{m_k+m_l}\Big[\frac{5}{4}\frac{\big[(\Delta\epsilon_k\Delta\epsilon_{kl}\big]_{kl}}{\Omega_{kl}^{(1,1)}} + \frac{5}{2}\frac{m_l}{m_k}\frac{\big[(\epsilon_{kK}^0(\gamma^2-\gamma\gamma'\cos\chi)\big]_{kl}}{\Omega_{kl}^{(1,1)}}$$

$$- \frac{m_l}{m_k}\frac{\big[(\epsilon_{kK}^0(\gamma^4-\gamma\gamma'^3\cos\chi)\big]_{kl}}{\Omega_{kl}^{(1,1)}}\Big] - \frac{5}{8}\frac{X_k}{\mathcal{D}_{kk}}\frac{\big[(\Delta\epsilon_{kk})^2\big]_{kk}}{\Omega_{kk}^{(1,1)}}, \qquad k\in\mathcal{P}, \tag{3.4.27}$$

$$\widetilde{L}_{kl}^{0110} = \frac{X_l}{\mathcal{D}_{kl}} \frac{m_k}{m_k + m_l} \Big[-\frac{5}{4} \frac{\big[(\Delta\epsilon_k \Delta\epsilon_{kl}\big]_{kl}}{\Omega_{kl}^{(1,1)}} + \frac{5}{2} \frac{\big[(\epsilon_{kK}^0 (\gamma^2 - \gamma\gamma' \cos\chi)\big]_{kl}}{\Omega_{kl}^{(1,1)}}$$

$$- \frac{\big[(\epsilon_{kK}^0 (\gamma^4 - \gamma\gamma'^3 \cos\chi)\big]_{kl}}{\Omega_{kl}^{(1,1)}} \Big], \qquad k \in \mathcal{P}, \quad l \in \mathcal{S}, \quad k \neq l, \tag{3.4.28}$$

$$\widetilde{L}_{kk}^{0101} = \sum_{\substack{l \in \mathcal{S} \\ l \neq k}} X_l \Big[\frac{c_k^{\mathrm{int}}}{k_{\mathrm{B}} \mathcal{D}_{k\,\mathrm{int},l}} + \frac{3}{4} \frac{m_k}{m_l} \frac{\big[(\Delta\epsilon_k)^2\big]_{kl}}{\Omega_{kl}^{(1,1)} \mathcal{D}_{kl}} \Big]$$

$$+ X_k \Big[\frac{c_k^{\mathrm{int}}}{k_{\mathrm{B}} \mathcal{D}_{k\,\mathrm{int},k}} + \frac{3}{8} \frac{\big[(\Delta\epsilon_{kk})^2\big]_{kk}}{\Omega_{kk}^{(1,1)} \mathcal{D}_{kk}} \Big], \qquad k \in \mathcal{P}, \tag{3.4.29}$$

$$\widetilde{L}_{kl}^{0101} = -X_l \Big[\frac{\big[\epsilon_{kK}^0 (\epsilon_{lL}^0 \gamma^2 - \epsilon_{lL'}^0 \gamma\gamma' \cos\chi)\big]_{kl}}{\Omega_{kl}^{(1,1)} \mathcal{D}_{kl}}$$

$$- \frac{3}{4} \frac{\big[\Delta\epsilon_k \widetilde{\Delta}\epsilon_l\big]_{kl}}{\Omega_{kl}^{(1,1)} \mathcal{D}_{kl}} \Big], \qquad k, l \in \mathcal{P}, \quad k \neq l. \tag{3.4.30}$$

Denoting then by $\widetilde{\alpha}^{D_l}$, $l \in \mathcal{S}$, the solution of the system

$$\begin{cases} \widetilde{L}\widetilde{\alpha}^{D_l} = \widetilde{\beta}^{D_l}, \\ \langle \mathcal{L}, \widetilde{\alpha}^{D_l} \rangle = 0, \end{cases} \tag{3.4.31}$$

where the constraint vector $\mathcal{L}$ has already been introduced in Section 2.6.1, the flux diffusion matrix $\widetilde{D}$ is defined by

$$\widetilde{D}_{kl} = \langle \widetilde{\alpha}^{D_k}, \beta^{D_l} \rangle = \langle \widetilde{\alpha}^{00D_k}, \beta^{00D_l} \rangle, \qquad k, l \in \mathcal{S}, \tag{3.4.32}$$

which can be simplified into

$$\widetilde{D}_{kl} = \widetilde{\alpha}_l^{00D_k}, \qquad k, l \in \mathcal{S}, \tag{3.4.33}$$

by explicitly using the constraint in (3.4.31). When the mass fractions are positive, the matrix $\mathcal{X}^D$ is then invertible, and we have the relations $\widetilde{L} = (\mathcal{X}^D)^{-1} L$ and $\widetilde{\beta}^{D_k} = Y_k (\mathcal{X}^D)^{-1} \beta^{D_k}$. As a consequence, we deduce that the quantity $x = (1/Y_k)\widetilde{\alpha}^{D_k}$ satisfies $Lx = \beta^{D_k}$ and $\langle \mathcal{L}, x \rangle = 0$ so that $x = \alpha^{D_k}$ and $\widetilde{\alpha}^{D_k} = Y_k \alpha^{D_k}$. Therefore, we have $\widetilde{D}_{kl} = Y_k D_{kl}$, and both definitions (3.4.32)(3.4.33) and (2.6.23)(2.6.24) are then equivalent.

3.4.3 Mathematical Structure of the Rescaled System $\widetilde{L}\widetilde{\alpha}^{D_l} = \widetilde{\beta}^{D_l}$

Properties of $\mathcal{A}^D$. The functional space $\mathcal{A}^D$ introduced in Section 2.6 to approximate ϕ^{D_k} satisfies the perpendicularity property (3.1.1), the species orthogonality property

(3.1.2), and the relation (3.1.60) from (2.3.51). In addition, the basis functions satisfy the species localization property (3.1.3). As a consequence, all the results established in Section 3.1.6 apply to the matrices L and $\widetilde{L}$. Note, however, that these results do not apply to the system $L\alpha^{D_l} = \beta^{D_l}$ since Ψ^{D_l} is not defined for $n_l = 0$.

Properties of L. Denoting by Γ^D the permutation matrix associated with the re-ordering of $\mathcal{B}^D$ into $(\mathcal{B}^{D+}, \mathcal{B}^{D-})$, the matrix L admits the block-decomposition

$$(\Gamma^D)^t L \Gamma^D = \begin{bmatrix} L^{++} & 0 \\ 0 & 0 \end{bmatrix}.$$

The matrix L^{++} is also symmetric positive semi-definite, $N(L^{++}) = \mathbb{R}\mathcal{U}^+$, where $\mathcal{U}_k^{00} = 1$, $k \in \mathcal{S}$, $\mathcal{U}_k^{10} = 0$, $k \in \mathcal{S}$, and $\mathcal{U}_k^{01} = 0$, $k \in \mathcal{P}$. The matrix L^{++} is positive definite on the constrained space $(\mathcal{L}^+)^\perp$ where $\mathcal{L}$ is given by $\mathcal{L}_k^{00} = Y_k$, $k \in \mathcal{S}$, $\mathcal{L}_k^{10} = 0$, $k \in \mathcal{S}$, and $\mathcal{L}_k^{01} = 0$, $k \in \mathcal{P}$. Moreover, we have $\langle \mathcal{L}^+, \mathcal{U}^+ \rangle \neq 0$. An explicit calculation also yields that $\beta^{D_k+} \in R(L^{++})$, that is, $\langle \beta^{D_k+}, \mathcal{U}^+ \rangle = 0$.

The matrix $db(L^{++})$ is positive definite for any $n^+ \geq 2$, whereas for $n^+ = 1$, we have $db(L^{++}) = L^{++}$ and $N\big(db(L^{++})\big) = \mathbb{R}\mathcal{U}^+$. Finally, the matrix $2db(L^{++}) - L^{++}$ is positive definite if $n^+ \geq 3$, whereas its nullspace is $N\big(2db(L^{++}) - L^{++}\big) = \mathbb{R}\mathcal{U}^{*+}$ for $n^+ = 2$, $\mathcal{S}^+ = \{1, 2\}$, where $\mathcal{U}^{*+}$ is defined by $\mathcal{U}_1^{00*+} = 1$, $\mathcal{U}_1^{r*+} = 0$, $r \in \mathcal{F}_1$, $r \neq 00$, and $\mathcal{U}_2^{00*+} = -1$, $\mathcal{U}_2^{r*+} = 0$, $r \in \mathcal{F}_2$, $r \neq 00$, and its nullspace is $N\big(2db(L^{++}) - L^{++}\big) = \mathbb{R}\mathcal{U}^+$ if $n^+ = 1$.

Properties of $\widetilde{L}$. The matrix $\widetilde{L}$ admits the block-decomposition

$$(\Gamma^D)^t \widetilde{L} \Gamma^D = \begin{bmatrix} \widetilde{L}^{++} & 0 \\ \widetilde{L}^{-+} & \widetilde{L}^{--} \end{bmatrix}.$$

The matrix $\widetilde{L}^{++}$ satisfies $N(\widetilde{L}^{++}) = \mathbb{R}\mathcal{U}^+$ and $R(\widetilde{L}^{++}) = (\widetilde{\mathcal{U}}^+)^\perp$, where we have introduced the notation

$$\widetilde{\mathcal{U}} = \mathcal{X}^D \mathcal{U},$$

so that $\widetilde{\mathcal{U}}_k^{00} = X_k$, $k \in \mathcal{S}$, $\widetilde{\mathcal{U}}_k^{10} = 0$, $k \in \mathcal{S}$, and $\widetilde{\mathcal{U}}_k^{01} = 0$, $k \in \mathcal{P}$. An explicit calculation also yields that $\widetilde{\beta}^{D_k+} \in R(\widetilde{L}^{++})$, that is, $\langle \widetilde{\beta}^{D_k+}, \widetilde{\mathcal{U}}^+ \rangle = 0$. In addition, the matrix $\widetilde{L}^{--}$ is block-diagonal, i.e., $\widetilde{L}^{--} = db(\widetilde{L}^{--})$, and is symmetric positive definite for any $n^+ \geq 1$. Finally, we also have $R(\widetilde{L}) = \widetilde{\mathcal{U}}^\perp$, $N(\widetilde{L}) = \mathbb{R}\mathcal{U}$, and an explicit calculation yields that $\widetilde{\beta}^{D_l} \in R(\widetilde{L})$.

The matrix $db(\widetilde{L})$ admits the block-decomposition

$$(\Gamma^D)^t \, db(\widetilde{L}) \Gamma^D = \begin{bmatrix} db(\widetilde{L}^{++}) & 0 \\ 0 & db(\widetilde{L}^{--}) \end{bmatrix}.$$

The matrix $db(\widetilde{L}^{++})$ is also symmetric, positive semi-definite, positive definite if $n^+ \geq 2$, and its nullspace is given by $N(db(\widetilde{L}^{++})) = \mathbb{R}\mathcal{U}^+$ if $n^+ = 1$. In addition, the matrix $db(\widetilde{L}^{--})$ is symmetric positive definite for any $n^+ \geq 1$. As a consequence, $db(\widetilde{L})$ is symmetric positive definite for $n^+ \geq 2$, whereas it is symmetric positive semi-definite for $n^+ = 1$ with nullspace $N(db(\widetilde{L})) = \mathbb{R}\Gamma^D(\mathcal{U}^+, 0)$.

3.4.4 The Rescaled System $\widetilde{L}_{[e]}\widetilde{\alpha}_{[e]}^{D_l} = \widetilde{\beta}_{[e]}^{D_l}$

The rescaled matrix associated with $L_{[e]}$ is denoted by $\widetilde{L}_{[e]}$ and is defined, for convenience, by $\widetilde{L}_{[e]kl}^{rs} = (2\bar{n}\sqrt{m_k m_l}/3\bar{p})\langle\!\langle\!\langle \phi^{10rk}, \Im(\phi^{10sl})\rangle\!\rangle\!\rangle$ for $(r,k),(s,l) \in \mathcal{B}_{[e]}^D$. Denoting by $\mathcal{X}_{[e]}^D$ the diagonal matrix $\mathcal{X}_{[e]}^D = \mathrm{diag}\big((X_k)_{(r,k)\in\mathcal{B}_{[e]}^D}\big)$

$$\mathcal{X}_{[e]}^D = \mathrm{diag}(X_1, \ldots, X_n, X_1, \ldots, X_n), \tag{3.4.34}$$

the matrix $\widetilde{L}_{[e]}$ is such that

$$\mathcal{X}_{[e]}^D \widetilde{L}_{[e]} = L_{[e]}. \tag{3.4.35}$$

In addition, we denote by $\mathcal{R} = \big(\mathcal{R}_{kl}^{rs}\big)_{(r,k)\in\mathcal{B}_{[e]}^D,(s,l)\in\mathcal{B}^D}$ the corresponding transformation matrix already given in (2.6.28). From the definition of $\widetilde{L}$ and $\widetilde{L}_{[e]}$ and from the linear relations between the basis functions of the approximation spaces $\mathcal{A}^D$ and $\mathcal{A}_{[e]}^D$, it is readily seen that we have the relation

$$\widetilde{L}_{[e]} = \mathcal{R}\widetilde{L}\mathcal{R}^t. \tag{3.4.36}$$

As a consequence, we have

$$\widetilde{L}_{[e]kl}^{0000} = \widetilde{L}_{kl}^{0000}, \qquad k,l \in \mathcal{S}, \tag{3.4.37}$$

$$\widetilde{L}_{[e]kl}^{00e} = \widetilde{L}_{kl}^{0010} + \delta_{\mathcal{P}}(l)\widetilde{L}_{kl}^{0001}, \qquad k,l \in \mathcal{S}, \tag{3.4.38}$$

$$\widetilde{L}_{[e]kl}^{e00} = \widetilde{L}_{kl}^{1000} + \delta_{\mathcal{P}}(k)\widetilde{L}_{kl}^{0100}, \qquad k,l \in \mathcal{S}, \tag{3.4.39}$$

$$\widetilde{L}_{[e]kl}^{ee} = \widetilde{L}_{kl}^{1010} + \delta_{\mathcal{P}}(k)\widetilde{L}_{kl}^{0110} + \delta_{\mathcal{P}}(l)\widetilde{L}_{kl}^{1001} + \delta_{\mathcal{P}}(k)\delta_{\mathcal{P}}(l)\widetilde{L}_{kl}^{0101}, \qquad k,l \in \mathcal{S}, \tag{3.4.40}$$

from which the matrix $\widetilde{L}_{[e]}$ is easily evaluated. In addition, we define new quantities $\widetilde{\beta}_{[e]}^{D_l}$ by

$$\widetilde{\beta}_{[e]}^{D_l} = \mathcal{R}\widetilde{\beta}^{D_l}, \qquad l \in \mathcal{S}. \tag{3.4.41}$$

Noting that $\mathcal{X}_{[e]}^D\mathcal{R} = \mathcal{R}\mathcal{X}^D$, we have $\mathcal{X}_{[e]}^D\widetilde{\beta}_{[e]}^{D_l} = \mathcal{X}_{[e]}^D\mathcal{R}\widetilde{\beta}^{D_l} = \mathcal{R}\mathcal{X}^D\widetilde{\beta}^{D_l}$. Using then the relations $\mathcal{X}^D\widetilde{\beta}^{D_l} = Y_l\beta^{D_l}$ and $\beta_{[e]}^{D_l} = \mathcal{R}\beta^{D_l}$, obtained from (3.4.9) and (2.6.29), we thus conclude that

$$\mathcal{X}_{[e]}^D\widetilde{\beta}_{[e]}^{D_l} = Y_l\beta_{[e]}^{D_l}, \qquad l \in \mathcal{S}. \tag{3.4.42}$$

An explicit calculation also yields

$$\widetilde{\beta}^{00D_l}_{[e]k} = \widetilde{\beta}^{00D_l}_k = \frac{m_k}{m}\delta_{kl} - \frac{m_k}{m}Y_l, \qquad k,l \in \mathcal{S}, \tag{3.4.43}$$

$$\widetilde{\beta}^{eD_l}_{[e]k} = \widetilde{\beta}^{10D_l}_k + \delta_{\mathcal{P}}(k)\widetilde{\beta}^{01D_l}_k = 0, \qquad k,l \in \mathcal{S}. \tag{3.4.44}$$

Denoting by $\widetilde{\alpha}^{D_l}_{[e]}$, $l \in \mathcal{S}$, the solution of the rescaled transport linear system

$$\begin{cases} \widetilde{L}_{[e]}\widetilde{\alpha}^{D_l}_{[e]} = \widetilde{\beta}^{D_l}_{[e]}, \\ \langle \mathcal{L}_{[e]}, \widetilde{\alpha}^{D_l}_{[e]} \rangle = 0, \end{cases} \tag{3.4.45}$$

where the constraint vector $\mathcal{L}_{[e]} = \mathcal{R}\mathcal{L}$ has already been introduced in Section 2.6.3, the flux diffusion matrix $\widetilde{D}_{[e]}$ is defined by

$$\widetilde{D}_{[e]kl} = \langle \widetilde{\alpha}^{D_k}_{[e]}, \beta^{D_l}_{[e]} \rangle = \langle \widetilde{\alpha}^{00D_k}_{[e]}, \beta^{00D_l}_{[e]} \rangle, \qquad k,l \in \mathcal{S}, \tag{3.4.46}$$

which can be simplified into

$$\widetilde{D}_{[e]kl} = \widetilde{\alpha}^{00D_k}_{[e]l}, \qquad k,l \in \mathcal{S}, \tag{3.4.47}$$

by explicitly using the constraint in (3.4.45). When the mass fractions are positive, the matrix $\mathcal{X}^D_{[e]}$ is then invertible, and we have the relations $\widetilde{L}_{[e]} = (\mathcal{X}^D_{[e]})^{-1}L_{[e]}$ and $\widetilde{\beta}^{D_k}_{[e]} = Y_k(\mathcal{X}^D_{[e]})^{-1}\beta^{D_k}_{[e]}$. As a consequence, we deduce that the quantity $x = (1/Y_k)\widetilde{\alpha}^{D_k}_{[e]}$ satisfies $L_{[e]}x = \beta^{D_k}_{[e]}$ and $\langle \mathcal{L}_{[e]}, x \rangle = 0$ so that $x = \alpha^{D_k}_{[e]}$ and $\widetilde{\alpha}^{D_k}_{[e]} = Y_k\alpha^{D_k}_{[e]}$. Therefore, we have $\widetilde{D}_{[e]kl} = Y_k D_{[e]kl}$, and both definitions (3.4.46)(3.4.47) and (2.6.36)(2.6.37) are then equivalent.

3.4.5 Mathematical Structure of the Rescaled System $\widetilde{L}_{[e]}\widetilde{\alpha}^{D_l}_{[e]} = \widetilde{\beta}^{D_l}_{[e]}$

Properties of $\mathcal{A}^D_{[e]}$. The functional space $\mathcal{A}^D_{[e]}$ introduced in Section 2.6 to approximate ϕ^{D_k} satisfies the perpendicularity property (3.1.1), the species orthogonality property (3.1.2), and the relation (3.1.60) from (2.3.51). In addition, the basis functions satisfy the species localization property (3.1.3). As a consequence, all the results established in Section 3.1.6 apply to the matrices $L_{[e]}$ and $\widetilde{L}_{[e]}$. Note, however, that these results do not apply to the system $L_{[e]}\alpha^{D_l}_{[e]} = \beta^{D_l}_{[e]}$ since Ψ^{D_l} is not defined for $n_l = 0$.

Properties of $L_{[e]}$. Denoting by $\Gamma^D_{[e]}$ the permutation matrix associated with the reordering of the set $\mathcal{B}^D_{[e]}$ into $(\mathcal{B}^{D+}_{[e]}, \mathcal{B}^{D-}_{[e]})$, the matrix $L_{[e]}$ admits the block-decomposition

$$(\Gamma^D_{[e]})^t L_{[e]} \Gamma^D_{[e]} = \begin{bmatrix} L^{++}_{[e]} & 0 \\ 0 & 0 \end{bmatrix}.$$

The matrix $L^{++}_{[e]}$ is also symmetric positive semi-definite and $N(L^{++}_{[e]}) = \mathbb{R}\mathcal{U}^{+}_{[e]}$, where $\mathcal{U}^{00}_{[e]k} = 1$, $k \in \mathcal{S}$, and $\mathcal{U}^{e}_{[e]k} = 0$, $k \in \mathcal{S}$. The matrix $L^{++}_{[e]}$ is positive definite on the constrained space $(\mathcal{L}^{+}_{[e]})^{\perp}$ where $\mathcal{L}_{[e]}$ is given by $\mathcal{L}^{00}_{[e]k} = Y_k$, $k \in \mathcal{S}$, and $\mathcal{L}^{e}_{[e]k} = 0$, $k \in \mathcal{S}$. Moreover, we have $\langle \mathcal{L}^{+}_{[e]}, \mathcal{U}^{+}_{[e]} \rangle \neq 0$. An explicit calculation also yields that $\beta^{D_k+}_{[e]} \in R(L^{++}_{[e]})$, that is, $\langle \beta^{D_k+}_{[e]}, \mathcal{U}^{+}_{[e]} \rangle = 0$.

The matrix $db(L^{++}_{[e]})$ is positive definite for any $n^+ \geq 2$, whereas for $n^+ = 1$, we have $db(L^{++}_{[e]}) = L^{++}_{[e]}$ and $N\big(db(L^{++}_{[e]})\big) = \mathbb{R}\mathcal{U}^{+}_{[e]}$. Finally, the matrix $2db(L^{++}_{[e]}) - L^{++}_{[e]}$ is positive definite if $n^+ \geq 3$, whereas its nullspace is $N\big(2db(L^{++}_{[e]}) - L^{++}_{[e]}\big) = \mathbb{R}\mathcal{U}^{*+}_{[e]}$ for $n^+ = 2$, $\mathcal{S}^+ = \{1, 2\}$, where $\mathcal{U}^{*+}_{[e]}$ is defined by $\mathcal{U}^{00*+}_{[e]1} = 1$, $\mathcal{U}^{e*+}_{[e]1} = 0$, and $\mathcal{U}^{00*+}_{[e]2} = -1$, $\mathcal{U}^{e*+}_{[e]2} = 0$, i.e., $\mathcal{U}^{*+}_{[e]} = (1, -1, 0, 0)$, and its nullspace is $N\big(2db(L^{++}_{[e]}) - L^{++}_{[e]}\big) = \mathbb{R}\mathcal{U}^{+}_{[e]}$ if $n^+ = 1$.

Properties of $\widetilde{L}_{[e]}$. The matrix $\widetilde{L}_{[e]}$ admits the block-decomposition

$$(\Gamma^{D}_{[e]})^t \widetilde{L}_{[e]} \Gamma^{D}_{[e]} = \begin{bmatrix} \widetilde{L}^{++}_{[e]} & 0 \\ \widetilde{L}^{-+}_{[e]} & \widetilde{L}^{--}_{[e]} \end{bmatrix}.$$

The matrix $\widetilde{L}^{++}_{[e]}$ also satisfies $N(\widetilde{L}^{++}_{[e]}) = \mathbb{R}\mathcal{U}^{+}_{[e]}$ and $R(\widetilde{L}^{++}_{[e]}) = (\widetilde{\mathcal{U}}^{+}_{[e]})^{\perp}$, where we have introduced the notation

$$\widetilde{\mathcal{U}}_{[e]} = \mathcal{X}^{D}_{[e]} \mathcal{U}_{[e]},$$

so that $\widetilde{\mathcal{U}}^{00}_{[e]k} = X_k$, $k \in \mathcal{S}$, and $\widetilde{\mathcal{U}}^{e}_{[e]k} = 0$, $k \in \mathcal{S}$. An explicit calculation also yields that $\widetilde{\beta}^{D_k+}_{[e]} \in R(\widetilde{L}^{++}_{[e]})$, that is, $\langle \widetilde{\beta}^{D_k+}_{[e]}, \widetilde{\mathcal{U}}^{+}_{[e]} \rangle = 0$. In addition, the matrix $\widetilde{L}^{--}_{[e]}$ is block-diagonal, i.e., $\widetilde{L}^{--}_{[e]} = db(\widetilde{L}^{--}_{[e]})$, and is symmetric positive definite for any $n^+ \geq 1$. Finally, we have $R(\widetilde{L}_{[e]}) = (\widetilde{\mathcal{U}}_{[e]})^{\perp}$, $N(\widetilde{L}_{[e]}) = \mathbb{R}\mathcal{U}_{[e]}$, and an explicit calculation yields that $\widetilde{\beta}^{D_l}_{[e]} \in R(\widetilde{L}_{[e]})$.

The matrix $db(\widetilde{L}_{[e]})$ admits the block-decomposition

$$(\Gamma^{D}_{[e]})^t \, db(\widetilde{L}_{[e]}) \Gamma^{D}_{[e]} = \begin{bmatrix} db(\widetilde{L}^{++}_{[e]}) & 0 \\ 0 & db(\widetilde{L}^{--}_{[e]}) \end{bmatrix}.$$

The matrix $db(\widetilde{L}^{++}_{[e]})$ is also symmetric, positive semi-definite, positive definite if $n^+ \geq 2$, and its nullspace is given by $N\big(db(\widetilde{L}^{++}_{[e]})\big) = \mathbb{R}\mathcal{U}^{+}_{[e]}$ if $n^+ = 1$. In addition, the matrix $db(\widetilde{L}^{--}_{[e]})$ is symmetric positive definite for any $n^+ \geq 1$. As a consequence, $db(\widetilde{L}_{[e]})$ is symmetric positive definite for $n^+ \geq 2$, whereas it is symmetric positive semi-definite with nullspace $N\big(db(\widetilde{L}_{[e]})\big) = \mathbb{R}\Gamma^{D}_{[e]}(\mathcal{U}^{+}_{[e]}, 0)$ for $n^+ = 1$.

3.4.6 The Rescaled System $\widetilde{L}_{[00]}\widetilde{\alpha}_{[00]}^{D_l} = \widetilde{\beta}_{[00]}^{D_l}$

The rescaled matrix associated with $L_{[00]}$ is denoted by $\widetilde{L}_{[00]}$ and is defined, for convenience, by $\widetilde{L}_{[00]kl}^{rs} = (2\bar{n}\sqrt{m_k m_l}/3\bar{p})\langle\!\langle\!\langle \phi^{10rk}, \Im(\phi^{10sl})\rangle\!\rangle\!\rangle$ for $(r,k),(s,l) \in \mathcal{B}_{[00]}^{D}$. Denoting by $\mathcal{X}_{[00]}^{D}$ the diagonal matrix $\mathcal{X}_{[00]}^{D} = \operatorname{diag}\big((X_k)_{(r,k)\in\mathcal{B}_{[00]}^{D}}\big)$,

$$\mathcal{X}_{[00]}^{D} = \operatorname{diag}(X_1,\dots,X_n), \tag{3.4.48}$$

the matrix $\widetilde{L}_{[00]}$ is then such that

$$\mathcal{X}_{[00]}^{D}\widetilde{L}_{[00]} = L_{[00]}. \tag{3.4.49}$$

In addition, we denote by $\mathcal{R} = \big(\mathcal{R}_{kl}^{rs}\big)_{(r,k)\in\mathcal{B}_{[00]}^{D},(s,l)\in\mathcal{B}^{D}}$ the corresponding transformation matrix already given in (2.6.40). From the definition of $\widetilde{L}$ and $\widetilde{L}_{[00]}$ and from the linear relations between the basis functions of the approximation spaces $\mathcal{A}^{D}$ and $\mathcal{A}_{[00]}^{D}$, it is readily seen that we also have the relation

$$\widetilde{L}_{[00]} = \mathcal{R}\widetilde{L}\mathcal{R}^{t}, \tag{3.4.50}$$

which yields

$$\widetilde{L}_{[00]} = \widetilde{L}^{0000}. \tag{3.4.51}$$

In addition, we define new quantities $\widetilde{\beta}_{[00]}^{D_l}$, $l \in \mathcal{S}$, by

$$\widetilde{\beta}_{[00]}^{D_l} = \mathcal{R}\widetilde{\beta}^{D_l}. \tag{3.4.52}$$

Noting that $\mathcal{X}_{[00]}^{D}\mathcal{R} = \mathcal{R}\mathcal{X}^{D}$, we have $\mathcal{X}_{[00]}^{D}\widetilde{\beta}_{[00]}^{D_l} = \mathcal{X}_{[00]}^{D}\mathcal{R}\widetilde{\beta}^{D_l} = \mathcal{R}\mathcal{X}^{D}\widetilde{\beta}^{D_l}$. Using then the relations $\mathcal{X}^{D}\widetilde{\beta}^{D_l} = Y_l\beta^{D_l}$ and $\beta_{[00]}^{D_l} = \mathcal{R}\beta^{D_l}$, obtained from (3.4.9) and (2.6.36), we thus conclude that

$$\mathcal{X}_{[00]}^{D}\widetilde{\beta}_{[00]}^{D_l} = Y_l\beta_{[00]}^{D_l}, \qquad l \in \mathcal{S}. \tag{3.4.53}$$

We also obtain from (3.4.52) that

$$\widetilde{\beta}_{[00]k}^{D_l} = \widetilde{\beta}_k^{00D_l} = \frac{m_k}{m}\delta_{kl} - \frac{m_k}{m}Y_l, \qquad k,l \in \mathcal{S}. \tag{3.4.54}$$

Denoting by $\widetilde{\alpha}_{[00]}^{D_l}$, $l \in \mathcal{S}$, the solution of the rescaled transport linear system

$$\begin{cases} \widetilde{L}_{[00]}\widetilde{\alpha}_{[00]}^{D_l} = \widetilde{\beta}_{[00]}^{D_l}, \\ \langle Y, \widetilde{\alpha}_{[00]}^{D_l}\rangle = 0, \end{cases} \tag{3.4.55}$$

where the constraint vector $Y = \mathcal{R}\mathcal{L}$ has already been introduced in Section 2.6.5, the flux diffusion matrix $\widetilde{D}_{[00]}$ is defined by

$$\widetilde{D}_{[00]kl} = \langle \widetilde{\alpha}^{D_k}_{[00]}, \beta^{D_l}_{[00]} \rangle = \langle \widetilde{\alpha}^{00D_k}_{[00]}, \beta^{00D_l}_{[00]} \rangle, \qquad k,l \in \mathcal{S}, \tag{3.4.56}$$

which can be simplified into

$$\widetilde{D}_{[00]kl} = \widetilde{\alpha}^{00D_k}_{[00]l}, \qquad k,l \in \mathcal{S}, \tag{3.4.57}$$

by explicitly using the constraint in (3.4.55). When the mass fractions are positive, the matrix $\mathcal{X}^D_{[00]}$ is then invertible and we have the relations $\widetilde{L}_{[00]} = (\mathcal{X}^D_{[00]})^{-1} L_{[00]}$ and $\widetilde{\beta}^{D_k}_{[00]} = Y_k (\mathcal{X}^D_{[00]})^{-1} \beta^{D_k}_{[00]}$. As a consequence, we deduce that the quantity $x = (1/Y_k)\widetilde{\alpha}^{D_k}_{[00]}$ satisfies $L_{[00]}x = \beta^{D_k}_{[00]}$ and $\langle Y, x\rangle = 0$ so that $x = \alpha^{D_k}_{[00]}$ and $\widetilde{\alpha}^{D_k}_{[00]} = Y_k \alpha^{D_k}_{[00]}$. Therefore, we have $\widetilde{D}_{[00]kl} = Y_k D_{[00]kl}$, and definitions (3.4.56)(3.4.57) and (2.6.45)(2.6.46) are then equivalent.

3.4.7 Mathematical Structure of the Rescaled System $\widetilde{L}_{[00]}\widetilde{\alpha}^{D_l}_{[00]} = \widetilde{\beta}^{D_l}_{[00]}$

Properties of $\mathcal{A}^D_{[00]}$. The functional space $\mathcal{A}^D_{[00]}$ introduced in Section 2.6 to approximate ϕ^{D_k} satisfies the perpendicularity property (3.1.1), the species orthogonality property (3.1.2), and the relation (3.1.60) from (2.3.51). In addition, the basis functions satisfy the species localization property (3.1.3). As a consequence, all the results established in Section 3.1.6 apply to the matrices $L_{[00]}$ and $\widetilde{L}_{[00]}$. Note, however, that these results do not apply to the system $L_{[00]}\alpha^{D_l}_{[00]} = \beta^{D_l}_{[00]}$ since Ψ^{D_l} is not defined for $n_l = 0$.

Properties of $L_{[00]}$. Denoting by $\Gamma^D_{[00]}$ the permutation matrix associated with the reordering of $\mathcal{B}^D_{[00]}$ into $(\mathcal{B}^{D+}_{[00]}, \mathcal{B}^{D-}_{[00]})$, the matrix $L_{[00]}$ admits the block-decomposition

$$(\Gamma^D_{[00]})^t L_{[00]} \Gamma^D_{[00]} = \begin{bmatrix} L^{++}_{[00]} & 0 \\ 0 & 0 \end{bmatrix}.$$

The matrix $L^{++}_{[00]}$ is symmetric positive semi-definite and $N(L^{++}_{[00]}) = \mathbb{R}U^+$, where $U_k = 1$, $k \in \mathcal{S}$. The matrix $L^{++}_{[00]}$ is positive definite on the constrained space $(Y^+)^\perp$ where we have $Y = (Y_k)_{k\in\mathcal{S}}$. Moreover, we have $\langle Y^+, U^+\rangle \neq 0$. We also have $\beta^{D_l+}_{[00]} \in R(L^{++}_{[00]})$, since an explicit calculation yields $\langle \beta^{D_l+}_{[00]}, U^+\rangle = 0$.

The matrix $db(L^{++}_{[00]})$ is positive definite for any $n^+ \geq 2$, whereas for $n^+ = 1$, $db(L^{++}_{[00]}) = L^{++}_{[00]} = [0]$ and $N\big(db(L^{++}_{[00]})\big) = \mathbb{R}U^+$. Finally, the matrix $2db(L^{++}_{[00]}) - L^{++}_{[00]}$

is positive definite if $n^+ \geq 3$, whereas its nullspace is $N\big(2db(L^{++}_{[00]}) - L^{++}_{[00]}\big) = \mathbb{R}U^{*+}$ for $n^+ = 2$, $\mathcal{S}^+ = \{1,2\}$, where U^{*+} is defined by $U^{*+}_1 = 1$, and $U^{*+}_2 = -1$, i.e., $U^{*+} = (1,-1)$, and its nullspace is $N\big(2db(L^{++}_{[00]}) - L^{++}_{[00]}\big) = \mathbb{R}U^+$ if $n^+ = 1$.

Properties of $\widetilde{L}_{[00]}$. The matrix $\widetilde{L}_{[00]}$ admits the block-decomposition

$$(\Gamma^D_{[00]})^t \widetilde{L}_{[00]} \Gamma^D_{[00]} = \begin{bmatrix} \widetilde{L}^{++}_{[00]} & 0 \\ \widetilde{L}^{-+}_{[00]} & \widetilde{L}^{--}_{[00]} \end{bmatrix}.$$

The matrix $\widetilde{L}^{++}_{[00]}$ also satisfies $N(\widetilde{L}^{++}_{[00]}) = \mathbb{R}U^+$ and $R(\widetilde{L}^{++}_{[00]}) = (\widetilde{U}^+)^\perp$, where we have introduced the notation

$$\widetilde{U} = \mathcal{X}^D_{[00]} U,$$

so that $\widetilde{U}_k = X_k$, $k \in \mathcal{S}$. We also have $\widetilde{\beta}^{D_k+}_{[00]} \in R(\widetilde{L}^{++}_{[00]})$ since an explicit calculation yields that $\langle \widetilde{\beta}^{D_k+}_{[00]}, \widetilde{U}^+ \rangle = 0$. In addition, the matrix $\widetilde{L}^{--}_{[00]}$ is block-diagonal, i.e., $\widetilde{L}^{--}_{[00]} = db(\widetilde{L}^{--}_{[00]})$, and is symmetric positive definite for any $n^+ \geq 1$. Finally, $R(\widetilde{L}_{[00]}) = \widetilde{U}^\perp$, $N(\widetilde{L}_{[00]}) = \mathbb{R}U$, and an explicit calculation yields that $\widetilde{\beta}^{D_l}_{[00]} \in R(\widetilde{L}_{[00]})$.

The matrix $db(\widetilde{L}_{[00]})$ admits the block-decomposition

$$(\Gamma^D_{[00]})^t \, db(\widetilde{L}_{[00]}) \Gamma^D_{[00]} = \begin{bmatrix} db(\widetilde{L}^{++}_{[00]}) & 0 \\ 0 & db(\widetilde{L}^{--}_{[00]}) \end{bmatrix}.$$

The matrix $db(\widetilde{L}^{++}_{[00]})$ is also symmetric, positive semi-definite, positive definite if $n^+ \geq 2$, and its nullspace is given by $N\big(db(\widetilde{L}^{++}_{[00]})\big) = \mathbb{R}U^+$ if $n^+ = 1$. In addition, the matrix $db(\widetilde{L}^{--}_{[00]})$ is symmetric positive definite for any $n^+ \geq 1$. As a consequence, $db(\widetilde{L})$ is symmetric positive definite for $n^+ \geq 2$, whereas it is symmetric positive semi-definite with nullspace $N\big(db(\widetilde{L}_{[00]})\big) = \mathbb{R}\Gamma^D_{[00]}(U^+, 0)$ for $n^+ = 1$.

3.4.8 Alternative Definition of the Flux Diffusion Matrix

In Section 3.4.2 the flux diffusion coefficients $\widetilde{D}_{kl}$, $k, l \in \mathcal{S}$, have been defined in terms of the matrix $\widetilde{L}$ which is a left rescaled version of the matrix of L since $\mathcal{X}^D \widetilde{L} = L$. It is also possible, however, to use right rescaled matrices. Indeed, let $\overline{L}$ be defined such that $\overline{L}^{rs}_{kl} = (2\rho\sqrt{m_k}/3\overline{p}\sqrt{m_l})\langle\!\langle\!\langle \Im(\phi^{10rk}), \phi^{10sl} \rangle\!\rangle\!\rangle$. Denoting then by $\mathcal{Y}^D \in \mathbb{R}^{2n+p,2n+p}$ the diagonal matrix defined by $\mathcal{Y}^D = \big((Y_k)_{(r,k)\in\mathcal{B}^D}\big)$

$$\mathcal{Y}^D = \mathrm{diag}(Y_1, \ldots, Y_n, Y_1, \ldots, Y_n, Y_{i_1}, \ldots, Y_{i_p}), \tag{3.4.58}$$

where $\{i_1, \ldots, i_p\} = \mathcal{P}$ are the polyatomic species, we have

$$\overline{L}\mathcal{Y}^D = L, \tag{3.4.59}$$

so that $\overline{L}$ is a right rescaled version of the matrix L. The choice of the matrix $\mathcal{Y}^D$ instead of $\mathcal{X}^D$ is made for convenience since we want to evaluate the fluxes $Y_k V_k$, $k \in \mathcal{S}$. We also have the relation

$$\overline{L}^t = (\mathcal{W}^D)^{-1}\widetilde{L}, \tag{3.4.60}$$

where $\mathcal{W}^D$ is given by $\mathcal{W}^D = \mathrm{diag}\big((m_k/m)_{(r,k)\in\mathcal{B}^D}\big)$ and where $m = \sum_{i\in\mathcal{S}} X_i m_i$. The rescaled matrix $\overline{L}$ can then be evaluated from (3.4.60) and (3.4.10)–(3.4.27).

In order to derive the properties of the matrix $\overline{L}$, a theory of right rescaled systems could easily be developed, as for left rescaled systems. However, using (3.4.60), the results already obtained for the matrix $\widetilde{L}$ can readily be rewritten in terms of the matrix $\overline{L}$. In particular, the nullspace and range of the matrix $\overline{L}$ are given by $N(\overline{L}) = \mathbb{R}\mathcal{L}$ and $R(\overline{L}) = \mathcal{U}^\perp$, and the linear system

$$\begin{cases} \overline{L}\overline{\alpha}^{D_l} = \beta^{D_l}, \\ \langle \mathcal{U}, \overline{\alpha}^{D_l}\rangle = 0, \end{cases} \tag{3.4.61}$$

admits a unique solution $\overline{\alpha}^{D_l}$. We further introduce the new quantities $\overline{\beta}^{D_l} \in \mathbb{R}^{2n+p}$ defined by

$$\overline{\beta}_k^{00D_l} = \delta_{kl} - Y_l, \qquad k,l \in \mathcal{S}, \tag{3.4.62}$$

$$\overline{\beta}_k^{10D_l} = 0, \qquad k,l \in \mathcal{S}, \tag{3.4.63}$$

$$\overline{\beta}_k^{01D_l} = 0, \qquad k \in \mathcal{P}, \quad l \in \mathcal{S}, \tag{3.4.64}$$

and the flux diffusion coefficients are then defined by

$$\widetilde{D}_{kl} = \langle \overline{\alpha}^{D_l}, \overline{\beta}^{D_k}\rangle = \langle \overline{\alpha}^{00D_l}, \overline{\beta}^{00D_k}\rangle, \qquad k,l \in \mathcal{S}, \tag{3.4.65}$$

which can be simplified into

$$\widetilde{D}_{kl} = \overline{\alpha}_k^{00D_l}, \qquad k,l \in \mathcal{S}, \tag{3.4.66}$$

by explicitly using the constraint in (3.4.61). When the mass fractions are positive, the matrix $\mathcal{Y}^D$ is invertible, and $\overline{L} = L(\mathcal{Y}^D)^{-1}$. Introducing the quantity $x = (\mathcal{Y}^D)^{-1}\overline{\alpha}^{D_l}$, we obtain that $Lx = \beta^{D_l}$ and $\langle \mathcal{L}, x\rangle = \langle \mathcal{Y}^D\mathcal{U}, x\rangle = \langle \mathcal{U}, \overline{\alpha}^{D_l}\rangle = 0$. As a consequence, we have $x = \alpha^{D_l}$ so that $\overline{\alpha}^{D_l} = \mathcal{Y}^D\alpha^{D_l}$ and $\widetilde{D}_{kl} = \overline{\alpha}_k^{00D_l} = Y_k\alpha_k^{00D_l} = Y_k D_{kl}$, and the definitions (3.4.65)(3.4.66) and (2.6.23)(2.6.24) are equivalent.

3.5 The Partial Thermal Conductivity and the Thermal Diffusion Vector

3.5.1 The Rescaled System $\widetilde{L}\alpha^{\lambda'} = \widetilde{\beta}^{\lambda'}$

The rescaled system associated with $L\alpha^{\lambda'} = \beta^{\lambda'}$ is denoted by $\widetilde{L}\alpha^{\lambda'} = \widetilde{\beta}^{\lambda'}$. The matrix $\widetilde{L}$ has already been introduced in Section 3.4.2, and its mathematical structure is given in Section 3.4.3. In particular, denoting by $\mathcal{X}^{\lambda'} \in \mathbb{R}^{2n+p,2n+p}$ the diagonal matrix $\mathcal{X}^{\lambda'} = \operatorname{diag}\big((X_k)_{(r,k)\in\mathcal{B}^{\lambda'}}\big) = \mathcal{X}^D$, we have the relation $\mathcal{X}^{\lambda'}\widetilde{L} = L$. On the other hand, the rescaled right members $\widetilde{\beta}^{\lambda'} \in \mathbb{R}^{2n+p}$ are easily evaluated to be

$$\widetilde{\beta}_k^{00\lambda'} = 0, \qquad k \in \mathcal{S}, \tag{3.5.1}$$

$$\widetilde{\beta}_k^{10\lambda'} = \frac{c_{\mathrm{p}}^{\mathrm{tr}}}{k_{\mathrm{B}}}, \qquad k \in \mathcal{S}, \tag{3.5.2}$$

$$\widetilde{\beta}_k^{10\lambda'} = \frac{c_k^{\mathrm{int}}}{k_{\mathrm{B}}}, \qquad k \in \mathcal{P}, \tag{3.5.3}$$

so that we have

$$\mathcal{X}^{\lambda'}\widetilde{\beta}^{\lambda'} = \beta^{\lambda'}. \tag{3.5.4}$$

Denoting by $\alpha^{\lambda'}$ the solution of the system

$$\begin{cases} \widetilde{L}\alpha^{\lambda'} = \widetilde{\beta}^{\lambda'}, \\ \langle \mathcal{L}, \alpha^{\lambda'} \rangle = 0, \end{cases} \tag{3.5.5}$$

the partial thermal conductivity λ' is defined by

$$\begin{aligned} \lambda' &= \frac{\overline{p}}{\overline{T}}\langle \alpha^{\lambda'}, \beta^{\lambda'} \rangle = \frac{\overline{p}}{\overline{T}}\langle \alpha^{10\lambda'}, \beta^{10\lambda'} \rangle + \frac{\overline{p}}{\overline{T}}\langle \alpha^{01\lambda'}, \beta^{01\lambda'} \rangle \\ &= \frac{\overline{p}}{k_{\mathrm{B}}\overline{T}}\Big(\sum_{k\in\mathcal{S}} X_k c_{\mathrm{p}}^{\mathrm{tr}} \alpha_k^{10\lambda'} + \sum_{k\in\mathcal{P}} X_k c_k^{\mathrm{int}} \alpha_k^{01\lambda'} \Big), \end{aligned} \tag{3.5.6}$$

and the thermal diffusion vector θ by

$$\theta_k = -\langle \alpha^{\lambda'}, \beta^{D_k} \rangle = -\langle \alpha^{00\lambda'}, \beta^{00D_k} \rangle, \qquad k \in \mathcal{S}, \tag{3.5.7}$$

which can be simplified into

$$\theta_k = -\alpha_k^{00\lambda'}, \qquad k \in \mathcal{S}, \tag{3.5.8}$$

by explicitly using the constraint in (3.5.5). When the mass fractions are positive, the matrix $\mathcal{X}^{\lambda'}$ is then invertible, and we have the relations $\widetilde{L} = (\mathcal{X}^{\lambda'})^{-1}L$ and $\widetilde{\beta}^{\lambda'} =$

$(\mathcal{X}^{\lambda'})^{-1}\beta^{\lambda'}$ so that $\widetilde{L}\alpha^{\lambda'} = \widetilde{\beta}^{\lambda'}$ if and only if $L\alpha^{\lambda'} = \beta^{\lambda'}$. As a consequence, the definitions (2.7.11) and (2.7.12) of the partial thermal conductivity and of the thermal diffusion vector are equivalent to the definitions (3.5.6) and (3.5.7).

3.5.2 The Rescaled System $\widetilde{L}_{[e]}\alpha^{\lambda'}_{[e]} = \widetilde{\beta}^{\lambda'}_{[e]}$

The rescaled system associated with $L_{[e]}\alpha^{\lambda'}_{[e]} = \beta^{\lambda'}_{[e]}$ is denoted by $\widetilde{L}_{[e]}\alpha^{\lambda'}_{[e]} = \widetilde{\beta}^{\lambda'}_{[e]}$. The matrix $\widetilde{L}_{[e]}$ has already been described in Section 3.4.4, and its mathematical properties are given in Section 3.4.5. In particular, denoting by $\mathcal{X}^{\lambda'}_{[e]} \in \mathbb{R}^{2n,2n}$ the diagonal matrix $\mathcal{X}^{\lambda'}_{[e]} = \operatorname{diag}\big((X_k)_{(r,k)\in\mathcal{B}^{\lambda'}_{[e]}}\big) = \mathcal{X}^{D}_{[e]}$, we have the relation $\mathcal{X}^{\lambda'}_{[e]}\widetilde{L}_{[e]} = L_{[e]}$, and $\widetilde{L}_{[e]}$ is given by $\mathcal{R}\widetilde{L}\mathcal{R}^t$ where $\mathcal{R}$ is given in (2.6.28). Similarly, the right member $\widetilde{\beta}^{\lambda'}_{[e]} = \mathcal{R}\widetilde{\beta}^{\lambda'}$ is easily evaluated to be

$$\widetilde{\beta}^{00\lambda'}_{[e]k} = 0, \qquad k \in \mathcal{S}, \tag{3.5.9}$$

$$\widetilde{\beta}^{e\lambda'}_{[e]k} = \frac{c_{\mathrm{p}}^{\mathrm{tr}} + c_k^{\mathrm{int}}}{k_{\mathrm{B}}}, \qquad k \in \mathcal{S}, \tag{3.5.10}$$

and we also have

$$\mathcal{X}^{\lambda'}_{[e]}\widetilde{\beta}^{\lambda'}_{[e]} = \beta^{\lambda'}_{[e]}. \tag{3.5.11}$$

Denoting by $\alpha^{\lambda'}_{[e]}$ the solution of the system

$$\begin{cases} \widetilde{L}_{[e]}\alpha^{\lambda'}_{[e]} = \widetilde{\beta}^{\lambda'}_{[e]}, \\ \langle \mathcal{L}_{[e]}, \alpha^{\lambda'}_{[e]} \rangle = 0, \end{cases} \tag{3.5.12}$$

the partial thermal conductivity $\lambda'_{[e]}$ is defined by

$$\begin{aligned} \lambda'_{[e]} &= \frac{\overline{p}}{\overline{T}}\langle \alpha^{\lambda'}_{[e]}, \beta^{\lambda'}_{[e]} \rangle = \frac{\overline{p}}{\overline{T}}\langle \alpha^{e\lambda'}_{[e]}, \beta^{e\lambda'}_{[e]} \rangle \\ &= \frac{\overline{p}}{k_{\mathrm{B}}\overline{T}} \sum_{k\in\mathcal{S}} X_k (c_{\mathrm{p}}^{\mathrm{tr}} + c_k^{\mathrm{int}}) \alpha^{e\lambda'}_{[e]k}, \end{aligned} \tag{3.5.13}$$

and the thermal diffusion vector $\theta_{[e]}$ by

$$\theta_{[e]k} = -\langle \alpha^{\lambda'}_{[e]}, \beta^{D_k}_{[e]} \rangle = -\langle \alpha^{00\lambda'}_{[e]}, \beta^{00D_k}_{[e]} \rangle, \qquad k \in \mathcal{S}, \tag{3.5.14}$$

which can be simplified into

$$\theta_{[e]k} = -\alpha^{00\lambda'}_{[e]k}, \qquad k \in \mathcal{S}, \tag{3.5.15}$$

making use of the constraint in (3.5.12). When the mass fractions are positive, the matrix $\mathcal{X}^{\lambda'}_{[e]}$ is then invertible, and we have the relations $\widetilde{L}_{[e]} = (\mathcal{X}^{\lambda'}_{[e]})^{-1} L_{[e]}$ and $\widetilde{\beta}^{\lambda'}_{[e]} = (\mathcal{X}^{\lambda'}_{[e]})^{-1} \beta^{\lambda'}_{[e]}$ so that $\widetilde{L}_{[e]}\alpha^{\lambda'}_{[e]} = \widetilde{\beta}^{\lambda'}_{[e]}$ if and only if $L_{[e]}\alpha^{\lambda'}_{[e]} = \beta^{\lambda'}_{[e]}$. As a consequence, the definitions (2.7.22) and (2.7.23) of the partial thermal conductivity and of the thermal diffusion vector are equivalent to the definitions (3.5.13) and (3.5.14).

3.6 The Thermal Conductivity and the Thermal Diffusion Ratios

3.6.1 The Rescaled System $\widetilde{\Lambda}\alpha^\lambda = \widetilde{\beta}^\lambda$

The rescaled system associated with $\Lambda\alpha^\lambda = \beta^\lambda$ is denoted by $\widetilde{\Lambda}\alpha^\lambda = \widetilde{\beta}^\lambda$ and is defined, for convenience, by $\widetilde{\Lambda}^{rs}_{kl} = (2\overline{n}\sqrt{m_k m_l}/3\overline{p})\langle\!\langle\!\langle \phi^{10rk}, \mathfrak{I}(\phi^{10sl})\rangle\!\rangle\!\rangle$, $(r,k),(s,l) \in \mathcal{B}^\lambda$, and $\widetilde{\beta}^{r\lambda}_k = (\overline{n}\sqrt{2m_k}/3\overline{p}\sqrt{k_B\overline{T}})\langle\!\langle\!\langle \phi^{10rk}, \Psi^{\lambda'}\rangle\!\rangle\!\rangle$, $(r,k) \in \mathcal{B}^\lambda$. Note here that we have used the fundamental property that $\langle\!\langle \phi^{10rk}, \Psi^{\lambda}\rangle\!\rangle = \langle\!\langle \phi^{10rk}, \Psi^{\lambda'}\rangle\!\rangle$, $(r,k) \in \mathcal{B}^\lambda$, for rescaling the right member. Denoting by $\mathcal{X}^\lambda$ the diagonal matrix $\mathcal{X}^\lambda = \mathrm{diag}\big((X_k)_{(r,k)\in\mathcal{B}^\lambda}\big)$

$$\mathcal{X}^\lambda = \mathrm{diag}(X_1, \ldots, X_n, X_{i_1}, \ldots, X_{i_p}), \tag{3.6.1}$$

where $\{i_1, \ldots, i_p\} = \mathcal{P}$ are the polyatomic species, we then have

$$\mathcal{X}^\lambda \widetilde{\Lambda} = \Lambda, \tag{3.6.2}$$

and

$$\mathcal{X}^\lambda \widetilde{\beta}^\lambda = \beta^\lambda. \tag{3.6.3}$$

It is also straightforward to establish that $\widetilde{\Lambda}$ is actually a submatrix of $\widetilde{L}$

$$\widetilde{\Lambda}^{1010}_{kl} = \widetilde{L}^{1010}_{kl}, \quad k \in \mathcal{S}, \quad l \in \mathcal{S}, \tag{3.6.4}$$

$$\widetilde{\Lambda}^{1001}_{kl} = \widetilde{L}^{1001}_{kl}, \quad k \in \mathcal{S}, \quad l \in \mathcal{P}, \tag{3.6.5}$$

$$\widetilde{\Lambda}^{0110}_{kl} = \widetilde{L}^{0110}_{kl}, \quad k \in \mathcal{P}, \quad l \in \mathcal{S}, \tag{3.6.6}$$

$$\widetilde{\Lambda}^{0101}_{kl} = \widetilde{L}^{0101}_{kl}, \quad k \in \mathcal{P}, \quad l \in \mathcal{P}, \tag{3.6.7}$$

and that $\widetilde{\beta}^\lambda$ is also a subvector of $\widetilde{\beta}^{\lambda'}$

$$\widetilde{\beta}^{10\lambda} = \widetilde{\beta}^{10\lambda'}, \tag{3.6.8}$$

$$\widetilde{\beta}^{01\lambda} = \widetilde{\beta}^{01\lambda'}. \tag{3.6.9}$$

Denoting by α^λ the solution of the system

$$\widetilde{\Lambda}\alpha^\lambda = \widetilde{\beta}^\lambda, \tag{3.6.10}$$

the thermal conductivity λ is then defined by

$$\begin{aligned}\lambda &= \frac{\overline{p}}{\overline{\overline{T}}}\langle\alpha^\lambda, \beta^\lambda\rangle = \frac{\overline{p}}{\overline{\overline{T}}}\langle\alpha^{10\lambda}, \beta^{10\lambda}\rangle + \frac{\overline{p}}{\overline{\overline{T}}}\langle\alpha^{01\lambda}, \beta^{01\lambda}\rangle \\ &= \frac{\overline{p}}{k_B\overline{\overline{T}}}\Big(\sum_{k\in S} X_k c_{\mathrm{p}}^{\mathrm{tr}} \alpha_k^{10\lambda} + \sum_{k\in \mathcal{P}} X_k c_k^{\mathrm{int}} \alpha_k^{01\lambda}\Big),\end{aligned} \tag{3.6.11}$$

and the thermal diffusion ratios χ are defined in terms of the matrix L and the vector α^λ with the relation

$$\chi = [L^{0010}, L^{0001}]\alpha^\lambda = L^{0010}\alpha^{10\lambda} + L^{0001}\alpha^{01\lambda}. \tag{3.6.12}$$

When the mass fractions are positive, the matrix $\mathcal{X}^\lambda$ is invertible, and we have $\widetilde{\Lambda}\alpha^\lambda = \widetilde{\beta}^\lambda$ if and only if $\Lambda\alpha^\lambda = \beta^\lambda$, since $\widetilde{\Lambda} = (\mathcal{X}^\lambda)^{-1}\Lambda$ and $\widetilde{\beta}^\lambda = (\mathcal{X}^\lambda)^{-1}\beta^\lambda$. As a consequence, the definitions (2.8.14) and (2.8.15) of the thermal conductivity and the thermal diffusion ratios are equivalent to the definitions (3.6.11) and (3.6.12).

Remark. We will establish in Section 4.6.3 that the thermal diffusion ratios are equivalently defined in terms of the thermal diffusion vector and the flux diffusion coefficients by the solution of the constrained linear system

$$\begin{cases} \widetilde{D}\chi = \mathcal{Y}\theta, \\ \langle U, \chi\rangle = 0, \end{cases} \tag{3.6.13}$$

where $\mathcal{Y}$ denotes the diagonal matrix $\mathcal{Y} = \mathrm{diag}(Y_1, \dots, Y_n)$. We will also establish that the relation $\lambda = \lambda' - (\overline{p}/\overline{T})\sum_{k\in S}\theta_k\chi_k$ is still valid for nonnegative mass fractions.

3.6.2 Mathematical Structure of the Rescaled System $\widetilde{\Lambda}\alpha^\lambda = \widetilde{\beta}^\lambda$

Properties of $\mathcal{A}^\lambda$. In Section 2.8 we have seen that the matrix Λ corresponds to a functional space $\mathcal{A}^\lambda$ satisfying the perpendicularity property (3.1.1), the species orthogonality property (3.1.2), and the relation (3.1.56). In addition, the basis functions satisfy the species localization property (3.1.3). As a consequence, the results (3.1.58)(3.1.59) established in Section 3.1.5 apply to the matrices Λ and $\widetilde{\Lambda}$.

Properties of Λ. Denoting by Γ^λ the permutation matrix associated with the reordering of $\mathcal{B}^\lambda$ into $(\mathcal{B}^{\lambda+}, \mathcal{B}^{\lambda-})$, the matrix Λ admits the block-decomposition

$$(\Gamma^\lambda)^t \Lambda \Gamma^\lambda = \begin{bmatrix} \Lambda^{++} & 0 \\ 0 & 0 \end{bmatrix}.$$

In addition, the matrices Λ^{++}, $db(\Lambda^{++})$, and $2db(\Lambda^{++}) - \Lambda^{++}$ are symmetric positive definite for any $n^+ \geq 1$.

Properties of $\widetilde{\Lambda}$. The matrix $\widetilde{\Lambda}$ admits the block-decomposition

$$(\Gamma^\lambda)^t \widetilde{\Lambda} \Gamma^\lambda = \begin{bmatrix} \widetilde{\Lambda}^{++} & 0 \\ \widetilde{\Lambda}^{-+} & \widetilde{\Lambda}^{--} \end{bmatrix}.$$

The matrix $\widetilde{\Lambda}^{++}$ is nonsingular and the matrix $\widetilde{\Lambda}^{--}$ is block-diagonal, i.e., $\widetilde{\Lambda}^{--} = db(\widetilde{\Lambda}^{--})$, and is symmetric positive definite for any $n^+ \geq 1$. As a consequence, $\widetilde{\Lambda}$ is nonsingular for any $n^+ \geq 1$.

The matrix $db(\widetilde{\Lambda})$ admits the block-decomposition

$$(\Gamma^\lambda)^t \, db(\widetilde{\Lambda}) \, \Gamma^\lambda = \begin{bmatrix} db(\widetilde{\Lambda}^{++}) & 0 \\ 0 & db(\widetilde{\Lambda}^{--}) \end{bmatrix}.$$

The matrix $db(\widetilde{\Lambda}^{++})$ is symmetric positive definite, and so is the matrix $db(\widetilde{\Lambda}^{--})$. As a consequence, $db(\widetilde{\Lambda})$ is symmetric positive definite for any $n^+ \geq 1$.

3.6.3 The Rescaled System $\widetilde{\Lambda}_{[e]}\alpha^\lambda_{[e]} = \widetilde{\beta}^\lambda_{[e]}$

The rescaled system associated with $\Lambda_{[e]}\alpha^\lambda_{[e]} = \beta^\lambda_{[e]}$ is denoted by $\widetilde{\Lambda}_{[e]}\alpha^\lambda_{[e]} = \widetilde{\beta}^\lambda_{[e]}$ and is defined, for convenience, by $\widetilde{\Lambda}^{rs}_{[e]kl} = (2\bar{n}\sqrt{m_k m_l}/3\bar{p})\langle\!\langle\!\langle \phi^{10rk}, \Im(\phi^{10sl})\rangle\!\rangle\!\rangle$, $(r,k),(s,l) \in \mathcal{B}^\lambda_{[e]}$, and $\widetilde{\beta}^{r\lambda}_{[e]k} = (\bar{n}\sqrt{2m_k}/3\bar{p}\sqrt{k_{\mathrm{B}}\overline{T}})\langle\!\langle\!\langle \phi^{10rk}, \Psi^{\lambda'}\rangle\!\rangle\!\rangle$, $(r,k) \in \mathcal{B}^\lambda_{[e]}$. Note here that we have used the fundamental property that $\langle\!\langle \phi^{10rk}, \Psi^\lambda \rangle\!\rangle = \langle\!\langle \phi^{10rk}, \Psi^{\lambda'} \rangle\!\rangle$, $(r,k) \in \mathcal{B}^\lambda_{[e]}$, for rescaling the right member. Denoting then by $\mathcal{X}^\lambda_{[e]}$ the diagonal matrix $\mathcal{X}^\lambda_{[e]} = \mathrm{diag}\big((X_k)_{(r,k)\in\mathcal{B}^\lambda_{[e]}}\big)$

$$\mathcal{X}^\lambda_{[e]} = \mathrm{diag}(X_1, \ldots, X_n), \tag{3.6.14}$$

we have

$$\mathcal{X}^\lambda_{[e]}\widetilde{\Lambda}_{[e]} = \Lambda_{[e]}, \tag{3.6.15}$$

and

$$\mathcal{X}^\lambda_{[e]}\widetilde{\beta}^\lambda_{[e]} = \beta^\lambda_{[e]}. \tag{3.6.16}$$

It is then straightforward to establish that the matrix $\widetilde{\Lambda}_{[e]}$ and the right member $\widetilde{\beta}^{\lambda}_{[e]}$ are actually a submatrix of $\widetilde{L}_{[e]}$ and a subvector of $\widetilde{\beta}^{\lambda'}_{[e]}$, respectively, since

$$\widetilde{\Lambda}_{[e]} = \widetilde{L}^{ee}_{[e]}, \tag{3.6.17}$$

and

$$\widetilde{\beta}^{\lambda}_{[e]} = \widetilde{\beta}^{e\lambda'}_{[e]}. \tag{3.6.18}$$

Denoting by $\alpha^{\lambda}_{[e]}$ the solution of the system

$$\widetilde{\Lambda}_{[e]}\alpha^{\lambda}_{[e]} = \widetilde{\beta}^{\lambda}_{[e]}, \tag{3.6.19}$$

the thermal conductivity $\lambda_{[e]}$ is defined by

$$\begin{aligned} \lambda_{[e]} &= \frac{\bar{p}}{\overline{T}}\langle \alpha^{\lambda}_{[e]}, \beta^{\lambda}_{[e]} \rangle = \frac{\bar{p}}{\overline{T}}\langle \alpha^{e\lambda}_{[e]}, \beta^{e\lambda}_{[e]} \rangle \\ &= \frac{\bar{p}}{k_B \overline{T}} \sum_{k\in\mathcal{S}} X_k (c^{tr}_p + c^{int}_k) \alpha^{e\lambda}_{[e]k}, \end{aligned} \tag{3.6.20}$$

and the thermal diffusion ratios $\chi_{[e]}$ are expressed in terms of the matrix $L_{[e]}$ and the vector $\alpha^{\lambda}_{[e]}$ with the relation

$$\chi_{[e]} = [L^{00e}_{[e]}]\alpha^{\lambda}_{[e]} = L^{00e}_{[e]}\alpha^{e\lambda}_{[e]}. \tag{3.6.21}$$

When the mass fractions are positive, the matrix $\mathcal{X}^{\lambda}_{[e]}$ is invertible, and we have $\widetilde{\Lambda}_{[e]}\alpha^{\lambda}_{[e]} = \widetilde{\beta}^{\lambda}_{[e]}$ if and only if $\Lambda_{[e]}\alpha^{\lambda}_{[e]} = \beta^{\lambda}_{[e]}$, since $\widetilde{\Lambda}_{[e]} = (\mathcal{X}^{\lambda}_{[e]})^{-1}\Lambda_{[e]}$ and $\widetilde{\beta}^{\lambda}_{[e]} = (\mathcal{X}^{\lambda}_{[e]})^{-1}\beta^{\lambda}_{[e]}$. As a consequence, the definitions (2.8.22) and (2.8.23) of the thermal conductivity and the thermal diffusion ratios are equivalent to the definitions (3.6.20) and (3.6.21) in this situation.

Remark. We will establish in Section 4.6.5 that the thermal diffusion ratios $\chi_{[e]}$ are equivalently defined in terms of the thermal diffusion vector $\theta_{[e]}$ and the flux diffusion coefficients $\widetilde{D}_{[e]}$ by the solution of the constrained linear system

$$\begin{cases} \widetilde{D}_{[e]}\chi_{[e]} = \mathcal{Y}\theta_{[e]}, \\ \langle U, \chi_{[e]} \rangle = 0, \end{cases} \tag{3.6.22}$$

where $\mathcal{Y} = \mathrm{diag}(Y_1, \dots, Y_n)$. We will also establish that the relation $\lambda_{[e]} = \lambda'_{[e]} - (\bar{p}/\overline{T})\sum_{k\in\mathcal{S}} \theta_{[e]k}\chi_{[e]k}$ is still valid for nonnegative mass fractions.

3.6.4 Mathematical Structure of the Rescaled System $\widetilde{\Lambda}_{[e]}\alpha^{\lambda}_{[e]} = \widetilde{\beta}^{\lambda}_{[e]}$

Properties of $\mathcal{A}^{\lambda}_{[e]}$. In Section 2.8.3 we have seen that the matrix $\Lambda_{[e]}$ corresponds to a functional space $\mathcal{A}^{\lambda}_{[e]}$ satisfying the perpendicularity property (3.1.1), the species orthogonality property (3.1.2), and the relation (3.1.56). In addition, the basis functions satisfy the species localization property (3.1.3). As a consequence, all the results (3.1.58)(3.1.59) established in Section 3.1.5 apply to the matrices $\Lambda_{[e]}$ and $\widetilde{\Lambda}_{[e]}$.

Properties of $\Lambda_{[e]}$. Denoting by $\Gamma^{\lambda}_{[e]}$ the permutation matrix associated with the reordering of $\mathcal{B}^{\lambda}_{[e]}$ into $(\mathcal{B}^{\lambda+}_{[e]}, \mathcal{B}^{\lambda-}_{[e]})$, the matrix $\Lambda_{[e]}$ admits the block-decomposition

$$(\Gamma^{\lambda}_{[e]})^t \,\Lambda_{[e]}\, \Gamma^{\lambda}_{[e]} = \begin{bmatrix} \Lambda^{++}_{[e]} & 0 \\ 0 & 0 \end{bmatrix}.$$

In addition, the matrices $\Lambda^{++}_{[e]}$, $db(\Lambda^{++}_{[e]})$, and $2db(\Lambda^{++}_{[e]}) - \Lambda^{++}_{[e]}$ are positive definite for any $n^+ \geq 1$.

Properties of $\widetilde{\Lambda}_{[e]}$. The matrix $\widetilde{\Lambda}_{[e]}$ admits the following block-decomposition

$$(\Gamma^{\lambda}_{[e]})^t \,\widetilde{\Lambda}_{[e]}\, \Gamma^{\lambda}_{[e]} = \begin{bmatrix} \widetilde{\Lambda}^{++}_{[e]} & 0 \\ \widetilde{\Lambda}^{-+}_{[e]} & \widetilde{\Lambda}^{--}_{[e]} \end{bmatrix}.$$

The matrix $\widetilde{\Lambda}^{++}_{[e]}$ is nonsingular, and the matrix $\widetilde{\Lambda}^{--}_{[e]}$ is block-diagonal, i.e., $\widetilde{\Lambda}^{--}_{[e]} = db(\widetilde{\Lambda}^{--}_{[e]})$, and is symmetric positive definite for any $n^+ \geq 1$. As a consequence, $\widetilde{\Lambda}_{[e]}$ is nonsingular for any $n^+ \geq 1$.

The matrix $db(\widetilde{\Lambda}_{[e]})$ admits the block-decomposition

$$(\Gamma^{\lambda}_{[e]})^t \, db(\widetilde{\Lambda}_{[e]})\, \Gamma^{\lambda}_{[e]} = \begin{bmatrix} db(\widetilde{\Lambda}^{++}_{[e]}) & 0 \\ 0 & db(\widetilde{\Lambda}^{--}_{[e]}) \end{bmatrix}.$$

The matrix $db(\widetilde{\Lambda}^{++}_{[e]})$ is symmetric positive definite, and so is the matrix $db(\widetilde{\Lambda}^{--}_{[e]})$. As a consequence, $db(\widetilde{\Lambda}_{[e]})$ is symmetric positive definite for any $n^+ \geq 1$.

3.7 The Mason and Monchick Approximations

We give here the expressions of the matrices $\widetilde{H}$, $\widetilde{K}$, $\widetilde{L}$, and $\widetilde{L}_{[e]}$, under the approximations [MM62] [MPM65] described in Section 2.10 since they are often used in practice and will be used in the following sections. Note that the matrices $\widetilde{K}_{[01]}$, $\widetilde{L}_{[00]}$, $\widetilde{\Lambda}$, and $\widetilde{\Lambda}_{[e]}$ are then easily obtained as blocks of $\widetilde{K}$, $\widetilde{L}$, and $\widetilde{L}_{[e]}$.

Under these approximations, the matrix $\widetilde{H}$ is given by

$$\widetilde{H}_{kk}^{0000} = \sum_{\substack{l\in\mathcal{S}\\ l\neq k}} 2\frac{X_l}{\eta_{kl}}\Big[\frac{5}{3\bar{A}_{kl}}\frac{m_k m_l}{(m_k+m_l)^2} + \frac{m_l^2}{(m_k+m_l)^2}\Big] + \frac{X_k}{\eta_k}, \qquad k\in\mathcal{S}, \quad (3.7.1)$$

$$\widetilde{H}_{kl}^{0000} = 2\frac{X_l}{\eta_{kl}}\frac{m_k m_l}{(m_k+m_l)^2}\Big[-\frac{5}{3\bar{A}_{kl}}+1\Big], \qquad k,l\in\mathcal{S}, \qquad k\neq l, \quad (3.7.2)$$

the matrix $\widetilde{K}$ is given by

$$\widetilde{K}_{kk}^{1010} = \sum_{\substack{l\in\mathcal{S}\\ l\neq k}} 5\frac{X_l}{\bar{A}_{kl}\eta_{kl}}\frac{m_k m_l}{(m_k+m_l)^2} + \frac{4}{k_B\pi}\frac{X_l}{\eta_{kl}}\frac{m_l^2}{(m_k+m_l)^2}\Big(\frac{c_k^{\rm int}}{\xi_k^{\rm int}}+\frac{c_l^{\rm int}}{\xi_l^{\rm int}}\Big)$$
$$+\frac{4}{k_B\pi}\frac{X_k}{\eta_k}\frac{c_k^{\rm int}}{\xi_k^{\rm int}}, \qquad k\in\mathcal{S}, \quad (3.7.3)$$

$$\widetilde{K}_{kl}^{1010} = -5\frac{X_l}{\bar{A}_{kl}\eta_{kl}}\frac{m_k m_l}{(m_k+m_l)^2}$$
$$+\frac{4}{k_B\pi}\frac{X_l}{\eta_{kl}}\frac{m_k m_l}{(m_k+m_l)^2}\Big(\frac{c_k^{\rm int}}{\xi_k^{\rm int}}+\frac{c_l^{\rm int}}{\xi_l^{\rm int}}\Big), \qquad k,l\in\mathcal{S}, \quad k\neq l, \quad (3.7.4)$$

$$\widetilde{K}_{kk}^{1001} = \sum_{\substack{l\in\mathcal{S}\\ l\neq k}} -\frac{4}{k_B\pi}\frac{X_l}{\eta_{kl}}\frac{m_l}{(m_k+m_l)}\frac{c_k^{\rm int}}{\xi_k^{\rm int}} - \frac{4}{k_B\pi}\frac{X_k}{\eta_k}\frac{c_k^{\rm int}}{\xi_k^{\rm int}}, \qquad k\in\mathcal{P}, \quad (3.7.5)$$

$$\widetilde{K}_{kl}^{1001} = -\frac{4}{k_B\pi}\frac{X_l}{\eta_{kl}}\frac{m_l}{(m_k+m_l)}\frac{c_l^{\rm int}}{\xi_l^{\rm int}}, \qquad k\in\mathcal{S}, \quad l\in\mathcal{P}, \quad k\neq l, \quad (3.7.6)$$

$$\widetilde{K}_{kk}^{0110} = \sum_{\substack{l\in\mathcal{S}\\ l\neq k}} -\frac{4}{k_B\pi}\frac{X_l}{\eta_{kl}}\frac{m_l}{(m_k+m_l)}\frac{c_k^{\rm int}}{\xi_k^{\rm int}} - \frac{4}{k_B\pi}\frac{X_k}{\eta_k}\frac{c_k^{\rm int}}{\xi_k^{\rm int}}, \qquad k\in\mathcal{P}, \quad (3.7.7)$$

$$\widetilde{K}_{kl}^{0110} = -\frac{4}{k_B\pi}\frac{X_l}{\eta_{kl}}\frac{m_k}{(m_k+m_l)}\frac{c_k^{\rm int}}{\xi_k^{\rm int}}, \qquad k\in\mathcal{P}, \quad l\in\mathcal{S}, \quad k\neq l, \quad (3.7.8)$$

$$\widetilde{K}_{kk}^{0101} = \sum_{\substack{l\in\mathcal{S}\\ l\neq k}} \frac{4}{k_B\pi}\frac{X_l}{\eta_{kl}}\frac{c_k^{\rm int}}{\xi_k^{\rm int}} + \frac{4}{k_B\pi}\frac{X_k}{\eta_k}\frac{c_k^{\rm int}}{\xi_k^{\rm int}}, \qquad k\in\mathcal{P}, \quad (3.7.9)$$

$$\widetilde{K}_{kl}^{0101} = 0, \qquad k,l\in\mathcal{P}, \quad k\neq l, \quad (3.7.10)$$

the matrix $\widetilde{L}$ is given by

$$\widetilde{L}_{kk}^{0000} = \sum_{\substack{l\in\mathcal{S}\\ l\neq k}} \frac{X_l}{\mathcal{D}_{kl}}, \qquad k\in\mathcal{S}, \quad (3.7.11)$$

$$\widetilde{L}_{kl}^{0000} = -\frac{X_l}{\mathcal{D}_{kl}}, \qquad k,l \in \mathcal{S}, \quad k \neq l, \tag{3.7.12}$$

$$\widetilde{L}_{kk}^{0010} = -\sum_{\substack{l\in\mathcal{S}\\ l\neq k}} \frac{X_l}{2\mathcal{D}_{kl}} \frac{m_l}{m_k+m_l}(6\bar{c}_{kl}-5), \qquad k \in \mathcal{S}, \tag{3.7.13}$$

$$\widetilde{L}_{kl}^{0010} = \frac{X_l}{2\mathcal{D}_{kl}} \frac{m_k}{m_k+m_l}(6\bar{c}_{kl}-5), \qquad k,l \in \mathcal{S}, \quad k \neq l, \tag{3.7.14}$$

$$\widetilde{L}_{kl}^{0001} = 0, \qquad k \in \mathcal{S}, \quad l \in \mathcal{P}, \tag{3.7.15}$$

$$\widetilde{L}_{kk}^{1000} = -\sum_{\substack{l\in\mathcal{S}\\ l\neq k}} \frac{X_l}{2\mathcal{D}_{kl}} \frac{m_l}{m_k+m_l}(6\bar{c}_{kl}-5), \qquad k \in \mathcal{S}, \tag{3.7.16}$$

$$\widetilde{L}_{kl}^{1000} = \frac{X_l}{2\mathcal{D}_{kl}} \frac{m_l}{m_k+m_l}(6\bar{c}_{kl}-5), \qquad k,l \in \mathcal{S}, \quad k \neq l, \tag{3.7.17}$$

$$\begin{aligned}\widetilde{L}_{kk}^{1010} = &\sum_{\substack{l\in\mathcal{S}\\ l\neq k}} \frac{X_l}{\mathcal{D}_{kl}} \frac{m_k m_l}{(m_k+m_l)^2}\Big[\frac{15}{2}\frac{m_k}{m_l} + \frac{25}{4}\frac{m_l}{m_k} - 3\frac{m_l}{m_k}\bar{B}_{kl} + 4\bar{A}_{kl} \\ &+ \frac{20}{3}\frac{\bar{A}_{kl}}{k_{\mathrm{B}}\pi}\Big(\frac{c_k^{\mathrm{int}}}{\xi_k^{\mathrm{int}}} + \frac{c_l^{\mathrm{int}}}{\xi_l^{\mathrm{int}}}\Big)\Big] + \frac{X_k}{\mathcal{D}_{kk}}\Big[2\bar{A}_{kk} + \frac{20}{3}\frac{\bar{A}_{kk}}{k_{\mathrm{B}}\pi}\frac{c_k^{\mathrm{int}}}{\xi_k^{\mathrm{int}}}\Big], \qquad k \in \mathcal{S},\end{aligned} \tag{3.7.18}$$

$$\begin{aligned}\widetilde{L}_{kl}^{1010} = &-\frac{X_l}{\mathcal{D}_{kl}} \frac{m_k m_l}{(m_k+m_l)^2}\Big[\frac{55}{4} - 3\bar{B}_{kl} - 4\bar{A}_{kl} \\ &- \frac{20}{3}\frac{\bar{A}_{kl}}{k_{\mathrm{B}}\pi}\Big(\frac{c_k^{\mathrm{int}}}{\xi_k^{\mathrm{int}}} + \frac{c_l^{\mathrm{int}}}{\xi_l^{\mathrm{int}}}\Big)\Big], \qquad k,l \in \mathcal{S}, \quad k \neq l,\end{aligned} \tag{3.7.19}$$

$$\widetilde{L}_{kk}^{1001} = -\sum_{\substack{l\in\mathcal{S}\\ l\neq k}} 4\frac{\bar{A}_{kl}}{k_{\mathrm{B}}\pi} \frac{X_l}{\mathcal{D}_{kl}} \frac{m_k}{m_k+m_l} \frac{c_k^{\mathrm{int}}}{\xi_k^{\mathrm{int}}} - 4\frac{\bar{A}_{kk}}{k_{\mathrm{B}}\pi}\frac{X_k}{\mathcal{D}_{kk}}\frac{c_k^{\mathrm{int}}}{\xi_k^{\mathrm{int}}}, \qquad k \in \mathcal{P}, \tag{3.7.20}$$

$$\widetilde{L}_{kl}^{1001} = -4\frac{\bar{A}_{kl}}{k_{\mathrm{B}}\pi} \frac{X_l}{\mathcal{D}_{kl}} \frac{m_l}{m_k+m_l} \frac{c_l^{\mathrm{int}}}{\xi_l^{\mathrm{int}}}, \qquad l \in \mathcal{S}, \quad l \in \mathcal{P}, \quad k \neq l, \tag{3.7.21}$$

$$\widetilde{L}_{kl}^{0100} = 0, \qquad k \in \mathcal{P}, \quad l \in \mathcal{S}, \tag{3.7.22}$$

$$\widetilde{L}_{kk}^{0110} = -\sum_{\substack{l\in\mathcal{S}\\ l\neq k}} 4\frac{\bar{A}_{kl}}{k_{\mathrm{B}}\pi} \frac{X_l}{\mathcal{D}_{kl}} \frac{m_k}{m_k+m_l} \frac{c_k^{\mathrm{int}}}{\xi_k^{\mathrm{int}}} - 4\frac{\bar{A}_{kk}}{k_{\mathrm{B}}\pi}\frac{X_k}{\mathcal{D}_{kk}}\frac{c_k^{\mathrm{int}}}{\xi_k^{\mathrm{int}}}, \qquad k \in \mathcal{P}, \tag{3.7.23}$$

$$\widetilde{L}_{kl}^{0110} = -4\frac{\bar{A}_{kl}}{k_{\mathrm{B}}\pi} \frac{X_l}{\mathcal{D}_{kl}} \frac{m_k}{m_k+m_l} \frac{c_k^{\mathrm{int}}}{\xi_k^{\mathrm{int}}}, \qquad k \in \mathcal{P}, \quad l \in \mathcal{S}, \quad k \neq l, \tag{3.7.24}$$

$$\widetilde{L}_{kk}^{0101} = \sum_{\substack{l\in\mathcal{S}\\ l\neq k}} X_l \frac{c_k^{\mathrm{int}}}{k_{\mathrm{B}}\mathcal{D}_{k\,\mathrm{int},l}} + X_k \frac{c_k^{\mathrm{int}}}{k_{\mathrm{B}}\mathcal{D}_{k\,\mathrm{int},k}}$$

$$+\sum_{\substack{l\in\mathcal{S}\\ l\neq k}} \frac{12}{5}\frac{\bar{A}_{kl}}{k_B\pi}\frac{X_k}{\mathcal{D}_{kl}}\frac{m_k}{m_l}\frac{c_k^{\rm int}}{\xi_k^{\rm int}} + \frac{12}{5}\frac{\bar{A}_{kk}}{k_B\pi}\frac{X_k}{\mathcal{D}_{kk}}\frac{c_k^{\rm int}}{\xi_k^{\rm int}}, \qquad k\in\mathcal{P}, \tag{3.7.25}$$

$$\widetilde{L}_{kl}^{0101} = 0, \qquad k,l\in\mathcal{P}, \quad k\neq l, \tag{3.7.26}$$

and, finally, the matrix $\widetilde{L}_{[e]}$ is given by

$$\widetilde{L}_{[e]kk}^{0000} = \sum_{\substack{l\in\mathcal{S}\\ l\neq k}} \frac{X_l}{\mathcal{D}_{kl}}, \qquad k\in\mathcal{S}, \tag{3.7.27}$$

$$\widetilde{L}_{[e]kl}^{0000} = -\frac{X_l}{\mathcal{D}_{kl}}, \qquad k,l\in\mathcal{S}, \quad k\neq l, \tag{3.7.28}$$

$$\widetilde{L}_{[e]kk}^{00e} = -\sum_{\substack{l\in\mathcal{S}\\ l\neq k}} \frac{X_l}{2\mathcal{D}_{kl}}\frac{m_l}{m_k+m_l}(6\bar{C}_{kl}-5), \qquad k\in\mathcal{S}, \tag{3.7.29}$$

$$\widetilde{L}_{[e]kl}^{00e} = \frac{X_l}{2\mathcal{D}_{kl}}\frac{m_k}{m_k+m_l}(6\bar{C}_{kl}-5), \qquad k,l\in\mathcal{S}, \quad k\neq l, \tag{3.7.30}$$

$$\widetilde{L}_{[e]kk}^{e00} = -\sum_{\substack{l\in\mathcal{S}\\ l\neq k}} \frac{X_l}{2\mathcal{D}_{kl}}\frac{m_l}{m_k+m_l}(6\bar{C}_{kl}-5), \qquad k\in\mathcal{S}, \tag{3.7.31}$$

$$\widetilde{L}_{[e]kl}^{e00} = \frac{X_l}{2\mathcal{D}_{kl}}\frac{m_l}{m_k+m_l}(6\bar{C}_{kl}-5), \qquad k,l\in\mathcal{S}, \quad k\neq l, \tag{3.7.32}$$

$$\begin{aligned}\widetilde{L}_{[e]kk}^{ee} = &\sum_{\substack{l\in\mathcal{S}\\ l\neq k}} \frac{X_l}{\mathcal{D}_{kl}}\frac{m_k m_l}{(m_k+m_l)^2}\Big[\frac{15}{2}\frac{m_k}{m_l} + \frac{25}{4}\frac{m_l}{m_k} - 3\frac{m_l}{m_k}\bar{B}_{kl} + 4\bar{A}_{kl} \\ &+ \frac{4}{15}\frac{(3m_k-2m_l)^2}{m_l^2}\frac{\bar{A}_{kl}}{k_B\pi}\frac{c_k^{\rm int}}{\xi_k^{\rm int}} + \frac{20}{3}\frac{\bar{A}_{kl}}{k_B\pi}\frac{c_l^{\rm int}}{\xi_l^{\rm int}} + \frac{(m_k+m_l)^2}{m_k m_l}\frac{c_k^{\rm int}\mathcal{D}_{kl}}{k_B\mathcal{D}_{k\,{\rm int},l}}\Big] \\ &+ \frac{X_k}{\mathcal{D}_{kk}}\Big[2\bar{A}_{kk} + \frac{16}{15}\frac{\bar{A}_{kk}}{k_B\pi}\frac{c_k^{\rm int}}{\xi_k^{\rm int}} + \frac{c_k^{\rm int}\mathcal{D}_{kk}}{k_B\mathcal{D}_{k\,{\rm int},k}}\Big], \qquad k\in\mathcal{S},\end{aligned} \tag{3.7.33}$$

$$\begin{aligned}\widetilde{L}_{[e]kl}^{ee} = &-\frac{X_l}{\mathcal{D}_{kl}}\frac{m_k m_l}{(m_k+m_l)^2}\Big[\frac{55}{4} - 3\bar{B}_{kl} - 4\bar{A}_{kl} + \frac{4}{3}\frac{\bar{A}_{kl}}{k_B\pi}\frac{3m_k-2m_l}{m_l}\frac{c_k^{\rm int}}{\xi_k^{\rm int}} \\ &+ \frac{4}{3}\frac{\bar{A}_{kl}}{k_B\pi}\frac{3m_l-2m_k}{m_k}\frac{c_l^{\rm int}}{\xi_l^{\rm int}}\Big], \qquad k,l\in\mathcal{S}, \quad k\neq l.\end{aligned} \tag{3.7.34}$$

4 Mathematical Properties

The kinetic theory investigations of Chapters 2 and 3 have yielded the structure properties of the transport linear systems of a given polyatomic gas mixture. In Chapters 4 and 5 these properties are recast into a set of assumptions written in a mathematical framework. The purpose of this framework is to establish several results that are valid for any system matrix that satisfies the corresponding mathematical assumptions. This approach is needed in order to investigate rigorously the well posedness of the transport linear systems and the singular limit of vanishing mass fractions. Furthermore, when collision integrals are not exact but estimated, it is then sufficient to verify that the approximate system matrix satisfies the corresponding mathematical assumptions in order for the theoretical results derived in Chapters 4 and 5 to apply. This will be verified systematically in the case of the practical approximations presented in Section 2.10 [MM62] [MPM65]. The mathematical framework will be fundamental in Chapter 5 also, in order to prove the convergence of various iterative algorithms.

For each transport coefficient, we first discuss the case of positive mass fractions and investigate the transport linear systems—including the simplified formulations—as they result from Chapter 2. We establish the well posedness of these systems and obtain various properties of the transport coefficients. We then consider the left rescaled transport linear systems—including the simplified formulations—as described in Chapter 3 and establish the well posedness of these systems. Furthermore, provided the diffusion matrix is replaced by the flux diffusion matrix, we prove that all the transport coefficients are smooth rational functions of the mass fractions and admit finite limits when some mass fractions become arbitrarily small. This establishes rigorously the validity of a common practice in numerical calculations, which consists in evaluating transport properties of a given gas mixture by first adding to all the species mass fractions a very small number, typically lower than the machine precision. The structure properties of the symmetric rescaled transport linear systems introduced in Section 3.1.8 are also presented for all the standard formulations, but they are omitted for the simplified formulations for the sake of brevity. These symmetric rescaled systems will be useful in Chapter 5 when deriving stabilized versions, for vanishing mass fractions, of preconditioned conjugate gradient methods.

Mathematical preliminaries are presented in Section 4.1. The state of the mixture, as given by the pressure, the temperature, and the species mass fractions is first discussed in Section 4.1.1. In Section 4.1.2 we next introduce some notation which completes the one given in Sections 2.3.1 and 3.1.2. In Section 4.1.3 we then restate several results on generalized inverses and constrained singular systems. For completeness, we present proofs of these results, which are generally imbedded in highly technical papers and dispersed in the literature.

In Sections 4.2 to 4.6 these results are then used to investigate the properties of all the transport linear systems and their various simplified formulations. Except for the first-order diffusion and flux diffusion matrices [Gi91], it is the first time, to the authors' knowledge, that such an analysis is made. The proofs of the mathematical results will be omitted for the simplified formulations since they are similar to the ones given for the standard formulations. We consider the shear viscosity in Section 4.2, the volume viscosity in Section 4.3, the diffusion matrix and the flux diffusion matrix in Section 4.4, the partial thermal conductivity and the thermal diffusion vector in Section 4.5, and the thermal conductivity and the thermal diffusion ratios in Section 4.6. Finally, the dilution—or pure species—limit is considered in Section 4.7 for all the transport coefficients.

4.1 Mathematical Preliminaries

4.1.1 State of the Mixture

General Assumptions. We consider a mixture of gases with molecules having internal degrees of freedom, $n \geq 1$ denotes the number of species and $\mathcal{S} = [1, n]$ the set of species indices. We denote by $p \geq 0$ the number of polyatomic species, i.e., the species which have at least two different energy levels and by $\mathcal{P}$ the set of polyatomic species indices. The state of a polyatomic gas mixture is given by the pressure $\overline{p}$, the temperature $\overline{T}$, and the species mass fractions $Y_1, \dots, Y_n$—or, equivalently, the species mole fractions $X_1, \dots, X_n$. We assume that the pressure $\overline{p}$ and the temperature $\overline{T}$ are given positive constants and that the mass and mole fractions are related by

$$X_k = Y_k \frac{m}{m_k}, \qquad k \in \mathcal{S}, \tag{4.1.1}$$

where the species masses m_k, $k \in S$, are given positive constants, and the mean species mass of the mixture, m, is given by [Gi90]

$$\frac{\sum_{k\in S} Y_k}{m} = \sum_{k\in S} \frac{Y_k}{m_k}. \tag{4.1.2}$$

The Mass and Mole Fraction Vectors. We denote $Y = (Y_k)_{k\in S}$ the mass fraction vector, i.e., the vector $Y \in \mathbb{R}^n$ whose components are the mass fractions $Y_1, \ldots, Y_n$. Similarly, we denote $X = (X_k)_{k\in S}$ the mole fraction vector. If all the mass fractions are positive (nonnegative) we will write $Y > 0$ ($Y \geq 0$). Finally, we deduce from (4.1.1) that the mass and mole fractions can only vanish simultaneously.

Positive Mass Fractions State of the Mixture. This state is obtained when all the mass and mole fractions are positive. In this case, the pressure, the temperature, and the mass fractions are in the domain

$$\mathfrak{D}_{Y>0} = \{\, \bar{p} > 0;\ \overline{T} > 0;\ Y > 0 \,\}, \tag{4.1.3}$$

and for later use, we also introduce the domain

$$\mathcal{D}_{Y>0} = \{\, \overline{T} > 0;\ Y > 0 \,\}. \tag{4.1.4}$$

Nonnegative Mass Fractions State of the Mixture. This state is obtained when the mass and mole fractions are nonnegative, but with at least one index $k \in S$ such that $Y_k > 0$ and $X_k > 0$. In this case, the pressure, the temperature, and the mass fractions are in the domain

$$\mathfrak{D}_{Y\geq 0, Y\neq 0} = \{\, \bar{p} > 0;\ \overline{T} > 0;\ Y \geq 0,\ Y \neq 0 \,\}, \tag{4.1.5}$$

and for later use, we also introduce the domains

$$\mathfrak{D}_{Y_k>0} = \{\, \bar{p} > 0;\ \overline{T} > 0;\ Y \geq 0,\ Y_k > 0 \,\}, \qquad k \in S, \tag{4.1.6}$$

and

$$\mathcal{D}_{Y\geq 0, Y\neq 0} = \{\, \overline{T} > 0;\ Y \geq 0,\ Y \neq 0 \,\}. \tag{4.1.7}$$

Remark. For both positive and nonnegative mass fractions states of the mixture, we always have the relation

$$\sum_{k\in S} Y_k > 0,$$

and we also deduce from (4.1.1)(4.1.2) that $\sum_{k\in S} Y_k = \sum_{k\in S} X_k$. In the previous chapters the mass fractions were assumed to satisfy the physical relation $\sum_{k\in S} Y_k = 1$. However, in numerical calculations where all the mass fractions are considered as independent unknowns, small deviations of $\sum_{k\in S} Y_k$ from unity can occur. The origin of these deviations may be, for instance, the iterative processes—such as Newton's method—used to solve the discrete governing equations. In the following chapters we will therefore assume that the mass fractions only satisfy $\sum_{k\in S} Y_k > 0$. As a consequence, the numerical factors $\sum_{k\in S} Y_k$ and $\sum_{k\in S} X_k$ will not be omitted from various explicit expressions. Note also that omitting such factors may modify Jacobian matrices of discrete systems when all the mass fractions are considered as independent unknowns [Gi90].

4.1.2 Notation

The general notation used in this book is presented in Section 2.3.1, and some additional notation is given in Section 3.1.2 for the singular case of nonnegative mass fractions. We recommend that the reader familiarize himself with these sections before reading the following chapters.

Notation Associated with the Variational Procedure. Referring to Section 2.3.1, we restate that the transport linear systems are obtained by applying a variational procedure with a finite dimensional functional space $\mathcal{A}^\mu = \mathrm{span}\{\ \xi^{rk},\ (r,k) \in \mathcal{B}^\mu\ \}$, where μ denotes a transport coefficient and ξ^{rk}, $(r,k) \in \mathcal{B}^\mu$, are basis functions. The indexing set $\mathcal{B}^\mu$ is a subset of $\mathcal{F}\times\mathcal{S}$, where $\mathcal{F}$ denotes a set of function type indices and $\mathcal{S}$ the set of species indices. Denoting by ω the number of elements in $\mathcal{B}^\mu$ and ordering $\mathcal{B}^\mu$ with the lexicographical order, we identify $\mathbb{R}^\omega$ with $\mathbb{R}^{\mathcal{B}^\mu}$. Therefore, a vector $x \in \mathbb{R}^\omega$ has components $x = (x_k^r)_{(r,k)\in\mathcal{B}^\mu}$, and a matrix $G \in \mathbb{R}^{\omega,\omega}$ has coefficients $G = (G_{kl}^{rs})_{(r,k),(s,l)\in\mathcal{B}^\mu}$. The corresponding block-decomposition of vectors and matrices is discussed in Section 2.3.1. In addition, we introduce the diagonal matrices $\mathcal{X}^\mu \in \mathbb{R}^{\omega,\omega}$ given by

$$\mathcal{X}^\mu = \mathrm{diag}\big((X_k)_{(r,k)\in\mathcal{B}^\mu}\big), \tag{4.1.8}$$

and similarly, for the reduced transport linear systems, the diagonal matrices $\mathcal{X}^\mu_{[\mathrm{red}]} \in \mathbb{R}^{\omega_{[\mathrm{red}]},\omega_{[\mathrm{red}]}}$ given by

$$\mathcal{X}^\mu_{[\mathrm{red}]} = \mathrm{diag}\big((X_k)_{(r,k)\in\mathcal{B}^\mu_{[\mathrm{red}]}}\big). \tag{4.1.9}$$

These matrices will be useful when considering rescaled versions of the transport linear systems, and examples of such matrices are given in Sections 3.2–3.6 for all the transport coefficients. We also define for later use the diagonal matrices

$$\begin{cases} \mathcal{W} = \mathrm{diag}(W_1, \dots, W_n), \\ \mathcal{X} = \mathrm{diag}(X_1, \dots, X_n), \\ \mathcal{Y} = \mathrm{diag}(Y_1, \dots, Y_n), \end{cases} \tag{4.1.10}$$

and the vector $W = (W_1, \dots, W_n) \in \mathbb{R}^n$ given by $W_k = m_k/m$, $k \in \mathcal{S}$. Finally, if S_1 and S_2 are two complementary subspaces of $\mathbb{R}^\omega$, i.e., $S_1 \oplus S_2 = \mathbb{R}^\omega$, we denote by P_{S_1,S_2} the oblique projector onto the subspace S_1 along the subspace S_2.

Partitioning for Nonnegative Mass Fractions. We denote by $n^+ \geq 1$ and $p^+ \geq 0$ the number of species and polyatomic species, respectively, such that $Y_k > 0$, and similarly by $n^- = n - n^+$ and $p^- = p - p^+$ the number of species and polyatomic species, respectively, such that $Y_k = 0$. We define the subsets $\mathcal{S}^+ = \{ k \in \mathcal{S},\ Y_k > 0 \}$, $\mathcal{S}^- = \{ k \in \mathcal{S},\ Y_k = 0 \}$, $\mathcal{P}^+ = \{ k \in \mathcal{P},\ Y_k > 0 \}$, and $\mathcal{P}^- = \{ k \in \mathcal{P},\ Y_k = 0 \}$, and the submixture formed by the positive mass fraction species only will be referred to as the $\mathcal{S}^+$ mixture.

Referring to Section 3.1.2 for more details, we denote by $\Upsilon \in \mathbb{R}^{n,n}$ the permutation matrix associated with the reordering of $\mathcal{S}$ into $(\mathcal{S}^+, \mathcal{S}^-)$. Consider then a transport coefficient μ and the corresponding indexing set $\mathcal{B}^\mu$. We define the subsets $\mathcal{B}^{\mu+} = \{ (r,k) \in \mathcal{B}^\mu,\ Y_k > 0 \}$ and $\mathcal{B}^{\mu-} = \{ (r,k) \in \mathcal{B}^\mu,\ Y_k = 0 \}$, and we denote by $\Gamma^\mu \in \mathbb{R}^{\omega,\omega}$ the permutation matrix associated with the reordering of $\mathcal{B}^\mu$ into $\left(\mathcal{B}^{\mu+}, \mathcal{B}^{\mu-}\right)$. We denote by ω^+ the number of elements of $\mathcal{B}^{\mu+}$ and ω^- the number of elements of $\mathcal{B}^{\mu-}$, so that $\omega = \omega^+ + \omega^-$. Using the matrix Γ^μ, we may decompose each vector $x \in \mathbb{R}^\omega$ into the vectors $x^+ \in \mathbb{R}^{\omega^+}$ and $x^- \in \mathbb{R}^{\omega^-}$ defined by $x^+ = (x_k^r)_{(r,k)\in\mathcal{B}^{\mu+}}$ and $x^- = (x_k^r)_{(r,k)\in\mathcal{B}^{\mu-}}$ so that $x = \Gamma^\mu(x^+, x^-)$. Correspondingly, we decompose each matrix $G \in \mathbb{R}^{\omega,\omega}$ into the blocks $G^{++} \in \mathbb{R}^{\omega^+,\omega^+}$, $G^{+-} \in \mathbb{R}^{\omega^+,\omega^-}$, $G^{-+} \in \mathbb{R}^{\omega^-,\omega^+}$, and $G^{--} \in \mathbb{R}^{\omega^-,\omega^-}$ so that

$$\left(\Gamma^\mu\right)^t G \Gamma^\mu = \begin{bmatrix} G^{++} & G^{+-} \\ G^{-+} & G^{--} \end{bmatrix}.$$

4.1.3 Generalized Inverses and Constrained Singular Systems

In this section we restate several results on generalized inverses and constrained singular systems. For completeness, we include proofs of these results, which are usually imbedded in highly technical papers and dispersed in the literature. These results will be systematically used in Sections 4.2–4.6 for all the transport linear systems.

The following proposition can be found in [BG74] [BP79] and characterizes generalized inverses with prescribed range and nullspace.

Proposition 4.1.1. *Let $G \in \mathbb{R}^{\omega,\omega}$ be a matrix and let $\mathcal{C}$ and S be two subspaces of $\mathbb{R}^\omega$ such that $N(G) \oplus \mathcal{C} = \mathbb{R}^\omega$ and $R(G) \oplus S = \mathbb{R}^\omega$. Then there exists a unique matrix Z such that $GZG = G$, $ZGZ = Z$, $N(Z) = S$, and $R(Z) = \mathcal{C}$. The matrix Z is called the generalized inverse of G with prescribed range $\mathcal{C}$ and nullspace S and is also such that $GZ = P_{R(G),S}$ and $ZG = P_{\mathcal{C},N(G)}$.*

Proof. We first show that there exists a matrix $M \in \mathbb{R}^{\omega,\omega}$ such that $GMG = G$ and $MGM = M$. Let f_i, $i = 1,\dots,\omega$, be a basis of $\mathbb{R}^\omega$ such that f_i, $i = 1,\dots,r$, is a basis of $N(G)$. Then, by construction, the vectors $g_i = Gf_i$, $i = r+1,\dots,\omega$ are linearly independent and may be completed to form a basis g_i, $i = 1,\dots,\omega$ of $\mathbb{R}^\omega$. Define then M such that $Mg_i = 0$, $1 \le i \le r$, and $Mg_i = f_i$, $r+1 \le i \le \omega$. One may then easily check that $GMGf_i = Gf_i$, $1 \le i \le \omega$, and $MGMg_i = Mg_i$, $1 \le i \le \omega$, so that $GMG = G$ and $MGM = M$. Defining now $Z = P_{\mathcal{C},N(G)}MP_{R(G),S}$, we have $GZG = GP_{\mathcal{C},N(G)}MP_{R(G),S}G = GMG$, since $G = GP_{\mathcal{C},N(G)}$ and $P_{R(G),S}G = G$, so that $GZG = G$. Similarly, from $ZGZ = P_{\mathcal{C},N(G)}MP_{R(G),S}GP_{\mathcal{C},N(G)}MP_{R(G),S}$, we obtain that $ZGZ = P_{\mathcal{C},N(G)}MGMP_{R(G),S}$ and thus that $ZGZ = Z$. By construction, we also have $R(Z) \subset \mathcal{C}$ and $S \subset N(Z)$, and from $GZG = G$ and $ZGZ = Z$, we obtain that $\mathrm{rank}(G) = \mathrm{rank}(Z)$. Since $\dim(\mathcal{C}) = \mathrm{rank}(G)$ and $\dim(S) = \omega - \mathrm{rank}(G)$ by assumption, we conclude that $R(Z) = \mathcal{C}$ and $N(Z) = S$. From the relations $GZG = G$ and $ZGZ = Z$, we also deduce that GZ and ZG are projectors and that $\mathrm{rank}(GZ) = \mathrm{rank}(G) = \mathrm{rank}(Z) = \mathrm{rank}(ZG)$. Since we also have $R(ZG) \subset R(Z) = \mathcal{C}$ and $N(G) \subset N(ZG)$, we obtain that $R(ZG) = \mathcal{C}$ and $N(ZG) = N(G)$ and similarly that $R(GZ) = R(G)$ and $N(GZ) = S$ so that $GZ = P_{R(G),S}$ and $ZG = P_{\mathcal{C},N(G)}$. Finally, if there are two such generalized inverses Z_1 and Z_2, we have $Z_i = Z_iGZ_i$, $GZ_i = P_{R(G),S}$ and $Z_iG = P_{\mathcal{C},N(G)}$, $i = 1,2$, so that $Z_1 = Z_1GZ_1 = P_{\mathcal{C},N(G)}Z_1 = Z_2GZ_1 = Z_2P_{R(G),S} = Z_2GZ_2 = Z_2$. □

Similarly, in the case where the nullspace and range of the matrix G are complementary spaces, G admits a group inverse which is characterized by the following result.

Proposition 4.1.2. *Let $G \in \mathbb{R}^{\omega,\omega}$ be a matrix such that $N(G) \oplus R(G) = \mathbb{R}^\omega$. Then there exists a unique matrix Z such that $GZG = G$, $ZGZ = Z$, and $GZ = ZG$. The matrix Z is called the group inverse of G and is denoted by $G^\sharp$. The group inverse is also the generalized inverse with prescribed range $R(G)$ and nullspace $N(G)$ and is also such that $GZ = ZG = P_{R(G),N(G)}$.*

Proof. From the relations $GZG = G$, $ZGZ = Z$, and $GZ = ZG$, we easily deduce that

$R(G) = R(Z)$ and $N(G) = N(Z)$. Therefore, Z is necessarily the generalized inverse of G with prescribed range $R(G)$ and nullspace $N(G)$. Since $N(G) \oplus R(G) = \mathbb{R}^\omega$, we deduce from Proposition 4.1.1 that the matrix Z exists and is unique. Hence, we only have to show that $GZ = ZG$, but this directly follows from $GZ = P_{R(G),N(G)}$ and $ZG = P_{R(G),N(G)}$ established in Proposition 4.1.1 with $\mathcal{C} = R(G)$ and $S = N(G)$. □

We now consider the well posedness of constrained singular systems and relate their solution to generalized inverses.

Proposition 4.1.3. *Let $G \in \mathbb{R}^{\omega,\omega}$ be a singular matrix, let $\mathcal{C}$ be a subspace of $\mathbb{R}^\omega$, and consider the constrained singular system*

$$\begin{cases} G\alpha^\mu = \beta^\mu, \\ \alpha^\mu \in \mathcal{C}. \end{cases} \tag{4.1.11}$$

Then (4.1.11) is well posed, i.e., admits a unique solution α^μ for any $\beta^\mu \in R(G)$, if and only if $N(G) \oplus \mathcal{C} = \mathbb{R}^\omega$. In this situation, for any subspace S such that $R(G) \oplus S = \mathbb{R}^\omega$, the solution α^μ can be written $\alpha^\mu = Z\beta^\mu$, where Z is the generalized inverse of G with prescribed range $\mathcal{C}$ and nullspace S.

Proof. Assume first that the system (4.1.11) is well posed and let $x \in \mathbb{R}^\omega$. Then there exists a unique solution $y \in \mathcal{C}$ to the system $Gy = Gx$, and hence $x - y \in N(G)$ so that $N(G) + \mathcal{C} = \mathbb{R}^\omega$. Furthermore, for any $z \in N(G) \cap \mathcal{C}$, $y + z$ also satisfies $G(y + z) = Gx$ and $y + z \in \mathcal{C}$, so that we must have $N(G) \cap \mathcal{C} = \{0\}$ by uniqueness. Conversely, if $N(G) \oplus \mathcal{C} = \mathbb{R}^\omega$ and $\beta^\mu \in R(G)$, there exists $x \in \mathbb{R}^\omega$ such that $Gx = \beta^\mu$, and we may write $x = y + z$ where $y \in N(G)$ and $z \in \mathcal{C}$. Therefore, we have $Gz = \beta^\mu$ and $z \in \mathcal{C}$ so that (4.1.11) has at least one solution which is also unique since the difference between any two solutions is in $N(G) \cap \mathcal{C} = \{0\}$. Let now S be a subspace such that $R(G) \oplus S = \mathbb{R}^\omega$. The generalized inverse Z then exists by Proposition 4.1.1 since $N(G) \oplus \mathcal{C} = R(G) \oplus S = \mathbb{R}^\omega$. Moreover, the vector $Z\beta^\mu$ satisfies $GZ\beta^\mu = P_{R(G),S}\beta^\mu = \beta^\mu$ since $\beta^\mu \in R(G)$, and we also have $Z\beta^\mu \in \mathcal{C}$ since $R(Z) = \mathcal{C}$, so that $\alpha^\mu = Z\beta^\mu$. □

In Sections 4.2 to 4.6, only the particular case where the nullspace of the matrix G is one-dimensional will be considered. In this situation, an explicit expression of the generalized inverse is given in the next proposition. A similar result can also be stated in the more general case where the dimension of $N(G)$ is larger than one, but is omitted for brevity.

Proposition 4.1.4. *Let $G \in \mathbb{R}^{\omega,\omega}$ be a symmetric positive semi-definite matrix and let $\mathcal{G}$ and $\mathcal{Z} \in \mathbb{R}^\omega$ be two vectors such that $\langle \mathcal{G}, \mathcal{Z} \rangle \neq 0$. Assume that the nullspace* *and range of G are given by $N(G) = \mathbb{R}\mathcal{Z}$ and $R(G) = \mathcal{Z}^\perp$, respectively, so that G is positive definite on the hyperplane $\mathcal{G}^\perp$. Consider further the generalized inverse $Z \in \mathbb{R}^{\omega,\omega}$ of G with prescribed nullspace $N(Z) = \mathbb{R}\mathcal{G}$ and range $R(Z) = \mathcal{G}^\perp$. Then the matrix Z is symmetric, positive semi-definite, and positive definite on the hyperplane $\mathcal{Z}^\perp$. Furthermore, for any positive real numbers a and b such that $ab\langle \mathcal{G}, \mathcal{Z}\rangle^2 = 1$, the matrices $G + a\,\mathcal{G}\otimes\mathcal{G}$ and $Z + b\,\mathcal{Z}\otimes\mathcal{Z}$ are symmetric, positive definite, inverse of each other, and coincide with G and Z on the hyperplanes $\mathcal{G}^\perp$ and $\mathcal{Z}^\perp$, respectively. In particular, the matrix Z is given by*

$$Z = (G + a\,\mathcal{G}\otimes\mathcal{G})^{-1} - b\,\mathcal{Z}\otimes\mathcal{Z}. \tag{4.1.12}$$

Proof. We have $\mathbb{R}\mathcal{Z} \oplus \mathcal{G}^\perp = \mathbb{R}\mathcal{G} \oplus \mathcal{Z}^\perp = \mathbb{R}^\omega$ since $\langle \mathcal{G}, \mathcal{Z}\rangle \neq 0$, so that, by Proposition 4.1.1, the generalized inverse Z exists. Furthermore, from $GZG = G$, $ZGZ = Z$, $N(Z) = \mathbb{R}\mathcal{G}$, $R(Z) = \mathcal{G}^\perp$, and $G^t = G$, we first deduce that $GZ^tG = G$, $Z^tGZ^t = Z^t$, $N(Z^t) = \mathbb{R}\mathcal{G}$, and $R(Z^t) = \mathcal{G}^\perp$, since $R(Z^t) = \big(N(Z)\big)^\perp$ and $N(Z^t) = \big(R(Z)\big)^\perp$. From the uniqueness of the generalized inverse with prescribed range and nullspace, we deduce that $Z = Z^t$, i.e., Z is symmetric. Moreover, Z is positive semi-definite since for $y \in \mathbb{R}^\omega$, we have $\langle y, Zy\rangle = \langle Zy, GZy\rangle \geq 0$ because $Z = ZGZ$, Z is symmetric, and G is positive semi-definite. We then deduce that Z is positive definite on $\mathcal{Z}^\perp$ since $\langle \mathcal{G}, \mathcal{Z}\rangle \neq 0$ and $N(Z) = \mathbb{R}\mathcal{G}$. Consider now the matrices $G + a\,\mathcal{G}\otimes\mathcal{G}$ and $Z + b\,\mathcal{Z}\otimes\mathcal{Z}$. First, these matrices are clearly symmetric. Furthermore, since a is positive, the quadratic forms defined by G and $a\,\mathcal{G}\otimes\mathcal{G}$ are both nonnegative and positive definite on $\mathcal{G}^\perp$ and $\mathbb{R}\mathcal{Z}$, respectively, so that their sum is positive definite on $\mathbb{R}^\omega$. The same argument holds for Z and $b\,\mathcal{Z}\otimes\mathcal{Z}$ since Z is positive definite on $\mathcal{Z}^\perp$ and b is positive. In addition, the formulas $(G + a\,\mathcal{G}\otimes\mathcal{G})(Z + b\,\mathcal{Z}\otimes\mathcal{Z}) = I$ and $(Z + b\,\mathcal{Z}\otimes\mathcal{Z})(G + a\,\mathcal{G}\otimes\mathcal{G}) = I$ are easily obtained since $GZ = I - \mathcal{G}\otimes\mathcal{Z}/\langle \mathcal{G}, \mathcal{Z}\rangle$ and $ZG = I - \mathcal{Z}\otimes\mathcal{G}/\langle \mathcal{G}, \mathcal{Z}\rangle$, and since for $x, y, x', y' \in \mathbb{R}^\omega$ and $G \in \mathbb{R}^{\omega,\omega}$, one has $x\otimes y\; x'\otimes y' = \langle y, x'\rangle\, x\otimes y'$, $x\otimes y\, G = x\otimes(G^t y)$ and $G\, x\otimes y = (Gx)\otimes y$. Finally, the matrices $G + a\,\mathcal{G}\otimes\mathcal{G}$ and $Z + b\,\mathcal{Z}\otimes\mathcal{Z}$ coincide with G and Z on the hyperplanes $\mathcal{G}^\perp$ and $\mathcal{Z}^\perp$, respectively, since for $x, y, x' \in \mathbb{R}^\omega$, one has $(x\otimes y)x' = \langle y, x'\rangle x$. □

The next proposition generalizes Proposition 4.1.4 to nonsymmetric matrices, such as the ones arising in the left rescaled transport linear systems.

Proposition 4.1.5. *Let $\widetilde{G} \in \mathbb{R}^{\omega,\omega}$ be a matrix and let $\mathcal{G}, \mathcal{Z}, \widetilde{\mathcal{G}}, \widetilde{\mathcal{Z}} \in \mathbb{R}^\omega$ be four vectors such that $\langle \mathcal{G}, \mathcal{Z}\rangle \neq 0$ and $\langle \widetilde{\mathcal{G}}, \widetilde{\mathcal{Z}}\rangle \neq 0$. Assume that the nullspace and range of $\widetilde{G}$ are given by $N(\widetilde{G}) = \mathbb{R}\mathcal{Z}$ and $R(\widetilde{G}) = \widetilde{\mathcal{Z}}^\perp$, respectively. Consider further the generalized* *inverse $\widetilde{Z}$ of $\widetilde{G}$ with prescribed nullspace $N(\widetilde{Z}) = \mathbb{R}\widetilde{\mathcal{G}}$ and range $R(\widetilde{Z}) = \mathcal{G}^{\perp}$. Then for any positive real numbers a and b such that $ab\langle \mathcal{G}, \mathcal{Z}\rangle\langle \widetilde{\mathcal{G}}, \widetilde{\mathcal{Z}}\rangle = 1$, the matrices $\widetilde{G}+a\,\widetilde{\mathcal{G}}\otimes\mathcal{G}$ and $\widetilde{Z}+b\,\mathcal{Z}\otimes\widetilde{\mathcal{Z}}$ are inverse of each other and coincide with $\widetilde{G}$ and $\widetilde{Z}$ on the hyperplanes $\mathcal{G}^{\perp}$ and $\widetilde{\mathcal{Z}}^{\perp}$, respectively. In particular, the matrix $\widetilde{Z}$ is given by*

$$\widetilde{Z} = [\widetilde{G} + a\,\widetilde{\mathcal{G}}\otimes\mathcal{G}]^{-1} - b\,\mathcal{Z}\otimes\widetilde{\mathcal{Z}}. \tag{4.1.13}$$

Proof. It is similar to the proof of Proposition 4.1.4. □

4.2 The Shear Viscosity

4.2.1 Mathematical Properties of the System $H\alpha^{\eta} = \beta^{\eta}$

We consider the linear system

$$H\alpha^{\eta} = \beta^{\eta}, \tag{4.2.1}$$

associated with the evaluation of the shear viscosity

$$\eta = \langle \alpha^{\eta}, \beta^{\eta}\rangle, \tag{4.2.2}$$

as described in Section 2.4.1. We have $H \in \mathbb{R}^{n,n}$, α^{η} and $\beta^{\eta} \in \mathbb{R}^{n}$, and the corresponding indexing set is given by

$$\mathcal{B}^{\eta} = \{00\}\times S. \tag{4.2.3}$$

Furthermore, from the kinetic theory results obtained in Sections 2.3 and 2.4, we can make the following assumptions.

($H0$) The coefficients of the matrix H are functions of the temperature and the mass fractions and can be expressed as quadratic functions of the mole fractions in the form

$$\begin{cases} H_{kk}^{rs} = \sum_{l\in S} X_k X_l H_{kl}^{\prime rs} + X_k^2 H_{kk}^{\prime\prime rs}, & (r,k),(s,k) \in \mathcal{B}^{\eta}, \\ H_{kl}^{rs} = X_k X_l H_{kl}^{\prime\prime rs}, \quad (r,k),(s,l) \in \mathcal{B}^{\eta}, \quad k \neq l, \end{cases} \tag{4.2.4}$$

where the matrices H' and H'' solely depend on the temperature.

($H1$) H is symmetric.

($H2$) H is positive definite.

Moreover, following Section 2.4.1, we assume that the right-hand side β^η is given by

$$\beta_k^{00\eta} = X_k, \qquad k \in \mathcal{S}. \tag{4.2.5}$$

Theorem 4.2.1. *Let $H \in \mathbb{R}^{n,n}$ be a matrix satisfying the properties ($H0$)–($H2$) and let $\beta^\eta \in \mathbb{R}^n$ be given by (4.2.5). Then the linear system (4.2.1) admits a unique solution α^η. Furthermore, the quantity η is positive.*

Proof. We deduce from ($H2$) that the matrix H is nonsingular so that the linear system (4.2.1) admits a unique solution α^η. Furthermore, we have $\eta = \langle \alpha^\eta, \beta^\eta \rangle = \langle \alpha^\eta, H\alpha^\eta \rangle$, and since the mole fractions are positive, $\beta^\eta \neq 0$ so that $\alpha^\eta \neq 0$. Hence, the quantity η is positive by ($H2$). □

Finally, we consider the matrix H resulting from the practical approximations presented in Section 2.10.

Proposition 4.2.2. *Let $\bar{A}_{kl}$ and η_{kl}, $k, l \in \mathcal{S}$, be symmetric and positive coefficients, let η_k, $k \in \mathcal{S}$, be positive coefficients, and assume that $Y > 0$. Then the matrix H given by (2.10.7)(2.10.8) satisfies ($H0$)–($H2$).*

Proof. The proof of ($H0$) directly results from (2.10.7)(2.10.8), and from the symmetry of $\bar{A}_{kl}$ and η_{kl}, we deduce that the matrix H is symmetric. Furthermore, an explicit calculation yields for $x \in \mathbb{R}^n$

$$\begin{aligned}\langle x, Hx \rangle = {} & \sum_{k \in \mathcal{S}} \frac{X_k^2}{\eta_k} x_k^2 \\ & + \sum_{\substack{k,l \in \mathcal{S} \\ l \neq k}} \frac{X_k X_l}{\eta_{kl}} \Big(\frac{5}{3\bar{A}_{kl}} \frac{m_k m_l}{(m_k + m_l)^2} (x_k - x_l)^2 + \big(\frac{m_l x_k + m_k x_l}{m_k + m_l}\big)^2 \Big).\end{aligned}$$

Since $Y > 0$ implies $X > 0$, we easily deduce from the assumptions and the above explicit expression that $\langle x, Hx \rangle = 0$ if and only if $x = 0$. □

4.2.2 Mathematical Properties of the System $\widetilde{H}\alpha^\eta = \widetilde{\beta}^\eta$

We consider the linear system

$$\widetilde{H}\alpha^\eta = \widetilde{\beta}^\eta, \tag{4.2.6}$$

associated with the evaluation of the shear viscosity

$$\eta = \langle \alpha^\eta, \beta^\eta \rangle, \tag{4.2.7}$$

as described in Section 3.2.1. We have $\widetilde{H} \in \mathbb{R}^{n,n}$, α^η and $\widetilde{\beta}^\eta \in \mathbb{R}^n$, and the indexing set is $\mathcal{B}^\eta = \{00\} \times \mathcal{S}$. Furthermore, from the kinetic theory results obtained in Sections 3.1 and 3.2, we can make the following assumptions.

($\widetilde{H}0$) The coefficients of the matrix $\widetilde{H}$ are functions of the temperature and the mass fractions and can be expressed as linear functions of the mole fractions in the form

$$\begin{cases} \widetilde{H}_{kk}^{rs} = \sum_{l \in \mathcal{S}} X_l H_{kl}^{\prime rs} + X_k H_{kk}^{\prime\prime rs}, & (r,k),(s,k) \in \mathcal{B}^\eta, \\ \widetilde{H}_{kl}^{rs} = X_l H_{kl}^{\prime\prime rs}, \quad (r,k),(s,l) \in \mathcal{B}^\eta, \quad k \neq l, \end{cases} \tag{4.2.8}$$

where the matrices H' and H'' coincide with the ones in (4.2.4).

($\widetilde{H}1$) The matrix $\widetilde{H}$ admits the block-decomposition

$$(\Gamma^\eta)^t \widetilde{H} \Gamma^\eta = \begin{bmatrix} \widetilde{H}^{++} & 0 \\ \widetilde{H}^{-+} & \widetilde{H}^{--} \end{bmatrix}, \tag{4.2.9}$$

where Γ^η is the permutation matrix associated with the reordering of $\mathcal{B}^\eta$ into $(\mathcal{B}^{\eta+}, \mathcal{B}^{\eta-})$.

($\widetilde{H}2$) $H^{++} = \mathcal{X}^{\eta++} \widetilde{H}^{++}$ corresponds to the $\mathcal{S}^+$ mixture and, in particular, is symmetric positive definite.

($\widetilde{H}3$) $\widetilde{H}^{--}$ is symmetric positive definite and we have $db(\widetilde{H}^{--}) = \widetilde{H}^{--}$.

($\widetilde{H}4$) $\widetilde{H}$ is nonsingular.

Moreover, following Section 3.2.1, we assume that the right-hand side $\widetilde{\beta}^\eta$ is given by

$$\widetilde{\beta}_k^{00\eta} = 1, \qquad k \in \mathcal{S}. \tag{4.2.10}$$

As a consequence of (4.2.4)(4.2.8) and (4.2.5)(4.2.10), the matrix $\widetilde{H}$ and the right-hand side $\widetilde{\beta}^\eta$ are rescaled versions of H and β^η, respectively,

$$\mathcal{X}^\eta \widetilde{H} = H, \qquad \mathcal{X}^\eta \widetilde{\beta}^\eta = \beta^\eta, \tag{4.2.11}$$

where $\mathcal{X}^\eta = \operatorname{diag}\big((X_k)_{(r,k) \in \mathcal{B}^\eta}\big)$.

Theorem 4.2.3. *Let $\widetilde{H} \in \mathbb{R}^{n,n}$ be a matrix satisfying the properties ($\widetilde{H}0$)–($\widetilde{H}4$) and let β^η and $\widetilde{\beta}^\eta \in \mathbb{R}^n$ be given by (4.2.5) and (4.2.10), respectively. Then the linear system (4.2.6) admits a unique solution α^η, and this solution is such that $H^{++}\alpha^{\eta+} = \beta^{\eta+}$. Furthermore, the quantity η is positive and is the shear viscosity of the $\mathcal{S}^+$ mixture.*

Proof. First, the matrix $\widetilde{H}$ is nonsingular by ($\widetilde{H}4$) so that the linear system (4.2.6) admits a unique solution α^η. Furthermore, we have $\widetilde{H}^{++}\alpha^{\eta+} = \widetilde{\beta}^{\eta+}$ from ($\widetilde{H}1$) and

$\mathcal{X}^{\eta++}\widetilde{\beta}^{\eta+} = \beta^{\eta+}$ from (4.2.11), and multiplying by $\mathcal{X}^{\eta++}$ then yields $H^{++}\alpha^{\eta+} = \beta^{\eta+}$. Using $\beta^{\eta-} = 0$, we next obtain $\eta = \langle \alpha^{\eta+}, \beta^{\eta+} \rangle = \langle \alpha^{\eta+}, H^{++}\alpha^{\eta+} \rangle$ so that η is the shear viscosity of the S^+ mixture, since H^{++} is the corresponding system matrix. Finally, since at least one mole fraction is positive, $\beta^{\eta+} \neq 0$ and hence $\alpha^{\eta+} \neq 0$ so that the quantity η is positive by $(\widetilde{H}2)$. □

We now prove that the shear viscosity η is a smooth function of the temperature and the mass fractions in $\mathcal{D}_{Y \geq 0, Y \neq 0}$ and that it is independent of the pressure.

Theorem 4.2.4. *Let H and $\widetilde{H} \in \mathbb{R}^{n,n}$ be such that H satisfies $(H0)$–$(H2)$ in $\mathcal{D}_{Y>0}$ and $\widetilde{H}$ satisfies $(\widetilde{H}0)$–$(\widetilde{H}4)$ in $\mathcal{D}_{Y \geq 0, Y \neq 0}$. Assume also that the matrices H' and H'' in (4.2.4)(4.2.8) are smooth functions of the temperature. Let α^η and η be the solution and the shear viscosity evaluated from H for $Y > 0$ and let $\widetilde{\alpha}^\eta$ and $\widetilde{\eta}$ be the solution and the shear viscosity evaluated from $\widetilde{H}$ for $Y \geq 0$, $Y \neq 0$. Then $\alpha^\eta = \widetilde{\alpha}^\eta$ and $\eta = \widetilde{\eta}$ for $Y > 0$. Furthermore, $\widetilde{\alpha}^\eta$ and $\widetilde{\eta}$ are independent of the pressure and are smooth functions of $(\overline{T}, Y)$ in $\mathcal{D}_{Y \geq 0, Y \neq 0}$. In particular, for $Z \in \mathbb{R}^n$, $Z \geq 0$, $Z \neq 0$, we have the following limit*

$$\lim_{\substack{Y \to Z \\ Y > 0}} \eta(\overline{T}, Y) = \widetilde{\eta}(\overline{T}, Z). \tag{4.2.12}$$

Proof. When all the mass fractions are positive, the matrix $\mathcal{X}^\eta$ is invertible and we deduce from (4.2.11) that the linear systems (4.2.1)(4.2.6) admit the same solution. Therefore, $\alpha^\eta = \widetilde{\alpha}^\eta$ and $\eta = \widetilde{\eta}$ for $Y > 0$. Furthermore, the components of $\widetilde{\alpha}^\eta$ can be written, using Cramer's rule, as rational functions of the coefficients of $\widetilde{H}$ and $\widetilde{\beta}^\eta$. These latter coefficients are, by assumption, independent of the pressure and smooth functions of the temperature and the mass fractions, keeping in mind that the mole fractions are smooth rational functions of the mass fractions by (4.1.1)(4.1.2). Therefore, $\widetilde{\alpha}^\eta$ and $\widetilde{\eta} = \langle \widetilde{\alpha}^\eta, \beta^\eta \rangle$ are independent of the pressure and are smooth functions of $(\overline{T}, Y)$ in $\mathcal{D}_{Y \geq 0, Y \neq 0}$. In particular, we obtain $\lim_{Y \to Z, Y>0} \eta(\overline{T}, Y) = \widetilde{\eta}(\overline{T}, Z)$. □

Remark. As a consequence of the previous theorem, the solutions of the linear systems (4.2.1)(4.2.6) coincide when $Y > 0$. This justifies the use of the same notation α^η and η in Sections 4.2.1 and 4.2.2.

Finally, we consider the matrix $\widetilde{H}$ resulting from the practical approximations presented in Section 2.10.

Proposition 4.2.5. *Let $\bar{\mathcal{A}}_{kl}$ and η_{kl}, $k, l \in S$, be symmetric and positive coefficients, let η_k, $k \in S$, be positive coefficients, and assume that $Y \geq 0$, $Y \neq 0$. Then the matrix $\widetilde{H}$ given by (3.7.1)(3.7.2) satisfies $(\widetilde{H}0)$–$(\widetilde{H}4)$.*

Proof. The properties $(\widetilde{H}0)$–$(\widetilde{H}1)$ directly result from (3.7.1)(3.7.2), and the proof of $(\widetilde{H}2)$ is similar to the one of Proposition 4.2.2. Next, the matrix $\widetilde{H}^{--}$ is symmetric, diagonal, and given by

$$\widetilde{H}_{kk}^{--} = \sum_{l \in \mathcal{S}^+} 2\frac{X_l}{\eta_{kl}}\Big[\frac{5}{3\bar{A}_{kl}}\frac{m_k m_l}{(m_k+m_l)^2} + \frac{m_l^2}{(m_k+m_l)^2}\Big], \qquad k \in \mathcal{S}^-.$$

Since $X_l > 0$, $l \in \mathcal{S}^+$, $(\widetilde{H}3)$ is proven, and finally, $(\widetilde{H}4)$ is a direct consequence of $(\widetilde{H}1)$–$(\widetilde{H}3)$. □

4.2.3 Mathematical Properties of the System $\widehat{H}\widehat{\alpha}^\eta = \widehat{\beta}^\eta$

We consider the linear system

$$\widehat{H}\widehat{\alpha}^\eta = \widehat{\beta}^\eta, \tag{4.2.13}$$

associated with the evaluation of the shear viscosity

$$\eta = \langle \widehat{\alpha}^\eta, \widehat{\beta}^\eta \rangle. \tag{4.2.14}$$

We have $\widehat{H} \in \mathbb{R}^{n,n}$, $\widehat{\alpha}^\eta$ and $\widehat{\beta}^\eta \in \mathbb{R}^n$, and following Section 3.1.8, we can make the following assumptions.

$(\widehat{H}0)$ The coefficients of the matrix $\widehat{H}$ are functions of the temperature and the mass fractions and can be expressed as

$$\begin{cases} \widehat{H}_{kk}^{rs} = \sum_{l \in \mathcal{S}} X_l H_{kl}^{\prime rs} + X_k H_{kk}^{\prime\prime rs}, & (r,k),(s,k) \in \mathcal{B}^\eta, \\ \widehat{H}_{kl}^{rs} = \sqrt{X_k X_l} H_{kl}^{\prime\prime rs}, & (r,k),(s,l) \in \mathcal{B}^\eta, \quad k \neq l, \end{cases} \tag{4.2.15}$$

where the matrices H' and H'' coincide with the ones in (4.2.4).

$(\widehat{H}1)$ $\widehat{H}$ is symmetric.

$(\widehat{H}2)$ $\widehat{H}$ is positive definite.

In addition, we assume that the right-hand side $\widehat{\beta}^\eta$ is given by

$$\widehat{\beta}_k^{00\eta} = \sqrt{X_k}, \qquad k \in \mathcal{S}. \tag{4.2.16}$$

The matrix $\widehat{H}$ and the right-hand side $\widehat{\beta}^\eta$ are then rescaled versions of H and β^η, respectively,

$$(\mathcal{X}^\eta)^{1/2}\widehat{H}(\mathcal{X}^\eta)^{1/2} = H, \qquad (\mathcal{X}^\eta)^{1/2}\widehat{\beta}^\eta = \beta^\eta. \tag{4.2.17}$$

Using $(\widehat{H}0)$–$(\widehat{H}2)$, one can prove that the system (4.2.13) is well posed, i.e., admits a unique solution and that η is positive. Finally, we deduce from (4.2.17) that the definitions (4.2.2)(4.2.14) of the shear viscosity coincide for $Y > 0$.

4.3 The Volume Viscosity

In this section we investigate the mathematical properties of the linear systems associated with the evaluation of the volume viscosity. The simplified transport linear systems $K_{[10]}\alpha^{\kappa}_{[10]} = \beta^{\kappa}_{[10]}$ and $K_{[\mathrm{d}]}\alpha^{\kappa}_{[\mathrm{d}]} = \beta^{\kappa}_{[\mathrm{d}]}$ will not be considered since we will see in Chapter 6 that the volume viscosities $\kappa_{[10]}$ and $\kappa_{[\mathrm{d}]}$ are not accurate.

4.3.1 Mathematical Properties of the System $K\alpha^{\kappa} = \beta^{\kappa}$

We consider the constrained linear system

$$\begin{cases} K\alpha^{\kappa} = \beta^{\kappa}, \\ \langle \mathcal{K}, \alpha^{\kappa} \rangle = 0, \end{cases} \tag{4.3.1}$$

associated with the evaluation of the volume viscosity

$$\kappa = \langle \alpha^{\kappa}, \beta^{\kappa} \rangle, \tag{4.3.2}$$

as described in Section 2.5.1. We have $K \in \mathbb{R}^{n+p,n+p}$, α^{κ}, β^{κ}, and $\mathcal{K} \in \mathbb{R}^{n+p}$, and the corresponding indexing set is given by

$$\mathcal{B}^{\kappa} = \{10\}\times\mathcal{S} \cup \{01\}\times\mathcal{P}, \tag{4.3.3}$$

yielding the block-decomposition

$$\begin{cases} \begin{bmatrix} K^{1010} & K^{1001} \\ K^{0110} & K^{0101} \end{bmatrix} \begin{bmatrix} \alpha^{10\kappa} \\ \alpha^{01\kappa} \end{bmatrix} = \begin{bmatrix} \beta^{10\kappa} \\ \beta^{01\kappa} \end{bmatrix}, \\ \langle \mathcal{K}^{10}, \alpha^{10\kappa} \rangle + \langle \mathcal{K}^{01}, \alpha^{01\kappa} \rangle = 0. \end{cases} \tag{4.3.4}$$

Furthermore, from the kinetic theory results obtained in Sections 2.3 and 2.5, we can make the following assumptions.

$(K0)$ The coefficients of the matrix K are functions of the temperature and the mass fractions and can be expressed as quadratic functions of the mole fractions in the form

$$\begin{cases} K_{kk}^{rs} = \sum_{l \in \mathcal{S}} X_k X_l K_{kl}'^{rs} + X_k^2 K_{kk}''^{rs}, & (r,k),(s,k) \in \mathcal{B}^\kappa, \\ K_{kl}^{rs} = X_k X_l K_{kl}''^{rs}, \quad (r,k),(s,l) \in \mathcal{B}^\kappa, \quad k \neq l, \end{cases} \tag{4.3.5}$$

where the matrices K' and K'' solely depend on the temperature.

$(K1)$ K is symmetric.

$(K2)$ K is positive semi-definite.

$(K3)$ K is positive definite on the hyperplane $\mathcal{K}^\perp$.

$(K4)$ $N(K) = \mathbb{R}\mathcal{V}$.

$(K5)$ $R(K) = \mathcal{V}^\perp$.

Following Section 2.5.1, we assume that the vectors $\mathcal{K}$ and $\mathcal{V} \in \mathbb{R}^{n+p}$ are given by

$$\begin{cases} \mathcal{K}_k^{10} = c_v^{tr} X_k, & k \in \mathcal{S}, \\ \mathcal{K}_k^{01} = c_k^{int} X_k, & k \in \mathcal{P}, \end{cases} \tag{4.3.6}$$

and

$$\begin{cases} \mathcal{V}_k^{10} = 1, & k \in \mathcal{S}, \\ \mathcal{V}_k^{01} = 1, & k \in \mathcal{P}. \end{cases} \tag{4.3.7}$$

The quantity c_v^{tr} is a given positive constant, and the quantities c_k^{int}, $k \in \mathcal{P}$, are assumed to be smooth positive functions of the temperature. Letting $c_v = \sum_{k \in \mathcal{S}} X_k c_v^{tr} + \sum_{k \in \mathcal{P}} X_k c_k^{int}$, we then deduce from (4.3.6)(4.3.7) that we have

$$\langle \mathcal{K}, \mathcal{V} \rangle = c_v > 0, \tag{4.3.8}$$

and consequently, $\mathcal{V}^\perp \oplus \mathbb{R}\mathcal{K} = \mathbb{R}\mathcal{V} \oplus \mathcal{K}^\perp = \mathbb{R}^{n+p}$. In addition, following Section 2.5.1, we assume that the right-hand side β^κ is given by

$$\begin{cases} \beta_k^{10\kappa} = \dfrac{c^{int}}{c_v} X_k, & k \in \mathcal{S}, \\ \beta_k^{01\kappa} = -\dfrac{c_k^{int}}{c_v} X_k, & k \in \mathcal{P}, \end{cases} \tag{4.3.9}$$

where $c^{int} = \left(\sum_{k \in \mathcal{P}} X_k c_k^{int}\right) / \left(\sum_{k \in \mathcal{S}} X_k\right)$, so that we obtain

$$\beta^\kappa \in \mathcal{V}^\perp. \tag{4.3.10}$$

Theorem 4.3.1. *Let $K \in \mathbb{R}^{n+p,n+p}$ be a matrix satisfying the properties $(K0)$–$(K5)$ and let $\mathcal{K}$ and $\beta^\kappa \in \mathbb{R}^{n+p}$ be given by (4.3.6) and (4.3.9), respectively. Then the constrained linear system (4.3.1) admits a unique solution α^κ. Furthermore, the quantity κ may also be written*

$$\kappa = \sum_{k\in\mathcal{S}} X_k \alpha_k^{10\kappa}, \tag{4.3.11}$$

or, alternatively,

$$\kappa = -\sum_{k\in\mathcal{P}} \frac{c_k^{\mathrm{int}}}{c_{\mathrm{v}}^{\mathrm{tr}}} X_k \alpha_k^{01\kappa}. \tag{4.3.12}$$

Finally, if $p \geq 1$, the quantity κ is positive and vanishes otherwise.

Proof. Since $\beta^\kappa \in \mathcal{V}^\perp = R(K)$ by (4.3.10) and $(K5)$ and since $\mathcal{K}^\perp \oplus \mathbb{R}\mathcal{V} = \mathbb{R}^{n+p}$ by (4.3.8), Proposition 4.1.3 applies with $C = \mathcal{K}^\perp$ and $S = \mathbb{R}\mathcal{V}$, and the constrained linear system (4.3.1) admits therefore a unique solution α^κ. On the other hand, we deduce from (4.3.6)(4.3.9) that $\beta^{10\kappa} = (c^{\mathrm{int}}/(c_{\mathrm{v}} c_{\mathrm{v}}^{\mathrm{tr}}))\mathcal{K}^{10}$ and $\beta^{01\kappa} = -(1/c_{\mathrm{v}})\mathcal{K}^{01}$, and using (4.3.2) and the constraint $\langle \mathcal{K}, \alpha^\kappa \rangle = 0$, we easily obtain (4.3.11)(4.3.12). Finally, we have $\kappa = \langle \alpha^\kappa, \beta^\kappa \rangle = \langle \alpha^\kappa, K\alpha^\kappa \rangle$ and $\alpha^\kappa \in \mathcal{K}^\perp$. If $p \geq 1$, we have $\beta^\kappa \neq 0$ so that $\alpha^\kappa \neq 0$ and the quantity κ is positive by $(K3)$, whereas if $p = 0$, we get $\beta^\kappa = 0$ and $\alpha^\kappa = 0$ so that $\kappa = 0$. □

Remark. If there are no polyatomic species, $p = 0$ so that $\kappa = 0$ and the classical result of monatomic gas mixtures [FK72] is recovered. Notice also that for $p \geq 1$, (4.3.11) and (4.3.12) may be rewritten in the form $\kappa = (c_{\mathrm{v}}/c^{\mathrm{int}})\langle \alpha^{10\kappa}, \beta^{10\kappa} \rangle$ and $\kappa = (c_{\mathrm{v}}/c_{\mathrm{v}}^{\mathrm{tr}})\langle \alpha^{01\kappa}, \beta^{01\kappa} \rangle$, respectively.

In the next proposition, we express the volume viscosity κ in terms of either a generalized inverse of the matrix K or the inverse of a symmetric positive definite matrix.

Proposition 4.3.2. *Let $K^\dagger$ be the generalized inverse of K with prescribed range $\mathcal{K}^\perp$ and nullspace $\mathbb{R}\mathcal{K}$. Then the solution of the constrained linear system (4.3.1) is given by $\alpha^\kappa = K^\dagger \beta^\kappa$. Furthermore, if a and b are two positive real numbers such that $ab\langle \mathcal{K}, \mathcal{V} \rangle^2 = 1$, then the matrices $K + a\,\mathcal{K}\otimes\mathcal{K}$ and $K^\dagger + b\,\mathcal{V}\otimes\mathcal{V}$ are symmetric, positive definite, inverse of each other, and coincide with K and $K^\dagger$ on the hyperplanes $\mathcal{K}^\perp$ and $\mathcal{V}^\perp$, respectively. Therefore, we have*

$$\begin{cases} \alpha^\kappa = [K + a\,\mathcal{K}\otimes\mathcal{K}]^{-1}\beta^\kappa, \\ \kappa = \langle\, [K + a\,\mathcal{K}\otimes\mathcal{K}]^{-1}\beta^\kappa, \beta^\kappa \,\rangle. \end{cases} \tag{4.3.13}$$

Proof. The proof directly follows from Proposition 4.1.3 with $\mathcal{C} = \mathcal{K}^\perp$ and $S = \mathbb{R}\mathcal{V}$ and Proposition 4.1.4, keeping in mind that $\langle \mathcal{K}, \mathcal{V} \rangle \neq 0$ and $\beta^\kappa \in \mathcal{V}^\perp$. □

Finally, we consider the matrix K resulting from the practical approximations presented in Section 2.10.

Proposition 4.3.3. *Let $\bar{A}_{kl}$ and η_{kl}, $k,l \in \mathcal{S}$, be symmetric and positive coefficients, let η_k, $k \in \mathcal{S}$, $c_l^{\rm int}$ and $\xi_l^{\rm int}$, $l \in \mathcal{P}$, be positive quantities, and assume that $Y > 0$. Then the matrix K given by (2.10.9)–(2.10.14) satisfies ($K0$)–($K5$).*

Proof. The proof of ($K0$) directly results from (2.10.9)–(2.10.14), and it is clear from the symmetry of $\bar{A}_{kl}$ and η_{kl} that the matrix K is symmetric and hence satisfies ($K1$). Furthermore, an explicit calculation yields for $x \in \mathbb{R}^{n+p}$

$$\langle x, Kx \rangle = \sum_{k \in \mathcal{P}} \frac{4}{k_{\rm B}\pi} \frac{X_k^2}{\eta_k} \frac{c_k^{\rm int}}{\xi_k^{\rm int}} (x_k^{10} - x_k^{01})^2$$

$$+ \sum_{\substack{k,l \in \mathcal{S} \\ l \neq k}} \frac{5}{2} \frac{X_k X_l}{\bar{A}_{kl}\eta_{kl}} \frac{m_k m_l}{(m_k + m_l)^2} (x_k^{10} - x_l^{10})^2$$

$$+ \sum_{\substack{k \in \mathcal{S}, l \in \mathcal{P} \\ k \neq l}} \frac{4}{k_{\rm B}\pi} \frac{X_k X_l}{\eta_{kl}} \frac{c_l^{\rm int}}{\xi_l^{\rm int}} \Big(\frac{m_l x_k^{10} + m_k x_l^{10}}{m_k + m_l} - x_l^{01} \Big)^2 .$$

This shows that K is positive semi-definite and, therefore, ($K2$) is proven. Moreover, since $Y > 0$ implies $X > 0$, we deduce from the assumptions and the above explicit expression that $\langle x, Kx \rangle = 0$ if and only if $x \in \mathbb{R}\mathcal{V}$. This implies that K is positive definite on the hyperplane $\mathcal{K}^\perp$ since $\langle \mathcal{K}, \mathcal{V} \rangle \neq 0$ and also that $N(K) = \mathbb{R}\mathcal{V}$. Finally, $R(K) = \mathcal{V}^\perp$ since K is symmetric. □

4.3.2 Mathematical Properties of the System $K_{[01]}\alpha^\kappa_{[01]} = \beta^\kappa_{[01]}$

In this section we assume that $p \geq 1$ since we have seen in the previous section that the volume viscosity of monatomic gas mixtures vanishes. We consider the linear system

$$K_{[01]}\alpha^\kappa_{[01]} = \beta^\kappa_{[01]}, \tag{4.3.14}$$

associated with the evaluation of the volume viscosity

$$\kappa_{[01]} = \langle \alpha^\kappa_{[01]}, \beta^\kappa_{[01]} \rangle, \tag{4.3.15}$$

as described in Section 2.5.3. We have $K_{[01]} \in \mathbb{R}^{p,p}$, $\alpha^{\kappa}_{[01]}$ and $\beta^{\kappa}_{[01]} \in \mathbb{R}^{p}$, and the corresponding indexing set is given by

$$\mathcal{B}^{\kappa}_{[01]} = \{01\} \times \mathcal{P}. \tag{4.3.16}$$

Furthermore, from the kinetic theory results obtained in Sections 2.3 and 2.5, we can make the following assumptions.

($K_{[01]}0$) The coefficients of the matrix $K_{[01]}$ are functions of the temperature and the mass fractions and can be expressed as quadratic functions of the mole fractions in the form

$$\begin{cases} K^{rs}_{[01]kk} = \sum_{l \in \mathcal{S}} X_k X_l K'^{rs}_{[01]kl} + X_k^2 K''^{rs}_{[01]kk}, & (r,k),(s,k) \in \mathcal{B}^{\kappa}_{[01]}, \\ K^{rs}_{[01]kl} = X_k X_l K''^{rs}_{[01]kl}, \qquad (r,k),(s,l) \in \mathcal{B}^{\kappa}_{[01]}, & k \neq l, \end{cases} \tag{4.3.17}$$

where the matrices $K'_{[01]}$ and $K''_{[01]}$ solely depend on the temperature.

($K_{[01]}1$) $K_{[01]}$ is symmetric.

($K_{[01]}2$) $K_{[01]}$ is positive definite.

In addition, following Section 2.5.3, we assume that the right-hand side $\beta^{\kappa}_{[01]}$ is given by

$$\beta^{01\kappa}_{[01]k} = -\frac{c_k^{\text{int}}}{c_v} X_k, \qquad k \in \mathcal{P}. \tag{4.3.18}$$

Theorem 4.3.4. *Let $K_{[01]} \in \mathbb{R}^{p,p}$ be a matrix satisfying the properties ($K_{[01]}0$)–($K_{[01]}2$) and let $\beta^{\kappa}_{[01]} \in \mathbb{R}^{p}$ be given by (4.3.18). Then the linear system (4.3.14) admits a unique solution $\alpha^{\kappa}_{[01]}$. Furthermore, the quantity $\kappa_{[01]}$ is positive.*

Finally, we consider the matrix $K_{[01]}$ resulting from the practical approximations presented in Section 2.10.

Proposition 4.3.5. *Let $\bar{\lambda}_{kl}$ and η_{kl}, $k,l \in \mathcal{S}$, be symmetric and positive coefficients, let η_k, c_k^{int} and ξ_k^{int}, $k \in \mathcal{P}$, be positive quantities, and assume that $Y > 0$. Then the matrix $K_{[01]}$ given by (2.10.13)(2.10.14) satisfies ($K_{[01]}0$)–($K_{[01]}2$).*

4.3.3 Mathematical Properties of the System $\widetilde{K}\alpha^{\kappa} = \widetilde{\beta}^{\kappa}$

We consider the constrained linear system

$$\begin{cases} \widetilde{K}\alpha^{\kappa} = \widetilde{\beta}^{\kappa}, \\ \langle \mathcal{K}, \alpha^{\kappa} \rangle = 0, \end{cases} \tag{4.3.19}$$

associated with the evaluation of the volume viscosity

$$\kappa = \langle \alpha^\kappa, \beta^\kappa \rangle, \tag{4.3.20}$$

as described in Section 3.3.1. We have $\widetilde{K} \in \mathbb{R}^{n+p,n+p}$, α^κ, $\widetilde{\beta}^\kappa$, and $\mathcal{K} \in \mathbb{R}^{n+p}$, and the indexing set is $\mathcal{B}^\kappa = \{10\}\times\mathcal{S} \cup \{01\}\times\mathcal{P}$, yielding the block-decomposition

$$\begin{cases} \begin{bmatrix} \widetilde{K}^{1010} & \widetilde{K}^{1001} \\ \widetilde{K}^{0110} & \widetilde{K}^{0101} \end{bmatrix} \begin{bmatrix} \alpha^{10\kappa} \\ \alpha^{01\kappa} \end{bmatrix} = \begin{bmatrix} \widetilde{\beta}^{10\kappa} \\ \widetilde{\beta}^{01\kappa} \end{bmatrix}, \\ \langle \mathcal{K}^{10}, \alpha^{10\kappa} \rangle + \langle \mathcal{K}^{01}, \alpha^{01\kappa} \rangle = 0. \end{cases} \tag{4.3.21}$$

Furthermore, from the kinetic theory results obtained in Sections 3.1 and 3.3, we can make the following assumptions.

($\widetilde{K}0$) The coefficients of the matrix $\widetilde{K}$ are functions of the temperature and the mass fractions and can be expressed as linear functions of the mole fractions in the form

$$\begin{cases} \widetilde{K}_{kk}^{rs} = \sum_{l\in\mathcal{S}} X_l K_{kl}^{\prime rs} + X_k K_{kk}^{\prime\prime rs}, & (r,k),(s,k) \in \mathcal{B}^\kappa, \\ \widetilde{K}_{kl}^{rs} = X_l K_{kl}^{\prime\prime rs}, \quad (r,k),(s,l) \in \mathcal{B}^\kappa, & k \neq l, \end{cases} \tag{4.3.22}$$

where the matrices K' and K'' coincide with the ones in (4.3.5).

($\widetilde{K}1$) The matrix $\widetilde{K}$ admits the block-decomposition

$$(\Gamma^\kappa)^t \widetilde{K} \Gamma^\kappa = \begin{bmatrix} \widetilde{K}^{++} & 0 \\ \widetilde{K}^{-+} & \widetilde{K}^{--} \end{bmatrix}, \tag{4.3.23}$$

where Γ^κ is the permutation matrix associated with the reordering of $\mathcal{B}^\kappa$ into $(\mathcal{B}^{\kappa+}, \mathcal{B}^{\kappa-})$.

($\widetilde{K}2$) The matrix $K^{++} = \mathcal{X}^{\kappa++}\widetilde{K}^{++}$ corresponds to the $\mathcal{S}^+$ mixture and, in particular, is symmetric, positive semi-definite, positive definite on the hyperplane $(\mathcal{K}^+)^\perp$, and has nullspace $N(K^{++}) = \mathbb{R}\mathcal{V}^+$ and range $R(K^{++}) = (\mathcal{V}^+)^\perp$.

($\widetilde{K}3$) $\widetilde{K}^{--}$ is symmetric positive definite and we have $db(\widetilde{K}^{--}) = \widetilde{K}^{--}$.

($\widetilde{K}4$) $N(\widetilde{K}) = \mathbb{R}\mathcal{V}$.

($\widetilde{K}5$) $R(\widetilde{K}) = \widetilde{\mathcal{V}}^\perp$.

In addition, following Section 3.3.1, we assume that the right-hand side $\widetilde{\beta}^\kappa$ is given by

$$\begin{cases} \widetilde{\beta}_k^{10\kappa} = \dfrac{c^{\mathrm{int}}}{c_v}, & k \in \mathcal{S}, \\ \widetilde{\beta}_k^{01\kappa} = -\dfrac{c_k^{\mathrm{int}}}{c_v}, & k \in \mathcal{P}. \end{cases} \tag{4.3.24}$$

From (4.3.5)(4.3.22) and (4.3.9)(4.3.24), we deduce that the matrix $\widetilde{K}$ and the right-hand side $\widetilde{\beta}^\kappa$ are rescaled versions of K and β^κ, respectively,

$$\mathcal{X}^\kappa \widetilde{K} = K, \qquad \mathcal{X}^\kappa \widetilde{\beta}^\kappa = \beta^\kappa, \tag{4.3.25}$$

where $\mathcal{X}^\kappa = \mathrm{diag}\big((X_k)_{(r,k)\in\mathcal{B}^\kappa}\big)$. Finally, the vector $\widetilde{\mathcal{V}} \in \mathbb{R}^{n+p}$ is given by $\widetilde{\mathcal{V}} = \mathcal{X}^\kappa \mathcal{V}$, or, more explicitly,

$$\begin{cases} \widetilde{\mathcal{V}}_k^{10} = X_k, & k \in \mathcal{S}, \\ \widetilde{\mathcal{V}}_k^{01} = X_k, & k \in \mathcal{P}, \end{cases} \tag{4.3.26}$$

and since $\langle \widetilde{\beta}^\kappa, \widetilde{\mathcal{V}} \rangle = \langle \widetilde{\beta}^\kappa, \mathcal{X}^\kappa \mathcal{V} \rangle = \langle \mathcal{X}^\kappa \widetilde{\beta}^\kappa, \mathcal{V} \rangle = \langle \beta^\kappa, \mathcal{V} \rangle = 0$, we obtain

$$\widetilde{\beta}^\kappa \in \widetilde{\mathcal{V}}^\perp. \tag{4.3.27}$$

Theorem 4.3.6. *Let $\widetilde{K} \in \mathbb{R}^{n+p,n+p}$ be a matrix satisfying the properties $(\widetilde{K}0)$–$(\widetilde{K}5)$ and let $\mathcal{K}$, β^κ, and $\widetilde{\beta}^\kappa \in \mathbb{R}^{n+p}$ be given by (4.3.6), (4.3.9), and (4.3.24), respectively. Then the constrained linear system (4.3.19) admits a unique solution α^κ, and this solution is such that $K^{++}\alpha^{\kappa+} = \beta^{\kappa+}$ and $\langle \mathcal{K}^+, \alpha^{\kappa+} \rangle = 0$. Furthermore, the quantity κ may also be written*

$$\kappa = \sum_{k\in\mathcal{S}^+} X_k \alpha_k^{10\kappa}, \tag{4.3.28}$$

or, alternatively,

$$\kappa = -\sum_{k\in\mathcal{P}^+} \frac{c_k^{\mathrm{int}}}{c_{\mathrm{v}}^{\mathrm{tr}}} X_k \alpha_k^{01\kappa}, \tag{4.3.29}$$

and is the volume viscosity of the $\mathcal{S}^+$ mixture. Finally, the quantity κ is positive if $p^+ \geq 1$, and vanishes otherwise.

Proof. Since $\widetilde{\beta}^\kappa \in \widetilde{\mathcal{V}}^\perp = R(\widetilde{K})$ from (4.3.27) and $(\widetilde{K}5)$ and since $\mathcal{K}^\perp \oplus \mathbb{R}\mathcal{V} = \mathbb{R}^{n+p}$, Proposition 4.1.3 applies, and the constrained linear system (4.3.19) admits a unique solution α^κ. In addition, we deduce from $(\widetilde{K}1)$ and (4.3.25) that $K^{++}\alpha^{\kappa+} = \beta^{\kappa+}$, and we have $\langle \mathcal{K}^+, \alpha^{\kappa+} \rangle = 0$ since $\mathcal{K}^- = 0$. The proof of (4.3.28)(4.3.29) is the same as the one given in Theorem 4.3.1. On the other hand, using $\beta^{\kappa-} = 0$, we obtain $\kappa = \langle \alpha^{\kappa+}, \beta^{\kappa+} \rangle = \langle \alpha^{\kappa+}, K^{++}\alpha^{\kappa+} \rangle$ so that κ is the volume viscosity of the $\mathcal{S}^+$ mixture, since K^{++} is the corresponding system matrix. Finally, if $p^+ \geq 1$, then $\beta^{\kappa+} \neq 0$ so that $\alpha^{\kappa+} \neq 0$ and since $\alpha^{\kappa+} \in (\mathcal{K}^+)^\perp$, the quantity κ is positive by $(\widetilde{K}2)$, whereas if $p^+ = 0$, we have $\beta^{\kappa+} = 0$ and $\alpha^{\kappa+} = 0$ so that $\kappa = 0$. □

Remark. If there are no polyatomic species, $p^+ = 0$ so that $\kappa = 0$ and the classical result of monatomic gas mixtures [FK72] is again recovered. Furthermore, for $p^+ \geq 1$,

(4.3.28) and (4.3.29) can be rewritten in the form $\kappa = (c_v/c^{\text{int}})\langle \alpha^{10\kappa}, \beta^{10\kappa}\rangle$ and $\kappa = (c_v/c_v^{\text{tr}})\langle \alpha^{01\kappa}, \beta^{01\kappa}\rangle$, respectively.

In the next proposition, we express the volume viscosity κ in terms of a generalized inverse of the matrix $\widetilde{K}$ and also in terms of the solution of a nonsingular linear system.

Proposition 4.3.7. *Let $\widetilde{\mathcal{K}} \in \mathbb{R}^{n+p}$ be given by $\widetilde{\mathcal{K}}_k^{10} = c_v^{\text{tr}}$, $k \in \mathcal{S}$, and $\widetilde{\mathcal{K}}_k^{01} = c_k^{\text{int}}$, $k \in \mathcal{P}$, and let $\widetilde{K}^\dagger$ be the generalized inverse of $\widetilde{K}$ with prescribed range $\mathcal{K}^\perp$ and nullspace $\mathbb{R}\widetilde{\mathcal{K}}$. Then the solution of the constrained linear system (4.3.19) is given by $\alpha^\kappa = \widetilde{K}^\dagger\widetilde{\beta}^\kappa$. Furthermore, if a and b are two positive real numbers such that $ab\langle\mathcal{K},\mathcal{V}\rangle\langle\widetilde{\mathcal{K}},\widetilde{\mathcal{V}}\rangle = 1$, then the matrices $\widetilde{K} + a\,\widetilde{\mathcal{K}}\otimes\mathcal{K}$ and $\widetilde{K}^\dagger + b\,\mathcal{V}\otimes\widetilde{\mathcal{V}}$ are inverse of each other and coincide with $\widetilde{K}$ and $\widetilde{K}^\dagger$ on the hyperplanes $\mathcal{K}^\perp$ and $\widetilde{\mathcal{V}}^\perp$, respectively. Therefore, we have*

$$\begin{cases} \alpha^\kappa = [\widetilde{K} + a\,\widetilde{\mathcal{K}}\otimes\mathcal{K}]^{-1}\widetilde{\beta}^\kappa, \\ \kappa = \langle\, [\widetilde{K} + a\,\widetilde{\mathcal{K}}\otimes\mathcal{K}]^{-1}\widetilde{\beta}^\kappa, \beta^\kappa \,\rangle. \end{cases} \tag{4.3.30}$$

Proof. It directly follows from Propositions 4.1.3 and 4.1.5 since $\langle\mathcal{K},\mathcal{V}\rangle = \langle\widetilde{\mathcal{K}},\widetilde{\mathcal{V}}\rangle > 0$ and $\widetilde{\beta}^\kappa \in \widetilde{\mathcal{V}}^\perp$. □

We now prove that the volume viscosity κ is a smooth function of the temperature and the mass fractions in $\mathcal{D}_{Y\geq 0, Y\neq 0}$ and that it is independent of the pressure.

Theorem 4.3.8. *Let K and $\widetilde{K} \in \mathbb{R}^{n+p,n+p}$ be such that K satisfies $(K0)$–$(K5)$ in $\mathcal{D}_{Y>0}$ and $\widetilde{K}$ satisfies $(\widetilde{K}0)$–$(\widetilde{K}5)$ in $\mathcal{D}_{Y\geq 0, Y\neq 0}$. Assume also that the matrices K' and K'' in (4.3.5)(4.3.22) are smooth functions of the temperature. Let α^κ and κ be the solution and the volume viscosity evaluated from the matrix K for $Y > 0$ and let $\widetilde{\alpha}^\kappa$ and $\widetilde{\kappa}$ be the solution and the volume viscosity evaluated from the matrix $\widetilde{K}$ for $Y \geq 0$, $Y \neq 0$. Then $\alpha^\kappa = \widetilde{\alpha}^\kappa$ and $\kappa = \widetilde{\kappa}$ for $Y > 0$. Furthermore, $\widetilde{\alpha}^\kappa$ and $\widetilde{\kappa}$ are independent of the pressure and are smooth functions of $(\overline{T}, Y)$ in $\mathcal{D}_{Y\geq 0, Y\neq 0}$. In particular, for $Z \in \mathbb{R}^n$, $Z \geq 0$, $Z \neq 0$, we have the following limit*

$$\lim_{\substack{Y\to Z \\ Y>0}} \kappa(\overline{T}, Y) = \widetilde{\kappa}(\overline{T}, Z). \tag{4.3.31}$$

Proof. When all the mass fractions are positive, the matrix $\mathcal{X}^\kappa$ is invertible, and we deduce from (4.3.25) that the constrained linear systems (4.3.1)(4.3.19) admit the same solution and hence $\alpha^\kappa = \widetilde{\alpha}^\kappa$ and $\kappa = \widetilde{\kappa}$ for $Y > 0$. Furthermore, using (4.3.30) and Cramer's rule, we can write the vector $\widetilde{\alpha}^\kappa$ and the quantity $\widetilde{\kappa}$ as rational functions of the coefficients of $\widetilde{K}$ and β^κ. These coefficients are, by assumption, independent of the pressure and are smooth functions of the temperature and the mass fractions, keeping in mind that the mole fractions are smooth rational functions of the mass fractions by (4.1.1)(4.1.2). Consequently, $\widetilde{\alpha}^\kappa$ and $\widetilde{\kappa}$ are independent of the pressure and are smooth functions of $(\overline{T}, Y)$ in $\mathcal{D}_{Y\geq 0, Y\neq 0}$. In particular, we obtain $\lim_{Y\to Z, Y>0} \kappa(\overline{T}, Y) = \widetilde{\kappa}(\overline{T}, Z)$. □

Remark. As a consequence of the previous theorem, the solutions of the linear systems (4.3.1)(4.3.19) coincide when $Y > 0$. This justifies the use of the same notation α^κ and κ in Sections 4.3.1 and 4.3.3.

Finally, we consider the matrix $\widetilde{K}$ resulting from the practical approximations presented in Section 2.10.

Proposition 4.3.9. *Let $\bar{A}_{kl}$ and η_{kl}, $k, l \in \mathcal{S}$, be symmetric and positive coefficients, let η_k, $k \in \mathcal{S}$, $c_l^{\rm int}$ and $\xi_l^{\rm int}$, $l \in \mathcal{P}$, be positive quantities, and assume that $Y \geq 0$, $Y \neq 0$. Then the matrix $\widetilde{K}$ given by (3.7.3)–(3.7.10) satisfies $(\widetilde{K}0)$–$(\widetilde{K}5)$.*

Proof. The properties $(\widetilde{K}0)$–$(\widetilde{K}1)$ directly result from (3.7.3)–(3.7.10), and the proof of $(\widetilde{K}2)$ is exactly the same as the one of Proposition 4.3.3. Next, the matrix $\widetilde{K}^{--}$ is symmetric, block-diagonal, and an explicit calculation yields for $x = \Gamma^\kappa(0, x^-) \in \mathbb{R}^{n+p}$

$$\langle x^-, \widetilde{K}^{--} x^- \rangle = \sum_{\substack{k\in\mathcal{S}^-\\ l\in\mathcal{S}^+}} 5 \frac{X_l}{\bar{A}_{kl}\eta_{kl}} \frac{m_k m_l}{(m_k + m_l)^2} \left(x_k^{10}\right)^2$$

$$+ \sum_{\substack{k\in\mathcal{S}^-\\ l\in\mathcal{P}^+}} \frac{4}{k_{\rm B}\pi} \frac{X_l}{\eta_{kl}} \frac{m_l^2}{(m_k + m_l)^2} \frac{c_l^{\rm int}}{\xi_l^{\rm int}} \left(x_k^{10}\right)^2$$

$$+ \sum_{\substack{k\in\mathcal{P}^-\\ l\in\mathcal{S}^+}} \frac{4}{k_{\rm B}\pi} \frac{X_l}{\eta_{kl}} \frac{c_k^{\rm int}}{\xi_k^{\rm int}} \left(\frac{m_l}{m_k + m_l} x_k^{10} - x_k^{01}\right)^2.$$

Since $X_l > 0$, $l \in \mathcal{S}^+$, it is readily seen that $\langle x^-, \widetilde{K}^{--} x^- \rangle = 0$ if and only if $x^- = 0$ so that $(\widetilde{K}3)$ is proven. Finally, we can explicitly verify from (3.7.3)–(3.7.10) that $\mathcal{V}^t \widetilde{K} = 0$, i.e., $R(\widetilde{K}) \subset \mathcal{V}^\perp$, and the remaining properties $(\widetilde{K}4)$–$(\widetilde{K}5)$ are then easily obtained. □

4.3.4 Mathematical Properties of the System $\widehat{K}\widehat{\alpha}^\kappa = \widehat{\beta}^\kappa$

We consider the constrained linear system

$$\begin{cases} \widehat{K}\widehat{\alpha}^\kappa = \widehat{\beta}^\kappa, \\ \langle \widehat{\mathcal{K}}, \widehat{\alpha}^\kappa \rangle = 0, \end{cases} \tag{4.3.32}$$

associated with the evaluation of the volume viscosity

$$\kappa = \langle \widehat{\alpha}^{\kappa}, \widehat{\beta}^{\kappa} \rangle. \tag{4.3.33}$$

We have $\widehat{K} \in \mathbb{R}^{n+p,n+p}$, $\widehat{\alpha}^{\kappa}$, $\widehat{\beta}^{\kappa}$, and $\widehat{\mathcal{K}} \in \mathbb{R}^{n+p}$, and following Section 3.1.8, we can make the following assumptions.

($\widehat{K}0$) The coefficients of the matrix $\widehat{K}$ are functions of the temperature and the mass fractions and can be expressed as

$$\begin{cases} \widehat{K}_{kk}^{rs} = \sum_{l \in \mathcal{S}} X_l K_{kl}^{\prime rs} + X_k K_{kk}^{\prime\prime rs}, & (r,k),(s,k) \in \mathcal{B}^{\kappa}, \\ \widehat{K}_{kl}^{rs} = \sqrt{X_k X_l} K_{kl}^{\prime\prime rs}, & (r,k),(s,l) \in \mathcal{B}^{\kappa}, \quad k \neq l, \end{cases} \tag{4.3.34}$$

where the matrices K' and K'' coincide with the ones in (4.3.5).

($\widehat{K}1$) $\widehat{K}$ is symmetric.

($\widehat{K}2$) $\widehat{K}$ is positive semi-definite.

($\widehat{K}3$) $\widehat{K}$ is positive definite on the hyperplane $\widehat{\mathcal{K}}^{\perp}$.

($\widehat{K}4$) $N(\widehat{K}) = \mathbb{R}\widehat{\mathcal{V}}$.

($\widehat{K}5$) $R(\widehat{K}) = \widehat{\mathcal{V}}^{\perp}$.

In addition, we assume that the constraint vector $\widehat{\mathcal{K}}$ is given by

$$\begin{cases} \widehat{\mathcal{K}}_k^{10} = c_{\mathrm{v}}^{\mathrm{tr}} \sqrt{X_k}, & k \in \mathcal{S}, \\ \widehat{\mathcal{K}}_k^{01} = c_k^{\mathrm{int}} \sqrt{X_k}, & k \in \mathcal{P}, \end{cases} \tag{4.3.35}$$

the vector $\widehat{\mathcal{V}} \in \mathbb{R}^{n+p}$ by $\widehat{\mathcal{V}} = (\mathcal{X}^{\kappa})^{1/2} \mathcal{V}$, or, more explicitly,

$$\begin{cases} \widehat{\mathcal{V}}_k^{10} = \sqrt{X_k}, & k \in \mathcal{S}, \\ \widehat{\mathcal{V}}_k^{01} = \sqrt{X_k}, & k \in \mathcal{P}, \end{cases} \tag{4.3.36}$$

and the right-hand side $\widehat{\beta}^{\kappa}$ by

$$\begin{cases} \widehat{\beta}_k^{10\kappa} = \dfrac{c^{\mathrm{int}}}{c_{\mathrm{v}}} \sqrt{X_k}, & k \in \mathcal{S}, \\ \widehat{\beta}_k^{01\kappa} = -\dfrac{c_k^{\mathrm{int}}}{c_{\mathrm{v}}} \sqrt{X_k}, & k \in \mathcal{P}. \end{cases} \tag{4.3.37}$$

The matrix $\widehat{K}$, the right-hand side $\widehat{\beta}^{\kappa}$, and the constraint vector $\widehat{\mathcal{K}} \in \mathbb{R}^{n+p}$ are then rescaled versions of K, β^{κ}, and $\mathcal{K}$, respectively,

$$(\mathcal{X}^{\kappa})^{1/2} \widehat{K} (\mathcal{X}^{\kappa})^{1/2} = K, \qquad (\mathcal{X}^{\kappa})^{1/2} \widehat{\beta}^{\kappa} = \beta^{\kappa}, \qquad (\mathcal{X}^{\kappa})^{1/2} \widehat{\mathcal{K}} = \mathcal{K}. \tag{4.3.38}$$

Using $(\widehat{K}0)$–$(\widehat{K}5)$ and (4.3.35)–(4.3.37), one can prove that the system (4.3.32) is well posed, i.e., admits a unique solution and that κ is positive if $p \geq 1$ and vanishes otherwise. Finally, from (4.3.38) we deduce that the definitions (4.3.2)(4.3.33) of the volume viscosity coincide for $Y > 0$.

4.3.5 Mathematical Properties of the System $\widetilde{K}_{[01]}\alpha^{\kappa}_{[01]} = \widetilde{\beta}^{\kappa}_{[01]}$

In this section we assume that $p^+ \geq 1$ since we have already seen that the volume viscosity of monatomic gas mixtures vanishes. We consider the linear system

$$\widetilde{K}_{[01]}\alpha^{\kappa}_{[01]} = \widetilde{\beta}^{\kappa}_{[01]}, \tag{4.3.39}$$

associated with the evaluation of the volume viscosity

$$\kappa_{[01]} = \langle \alpha^{\kappa}_{[01]}, \beta^{\kappa}_{[01]} \rangle, \tag{4.3.40}$$

as described in Section 3.3.3. We have $\widetilde{K}_{[01]} \in \mathbb{R}^{p,p}$, $\alpha^{\kappa}_{[01]}$ and $\widetilde{\beta}^{\kappa}_{[01]} \in \mathbb{R}^{p}$, and the indexing set is $\mathcal{B}^{\kappa}_{[01]} = \{01\} \times \mathcal{P}$. Furthermore, from the kinetic theory results obtained in Sections 3.1 and 3.3, we can make the following assumptions.

$(\widetilde{K}_{[01]}0)$ The coefficients of the matrix $\widetilde{K}_{[01]}$ are functions of the temperature and the mass fractions and can be expressed as linear functions of the mole fractions in the form

$$\begin{cases} \widetilde{K}^{rs}_{[01]kk} = \sum\limits_{l \in S} X_l K'^{rs}_{[01]kl} + X_k K''^{rs}_{[01]kk}, & (r,k),(s,k) \in \mathcal{B}^{\kappa}_{[01]}, \\ \widetilde{K}^{rs}_{[01]kl} = X_l K''^{rs}_{[01]kl}, \qquad (r,k),(s,l) \in \mathcal{B}^{\kappa}_{[01]}, & k \neq l, \end{cases} \tag{4.3.41}$$

where the matrices $K'_{[01]}$ and $K''_{[01]}$ coincide with the ones in (4.3.17).

$(\widetilde{K}_{[01]}1)$ The matrix $\widetilde{K}_{[01]}$ admits the block-decomposition

$$\left(\Gamma^{\kappa}_{[01]}\right)^t \widetilde{K}_{[01]} \Gamma^{\kappa}_{[01]} = \begin{bmatrix} \widetilde{K}^{++}_{[01]} & 0 \\ \widetilde{K}^{-+}_{[01]} & \widetilde{K}^{--}_{[01]} \end{bmatrix}, \tag{4.3.42}$$

where $\Gamma^{\kappa}_{[01]}$ is the permutation matrix associated with the reordering of $\mathcal{B}^{\kappa}_{[01]}$ into $(\mathcal{B}^{\kappa+}_{[01]}, \mathcal{B}^{\kappa-}_{[01]})$.

$(\widetilde{K}_{[01]}2)$ $K^{++}_{[01]} = \mathcal{X}^{\kappa++}_{[01]} \widetilde{K}^{++}_{[01]}$ corresponds to the S^+ mixture and, in particular, is symmetric positive definite.

$(\widetilde{K}_{[01]}3)$ $K^{--}_{[01]}$ is symmetric positive definite and we have $db(K^{--}_{[01]}) = K^{--}_{[01]}$.

$(\widetilde{K}_{[01]}4)$ $K_{[01]}$ is nonsingular.

In addition, following Section 3.3.3, we assume that the right-hand side $\widetilde{\beta}^{\kappa}_{[01]}$ is given by

$$\widetilde{\beta}^{01\kappa}_{[01]k} = -\frac{c_k^{\text{int}}}{c_v}, \qquad k \in \mathcal{P}, \tag{4.3.43}$$

and we deduce from (4.3.17)(4.3.41) and (4.3.18)(4.3.43) that the matrix $\widetilde{K}_{[01]}$ and the right-hand side $\widetilde{\beta}^{\kappa}_{[01]}$ are rescaled versions of $K_{[01]}$ and $\beta^{\kappa}_{[01]}$, respectively,

$$\mathcal{X}^{\kappa}_{[01]}\widetilde{K}_{[01]} = K_{[01]}, \qquad \mathcal{X}^{\kappa}_{[01]}\widetilde{\beta}^{\kappa}_{[01]} = \beta^{\kappa}_{[01]}, \tag{4.3.44}$$

where $\mathcal{X}^{\kappa}_{[01]} = \text{diag}\big((X_k)_{(r,k)\in\mathcal{B}^{\kappa}_{[01]}}\big)$.

Theorem 4.3.10. *Let $\widetilde{K}_{[01]} \in \mathbb{R}^{p,p}$ be a matrix satisfying the properties $(\widetilde{K}_{[01]}0)$–$(\widetilde{K}_{[01]}4)$ and let $\beta^{\kappa}_{[01]}$ and $\widetilde{\beta}^{\kappa}_{[01]} \in \mathbb{R}^p$ be given by (4.3.18) and (4.3.43), respectively. Then the linear system (4.3.39) admits a unique solution $\alpha^{\kappa}_{[01]}$, and this solution is such that $K^{++}_{[01]}\alpha^{\kappa+}_{[01]} = \beta^{\kappa+}_{[01]}$. Furthermore, the quantity $\kappa_{[01]}$ is positive and is the volume viscosity of the $\mathcal{S}^+$ mixture.*

We now establish that the volume viscosity $\kappa_{[01]}$ is a smooth function of the temperature and the mass fractions in $\mathcal{D}_{Y\geq 0, Y\neq 0}$ and that it is independent of the pressure.

Theorem 4.3.11. *Let $K_{[01]}$ and $\widetilde{K}_{[01]} \in \mathbb{R}^{p,p}$ be such that $K_{[01]}$ satisfies $(K_{[01]}0)$–$(K_{[01]}2)$ in $\mathcal{D}_{Y>0}$ and $\widetilde{K}_{[01]}$ satisfies $(\widetilde{K}_{[01]}0)$–$(\widetilde{K}_{[01]}4)$ in $\mathcal{D}_{Y\geq 0, Y\neq 0}$. Assume also that the matrices $K'_{[01]}$ and $K''_{[01]}$ in (4.3.17)(4.3.41) are smooth functions of the temperature. Let $\alpha^{\kappa}_{[01]}$ and $\kappa_{[01]}$ be the solution and the volume viscosity evaluated from the matrix $K_{[01]}$ for $Y > 0$ and let $\widetilde{\alpha}^{\kappa}_{[01]}$ and $\widetilde{\kappa}_{[01]}$ be the solution and the volume viscosity evaluated from the matrix $\widetilde{K}_{[01]}$ for $Y \geq 0$, $Y \neq 0$. Then $\alpha^{\kappa}_{[01]} = \widetilde{\alpha}^{\kappa}_{[01]}$ and $\kappa_{[01]} = \widetilde{\kappa}_{[01]}$ for $Y > 0$. Furthermore, $\widetilde{\alpha}^{\kappa}_{[01]}$ and $\widetilde{\kappa}_{[01]}$ are independent of the pressure and are smooth functions of $(\overline{T}, Y)$ in $\mathcal{D}_{Y\geq 0, Y\neq 0}$. In particular, for $Z \in \mathbb{R}^n$, $Z \geq 0$, $Z \neq 0$, we have the following limit*

$$\lim_{\substack{Y\to Z\\ Y>0}} \kappa_{[01]}(\overline{T}, Y) = \widetilde{\kappa}_{[01]}(\overline{T}, Z). \tag{4.3.45}$$

Remark. As a consequence of the previous theorem, the solutions of the linear systems (4.3.14)(4.3.39) coincide when $Y > 0$. This justifies the use of the same notation $\alpha^{\kappa}_{[01]}$ and $\kappa_{[01]}$ in Sections 4.3.2 and 4.3.5.

Finally, we consider the matrix $\widetilde{K}_{[01]}$ resulting from the practical approximations presented in Section 2.10.

Proposition 4.3.12. *Let $\bar{A}_{kl}$ and η_{kl}, $k,l \in \mathcal{S}$, be symmetric and positive coefficients, let η_k, $c_k^{\rm int}$ and $\xi_k^{\rm int}$, $k \in \mathcal{P}$, be positive quantities, and assume that $Y \geq 0$, $Y \neq 0$. Then the matrix $\widetilde{K}_{[01]}$ given by (3.7.9)(3.7.10) satisfies $(\widetilde{K}_{[01]}0)$–$(\widetilde{K}_{[01]}4)$.*

4.4 The Diffusion Matrix and the Flux Diffusion Matrix

We first investigate the mathematical properties of the linear systems associated with the evaluation of the diffusion matrix, the flux diffusion matrix, and the symmetric flux diffusion matrix in Sections 4.4.1–4.4.7. Several generalized Stefan-Maxwell-Boltzmann equations are presented in Section 4.4.8, and diagonal diffusion processes are considered in Section 4.4.9. An alternative definition of the flux diffusion matrix is given in Section 4.4.10, and finally, singularities arising in discretized conservation equations due to diffusion matrices are briefly discussed in Section 4.4.11.

4.4.1 Mathematical Properties of the System $L\alpha^{D_l} = \beta^{D_l}$

We consider the n constrained linear systems indexed by $l \in \mathcal{S}$

$$\begin{cases} L\alpha^{D_l} = \beta^{D_l}, \\ \langle \mathcal{L}, \alpha^{D_l} \rangle = 0, \end{cases} \tag{4.4.1}$$

associated with the evaluation of the diffusion matrix $D = (D_{kl})_{k,l \in \mathcal{S}}$

$$D_{kl} = \langle \alpha^{D_k}, \beta^{D_l} \rangle, \qquad k,l \in \mathcal{S}, \tag{4.4.2}$$

as described in Section 2.6.1. We have $L \in \mathbb{R}^{2n+p,2n+p}$, α^{D_l}, β^{D_l}, and $\mathcal{L} \in \mathbb{R}^{2n+p}$, and the corresponding indexing set is given by

$$\mathcal{B}^D = \{00, 10\} \times \mathcal{S} \ \cup \ \{01\} \times \mathcal{P}, \tag{4.4.3}$$

yielding the block-decomposition

$$\begin{cases} \begin{bmatrix} L^{0000} & L^{0010} & L^{0001} \\ L^{1000} & L^{1010} & L^{1001} \\ L^{0100} & L^{0110} & L^{0101} \end{bmatrix} \begin{bmatrix} \alpha^{00D_l} \\ \alpha^{10D_l} \\ \alpha^{01D_l} \end{bmatrix} = \begin{bmatrix} \beta^{00D_l} \\ \beta^{10D_l} \\ \beta^{01D_l} \end{bmatrix}, \\ \langle \mathcal{L}^{00}, \alpha^{00D_l} \rangle + \langle \mathcal{L}^{10}, \alpha^{10D_l} \rangle + \langle \mathcal{L}^{01}, \alpha^{01D_l} \rangle = 0. \end{cases} \tag{4.4.4}$$

Furthermore, from the kinetic theory results obtained in Sections 2.3 and 2.6, we can make the following assumptions.

($L0$) The coefficients of the matrix L are functions of the pressure, the temperature, and the mass fractions, they are proportional to the pressure, and they can be expressed as quadratic functions of the mole fractions in the form

$$\begin{cases} L_{kk}^{rs} = \sum_{l\in\mathcal{S}} X_k X_l L_{kl}^{\prime rs} + X_k^2 L_{kk}^{\prime\prime rs}, & (r,k),(s,k) \in \mathcal{B}^D, \\ L_{kl}^{rs} = X_k X_l L_{kl}^{\prime\prime rs}, \quad (r,k),(s,l) \in \mathcal{B}^D, \quad k \neq l, \end{cases} \tag{4.4.5}$$

where the matrices L' and L'' solely depend on the pressure and the temperature.

($L1$) L is symmetric.

($L2$) L is positive semi-definite.

($L3$) L is positive definite on the hyperplane $\mathcal{L}^\perp$.

($L4$) $N(L) = \mathbb{R}\mathcal{U}$.

($L5$) $R(L) = \mathcal{U}^\perp$.

Following Section 2.6.1, we assume that the vectors $\mathcal{L}$ and $\mathcal{U} \in \mathbb{R}^{2n+p}$ are given by the block-decompositions

$$\mathcal{L}^{00} = Y, \qquad \mathcal{L}^{10} = 0, \qquad \mathcal{L}^{01} = 0, \tag{4.4.6}$$

and

$$\mathcal{U}^{00} = U, \qquad \mathcal{U}^{10} = 0, \qquad \mathcal{U}^{01} = 0, \tag{4.4.7}$$

where $Y = (Y_k)_{k\in\mathcal{S}} \in \mathbb{R}^n$ and $U = (1)_{k\in\mathcal{S}} \in \mathbb{R}^n$. Since the mass fractions are positive, we have $\langle Y, U\rangle = \sum_{k\in\mathcal{S}} Y_k > 0$, thus yielding

$$\langle \mathcal{L}, \mathcal{U}\rangle > 0, \tag{4.4.8}$$

from which we deduce that $\mathcal{U}^\perp \oplus \mathbb{R}\mathcal{L} = \mathbb{R}\mathcal{U} \oplus \mathcal{L}^\perp = \mathbb{R}^{2n+p}$. In addition, following Section 2.6.1, we assume that the right-hand sides β^{D_l}, $l \in \mathcal{S}$, are given by

$$\begin{cases} \beta_k^{00D_l} = \delta_{kl} - \dfrac{Y_k}{\langle Y, U\rangle}, & k \in \mathcal{S}, \\ \beta_k^{10D_l} = 0, \quad k \in \mathcal{S}, \\ \beta_k^{01D_l} = 0, \quad k \in \mathcal{P}, \end{cases} \tag{4.4.9}$$

and since $\sum_{k\in S} \beta_k^{00D_l} = 0$, we obtain

$$\beta^{D_l} \in \mathcal{U}^{\perp}, \qquad l \in S. \tag{4.4.10}$$

Theorem 4.4.1. *Let $L \in \mathbb{R}^{2n+p,2n+p}$ be a matrix satisfying the properties $(L0)$–$(L5)$ and let $\mathcal{L}$ and $\beta^{D_l} \in \mathbb{R}^{2n+p}$, $l \in S$, be given by (4.4.6) and (4.4.9), respectively. Then the constrained linear systems (4.4.1) admit a unique solution α^{D_l}, $l \in S$. Furthermore, the matrix D may also be written*

$$D_{kl} = \alpha_l^{00D_k}, \qquad k, l \in S. \tag{4.4.11}$$

This matrix is symmetric, positive semi-definite, positive definite on the hyperplane $U^{\perp}$, and has nullspace $N(D) = \mathbb{R}Y$ and range $R(D) = Y^{\perp}$.

Proof. Since $\beta^{D_l} \in \mathcal{U}^{\perp} = R(L)$ from (4.4.10) and $(L5)$ and since $\mathbb{R}\mathcal{U} \oplus \mathcal{L}^{\perp} = \mathbb{R}^{2n+p}$, Proposition 4.1.3 applies, and the constrained linear systems (4.4.1) admit a unique solution α^{D_l}, $l \in S$. On the other hand, since $\beta^{D_l} = e^{00l} - (1/\langle \mathcal{L}, \mathcal{U}\rangle)\mathcal{L}$, $l \in S$, we immediately deduce from the constraints $\langle \mathcal{L}, \alpha^{D_k}\rangle = 0$, $k \in S$, that $D_{kl} = \langle \alpha^{D_k}, \beta^{D_l}\rangle = \alpha_l^{00D_k}$ and (4.4.11) is proven. Furthermore, we can write $D_{kl} = \langle \alpha^{D_k}, \beta^{D_l}\rangle = \langle L\alpha^{D_k}, \alpha^{D_l}\rangle$ so that the matrix D is symmetric from $(L1)$. Moreover, for $x \in \mathbb{R}^n$, we have

$$\langle x, Dx\rangle = \Big\langle L \sum_{k\in S} x_k \alpha^{D_k}, \sum_{k\in S} x_k \alpha^{D_k} \Big\rangle,$$

so that D is positive semi-definite by $(L2)$. From the properties of the matrix L, we then deduce that $\langle x, Dx\rangle = 0$ if and only if $\sum_{k\in S} x_k \alpha^{D_k} \in N(L)$ which, in turn, is equivalent to $\sum_{k\in S} x_k \beta^{D_k} = 0$ and this immediately yields $x = \big(\langle x, U\rangle / \langle Y, U\rangle\big) Y$, i.e., $x \in \mathbb{R}Y$. We have thus proven that $\langle x, Dx\rangle = 0$ if and only if $x \in \mathbb{R}Y$ from which we deduce that $N(D) = \mathbb{R}Y$ and that D is positive definite on the hyperplane $U^{\perp}$ since $\langle Y, U\rangle \neq 0$. Finally, we have $R(D) = Y^{\perp}$ since D is symmetric. □

Remark. We recover the mathematical properties of the matrix D that are important from a thermodynamic viewpoint. Symmetric diffusion coefficients have been considered by [Wa58] [Cu68] [CC70] [FK72] [Gi91] and are consistent with Onsager reciprocal relations of thermodynamics of irreversible processes [Va67]. Furthermore, the property $R(D) = Y^{\perp}$ corresponds to the mass conservation constraint. Finally, as detailed in Section 2.1.6, the positive definiteness of the diffusion matrix on the hyperplane $U^{\perp}$ corresponds to a positive entropy production on the physical hyperplane of zero sum gradients [Wa58] [FK72] [Gi91].

In the next proposition, we express the diffusion matrix D in terms of either a generalized inverse of the matrix L or the inverse of a symmetric positive definite matrix.

Proposition 4.4.2. *Let $L^\dagger$ be the generalized inverse of L with prescribed range $\mathcal{L}^\perp$ and nullspace $\mathbb{R}\mathcal{L}$. Then the solution of the constrained linear systems (4.4.1) is given by $\alpha^{D_l} = L^\dagger \beta^{D_l}$, $l \in \mathcal{S}$. Furthermore, if a and b are two positive real numbers such that $ab\langle \mathcal{L},\mathcal{U}\rangle^2 = 1$, then the matrices $L + a\,\mathcal{L}\otimes\mathcal{L}$ and $L^\dagger + b\mathcal{U}\otimes\mathcal{U}$ are symmetric, positive definite, inverse of each other, and coincide with L and $L^\dagger$ on the hyperplanes $\mathcal{L}^\perp$ and $\mathcal{U}^\perp$, respectively. Therefore, we have*

$$\begin{cases} \alpha^{D_l} = [L + a\,\mathcal{L}\otimes\mathcal{L}]^{-1}\beta^{D_l}, & l \in \mathcal{S}, \\ D_{kl} = \langle\, [L + a\,\mathcal{L}\otimes\mathcal{L}]^{-1}\beta^{D_k}, \beta^{D_l}\,\rangle, & k,l \in \mathcal{S}. \end{cases} \tag{4.4.12}$$

Proof. The proof directly follows from Propositions 4.1.3 and 4.1.4 since $\langle \mathcal{L},\mathcal{U}\rangle \neq 0$ and $\beta^{D_l} \in \mathcal{U}^\perp$, $l \in \mathcal{S}$. □

We now consider the submatrix $\Lambda \in \mathbb{R}^{n+p,n+p}$ given by

$$\begin{cases} \Lambda_{kl}^{1010} = L_{kl}^{1010}, & k \in \mathcal{S}, \quad l \in \mathcal{S}, \\ \Lambda_{kl}^{1001} = L_{kl}^{1001}, & k \in \mathcal{S}, \quad l \in \mathcal{P}, \\ \Lambda_{kl}^{0110} = L_{kl}^{0110}, & k \in \mathcal{P}, \quad l \in \mathcal{S}, \\ \Lambda_{kl}^{0101} = L_{kl}^{0101}, & k \in \mathcal{P}, \quad l \in \mathcal{P}, \end{cases} \tag{4.4.13}$$

and prove in the following lemma that this matrix is symmetric positive definite.

Lemma 4.4.3. *Let $L \in \mathbb{R}^{2n+p,2n+p}$ be a matrix satisfying the properties ($L0$)–($L5$). Then the matrix $\Lambda \in \mathbb{R}^{n+p,n+p}$ given by (4.4.13) is symmetric positive definite.*

Proof. The matrix Λ is symmetric positive semi-definite by ($L1$)–($L2$). Furthermore, since L is positive definite on $\mathcal{L}^\perp$ and the vector $\mathcal{L}$ is such that $\mathcal{L}^{10} = 0$ and $\mathcal{L}^{01} = 0$, the matrix Λ is positive definite. □

As a consequence of Lemma 4.4.3, the matrix Λ is nonsingular and, therefore, the Schur complement of this matrix exists. In the next proposition, which generalizes the results from [Gi91], we investigate its mathematical properties.

Proposition 4.4.4. *Let $L \in \mathbb{R}^{2n+p,2n+p}$ be a matrix satisfying the properties ($L0$)–($L5$). Then the matrix $\Delta \in \mathbb{R}^{n,n}$ given by*

$$\Delta = L^{0000} - [L^{0010}, L^{0001}]\Lambda^{-1}\begin{bmatrix} L^{1000} \\ L^{0100} \end{bmatrix}, \tag{4.4.14}$$

is the generalized inverse of the diffusion matrix D with prescribed nullspace $N(\Delta) = \mathbb{R}U$ and range $R(\Delta) = U^\perp$. The matrix Δ is symmetric, positive semi-definite, and positive definite on the hyperplane $Y^\perp$.

Proof. First, we deduce from ($L1$) and Lemma 4.4.3 that the matrix Δ given by (4.4.14) is symmetric. Furthermore, since Λ is nonsingular, for any $x^{00} \in \mathbb{R}^n$ we can consider the vector x given by the block-decomposition $x = (x^{00}, \Pi x^{00})$ where $\Pi x^{00} = -\Lambda^{-1}(L^{1000}x^{00}, L^{0100}x^{00}) \in \mathbb{R}^{n+p}$. An explicit calculation then yields that $\langle x^{00}, \Delta x^{00}\rangle = \langle x, Lx\rangle$ and hence Δ is positive semi-definite. Moreover, $\langle x^{00}, \Delta x^{00}\rangle = 0$ if and only if $x \in \mathbb{R}\mathcal{U}$ which, in turn, is equivalent to $x^{00} \in \mathbb{R}U$. This shows that $N(\Delta) = \mathbb{R}U$ and that Δ is positive definite on the hyperplane $Y^\perp$ since $\langle U, Y\rangle \neq 0$. Since Δ is symmetric, we next obtain that $R(\Delta) = U^\perp$. Finally, by eliminating the last $n+p$ rows in the linear systems (4.4.1), we obtain $\Delta\alpha^{00D_l} = \beta^{00D_l}$ for all $l \in \mathcal{S}$ since $\beta^{10D_l} = \beta^{01D_l} = 0$. Using (4.4.9)(4.4.11), this relation can be rewritten in the form $\Delta D = P_{U^\perp,\mathbb{R}Y}$, where $P_{U^\perp,\mathbb{R}Y}$ is the oblique projector onto the hyperplane $U^\perp$ along $\mathbb{R}Y$. This implies that $\Delta D \Delta = \Delta$ and $D\Delta D = D$ since $N(D) = \mathbb{R}Y$ and $R(\Delta) = U^\perp$. Hence, Δ is the generalized inverse of D with prescribed range $R(\Delta) = U^\perp$ and nullspace $N(\Delta) = \mathbb{R}U$, and the proof is complete. □

In the next proposition, we express the matrix D in terms of the inverse of a symmetric positive definite matrix obtained from the matrix Δ.

Proposition 4.4.5. *Let Δ be given by (4.4.14) and let a and b be two positive real numbers such that $ab\langle Y, U\rangle^2 = 1$. Then the matrices $\Delta + a\, Y\otimes Y$ and $D + b\, U\otimes U$ are symmetric, positive definite, inverse of each other, and coincide with Δ and D on the hyperplanes $Y^\perp$ and $U^\perp$, respectively. As a consequence, the diffusion matrix D can be written*

$$D = [\Delta + a\, Y\otimes Y]^{-1} - b\, U\otimes U. \tag{4.4.15}$$

Proof. It follows from Propositions 4.1.4 and 4.4.4. □

Remark. By using the matrix $D + b\, U\otimes U$ instead of just D, one can suppress artificial singularities arising in Jacobian matrices of discretized conservation equations when all the mass fractions are considered as independent unknowns [Gi90]. This issue will be further discussed in Section 4.4.11.

Finally, we consider the matrix L resulting from the practical approximations presented in Section 2.10.

Proposition 4.4.6. *Let $\mathcal{D}_{k\,\mathrm{int},l}$, c_k^{int}, ξ_k^{int}, and m_l, $k \in \mathcal{P}$, $l \in \mathcal{S}$, be positive quantities, let $\mathcal{D}_{kl}$, $\bar{A}_{kl}$, $\bar{B}_{kl}$, and $\bar{C}_{kl}$, $k, l \in \mathcal{S}$, be symmetric and positive coefficients such that*

$$\frac{25}{4} - 3\bar{B}_{kl} > 0, \qquad 15\bar{C}_{kl} - 3\bar{B}_{kl} - 9\bar{C}_{kl}^2 > 0, \qquad k, l \in \mathcal{S}, \tag{4.4.16}$$

and assume that $Y > 0$. Then the matrix L given by (2.10.15)–(2.10.25) satisfies (L0)–(L5).

Proof. The property $(L0)$ is a consequence of the expressions (2.10.15)–(2.10.25) since the coefficients $\mathcal{D}_{kl}$, $k,l \in \mathcal{S}$, are proportional to the pressure, and the symmetry of L directly results from the symmetry of the coefficients $\mathcal{D}_{kl}$, $\bar{A}_{kl}$, $\bar{B}_{kl}$, and $\bar{C}_{kl}$. Moreover, an explicit calculation yields for $x \in \mathbb{R}^{2n+p}$

$$\langle x, Lx\rangle = \sum_{\substack{k,l\in\mathcal{S}\\ l\neq k}} \frac{X_k X_l}{2\mathcal{D}_{kl}}\Big[(x_k^{00}-x_l^{00})^2 - (6\bar{C}_{kl}-5)(x_k^{00}-x_l^{00})(\frac{m_l x_k^{10}-m_k x_l^{10}}{m_k+m_l})$$

$$+(\frac{25}{4}-3\bar{B}_{kl})(\frac{m_l x_k^{10}-m_k x_l^{10}}{m_k+m_l})^2\Big]$$

$$+\sum_{\substack{k,l\in\mathcal{S}\\ l\neq k}} \frac{X_k X_l}{\mathcal{D}_{kl}}\Big[\frac{15}{4}(\frac{m_k x_k^{10}-m_l x_l^{10}}{m_k+m_l})^2 + 2\bar{A}_{kl}\frac{m_k m_l}{(m_k+m_l)^2}(x_k^{10}+x_l^{10})^2\Big]$$

$$+\sum_{k\in\mathcal{S}} 2\bar{A}_{kk}\frac{X_k^2}{\mathcal{D}_{kk}}\left(x_k^{10}\right)^2 + \sum_{k\in\mathcal{P}}\Big(\sum_{\substack{l\in\mathcal{S}\\ l\neq k}} X_k X_l \frac{c_k^{\mathrm{int}}}{k_{\mathrm{B}}\mathcal{D}_{k\,\mathrm{int},l}} + X_k^2\frac{c_k^{\mathrm{int}}}{k_{\mathrm{B}}\mathcal{D}_{k\,\mathrm{int},k}}\Big)\left(x_k^{01}\right)^2$$

$$+\sum_{k\in\mathcal{P}} \frac{20}{3}\frac{\bar{A}_{kk}}{k_{\mathrm{B}}\pi}\frac{X_k^2}{\mathcal{D}_{kk}}\frac{c_k^{\mathrm{int}}}{\xi_k^{\mathrm{int}}}(x_k^{10}-\frac{3}{5}x_k^{01})^2$$

$$+\sum_{\substack{k\in\mathcal{S},l\in\mathcal{P}\\ l\neq k}} \frac{20}{3}\frac{\bar{A}_{kl}}{k_{\mathrm{B}}\pi}\frac{X_k X_l}{\mathcal{D}_{kl}}\frac{c_l^{\mathrm{int}}}{\xi_l^{\mathrm{int}}}\frac{m_l}{m_k}\Big(\frac{m_k}{m_k+m_l}(x_k^{10}+x_l^{10})-\frac{3}{5}x_l^{01}\Big)^2.$$

We note that the first term is nonnegative since $1/4(6\bar{C}_{kl}-5)^2 - (25/4 - 3\bar{B}_{kl}) = 9\bar{C}_{kl}^2 - 15\bar{C}_{kl} + 3\bar{B}_{kl} < 0$ and $25/4 - 3\bar{B}_{kl} > 0$ by (4.4.16). The remaining terms in $\langle x, Lx\rangle$ are obviously nonnegative. This shows that the matrix L is positive semi-definite, i.e., satisfies $(L2)$. Furthermore, since $Y>0$ implies $X>0$, we easily deduce from the assumptions and the above explicit expression that $\langle x, Lx\rangle = 0$ if and only if $x_k^{00} = x_l^{00}$, $k,l\in\mathcal{S}$, $k\neq l$, $x_k^{10}=0$, $k\in\mathcal{S}$, and $x_k^{01}=0$, $k\in\mathcal{P}$, i.e., if and only if $x\in\mathbb{R}\mathcal{U}$. The remaining properties $(L3)$–$(L5)$ are then easily obtained. □

4.4.2 Mathematical Properties of the System $L_{[\mathrm{e}]}\alpha_{[\mathrm{e}]}^{D_l} = \beta_{[\mathrm{e}]}^{D_l}$

We consider the n constrained linear systems indexed by $l \in \mathcal{S}$

$$\begin{cases} L_{[\mathrm{e}]}\alpha_{[\mathrm{e}]}^{D_l} = \beta_{[\mathrm{e}]}^{D_l}, \\ \langle \mathcal{L}_{[\mathrm{e}]}, \alpha_{[\mathrm{e}]}^{D_l}\rangle = 0, \end{cases} \tag{4.4.17}$$

associated with the evaluation of the diffusion matrix $D_{[e]} = (D_{[e]kl})_{k,l \in \mathcal{S}}$

$$D_{[e]kl} = \langle \alpha_{[e]}^{D_k}, \beta_{[e]}^{D_l} \rangle, \qquad k,l \in \mathcal{S}, \tag{4.4.18}$$

as described in Section 2.6.3. We have $L_{[e]} \in \mathbb{R}^{2n,2n}$, $\alpha_{[e]}^{D_l}$, $\beta_{[e]}^{D_l}$, and $\mathcal{L}_{[e]} \in \mathbb{R}^{2n}$, and the corresponding indexing set is given by

$$\mathcal{B}_{[e]}^{D} = \{00, \mathrm{e}\} \times \mathcal{S}, \tag{4.4.19}$$

yielding the block-decomposition

$$\begin{cases} \begin{bmatrix} L_{[e]}^{0000} & L_{[e]}^{00\mathrm{e}} \\ L_{[e]}^{\mathrm{e}00} & L_{[e]}^{\mathrm{ee}} \end{bmatrix} \begin{bmatrix} \alpha_{[e]}^{00D_l} \\ \alpha_{[e]}^{\mathrm{e}D_l} \end{bmatrix} = \begin{bmatrix} \beta_{[e]}^{00D_l} \\ \beta_{[e]}^{\mathrm{e}D_l} \end{bmatrix}, \\ \langle \mathcal{L}_{[e]}^{00}, \alpha_{[e]}^{00D_l} \rangle + \langle \mathcal{L}_{[e]}^{\mathrm{e}}, \alpha_{[e]}^{\mathrm{e}D_l} \rangle = 0. \end{cases} \tag{4.4.20}$$

Furthermore, from the kinetic theory results obtained in Sections 2.3 and 2.6, we can make the following assumptions.

($L_{[e]}0$) The coefficients of the matrix $L_{[e]}$ are functions of the pressure, the temperature, and the mass fractions, they are proportional to the pressure, and they can be expressed as quadratic functions of the mole fractions in the form

$$\begin{cases} L_{[e]kk}^{rs} = \sum_{l \in \mathcal{S}} X_k X_l L_{[e]kl}^{\prime rs} + X_k^2 L_{[e]kk}^{\prime\prime rs}, & (r,k),(s,k) \in \mathcal{B}_{[e]}^{D}, \\ L_{[e]kl}^{rs} = X_k X_l L_{[e]kl}^{\prime\prime rs}, \qquad (r,k),(s,l) \in \mathcal{B}_{[e]}^{D}, & k \neq l, \end{cases} \tag{4.4.21}$$

where the matrices $L'_{[e]}$ and $L''_{[e]}$ solely depend on the pressure and the temperature.

($L_{[e]}1$) $L_{[e]}$ is symmetric.

($L_{[e]}2$) $L_{[e]}$ is positive semi-definite.

($L_{[e]}3$) $L_{[e]}$ is positive definite on the hyperplane $\mathcal{L}_{[e]}^{\perp}$.

($L_{[e]}4$) $N(L_{[e]}) = \mathbb{R}\mathcal{U}_{[e]}$.

($L_{[e]}5$) $R(L_{[e]}) = \mathcal{U}_{[e]}^{\perp}$.

Following Section 2.6.3, we assume that the vectors $\mathcal{L}_{[e]}$ and $\mathcal{U}_{[e]} \in \mathbb{R}^{2n}$ are given by the block-decompositions

$$\mathcal{L}_{[e]}^{00} = Y, \qquad \mathcal{L}_{[e]}^{\mathrm{e}} = 0, \tag{4.4.22}$$

and

$$\mathcal{U}_{[e]}^{00} = U, \qquad \mathcal{U}_{[e]}^{\mathrm{e}} = 0, \tag{4.4.23}$$

so that

$$\langle \mathcal{L}_{[e]}, \mathcal{U}_{[e]} \rangle > 0, \tag{4.4.24}$$

which yields $\mathcal{U}_{[e]}^{\perp} \oplus \mathbb{R}\mathcal{L}_{[e]} = \mathbb{R}\mathcal{U}_{[e]} \oplus \mathcal{L}_{[e]}^{\perp} = \mathbb{R}^{2n}$. In addition, following Section 2.6.3, we assume that the right-hand sides $\beta_{[e]}^{D_l}$, $l \in \mathcal{S}$, are given by

$$\begin{cases} \beta_{[e]k}^{00D_l} = \delta_{kl} - \dfrac{Y_k}{\langle Y, U \rangle}, & k \in \mathcal{S}, \\ \beta_{[e]k}^{eD_l} = 0, \quad k \in \mathcal{S}, & \end{cases} \tag{4.4.25}$$

and since $\sum_{k \in \mathcal{S}} \beta_{[e]k}^{00D_l} = 0$, we obtain

$$\beta_{[e]}^{D_l} \in \mathcal{U}_{[e]}^{\perp}, \qquad l \in \mathcal{S}. \tag{4.4.26}$$

Theorem 4.4.7. *Let $L_{[e]} \in \mathbb{R}^{2n,2n}$ be a matrix satisfying the properties $(L_{[e]}0)$–$(L_{[e]}5)$ and let $\mathcal{L}_{[e]}$ and $\beta_{[e]}^{D_l} \in \mathbb{R}^{2n}$, $l \in \mathcal{S}$, be given by (4.4.22) and (4.4.25), respectively. Then the constrained linear systems (4.4.17) admit a unique solution $\alpha_{[e]}^{D_l}$, $l \in \mathcal{S}$. Furthermore, the matrix $D_{[e]}$ may also be written*

$$D_{[e]kl} = \alpha_{[e]l}^{00D_k}, \qquad k, l \in \mathcal{S}. \tag{4.4.27}$$

This matrix is symmetric, positive semi-definite, positive definite on the hyperplane $U^{\perp}$, and has nullspace $N(D_{[e]}) = \mathbb{R}Y$ and range $R(D_{[e]}) = Y^{\perp}$.

Remark. We recover the mathematical properties of the matrix $D_{[e]}$ that are important from a thermodynamic viewpoint. Indeed, the matrix $D_{[e]}$ is symmetric, conserves mass since $R(D_{[e]}) = Y^{\perp}$, and yields a positive entropy production on the physical hyperplane of zero sum gradients since $D_{[e]}$ is positive definite on $U^{\perp}$.

In the next proposition, we express the diffusion matrix $D_{[e]}$ in terms of either a generalized inverse of the matrix $L_{[e]}$ or the inverse of a symmetric positive definite matrix.

Proposition 4.4.8. *Let $L_{[e]}^{\dagger}$ be the generalized inverse of $L_{[e]}$ with prescribed range $\mathcal{L}_{[e]}^{\perp}$ and nullspace $\mathbb{R}\mathcal{L}_{[e]}$. Then the solution of the constrained linear systems (4.4.17) is given by $\alpha_{[e]}^{D_l} = L_{[e]}^{\dagger}\beta_{[e]}^{D_l}$, $l \in \mathcal{S}$. Furthermore, if a and b are two positive real numbers such that $ab\langle \mathcal{L}_{[e]}, \mathcal{U}_{[e]} \rangle^2 = 1$, then the matrices $L_{[e]} + a\,\mathcal{L}_{[e]} \otimes \mathcal{L}_{[e]}$ and $L_{[e]}^{\dagger} + b\,\mathcal{U}_{[e]} \otimes \mathcal{U}_{[e]}$ are symmetric, positive definite, inverse of each other, and coincide with $L_{[e]}$ and $L_{[e]}^{\dagger}$ on the hyperplanes $\mathcal{L}_{[e]}^{\perp}$ and $\mathcal{U}_{[e]}^{\perp}$, respectively. Therefore, we have*

$$\begin{cases} \alpha_{[e]}^{D_l} = [L_{[e]} + a\,\mathcal{L}_{[e]} \otimes \mathcal{L}_{[e]}]^{-1} \beta_{[e]}^{D_l}, & l \in \mathcal{S}, \\ D_{[e]kl} = \langle\, [L_{[e]} + a\,\mathcal{L}_{[e]} \otimes \mathcal{L}_{[e]}]^{-1} \beta_{[e]}^{D_k}, \beta_{[e]}^{D_l} \,\rangle, & k, l \in \mathcal{S}. \end{cases} \tag{4.4.28}$$

We now consider the submatrix $\Lambda_{[e]} \in \mathbb{R}^{n,n}$ given by

$$\Lambda_{[e]} = L_{[e]}^{ee}, \tag{4.4.29}$$

and establish in the following lemma that this matrix is symmetric positive definite.

Lemma 4.4.9. *Let $L_{[e]} \in \mathbb{R}^{2n,2n}$ be a matrix satisfying the properties $(L_{[e]}0)$–$(L_{[e]}5)$. Then the matrix $\Lambda_{[e]} \in \mathbb{R}^{n,n}$ given by (4.4.29) is symmetric positive definite.*

As a consequence of Lemma 4.4.9, the matrix $\Lambda_{[e]}$ is nonsingular and, therefore, the Schur complement of this matrix exists. In the next proposition, we investigate its mathematical properties.

Proposition 4.4.10. *Let $L_{[e]} \in \mathbb{R}^{2n,2n}$ be a matrix satisfying the properties $(L_{[e]}0)$–$(L_{[e]}5)$. Then the matrix $\Delta_{[e]} \in \mathbb{R}^{n,n}$ given by*

$$\Delta_{[e]} = L_{[e]}^{0000} - L_{[e]}^{00e} \Lambda_{[e]}^{-1} L_{[e]}^{e00}, \tag{4.4.30}$$

is the generalized inverse of the matrix $D_{[e]}$ with prescribed nullspace $N(\Delta_{[e]}) = \mathbb{R}U$ and range $R(\Delta_{[e]}) = U^{\perp}$. The matrix $\Delta_{[e]}$ is symmetric, positive semi-definite, and positive definite on the hyperplane $Y^{\perp}$.

In the next proposition, we express the matrix $D_{[e]}$ in terms of the inverse of a symmetric positive definite matrix obtained from the matrix $\Delta_{[e]}$.

Proposition 4.4.11. *Let $\Delta_{[e]}$ be given by (4.4.30) and let a and b be two positive real numbers such that $ab\langle Y, U\rangle^2 = 1$. Then the matrices $\Delta_{[e]} + a\,Y\otimes Y$ and $D_{[e]} + b\,U\otimes U$ are symmetric, positive definite, inverse of each other, and coincide with $\Delta_{[e]}$ and $D_{[e]}$ on the hyperplanes $Y^{\perp}$ and $U^{\perp}$, respectively. As a consequence, the diffusion matrix $D_{[e]}$ can be written*

$$D_{[e]} = [\Delta_{[e]} + a\,Y\otimes Y]^{-1} - b\,U\otimes U. \tag{4.4.31}$$

Finally, we consider the matrix $L_{[e]}$ resulting from the practical approximations presented in Section 2.10.

Proposition 4.4.12. *Let $\mathcal{D}_{k\,\mathrm{int},l}$, c_k^{int}, ξ_k^{int}, and m_l, $k \in \mathcal{P}$, $l \in \mathcal{S}$, be positive quantities, let $\mathcal{D}_{kl}$, $\bar{\mathcal{A}}_{kl}$, $\bar{\mathcal{B}}_{kl}$, and $\bar{\mathcal{C}}_{kl}$, $k,l \in \mathcal{S}$, be symmetric and positive coefficients such that $25/4 - 3\bar{\mathcal{B}}_{kl} > 0$ and $15\bar{\mathcal{C}}_{kl} - 3\bar{\mathcal{B}}_{kl} - 9\bar{\mathcal{C}}_{kl}^2 > 0$, $k,l \in \mathcal{S}$, and assume that $Y > 0$. Then the matrix $L_{[e]}$ given by (2.10.26)–(2.10.31) satisfies $(L_{[e]}0)$–$(L_{[e]}5)$.*

4.4.3 Mathematical Properties of the System $L_{[00]}\alpha_{[00]}^{D_l} = \beta_{[00]}^{D_l}$

The mathematical properties of the matrix $D_{[00]}$ have already been studied in [Gi91], but we include the results in this section for completeness. We consider the n constrained linear systems indexed by $l \in \mathcal{S}$

$$\begin{cases} L_{[00]}\alpha_{[00]}^{D_l} = \beta_{[00]}^{D_l}, \\ \langle Y, \alpha_{[00]}^{D_l} \rangle = 0, \end{cases} \tag{4.4.32}$$

associated with the evaluation of the diffusion matrix $D_{[00]} = (D_{[00]kl})_{k,l\in\mathcal{S}}$

$$D_{[00]kl} = \langle \alpha_{[00]}^{D_k}, \beta_{[00]}^{D_l} \rangle, \qquad k, l \in \mathcal{S}, \tag{4.4.33}$$

as described in Section 2.6.5. We have $L_{[00]} \in \mathbb{R}^{n,n}$, $\alpha_{[00]}^{D_l}$ and $\beta_{[00]}^{D_l} \in \mathbb{R}^n$, and the corresponding indexing set is given by

$$\mathcal{B}_{[00]}^{D} = \{00\}\times\mathcal{S}. \tag{4.4.34}$$

Furthermore, from the kinetic theory results obtained in Sections 2.3 and 2.6, we can make the following assumptions.

($L_{[00]}0$) The coefficients of the matrix $L_{[00]}$ are functions of the pressure, the temperature, and the mass fractions, they are proportional to the pressure, and they can be expressed as quadratic functions of the mole fractions in the form

$$\begin{cases} L_{[00]kk}^{rs} = \sum_{l\in\mathcal{S}} X_k X_l L_{[00]kl}^{\prime rs} + X_k^2 L_{[00]kk}^{\prime\prime rs}, \qquad (r,k),(s,k) \in \mathcal{B}_{[00]}^{D}, \\ L_{[00]kl}^{rs} = X_k X_l L_{[00]kl}^{\prime\prime rs}, \qquad (r,k),(s,l) \in \mathcal{B}_{[00]}^{D}, \quad k \neq l, \end{cases} \tag{4.4.35}$$

where the matrices $L'_{[00]}$ and $L''_{[00]}$ solely depend on the pressure and the temperature.

($L_{[00]}1$) $L_{[00]}$ is symmetric.

($L_{[00]}2$) $L_{[00]}$ is positive semi-definite.

($L_{[00]}3$) $L_{[00]}$ is positive definite on the hyperplane $Y^\perp$.

($L_{[00]}4$) $N(L_{[00]}) = \mathbb{R}U$.

($L_{[00]}5$) $R(L_{[00]}) = U^\perp$.

Moreover, following Section 2.6.5, we assume that the right-hand sides $\beta_{[00]}^{D_l}$, $l \in \mathcal{S}$, are given by

$$\beta_{[00]k}^{00D_l} = \delta_{kl} - \frac{Y_k}{\langle Y, U \rangle}, \qquad k \in \mathcal{S}, \tag{4.4.36}$$

and since $\sum_{k\in S} \beta^{00D_l}_{[00]k} = 0$, we obtain

$$\beta^{D_l}_{[00]} \in U^{\perp}, \qquad l \in S. \tag{4.4.37}$$

Theorem 4.4.13. *Let $L_{[00]} \in \mathbb{R}^{n,n}$ be a matrix satisfying the properties $(L_{[00]}0)$–$(L_{[00]}5)$ and let $\beta^{D_l}_{[00]} \in \mathbb{R}^n$, $l \in S$, be given by (4.4.36). Then the constrained linear systems (4.4.32) admit a unique solution $\alpha^{D_l}_{[00]}$, $l \in S$. Furthermore, the matrix $D_{[00]}$ may also be written*

$$D_{[00]kl} = \alpha^{00D_k}_{[00]l}, \qquad k, l \in S. \tag{4.4.38}$$

This matrix is symmetric, positive semi-definite, positive definite on the hyperplane $U^{\perp}$, and has nullspace $N(D_{[00]}) = \mathbb{R}Y$ and range $R(D_{[00]}) = Y^{\perp}$.

Remark. We recover the mathematical properties of the matrix $D_{[00]}$ which are important from a thermodynamic viewpoint. Indeed, the matrix $D_{[00]}$ is symmetric, conserves mass since $R(D_{[00]}) = Y^{\perp}$, and yields a positive entropy production on the physical hyperplane of zero sum gradients since $D_{[00]}$ is positive definite on $U^{\perp}$.

In the next proposition, we clarify the relation between the diffusion matrix $D_{[00]}$ and the matrix $L_{[00]}$ and also express the matrix $D_{[00]}$ in terms of the inverse of a symmetric positive definite matrix [Gi91].

Proposition 4.4.14. *Let $L_{[00]} \in \mathbb{R}^{n,n}$ be a matrix satisfying the properties $(L_{[00]}0)$–$(L_{[00]}5)$ and consider the diffusion matrix $D_{[00]}$ given by (4.4.33). Then the matrix $D_{[00]}$ is the generalized inverse of $L_{[00]}$ with prescribed nullspace $N(D_{[00]}) = \mathbb{R}Y$ and range $R(D_{[00]}) = Y^{\perp}$. In addition, the solution of the constrained linear systems (4.4.32) is given by $\alpha^{D_l}_{[00]} = D_{[00]}\beta^{D_l}_{[00]}$, $l \in S$. Furthermore, if a and b are two positive real numbers such that $ab\langle Y, U\rangle^2 = 1$, then the matrices $L_{[00]} + a\,Y\otimes Y$ and $D_{[00]} + b\,U\otimes U$ are symmetric, positive definite, inverse of each other, and coincide with $L_{[00]}$ and $D_{[00]}$ on the hyperplanes $Y^{\perp}$ and $U^{\perp}$, respectively. Therefore, we have*

$$\begin{cases} \alpha^{D_l}_{[00]} = [L_{[00]} + a\,Y\otimes Y]^{-1}\beta^{D_l}_{[00]}, & l \in S, \\ D_{[00]} = [L_{[00]} + a\,Y\otimes Y]^{-1} - b\,U\otimes U. \end{cases} \tag{4.4.39}$$

Finally, we consider the matrix $L_{[00]}$ resulting from the practical approximations presented in Section 2.10.

Proposition 4.4.15. *Let $\mathcal{D}_{kl}$, $k, l \in S$, be symmetric and positive coefficients and assume that $Y > 0$. Then the matrix $L_{[00]}$ given by (2.10.15)(2.10.16) satisfies $(L_{[00]}0)$–$(L_{[00]}5)$.*

4.4.4 Mathematical Properties of the System $\widetilde{L}\widetilde{\alpha}^{D_l} = \widetilde{\beta}^{D_l}$

We consider the n constrained linear systems indexed by $l \in \mathcal{S}$

$$\begin{cases} \widetilde{L}\widetilde{\alpha}^{D_l} = \widetilde{\beta}^{D_l}, \\ \langle \mathcal{L}, \widetilde{\alpha}^{D_l} \rangle = 0, \end{cases} \tag{4.4.40}$$

associated with the evaluation of the flux diffusion matrix $\widetilde{D} = (\widetilde{D}_{kl})_{k,l\in\mathcal{S}}$

$$\widetilde{D}_{kl} = \langle \widetilde{\alpha}^{D_k}, \beta^{D_l} \rangle, \qquad k,l \in \mathcal{S}, \tag{4.4.41}$$

as described in Section 3.4.2. We have $\widetilde{L} \in \mathbb{R}^{2n+p,2n+p}$, $\widetilde{\alpha}^{D_l}$, $\widetilde{\beta}^{D_l}$, and $\mathcal{L} \in \mathbb{R}^{2n+p}$, and the indexing set is $\mathcal{B}^D = \{00,10\}\times\mathcal{S} \cup \{01\}\times\mathcal{P}$, yielding the block-decomposition

$$\begin{cases} \begin{bmatrix} \widetilde{L}^{0000} & \widetilde{L}^{0010} & \widetilde{L}^{0001} \\ \widetilde{L}^{1000} & \widetilde{L}^{1010} & \widetilde{L}^{1001} \\ \widetilde{L}^{0100} & \widetilde{L}^{0110} & \widetilde{L}^{0101} \end{bmatrix} \begin{bmatrix} \widetilde{\alpha}^{00D_l} \\ \widetilde{\alpha}^{10D_l} \\ \widetilde{\alpha}^{01D_l} \end{bmatrix} = \begin{bmatrix} \widetilde{\beta}^{00D_l} \\ \widetilde{\beta}^{10D_l} \\ \widetilde{\beta}^{01D_l} \end{bmatrix}, \\ \langle \mathcal{L}^{00}, \widetilde{\alpha}^{00D_l} \rangle + \langle \mathcal{L}^{10}, \widetilde{\alpha}^{10D_l} \rangle + \langle \mathcal{L}^{01}, \widetilde{\alpha}^{01D_l} \rangle = 0. \end{cases} \tag{4.4.42}$$

Furthermore, from the kinetic theory results obtained in Sections 3.1 and 3.4, we can make the following assumptions.

($\widetilde{L}0$) The coefficients of the matrix $\widetilde{L}$ are functions of the pressure, the temperature, and the mass fractions, they are proportional to the pressure, and they can be expressed as linear functions of the mole fractions in the form

$$\begin{cases} \widetilde{L}_{kk}^{rs} = \sum_{l\in\mathcal{S}} X_l L_{kl}^{\prime rs} + X_k L_{kk}^{\prime\prime rs}, & (r,k),(s,k) \in \mathcal{B}^D, \\ \widetilde{L}_{kl}^{rs} = X_l L_{kl}^{\prime\prime rs}, \quad (r,k),(s,l) \in \mathcal{B}^D, & k \neq l, \end{cases} \tag{4.4.43}$$

where the matrices L' and L'' coincide with the ones in (4.4.5).

($\widetilde{L}1$) The matrix $\widetilde{L}$ admits the block-decomposition

$$(\Gamma^D)^t \widetilde{L} \Gamma^D = \begin{bmatrix} \widetilde{L}^{++} & 0 \\ \widetilde{L}^{-+} & \widetilde{L}^{--} \end{bmatrix}, \tag{4.4.44}$$

where Γ^D is the permutation matrix associated with the reordering of $\mathcal{B}^D$ into $(\mathcal{B}^{D+}, \mathcal{B}^{D-})$.

($\widetilde{L}2$) The matrix $L^{++} = \mathcal{X}^{D++}\widetilde{L}^{++}$ corresponds to the $\mathcal{S}^+$ mixture and, in particular, is symmetric, positive semi-definite, positive definite on the hyperplane $(\mathcal{L}^+)^\perp$, and has nullspace $N(L^{++}) = \mathbb{R}\mathcal{U}^+$ and range $R(L^{++}) = (\mathcal{U}^+)^\perp$.

($\widetilde{L}3$) $\widetilde{L}^{--}$ is symmetric positive definite and we have $db(\widetilde{L}^{--}) = \widetilde{L}^{--}$.

($\widetilde{L}4$) $N(\widetilde{L}) = \mathbb{R}\mathcal{U}$.

($\widetilde{L}5$) $R(\widetilde{L}) = \widetilde{\mathcal{U}}^{\perp}$.

Moreover, following Section 3.4.2, we assume that the right-hand sides $\widetilde{\beta}^{D_l}$, $l \in \mathcal{S}$, are given by

$$\begin{cases} \widetilde{\beta}_k^{00D_l} = \dfrac{m_k}{m}\delta_{kl} - \dfrac{m_k}{m}\dfrac{Y_l}{\langle Y, U\rangle}, & k \in \mathcal{S}, \\ \widetilde{\beta}_k^{10D_l} = 0, \quad k \in \mathcal{S}, \\ \widetilde{\beta}_k^{01D_l} = 0, \quad k \in \mathcal{P}. \end{cases} \tag{4.4.45}$$

We deduce from (4.4.5)(4.4.43) and (4.4.9)(4.4.45) that the matrix $\widetilde{L}$ and the right-hand sides $\widetilde{\beta}^{D_l}$ are rescaled versions of L and β^{D_l}, respectively,

$$\mathcal{X}^D \widetilde{L} = L, \qquad \mathcal{X}^D \widetilde{\beta}^{D_l} = Y_l \beta^{D_l}, \ l \in \mathcal{S}, \tag{4.4.46}$$

where $\mathcal{X}^D = \mathrm{diag}\big((X_k)_{(r,k)\in\mathcal{B}^D}\big)$. Finally, the vector $\widetilde{\mathcal{U}} \in \mathbb{R}^{2n+p}$ is given by the block-decomposition

$$\widetilde{\mathcal{U}}^{00} = X, \qquad \widetilde{\mathcal{U}}^{10} = 0, \qquad \widetilde{\mathcal{U}}^{01} = 0, \tag{4.4.47}$$

and since $\sum_{k\in\mathcal{S}} X_k \widetilde{\beta}_k^{00D_l} = 0$, we obtain

$$\widetilde{\beta}^{D_l} \in \widetilde{\mathcal{U}}^{\perp}, \qquad l \in \mathcal{S}. \tag{4.4.48}$$

Theorem 4.4.16. *Let $\widetilde{L} \in \mathbb{R}^{2n+p,2n+p}$ be a matrix satisfying the properties ($\widetilde{L}0$)–($\widetilde{L}5$) and let $\mathcal{L}$, β^{D_l}, and $\widetilde{\beta}^{D_l} \in \mathbb{R}^{2n+p}$, $l \in \mathcal{S}$, be given by (4.4.6), (4.4.9), and (4.4.45), respectively. Then the constrained linear systems (4.4.40) admit a unique solution $\widetilde{\alpha}^{D_l}$, $l \in \mathcal{S}$. In addition, the matrix $\widetilde{D}$ may also be written*

$$\widetilde{D}_{kl} = \widetilde{\alpha}_l^{00D_k}, \qquad k, l \in \mathcal{S}, \tag{4.4.49}$$

and admits the block-decomposition

$$\Upsilon^t \widetilde{D} \Upsilon = \begin{bmatrix} \widetilde{D}^{++} & \widetilde{D}^{+-} \\ 0 & \widetilde{D}^{--} \end{bmatrix}, \tag{4.4.50}$$

where Υ is the permutation matrix associated with the reordering of $\mathcal{S}$ into $(\mathcal{S}^+, \mathcal{S}^-)$. The matrix $\widetilde{D}^{--}$ is diagonal with positive entries, and we have $\widetilde{D}^{++} = \mathcal{Y}^{++} D^{++}$, where D^{++} is the diffusion matrix of the $\mathcal{S}^+$ mixture. Finally, the matrix $\widetilde{D}$ has nullspace $N(\widetilde{D}) = \mathbb{R}Y$ and range $R(\widetilde{D}) = U^{\perp}$.

Proof. Since $\widetilde{\beta}^{D_l} \in \widetilde{\mathcal{U}}^{\perp} = R(\widetilde{L})$ from (4.4.48) and ($\widetilde{L}5$) and since $\mathcal{L}^{\perp} \oplus \mathbb{R}\mathcal{U} = \mathbb{R}^{2n+p}$, Proposition 4.1.3 applies, and the constrained linear systems (4.4.40) admit a unique solution $\widetilde{\alpha}^{D_l}$, $l \in \mathcal{S}$. The proof of (4.4.49) is the same as the one of (4.4.11) in the proof of Theorem 4.4.1 since $\widetilde{\alpha}^{D_k} \in \mathcal{L}^{\perp}$ and $\beta^{D_l} = e^{00l} - (1/\langle \mathcal{L}, \mathcal{U}\rangle)\mathcal{L}$. We now establish the block-decomposition (4.4.50).

1. For $k, l \in \mathcal{S}^+$, we have $\beta^{D_l+} \in (\mathcal{U}^+)^{\perp}$ and we deduce from $(\widetilde{L}2)$ that the constrained linear system
$$\begin{cases} L^{++}\alpha^{D_l+} = \beta^{D_l+}, \\ \langle \mathcal{L}^+, \alpha^{D_l+}\rangle = 0, \end{cases}$$
is well posed and hence admits a unique solution α^{D_l+}, $l \in \mathcal{S}^+$. From $(\widetilde{L}2)$ we also deduce that $D_{kl}^{++} = \langle \alpha^{D_k+}, \beta^{D_l+}\rangle$ is the diffusion matrix of the $\mathcal{S}^+$ mixture. From $(\widetilde{L}1)$ we next obtain that the solution $\widetilde{\alpha}^{D_l}$ of (4.4.40) is such that $\widetilde{L}^{++}\widetilde{\alpha}^{D_l+} = \widetilde{\beta}^{D_l+}$, and using (4.4.46) we then deduce that $\widetilde{\alpha}^{D_l+} = Y_l\alpha^{D_l+}$, $l \in \mathcal{S}^+$. Since $\beta^{D_l-} = 0$, $l \in \mathcal{S}^+$, this yields
$$\widetilde{D}_{kl}^{++} = \langle \widetilde{\alpha}^{D_k}, \beta^{D_l}\rangle = \langle \widetilde{\alpha}^{D_k+}, \beta^{D_l+}\rangle = Y_k D_{kl}^{++}.$$
2. For $k \in \mathcal{S}^-$ and $l \in \mathcal{S}^+$, we deduce from $(\widetilde{L}1)$ that $\widetilde{L}^{++}\widetilde{\alpha}^{D_k+} = 0$ since $\widetilde{\beta}^{D_k+} = 0$ for $k \in \mathcal{S}^-$, so that $\widetilde{\alpha}^{D_k+} = 0$. Keeping in mind that $\beta^{D_l-} = 0$ for $l \in \mathcal{S}^+$, we then obtain $\widetilde{D}_{kl}^{-+} = \langle \widetilde{\alpha}^{D_k+}, \beta^{D_l+}\rangle = 0$.
3. For $k, l \in \mathcal{S}^-$, we have already seen that $\widetilde{\alpha}^{D_k+} = 0$ for $k \in \mathcal{S}^-$, so that $\widetilde{\alpha}^{D_k-} = (\widetilde{L}^{--})^{-1}\widetilde{\beta}^{D_k-}$ from $(\widetilde{L}1)$. Using $\widetilde{\beta}^{D_k-} = (m_k/m)e^{00k-}$ and $\beta^{D_l-} = e^{00l-}$ then yields $\widetilde{D}_{kl}^{--} = (m_k/m)\langle (\widetilde{L}^{--})^{-1}e^{00k-}, e^{00l-}\rangle$. Since the matrix $(\widetilde{L}^{--})^{-1}$ is block-diagonal and positive definite, the matrix $\widetilde{D}^{--}$ is diagonal with positive entries.

Using now the block-decomposition (4.4.50), the positive definiteness of $\widetilde{D}^{--}$, and keeping in mind that $N(D^{++}) = \mathbb{R}Y^+$ since D^{++} is the diffusion matrix of the $\mathcal{S}^+$ mixture, we obtain $N(\widetilde{D}) = \mathbb{R}\Upsilon(Y^+, 0) = \mathbb{R}Y$. On the other hand, the vector $\widetilde{\alpha} = \sum_{k\in\mathcal{S}} \widetilde{\alpha}^{D_k}$ satisfies $\widetilde{L}\widetilde{\alpha} = \sum_{k\in\mathcal{S}} \widetilde{\beta}^{D_k} = 0$ and $\langle \mathcal{L}, \widetilde{\alpha}\rangle = 0$ so that $\widetilde{\alpha} = 0$ which, in turn, implies that $\sum_{k\in\mathcal{S}} \widetilde{D}_{kl} = 0$, i.e., $R(\widetilde{D}) \subset U^{\perp}$. Since the dimensions of the nullspace and range of the matrix $\widetilde{D}$ must sum up to n, we have $R(\widetilde{D}) = U^{\perp}$, and the proof is complete. □

In the next proposition, we express the flux diffusion matrix $\widetilde{D}$ in terms of a generalized inverse of the matrix $\widetilde{L}$ and also in terms of the solution of a nonsingular linear system.

Proposition 4.4.17. *Let $\widetilde{\mathcal{L}} \in \mathbb{R}^{2n+p}$ be given by the block-decomposition $\widetilde{\mathcal{L}}^{00} = W$, $\widetilde{\mathcal{L}}^{10} = 0$, $\widetilde{\mathcal{L}}^{01} = 0$, and let $\widetilde{L}^{\dagger}$ be the generalized inverse of $\widetilde{L}$ with prescribed range $\mathcal{L}^{\perp}$ and nullspace $\mathbb{R}\widetilde{\mathcal{L}}$. Then the solution of the constrained linear systems (4.4.40) is given* *by* $\widetilde{\alpha}^{D_l} = \widetilde{L}^\dagger \widetilde{\beta}^{D_l}$, $l \in \mathcal{S}$. *Furthermore, if* a *and* b *are two positive real numbers such that* $ab\langle \mathcal{L}, \mathcal{U}\rangle\langle \widetilde{\mathcal{L}}, \widetilde{\mathcal{U}}\rangle = 1$, *then the matrices* $\widetilde{L} + a\,\widetilde{\mathcal{L}}\otimes\mathcal{L}$ *and* $\widetilde{L}^\dagger + b\,\mathcal{U}\otimes\widetilde{\mathcal{U}}$ *are inverse of each other and coincide with* $\widetilde{L}$ *and* $\widetilde{L}^\dagger$ *on the hyperplanes* $\mathcal{L}^\perp$ *and* $\widetilde{\mathcal{U}}^\perp$, *respectively. Therefore, we have*

$$\begin{cases} \widetilde{\alpha}^{D_l} = [\widetilde{L} + a\,\widetilde{\mathcal{L}}\otimes\mathcal{L}]^{-1}\widetilde{\beta}^{D_l}, & l \in \mathcal{S}, \\ \widetilde{D}_{kl} = \langle\, [\widetilde{L} + a\,\widetilde{\mathcal{L}}\otimes\mathcal{L}]^{-1}\widetilde{\beta}^{D_k}, \beta^{D_l}\,\rangle, & k,l \in \mathcal{S}. \end{cases} \tag{4.4.51}$$

Proof. The proof directly follows from Propositions 4.1.3 and 4.1.5 since $\langle \mathcal{L}, \mathcal{U}\rangle \neq 0$, $\langle \widetilde{\mathcal{L}}, \widetilde{\mathcal{U}}\rangle \neq 0$, and $\widetilde{\beta}^{D_l} \in \widetilde{\mathcal{U}}^\perp$, $l \in \mathcal{S}$. □

We now consider the submatrix $\widetilde{\Lambda} \in \mathbb{R}^{n+p,n+p}$ given by

$$\begin{cases} \widetilde{\Lambda}_{kl}^{1010} = \widetilde{L}_{kl}^{1010}, & k \in \mathcal{S}, \quad l \in \mathcal{S}, \\ \widetilde{\Lambda}_{kl}^{1001} = \widetilde{L}_{kl}^{1001}, & k \in \mathcal{S}, \quad l \in \mathcal{P}, \\ \widetilde{\Lambda}_{kl}^{0110} = \widetilde{L}_{kl}^{0110}, & k \in \mathcal{P}, \quad l \in \mathcal{S}, \\ \widetilde{\Lambda}_{kl}^{0101} = \widetilde{L}_{kl}^{0101}, & k \in \mathcal{P}, \quad l \in \mathcal{P}, \end{cases} \tag{4.4.52}$$

and prove in the following lemma that this matrix is nonsingular.

Lemma 4.4.18. *Let* $\widetilde{L} \in \mathbb{R}^{2n+p,2n+p}$ *be a matrix satisfying the properties* $(\widetilde{L}0)$–$(\widetilde{L}5)$. *Then the matrix* $\widetilde{\Lambda} \in \mathbb{R}^{n+p,n+p}$ *given by (4.4.52) is nonsingular.*

Proof. We first obtain, as in the proof of Lemma 4.4.3, that the submatrix Λ^{++} is symmetric positive definite. Since $\widetilde{\Lambda}^{++}$ is a left rescaled version of Λ^{++}, we deduce that $\widetilde{\Lambda}^{++}$ is nonsingular. Furthermore, we obtain from $(\widetilde{L}3)$ that the submatrix $\widetilde{\Lambda}^{--}$ is symmetric positive definite, and hence we conclude by $(\widetilde{L}1)$ that the matrix $\widetilde{\Lambda}$ is nonsingular. □

As a consequence of Lemma 4.4.18, the Schur complement of the matrix $\widetilde{\Lambda}$ exists, and we investigate its mathematical properties in the next proposition.

Proposition 4.4.19. *Let* $\widetilde{L} \in \mathbb{R}^{2n+p,2n+p}$ *be a matrix satisfying the properties* $(\widetilde{L}0)$–$(\widetilde{L}5)$ *and let* $\widetilde{\Delta} \in \mathbb{R}^{n,n}$ *be given by*

$$\widetilde{\Delta} = \widetilde{L}^{0000} - [\widetilde{L}^{0010}, \widetilde{L}^{0001}]\widetilde{\Lambda}^{-1}\begin{bmatrix} \widetilde{L}^{1000} \\ \widetilde{L}^{0100} \end{bmatrix}. \tag{4.4.53}$$

Then we have $N(\widetilde{\Delta}) = \mathbb{R}U$, $R(\widetilde{\Delta}) = X^\perp$, *and if* a *and* b *are two positive real numbers such that* $ab\langle Y, U\rangle^2 = 1$, *we have*

$$\widetilde{D} = \mathcal{W}\big[(\widetilde{\Delta})^t + a\,Y\otimes W\big]^{-1} - b\,Y\otimes U. \tag{4.4.54}$$

Proof. Let $x^{00} \in \mathbb{R}^n$ and let $x \in \mathbb{R}^{2n+p}$ be given by the block-decomposition $x = (x^{00}, \widetilde{\Pi} x^{00})$ where $\widetilde{\Pi} x^{00} = -\widetilde{\Lambda}^{-1}(\widetilde{L}^{1000} x^{00}, \widetilde{L}^{0100} x^{00}) \in \mathbb{R}^{n+p}$. We then have $x^{00} \in N(\widetilde{\Delta})$ if and only if $x \in N(\widetilde{L}) = \mathbb{R}\mathcal{U}$, which is, in turn, equivalent to $x^{00} \in \mathbb{R}U$. Moreover, from ($\widetilde{L}4$) and (4.4.47) we obtain $X^t \widetilde{L}^{0000} = 0$ and $X^t[\widetilde{L}^{0010}, \widetilde{L}^{0001}] = 0$ so that $R(\widetilde{\Delta}) \subset X^\perp$. Since the dimensions of $N(\widetilde{\Delta})$ and $R(\widetilde{\Delta})$ must sum up to n, we then obtain $R(\widetilde{\Delta}) = X^\perp$. On the other hand, we easily obtain from (4.4.45), (4.4.49), and (4.4.53) that $\mathcal{W}^{-1}\widetilde{\Delta}\widetilde{D}^t = I - (1/\langle \mathcal{L}, \mathcal{U}\rangle) U \otimes Y$, from which we easily deduce that $\widetilde{D}^t$ is the generalized inverse of $\mathcal{W}^{-1}\widetilde{\Delta}$ with prescribed nullspace $N(\widetilde{D}^t) = \mathbb{R}U$ and range $R(\widetilde{D}^t) = Y^\perp$. Proposition 4.1.5 then yields $\widetilde{D}^t = [\mathcal{W}^{-1}\widetilde{\Delta} + a\, U \otimes Y]^{-1} - b\, U \otimes Y$, and transposing yields (4.4.54) since $\mathcal{W}U = W$. □

We now prove that all the flux diffusion coefficients $\widetilde{D}_{kl}$, $k, l \in \mathcal{S}$, are smooth functions of the pressure, the temperature, and the mass fractions in $\mathfrak{D}_{Y \geq 0, Y \neq 0}$ and are inversely proportional to the pressure.

Theorem 4.4.20. *Let L and $\widetilde{L} \in \mathbb{R}^{2n+p, 2n+p}$ be such that L satisfies ($L0$)–($L5$) in $\mathfrak{D}_{Y>0}$ and $\widetilde{L}$ satisfies ($\widetilde{L}0$)–($\widetilde{L}5$) in $\mathfrak{D}_{Y \geq 0, Y \neq 0}$. Assume also that the matrices L' and L'' in (4.4.5)(4.4.43) are smooth functions of the pressure and the temperature. Then for $Y > 0$ we have $\widetilde{\alpha}^{D_l} = Y_l \alpha^{D_l}$, $l \in \mathcal{S}$, and $\widetilde{D}_{kl} = Y_k D_{kl}$, $k, l \in \mathcal{S}$. Furthermore, all the vectors $\widetilde{\alpha}^{D_l}$, $l \in \mathcal{S}$, and all the flux diffusion coefficients $\widetilde{D}_{kl}$, $k, l \in \mathcal{S}$, are smooth functions of $(\overline{p}, \overline{T}, Y)$ in $\mathfrak{D}_{Y \geq 0, Y \neq 0}$ and are inversely proportional to the pressure. Moreover, for $Z \in \mathbb{R}^n$, $Z \geq 0$, $Z \neq 0$, we have the following limits*

$$
\begin{cases}
\displaystyle \lim_{\substack{Y \to Z \\ Y > 0}} D_{kl}(\overline{p}, \overline{T}, Y) = \frac{\widetilde{D}_{kl}(\overline{p}, \overline{T}, Z)}{Z_k}, & Z_k > 0, \quad l \in \mathcal{S}, \\
\displaystyle \lim_{\substack{Y \to Z \\ Y > 0}} D_{kl}(\overline{p}, \overline{T}, Y) = \frac{\partial \widetilde{D}_{kl}(\overline{p}, \overline{T}, Z)}{\partial Z_k}, & Z_k = 0, \quad l \in \mathcal{S}, \quad l \neq k, \\
\displaystyle \lim_{\substack{Y \to Z \\ Y > 0}} Y_k D_{kk}(\overline{p}, \overline{T}, Y) = \widetilde{D}_{kk}(\overline{p}, \overline{T}, Z) > 0, & Z_k = 0.
\end{cases}
\tag{4.4.55}
$$

Consequently, the diffusion coefficients D_{kk}, $k \in \mathcal{S}$, admit a smooth extension to $\mathfrak{D}_{Y_k > 0}$, and the coefficients D_{kl}, $k, l \in \mathcal{S}$, $k \neq l$, admit a smooth extension to $\mathfrak{D}_{Y \geq 0, Y \neq 0}$.

Proof. When all the mass fractions are positive, the matrix $\mathcal{X}^D$ is invertible, and we deduce from (4.4.46) that $\widetilde{\alpha}^{D_l} = Y_l \alpha^{D_l}$, $l \in \mathcal{S}$, and $\widetilde{D}_{kl} = Y_k D_{kl}$, $k, l \in \mathcal{S}$. Furthermore, using (4.4.51) and Cramer's rule, we can write the vectors $\widetilde{\alpha}^{D_l}$ and the matrix $\widetilde{D}$ as rational functions of the coefficients of $\widetilde{L}$, β^{D_l}, and $\widetilde{\beta}^{D_l}$, $l \in \mathcal{S}$, which, in turn, are smooth functions of the pressure, the temperature, and the mass fractions. Therefore, $\widetilde{\alpha}^{D_l}$, $l \in \mathcal{S}$, and $\widetilde{D}_{kl}$, $k, l \in \mathcal{S}$, are smooth functions of $(\overline{p}, \overline{T}, Y)$ in $\mathfrak{D}_{Y \geq 0, Y \neq 0}$ and are clearly inversely proportional to the pressure since the coefficients of the matrix $\widetilde{L}$ are proportional to the pressure by ($\widetilde{L}$0). Furthermore, the limits (4.4.55) directly result from the relation $D_{kl}(\bar{p},\overline{T},Y) = \widetilde{D}_{kl}(\bar{p},\overline{T},Y)/Y_k$, $k,l \in S$, valid for $Y > 0$, the smoothness of $\widetilde{D}(\bar{p},\overline{T},Y)$, and the block-decomposition (4.4.50) obtained in Theorem 4.4.16. Finally, we deduce from the limits (4.4.55) that all the diffusion coefficients admit finite limits when some mass fractions are allowed to vanish, except for the diagonal coefficients D_{kk} which blow up when $Y_k \to 0$. □

Finally, we consider the matrix $\widetilde{L}$ resulting from the practical approximations presented in Section 2.10.

Proposition 4.4.21. *Let $\mathcal{D}_{k\,\mathrm{int},l}$, c_k^{int}, ξ_k^{int}, and m_l, $k \in \mathcal{P}$, $l \in \mathcal{S}$, be positive quantities, let $\mathcal{D}_{kl}$, $\bar{\mathcal{A}}_{kl}$, $\bar{\mathcal{B}}_{kl}$, and $\bar{\mathcal{C}}_{kl}$, $k,l \in \mathcal{S}$, be symmetric and positive coefficients such that $25/4 - 3\bar{\mathcal{B}}_{kl} > 0$ and $15\bar{\mathcal{C}}_{kl} - 3\bar{\mathcal{B}}_{kl} - 9\bar{\mathcal{C}}_{kl}^2 > 0$, $k,l \in \mathcal{S}$, and assume that $Y \geq 0$, $Y \neq 0$. Then the matrix $\widetilde{L}$ given by (3.7.11)–(3.7.26) satisfies ($\widetilde{L}$0)–($\widetilde{L}$5).*

Proof. The properties ($\widetilde{L}$0)–($\widetilde{L}$1) directly result from (3.7.11)–(3.7.26), and the proof of ($\widetilde{L}$2) is similar to the one of Proposition 4.4.6. Next, the matrix $\widetilde{L}^{--}$ is symmetric, block-diagonal, and an explicit calculation yields for $x = \Gamma^D(0,x^-) \in \mathbb{R}^{2n+p}$

$$\begin{aligned}
\langle x^-,\widetilde{L}^{--}x^-\rangle = &\sum_{\substack{k\in\mathcal{S}^-\\ l\in\mathcal{S}^+}} \frac{X_l}{\mathcal{D}_{kl}}\Big[\big(x_k^{00}\big)^2 - (6\bar{\mathcal{C}}_{kl}-5)x_k^{00}\frac{m_l x_k^{10}}{m_k+m_l} \\
&\qquad + \big(\frac{25}{4} - 3\bar{\mathcal{B}}_{kl}\big)\big(\frac{m_l x_k^{10}}{m_k+m_l}\big)^2\Big] \\
&+ \sum_{\substack{k\in\mathcal{S}^-\\ l\in\mathcal{S}^+}} \frac{X_l}{\mathcal{D}_{kl}}\frac{1}{(m_k+m_l)^2}\Big[\frac{15}{2}m_k^2 + 4\bar{\mathcal{A}}_{kl}m_k m_l\Big]\big(x_k^{10}\big)^2 \\
&+ \sum_{\substack{k\in\mathcal{P}^-\\ l\in\mathcal{S}^+}} X_l\frac{c_k^{\mathrm{int}}}{k_{\mathrm{B}}\mathcal{D}_{k\,\mathrm{int},l}}\big(x_k^{01}\big)^2 + \sum_{\substack{k\in\mathcal{S}^-\\ l\in\mathcal{P}^+}} \frac{20}{3}\frac{\bar{\mathcal{A}}_{kl}}{k_{\mathrm{B}}\pi}\frac{X_l}{\mathcal{D}_{kl}}\frac{c_l^{\mathrm{int}}}{\xi_l^{\mathrm{int}}}\frac{m_k m_l}{(m_k+m_l)^2}\big(x_k^{10}\big)^2 \\
&+ \sum_{\substack{k\in\mathcal{P}^-\\ l\in\mathcal{S}^+}} \frac{20}{3}\frac{\bar{\mathcal{A}}_{kl}}{k_{\mathrm{B}}\pi}\frac{X_l}{\mathcal{D}_{kl}}\frac{c_k^{\mathrm{int}}}{\xi_k^{\mathrm{int}}}\frac{m_k}{m_l}\Big(\frac{m_l}{m_k+m_l}x_k^{10} - \frac{3}{5}x_k^{01}\Big)^2.
\end{aligned}$$

We deduce from $25/4-3\bar{\mathcal{B}}_{kl} > 0$ and $15\bar{\mathcal{C}}_{kl}-3\bar{\mathcal{B}}_{kl}-9\bar{\mathcal{C}}_{kl}^2 > 0$, $k,l \in \mathcal{S}$, that the first term in $\langle x^-,\widetilde{L}^{--}x^-\rangle$ is nonnegative and the remaining terms are obviously nonnegative so that $\widetilde{L}^{--}$ is positive semi-definite. Moreover, since $X_l > 0$, $l \in \mathcal{S}^+$, it is readily seen that $\langle x^-, \widetilde{L}^{--}x^-\rangle = 0$ if and only if $x^- = 0$ so that ($\widetilde{L}3$) is proven. Finally, we can explicitly verify from (3.7.11)–(3.7.26) that $\mathcal{U}^t\widetilde{L} = 0$, i.e., $R(\widetilde{L}) \subset \mathcal{U}^\perp$, and ($\widetilde{L}4$)–($\widetilde{L}5$) are then easily obtained. □

4.4.5 Mathematical Properties of the System $\widehat{L}\widehat{\alpha}^{D_l} = \widehat{\beta}^{D_l}$

We consider the n constrained linear systems indexed by $l \in \mathcal{S}$

$$\begin{cases} \widehat{L}\widehat{\alpha}^{D_l} = \widehat{\beta}^{D_l}, \\ \langle \widehat{\mathcal{L}}, \widehat{\alpha}^{D_l}\rangle = 0, \end{cases} \tag{4.4.56}$$

associated with the evaluation of the symmetric flux diffusion matrix $\widehat{D} = (\widehat{D}_{kl})_{k,l\in\mathcal{S}}$

$$\widehat{D}_{kl} = \langle \widehat{\alpha}^{D_k}, \widehat{\beta}^{D_l}\rangle, \qquad k, l \in \mathcal{S}. \tag{4.4.57}$$

We have $\widehat{L} \in \mathbb{R}^{2n+p,2n+p}$, $\widehat{\alpha}^{D_l}$, $\widehat{\beta}^{D_l}$, and $\widehat{\mathcal{L}} \in \mathbb{R}^{2n+p}$, and following Section 3.1.8, we can make the following assumptions.

($\widehat{L}0$) The coefficients of the matrix $\widehat{L}$ are functions of the pressure, the temperature, and the mass fractions, they are proportional to the pressure, and they can be expressed as

$$\begin{cases} \widehat{L}_{kk}^{rs} = \sum_{l\in\mathcal{S}} X_l L_{kl}^{\prime rs} + X_k L_{kk}^{\prime\prime rs}, & (r,k),(s,k) \in \mathcal{B}^D, \\ \widehat{L}_{kl}^{rs} = \sqrt{X_k X_l} L_{kl}^{\prime\prime rs}, & (r,k),(s,l) \in \mathcal{B}^D, \quad k \neq l, \end{cases} \tag{4.4.58}$$

where the matrices L' and L'' coincide with the ones in (4.4.5).

($\widehat{L}1$) $\widehat{L}$ is symmetric.

($\widehat{L}2$) $\widehat{L}$ is positive semi-definite.

($\widehat{L}3$) $\widehat{L}$ is positive definite on the hyperplane $\widehat{\mathcal{L}}^\perp$.

($\widehat{L}4$) $N(\widehat{L}) = \mathbb{R}\widehat{\mathcal{U}}$.

($\widehat{L}5$) $R(\widehat{L}) = \widehat{\mathcal{U}}^\perp$.

In addition, we assume that the constraint vector $\widehat{\mathcal{L}}$ is given by

$$\begin{cases} \widehat{\mathcal{L}}_k^{00} = \sqrt{\frac{m_k}{m}Y_k}, & k \in \mathcal{S}, \\ \widehat{\mathcal{L}}_k^{10} = 0, & k \in \mathcal{S}, \\ \widehat{\mathcal{L}}_k^{01} = 0, & k \in \mathcal{P}, \end{cases} \tag{4.4.59}$$

the vector $\widehat{\mathcal{U}} \in \mathbb{R}^{2n+p}$ by $\widehat{\mathcal{U}} = (\mathcal{X}^D)^{1/2}\mathcal{U}$ or, more explicitly,

$$\begin{cases} \widehat{\mathcal{U}}_k^{00} = \sqrt{X_k}, & k \in \mathcal{S}, \\ \widehat{\mathcal{U}}_k^{10} = 0, & k \in \mathcal{S}, \\ \widehat{\mathcal{U}}_k^{01} = 0, & k \in \mathcal{P}, \end{cases} \tag{4.4.60}$$

and the right-hand sides $\widehat{\beta}^{D_l}$, $l \in \mathcal{S}$, are given by

$$\begin{cases} \widehat{\beta}_k^{00D_l} = \sqrt{\frac{m_k}{m}}\delta_{kl} - \sqrt{\frac{m_k}{m}}\frac{\sqrt{Y_kY_l}}{\langle Y, U\rangle}, & k \in \mathcal{S}, \\ \widehat{\beta}_k^{10D_l} = 0, & k \in \mathcal{S}, \\ \widehat{\beta}_k^{01D_l} = 0, & k \in \mathcal{P}. \end{cases} \tag{4.4.61}$$

The matrix $\widehat{L}$, the right-hand sides $\widehat{\beta}^{D_l}$, and the constraint vector $\widehat{\mathcal{L}}$ are then rescaled versions of L, β^{D_l}, and $\mathcal{L}$, respectively,

$$(\mathcal{X}^D)^{1/2}\widehat{L}(\mathcal{X}^D)^{1/2} = L, \qquad (\mathcal{X}^D)^{1/2}\widehat{\beta}^{D_l} = \sqrt{Y_l}\beta^{D_l},\ l \in \mathcal{S}, \qquad (\mathcal{X}^D)^{1/2}\widehat{\mathcal{L}} = \mathcal{L}. \tag{4.4.62}$$

Using ($\widehat{L}$0)–($\widehat{L}$5) and (4.4.59)–(4.4.61), one can prove that the system (4.4.56) is well posed, i.e., admits a unique solution and that the symmetric flux diffusion matrix is also given by $\widehat{D}_{kl} = \sqrt{m_l/m}\,\widehat{\alpha}_l^{00D_k}$, $k, l \in \mathcal{S}$. Finally, we deduce from (4.4.62) that for positive mass fractions, we have

$$\widehat{D}_{kl} = \sqrt{Y_kY_l}D_{kl}, \qquad k, l \in \mathcal{S}. \tag{4.4.63}$$

4.4.6 Mathematical Properties of the System $\widetilde{L}_{[e]}\widetilde{\alpha}_{[e]}^{D_l} = \widetilde{\beta}_{[e]}^{D_l}$

We consider the n constrained linear systems indexed by $l \in \mathcal{S}$

$$\begin{cases} \widetilde{L}_{[e]}\widetilde{\alpha}_{[e]}^{D_l} = \widetilde{\beta}_{[e]}^{D_l}, \\ \langle \mathcal{L}_{[e]}, \widetilde{\alpha}_{[e]}^{D_l}\rangle = 0, \end{cases} \tag{4.4.64}$$

associated with the evaluation of the flux diffusion matrix $\widetilde{D}_{[e]} = (\widetilde{D}_{[e]kl})_{k,l\in\mathcal{S}}$

$$\widetilde{D}_{[e]kl} = \langle \widetilde{\alpha}_{[e]}^{D_k}, \beta_{[e]}^{D_l}\rangle, \qquad k, l \in \mathcal{S}, \tag{4.4.65}$$

as described in Section 3.4.4. We have $\widetilde{L}_{[e]} \in \mathbb{R}^{2n,2n}$, $\widetilde{\alpha}_{[e]}^{D_l}$, $\widetilde{\beta}_{[e]}^{D_l}$, and $\mathcal{L}_{[e]} \in \mathbb{R}^{2n}$, and the indexing set is $\mathcal{B}_{[e]}^D = \{00, \mathrm{e}\}\times\mathcal{S}$, yielding the block-decomposition

$$\begin{cases} \begin{bmatrix} \widetilde{L}_{[e]}^{0000} & \widetilde{L}_{[e]}^{00e} \\ \widetilde{L}_{[e]}^{e00} & \widetilde{L}_{[e]}^{ee} \end{bmatrix} \begin{bmatrix} \widetilde{\alpha}_{[e]}^{00D_l} \\ \widetilde{\alpha}_{[e]}^{eD_l} \end{bmatrix} = \begin{bmatrix} \widetilde{\beta}_{[e]}^{00D_l} \\ \widetilde{\beta}_{[e]}^{eD_l} \end{bmatrix}, \\ \langle \mathcal{L}_{[e]}^{00}, \widetilde{\alpha}_{[e]}^{00D_l}\rangle + \langle \mathcal{L}_{[e]}^{e}, \widetilde{\alpha}_{[e]}^{eD_l}\rangle = 0. \end{cases} \tag{4.4.66}$$

Furthermore, from the kinetic theory results obtained in Sections 3.1 and 3.4, we can make the following assumptions.

($\widetilde{L}_{[e]}0$) The coefficients of the matrix $\widetilde{L}_{[e]}$ are functions of the pressure, the temperature, and the mass fractions, they are proportional to the pressure, and they can be expressed as linear functions of the mole fractions in the form

$$\begin{cases} \widetilde{L}^{rs}_{[e]kk} = \sum_{l\in\mathcal{S}} X_l L'^{rs}_{[e]kl} + X_k L''^{rs}_{[e]kk}, & (r,k),(s,k) \in \mathcal{B}^D_{[e]}, \\ \widetilde{L}^{rs}_{[e]kl} = X_l L''^{rs}_{[e]kl}, \quad (r,k),(s,l) \in \mathcal{B}^D_{[e]}, & k \neq l, \end{cases} \tag{4.4.67}$$

where the matrices $L'_{[e]}$ and $L''_{[e]}$ coincide with the ones in (4.4.21).

($\widetilde{L}_{[e]}1$) The matrix $\widetilde{L}_{[e]}$ admits the block-decomposition

$$(\Gamma^D_{[e]})^t \widetilde{L}_{[e]} \Gamma^D_{[e]} = \begin{bmatrix} \widetilde{L}^{++}_{[e]} & 0 \\ \widetilde{L}^{-+}_{[e]} & \widetilde{L}^{--}_{[e]} \end{bmatrix}, \tag{4.4.68}$$

where $\Gamma^D_{[e]}$ is the permutation matrix associated with the reordering of $\mathcal{B}^D_{[e]}$ into $(\mathcal{B}^{D+}_{[e]}, \mathcal{B}^{D-}_{[e]})$.

($\widetilde{L}_{[e]}2$) The matrix $L^{++}_{[e]} = \mathcal{X}^{D++}_{[e]} \widetilde{L}^{++}_{[e]}$ corresponds to the $\mathcal{S}^+$ mixture and, in particular, is symmetric, positive semi-definite, positive definite on the hyperplane $(\mathcal{L}^+_{[e]})^\perp$, and has nullspace $N(L^{++}_{[e]}) = \mathbb{R}\mathcal{U}^+_{[e]}$ and range $R(L^{++}_{[e]}) = (\mathcal{U}^+_{[e]})^\perp$.

($\widetilde{L}_{[e]}3$) $\widetilde{L}^{--}_{[e]}$ is symmetric positive definite and we have $db(\widetilde{L}^{--}_{[e]}) = \widetilde{L}^{--}_{[e]}$.

($\widetilde{L}_{[e]}4$) $N(\widetilde{L}_{[e]}) = \mathbb{R}\mathcal{U}_{[e]}$.

($\widetilde{L}_{[e]}5$) $R(\widetilde{L}_{[e]}) = \widetilde{\mathcal{U}}^\perp_{[e]}$.

Moreover, following Section 3.4.4, we assume that the right-hand sides $\widetilde{\beta}^{D_l}_{[e]}$, $l \in \mathcal{S}$, are given by

$$\begin{cases} \widetilde{\beta}^{00D_l}_{[e]k} = \dfrac{m_k}{m}\delta_{kl} - \dfrac{m_k}{m}\dfrac{Y_l}{\langle Y, U\rangle}, & k \in \mathcal{S}, \\ \widetilde{\beta}^{eD_l}_{[e]k} = 0, \quad k \in \mathcal{S}, & \end{cases} \tag{4.4.69}$$

and we deduce from (4.4.21)(4.4.67) and (4.4.25)(4.4.69) that the matrix $\widetilde{L}_{[e]}$ and the right-hand sides $\widetilde{\beta}^{D_l}_{[e]}$ are rescaled versions of $L_{[e]}$ and $\beta^{D_l}_{[e]}$, respectively,

$$\mathcal{X}^D_{[e]} \widetilde{L}_{[e]} = L_{[e]}, \qquad \mathcal{X}^D_{[e]} \widetilde{\beta}^{D_l}_{[e]} = Y_l \beta^{D_l}_{[e]}, \ l \in \mathcal{S}, \tag{4.4.70}$$

where $\mathcal{X}^D_{[e]} = \operatorname{diag}\big((X_k)_{(r,k)\in\mathcal{B}^D_{[e]}}\big)$. Finally, the vector $\widetilde{\mathcal{U}}_{[e]} \in \mathbb{R}^{2n}$ is given by the block-decomposition

$$\widetilde{\mathcal{U}}^{00}_{[e]} = X, \qquad \widetilde{\mathcal{U}}^{e}_{[e]} = 0, \tag{4.4.71}$$

and since $\sum_{k\in S} X_k \widetilde{\beta}^{00D_l}_{[e]k} = 0$, we obtain

$$\widetilde{\beta}^{D_l}_{[e]} \in \widetilde{\mathcal{U}}^{\perp}_{[e]}, \qquad l \in S. \tag{4.4.72}$$

Theorem 4.4.22. *Let $\widetilde{L}_{[e]} \in \mathbb{R}^{2n,2n}$ be a matrix satisfying the properties $(\widetilde{L}_{[e]}0)$–$(\widetilde{L}_{[e]}5)$, and let $\mathcal{L}_{[e]}$, $\beta^{D_l}_{[e]}$, and $\widetilde{\beta}^{D_l}_{[e]} \in \mathbb{R}^{2n}$, $l \in S$, be given by (4.4.21), (4.4.25), and (4.4.69), respectively. Then the constrained linear systems (4.4.64) admit a unique solution $\widetilde{\alpha}^{D_l}_{[e]}$, $l \in S$. In addition, the matrix $\widetilde{D}_{[e]}$ may also be written*

$$\widetilde{D}_{[e]kl} = \widetilde{\alpha}^{00D_k}_{[e]l}, \qquad k, l \in S, \tag{4.4.73}$$

and admits the block-decomposition

$$\Upsilon^t \widetilde{D}_{[e]} \Upsilon = \begin{bmatrix} \widetilde{D}^{++}_{[e]} & \widetilde{D}^{+-}_{[e]} \\ 0 & \widetilde{D}^{--}_{[e]} \end{bmatrix}, \tag{4.4.74}$$

where Υ is the permutation matrix associated with the reordering of S into (S^+, S^-). The matrix $\widetilde{D}^{--}_{[e]}$ is diagonal with positive entries, and we have $\widetilde{D}^{++}_{[e]} = \mathcal{Y}^{++} D^{++}_{[e]}$, where $D^{++}_{[e]}$ is the diffusion matrix of the S^+ mixture. Finally, the matrix $\widetilde{D}_{[e]}$ has nullspace $N(\widetilde{D}_{[e]}) = \mathbb{R}Y$ and range $R(\widetilde{D}_{[e]}) = U^{\perp}$.

In the next proposition, we express the flux diffusion matrix $\widetilde{D}_{[e]}$ in terms of a generalized inverse of the matrix $\widetilde{L}_{[e]}$ and also in terms of the solution of a nonsingular linear system.

Proposition 4.4.23. *Let $\widetilde{\mathcal{L}}_{[e]} \in \mathbb{R}^{2n}$ be given by the block-decomposition $\widetilde{\mathcal{L}}^{00}_{[e]} = W$ and $\widetilde{\mathcal{L}}^{e}_{[e]} = 0$, and let $\widetilde{L}^{\dagger}_{[e]}$ be the generalized inverse of $\widetilde{L}_{[e]}$ with prescribed range $\mathcal{L}^{\perp}_{[e]}$ and nullspace $\mathbb{R}\widetilde{\mathcal{L}}_{[e]}$. Then the solution of the constrained linear systems (4.4.64) is given by $\widetilde{\alpha}^{D_l}_{[e]} = \widetilde{L}^{\dagger}_{[e]} \widetilde{\beta}^{D_l}_{[e]}$, $l \in S$. Furthermore, if a and b are two positive real numbers such that $ab\langle \mathcal{L}_{[e]}, \mathcal{U}_{[e]} \rangle \langle \widetilde{\mathcal{L}}_{[e]}, \widetilde{\mathcal{U}}_{[e]} \rangle = 1$, then the matrices $\widetilde{L}_{[e]} + a\,\widetilde{\mathcal{L}}_{[e]} \otimes \mathcal{L}_{[e]}$ and $\widetilde{L}^{\dagger}_{[e]} + b\,\mathcal{U}_{[e]} \otimes \widetilde{\mathcal{U}}_{[e]}$ are inverse of each other and coincide with $\widetilde{L}_{[e]}$ and $\widetilde{L}^{\dagger}_{[e]}$ on the hyperplanes $\mathcal{L}^{\perp}_{[e]}$ and $\widetilde{\mathcal{U}}^{\perp}_{[e]}$, respectively. Therefore, we have*

$$\begin{cases} \widetilde{\alpha}^{D_l}_{[e]} = [\widetilde{L}_{[e]} + a\,\widetilde{\mathcal{L}}_{[e]} \otimes \mathcal{L}_{[e]}]^{-1} \widetilde{\beta}^{D_l}_{[e]}, & l \in S, \\ \widetilde{D}_{[e]kl} = \langle\, [\widetilde{L}_{[e]} + a\widetilde{\mathcal{L}}_{[e]} \otimes \mathcal{L}_{[e]}]^{-1} \widetilde{\beta}^{D_k}_{[e]}, \beta^{D_l}_{[e]} \,\rangle, & k, l \in S. \end{cases} \tag{4.4.75}$$

We now consider the submatrix $\widetilde{\Lambda}_{[e]} \in \mathbb{R}^{n,n}$ given by

$$\widetilde{\Lambda}_{[e]} = \widetilde{L}^{ee}_{[e]}, \tag{4.4.76}$$

and establish in the following lemma that this matrix is nonsingular.

Lemma 4.4.24. *Let $\widetilde{L}_{[e]} \in \mathbb{R}^{2n,2n}$ be a matrix satisfying the properties $(\widetilde{L}_{[e]}0)$–$(\widetilde{L}_{[e]}5)$. Then the matrix $\widetilde{\Lambda}_{[e]} \in \mathbb{R}^{n,n}$ given by (4.4.76) is nonsingular.*

As a consequence of Lemma 4.4.24, the Schur complement of the matrix $\widetilde{\Lambda}_{[e]}$ is well defined, and we investigate its mathematical properties in the next proposition.

Proposition 4.4.25. *Let $\widetilde{L}_{[e]} \in \mathbb{R}^{2n,2n}$ be a matrix satisfying the properties $(\widetilde{L}_{[e]}0)$–$(\widetilde{L}_{[e]}5)$ and let $\widetilde{\Delta}_{[e]} \in \mathbb{R}^{n,n}$ be given by*

$$\widetilde{\Delta}_{[e]} = \widetilde{L}_{[e]}^{0000} - \widetilde{L}_{[e]}^{00e} \widetilde{\Lambda}_{[e]}^{-1} \widetilde{L}_{[e]}^{e00}. \tag{4.4.77}$$

Then we have $N(\widetilde{\Delta}_{[e]}) = \mathbb{R}U$, $R(\widetilde{\Delta}_{[e]}) = X^{\perp}$, and if a and b are two positive real numbers such that $ab\langle Y, U\rangle^2 = 1$, we have

$$\widetilde{D}_{[e]} = \mathcal{W}\big[(\widetilde{\Delta}_{[e]})^t + a\, Y{\otimes}W\big]^{-1} - b\, Y{\otimes}U. \tag{4.4.78}$$

We now prove that all the flux diffusion coefficients $\widetilde{D}_{[e]kl}$, $k,l \in \mathcal{S}$, are smooth functions of the pressure, the temperature, and the mass fractions in $\mathfrak{D}_{Y\geq 0, Y\neq 0}$ and are inversely proportional to the pressure.

Theorem 4.4.26. *Let $L_{[e]}$ and $\widetilde{L}_{[e]} \in \mathbb{R}^{2n,2n}$ be such that $L_{[e]}$ satisfies $(L_{[e]}0)$–$(L_{[e]}5)$ in $\mathfrak{D}_{Y>0}$ and $\widetilde{L}_{[e]}$ satisfies $(\widetilde{L}_{[e]}0)$–$(\widetilde{L}_{[e]}5)$ in $\mathfrak{D}_{Y\geq 0, Y\neq 0}$. Assume also that the matrices $L'_{[e]}$ and $L''_{[e]}$ in (4.4.21)(4.4.67) are smooth functions of the pressure and the temperature. Then for $Y > 0$ we have $\widetilde{\alpha}_{[e]}^{D_l} = Y_l \alpha_{[e]}^{D_l}$, $l \in \mathcal{S}$, and $\widetilde{D}_{[e]kl} = Y_k D_{[e]kl}$, $k,l \in \mathcal{S}$. Furthermore, all the vectors $\widetilde{\alpha}_{[e]}^{D_l}$, $l \in \mathcal{S}$, and all the flux diffusion coefficients $\widetilde{D}_{[e]kl}$, $k,l \in \mathcal{S}$, are smooth functions of $(\bar{p}, \overline{T}, Y)$ in $\mathfrak{D}_{Y\geq 0, Y\neq 0}$ and are inversely proportional to the pressure. Moreover, for $Z \in \mathbb{R}^n$, $Z \geq 0$, $Z \neq 0$, we have the following limits*

$$\begin{cases} \displaystyle\lim_{\substack{Y\to Z\\ Y>0}} D_{[e]kl}(\bar{p}, \overline{T}, Y) = \frac{\widetilde{D}_{[e]kl}(\bar{p}, \overline{T}, Z)}{Z_k}, & Z_k > 0, \quad l \in \mathcal{S}, \\[2ex] \displaystyle\lim_{\substack{Y\to Z\\ Y>0}} D_{[e]kl}(\bar{p}, \overline{T}, Y) = \frac{\partial \widetilde{D}_{[e]kl}(\bar{p}, \overline{T}, Z)}{\partial Z_k}, & Z_k = 0, \quad l \in \mathcal{S}, \quad l \neq k, \\[2ex] \displaystyle\lim_{\substack{Y\to Z\\ Y>0}} Y_k D_{[e]kk}(\bar{p}, \overline{T}, Y) = \widetilde{D}_{[e]kk}(\bar{p}, \overline{T}, Z) > 0, & Z_k = 0. \end{cases} \tag{4.4.79}$$

Consequently, the diffusion coefficients $D_{[e]kk}$, $k \in \mathcal{S}$, admit a smooth extension to $\mathfrak{D}_{Y_k>0}$, and the coefficients $D_{[e]kl}$, $k,l \in \mathcal{S}$, $k \neq l$, admit a smooth extension to $\mathfrak{D}_{Y\geq 0, Y\neq 0}$.

Finally, we consider the matrix $\widetilde{L}_{[e]}$ resulting from the practical approximations presented in Section 2.10.

Proposition 4.4.27. *Let $\mathcal{D}_{k\,\mathrm{int},l}$, c_k^{int}, ξ_k^{int}, and m_l, $k \in \mathcal{P}$, $l \in \mathcal{S}$, be positive quantities, let $\mathcal{D}_{kl}$, $\bar{A}_{kl}$, $\bar{B}_{kl}$, and $\bar{C}_{kl}$, $k,l \in \mathcal{S}$, be symmetric and positive coefficients such that $25/4 - 3\bar{B}_{kl} > 0$ and $15\bar{C}_{kl} - 3\bar{B}_{kl} - 9\bar{C}_{kl}^2 > 0$, $k,l \in \mathcal{S}$, and assume that $Y \geq 0$, $Y \neq 0$. Then the matrix $\widetilde{L}_{[e]}$ given by (3.7.27)–(3.7.34) satisfies $(\widetilde{L}_{[e]}0)$–$(\widetilde{L}_{[e]}5)$.*

4.4.7 Mathematical Properties of the System $\widetilde{L}_{[00]}\widetilde{\alpha}_{[00]}^{D_l} = \widetilde{\beta}_{[00]}^{D_l}$

The mathematical properties of the matrix $\widetilde{D}_{[00]}$ have already been studied in [Gi91], using the right rescaled system $\overline{L}_{[00]}\overline{\alpha}_{[00]}^{D_l} = \overline{\beta}_{[00]}^{D_l}$. The definition of the standard flux diffusion matrix, $\widetilde{D}$, in terms of a right rescaled transport linear system is given in Section 4.4.10, and the corresponding definition of the matrix $\widetilde{D}_{[00]}$ can then easily be deduced. In this section we consider instead the n constrained linear systems indexed by $l \in \mathcal{S}$

$$\begin{cases} \widetilde{L}_{[00]}\widetilde{\alpha}_{[00]}^{D_l} = \widetilde{\beta}_{[00]}^{D_l}, \\ \langle Y, \widetilde{\alpha}_{[00]}^{D_l} \rangle = 0, \end{cases} \tag{4.4.80}$$

associated with the evaluation of the flux diffusion matrix $\widetilde{D}_{[00]} = (\widetilde{D}_{[00]kl})_{k,l\in\mathcal{S}}$

$$\widetilde{D}_{[00]kl} = \langle \widetilde{\alpha}_{[00]}^{D_k}, \beta_{[00]}^{D_l} \rangle, \qquad k,l \in \mathcal{S}, \tag{4.4.81}$$

as described in Section 3.4.6. We have $\widetilde{L}_{[00]} \in \mathbb{R}^{n,n}$, $\widetilde{\alpha}_{[00]}^{D_l}$ and $\widetilde{\beta}_{[00]}^{D_l} \in \mathbb{R}^n$, and the indexing set is $\mathcal{B}_{[00]}^{D} = \{00\}\times\mathcal{S}$. Furthermore, from the kinetic theory results obtained in Sections 3.1 and 3.4, we can make the following assumptions.

$(\widetilde{L}_{[00]}0)$ The coefficients of the matrix $\widetilde{L}_{[00]}$ are functions of the pressure, the temperature, and the mass fractions, they are proportional to the pressure, and they can be expressed as linear functions of the mole fractions in the form

$$\begin{cases} \widetilde{L}_{[00]kk}^{rs} = \sum_{l\in\mathcal{S}} X_l L_{[00]kl}^{\prime rs} + X_k L_{[00]kk}^{\prime\prime rs}, & (r,k) \in \mathcal{B}_{[00]}^{D}, \\ \widetilde{L}_{[00]kl}^{rs} = X_l L_{[00]kl}^{\prime\prime rs}, & (r,k),(s,l) \in \mathcal{B}_{[00]}^{D}, \quad k \neq l, \end{cases} \tag{4.4.82}$$

where the matrices $L'_{[00]}$ and $L''_{[00]}$ coincide with the ones in (4.4.35).

$(\widetilde{L}_{[00]}1)$ The matrix $\widetilde{L}_{[00]}$ admits the block-decomposition

$$(\Gamma_{[00]}^{D})^t \widetilde{L}_{[00]} \Gamma_{[00]}^{D} = \begin{bmatrix} \widetilde{L}_{[00]}^{++} & 0 \\ \widetilde{L}_{[00]}^{-+} & \widetilde{L}_{[00]}^{--} \end{bmatrix}, \tag{4.4.83}$$

where $\Gamma_{[00]}^{D}$ is the permutation matrix associated with the reordering of $\mathcal{B}_{[00]}^{D}$ into $(\mathcal{B}_{[00]}^{D+}, \mathcal{B}_{[00]}^{D-})$.

($\widetilde{L}_{[00]}2$) The matrix $L_{[00]}^{++} = \mathcal{X}_{[00]}^{D++}\widetilde{L}_{[00]}^{++}$ corresponds to the $\mathcal{S}^+$ mixture and, in particular, is symmetric, positive semi-definite, positive definite on the hyperplane $(Y^+)^\perp$, and has nullspace $N(L_{[00]}^{++}) = \mathbb{R}U^+$ and range $R(L_{[00]}^{++}) = (U^+)^\perp$.

($\widetilde{L}_{[00]}3$) $\widetilde{L}_{[00]}^{--}$ is symmetric positive definite and we have $db(\widetilde{L}_{[00]}^{--}) = \widetilde{L}_{[00]}^{--}$.

($\widetilde{L}_{[00]}4$) $N(\widetilde{L}_{[00]}) = \mathbb{R}U$.

($\widetilde{L}_{[00]}5$) $R(\widetilde{L}_{[00]}) = X^\perp$.

Moreover, following Section 3.4.6, we assume that the right-hand sides $\widetilde{\beta}_{[00]}^{D_l}$, $l \in \mathcal{S}$, are given by

$$\widetilde{\beta}_{[00]k}^{00D_l} = \frac{m_k}{m}\delta_{kl} - \frac{m_k}{m}\frac{Y_l}{\langle Y, U\rangle}, \qquad k \in \mathcal{S}, \tag{4.4.84}$$

and we deduce from (4.4.35)(4.4.82) and (4.4.36)(4.4.84) that the matrix $\widetilde{L}_{[00]}$ and the right-hand sides $\widetilde{\beta}_{[00]}^{D_l}$ are rescaled versions of $L_{[00]}$ and $\beta_{[00]}^{D_l}$, respectively,

$$\mathcal{X}_{[00]}^{D}\widetilde{L}_{[00]} = L_{[00]}, \qquad \mathcal{X}_{[00]}^{D}\widetilde{\beta}_{[00]}^{D_l} = Y_l\beta_{[00]}^{D_l}, \ l \in \mathcal{S}, \tag{4.4.85}$$

where $\mathcal{X}_{[00]}^{D} = \operatorname{diag}\big((X_k)_{(r,k)\in\mathcal{B}_{[00]}^{D}}\big)$. Since $\sum_{k\in\mathcal{S}} X_k\widetilde{\beta}_{[00]k}^{00D_l} = 0$, we obtain

$$\widetilde{\beta}_{[00]}^{D_l} \in X^\perp, \qquad l \in \mathcal{S}. \tag{4.4.86}$$

Theorem 4.4.28. *Let $\widetilde{L}_{[00]} \in \mathbb{R}^{n,n}$ be a matrix satisfying the properties ($\widetilde{L}_{[00]}0$)–($\widetilde{L}_{[00]}5$), and let $\beta_{[00]}^{D_l}$ and $\widetilde{\beta}_{[00]}^{D_l} \in \mathbb{R}^n$, $l \in \mathcal{S}$, be given by (4.4.36) and (4.4.84), respectively. Then the constrained linear systems (4.4.80) admit a unique solution $\widetilde{\alpha}_{[00]}^{D_l}$, $l \in \mathcal{S}$. In addition, the matrix $\widetilde{D}_{[00]}$ may also be written*

$$\widetilde{D}_{[00]kl} = \widetilde{\alpha}_{[00]l}^{00D_k}, \qquad k, l \in \mathcal{S}, \tag{4.4.87}$$

and admits the block-decomposition

$$\Upsilon^t\widetilde{D}_{[00]}\Upsilon = \begin{bmatrix} \widetilde{D}_{[00]}^{++} & \widetilde{D}_{[00]}^{+-} \\ 0 & \widetilde{D}_{[00]}^{--} \end{bmatrix}, \tag{4.4.88}$$

where Υ is the permutation matrix associated with the reordering of $\mathcal{S}$ into $(\mathcal{S}^+, \mathcal{S}^-)$. The matrix $\widetilde{D}_{[00]}^{--}$ is diagonal with positive entries, and we have $\widetilde{D}_{[00]}^{++} = \mathcal{Y}^{++}D_{[00]}^{++}$, where $D_{[00]}^{++}$ is the diffusion matrix of the $\mathcal{S}^+$ mixture. Finally, the matrix $\widetilde{D}_{[00]}$ has nullspace $N(\widetilde{D}_{[00]}) = \mathbb{R}Y$ and range $R(\widetilde{D}_{[00]}) = U^\perp$.

In the next proposition, we clarify the relation between the flux diffusion matrix $\widetilde{D}_{[00]}$ and the matrix $\widetilde{L}_{[00]}$ and also express the matrix $\widetilde{D}_{[00]}$ in terms of the inverse of a nonsingular matrix [Gi91].

Proposition 4.4.29. *Let $\widetilde{L}_{[00]} \in \mathbb{R}^{n,n}$ be a matrix satisfying the properties $(\widetilde{L}_{[00]}0)$–$(\widetilde{L}_{[00]}5)$ and consider the flux diffusion matrix $\widetilde{D}_{[00]}$ given by (4.4.81). Then the matrix $\widetilde{D}_{[00]}$ is the generalized inverse of $(\widetilde{L}_{[00]})^t \mathcal{W}^{-1}$ with prescribed nullspace $N(\widetilde{D}_{[00]}) = \mathbb{R}Y$ and range $R(\widetilde{D}_{[00]}) = U^{\perp}$. In addition, the solution of the constrained linear systems (4.4.80) is given by $\widetilde{\alpha}^{D_l}_{[00]} = (\widetilde{D}_{[00]})^t \mathcal{W}^{-1} \widetilde{\beta}^{D_l}_{[00]}$, $l \in S$. Furthermore, if a and b are two positive real numbers such that $ab\langle Y, U\rangle^2 = 1$, we have*

$$\widetilde{D}_{[00]} = \mathcal{W}\big[(\widetilde{L}_{[00]})^t + a\,Y \otimes W\big]^{-1} - b\,Y \otimes U. \tag{4.4.89}$$

We now prove that all the flux diffusion coefficients $\widetilde{D}_{[00]kl}$, $k, l \in S$, are smooth functions of the pressure, the temperature, and the mass fractions in $\mathfrak{D}_{Y\geq 0, Y\neq 0}$ and are inversely proportional to the pressure.

Theorem 4.4.30. *Let $L_{[00]}$ and $\widetilde{L}_{[00]} \in \mathbb{R}^{n,n}$ be such that $L_{[00]}$ satisfies $(L_{[00]}0)$–$(L_{[00]}5)$ in $\mathfrak{D}_{Y>0}$ and $\widetilde{L}_{[00]}$ satisfies $(\widetilde{L}_{[00]}0)$–$(\widetilde{L}_{[00]}5)$ in $\mathfrak{D}_{Y\geq 0, Y\neq 0}$. Assume also that the matrices $L'_{[00]}$ and $L''_{[00]}$ in (4.4.35)(4.4.82) are smooth functions of the pressure and the temperature. Then for $Y > 0$ we have $\widetilde{\alpha}^{D_l}_{[00]} = Y_l \alpha^{D_l}_{[00]}$, $l \in S$, and $\widetilde{D}_{[00]kl} = Y_k D_{[00]kl}$, $k, l \in S$. Furthermore, all the vectors $\widetilde{\alpha}^{D_l}_{[00]}$, $l \in S$, and all the flux diffusion coefficients $\widetilde{D}_{[00]kl}$, $k, l \in S$, are smooth functions of $(\overline{p}, \overline{T}, Y)$ in $\mathfrak{D}_{Y\geq 0, Y\neq 0}$ and are inversely proportional to the pressure. Moreover, for $Z \in \mathbb{R}^n$, $Z \geq 0$, $Z \neq 0$, we have the following limits*

$$\begin{cases} \lim\limits_{\substack{Y\to Z\\ Y>0}} D_{[00]kl}(\overline{p}, \overline{T}, Y) = \dfrac{\widetilde{D}_{[00]kl}(\overline{p}, \overline{T}, Z)}{Z_k}, & Z_k > 0, \quad l \in S, \\[2ex] \lim\limits_{\substack{Y\to Z\\ Y>0}} D_{[00]kl}(\overline{p}, \overline{T}, Y) = \dfrac{\partial \widetilde{D}_{[00]kl}(\overline{p}, \overline{T}, Z)}{\partial Z_k}, & Z_k = 0, \quad l \in S, \quad l \neq k, \\[2ex] \lim\limits_{\substack{Y\to Z\\ Y>0}} Y_k D_{[00]kk}(\overline{p}, \overline{T}, Y) = \widetilde{D}_{[00]kk}(\overline{p}, \overline{T}, Z) > 0, & Z_k = 0. \end{cases} \tag{4.4.90}$$

Consequently, the diffusion coefficients $D_{[00]kk}$, $k \in S$, admit a smooth extension to $\mathfrak{D}_{Y_k>0}$, and the coefficients $D_{[00]kl}$, $k, l \in S$, $k \neq l$, admit a smooth extension to $\mathfrak{D}_{Y\geq 0, Y\neq 0}$.

Finally, we consider the matrix $\widetilde{L}_{[00]}$ resulting from the practical approximations presented in Section 2.10.

Proposition 4.4.31. *Let $\mathcal{D}_{kl}$, $k, l \in S$, be symmetric and positive coefficients and assume that $Y \geq 0$, $Y \neq 0$. Then the matrix $\widetilde{L}_{[00]}$ given by (3.7.11)(3.7.12) satisfies $(\widetilde{L}_{[00]}0)$–$(\widetilde{L}_{[00]}5)$.*

4.4.8 Generalized Stefan-Maxwell-Boltzmann Equations

In this section we consider the linear relations that relate the diffusion velocities V_i, $i \in \mathcal{S}$, to the diffusion driving forces d_i, $i \in \mathcal{S}$. For isotropic mixtures, the same relation is obtained in all directions of the physical three-dimensional space $\mathbb{R}^3$ so that, without loss of generality, we can restrict ourselves to the case where V_i and d_i, $i \in \mathcal{S}$, are scalars. Furthermore, introducing the vectors $V = (V_i)_{i\in\mathcal{S}}$ and $d = (d_i)_{i\in\mathcal{S}}$, the mass constraint relations $\sum_{i\in\mathcal{S}} Y_i V_i = 0$ and $\sum_{i\in\mathcal{S}} d_i = 0$ can be rewritten as $V \in Y^\perp$ and $d \in U^\perp$, respectively, where $Y = (Y_i)_{i\in\mathcal{S}}$ and $U = (1)_{i\in\mathcal{S}}$.

We first consider the standard—or second-order—formulation for the diffusion matrix D. In this case, the diffusion velocities are given by

$$V = -Dd. \tag{4.4.91}$$

Multiplying by the matrix Δ given by Proposition 4.4.4, we obtain $d = -\Delta V$ since we have $\Delta D = P_{U^\perp,\mathbb{R}Y}$, where $P_{U^\perp,\mathbb{R}Y}$ is the oblique projector onto the hyperplane $U^\perp$ along $\mathbb{R}Y$ and $d \in U^\perp$. Keeping in mind that $N(\Delta) = \mathbb{R}U$, this yields the following relations

$$\sum_{\substack{j\in\mathcal{S}\\ j\neq i}} \Delta_{ij}(V_i - V_j) = d_i, \qquad i \in \mathcal{S}, \tag{4.4.92}$$

which are a generalized form of the Stefan-Maxwell-Boltzmann equations. Note that a similar equation has also been considered in [MMM66] for monatomic gas mixtures while (4.4.92) is also valid for polyatomic gas mixtures.

Similarly, the simplified diffusion velocities $V_{[\mathrm{e}]} = (V_{[\mathrm{e}]i})_{i\in\mathcal{S}}$ are given by $V_{[\mathrm{e}]} = -D_{[\mathrm{e}]}d$ and are such that $V_{[\mathrm{e}]} \in Y^\perp$. We next obtain that $d = -\Delta_{[\mathrm{e}]} V_{[\mathrm{e}]}$ since $d \in U^\perp$ and $\Delta_{[\mathrm{e}]} D_{[\mathrm{e}]} = P_{U^\perp,\mathbb{R}Y}$. Using $N(\Delta_{[\mathrm{e}]}) = \mathbb{R}U$ then yields the relations

$$\sum_{\substack{j\in\mathcal{S}\\ j\neq i}} \Delta_{[\mathrm{e}]ij}(V_{[\mathrm{e}]i} - V_{[\mathrm{e}]j}) = d_i, \qquad i \in \mathcal{S}, \tag{4.4.93}$$

which take on the form of generalized Stefan-Maxwell-Boltzmann equations.

Finally, the simplified—or first-order—diffusion velocities $V_{[00]} = (V_{[00]i})_{i\in\mathcal{S}}$ are given by $V_{[00]} = -D_{[00]}d$ and are such that $V_{[00]} \in Y^\perp$. We next obtain that $d = -L_{[00]}V_{[00]}$ since $d \in U^\perp$ and $L_{[00]}D_{[00]} = P_{U^\perp,\mathbb{R}Y}$. Using $N(L_{[00]}) = \mathbb{R}U$ then yields the relations

$$\sum_{\substack{j\in\mathcal{S}\\ j\neq i}} L_{[00]ij}(V_{[00]i} - V_{[00]j}) = \sum_{\substack{j\in\mathcal{S}\\ j\neq i}} \frac{X_i X_j}{\mathcal{D}_{ij}}(V_{[00]j} - V_{[00]i}) = d_i, \qquad i \in \mathcal{S}, \tag{4.4.94}$$

which are referred to as the Stefan-Maxwell-Boltzmann equations. An elementary derivation of these equations has been given by Williams [Wi58].

Remark. Notice that iterative algorithms for diffusion velocities using Stefan-Maxwell-Boltzmann equations [JB81] are only interesting for the first-order diffusion velocities $V_{[00]}$. For the simplified diffusion velocities $V_{[e]}$ and the second-order diffusion velocities V, indeed, the matrices $\Delta_{[e]}$ and Δ are not explicitly known.

4.4.9 Diagonal Diffusion Processes

Because of the considerable simplifications that may result, we now investigate the situations where the diffusion process can be represented by a diagonal matrix, that is, when the diffusion matrix D coincides with a diagonal matrix on the hyperplane $U^\perp$. The following proposition generalizes a result from [Gi91].

Proposition 4.4.32. *Let $L \in \mathbb{R}^{2n+p,2n+p}$ be a matrix satisfying the properties $(L0)$–$(L5)$ and let the matrix $\Delta \in \mathbb{R}^{n,n}$ be given by (4.4.14). Then the diffusion matrix D coincides with a diagonal matrix on the hyperplane $U^\perp$ if and only if the matrix Δ may be written*

$$\Delta = \frac{1}{\mathcal{D}}\Big(\mathcal{Y} - \frac{Y \otimes Y}{\langle Y, U\rangle}\Big), \tag{4.4.95}$$

where $\mathcal{Y} = \mathrm{diag}(Y_1, \ldots, Y_n)$ and the quantity $\mathcal{D}$ is a constant independent of the species. In this situation, we have

$$D = \mathcal{D}\Big(\mathcal{Y}^{-1} - \frac{U \otimes U}{\langle Y, U\rangle}\Big). \tag{4.4.96}$$

Proof. Note first that (4.4.95) and (4.4.96) are equivalent. Indeed, this follows from Proposition 4.4.5 with the coefficients a and b taken equal to $a = 1/(\mathcal{D}\langle Y, U\rangle)$ and $b = \mathcal{D}/\langle Y, U\rangle$, respectively. Assume now that the matrix D coincides with a diagonal matrix on the hyperplane $U^\perp$. We then have $U^\perp \subset N(D - \Phi)$, where Φ is a diagonal matrix $\Phi = \mathrm{diag}(\phi_1, \ldots, \phi_n) \in \mathbb{R}^{n,n}$ and this yields $D = \Phi + C \otimes U$ where $C \in \mathbb{R}^n$ is a vector. By Theorem 4.4.1, we have $N(D) = \mathbb{R}Y$ and $R(D) = Y^\perp$, i.e., $DY = 0$ and $Y^t D = 0$. By combining these relations, we obtain for all $k \in S$ that $\phi_k Y_k = -C_k \langle Y, U\rangle$ and $\phi_k Y_k = -\langle Y, C\rangle$ since for a, b, and $c \in \mathbb{R}^n$ one has $(a \otimes b)c = \langle b, c\rangle a$. Denoting by $\mathcal{D}$ the constant $-\langle Y, C\rangle$, we deduce from the previous relations after some algebra that (4.4.96) holds. Conversely, assuming that either (4.4.95) or (4.4.96) hold, the matrix D obviously coincides with the diagonal matrix $\mathcal{D}\mathcal{Y}^{-1}$ on the hyperplane $U^\perp$. □

The diffusion matrix D relates the diffusion velocities to the diffusion driving forces, which involve, in particular, the gradients of the mole fractions ∇X_k, $k \in S$.

However, the mass fractions are often chosen as the fundamental unknowns—together with the density ρ, the mass averaged flow velocity v, and the temperature T—for solving the equations governing multicomponent reacting flows. An especially interesting simplification occurs when the diffusion driving forces d_k, $k \in \mathcal{S}$, are approximated by $d_k = \nabla X_k$. In this case, it becomes possible to express the diffusion velocities in terms of the gradients of the mass fractions ∇Y_k, $k \in \mathcal{S}$, by using the matrix DE, where the matrix $E = (E_{kl})_{k,l\in\mathcal{S}}$ is given by

$$\begin{cases} E_{kk} = \dfrac{m}{m_k} + \dfrac{X_k}{\langle Y,U\rangle}\Big(1 - \dfrac{m}{m_k}\Big), & k \in \mathcal{S}, \\ E_{kl} = \dfrac{X_k}{\langle Y,U\rangle}\Big(1 - \dfrac{m}{m_l}\Big), & k,l \in \mathcal{S}, \quad k \neq l. \end{cases} \tag{4.4.97}$$

The matrix E is nonsingular and is such that $\nabla X_k = \sum_{l\in\mathcal{S}} E_{kl}\nabla Y_l$ for all $k \in \mathcal{S}$ [Gi90]. In the following proposition, we examine the cases where the matrix DE coincides with a diagonal matrix on the hyperplane $U^\perp$.

Proposition 4.4.33. *With the assumptions and notation of Proposition 4.4.32, the matrix DE coincides with a diagonal matrix on the hyperplane $U^\perp$ if and only if the matrix Δ may be written*

$$\Delta = \frac{1}{\mathcal{D}}\Big(\mathcal{X} - \frac{X\otimes X}{\langle X,U\rangle}\Big), \tag{4.4.98}$$

where $\mathcal{X} = \mathrm{diag}(X_1,\ldots,X_n)$ and the quantity $\mathcal{D}$ is a constant independent of the species. In this situation, we have

$$DE = \mathcal{D}\Big(\mathcal{Y}^{-1} - \frac{Z\otimes U}{\langle X,U\rangle}\Big), \tag{4.4.99}$$

where the vector Z is given by $Z = \mathcal{Y}^{-1}E^{-1}Y$.

Proof. It is similar to the previous one, and we refer to [Gi91] for more details. □

With the same motivations as for Propositions 4.4.32 and 4.4.33, we now investigate the cases where the matrices $\widetilde{D}$ and $\widetilde{D}E$ coincide with a diagonal matrix on the physical hyperplane $U^\perp$. The proofs are omitted since they are similar to the previous ones.

Proposition 4.4.34. *Let $\widetilde{L} \in \mathbb{R}^{2n+p,2n+p}$ be a matrix satisfying the properties ($\widetilde{L}0$)–($\widetilde{L}5$) and let the matrix $\widetilde{\Delta} \in \mathbb{R}^{n,n}$ be given by (4.4.53). Then the matrix $\widetilde{D}$ coincides with a diagonal matrix on the hyperplane $U^\perp$ if and only if the matrix $\widetilde{\Delta}$ may be written*

$$\widetilde{\Delta} = \frac{1}{\widetilde{\mathcal{D}}}\Big(\mathcal{W} - \frac{W\otimes Y}{\langle Y,U\rangle}\Big), \tag{4.4.100}$$

where $\mathcal{W} = \mathrm{diag}(W_1, \ldots, W_n)$, $W_k = m_k/m$, $k \in \mathcal{S}$, *and the quantity* $\widetilde{\mathcal{D}}$ *is a constant independent of the species. In this situation, we have*

$$\widetilde{D} = \widetilde{\mathcal{D}}\Big(I - \frac{Y \otimes U}{\langle Y, U \rangle}\Big). \qquad (4.4.101)$$

Proposition 4.4.35. *Keeping the assumptions of Proposition 4.4.34, the matrix* $\widetilde{D}E$ *coincides with a diagonal matrix on the hyperplane* $U^{\perp}$ *if and only if the matrix* $\widetilde{\Delta}$ *may be written*

$$\widetilde{\Delta} = \frac{1}{\widetilde{\mathcal{D}}}\Big(I - \frac{U \otimes X}{\langle X, U \rangle}\Big), \qquad (4.4.102)$$

where the quantity $\widetilde{\mathcal{D}}$ *is a constant independent of the species. In this situation, we have*

$$\widetilde{D}E = \widetilde{\mathcal{D}}\Big(I - \frac{(E^{-1})Y \otimes U}{\langle X, U \rangle}\Big). \qquad (4.4.103)$$

Finally, similar results can also be stated for the simplified flux diffusion matrices $\widetilde{D}_{[\mathrm{e}]}$ and $\widetilde{D}_{[00]}$, but are omitted for brevity.

The previous propositions show that the use of generalized Fick laws of the form $F_k = -\alpha_k \nabla Y_k$, $k \in \mathcal{S}$, for all the species are possible only if all the scalar coefficients α_k, $k \in \mathcal{S}$, are equal, and this generalizes a result from [Gi91]. In particular, generalized Fick laws for the simplified diffusion fluxes $F_{[00]k} = -\alpha_k \nabla Y_k$, $k \in \mathcal{S}$, cannot be used for all the species unless all the binary diffusion coefficients $\mathcal{D}_{kl}$, $k, l \in \mathcal{S}$, $k \neq l$, are equal [Gi91].

4.4.10 Alternative Definition of the Flux Diffusion Matrix

In Sections 4.4.4, 4.4.6, and 4.4.7, we have expressed the flux diffusion matrices in terms of left rescaled system matrices using the mole fractions. This is indeed convenient since the same system matrices will then be considered in Section 4.5 for the partial thermal conductivity and the thermal diffusion vector. Nevertheless, when considering flux diffusion matrices only, it may be more convenient to introduce right rescaled system matrices using the mass fractions. All the results obtained in the previous sections can be rewritten in terms of these matrices, but, for brevity, most of the details are omitted. In particular, we only consider the standard formulation for the flux diffusion matrix $\widetilde{D}$ presented in Section 4.4.4.

Following Section 3.4.8, we introduce the matrix $\overline{L} \in \mathbb{R}^{2n+p,2n+p}$ given by

$$\overline{L} = (\widetilde{L})^t (\mathcal{W}^D)^{-1}, \qquad (4.4.104)$$

where the matrix $\widetilde{L}$ is assumed to satisfy the properties ($\widetilde{L}0$)–($\widetilde{L}5$) given in Section 4.4.4 and $\mathcal{W}^D = \operatorname{diag}\big((m_k/m)_{(r,k)\in\mathcal{B}^D}\big)$. In particular, we deduce that the coefficients of the matrix $\overline{L}$ are functions of the pressure, the temperature, and the mass fractions, and are proportional to the pressure. Moreover, they can be expressed as rational functions of the mass fractions since m is a rational function of the mass fractions by (4.1.2).

Remark. Transposing (4.4.104) and multiplying on the left by the matrix $\mathcal{Y}^D = \operatorname{diag}\big((Y_k)_{(r,k)\in\mathcal{B}^D}\big)$ yields $\mathcal{Y}^D\big(\overline{L}\big)^t = \mathcal{X}^D\widetilde{L}$. Since $\mathcal{X}^D\widetilde{L} = L$ and L is symmetric by ($L1$), we then obtain $\overline{L}\mathcal{Y}^D = \mathcal{X}^D\widetilde{L} = L$, so that the matrix $\overline{L}$ is indeed a right rescaled version of L using the mass fractions.

Theorem 4.4.36. *Let $\overline{L} \in \mathbb{R}^{2n+p,2n+p}$ be given by (4.4.104) and assume that $\widetilde{L}$ satisfies the properties ($\widetilde{L}0$)–($\widetilde{L}5$). Then the constrained linear systems*

$$\begin{cases} \overline{L}\overline{\alpha}^{D_l} = \beta^{D_l}, \\ \langle \mathcal{U}, \overline{\alpha}^{D_l}\rangle = 0, \end{cases} \tag{4.4.105}$$

admit a unique solution $\overline{\alpha}^{D_l}$, $l \in \mathcal{S}$. Furthermore, if the vectors $\overline{\beta}^{D_l} \in \mathbb{R}^{2n+p}$, $l \in \mathcal{S}$, are given by

$$\begin{cases} \overline{\beta}_k^{00D_l} = \delta_{kl} - \dfrac{Y_l}{\langle Y, U\rangle}, & k \in \mathcal{S}, \\ \overline{\beta}_k^{10D_l} = 0, \quad k \in \mathcal{S}, \\ \overline{\beta}_k^{01D_l} = 0, \quad k \in \mathcal{P}, \end{cases} \tag{4.4.106}$$

the flux diffusion matrix $\widetilde{D}$ can then be expressed as

$$\widetilde{D}_{kl} = \langle \overline{\alpha}^{D_l}, \overline{\beta}^{D_k}\rangle = \overline{\alpha}_k^{00D_l}, \qquad k,l \in \mathcal{S}. \tag{4.4.107}$$

Proof. We first obtain from (4.4.104) that the nullspace and range of the matrix $\overline{L}$ are given by $N(\overline{L}) = \mathbb{R}\mathcal{L}$ and $R(\overline{L}) = \mathcal{U}^\perp$, respectively, and since $\langle \mathcal{L}, \mathcal{U}\rangle \neq 0$, Proposition 4.1.3 applies, and the linear system (4.4.105) admits therefore a unique solution $\overline{\alpha}^{D_l}$, $l \in \mathcal{S}$. Furthermore, we have $\widetilde{D}_{kl} = \langle \widetilde{\alpha}^{D_k}, \beta^{D_l}\rangle = \langle \widetilde{\alpha}^{D_k}, \overline{L}\overline{\alpha}^{D_l}\rangle = \langle (\overline{L})^t\widetilde{\alpha}^{D_k}, \overline{\alpha}^{D_l}\rangle$ and $(\overline{L})^t\widetilde{\alpha}^{D_k} = (\mathcal{W}^D)^{-1}\widetilde{L}\widetilde{\alpha}^{D_k} = (\mathcal{W}^D)^{-1}\widetilde{\beta}^{D_k} = \overline{\beta}^{D_k}$ so that $\widetilde{D}_{kl} = \langle \overline{\alpha}^{D_l}, \overline{\beta}^{D_k}\rangle$. Finally, since $\overline{\beta}^{D_k} = e^{00k} - (Y_k/\langle \mathcal{L}, \mathcal{U}\rangle)\mathcal{U}$, $k \in \mathcal{S}$, and $\overline{\alpha}^{D_l} \in \mathcal{U}^\perp$, $l \in \mathcal{S}$, we obtain $\widetilde{D}_{kl} = \overline{\alpha}_k^{00D_l}$, $k,l \in \mathcal{S}$. □

In the next proposition, we express the flux diffusion matrix $\widetilde{D}$ in terms of a generalized inverse of the matrix $\overline{L}$ and also in terms of the solution of a nonsingular linear system.

Proposition 4.4.37. *Let $\overline{L}^\dagger$ be the group inverse of $\overline{L}$. Then the solution of the constrained linear systems (4.4.105) is given by $\overline{\alpha}^{D_l} = \overline{L}^\dagger \beta^{D_l}$, $l \in \mathcal{S}$. Furthermore, if a and b are two positive real numbers such that $ab\langle \mathcal{L},\mathcal{U}\rangle^2 = 1$, then the matrices $\overline{L} + a\,\mathcal{L}\otimes\mathcal{U}$ and $\overline{L}^\dagger + b\,\mathcal{L}\otimes\mathcal{U}$ are inverse of each other and coincide with $\overline{L}$ and $\overline{L}^\dagger$, respectively, on the hyperplane $\mathcal{U}^\perp$. Therefore, we have*

$$\begin{cases} \overline{\alpha}^{D_l} = [\overline{L} + a\,\mathcal{L}\otimes\mathcal{U}]^{-1}\beta^{D_l}, & l \in \mathcal{S}, \\ \widetilde{D}_{kl} = \langle\, [\overline{L} + a\,\mathcal{L}\otimes\mathcal{U}]^{-1}\beta^{D_l}, \overline{\beta}^{D_k} \,\rangle, & k,l \in \mathcal{S}. \end{cases} \tag{4.4.108}$$

Proof. Since $\langle \mathcal{L},\mathcal{U}\rangle \neq 0$ and $\overline{\beta}^{D_l} \in \mathcal{L}^\perp$, $l \in \mathcal{S}$, as may easily be verified, the proof directly results from Propositions 4.1.3 and 4.1.5. □

The following proposition generalizes a result from [Gi91].

Proposition 4.4.38. *Let $\widetilde{L} \in \mathbb{R}^{2n+p,2n+p}$ be a matrix satisfying the properties ($\widetilde{L}0$)–($\widetilde{L}5$) and let $\overline{\Delta} \in \mathbb{R}^{n,n}$ be given by*

$$\overline{\Delta} = (\widetilde{\Delta})^t \mathcal{W}^{-1}, \tag{4.4.109}$$

where $\widetilde{\Delta}$ is given by (4.4.53). Then we have $\overline{\Delta}\mathcal{Y} = \mathcal{X}\widetilde{\Delta}$, $N(\overline{\Delta}) = \mathbb{R}Y$, $R(\overline{\Delta}) = U^\perp$, and the matrix $\overline{\Delta}$ is the group inverse of the matrix $\widetilde{D}$. Furthermore, if a and b are two positive real numbers such that $ab\langle Y,U\rangle^2 = 1$, we have

$$\widetilde{D} = [\overline{\Delta} + b\,Y\otimes U]^{-1} - a\,Y\otimes U. \tag{4.4.110}$$

Proof. Transposing (4.4.109) and multiplying on the left by $\mathcal{Y}$ yields $\mathcal{Y}\overline{\Delta} = \mathcal{X}\widetilde{\Delta} = (\widetilde{\Delta})^t\mathcal{X}$, since the matrix $\mathcal{X}\widetilde{\Delta} = \Delta$ is symmetric. In addition, we deduce from Proposition 4.4.19 that $N(\overline{\Delta}) = \mathbb{R}Y$ and $R(\overline{\Delta}) = U^\perp$ since $N(\widetilde{\Delta}) = \mathbb{R}U$, $R(\widetilde{\Delta}) = X^\perp$, and $\mathcal{W}X = Y$. On the other hand, by eliminating the last $n+p$ rows in the linear systems (4.4.105), we obtain $\overline{\Delta}\widetilde{D} = P_{U^\perp,\mathbb{R}Y}$, from which we deduce that $\overline{\Delta}\widetilde{D}\overline{\Delta} = \overline{\Delta}$ and $\widetilde{D}\overline{\Delta}\widetilde{D} = \widetilde{D}$ since $R(\overline{\Delta}) = U^\perp$ and $N(\widetilde{D}) = \mathbb{R}Y$. From Proposition 4.1.2 we then obtain that $\overline{\Delta}$ is the group inverse of $\widetilde{D}$, and (4.4.110) results from Proposition 4.1.5 and (4.4.109). □

Finally, we note that when all the mass fractions are positive, we have the relation $\overline{\alpha}^{D_l} = \mathcal{Y}\alpha^{D_l}$, $l \in \mathcal{S}$, where α^{D_l} is the solution of (4.4.1), and, therefore, we recover the relation $\widetilde{D} = \mathcal{Y}D$ for $Y > 0$.

4.4.11 Singular Behavior of Discretized Conservation Equations

It is worthwhile to note that by using the matrix $D + a\,U\otimes U$ instead of just D, where a is a positive real number, one can suppress artificial singularities arising in Jacobian matrices of discretized conservation equations when all the mass fractions are considered as independent unknowns. Indeed, one can show that when the diffusion matrix is written as $D + a\,U\otimes U$, one obtains the following relation

$$\sum_{k\in\mathcal{S}} Y_k V_k = -\mathcal{D}\nabla\big(\sum_{k\in\mathcal{S}} Y_k\big), \tag{4.4.111}$$

and the following conservation equation for $\sum_{k\in\mathcal{S}} Y_k$—given in [Gi90] except for a sign misprint—,

$$\rho\partial_t\big(\sum_{k\in\mathcal{S}} Y_k\big) + \rho v\cdot\nabla\big(\sum_{k\in\mathcal{S}} Y_k\big) = \nabla\cdot\Big(\mathcal{D}\nabla\big(\sum_{k\in\mathcal{S}} Y_k\big)\Big), \tag{4.4.112}$$

and the new diffusion term $\mathcal{D} = a(\sum_{k\in\mathcal{S}} Y_k)$ suppresses the singular behavior [Gi90].

4.5 The Partial Thermal Conductivity and the Thermal Diffusion Vector

4.5.1 Mathematical Properties of the System $L\alpha^{\lambda'} = \beta^{\lambda'}$

We consider the constrained linear system

$$\begin{cases} L\alpha^{\lambda'} = \beta^{\lambda'}, \\ \langle \mathcal{L}, \alpha^{\lambda'}\rangle = 0, \end{cases} \tag{4.5.1}$$

associated with the evaluation of the partial thermal conductivity

$$\lambda' = \frac{\bar{p}}{T}\langle \alpha^{\lambda'}, \beta^{\lambda'}\rangle, \tag{4.5.2}$$

and the thermal diffusion vector $\theta = (\theta_k)_{k\in\mathcal{S}}$

$$\theta_k = -\langle \alpha^{\lambda'}, \beta^{D_k}\rangle, \qquad k\in\mathcal{S}, \tag{4.5.3}$$

as described in Section 2.7.1. We have $L \in \mathbb{R}^{2n+p,2n+p}$, $\alpha^{\lambda'}$, $\beta^{\lambda'}$, and $\mathcal{L} \in \mathbb{R}^{2n+p}$, and the corresponding indexing set is given by

$$\mathcal{B}^{\lambda'} = \{00, 10\}\times\mathcal{S} \,\cup\, \{01\}\times\mathcal{P}, \tag{4.5.4}$$

yielding the block-decomposition

$$\begin{cases} \begin{bmatrix} L^{0000} & L^{0010} & L^{0001} \\ L^{1000} & L^{1010} & L^{1001} \\ L^{0100} & L^{0110} & L^{0101} \end{bmatrix} \begin{bmatrix} \alpha^{00\lambda'} \\ \alpha^{10\lambda'} \\ \alpha^{01\lambda'} \end{bmatrix} = \begin{bmatrix} \beta^{00\lambda'} \\ \beta^{10\lambda'} \\ \beta^{01\lambda'} \end{bmatrix}, \\ \langle \mathcal{L}^{00}, \alpha^{00\lambda'} \rangle + \langle \mathcal{L}^{10}, \alpha^{10\lambda'} \rangle + \langle \mathcal{L}^{01}, \alpha^{01\lambda'} \rangle = 0. \end{cases} \tag{4.5.5}$$

In addition, following Section 2.7.1, we assume that the vectors β^{D_k}, $k \in \mathcal{S}$, are given by (4.4.9), that the matrix L satisfies the properties ($L0$)–($L5$) given in Section 4.4.1, and that the right-hand side $\beta^{\lambda'}$ is given by

$$\begin{cases} \beta_k^{00\lambda'} = 0, & k \in \mathcal{S}, \\ \beta_k^{10\lambda'} = \dfrac{c_{\mathrm{p}}^{\mathrm{tr}}}{k_{\mathrm{B}}} X_k, & k \in \mathcal{S}, \\ \beta_k^{01\lambda'} = \dfrac{c_k^{\mathrm{int}}}{k_{\mathrm{B}}} X_k, & k \in \mathcal{P}, \end{cases} \tag{4.5.6}$$

where $c_{\mathrm{p}}^{\mathrm{tr}}$ and k_{B} are positive constants and c_k^{int}, $k \in \mathcal{P}$, smooth positive functions of the temperature. The vectors $\mathcal{L}$ and $\mathcal{U}$ are, in turn, given by (4.4.6) and (4.4.7), respectively, and one can easily verify that

$$\beta^{\lambda'} \in \mathcal{U}^{\perp}. \tag{4.5.7}$$

Remark. We deduce from the symmetry of L that the vector θ can also be obtained from the solutions of the constrained linear systems (4.4.1) since

$$\theta_k = -\langle \alpha^{\lambda'}, \beta^{D_k} \rangle = -\langle \alpha^{\lambda'}, L\alpha^{D_k} \rangle = -\langle \beta^{\lambda'}, \alpha^{D_k} \rangle, \qquad k \in \mathcal{S}, \tag{4.5.8}$$

noticing that (4.5.8) involves the solution of n constrained linear systems instead of just one as in (4.5.3).

Theorem 4.5.1. *Let $L \in \mathbb{R}^{2n+p,2n+p}$ be a matrix satisfying the properties ($L0$)–($L5$) and let $\mathcal{L}$, β^{D_k}, $k \in \mathcal{S}$, and $\beta^{\lambda'} \in \mathbb{R}^{2n+p}$ be given by (4.4.6), (4.4.9), and (4.5.6), respectively. Then the constrained linear system (4.5.1) admits a unique solution $\alpha^{\lambda'}$. Furthermore, the quantity λ' is positive, and the vector θ satisfies the relation*

$$\langle \theta, Y \rangle = \sum_{k \in \mathcal{S}} Y_k \theta_k = 0, \tag{4.5.9}$$

and may also be written

$$\theta = -\alpha^{00\lambda'}. \tag{4.5.10}$$

Proof. Since $\beta^{\lambda'} \in \mathcal{U}^\perp = R(L)$ by (4.5.7) and ($L5$) and since $\langle \mathcal{L}, \mathcal{U} \rangle \neq 0$, Proposition 4.1.3 applies, and the constrained linear system (4.5.1) admits a unique solution $\alpha^{\lambda'}$. Furthermore, we have $\lambda' = (\bar{p}/\overline{T})\langle \alpha^{\lambda'}, L\alpha^{\lambda'} \rangle > 0$ since $\alpha^{\lambda'} \in \mathcal{L}^\perp$ and $\alpha^{\lambda'} \neq 0$ because $\beta^{\lambda'} \neq 0$. Finally, since $\beta^{D_k} = e^{00k} - (1/\langle \mathcal{L}, \mathcal{U} \rangle)\mathcal{L}$, $k \in \mathcal{S}$, we have $\sum_{k \in \mathcal{S}} Y_k \beta^{D_k} = 0$ thus yielding $\langle \theta, Y \rangle = -\langle \alpha^{\lambda'}, \sum_{k \in \mathcal{S}} Y_k \beta^{D_k} \rangle = 0$ and $\theta_k = -\langle \alpha^{\lambda'}, e^{00k} \rangle = -\alpha_k^{00\lambda'}$, $k \in \mathcal{S}$, since $\langle \mathcal{L}, \alpha^{\lambda'} \rangle = 0$. □

Remark. Note that (4.5.9) corresponds to the mass conservation constraint on the diffusion velocities V_k, $k \in \mathcal{S}$, that may be written $\sum_{k \in \mathcal{S}} Y_k V_k = 0$.

In the next proposition, we express the partial thermal conductivity λ' and the thermal diffusion vector θ in terms of either a generalized inverse of the matrix L or the inverse of a symmetric positive definite matrix.

Proposition 4.5.2. *Let $L^\dagger$ be the generalized inverse of L with prescribed range $\mathcal{L}^\perp$ and nullspace $\mathbb{R}\mathcal{L}$. Then the solution of the constrained linear system (4.5.1) is given by $\alpha^{\lambda'} = L^\dagger \beta^{\lambda'}$. Furthermore, if a and b are two positive real numbers such that $ab\langle \mathcal{L}, \mathcal{U} \rangle^2 = 1$, then the matrices $L + a\,\mathcal{L}\otimes\mathcal{L}$ and $L^\dagger + b\,\mathcal{U}\otimes\mathcal{U}$ are symmetric, positive definite, inverse of each other, and coincide with L and $L^\dagger$ on the hyperplanes $\mathcal{L}^\perp$ and $\mathcal{U}^\perp$, respectively. Therefore, we have*

$$\begin{cases} \alpha^{\lambda'} = [L + a\,\mathcal{L}\otimes\mathcal{L}]^{-1}\beta^{\lambda'}, \\ \lambda' = \dfrac{\bar{p}}{\overline{T}}\langle\, [L + a\,\mathcal{L}\otimes\mathcal{L}]^{-1}\beta^{\lambda'}, \beta^{\lambda'} \,\rangle, \\ \theta_k = -\langle\, [L + a\,\mathcal{L}\otimes\mathcal{L}]^{-1}\beta^{\lambda'}, \beta^{D_k} \,\rangle, \qquad k \in \mathcal{S}. \end{cases} \tag{4.5.11}$$

Proof. It follows from Propositions 4.1.3 and 4.1.4 since we have $\langle \mathcal{L}, \mathcal{U} \rangle \neq 0$ and $\beta^{\lambda'} \in \mathcal{U}^\perp$. □

4.5.2 Mathematical Properties of the System $L_{[e]}\alpha_{[e]}^{\lambda'} = \beta_{[e]}^{\lambda'}$

We consider the constrained linear system

$$\begin{cases} L_{[e]}\alpha_{[e]}^{\lambda'} = \beta_{[e]}^{\lambda'}, \\ \langle \mathcal{L}_{[e]}, \alpha_{[e]}^{\lambda'} \rangle = 0, \end{cases} \tag{4.5.12}$$

associated with the evaluation of the partial thermal conductivity

$$\lambda'_{[e]} = \frac{\bar{p}}{\overline{T}}\langle \alpha_{[e]}^{\lambda'}, \beta_{[e]}^{\lambda'} \rangle, \tag{4.5.13}$$

and the thermal diffusion vector $\theta_{[e]} = (\theta_{[e]k})_{k\in\mathcal{S}}$

$$\theta_{[e]k} = -\langle \alpha_{[e]}^{\lambda'}, \beta_{[e]}^{D_k} \rangle, \qquad k \in \mathcal{S}, \tag{4.5.14}$$

as described in Section 2.7.2. We have $L_{[e]} \in \mathbb{R}^{2n,2n}$, $\alpha_{[e]}^{\lambda'}$, $\beta_{[e]}^{\lambda'}$, and $\mathcal{L}_{[e]} \in \mathbb{R}^{2n}$, and the corresponding indexing set is given by

$$\mathcal{B}_{[e]}^{\lambda'} = \{00, \mathrm{e}\} \times \mathcal{S}, \tag{4.5.15}$$

yielding the block-decomposition

$$\begin{cases} \begin{bmatrix} L_{[e]}^{0000} & L_{[e]}^{00\mathrm{e}} \\ L_{[e]}^{\mathrm{e}00} & L_{[e]}^{\mathrm{ee}} \end{bmatrix} \begin{bmatrix} \alpha_{[e]}^{00\lambda'} \\ \alpha_{[e]}^{\mathrm{e}\lambda'} \end{bmatrix} = \begin{bmatrix} \beta_{[e]}^{00\lambda'} \\ \beta_{[e]}^{\mathrm{e}\lambda'} \end{bmatrix}, \\ \langle \mathcal{L}_{[e]}^{00}, \alpha_{[e]}^{00\lambda'} \rangle + \langle \mathcal{L}_{[e]}^{\mathrm{e}}, \alpha_{[e]}^{\mathrm{e}\lambda'} \rangle = 0. \end{cases} \tag{4.5.16}$$

In addition, following Section 2.7.2, we assume that the vectors $\beta_{[e]}^{D_k}$ are given by (4.4.25), that the matrix $L_{[e]}$ satisfies the properties ($L_{[e]}0$)–($L_{[e]}5$) given in Section 4.4.2, and that the right-hand side $\beta_{[e]}^{\lambda'}$ is given by

$$\begin{cases} \beta_{[e]k}^{00\lambda'} = 0, \qquad k \in \mathcal{S}, \\ \beta_{[e]k}^{\mathrm{e}\lambda'} = \dfrac{c_{\mathrm{p}}^{\mathrm{tr}} + c_k^{\mathrm{int}}}{k_{\mathrm{B}}} X_k, \qquad k \in \mathcal{S}. \end{cases} \tag{4.5.17}$$

The vectors $\mathcal{L}_{[e]}$ and $\mathcal{U}_{[e]}$ are, in turn, given by (4.4.22) and (4.4.23), respectively, and one can easily verify that

$$\beta_{[e]}^{\lambda'} \in \mathcal{U}_{[e]}^{\perp}. \tag{4.5.18}$$

Remark. We deduce from the symmetry of $L_{[e]}$ that the vector $\theta_{[e]}$ can also be obtained from the solutions of the constrained linear systems (4.4.17) since

$$\theta_{[e]k} = -\langle \alpha_{[e]}^{\lambda'}, \beta_{[e]}^{D_k} \rangle = -\langle \alpha_{[e]}^{\lambda'}, L_{[e]} \alpha_{[e]}^{D_k} \rangle = -\langle \beta_{[e]}^{\lambda'}, \alpha_{[e]}^{D_k} \rangle, \qquad k \in \mathcal{S}, \tag{4.5.19}$$

noticing that (4.5.19) involves the solution of n constrained linear systems instead of just one as in (4.5.14).

Theorem 4.5.3. *Let $L_{[e]} \in \mathbb{R}^{2n,2n}$ be a matrix satisfying the properties ($L_{[e]}0$)–($L_{[e]}5$) and let $\mathcal{L}_{[e]}$, $\beta_{[e]}^{D_k}$, $k \in \mathcal{S}$, and $\beta_{[e]}^{\lambda'} \in \mathbb{R}^{2n}$ be given by (4.4.22), (4.4.25), and (4.5.17), respectively. Then the constrained linear system (4.5.12) admits a unique solution $\alpha_{[e]}^{\lambda'}$. Furthermore, the quantity $\lambda'_{[e]}$ is positive, and the vector $\theta_{[e]}$ satisfies the relation*

$$\langle \theta_{[e]}, Y \rangle = \sum_{k\in\mathcal{S}} Y_k \theta_{[e]k} = 0, \tag{4.5.20}$$

and may also be written

$$\theta_{[e]} = -\alpha_{[e]}^{00\lambda'}. \tag{4.5.21}$$

Remark. Note that (4.5.20) corresponds to the mass conservation constraint on the simplified diffusion velocities $V_{[e]k}$, $k \in \mathcal{S}$, that may be written $\sum_{k\in\mathcal{S}} Y_k V_{[e]k} = 0$.

In the next proposition, we express the partial thermal conductivity $\lambda'_{[e]}$ and the thermal diffusion vector $\theta_{[e]}$ in terms of either a generalized inverse of the matrix $L_{[e]}$ or the inverse of a symmetric positive definite matrix.

Proposition 4.5.4. *Let $L_{[e]}^{\dagger}$ be the generalized inverse of $L_{[e]}$ with prescribed range $\mathcal{L}_{[e]}^{\perp}$ and nullspace $\mathbb{R}\mathcal{L}_{[e]}$. Then the solution of the constrained linear system (4.5.12) is given by $\alpha_{[e]}^{\lambda'} = L_{[e]}^{\dagger}\beta_{[e]}^{\lambda'}$. Furthermore, if a and b are two positive real numbers such that $ab\langle \mathcal{L}_{[e]}, \mathcal{U}_{[e]}\rangle^2 = 1$, then the matrices $L_{[e]} + a\,\mathcal{L}_{[e]}\otimes\mathcal{L}_{[e]}$ and $L_{[e]}^{\dagger} + b\,\mathcal{U}_{[e]}\otimes\mathcal{U}_{[e]}$ are symmetric, positive definite, inverse of each other, and coincide with $L_{[e]}$ and $L_{[e]}^{\dagger}$ on the hyperplanes $\mathcal{L}_{[e]}^{\perp}$ and $\mathcal{U}_{[e]}^{\perp}$, respectively. Therefore, we have*

$$\begin{cases} \alpha_{[e]}^{\lambda'} = [L_{[e]} + a\,\mathcal{L}_{[e]}\otimes\mathcal{L}_{[e]}]^{-1}\beta_{[e]}^{\lambda'}, \\ \lambda'_{[e]} = \frac{\overline{p}}{\overline{T}}\langle\, [L_{[e]} + a\,\mathcal{L}_{[e]}\otimes\mathcal{L}_{[e]}]^{-1}\beta_{[e]}^{\lambda'}, \beta_{[e]}^{\lambda'}\,\rangle, \\ \theta_{[e]k} = -\langle\, [L_{[e]} + a\,\mathcal{L}_{[e]}\otimes\mathcal{L}_{[e]}]^{-1}\beta_{[e]}^{\lambda'}, \beta_{[e]}^{D_k}\,\rangle, \qquad k \in \mathcal{S}. \end{cases} \tag{4.5.22}$$

4.5.3 Mathematical Properties of the System $\widetilde{L}\alpha^{\lambda'} = \widetilde{\beta}^{\lambda'}$

We consider the constrained linear system

$$\begin{cases} \widetilde{L}\alpha^{\lambda'} = \widetilde{\beta}^{\lambda'}, \\ \langle \mathcal{L}, \alpha^{\lambda'} \rangle = 0, \end{cases} \tag{4.5.23}$$

associated with the evaluation of the partial thermal conductivity

$$\lambda' = \frac{\overline{p}}{\overline{T}}\langle \alpha^{\lambda'}, \beta^{\lambda'} \rangle, \tag{4.5.24}$$

and the thermal diffusion vector $\theta = (\theta_k)_{k\in\mathcal{S}}$

$$\theta_k = -\langle \alpha^{\lambda'}, \beta^{D_k} \rangle, \qquad k \in \mathcal{S}, \tag{4.5.25}$$

as described in Section 3.5.1. We have $\widetilde{L} \in \mathbb{R}^{2n+p,2n+p}$, $\alpha^{\lambda'}$, $\widetilde{\beta}^{\lambda'}$, and $\mathcal{L} \in \mathbb{R}^{2n+p}$, and the indexing set is $\mathcal{B}^{\lambda'} = \{00, 10\}\times\mathcal{S} \cup \{01\}\times\mathcal{P}$, yielding the block-decomposition

$$\begin{cases} \begin{bmatrix} \widetilde{L}^{0000} & \widetilde{L}^{0010} & \widetilde{L}^{0001} \\ \widetilde{L}^{1000} & \widetilde{L}^{1010} & \widetilde{L}^{1001} \\ \widetilde{L}^{0100} & \widetilde{L}^{0110} & \widetilde{L}^{0101} \end{bmatrix} \begin{bmatrix} \alpha^{00\lambda'} \\ \alpha^{10\lambda'} \\ \alpha^{01\lambda'} \end{bmatrix} = \begin{bmatrix} \widetilde{\beta}^{00\lambda'} \\ \widetilde{\beta}^{10\lambda'} \\ \widetilde{\beta}^{01\lambda'} \end{bmatrix}, \\ \langle \mathcal{L}^{00}, \alpha^{00\lambda'} \rangle + \langle \mathcal{L}^{10}, \alpha^{10\lambda'} \rangle + \langle \mathcal{L}^{01}, \alpha^{01\lambda'} \rangle = 0. \end{cases} \tag{4.5.26}$$

Following Section 3.5.1, we assume that the matrix $\widetilde{L}$ satisfies the properties ($\widetilde{L}0$)–($\widetilde{L}5$) given in Section 4.4.4 and that the right-hand side $\widetilde{\beta}^{\lambda'}$ is given by

$$\begin{cases} \widetilde{\beta}_k^{00\lambda'} = 0, & k \in \mathcal{S}, \\ \widetilde{\beta}_k^{10\lambda'} = \dfrac{c_{\mathrm{p}}^{\mathrm{tr}}}{k_{\mathrm{B}}}, & k \in \mathcal{S}, \\ \widetilde{\beta}_k^{01\lambda'} = \dfrac{c_k^{\mathrm{int}}}{k_{\mathrm{B}}}, & k \in \mathcal{P}. \end{cases} \tag{4.5.27}$$

We deduce from (4.4.5)(4.4.43) and (4.5.6)(4.5.27) that the matrix $\widetilde{L}$ and the right-hand side $\widetilde{\beta}^{\lambda'}$ are rescaled versions of L and $\beta^{\lambda'}$, respectively,

$$\mathcal{X}^{\lambda'}\widetilde{L} = L, \qquad \mathcal{X}^{\lambda'}\widetilde{\beta}^{\lambda'} = \beta^{\lambda'}, \tag{4.5.28}$$

where $\mathcal{X}^{\lambda'} = \mathrm{diag}\big((X_k)_{(r,k)\in\mathcal{B}^{\lambda'}}\big)$. The vector $\widetilde{\mathcal{U}}$ is, in turn, given by (4.4.47), and from (4.4.47)(4.5.27) we obtain

$$\widetilde{\beta}^{\lambda'} \in \widetilde{\mathcal{U}}^{\perp}. \tag{4.5.29}$$

Theorem 4.5.5. *Let $\widetilde{L} \in \mathbb{R}^{2n+p,2n+p}$ be a matrix satisfying the properties ($\widetilde{L}0$)–($\widetilde{L}5$) and let $\mathcal{L}$, β^{D_k}, $k \in \mathcal{S}$, $\beta^{\lambda'}$, and $\widetilde{\beta}^{\lambda'} \in \mathbb{R}^{2n+p}$ be given by (4.4.6), (4.4.9), (4.5.6), and (4.5.27), respectively. Then the constrained linear system (4.5.23) admits a unique solution $\alpha^{\lambda'}$, and this solution is such that $L^{++}\alpha^{\lambda'+} = \beta^{\lambda'+}$ and $\langle \mathcal{L}^+, \alpha^{\lambda'+}\rangle = 0$. Furthermore, the quantity λ' is positive and is the partial thermal conductivity of the $\mathcal{S}^+$ mixture. The vector θ satisfies the relation*

$$\langle \theta, Y\rangle = \sum_{k\in\mathcal{S}} Y_k\theta_k = 0, \tag{4.5.30}$$

and may also be written

$$\theta = -\alpha^{00\lambda'}. \tag{4.5.31}$$

Finally, θ admits the block-decomposition $\theta = \Upsilon(\theta^+, \theta^-)$, where θ^+ is the thermal diffusion vector of the $\mathcal{S}^+$ mixture.

Proof. Since $\widetilde{\beta}^{\lambda'} \in \widetilde{\mathcal{U}}^{\perp} = R(\widetilde{L})$ by (4.5.29) and ($\widetilde{L}5$) and since $\langle \mathcal{L}, \mathcal{U}\rangle \neq 0$, Proposition 4.1.3 applies, and the constrained linear system (4.5.23) admits a unique solution $\alpha^{\lambda'}$. In addition, we have $L^{++}\alpha^{\lambda'+} = \beta^{\lambda'+}$ from ($\widetilde{L}1$) and (4.5.28), and $\langle \mathcal{L}^+, \alpha^{\lambda'+}\rangle = 0$ from $\mathcal{L}^- = 0$. Furthermore, from $\beta^{\lambda'-} = 0$, we deduce that $\lambda' = (\overline{p}/\overline{T})\langle \alpha^{\lambda'+}, L^{++}\alpha^{\lambda'+}\rangle$, and hence, λ' is positive and is the partial thermal conductivity of the $\mathcal{S}^+$ mixture. The proof of (4.5.30)(4.5.31) is similar to the one of (4.5.9)(4.5.10) given in Theorem

4.5.1. Finally, for $k \in \mathcal{S}^+$, $\beta^{D_k-} = 0$ so that $\theta_k = -\langle \alpha^{\lambda'+}, \beta^{D_k+} \rangle$, and hence, θ^+ is the thermal diffusion vector of the $\mathcal{S}^+$ mixture. □

Remark. It is interesting to point out that the components θ_k, $k \in \mathcal{S}^-$, are a priori nonzero. Their explicit expression is given in Section 4.7 in the practically important dilution limit.

In the next proposition, we express the partial thermal conductivity λ' and the thermal diffusion vector θ in terms of a generalized inverse of the matrix $\widetilde{L}$ and also in terms of the solution of a nonsingular linear system.

Proposition 4.5.6. *Let $\widetilde{\mathcal{L}} \in \mathbb{R}^{2n+p}$ be given by the block-decomposition $\widetilde{\mathcal{L}}^{00} = W$, $\widetilde{\mathcal{L}}^{10} = 0$, and $\widetilde{\mathcal{L}}^{01} = 0$, and let $\widetilde{L}^\dagger$ be the generalized inverse of $\widetilde{L}$ with prescribed range $\mathcal{L}^\perp$ and nullspace $\mathbb{R}\widetilde{\mathcal{L}}$. Then the solution of the constrained linear system (4.5.23) is given by $\alpha^{\lambda'} = \widetilde{L}^\dagger \widetilde{\beta}^{\lambda'}$. Furthermore, if a and b are two positive real numbers such that $ab\langle \mathcal{L},\mathcal{U}\rangle\langle \widetilde{\mathcal{L}},\widetilde{\mathcal{U}}\rangle = 1$, then the matrices $\widetilde{L} + a\,\widetilde{\mathcal{L}}\otimes\mathcal{L}$ and $\widetilde{L}^\dagger + b\mathcal{U}\otimes\widetilde{\mathcal{U}}$ are inverse of each other and coincide with $\widetilde{L}$ and $\widetilde{L}^\dagger$ on the hyperplanes $\mathcal{L}^\perp$ and $\widetilde{\mathcal{U}}^\perp$, respectively. Therefore, we have*

$$\begin{cases} \alpha^{\lambda'} = [\widetilde{L} + a\,\widetilde{\mathcal{L}}\otimes\mathcal{L}]^{-1}\widetilde{\beta}^{\lambda'}, \\ \lambda' = \dfrac{\overline{p}}{\overline{T}}\langle\, [\widetilde{L} + a\,\widetilde{\mathcal{L}}\otimes\mathcal{L}]^{-1}\widetilde{\beta}^{\lambda'}, \beta^{\lambda'}\,\rangle, \\ \theta_k = -\langle\, [\widetilde{L} + a\,\widetilde{\mathcal{L}}\otimes\mathcal{L}]^{-1}\widetilde{\beta}^{\lambda'}, \beta^{D_k}\,\rangle, \qquad k \in \mathcal{S}. \end{cases} \tag{4.5.32}$$

Proof. The proof directly results from Propositions 4.1.3 and 4.1.5 since $\langle \mathcal{L},\mathcal{U}\rangle \neq 0$, $\langle \widetilde{\mathcal{L}},\widetilde{\mathcal{U}}\rangle \neq 0$, and $\widetilde{\beta}^{\lambda'} \in \widetilde{\mathcal{U}}^\perp$. □

We now prove that the partial thermal conductivity λ' is a smooth function of the temperature and the mass fractions in $\mathcal{D}_{Y\geq 0, Y\neq 0}$ and is independent of the pressure, whereas the thermal diffusion vector θ is a smooth function of the pressure, the temperature, and the mass fractions in $\mathfrak{D}_{Y\geq 0, Y\neq 0}$ and is inversely proportional to the pressure.

Theorem 4.5.7. *Keeping the assumptions of Theorem 4.4.20, let $\alpha^{\lambda'}$, λ', and θ be the solution, the partial thermal conductivity, and the thermal diffusion vector evaluated from L for $Y > 0$ and let $\widetilde{\alpha}^{\lambda'}$, $\widetilde{\lambda}'$, and $\widetilde{\theta}$ be the solution, the partial thermal conductivity, and the thermal diffusion vector evaluated from $\widetilde{L}$ for $Y \geq 0$, $Y \neq 0$. Then $\alpha^{\lambda'} = \widetilde{\alpha}^{\lambda'}$, $\lambda' = \widetilde{\lambda}'$, and $\theta = \widetilde{\theta}$ for $Y > 0$. Furthermore, $\widetilde{\alpha}^{\lambda'}$ and $\widetilde{\theta}$ are smooth functions of $(\overline{p}, \overline{T}, Y)$ in $\mathfrak{D}_{Y\geq 0, Y\neq 0}$ and are inversely proportional to the pressure, whereas $\widetilde{\lambda}'$ is a smooth function of $(\overline{T}, Y)$ in $\mathcal{D}_{Y\geq 0, Y\neq 0}$ and is independent of the pressure. Moreover, for*

$Z \in \mathbb{R}^n$, $Z \geq 0$, $Z \neq 0$, *we have the following limits*

$$\begin{cases} \lim\limits_{\substack{Y \to Z \\ Y > 0}} \lambda'(\overline{T}, Y) = \widetilde{\lambda}'(\overline{T}, Z), \\ \lim\limits_{\substack{Y \to Z \\ Y > 0}} \theta(\overline{p}, \overline{T}, Y) = \widetilde{\theta}(\overline{p}, \overline{T}, Z). \end{cases} \tag{4.5.33}$$

Proof. When all the mass fractions are positive, the matrix $\mathcal{X}^{\lambda'}$ is invertible and the relations $\alpha^{\lambda'} = \widetilde{\alpha}^{\lambda'}$, $\lambda' = \widetilde{\lambda}'$, and $\theta = \widetilde{\theta}$ are then easily obtained from (4.5.28). Furthermore, we deduce from ($\widetilde{L}0$) that the vector $\widetilde{\alpha}^{\lambda'}$ is inversely proportional to the pressure so that the partial thermal conductivity is independent of the pressure, and the thermal diffusion vector is inversely proportional to the pressure. The rest of the proof is similar to the previous ones. □

Remark. As a consequence of the previous theorem, the solutions of the linear systems (4.5.1)(4.5.23) coincide when $Y > 0$. This justifies the use of the same notation $\alpha^{\lambda'}$, λ', and θ in Sections 4.5.1 and 4.5.3.

4.5.4 Mathematical Properties of the System $\widehat{L}\widehat{\alpha}^{\lambda'} = \widehat{\beta}^{\lambda'}$

We consider the constrained linear system

$$\begin{cases} \widehat{L}\widehat{\alpha}^{\lambda'} = \widehat{\beta}^{\lambda'}, \\ \langle \widehat{\mathcal{L}}, \widehat{\alpha}^{\lambda'} \rangle = 0, \end{cases} \tag{4.5.34}$$

associated with the evaluation of the partial thermal conductivity

$$\lambda' = \frac{\overline{p}}{\overline{T}} \langle \widehat{\alpha}^{\lambda'}, \widehat{\beta}^{\lambda'} \rangle, \tag{4.5.35}$$

and the rescaled thermal diffusion vector $\widehat{\theta} = (\widehat{\theta}_k)_{k \in S}$

$$\widehat{\theta}_k = -\langle \widehat{\alpha}^{\lambda'}, \widehat{\beta}^{D_k} \rangle, \qquad k \in S. \tag{4.5.36}$$

In addition to the notation and assumptions given in Section 4.4.5, we assume, following Section 3.1.8, that the right-hand side $\widehat{\beta}^{\lambda'} \in \mathbb{R}^{2n+p}$ is given by

$$\begin{cases} \widehat{\beta}_k^{00\lambda'} = 0, \qquad k \in S, \\ \widehat{\beta}_k^{10\lambda'} = \dfrac{c_{\mathrm{P}}^{\mathrm{tr}}}{k_{\mathrm{B}}} \sqrt{X_k}, \qquad k \in S, \\ \widehat{\beta}_k^{01\lambda'} = \dfrac{c_k^{\mathrm{int}}}{k_{\mathrm{B}}} \sqrt{X_k}, \qquad k \in \mathcal{P}, \end{cases} \tag{4.5.37}$$

so that $\widehat{\beta}^{\lambda'}$ is a rescaled version of $\beta^{\lambda'}$

$$(\mathcal{X}^{\lambda'})^{1/2}\widehat{\beta}^{\lambda'} = \beta^{\lambda'}. \tag{4.5.38}$$

One can then prove that the constrained linear system (4.5.34) is well posed, i.e., admits a unique solution and that the rescaled thermal diffusion vector is also given by $\widehat{\theta}_k = -\sqrt{m_k/m}\,\widehat{\alpha}_k^{00\lambda'}$, $k \in \mathcal{S}$. Finally, when all the mass fractions are positive, we deduce from (4.4.62)(4.5.38) that the definitions (4.5.2)(4.5.35) of the partial thermal conductivity coincide, whereas we have

$$\widehat{\theta}_k = \sqrt{Y_k}\theta_k, \qquad k \in \mathcal{S}. \tag{4.5.39}$$

4.5.5 Mathematical Properties of the System $\widetilde{L}_{[e]}\alpha_{[e]}^{\lambda'} = \widetilde{\beta}_{[e]}^{\lambda'}$

We consider the constrained linear system

$$\begin{cases} \widetilde{L}_{[e]}\alpha_{[e]}^{\lambda'} = \widetilde{\beta}_{[e]}^{\lambda'}, \\ \langle \mathcal{L}_{[e]}, \alpha_{[e]}^{\lambda'} \rangle = 0, \end{cases} \tag{4.5.40}$$

associated with the evaluation of the partial thermal conductivity

$$\lambda'_{[e]} = \frac{\bar{p}}{\bar{T}}\langle \alpha_{[e]}^{\lambda'}, \beta_{[e]}^{\lambda'} \rangle, \tag{4.5.41}$$

and the thermal diffusion vector $\theta_{[e]} = (\theta_{[e]k})_{k\in\mathcal{S}}$

$$\theta_{[e]k} = -\langle \alpha_{[e]}^{\lambda'}, \beta_{[e]}^{D_k} \rangle, \qquad k \in \mathcal{S}, \tag{4.5.42}$$

as described in Section 3.5.2. We have $\widetilde{L}_{[e]} \in \mathbb{R}^{2n,2n}$, $\alpha_{[e]}^{\lambda'}$, $\widetilde{\beta}_{[e]}^{\lambda'}$, and $\mathcal{L}_{[e]} \in \mathbb{R}^{2n}$, and the indexing set is $\mathcal{B}_{[e]}^{\lambda'} = \{00, \mathrm{e}\}\times\mathcal{S}$, yielding the block-decomposition

$$\begin{cases} \begin{bmatrix} \widetilde{L}_{[e]}^{0000} & \widetilde{L}_{[e]}^{00e} \\ \widetilde{L}_{[e]}^{e00} & \widetilde{L}_{[e]}^{ee} \end{bmatrix} \begin{bmatrix} \alpha_{[e]}^{00\lambda'} \\ \alpha_{[e]}^{e\lambda'} \end{bmatrix} = \begin{bmatrix} \widetilde{\beta}_{[e]}^{00\lambda'} \\ \widetilde{\beta}_{[e]}^{e\lambda'} \end{bmatrix}, \\ \langle \mathcal{L}_{[e]}^{00}, \alpha_{[e]}^{00\lambda'} \rangle + \langle \mathcal{L}_{[e]}^{e}, \alpha_{[e]}^{e\lambda'} \rangle = 0. \end{cases} \tag{4.5.43}$$

Following Section 3.5.2, we assume that the matrix $\widetilde{L}_{[e]}$ satisfies the properties $(\widetilde{L}_{[e]}0)$–$(\widetilde{L}_{[e]}5)$ given in Section 4.4.5 and that the right-hand side $\widetilde{\beta}_{[e]}^{\lambda'}$ is given by

$$\begin{cases} \widetilde{\beta}_{[e]k}^{00\lambda'} = 0, \qquad k \in \mathcal{S}, \\ \widetilde{\beta}_{[e]k}^{e\lambda'} = \dfrac{c_p^{tr} + c_k^{int}}{k_B}, \qquad k \in \mathcal{S}. \end{cases} \tag{4.5.44}$$

We deduce from (4.4.21)(4.4.67) and (4.5.17)(4.5.44) that the matrix $\widetilde{L}_{[e]}$ and the right-hand side $\widetilde{\beta}^{\lambda'}_{[e]}$ are rescaled versions of $L_{[e]}$ and $\beta^{\lambda'}_{[e]}$, respectively,

$$\mathcal{X}^{\lambda'}_{[e]}\widetilde{L}_{[e]} = L_{[e]}, \qquad \mathcal{X}^{\lambda'}_{[e]}\widetilde{\beta}^{\lambda'}_{[e]} = \beta^{\lambda'}_{[e]}, \tag{4.5.45}$$

where $\mathcal{X}^{\lambda'}_{[e]} = \mathrm{diag}\big((X_k)_{(r,k)\in\mathcal{B}^{\lambda'}_{[e]}}\big)$. The vector $\widetilde{\mathcal{U}}_{[e]}$ is, in turn, given by (4.4.71), and from (4.4.71)(4.5.44) we obtain

$$\widetilde{\beta}^{\lambda'}_{[e]} \in \widetilde{\mathcal{U}}^{\perp}_{[e]}. \tag{4.5.46}$$

Theorem 4.5.8. *Let $\widetilde{L}_{[e]} \in \mathbb{R}^{2n,2n}$ be a matrix satisfying the properties $(\widetilde{L}_{[e]}0)$–$(\widetilde{L}_{[e]}5)$ and let $\mathcal{L}_{[e]}$, $\beta^{D_k}_{[e]}$, $k \in \mathcal{S}$, $\beta^{\lambda'}_{[e]}$, and $\widetilde{\beta}^{\lambda'}_{[e]} \in \mathbb{R}^{2n}$ be given by (4.4.22), (4.4.25), (4.5.17), and (4.5.44), respectively. Then the constrained linear system (4.5.40) admits a unique solution $\alpha^{\lambda'}_{[e]}$, and this solution is such that $L^{++}_{[e]}\alpha^{\lambda'+}_{[e]} = \beta^{\lambda'+}_{[e]}$ and $\langle \mathcal{L}^{+}_{[e]}, \alpha^{\lambda'+}_{[e]}\rangle = 0$. Furthermore, the quantity $\lambda'_{[e]}$ is positive and is the partial thermal conductivity of the $\mathcal{S}^+$ mixture. The vector $\theta_{[e]}$ satisfies the relation*

$$\langle \theta_{[e]}, Y\rangle = \sum_{k\in\mathcal{S}} Y_k \theta_{[e]k} = 0, \tag{4.5.47}$$

and may also be written

$$\theta_{[e]} = -\alpha^{00\lambda'}_{[e]}. \tag{4.5.48}$$

Finally, $\theta_{[e]}$ admits the block-decomposition $\theta_{[e]} = \Upsilon(\theta^{+}_{[e]}, \theta^{-}_{[e]})$, where $\theta^{+}_{[e]}$ is the thermal diffusion vector of the $\mathcal{S}^+$ mixture.

Remark. It is interesting to point out that the components $\theta_{[e]k}$, $k \in \mathcal{S}^-$, are a priori nonzero. Their explicit expression is given in Section 4.7 in the practically important dilution limit.

In the next proposition, we express the partial thermal conductivity $\lambda'_{[e]}$ and the thermal diffusion vector $\theta_{[e]}$ in terms of a generalized inverse of the matrix $\widetilde{L}_{[e]}$ and also in terms of the solution of a nonsingular linear system.

Proposition 4.5.9. *Let $\widetilde{\mathcal{L}}_{[e]} \in \mathbb{R}^{2n}$ be given by the block-decomposition $\widetilde{\mathcal{L}}^{00}_{[e]} = W$ and $\widetilde{\mathcal{L}}^{e}_{[e]} = 0$, and let $\widetilde{L}^{\dagger}_{[e]}$ be the generalized inverse of $\widetilde{L}_{[e]}$ with prescribed range $\mathcal{L}^{\perp}_{[e]}$ and nullspace $\mathbb{R}\widetilde{\mathcal{L}}_{[e]}$. Then the solution of the constrained linear system (4.5.40) is given by $\alpha^{\lambda'}_{[e]} = \widetilde{L}^{\dagger}_{[e]}\widetilde{\beta}^{\lambda'}_{[e]}$. Furthermore, if a and b are two positive real numbers such that $ab\langle \mathcal{L}_{[e]}, \mathcal{U}_{[e]}\rangle\langle \widetilde{\mathcal{L}}_{[e]}, \widetilde{\mathcal{U}}_{[e]}\rangle = 1$, then the matrices $\widetilde{L}_{[e]} + a\,\widetilde{\mathcal{L}}_{[e]}\otimes\mathcal{L}_{[e]}$ and $\widetilde{L}^{\dagger}_{[e]} + b\,\mathcal{U}_{[e]}\otimes\widetilde{\mathcal{U}}_{[e]}$ are inverse of each other and coincide with $\widetilde{L}_{[e]}$ and $\widetilde{L}^{\dagger}_{[e]}$ on the hyperplanes $\mathcal{L}^{\perp}_{[e]}$ and $\widetilde{\mathcal{U}}^{\perp}_{[e]}$,* *respectively. Therefore, we have*

$$\begin{cases} \alpha^{\lambda'}_{[e]} = [\widetilde{L}_{[e]} + a\,\widetilde{\mathcal{L}}_{[e]}\otimes\mathcal{L}_{[e]}]^{-1}\widetilde{\beta}^{\lambda'}_{[e]}, \\ \lambda'_{[e]} = \frac{\bar{p}}{\overline{T}}\langle\, [\widetilde{L}_{[e]} + a\,\widetilde{\mathcal{L}}_{[e]}\otimes\mathcal{L}_{[e]}]^{-1}\widetilde{\beta}^{\lambda'}_{[e]}, \beta^{\lambda'}_{[e]}\,\rangle, \\ \theta_{[e]k} = -\langle\, [\widetilde{L}_{[e]} + a\,\widetilde{\mathcal{L}}_{[e]}\otimes\mathcal{L}_{[e]}]^{-1}\widetilde{\beta}^{\lambda'}_{[e]}, \beta^{D_k}_{[e]}\,\rangle, \qquad k\in\mathcal{S}. \end{cases} \tag{4.5.49}$$

We now prove that the partial thermal conductivity $\lambda'_{[e]}$ is a smooth function of the temperature and the mass fractions in $\mathcal{D}_{Y\geq 0, Y\neq 0}$ and is independent of the pressure, whereas the thermal diffusion vector $\theta_{[e]}$ is a smooth function of the pressure, the temperature, and the mass fractions in $\mathfrak{D}_{Y\geq 0, Y\neq 0}$ and is inversely proportional to the pressure.

Theorem 4.5.10. *Keeping the assumptions of Theorem 4.4.26, let $\alpha^{\lambda'}_{[e]}$, $\lambda'_{[e]}$, and $\theta_{[e]}$ be the solution, the partial thermal conductivity, and the thermal diffusion vector evaluated from $L_{[e]}$ for $Y > 0$ and let $\widetilde{\alpha}^{\lambda'}_{[e]}$, $\widetilde{\lambda}'_{[e]}$, and $\widetilde{\theta}_{[e]}$ be the solution, the partial thermal conductivity, and the thermal diffusion vector evaluated from $\widetilde{L}_{[e]}$ for $Y \geq 0$, $Y \neq 0$. Then $\alpha^{\lambda'}_{[e]} = \widetilde{\alpha}^{\lambda'}_{[e]}$, $\lambda'_{[e]} = \widetilde{\lambda}'_{[e]}$, and $\theta_{[e]} = \widetilde{\theta}_{[e]}$ for $Y > 0$. Furthermore, $\widetilde{\alpha}^{\lambda'}_{[e]}$ and $\widetilde{\theta}_{[e]}$ are smooth functions of $(\bar{p}, \overline{T}, Y)$ in $\mathfrak{D}_{Y\geq 0, Y\neq 0}$ and are inversely proportional to the pressure, whereas $\widetilde{\lambda}'_{[e]}$ is a smooth function of $(\overline{T}, Y)$ in $\mathcal{D}_{Y\geq 0, Y\neq 0}$ and is independent of the pressure. Moreover, for $Z \in \mathbb{R}^n$, $Z \geq 0$, $Z \neq 0$, we have the following limits*

$$\begin{cases} \lim\limits_{\substack{Y\to Z\\ Y>0}} \lambda'_{[e]}(\overline{T}, Y) = \widetilde{\lambda}'_{[e]}(\overline{T}, Z), \\ \lim\limits_{\substack{Y\to Z\\ Y>0}} \theta_{[e]}(\bar{p}, \overline{T}, Y) = \widetilde{\theta}_{[e]}(\bar{p}, \overline{T}, Z). \end{cases} \tag{4.5.50}$$

Remark. As a consequence of the previous theorem, the solutions of the linear systems (4.5.12)(4.5.40) coincide when $Y > 0$. This justifies the use of the same notation $\alpha^{\lambda'}_{[e]}$, $\lambda'_{[e]}$, and $\theta_{[e]}$ in Sections 4.5.2 and 4.5.5.

4.6 The Thermal Conductivity and the Thermal Diffusion Ratios

4.6.1 Mathematical Properties of the System $\Lambda\alpha^{\lambda} = \beta^{\lambda}$

We consider the linear system

$$\Lambda\alpha^{\lambda} = \beta^{\lambda}, \tag{4.6.1}$$

associated with the evaluation of the thermal conductivity

$$\lambda = \frac{\overline{p}}{\overline{T}} \langle \alpha^\lambda, \beta^\lambda \rangle, \tag{4.6.2}$$

and the thermal diffusion ratios $\chi = (\chi_k)_{k\in\mathcal{S}}$

$$\chi = [L^{0010}, L^{0001}]\alpha^\lambda, \tag{4.6.3}$$

as described in Section 2.8.1. We have $\Lambda \in \mathbb{R}^{n+p,n+p}$, α^λ and $\beta^\lambda \in \mathbb{R}^{n+p}$, and the corresponding indexing set is given by

$$\mathcal{B}^\lambda = \{10\}\times\mathcal{S} \cup \{01\}\times\mathcal{P}, \tag{4.6.4}$$

yielding the block-decomposition

$$\begin{bmatrix} \Lambda^{1010} & \Lambda^{1001} \\ \Lambda^{0110} & \Lambda^{0101} \end{bmatrix} \begin{bmatrix} \alpha^{10\lambda} \\ \alpha^{01\lambda} \end{bmatrix} = \begin{bmatrix} \beta^{10\lambda} \\ \beta^{01\lambda} \end{bmatrix}. \tag{4.6.5}$$

We assume that the blocks $L^{0010} \in \mathbb{R}^{n,n}$ and $L^{0001} \in \mathbb{R}^{n,p}$ forming the rectangular matrix $[L^{0010}, L^{0001}] \in \mathbb{R}^{n,n+p}$ are such that

$$R(L^{0010}) \subset U^\perp, \qquad R(L^{0001}) \subset U^\perp, \tag{4.6.6}$$

where $U = (1)_{k\in\mathcal{S}}$. Furthermore, from the kinetic theory results obtained in Sections 2.3 and 2.8, we can make the following assumptions.

(Λ0) The coefficients of the matrix Λ are functions of the pressure, the temperature, and the mass fractions, they are proportional to the pressure, and they can be expressed as quadratic functions of the mole fractions in the form

$$\begin{cases} \Lambda_{kk}^{rs} = \sum_{l\in\mathcal{S}} X_k X_l \Lambda_{kl}^{\prime rs} + X_k^2 \Lambda_{kk}^{\prime\prime rs}, & (r,k),(s,k) \in \mathcal{B}^\lambda, \\ \Lambda_{kl}^{rs} = X_k X_l \Lambda_{kl}^{\prime\prime rs}, & (r,k),(s,l) \in \mathcal{B}^\lambda, \quad k \neq l, \end{cases} \tag{4.6.7}$$

where the matrices Λ' and Λ'' solely depend on the pressure and the temperature.

(Λ1) Λ is symmetric.

(Λ2) Λ is positive definite.

Moreover, following Section 2.8.1, we assume that the right-hand side β^λ is given by

$$\begin{cases} \beta_k^{10\lambda} = \frac{c_{\mathrm{p}}^{\mathrm{tr}}}{k_{\mathrm{B}}} X_k, & k \in \mathcal{S}, \\ \beta_k^{01\lambda} = \frac{c_k^{\mathrm{int}}}{k_{\mathrm{B}}} X_k, & k \in \mathcal{P}, \end{cases} \tag{4.6.8}$$

where the quantities $c_{\mathrm{p}}^{\mathrm{tr}}$ and k_{B} are positive constants and c_k^{int}, $k \in \mathcal{P}$, smooth positive functions of the temperature.

Theorem 4.6.1. *Let $\Lambda \in \mathbb{R}^{n+p,n+p}$ be a matrix satisfying the properties (Λ0)–(Λ2), assume that the matrices L^{0010} and L^{0001} satisfy (4.6.6), and let $\beta^\lambda \in \mathbb{R}^{n+p}$ be given by (4.6.8). Then the linear system (4.6.1) admits a unique solution α^λ. Furthermore, the quantity λ is positive, and the vector χ satisfies the relation*

$$\langle \chi, U\rangle = \sum_{k\in S} \chi_k = 0. \tag{4.6.9}$$

Proof. We deduce from (Λ2) that the matrix Λ is nonsingular so that the linear system (4.6.1) admits a unique solution α^λ. Furthermore, we have $\lambda = (\bar{p}/\overline{T})\langle \alpha^\lambda, \Lambda\alpha^\lambda\rangle$ and $\alpha^\lambda \neq 0$ since $\beta^\lambda \neq 0$ so that by (Λ2), the quantity λ is positive. Finally, we have $\langle \chi, U\rangle = \langle L^{0010}\alpha^{10\lambda}, U\rangle + \langle L^{0001}\alpha^{01\lambda}, U\rangle = 0$ by (4.6.6). □

In the next proposition, we express the thermal conductivity and the thermal diffusion ratios in terms of the partial thermal conductivity and the thermal diffusion vector obtained in Section 4.5.1.

Proposition 4.6.2. *Keeping the assumptions of Theorems 4.4.1, 4.5.1, and 4.6.1, consider the resulting partial thermal conductivity λ', the thermal diffusion vector θ, the diffusion matrix D, and the matrix Δ given by (4.4.14). Assume also that the matrix Λ is a submatrix of the matrix L given by $\Lambda^{rs} = L^{rs}$, $r,s \in \{10, 01\}$, and that the rectangular matrix $[L^{0010}, L^{0001}]$ is formed by the corresponding blocks of the matrix L. Then the thermal conductivity may be written*

$$\lambda = \lambda' - \frac{\bar{p}}{\overline{T}}\langle \theta, \chi\rangle, \tag{4.6.10}$$

and the thermal diffusion ratios are given by

$$\chi = \Delta\theta = (D + a\, U\otimes U)^{-1}\theta, \tag{4.6.11}$$

for any positive real number a. Similarly, we have

$$\theta = D\chi = (\Delta + b\, Y\otimes Y)^{-1}\chi, \tag{4.6.12}$$

for any positive real number b.

Proof. The last $n+p$ rows in the linear system $L\alpha^{\lambda'} = \beta^{\lambda'}$ can be cast into the form

$$\Lambda^{-1}(L^{1000}\alpha^{00\lambda'}, L^{0100}\alpha^{00\lambda'}) + (\alpha^{10\lambda'}, \alpha^{01\lambda'}) = \Lambda^{-1}\beta^\lambda = \alpha^\lambda,$$

since $(\beta^{10\lambda'}, \beta^{01\lambda'}) = \beta^\lambda$. Multiplying by β^λ yields $\langle \alpha^\lambda, \beta^\lambda\rangle = \langle \alpha^{00\lambda'}, \chi\rangle + \langle \alpha^{\lambda'}, \beta^{\lambda'}\rangle$ since L is symmetric, $\chi = [L^{0010}, L^{0001}]\Lambda^{-1}\beta^\lambda$, and $\beta^{00\lambda'} = 0$. Therefore, we have

$\lambda = (\bar{p}/\bar{T})\langle\alpha^{\lambda},\beta^{\lambda}\rangle = \lambda' - (\bar{p}/\bar{T})\langle\theta,\chi\rangle$ since $\theta = -\alpha^{00\lambda'}$ from Theorem 4.5.1. Next, the first n rows in the linear system $L\alpha^{\lambda'} = \beta^{\lambda'}$ can be written using (4.4.14) in the form $\Delta\alpha^{00\lambda'} = -[L^{0010}, L^{0001}]\Lambda^{-1}\beta^{\lambda}$ and since $\theta = -\alpha^{00\lambda'}$, we obtain $\Delta\theta = \chi$. By multiplying this equation by D, we obtain $D\chi = D\Delta\theta = \theta$ since $D\Delta$ is the oblique projector onto $Y^{\perp}$ along $\mathbb{R}U$ and $\theta \in Y^{\perp}$. Finally, by Proposition 4.4.5, the matrices $D + a\,U\otimes U$ and $\Delta + b\,Y\otimes Y$, where $a > 0$, $b > 0$, and $ab\langle Y,U\rangle^2 = 1$, are inverse of each other. Since $\theta \in Y^{\perp}$, we then obtain $\chi = (\Delta + b\,Y\otimes Y)\theta = (D + a\,U\otimes U)^{-1}\theta$ and similarly $\theta = (D + b\,U\otimes U)\chi = (\Delta + b\,Y\otimes Y)^{-1}\chi$ since $\chi \in U^{\perp}$. □

Remark. Notice that $\lambda < \lambda'$ since $\langle\theta,\chi\rangle = \langle\theta,\Delta\theta\rangle > 0$ because $\theta \in Y^{\perp}$, $\theta \neq 0$, and Δ is positive definite on $Y^{\perp}$.

Finally, we consider the matrix Λ resulting from the practical approximations presented in Section 2.10.

Proposition 4.6.3. *Let $\mathcal{D}_{k\,\mathrm{int},l}$, c_k^{int}, ξ_k^{int}, and m_l, $k \in \mathcal{P}$, $l \in \mathcal{S}$, be positive quantities, let $\mathcal{D}_{kl}$, $\bar{A}_{kl}$, $\bar{B}_{kl}$, and $\bar{C}_{kl}$, $k,l \in \mathcal{S}$, be symmetric and positive coefficients such that*

$$\frac{25}{4} - 3\bar{B}_{kl} > 0, \qquad k,l \in \mathcal{S}, \tag{4.6.13}$$

and assume that $Y > 0$. Then the matrix Λ given by (2.10.20)–(2.10.25) satisfies (Λ0)–(Λ2).

Proof. The properties (Λ0)–(Λ1) directly result from (2.10.20)–(2.10.25), keeping in mind that the coefficients $\mathcal{D}_{kl}$, $k,l \in \mathcal{S}$, are proportional to the inverse of the pressure. Furthermore, an explicit calculation yields for $x \in \mathbb{R}^{n+p}$

$$\begin{aligned}
\langle x,\Lambda x\rangle = &\sum_{\substack{k,l\in\mathcal{S}\\ l\neq k}} \Big(\frac{25}{4} - 3\bar{B}_{kl}\Big)\Big(\frac{m_l x_k^{10} - m_k x_l^{10}}{m_k + m_l}\Big)^2\\
&+ \sum_{\substack{k,l\in\mathcal{S}\\ l\neq k}} \frac{X_k X_l}{\mathcal{D}_{kl}}\Big[\frac{15}{4}\Big(\frac{m_k x_k^{10} - m_l x_l^{10}}{m_k + m_l}\Big)^2 + 2\bar{A}_{kl}\frac{m_k m_l}{(m_k+m_l)^2}(x_k^{10} + x_l^{10})^2\Big]\\
&+ \sum_{k\in\mathcal{S}} 2\bar{A}_{kk}\frac{X_k^2}{\mathcal{D}_{kk}}\big(x_k^{10}\big)^2 + \sum_{k\in\mathcal{P}}\Big(\sum_{\substack{l\in\mathcal{S}\\ l\neq k}} X_k X_l \frac{c_k^{\mathrm{int}}}{k_{\mathrm{B}}\mathcal{D}_{k\,\mathrm{int},l}} + X_k^2 \frac{c_k^{\mathrm{int}}}{k_{\mathrm{B}}\mathcal{D}_{k\,\mathrm{int},k}}\Big)\big(x_k^{01}\big)^2\\
&+ \sum_{k\in\mathcal{P}} \frac{20}{3}\frac{\bar{A}_{kk}}{k_{\mathrm{B}}\pi}\frac{X_k^2}{\mathcal{D}_{kk}}\frac{c_k^{\mathrm{int}}}{\xi_k^{\mathrm{int}}}\Big(x_k^{10} - \frac{3}{5}x_k^{01}\Big)^2\\
&+ \sum_{\substack{k\in\mathcal{S},l\in\mathcal{P}\\ l\neq k}} \frac{20}{3}\frac{\bar{A}_{kl}}{k_{\mathrm{B}}\pi}\frac{X_k X_l}{\mathcal{D}_{kl}}\frac{c_l^{\mathrm{int}}}{\xi_l^{\mathrm{int}}}\frac{m_l}{m_k}\Big(\frac{m_k}{m_k+m_l}(x_k^{10} + x_l^{10}) - \frac{3}{5}x_l^{01}\Big)^2.
\end{aligned}$$

Since $Y > 0$ implies $X > 0$, we deduce that $\langle x, \Lambda x\rangle = 0$ if and only if $x = 0$ and, hence, the matrix Λ satisfies $(\Lambda 2)$. □

4.6.2 Mathematical Properties of the System $\Lambda_{[e]}\alpha^{\lambda}_{[e]} = \beta^{\lambda}_{[e]}$

We consider the linear system

$$\Lambda_{[e]}\alpha^{\lambda}_{[e]} = \beta^{\lambda}_{[e]}, \tag{4.6.14}$$

associated with the evaluation of the thermal conductivity

$$\lambda_{[e]} = \frac{\bar{p}}{\overline{T}}\langle \alpha^{\lambda}_{[e]}, \beta^{\lambda}_{[e]}\rangle, \tag{4.6.15}$$

and the thermal diffusion ratios $\chi_{[e]} = (\chi_{[e]k})_{k\in\mathcal{S}}$

$$\chi_{[e]} = [L^{00e}_{[e]}]\alpha^{\lambda}_{[e]}, \tag{4.6.16}$$

as described in Section 2.8.3. We have $\Lambda_{[e]} \in \mathbb{R}^{n,n}$, $\alpha^{\lambda}_{[e]}$ and $\beta^{\lambda}_{[e]} \in \mathbb{R}^n$, and the corresponding indexing set is given by

$$\mathcal{B}^{\lambda}_{[e]} = \{e\}\times\mathcal{S}. \tag{4.6.17}$$

We assume that the block $L^{00e}_{[e]}$ forming the matrix $[L^{00e}_{[e]}] \in \mathbb{R}^{n,n}$ is such that

$$R(L^{00e}_{[e]}) \subset U^{\perp}. \tag{4.6.18}$$

Furthermore, from the kinetic theory results obtained in Sections 2.3 and 2.8, we can make the following assumptions.

$(\Lambda_{[e]}0)$ The coefficients of the matrix $\Lambda_{[e]}$ are functions of the pressure, the temperature, and the mass fractions, they are proportional to the pressure, and they can be expressed as quadratic functions of the mole fractions in the form

$$\begin{cases} \Lambda^{rs}_{[e]kk} = \sum_{l\in\mathcal{S}} X_k X_l \Lambda'^{rs}_{[e]kl} + X_k^2 \Lambda''^{rs}_{[e]kk}, & (r,k),(s,k) \in \mathcal{B}^{\lambda}_{[e]}, \\ \Lambda^{rs}_{[e]kl} = X_k X_l \Lambda''^{rs}_{[e]kl}, & (r,k),(s,l) \in \mathcal{B}^{\lambda}_{[e]}, \quad k \neq l, \end{cases} \tag{4.6.19}$$

where the matrices $\Lambda'_{[e]}$ and $\Lambda''_{[e]}$ solely depend on the pressure and the temperature.

$(\Lambda_{[e]}1)$ $\Lambda_{[e]}$ is symmetric.

$(\Lambda_{[e]}2)$ $\Lambda_{[e]}$ is positive definite.

Moreover, following Section 2.8.3, we assume that the right-hand side $\beta_{[e]}^{\lambda}$ is given by

$$\beta_{[e]k}^{e\lambda} = \frac{c_p^{tr} + c_k^{int}}{k_B} X_k, \qquad k \in S. \tag{4.6.20}$$

Theorem 4.6.4. *Let $\Lambda_{[e]} \in \mathbb{R}^{n,n}$ be a matrix satisfying the properties $(\Lambda_{[e]}0)$–$(\Lambda_{[e]}2)$, assume that the matrix $[L_{[e]}^{00e}]$ satisfies (4.6.18), and let $\beta_{[e]}^{\lambda} \in \mathbb{R}^n$ be given by (4.6.20). Then the linear system (4.6.14) admits a unique solution $\alpha_{[e]}^{\lambda}$. Furthermore, the quantity $\lambda_{[e]}$ is positive, and the vector $\chi_{[e]}$ satisfies the relation*

$$\langle \chi_{[e]}, U \rangle = \sum_{k \in S} \chi_{[e]k} = 0. \tag{4.6.21}$$

In the next proposition, we express the thermal conductivity and the thermal diffusion ratios in terms of the partial thermal conductivity and the thermal diffusion vector obtained in Section 4.5.2.

Proposition 4.6.5. *Keeping the assumptions of Theorems 4.4.7, 4.5.3, and 4.6.4, consider the resulting partial thermal conductivity $\lambda'_{[e]}$, the thermal diffusion vector $\theta_{[e]}$, the diffusion matrix $D_{[e]}$, and the matrix $\Delta_{[e]}$ given by (4.4.30). Assume also that the matrix $\Lambda_{[e]}$ is given by $\Lambda_{[e]} = L_{[e]}^{ee}$ and that the matrix $[L_{[e]}^{00e}]$ is formed by the corresponding block of the matrix $L_{[e]}$. Then the thermal conductivity may be written*

$$\lambda_{[e]} = \lambda'_{[e]} - \frac{\bar{p}}{\overline{T}} \langle \theta_{[e]}, \chi_{[e]} \rangle, \tag{4.6.22}$$

and the thermal diffusion ratios are given by

$$\chi_{[e]} = \Delta_{[e]} \theta_{[e]} = (D_{[e]} + a\, U \otimes U)^{-1} \theta_{[e]}, \tag{4.6.23}$$

for any positive real number a. Similarly, we have

$$\theta_{[e]} = D_{[e]} \chi_{[e]} = (\Delta_{[e]} + b\, Y \otimes Y)^{-1} \chi_{[e]}, \tag{4.6.24}$$

for any positive real number b.

Finally, we consider the matrix $\Lambda_{[e]}$ resulting from the practical approximations presented in Section 2.10.

Proposition 4.6.6. *Let $\mathcal{D}_{k\,\mathrm{int},l}$, c_k^{int}, ξ_k^{int}, and m_l, $k \in \mathcal{P}$, $l \in S$, be positive quantities, let $\mathcal{D}_{kl}$, $\bar{A}_{kl}$, $\bar{B}_{kl}$, and $\bar{C}_{kl}$, $k, l \in S$, be symmetric and positive coefficients such that $25/4 - 3\bar{B}_{kl} > 0$, $k, l \in S$, and assume that $Y > 0$. Then the matrix $\Lambda_{[e]}$ given by (2.10.30)–(2.10.31) satisfies $(\Lambda_{[e]}0)$–$(\Lambda_{[e]}2)$.*

4.6.3 Mathematical Properties of the System $\widetilde{\Lambda}\alpha^\lambda = \widetilde{\beta}^\lambda$

We consider the linear system

$$\widetilde{\Lambda}\alpha^\lambda = \widetilde{\beta}^\lambda, \tag{4.6.25}$$

associated with the evaluation of the thermal conductivity

$$\lambda = \frac{\bar{p}}{\overline{T}}\langle \alpha^\lambda, \beta^\lambda \rangle, \tag{4.6.26}$$

and the thermal diffusion ratios $\chi = (\chi_k)_{k\in\mathcal{S}}$

$$\chi = [L^{0010}, L^{0001}]\alpha^\lambda, \tag{4.6.27}$$

as described in Section 3.6.1. We have $\widetilde{\Lambda} \in \mathbb{R}^{n+p,n+p}$, α^λ and $\widetilde{\beta}^\lambda \in \mathbb{R}^{n+p}$, and the indexing set is $\mathcal{B}^\lambda = \{10\}\times\mathcal{S} \cup \{01\}\times\mathcal{P}$, yielding the block-decomposition

$$\begin{bmatrix} \widetilde{\Lambda}^{1010} & \widetilde{\Lambda}^{1001} \\ \widetilde{\Lambda}^{0110} & \widetilde{\Lambda}^{0101} \end{bmatrix} \begin{bmatrix} \alpha^{10\lambda} \\ \alpha^{01\lambda} \end{bmatrix} = \begin{bmatrix} \widetilde{\beta}^{10\lambda} \\ \widetilde{\beta}^{01\lambda} \end{bmatrix}. \tag{4.6.28}$$

Furthermore, from the kinetic theory results obtained in Sections 3.1 and 3.6, we can make the following assumptions.

($\widetilde{\Lambda}$0) The coefficients of the matrix $\widetilde{\Lambda}$ are functions of the pressure, the temperature, and the mass fractions, they are proportional to the pressure, and they can be expressed as linear functions of the mole fractions in the form

$$\begin{cases} \widetilde{\Lambda}_{kk}^{rs} = \sum\limits_{l\in\mathcal{S}} X_l \Lambda_{kl}^{\prime rs} + X_k \Lambda_{kk}^{\prime\prime rs}, & (r,k),(s,k) \in \mathcal{B}^\lambda, \\ \widetilde{\Lambda}_{kl}^{rs} = X_l \Lambda_{kl}^{\prime\prime rs}, \qquad (r,k),(s,l) \in \mathcal{B}^\lambda, & k \neq l, \end{cases} \tag{4.6.29}$$

where the matrices Λ' and Λ'' coincide with the ones in (4.6.7).

($\widetilde{\Lambda}$1) The matrix $\widetilde{\Lambda}$ admits the block-decomposition

$$(\Gamma^\lambda)^t \widetilde{\Lambda} \Gamma^\lambda = \begin{bmatrix} \widetilde{\Lambda}^{++} & 0 \\ \widetilde{\Lambda}^{-+} & \widetilde{\Lambda}^{--} \end{bmatrix}, \tag{4.6.30}$$

where Γ^λ is the permutation matrix associated with the reordering of $\mathcal{B}^\lambda$ into $(\mathcal{B}^{\lambda+}, \mathcal{B}^{\lambda-})$.

($\widetilde{\Lambda}$2) The matrix $\Lambda^{++} = \mathcal{X}^{\lambda++}\widetilde{\Lambda}^{++}$ corresponds to the $\mathcal{S}^+$ mixture and, in particular, is symmetric positive definite.

($\widetilde{\Lambda}$3) $\widetilde{\Lambda}^{--}$ is symmetric positive definite and we have $db(\widetilde{\Lambda}^{--}) = \widetilde{\Lambda}^{--}$.

($\widetilde{\Lambda}$4) $\widetilde{\Lambda}$ is nonsingular.

Moreover, following Section 3.6.1, we assume that the right-hand side $\widetilde{\beta}^\lambda$ is given by

$$\begin{cases} \widetilde{\beta}_k^{10\lambda} = \dfrac{c_{\mathrm{P}}^{\mathrm{tr}}}{k_{\mathrm{B}}}, & k \in \mathcal{S}, \\ \widetilde{\beta}_k^{01\lambda} = \dfrac{c_k^{\mathrm{int}}}{k_{\mathrm{B}}}, & k \in \mathcal{P}. \end{cases} \tag{4.6.31}$$

We deduce from (4.6.7)(4.6.29) and (4.6.8)(4.6.31) that the matrix $\widetilde{\Lambda}$ and the right-hand side $\widetilde{\beta}^\lambda$ are rescaled versions of Λ and β^λ, respectively,

$$\mathcal{X}^\lambda \widetilde{\Lambda} = \Lambda, \qquad \mathcal{X}^\lambda \widetilde{\beta}^\lambda = \beta^\lambda, \tag{4.6.32}$$

where $\mathcal{X}^\lambda = \mathrm{diag}\big((X_k)_{(r,k)\in\mathcal{B}^\lambda}\big)$. Finally, we assume that

$$\Upsilon^t L^{0010} \Upsilon = \begin{bmatrix} L^{0010++} & 0 \\ 0 & 0 \end{bmatrix}, \qquad \Upsilon^t L^{0001} \Upsilon = \begin{bmatrix} L^{0001++} & 0 \\ 0 & 0 \end{bmatrix}, \tag{4.6.33}$$

where Υ is the permutation matrix associated with the reordering of $\mathcal{S}$ into $(\mathcal{S}^+, \mathcal{S}^-)$.

Theorem 4.6.7. *Let $\widetilde{\Lambda} \in \mathbb{R}^{n+p,n+p}$ be a matrix satisfying the properties $(\widetilde{\Lambda}0)$–$(\widetilde{\Lambda}4)$, assume that the matrices L^{0010} and L^{0001} satisfy (4.6.6)(4.6.33), and let β^λ and $\widetilde{\beta}^\lambda \in \mathbb{R}^{n+p}$ be given by (4.6.8) and (4.6.31), respectively. Then the linear system (4.6.25) admits a unique solution α^λ, and this solution is such that $\Lambda^{++}\alpha^{\lambda+} = \beta^{\lambda+}$. Furthermore, the quantity λ is positive and is the thermal conductivity of the $\mathcal{S}^+$ mixture. Finally, the vector χ satisfies the relation*

$$\langle \chi, U \rangle = \sum_{k\in\mathcal{S}} \chi_k = 0, \tag{4.6.34}$$

and admits the block-decomposition $\chi = \Upsilon(\chi^+, 0)$, where χ^+ are the thermal diffusion ratios of the $\mathcal{S}^+$ mixture.

Proof. Since the matrix $\widetilde{\Lambda}$ is nonsingular by $(\widetilde{\Lambda}4)$, the system (4.6.25) admits a unique solution α^λ. In addition, from $(\widetilde{\Lambda}1)$ and (4.6.32), we deduce that $\Lambda^{++}\alpha^{\lambda+} = \beta^{\lambda+}$. Furthermore, we have $\lambda = (\overline{p}/\overline{T})\langle \alpha^{\lambda+}, \Lambda^{++}\alpha^{\lambda+}\rangle$ since $\beta^{\lambda-} = 0$, and by $(\widetilde{\Lambda}2)$, λ is then positive and is the thermal conductivity of the $\mathcal{S}^+$ mixture. The relation (4.6.34) follows from (4.6.6), and we deduce from (4.6.33) that $\chi^- = 0$ and $\chi^+ = [L^{0010++}, L^{0001++}]\alpha^{\lambda+}$. Consequently, χ^+ are the thermal diffusion ratios of the $\mathcal{S}^+$ mixture. □

Remark. It is interesting to point out that the components χ_k, $k \in \mathcal{S}^-$, vanish. Nevertheless, the ratio χ_k/X_k admits a finite limit when X_k becomes arbitrarily small, as described in Section 4.7 in the case of the dilution limit.

In the next proposition, we express the thermal conductivity and the thermal diffusion ratios in terms of the partial thermal conductivity and the thermal diffusion vector obtained in Section 4.5.3.

Proposition 4.6.8. *Keeping the assumptions of Theorems 4.4.16, 4.5.5, and 4.6.7, consider the resulting partial thermal conductivity λ', the thermal diffusion vector θ, the flux diffusion matrix $\widetilde{D}$, the matrix $\widetilde{\Delta}$ given by (4.4.53), and let $\overline{\Delta} = (\widetilde{\Delta})^t \mathcal{W}^{-1}$. Assume also that the matrix $\widetilde{\Lambda}$ is a submatrix of the matrix $\widetilde{L}$ given by $\widetilde{\Lambda}^{rs} = \widetilde{L}^{rs}$, $r, s \in \{10, 01\}$, and that the rectangular matrix $[L^{0010}, L^{0001}]$ is formed by the corresponding blocks of the matrix L. Then the thermal conductivity may be written*

$$\lambda = \lambda' - \frac{\bar{p}}{\overline{T}} \langle \theta, \chi \rangle, \tag{4.6.35}$$

and the thermal diffusion ratios are given by

$$\chi = \mathcal{X} \widetilde{\Delta} \theta = \overline{\Delta}\,(\mathcal{Y}\theta) = (\widetilde{D} + a\, Y \otimes U)^{-1} \mathcal{Y}\theta, \tag{4.6.36}$$

for any positive real number a. Similarly, we have

$$\mathcal{Y}\theta = \widetilde{D}\chi = \mathcal{W}\big((\widetilde{\Delta})^t + b\, Y \otimes W\big)^{-1} \chi = (\overline{\Delta} + b\, Y \otimes U)^{-1} \chi, \tag{4.6.37}$$

for any positive real number b.

Proof. The last $n+p$ rows in the linear system $\widetilde{L}\alpha^{\lambda'} = \widetilde{\beta}^{\lambda'}$ can be cast into the form

$$\widetilde{\Lambda}^{-1}(\widetilde{L}^{1000}\alpha^{00\lambda'}, \widetilde{L}^{0100}\alpha^{00\lambda'}) + (\alpha^{10\lambda'}, \alpha^{01\lambda'}) = \widetilde{\Lambda}^{-1}\widetilde{\beta}^{\lambda} = \alpha^{\lambda},$$

since we have $(\widetilde{\beta}^{10\lambda'}, \widetilde{\beta}^{01\lambda'}) = \widetilde{\beta}^{\lambda}$. By multiplying both sides of this equation by β^{λ}, we obtain $\langle \alpha^{\lambda}, \beta^{\lambda} \rangle = \langle \alpha^{\lambda'}, \beta^{\lambda'} \rangle - \langle \theta, [(\widetilde{L}^{1000})^t, (\widetilde{L}^{0100})^t](\widetilde{\Lambda}^{-1})^t \beta^{\lambda} \rangle$, keeping in mind that $\beta^{00\lambda'} = 0$ and that $\theta = -\alpha^{00\lambda'}$. Considering $x = [(\widetilde{L}^{1000})^t, (\widetilde{L}^{0100})^t](\widetilde{\Lambda}^{-1})^t \beta^{\lambda}$, we obtain after some algebra that $x = \Upsilon(x^+, 0)$ where $x^+ = [L^{0010++}, L^{0001++}]\alpha^{\lambda+}$, and we deduce from the proof of Theorem 4.6.7 that $x = \chi$ so that $\lambda = \lambda' - (\bar{p}/\overline{T})\langle \theta, \chi \rangle$. Next, the first n rows in the linear system $\widetilde{L}\alpha^{\lambda'} = \widetilde{\beta}^{\lambda'}$ can be written using (4.4.53) in the form $\widetilde{\Delta}\alpha^{00\lambda'} = -[\widetilde{L}^{0010}, \widetilde{L}^{0001}]\widetilde{\Lambda}^{-1}\beta^{\lambda}$, and since $\theta = -\alpha^{00\lambda'}$ and $\mathcal{X}[\widetilde{L}^{0010}, \widetilde{L}^{0001}] = [L^{0010}, L^{0001}]$, we obtain $\chi = \mathcal{X}\widetilde{\Delta}\theta$ and $\chi = \overline{\Delta}\mathcal{Y}\theta$ because $\overline{\Delta}\mathcal{Y} = \mathcal{X}\widetilde{\Delta}$ as shown in Proposition 4.4.38. We then obtain $\widetilde{D}\chi = \widetilde{D}\overline{\Delta}\mathcal{Y}\theta = \mathcal{Y}\theta$ since $\widetilde{D}\overline{\Delta}$ is the oblique projector onto $U^{\perp}$ along $\mathbb{R}Y$ and $\mathcal{Y}\theta \in U^{\perp}$. Finally, the remaining expressions are obtained as in the proof of Proposition 4.6.2 using Propositions 4.4.19 and 4.4.38. □

We now prove that the thermal conductivity λ and the thermal diffusion ratios χ are independent of the pressure and are smooth functions of the temperature and the mass fractions in $\mathcal{D}_{Y \geq 0, Y \neq 0}$.

Theorem 4.6.9. *Let Λ and $\widetilde{\Lambda} \in \mathbb{R}^{n+p,n+p}$ be matrices such that Λ satisfies $(\Lambda 0)$–$(\Lambda 2)$ in $\mathfrak{D}_{Y>0}$ and $\widetilde{\Lambda}$ satisfies $(\widetilde{\Lambda}0)$–$(\widetilde{\Lambda}4)$ in $\mathfrak{D}_{Y\geq 0, Y\neq 0}$. Assume also that the matrices Λ' and Λ'' in (4.6.7)(4.6.29) are smooth functions of the temperature and that the matrices L^{0010} and L^{0001} are smooth functions of the pressure, the temperature, and the mass fractions in $\mathfrak{D}_{Y\geq 0, Y\neq 0}$ and are proportional to the pressure. Let α^{λ}, λ, and χ be the solution, the thermal conductivity, and the thermal diffusion ratios evaluated from the matrix Λ for $Y > 0$ and let $\widetilde{\alpha}^{\lambda}$, $\widetilde{\lambda}$, and $\widetilde{\chi}$ be the solution, the thermal conductivity, and the thermal diffusion ratios evaluated from the matrix $\widetilde{\Lambda}$ for $Y \geq 0$, $Y \neq 0$. Then $\alpha^{\lambda} = \widetilde{\alpha}^{\lambda}$, $\lambda = \widetilde{\lambda}$, and $\chi = \widetilde{\chi}$ for $Y > 0$. Furthermore, $\widetilde{\alpha}^{\lambda}$ is a smooth function of $(\overline{p}, \overline{T}, Y)$ in $\mathfrak{D}_{Y\geq 0, Y\neq 0}$ and is inversely proportional to the pressure, whereas $\widetilde{\lambda}$ and $\widetilde{\chi}$ are smooth functions of $(\overline{T}, Y)$ in $\mathcal{D}_{Y\geq 0, Y\neq 0}$ and are independent of the pressure. Moreover, for $Z \in \mathbb{R}^n$, $Z \geq 0$, $Z \neq 0$, we have the following limits*

$$\begin{cases} \lim\limits_{\substack{Y\to Z\\ Y>0}} \lambda(\overline{T}, Y) = \widetilde{\lambda}(\overline{T}, Z), \\ \lim\limits_{\substack{Y\to Z\\ Y>0}} \chi(\overline{T}, Y) = \widetilde{\chi}(\overline{T}, Z). \end{cases} \tag{4.6.38}$$

Proof. When all the mass fractions are positive, the matrix $\mathcal{X}^{\lambda}$ is invertible and we deduce from (4.6.32) that the linear systems (4.6.1)(4.6.25) admit the same solution and hence $\alpha^{\lambda} = \widetilde{\alpha}^{\lambda}$, $\lambda = \widetilde{\lambda}$, and $\chi = \widetilde{\chi}$ for $Y > 0$. Furthermore, we deduce from $(\widetilde{\Lambda}0)$ that the vector $\widetilde{\alpha}^{\lambda}$ is inversely proportional to the pressure so that $\widetilde{\lambda}$ is independent of the pressure. The same conclusion holds for $\widetilde{\chi}$ since the matrix $[L^{0010}, L^{0001}]$ is assumed to be proportional to the pressure. Moreover, $\widetilde{\alpha}^{\lambda}$, $\widetilde{\lambda}$, and $\widetilde{\chi}$ can be expressed, using Cramer's rule, as rational functions of the coefficients of the matrices $\widetilde{\Lambda}$, $[L^{0010}, L^{0001}]$ and the vector $\widetilde{\beta}^{\lambda}$. Therefore, $\widetilde{\alpha}^{\lambda}$ is a smooth function of $(\overline{p}, \overline{T}, Y)$ in $\mathfrak{D}_{Y\geq 0, Y\neq 0}$, whereas $\widetilde{\lambda}$ and $\widetilde{\chi}$ are smooth functions of $(\overline{T}, Y)$ in $\mathcal{D}_{Y\geq 0, Y\neq 0}$. The limits (4.6.38) are then easily obtained. □

Remark. As a consequence of the previous theorem, the solutions of the linear systems (4.6.1)(4.6.25) coincide when $Y > 0$. This justifies the use of the same notation α^{λ}, λ, and χ in Sections 4.6.1 and 4.6.3.

Finally, we consider the matrix $\widetilde{\Lambda}$ resulting from the practical approximations presented in Section 2.10.

Proposition 4.6.10. *Let $\mathcal{D}_{k\,\mathrm{int},l}$, c_k^{int}, ξ_k^{int}, and m_l, $k \in \mathcal{P}$, $l \in \mathcal{S}$, be positive quantities, let $\mathcal{D}_{kl}$, $\bar{A}_{kl}$, $\bar{B}_{kl}$, and $\bar{C}_{kl}$, $k, l \in \mathcal{S}$, be symmetric and positive coefficients such that $25/4 - 3\bar{B}_{kl} > 0$, $k, l \in \mathcal{S}$, and assume that $Y \geq 0$, $Y \neq 0$. Then the matrix $\widetilde{\Lambda}$ given by (3.7.18)–(3.7.26) satisfies $(\widetilde{\Lambda}0)$–$(\widetilde{\Lambda}4)$.*

Proof. The properties $(\widetilde{\Lambda}0)$–$(\widetilde{\Lambda}1)$ directly result from (3.7.18)–(3.7.26), and the proof of $(\widetilde{\Lambda}2)$ is exactly the same as the one of Proposition 4.6.3. Next, the matrix $\widetilde{\Lambda}^{--}$ is symmetric, block-diagonal, and an explicit calculation yields for $x = \Gamma^\lambda(0, x^-) \in \mathbb{R}^{n+p}$

$$\langle x^-, \widetilde{\Lambda}^{--} x^- \rangle = \sum_{\substack{k \in \mathcal{S}^- \\ l \in \mathcal{S}^+}} \frac{X_l}{\mathcal{D}_{kl}} \Big(\frac{25}{4} - 3\bar{B}_{kl}\Big) \Big(\frac{m_l x_k^{10}}{m_k + m_l}\Big)^2$$

$$+ \sum_{\substack{k \in \mathcal{S}^- \\ l \in \mathcal{S}^+}} \frac{X_l}{\mathcal{D}_{kl}} \frac{1}{(m_k + m_l)^2} \Big[\frac{15}{2} m_k^2 + 4\bar{A}_{kl} m_k m_l\Big] \big(x_k^{10}\big)^2$$

$$+ \sum_{\substack{k \in \mathcal{P}^- \\ l \in \mathcal{S}^+}} X_l \frac{c_k^{\mathrm{int}}}{k_B \mathcal{D}_{k\,\mathrm{int},l}} \big(x_k^{01}\big)^2 + \sum_{\substack{k \in \mathcal{S}^- \\ l \in \mathcal{P}^+}} \frac{20}{3} \frac{\bar{A}_{kl}}{k_B \pi} \frac{X_l}{\mathcal{D}_{kl}} \frac{c_l^{\mathrm{int}}}{\xi_l^{\mathrm{int}}} \frac{m_k m_l}{(m_k + m_l)^2} \big(x_k^{10}\big)^2$$

$$+ \sum_{\substack{k \in \mathcal{P}^- \\ l \in \mathcal{S}^+}} \frac{20}{3} \frac{\bar{A}_{kl}}{k_B \pi} \frac{X_l}{\mathcal{D}_{kl}} \frac{c_k^{\mathrm{int}}}{\xi_k^{\mathrm{int}}} \frac{m_k}{m_l} \Big(\frac{m_l}{m_k + m_l} x_k^{10} - \frac{3}{5} x_k^{01}\Big)^2.$$

Since $X_l > 0$, $l \in \mathcal{S}^+$, we deduce from the assumptions that all the terms in the previous expression are positive and hence $\widetilde{\Lambda}^{--}$ is positive definite. Finally, $(\widetilde{\Lambda}4)$ directly follows from $(\widetilde{\Lambda}1)$–$(\widetilde{\Lambda}3)$. □

4.6.4 Mathematical Properties of the System $\widehat{\Lambda}\widehat{\alpha}^\lambda = \widehat{\beta}^\lambda$

We consider the linear system

$$\widehat{\Lambda}\widehat{\alpha}^\lambda = \widehat{\beta}^\lambda, \tag{4.6.39}$$

associated with the evaluation of the thermal conductivity

$$\lambda = \frac{\bar{p}}{\bar{T}} \langle \widehat{\alpha}^\lambda, \widehat{\beta}^\lambda \rangle, \tag{4.6.40}$$

and the thermal diffusion ratios $\chi = (\chi_k)_{k \in \mathcal{S}}$

$$\chi = [\widehat{L}^{0010}, \widehat{L}^{0001}]\widehat{\alpha}^\lambda. \tag{4.6.41}$$

We have $\widehat{\Lambda} \in \mathbb{R}^{n+p,n+p}$, $\widehat{\alpha}^\lambda$ and $\widehat{\beta}^\lambda \in \mathbb{R}^{n+p}$, and following Section 3.1.8, we can make the following assumptions.

$(\widehat{\Lambda}0)$ The coefficients of the matrix $\widehat{\Lambda}$ are functions of the pressure, the temperature, and the mass fractions, they are proportional to the pressure, and they can be expressed as

$$\begin{cases} \widehat{\Lambda}_{kk}^{rs} = \sum_{l \in \mathcal{S}} X_l \Lambda_{kl}^{\prime rs} + X_k \Lambda_{kk}^{\prime\prime rs}, & (r,k),(s,k) \in \mathcal{B}^\lambda, \\ \widehat{\Lambda}_{kl}^{rs} = \sqrt{X_k X_l} \Lambda_{kl}^{\prime\prime rs}, \quad (r,k),(s,l) \in \mathcal{B}^\lambda, & k \neq l, \end{cases} \tag{4.6.42}$$

where the matrices Λ' and Λ'' coincide with the ones in (4.6.7).

($\widehat{\Lambda}1$) $\widehat{\Lambda}$ is symmetric.

($\widehat{\Lambda}2$) $\widehat{\Lambda}$ is positive definite.

In addition, we assume that the right-hand side $\widehat{\beta}^\lambda$ is given by

$$\begin{cases} \widehat{\beta}_k^{10\lambda} = \frac{c_{\mathrm{p}}^{\mathrm{tr}}}{k_{\mathrm{B}}} \sqrt{X_k}, & k \in \mathcal{S}, \\ \widehat{\beta}_k^{01\lambda} = \frac{c_k^{\mathrm{int}}}{k_{\mathrm{B}}} \sqrt{X_k}, & k \in \mathcal{P}. \end{cases} \tag{4.6.43}$$

The matrix $\widehat{\Lambda}$ and the right-hand side $\widehat{\beta}^\lambda$ are then rescaled versions of Λ and β^λ, respectively,

$$(\mathcal{X}^\lambda)^{1/2} \widehat{\Lambda} (\mathcal{X}^\lambda)^{1/2} = \Lambda, \qquad (\mathcal{X}^\lambda)^{1/2} \widehat{\beta}^\lambda = \beta^\lambda, \tag{4.6.44}$$

and we assume that

$$[\widehat{L}^{0010}, \widehat{L}^{0001}] (\mathcal{X}^\lambda)^{1/2} = [L^{0010}, L^{0001}]. \tag{4.6.45}$$

Using ($\widehat{\Lambda}0$)–($\widehat{\Lambda}2$), one can prove that the system (4.6.39) is well posed, i.e., admits a unique solution. Finally, when all the mass fractions are positive, we deduce from (4.6.44)(4.6.45) that the definitions of the thermal conductivity (4.6.2)(4.6.40) and of the thermal diffusion ratios (4.6.3)(4.6.41) coincide.

4.6.5 Mathematical Properties of the System $\widetilde{\Lambda}_{[e]} \alpha_{[e]}^\lambda = \widetilde{\beta}_{[e]}^\lambda$

We consider the linear system

$$\widetilde{\Lambda}_{[e]} \alpha_{[e]}^\lambda = \widetilde{\beta}_{[e]}^\lambda, \tag{4.6.46}$$

associated with the evaluation of the thermal conductivity

$$\lambda_{[e]} = \frac{\overline{p}}{\overline{T}} \langle \alpha_{[e]}^\lambda, \beta_{[e]}^\lambda \rangle, \tag{4.6.47}$$

and the thermal diffusion ratios $\chi_{[e]} = (\chi_{[e]k})_{k \in \mathcal{S}}$

$$\chi_{[e]} = [L_{[e]}^{00e}] \alpha_{[e]}^\lambda, \tag{4.6.48}$$

as described in Section 3.6.3. We have $\widetilde{\Lambda}_{[e]} \in \mathbb{R}^{n,n}$, $\alpha^{\lambda}_{[e]}$ and $\widetilde{\beta}^{\lambda}_{[e]} \in \mathbb{R}^{n}$, and the indexing set is $\mathcal{B}^{\lambda}_{[e]} = \{e\}\times\mathcal{S}$. Furthermore, from the kinetic theory results obtained in Sections 3.1 and 3.6, we can make the following assumptions.

($\widetilde{\Lambda}_{[e]}0$) The coefficients of the matrix $\widetilde{\Lambda}_{[e]}$ are functions of the pressure, the temperature, and the mass fractions, they are proportional to the pressure, and they can be expressed as linear functions of the mole fractions in the form

$$\begin{cases} \widetilde{\Lambda}^{rs}_{[e]kk} = \sum_{l\in\mathcal{S}} X_l \Lambda'^{rs}_{[e]kl} + X_k \Lambda''^{rs}_{[e]kk}, & (r,k),(s,k) \in \mathcal{B}^{\lambda}_{[e]}, \\ \widetilde{\Lambda}^{rs}_{[e]kl} = X_l \Lambda''^{rs}_{[e]kl}, \quad (r,k),(s,l) \in \mathcal{B}^{\lambda}_{[e]}, & k \neq l, \end{cases} \tag{4.6.49}$$

where the matrices $\Lambda'_{[e]}$ and $\Lambda''_{[e]}$ coincide with the ones in (4.6.19).

($\widetilde{\Lambda}_{[e]}1$) The matrix $\widetilde{\Lambda}_{[e]}$ admits the block-decomposition

$$(\Gamma^{\lambda}_{[e]})^t \widetilde{\Lambda}_{[e]} \Gamma^{\lambda}_{[e]} = \begin{bmatrix} \widetilde{\Lambda}^{++}_{[e]} & 0 \\ \widetilde{\Lambda}^{-+}_{[e]} & \widetilde{\Lambda}^{--}_{[e]} \end{bmatrix}, \tag{4.6.50}$$

where $\Gamma^{\lambda}_{[e]}$ is the permutation matrix associated with the reordering of $\mathcal{B}^{\lambda}_{[e]}$ into $(\mathcal{B}^{\lambda+}_{[e]}, \mathcal{B}^{\lambda-}_{[e]})$.

($\widetilde{\Lambda}_{[e]}2$) The matrix $\Lambda^{++}_{[e]} = \mathcal{X}^{\lambda++}_{[e]} \widetilde{\Lambda}^{++}_{[e]}$ corresponds to the $\mathcal{S}^+$ mixture and, in particular, is symmetric positive definite.

($\widetilde{\Lambda}_{[e]}3$) $\widetilde{\Lambda}^{--}_{[e]}$ is symmetric positive definite and we have $db(\widetilde{\Lambda}^{--}_{[e]}) = \widetilde{\Lambda}^{--}_{[e]}$.

($\widetilde{\Lambda}_{[e]}4$) $\widetilde{\Lambda}_{[e]}$ is nonsingular.

Moreover, following Section 3.6.3, we assume that the right-hand side $\widetilde{\beta}^{\lambda}_{[e]}$ is given by

$$\widetilde{\beta}^{e\lambda}_{[e]k} = \frac{c^{\mathrm{tr}}_{\mathrm{p}} + c^{\mathrm{int}}_k}{k_{\mathrm{B}}}, \qquad k \in \mathcal{S}, \tag{4.6.51}$$

and we deduce from (4.6.19)(4.6.49) and (4.6.20)(4.6.51) that the matrix $\widetilde{\Lambda}_{[e]}$ and the right-hand side $\widetilde{\beta}^{\lambda}_{[e]}$ are rescaled versions of $\Lambda_{[e]}$ and $\beta^{\lambda}_{[e]}$, respectively,

$$\mathcal{X}^{\lambda}_{[e]} \widetilde{\Lambda}_{[e]} = \Lambda_{[e]}, \qquad \mathcal{X}^{\lambda}_{[e]} \widetilde{\beta}^{\lambda}_{[e]} = \beta^{\lambda}_{[e]}, \tag{4.6.52}$$

where $\mathcal{X}^{\lambda}_{[e]} = \mathrm{diag}\big((X_k)_{(r,k)\in\mathcal{B}^{\lambda}_{[e]}}\big)$. Finally, we assume that

$$\Upsilon^t L^{00e}_{[e]} \Upsilon = \begin{bmatrix} L^{00e++}_{[e]} & 0 \\ 0 & 0 \end{bmatrix}, \tag{4.6.53}$$

where Υ is the permutation matrix associated with the reordering of $\mathcal{S}$ into $(\mathcal{S}^+, \mathcal{S}^-)$.

Theorem 4.6.11. *Let $\widetilde{\Lambda}_{[e]} \in \mathbb{R}^{n,n}$ be a matrix satisfying the properties $(\widetilde{\Lambda}_{[e]}0)$–$(\widetilde{\Lambda}_{[e]}4)$, assume that the matrix $[L^{00e}_{[e]}]$ satisfies (4.6.18)(4.6.53), and let $\beta^{\lambda}_{[e]}$ and $\widetilde{\beta}^{\lambda}_{[e]} \in \mathbb{R}^n$ be given by (4.6.20) and (4.6.51), respectively. Then the linear system (4.6.46) admits a unique solution $\alpha^{\lambda}_{[e]}$, and this solution is such that $\Lambda^{++}_{[e]}\alpha^{\lambda+}_{[e]} = \beta^{\lambda+}_{[e]}$. Furthermore, the quantity $\lambda_{[e]}$ is positive and is the thermal conductivity of the S^+ mixture. Finally, the vector $\chi_{[e]}$ satisfies the relation*

$$\langle \chi_{[e]}, U\rangle = \sum_{k\in S} \chi_{[e]k} = 0, \tag{4.6.54}$$

and admits the block-decomposition $\chi_{[e]} = \Upsilon(\chi^{+}_{[e]}, 0)$, where $\chi^{+}_{[e]}$ are the thermal diffusion ratios of the S^+ mixture.

Remark. It is interesting to point out that the components $\chi_{[e]k}$, $k \in S^-$, vanish. Nevertheless, the ratio $\chi_{[e]k}/X_k$ admits a finite limit when X_k becomes arbitrarily small, as described in Section 4.7 in the case of the dilution limit.

In the next proposition, we express the thermal conductivity and the thermal diffusion ratios in terms of the partial thermal conductivity and the thermal diffusion vector obtained in Section 4.5.5.

Proposition 4.6.12. *Keeping the assumptions of Theorems 4.4.22, 4.5.8, and 4.6.11, consider the resulting partial thermal conductivity $\lambda'_{[e]}$, the thermal diffusion vector $\theta_{[e]}$, the flux diffusion matrix $\widetilde{D}_{[e]}$, the matrix $\widetilde{\Delta}_{[e]}$ given by (4.4.77) and let $\overline{\Delta}_{[e]} = (\widetilde{\Delta}_{[e]})^t \mathcal{W}^{-1}$. Assume also that the matrix $\widetilde{\Lambda}_{[e]}$ is given by $\widetilde{\Lambda}_{[e]} = \widetilde{L}^{ee}_{[e]}$ and that the matrix $[L^{00e}_{[e]}]$ is formed by the corresponding block of the matrix $L_{[e]}$. Then the thermal conductivity may be written*

$$\lambda_{[e]} = \lambda'_{[e]} - \frac{\overline{p}}{\overline{T}}\langle \theta_{[e]}, \chi_{[e]}\rangle, \tag{4.6.55}$$

and the thermal diffusion ratios are given by

$$\chi_{[e]} = \mathcal{X}\widetilde{\Delta}_{[e]}\theta_{[e]} = \overline{\Delta}_{[e]}(\mathcal{Y}\theta_{[e]}) = (\widetilde{D}_{[e]} + a\,Y{\otimes}U)^{-1}\mathcal{Y}\theta_{[e]}, \tag{4.6.56}$$

for any positive real number a. Similarly, we have

$$\mathcal{Y}\theta_{[e]} = \widetilde{D}_{[e]}\chi_{[e]} = \mathcal{W}\big((\widetilde{\Delta}_{[e]})^t + b\,Y{\otimes}W\big)^{-1}\chi_{[e]} = (\overline{\Delta}_{[e]} + b\,Y{\otimes}U)^{-1}\chi_{[e]}, \tag{4.6.57}$$

for any positive real number b.

We now prove that the thermal conductivity $\lambda_{[e]}$ and the thermal diffusion ratios $\chi_{[e]}$ are independent of the pressure and are smooth functions of the temperature and the mass fractions in $\mathcal{D}_{Y\geq 0, Y\neq 0}$.

Theorem 4.6.13. *Let $\Lambda_{[e]}$ and $\widetilde{\Lambda}_{[e]} \in \mathbb{R}^{n,n}$ be such that $\Lambda_{[e]}$ satisfies $(\Lambda_{[e]}0)$–$(\Lambda_{[e]}2)$ in $\mathfrak{D}_{Y>0}$ and $\widetilde{\Lambda}_{[e]}$ satisfies $(\widetilde{\Lambda}_{[e]}0)$–$(\widetilde{\Lambda}_{[e]}4)$ in $\mathfrak{D}_{Y\geq 0, Y\neq 0}$. Assume also that the matrices $\Lambda'_{[e]}$ and $\Lambda''_{[e]}$ in (4.6.19)(4.6.49) are smooth functions of the temperature and that the matrix $L^{00e}_{[e]}$ is a smooth function of the pressure, the temperature, and the mass fractions in $\mathfrak{D}_{Y\geq 0, Y\neq 0}$ and is proportional to the pressure. Let $\alpha^{\lambda}_{[e]}$, $\lambda_{[e]}$, and $\chi_{[e]}$ be the solution, the thermal conductivity, and the thermal diffusion ratios evaluated from the matrix $\Lambda_{[e]}$ for $Y > 0$ and let $\widetilde{\alpha}^{\lambda}_{[e]}$, $\widetilde{\lambda}_{[e]}$, and $\widetilde{\chi}_{[e]}$ be the solution, the thermal conductivity, and the thermal diffusion ratios evaluated from the matrix $\widetilde{\Lambda}_{[e]}$ for $Y \geq 0$, $Y \neq 0$. Then $\alpha^{\lambda}_{[e]} = \widetilde{\alpha}^{\lambda}_{[e]}$, $\lambda_{[e]} = \widetilde{\lambda}_{[e]}$, and $\chi_{[e]} = \widetilde{\chi}_{[e]}$ for $Y > 0$. Furthermore, $\widetilde{\alpha}^{\lambda}_{[e]}$ is a smooth function of $(\overline{p}, \overline{T}, Y)$ in $\mathfrak{D}_{Y\geq 0, Y\neq 0}$ and is inversely proportional to the pressure, whereas $\widetilde{\lambda}_{[e]}$ and $\widetilde{\chi}_{[e]}$ are smooth functions of $(\overline{T}, Y)$ in $\mathcal{D}_{Y\geq 0, Y\neq 0}$ and are independent of the pressure. Moreover, for $Z \in \mathbb{R}^n$, $Z \geq 0$, $Z \neq 0$, we have the following limits*

$$\begin{cases} \lim\limits_{\substack{Y \to Z \\ Y>0}} \lambda_{[e]}(\overline{T}, Y) = \widetilde{\lambda}_{[e]}(\overline{T}, Z), \\ \lim\limits_{\substack{Y \to Z \\ Y>0}} \chi_{[e]}(\overline{T}, Y) = \widetilde{\chi}_{[e]}(\overline{T}, Z). \end{cases} \tag{4.6.58}$$

Remark. As a consequence of the previous theorem, the solutions of the linear systems (4.6.14)(4.6.46) coincide when $Y > 0$. This justifies the use of the same notation $\alpha^{\lambda}_{[e]}$, $\lambda_{[e]}$, and $\chi_{[e]}$ in Sections 4.6.2 and 4.6.5.

Finally, we consider the matrix $\widetilde{\Lambda}_{[e]}$ resulting from the practical approximations presented in Section 2.10.

Proposition 4.6.14. *Let $\mathcal{D}_{k\,\mathrm{int},l}$, c^{int}_k, ξ^{int}_k, and m_l, $k \in \mathcal{P}$, $l \in \mathcal{S}$, be positive quantities, let $\mathcal{D}_{kl}$, $\bar{A}_{kl}$, $\bar{B}_{kl}$, and $\bar{C}_{kl}$, $k, l \in \mathcal{S}$, be symmetric and positive coefficients such that $25/4 - 3\bar{B}_{kl} > 0$, $k, l \in \mathcal{S}$, and assume that $Y \geq 0$, $Y \neq 0$. Then the matrix $\widetilde{\Lambda}_{[e]}$ given by (3.7.33)–(3.7.34) satisfies $(\widetilde{\Lambda}_{[e]}0)$–$(\widetilde{\Lambda}_{[e]}4)$.*

4.7 The Dilution Limit

The dilution—or pure species—limit arises in several models of practical importance as, for instance, chemical vapor deposition models [Er94]. In this limit, the mixture is composed of an excess species while all the other species are in trace amounts. Because of the considerable simplifications that occur, we now present the explicit expressions of the transport coefficients of such mixtures. We will make use of the results obtained in the previous sections, and, in particular, the smoothness of all the transport coefficients for vanishing mass fractions.

We first introduce some notation. Without loss of generality, we assume that the pure species state of the mixture is given by the pressure $\overline{p}$, the temperature $\overline{T}$, and the mass fraction vector

$$Y_0 = (1, 0, \ldots, 0), \tag{4.7.1}$$

so that $\mathcal{S}^+ = \{1\}$ and $\mathcal{S}^- = [2, n]$. The corresponding mole fraction vector is then given by $X_0 = (1, 0, \ldots, 0)$. Furthermore, for an arbitrary $\epsilon > 0$, we introduce the domain

$$\mathcal{D}_\epsilon = \{\, Y \in \mathbb{R}^n;\ Y_1 > 0;\ 0 \le Y_k \le \epsilon,\ k \in [2, n] \,\}. \tag{4.7.2}$$

In Theorems 4.7.1–4.7.4 we first consider the simpler case of the shear viscosity, the volume viscosity, the partial thermal conductivity, and the thermal conductivity. Only the proof of Theorem 4.7.1 is given since the others are similar.

Theorem 4.7.1. *(Shear Viscosity) Keeping the assumptions of Theorem 4.2.4, we have*

$$\lim_{\substack{Y \in \mathcal{D}_\epsilon \\ \epsilon \to 0}} \eta(\overline{T}, Y) = \eta(\overline{T}, Y_0) = \eta_1(\overline{T}). \tag{4.7.3}$$

Proof. It directly follows from Theorems 4.2.3 and 4.2.4 since the shear viscosity of the $\mathcal{S}^+$ mixture is $\eta(\overline{T}, Y_0) = \eta_1(\overline{T})$. □

Theorem 4.7.2. *(Volume Viscosity) Keeping the assumptions of Theorem 4.3.8, we have*

$$\lim_{\substack{Y \in \mathcal{D}_\epsilon \\ \epsilon \to 0}} \kappa(\overline{T}, Y) = \kappa(\overline{T}, Y_0) = \kappa_1(\overline{T}). \tag{4.7.4}$$

Similarly, keeping the assumptions of Theorem 4.3.11, we have

$$\lim_{\substack{Y \in \mathcal{D}_\epsilon \\ \epsilon \to 0}} \kappa_{[01]}(\overline{T}, Y) = \kappa_{[01]}(\overline{T}, Y_0) = \kappa_{[01]1}(\overline{T}) = \kappa_1(\overline{T}). \tag{4.7.5}$$

Theorem 4.7.3. *(Partial Thermal Conductivity) Keeping the assumptions of Theorem 4.5.7, we have*

$$\lim_{\substack{Y \in \mathcal{D}_\epsilon \\ \epsilon \to 0}} \lambda'(\overline{T}, Y) = \lambda'(\overline{T}, Y_0) = \lambda'_1(\overline{T}). \tag{4.7.6}$$

Similarly, keeping the assumptions of Theorem 4.5.10, we have

$$\lim_{\substack{Y \in \mathcal{D}_\epsilon \\ \epsilon \to 0}} \lambda'_{[e]}(\overline{T}, Y) = \lambda'_{[e]}(\overline{T}, Y_0) = \lambda'_{[e]1}(\overline{T}). \tag{4.7.7}$$

Theorem 4.7.4. *(Thermal Conductivity) Keeping the assumptions of Theorem 4.6.9, we have*

$$\lim_{\substack{Y\in\mathcal{D}_\epsilon\\ \epsilon\to 0}} \lambda(\overline{T},Y) = \lambda(\overline{T},Y_0) = \lambda_1(\overline{T}). \tag{4.7.8}$$

Similarly, keeping the assumptions of Theorem 4.6.13, we have

$$\lim_{\substack{Y\in\mathcal{D}_\epsilon\\ \epsilon\to 0}} \lambda_{[e]}(\overline{T},Y) = \lambda_{[e]}(\overline{T},Y_0) = \lambda_{[e]1}(\overline{T}). \tag{4.7.9}$$

Remark. For mixtures in a pure species state, one can easily verify that $\lambda'(\overline{T},Y_0) = \lambda(\overline{T},Y_0)$ and $\lambda'_{[e]}(\overline{T},Y_0) = \lambda_{[e]}(\overline{T},Y_0)$.

We now consider the flux diffusion matrices $\widetilde{D}$, $\widetilde{D}_{[e]}$, and $\widetilde{D}_{[00]}$ in the dilution limit.

Theorem 4.7.5. *(Flux Diffusion Matrix) Keeping the assumptions of Theorem 4.4.20, we have*

$$\lim_{\substack{Y\in\mathcal{D}_\epsilon\\ \epsilon\to 0}} \widetilde{D}(\bar{p},\overline{T},Y) = \widetilde{D}(\bar{p},\overline{T},Y_0), \tag{4.7.10}$$

and $\widetilde{D}(\bar{p},\overline{T},Y_0)$ admits the block-decomposition

$$\widetilde{D}(\bar{p},\overline{T},Y_0) = \begin{bmatrix} 0 & -\widetilde{D}_{22}(\bar{p},\overline{T}) & \dots & \dots & -\widetilde{D}_{nn}(\bar{p},\overline{T}) \\ 0 & \widetilde{D}_{22}(\bar{p},\overline{T}) & 0 & \dots & 0 \\ \vdots & 0 & \ddots & & \vdots \\ \vdots & \vdots & & \ddots & 0 \\ 0 & 0 & \dots & 0 & \widetilde{D}_{nn}(\bar{p},\overline{T}) \end{bmatrix}, \tag{4.7.11}$$

where

$$\widetilde{D}_{kk}(\bar{p},\overline{T}) = \frac{m_k}{m_1}\frac{1}{\widetilde{\Delta}_{kk}}, \qquad k\in\mathcal{S}^-, \tag{4.7.12}$$

and $\widetilde{\Delta}^{--}$ is given by

$$\widetilde{\Delta}^{--} = \widetilde{L}^{0000--} - [\widetilde{L}^{0010--}, \widetilde{L}^{0001--}](\widetilde{\Lambda}^{--})^{-1}\begin{bmatrix}\widetilde{L}^{1000--}\\ \widetilde{L}^{0100--}\end{bmatrix}. \tag{4.7.13}$$

Similarly, keeping the assumptions of Theorem 4.4.26, we have

$$\lim_{\substack{Y\in\mathcal{D}_\epsilon\\ \epsilon\to 0}} \widetilde{D}_{[e]}(\bar{p},\overline{T},Y) = \widetilde{D}_{[e]}(\bar{p},\overline{T},Y_0), \tag{4.7.14}$$

and $\widetilde{D}_{[e]}(\bar{p},\overline{T},Y_0)$ admits a block-decomposition similar to (4.7.11), where

$$\widetilde{D}_{[e]kk}(\bar{p},\overline{T}) = \frac{m_k}{m_1}\frac{1}{\widetilde{\Delta}_{[e]kk}}, \qquad k\in\mathcal{S}^-, \tag{4.7.15}$$

and $\widetilde{\Delta}_{[e]kk}$, $k \in \mathcal{S}^-$, is given by

$$\widetilde{\Delta}_{[e]kk} = \widetilde{L}^{0000}_{[e]kk} - \frac{\left(\widetilde{L}^{00e}_{[e]kk}\right)^2}{\widetilde{L}^{ee}_{[e]kk}}, \qquad k \in \mathcal{S}^-. \tag{4.7.16}$$

Finally, keeping the assumptions of Theorem 4.4.30, we have

$$\lim_{\substack{Y \in \mathcal{D}_\epsilon \\ \epsilon \to 0}} \widetilde{D}_{[00]}(\bar{p}, \overline{T}, Y) = \widetilde{D}_{[00]}(\bar{p}, \overline{T}, Y_0), \tag{4.7.17}$$

and $\widetilde{D}_{[00]}(\bar{p}, \overline{T}, Y_0)$ admits a block-decomposition similar to (4.7.11), where

$$\widetilde{D}_{[00]kk}(\bar{p}, \overline{T}) = \frac{m_k}{m_1} \mathcal{D}_{1k}, \tag{4.7.18}$$

and $\mathcal{D}_{1k}$ is the binary diffusion coefficient of the species pair $(1, k)$.

Proof. We only consider the flux diffusion matrix $\widetilde{D}$ since the case of $\widetilde{D}_{[e]}$ and $\widetilde{D}_{[00]}$ is treated similarly. We first deduce from Theorem 4.4.20 that $\lim_{\substack{Y \in \mathcal{D}_\epsilon \\ \epsilon \to 0}} \widetilde{D}(\bar{p}, \overline{T}, Y) = \widetilde{D}(\bar{p}, \overline{T}, Y_0)$, and from Theorem 4.4.16 that the block-decomposition (4.7.11) holds. Consider now the matrix $\widetilde{\Delta}$ given by (4.4.53). We then obtain from ($\widetilde{L}1$) and ($\widetilde{L}3$) that $\widetilde{\Delta}^{+-} = 0$ and that $\widetilde{\Delta}^{--}$ is diagonal and given by (4.7.13). Furthermore, since $N(\widetilde{\Delta}) = \mathbb{R}U$ and $R(\widetilde{\Delta}) = X_0^\perp$ by Proposition 4.4.19, we have

$$\widetilde{\Delta} = \begin{bmatrix} 0 & 0 & \dots & \dots & 0 \\ -\widetilde{\Delta}_{22} & \widetilde{\Delta}_{22} & 0 & \dots & 0 \\ \vdots & 0 & \ddots & & \vdots \\ \vdots & \vdots & & \ddots & 0 \\ -\widetilde{\Delta}_{nn} & 0 & \dots & 0 & \widetilde{\Delta}_{nn} \end{bmatrix},$$

and finally, from $\widetilde{\Delta}(\widetilde{D})^t = \mathcal{W} - \big(1/\langle \mathcal{L}, \mathcal{U} \rangle\big) W \otimes Y$, we easily obtain (4.7.12). □

Finally, we consider the thermal diffusion vector and the thermal diffusion ratios. We first note that the thermal diffusion ratios of a mixture in a pure species state vanish, that is,

$$\chi(\overline{T}, Y_0) = \chi_{[e]}(\overline{T}, Y_0) = 0, \tag{4.7.19}$$

since $\chi^- = \chi^-_{[e]} = 0$ and $\sum_{k \in \mathcal{S}} \chi_k = \sum_{k \in \mathcal{S}} \chi_{[e]k} = 0$, as shown in Theorems 4.6.7 and 4.6.11. In this section we will therefore consider the rescaled thermal diffusion ratios

$$\begin{cases} \widetilde{\chi} = [\widetilde{L}^{0010}, \widetilde{L}^{0001}] \alpha^\lambda, \\ \widetilde{\chi}_{[e]} = [\widetilde{L}^{00e}_{[e]}] \alpha^\lambda_{[e]}, \end{cases} \tag{4.7.20}$$

which are such that

$$\begin{cases} X_k \widetilde{\chi}_k = \chi_k, & k \in \mathcal{S}, \\ X_k \widetilde{\chi}_{[e]k} = \chi_{[e]k}, & k \in \mathcal{S}, \end{cases} \tag{4.7.21}$$

and from $\langle \chi, U \rangle = \langle \chi_{[e]}, U \rangle = 0$, we also deduce that $\langle \widetilde{\chi}, X \rangle = \langle \widetilde{\chi}_{[e]}, X \rangle = 0$. In addition, using the notation of Section 2.3.1, we introduce the vectors $\widetilde{\beta}_k^\lambda = (\widetilde{\beta}_k^{r\lambda})_{r\in\mathcal{F}_k}$ and $\alpha_k^\lambda = (\alpha_k^{r\lambda})_{r\in\mathcal{F}_k}$, $k \in \mathcal{S}$, and the matrices $\widetilde{\Lambda}_{kl} = (\widetilde{\Lambda}_{kl}^{rs})_{r\in\mathcal{F}_k, s\in\mathcal{F}_l}$, $k,l \in \mathcal{S}$, where $\mathcal{F}_k = \{10, 01\}$ if $k \in \mathcal{P}$ and $\mathcal{F}_k = \{10\}$ otherwise.

Theorem 4.7.6. *(Thermal Diffusion) Keeping the assumptions of Proposition 4.6.8 and Theorems 4.4.20, 4.5.7, and 4.6.9, we have*

$$\begin{cases} \displaystyle\lim_{\substack{Y\in\mathcal{D}_\epsilon \\ \epsilon\to 0}} \theta(\overline{p}, \overline{T}, Y) = \theta(\overline{p}, \overline{T}, Y_0) = \begin{bmatrix} 0 \\ \theta^-(\overline{p}, \overline{T}) \end{bmatrix}, \\ \displaystyle\lim_{\substack{Y\in\mathcal{D}_\epsilon \\ \epsilon\to 0}} \widetilde{\chi}(\overline{T}, Y) = \widetilde{\chi}(\overline{T}, Y_0) = \begin{bmatrix} 0 \\ \widetilde{\chi}^-(\overline{T}) \end{bmatrix}, \end{cases} \tag{4.7.22}$$

where

$$\theta_k(\overline{p}, \overline{T}) = \frac{m_1}{m_k} \widetilde{D}_{kk}(\overline{p}, \overline{T}) \widetilde{\chi}_k(\overline{T}), \qquad k \in \mathcal{S}. \tag{4.7.23}$$

Furthermore, we have

$$\widetilde{\chi}_k = \sum_{s\in\mathcal{F}_1} \widetilde{L}_{k1}^{00s} \alpha_1^{s\lambda} + \sum_{s\in\mathcal{F}_k} \widetilde{L}_{kk}^{00s} \alpha_k^{s\lambda}, \qquad k \in \mathcal{S}^-, \tag{4.7.24}$$

where

$$\begin{cases} \alpha_1^\lambda = (\widetilde{\Lambda}_{11})^{-1} \widetilde{\beta}_1^\lambda, \\ \alpha_k^\lambda = (\widetilde{\Lambda}_{kk})^{-1} \big(\widetilde{\beta}_k^\lambda - \widetilde{\Lambda}_{k1} (\widetilde{\Lambda}_{11})^{-1} \widetilde{\beta}_1^\lambda \big), & k \in \mathcal{S}^-, \end{cases} \tag{4.7.25}$$

and we have

$$(\widetilde{\Lambda}_{kk})^{-1} = \frac{1}{\widetilde{\Lambda}_{kk}^{1010} \widetilde{\Lambda}_{kk}^{0101} - (\widetilde{\Lambda}_{kk}^{1001})^2} \begin{bmatrix} \widetilde{\Lambda}_{kk}^{0101} & -\widetilde{\Lambda}_{kk}^{1001} \\ -\widetilde{\Lambda}_{kk}^{0110} & \widetilde{\Lambda}_{kk}^{1010} \end{bmatrix}, \qquad k \in \mathcal{P}, \tag{4.7.26}$$

and

$$(\widetilde{\Lambda}_{kk})^{-1} = \frac{1}{\widetilde{\Lambda}_{kk}^{1010}}, \qquad k \in \mathcal{S} \setminus \mathcal{P}. \tag{4.7.27}$$

Similarly, keeping the assumptions of Proposition 4.6.12 and Theorems 4.4.26, 4.5.10, and 4.6.13, we have

$$\begin{cases} \displaystyle\lim_{\substack{Y\in\mathcal{D}_\epsilon \\ \epsilon\to 0}} \theta_{[e]}(\overline{p}, \overline{T}, Y) = \theta_{[e]}(\overline{p}, \overline{T}, Y_0) = \begin{bmatrix} 0 \\ \theta_{[e]}^-(\overline{p}, \overline{T}) \end{bmatrix}, \\ \displaystyle\lim_{\substack{Y\in\mathcal{D}_\epsilon \\ \epsilon\to 0}} \widetilde{\chi}_{[e]}(\overline{T}, Y) = \widetilde{\chi}_{[e]}(\overline{T}, Y_0) = \begin{bmatrix} 0 \\ \widetilde{\chi}_{[e]}^-(\overline{T}) \end{bmatrix}, \end{cases} \tag{4.7.28}$$

where

$$\theta_{[e]k}(\bar{p},\overline{T}) = \frac{m_1}{m_k}\widetilde{D}_{[e]kk}(\bar{p},\overline{T})\widetilde{\chi}_{[e]k}(\overline{T}), \qquad k \in \mathcal{S}. \tag{4.7.29}$$

Finally, we have

$$\widetilde{\chi}_{[e]k} = \widetilde{L}^{00e}_{[e]k1}\alpha^{e\lambda}_{[e]1} + \widetilde{L}^{00e}_{[e]kk}\alpha^{e\lambda}_{[e]k}, \qquad k \in \mathcal{S}^-, \tag{4.7.30}$$

where

$$\begin{cases} \alpha^{e\lambda}_{[e]1} = \big(\widetilde{\Lambda}^{ee}_{[e]11}\big)^{-1}\widetilde{\beta}^{e\lambda}_{[e]1}, \\ \alpha^{e\lambda}_{[e]k} = \big(\widetilde{\Lambda}^{ee}_{[e]kk}\big)^{-1}\Big(\widetilde{\beta}^{e\lambda}_{[e]k} - \widetilde{\Lambda}^{ee}_{[e]k1}\big(\widetilde{\Lambda}^{ee}_{[e]11}\big)^{-1}\widetilde{\beta}^{e\lambda}_{[e]1}\Big), \qquad k \in \mathcal{S}^-. \end{cases} \tag{4.7.31}$$

Proof. We only consider the thermal diffusion vector θ and the rescaled thermal diffusion ratios $\widetilde{\chi}$, since the case of $\theta_{[e]}$ and $\widetilde{\chi}_{[e]}$ is treated similarly. The limits (4.7.22) are directly obtained from Theorems 4.5.7 and 4.6.9, and from Theorem 4.5.5 we then deduce that $\theta_1(\bar{p},\overline{T},Y_0) = 0$ since $\langle\theta(\bar{p},\overline{T},Y_0),Y_0\rangle = 0$ and $Y_0 = (1,0,\dots,0)$. Similarly, $\widetilde{\chi}_1(\overline{T},Y_0) = 0$ since $\langle\widetilde{\chi}(\overline{T},Y_0),X_0\rangle = 0$ and $X_0 = (1,0,\dots,0)$. From Proposition 4.6.8 we next obtain that $\widetilde{\chi} = \widetilde{\Delta}\theta$, and from Theorem 4.7.5 we then easily obtain (4.7.23). Finally, the expressions (4.7.24)–(4.7.27) directly result from ($\widetilde{L}1$), ($\widetilde{L}3$), ($\widetilde{\Lambda}1$), and ($\widetilde{\Lambda}3$). □

5 Convergent Iterative Methods

In this chapter we present the convergence theorems for standard iterative and conjugate gradient methods for the solution of the transport linear systems. Using these results, all the transport coefficients are expressed as convergent series, and truncating these series then yields new, explicit, and rigorously derived expressions for the transport coefficients. With the exception of conjugate gradient methods for diffusion matrices, the partial sums also satisfy the mathematical properties that are important from a thermodynamic viewpoint, i.e., symmetry, mass conservation, and positive entropy production on the hyperplane of zero sum driving forces.

As in Chapter 4, all the convergence results are valid for any general system matrix satisfying a given set of mathematical properties. These properties include the ones used in Chapter 4 and additional properties extracted from the kinetic theory investigations of Chapters 2 and 3. As in the previous chapter, we systematically verify that these mathematical properties remain valid when collision integrals are estimated in the framework of the practical approximations presented in Section 2.10 [MM62] [MPM65].

All the convergence results for conjugate gradient methods are, to the authors' knowledge, new. On the other hand, standard iterative methods have been considered implicitly by [HCB54] [Br58] [Br64] when deriving approximate expressions for the shear viscosity and the thermal conductivity of monatomic gas mixtures. Standard iterative methods have also been considered by [JB81] [OB81] for evaluating the first-order diffusion velocities, which correspond to the simplified diffusion matrix $D_{[00]}$ in this book, but the convergence of the iterates was only verified numerically. The convergence of the Jones-Oran-Boris algorithm has been proven rigorously by Giovangigli [Gi91] who also established that the corresponding iteration matrix has a spectral radius unity. Additional algorithms for the simplified diffusion matrix $D_{[00]}$ and the simplified flux diffusion matrix $\widetilde{D}_{[00]}$ have been introduced in [Gi91] for which the iteration matrix has a spectral radius strictly lower than unity. All the remaining convergence results that are presented for standard iterative methods are, to the authors' knowledge, new.

In Section 5.1 we restate several mathematical results on convergent iterative methods for nonsingular and constrained singular systems. For completeness, we present proofs of these results, which are generally imbedded in highly technical papers and dispersed in the literature. Standard iterative methods are discussed in Section 5.1.1, and projected algorithms for constrained singular systems are described in Section 5.1.2. Convergence results for standard iterative methods applied to symmetric positive definite and positive semi-definite matrices are obtained in Section 5.1.3. The preconditioned conjugate gradient method is presented in Section 5.1.4, and its projected version for symmetric positive semi-definite systems is also introduced. Standard iterative and conjugate gradient methods for Schur complements are discussed in Section 5.1.5. Finally, in Section 5.1.6 stabilized versions, for vanishing mass fractions, of the iterative algorithms are obtained by considering appropriately rescaled system matrices, i.e., left rescaled matrices for standard iterative methods and symmetric rescaled matrices for conjugate gradient methods.

In Sections 5.2 to 5.6 we apply the convergence theorems obtained in Section 5.1 to all the transport linear systems described in the previous chapters. Only the most important proofs are given for the sake of brevity. We consider the shear viscosity in Section 5.2, the volume viscosity in Section 5.3, the diffusion matrix and the flux diffusion matrix in Section 5.4, the partial thermal conductivity and the thermal diffusion vector in Section 5.5, and the thermal conductivity and the thermal diffusion ratios in Section 5.6.

5.1 Iterative Methods for Constrained Singular Systems

5.1.1 Convergent Matrices and Standard Iterative Methods

We refer to Neumann and Plemmons [NP78], Bermann and Plemmons [BP79], and Keller [Ke65] for an introduction to the solution of singular consistent linear systems by standard iterative methods.

For a matrix $T \in \mathbb{R}^{\omega,\omega}$, $\sigma(T)$ and $\rho(T)$ denote the spectrum and the spectral radius of T, and we also define $\gamma(T) = \max\{\ |\lambda|;\ \lambda \in \sigma(T),\ \lambda \neq 1\ \}$. A matrix T is said to be convergent when $\lim_{i\to\infty} T^i$ exists, not necessarily being zero. This corresponds to the terminology of Neumann and Plemmons [NP78] as opposed to the more conventional one, where a matrix T is said to be convergent when $\lim_{i\to\infty} T^i = 0$. It is well known that the powers of a matrix T converge to zero if and only if $\rho(T) < 1$. More generally, we have the following result [Ol40].

Proposition 5.1.1. *A matrix $T \in \mathbb{R}^{\omega,\omega}$ is convergent if and only if either $\rho(T) < 1$ or $\rho(T) = 1$, $1 \in \sigma(T)$, $\gamma(T) < 1$, and $(I - T)^\sharp$ exists, i.e., T has only elementary divisors corresponding to the eigenvalue 1. Moreover, if T is convergent, we have $\rho(T) < 1$ if and only if $I - T$ is nonsingular.*

Proof. Let T be a convergent matrix and $\lambda \in \sigma(T)$. Then there exists $v \neq 0$ such that $Tv = \lambda v$ and hence such that $T^i v = \lambda^i v$, $i \geq 1$. Since $\lim_{i\to\infty} T^i$ exists, we deduce that $\lim_{i\to\infty} \lambda^i$ exists so that either $|\lambda| < 1$ or $\lambda = 1$, i.e., $\rho(T) < 1$ or $\rho(T) = 1$, $1 \in \sigma(T)$, and $\gamma(T) < 1$. Moreover, if T does not have only elementary divisors for the eigenvalue 1, then $N(I-T) \cap R(I-T) \neq \{0\}$, i.e., there exists $v \neq 0$ and $w \neq 0$ such that $Tv = v$ and $Tw = v + w$. This implies now that $T^i w = iv + w$, $i \geq 1$, which has no limit when $i \to \infty$. We have thus shown that either $\rho(T) < 1$ or $\rho(T) = 1$, $1 \in \sigma(T)$, $\gamma(T) < 1$, and T has only elementary divisors corresponding to the eigenvalue 1. Conversely, if T satisfies these assumptions, then there exists a nonsingular matrix $S \in \mathbb{R}^{\omega,\omega}$ and a matrix $\mathcal{T} \in \mathbb{R}^{\omega_1,\omega_1}$ with $\omega_1 \leq \omega$ and $\rho(\mathcal{T}) < 1$ such that

$$T = S^{-1}\begin{pmatrix} I & 0 \\ 0 & \mathcal{T} \end{pmatrix} S, \qquad \text{and} \qquad T^i = S^{-1}\begin{pmatrix} I & 0 \\ 0 & \mathcal{T}^i \end{pmatrix} S, \quad i \geq 1.$$

This implies that T is convergent since $\lim_{i\to\infty} \mathcal{T}^i = 0$. Finally, assuming that T is convergent, we immediately deduce from the first part of the proof that $\rho(T) < 1$ if and only if $I - T$ is nonsingular. □

Remark. We have

$$T^\infty = S^{-1}\begin{pmatrix} I & 0 \\ 0 & 0 \end{pmatrix} S,$$

and

$$I - T = S^{-1}\begin{pmatrix} 0 & 0 \\ 0 & I - \mathcal{T} \end{pmatrix} S, \qquad (I-T)^\sharp = S^{-1}\begin{pmatrix} 0 & 0 \\ 0 & (I - \mathcal{T})^{-1} \end{pmatrix} S,$$

so that $T^\infty = I - (I - T)(I - T)^\sharp$. Furthermore, in the course of the proof we have also shown that the asymptotic convergence rate is $-\log \gamma(T)$ since $\rho(\mathcal{T}) = \gamma(T)$.

Next, for a matrix $G \in \mathbb{R}^{\omega,\omega}$, the decomposition

$$G = M - Z, \tag{5.1.1}$$

is a splitting if the matrix M is invertible. In order to solve the linear system

$$G\alpha^\mu = \beta^\mu, \tag{5.1.2}$$

where $\beta^\mu \in \mathbb{R}^\omega$, the splitting (5.1.1) induces the iterative scheme

$$x_{i+1} = Tx_i + M^{-1}\beta^\mu, \qquad i = 0, 1, \ldots \tag{5.1.3}$$

where $T = M^{-1}Z$ is the iteration matrix. If G is nonsingular, the sequence of iterates (5.1.3) converges for every x_0 to the unique solution of (5.1.2) if and only if $\rho(T) < 1$. More generally, if G is singular and if the system (5.1.2) is consistent, i.e., $\beta^\mu \in R(G)$, we have $M^{-1}\beta^\mu \in R(I-T)$, and the limit of (5.1.3) depends on x_0. More precisely, this limit is given in the following lemma which can be found in [MP77] except for a misprint on the matrix E.

Lemma 5.1.2. *Let $T \in \mathbb{R}^{\omega,\omega}$ and let $z \in \mathbb{R}^\omega$ such that $z \in R(I-T)$. Then the iterative scheme $x_{i+1} = Tx_i + z$, $i \geq 0$, converges for any $x_0 \in \mathbb{R}^\omega$ if and only if T is convergent. In this situation, the limit $\lim_{i\to\infty} x_i = x_\infty$ is given by $x_\infty = (I-T)^\sharp z + Ex_0$ where $E = I - (I-T)\,(I-T)^\sharp$.*

Proof. Since $z \in R(I-T)$, there exists $c \in \mathbb{R}^\omega$ such that $(I-T)c = z$. Therefore, $x_{i+1} - c = T(x_i - c)$, $i \geq 0$, and thus $x_i - c = T^i(x_0 - c)$, $i \geq 0$. We then conclude that the iterative scheme is convergent for any x_0 if and only if $T^i(x_0 - c)$ has a limit when $i \to \infty$ for any x_0, and thus if and only if T is convergent. In this situation, $(I-T)^\sharp$ exists by Proposition 5.1.1, and we may take $c = (I-T)^\sharp z$ since then $(I-T)c = z$ because $(I-T)(I-T)^\sharp = P_{R(I-T),N(I-T)}$ and $z \in R(I-T)$. Moreover, we have seen in the proof of Proposition 5.1.1 that $T^\infty = E$ where $E = I - (I-T)\,(I-T)^\sharp$, so that $x_\infty = c + E(x_0 - c)$. Since $Ec = c - (I-T)\,(I-T)^\sharp c$, we deduce that $Ec = 0$ because $c \in R(I-T)^\sharp = R(I-T)$ and $(I-T)\,(I-T)^\sharp = P_{R(I-T),N(I-T)}$. □

Finally, with an eye towards the rescaled transport linear systems, we consider a matrix G that admits the block-decomposition

$$G = \begin{bmatrix} G^{11} & 0 \\ G^{21} & G^{22} \end{bmatrix}, \tag{5.1.4}$$

where $G^{11} \in \mathbb{R}^{\omega_1,\omega_1}$, $G^{21} \in \mathbb{R}^{\omega-\omega_1,\omega_1}$, and $G^{22} \in \mathbb{R}^{\omega-\omega_1,\omega-\omega_1}$ for some $0 < \omega_1 \leq \omega$. We also consider a splitting for the matrix G of the form $G = M - Z$ and we assume that the matrices M and Z admit the block-decomposition

$$M = \begin{bmatrix} M^{11} & 0 \\ M^{21} & M^{22} \end{bmatrix}, \qquad Z = \begin{bmatrix} Z^{11} & 0 \\ Z^{21} & Z^{22} \end{bmatrix}. \tag{5.1.5}$$

In the following lemma, we examine the cases where the matrix $T = M^{-1}Z$ is convergent.

Lemma 5.1.3. *Let G, M, and $Z \in \mathbb{R}^{\omega,\omega}$ be three matrices admitting the block-decompositions (5.1.4)(5.1.5), and consider the matrix $T = M^{-1}Z$. Then the matrix*

T admits the block-decomposition

$$T = \begin{bmatrix} T^{11} & 0 \\ T^{21} & T^{22} \end{bmatrix}. \tag{5.1.6}$$

Furthermore, if the matrix T^{11} is convergent and $\rho(T^{22}) < 1$, the matrix T is convergent, and we have

$$T^{\infty} = \begin{bmatrix} \left(T^{11}\right)^{\infty} & 0 \\ \left(I - T^{22}\right)^{-1} T^{21} \left(T^{11}\right)^{\infty} & 0 \end{bmatrix}. \tag{5.1.7}$$

Proof. The block-decomposition (5.1.6) directly results from (5.1.5), and we have

$$T^{11} = \left(M^{11}\right)^{-1} Z^{11}, \qquad T^{22} = \left(M^{22}\right)^{-1} Z^{22}.$$

One can then easily verify by induction that

$$T^{i} = \begin{bmatrix} \left(T^{11}\right)^{i} & 0 \\ \sum_{j=0}^{i-1} \left(T^{22}\right)^{j} T^{21} \left(T^{11}\right)^{i-j-1} & \left(T^{22}\right)^{i} \end{bmatrix}, \qquad i \geq 1.$$

Since the matrix T^{11} is convergent, we have $\lim_{i\to\infty}\left(T^{11}\right)^{i} = \left(T^{11}\right)^{\infty}$, and since $\rho(T^{22}) < 1$, we have $\lim_{i\to\infty}\left(T^{22}\right)^{i} = 0$. We thus obtain

$$\begin{aligned} \lim_{i\to\infty} \sum_{j=0}^{i-1} \left(T^{22}\right)^{j} T^{21} \left(T^{11}\right)^{i-j-1} &= \sum_{j=0}^{\infty} \left(T^{22}\right)^{j} T^{21} \left(T^{11}\right)^{\infty} \\ &= \left(I - T^{22}\right)^{-1} T^{21} \left(T^{11}\right)^{\infty}, \end{aligned}$$

so that the matrix T is convergent and (5.1.7) is proven. □

5.1.2 Generalized Inverses and Projected Iterative Algorithms

We are now interested in solving the constrained singular system

$$\begin{cases} G\alpha^{\mu} = \beta^{\mu}, \\ \alpha^{\mu} \in \mathcal{C}, \end{cases} \tag{5.1.8}$$

where the subspace $\mathcal{C}$ is complementary to $N(G)$, i.e., $N(G) \oplus \mathcal{C} = \mathbb{R}^{\omega}$. By Proposition 4.1.3, the constrained singular system (5.1.8) admits a unique solution, and this solution can be expressed in terms of a generalized inverse of the matrix G.

Consider now the splitting $G = M - Z$ and assume that the iteration matrix $T = M^{-1}Z$ is convergent. If the matrix G is invertible, we deduce from Proposition 5.1.1 that $\rho(T) < 1$ since $I - T = M^{-1}G$, whereas if the matrix G is singular, we have $\rho(T) = 1$ since $Tx = x$ for $x \in N(G)$. In order to obtain an iteration matrix of spectral radius strictly lower than unity, we now investigate the projected version of the iterative scheme (5.1.3) given by

$$y_{i+1} = PTy_i + PM^{-1}\beta^{\mu}, \qquad i = 0, 1, \ldots \tag{5.1.9}$$

where $P = P_{\mathcal{C},N(G)}$ is the projector on the subspace $\mathcal{C}$ along $N(G)$. Projected iterative schemes have been introduced in [Gi91] when studying convergent iterative methods for multicomponent diffusion coefficients. In the following theorem due to Neumann and Plemmons [NP78], we establish that the matrix PT has a spectral radius strictly lower than unity whenever the matrix T is convergent.

Theorem 5.1.4. *(Neumann-Plemmons) Let $G \in \mathbb{R}^{\omega,\omega}$ be a matrix and $G = M - Z$ be a splitting. Assume that the matrix $T = M^{-1}Z$ is convergent. Then for any projector matrix $P = P_{\mathcal{C},N(G)}$ on a subspace $\mathcal{C}$ complementary to $N(G)$, i.e., $N(G)\oplus\mathcal{C} = \mathbb{R}^{\omega}$, the product PT has a spectral radius strictly lower than unity given by $\rho\big(PT\big) = \gamma(T) < 1$.*

Proof. Let P' denote the projection on the joint of all root subspaces of T associated with the eigenvalues other than unity along the eigenspace of T associated with the eigenvalue 1, i.e., $N(G)$. By definition of $\gamma(T)$, we have the relation $\gamma(T) = \rho(TP')$ and it is well known that P' commutes with T. Denoting $P = P_{\mathcal{C},N(G)}$, one may also easily check that $PP' = P$ and $P'P = P'$. Keeping in mind that for any $A, B \in \mathbb{R}^{\omega,\omega}$, $\rho(AB) = \rho(BA)$, we now obtain

$$\gamma(T) = \rho(TP') = \rho(TP'P) = \rho(PTP') = \rho(PP'T) = \rho(PT),$$

so that $\rho(PT) = \gamma(T) < 1$ by Proposition 5.1.1. □

The properties of projected iterative schemes are given in the following theorem.

Theorem 5.1.5. *Let $G \in \mathbb{R}^{\omega,\omega}$ be a matrix, $G = M - Z$ be a splitting, and assume that the matrix $T = M^{-1}Z$ is convergent. Assume that the subspaces $\mathcal{C}$ and S are such that $N(G)\oplus\mathcal{C} = \mathbb{R}^{\omega}$ and $R(G)\oplus S = \mathbb{R}^{\omega}$, let P denote the oblique projector onto the subspace $\mathcal{C}$ along $N(G)$, and let Q denote the oblique projector onto $R(G)$ along the subspace S. Let $\beta^{\mu} \in R(G)$, $x_0 \in \mathbb{R}^{\omega}$, $y_0 = Px_0$, and consider for $i \geq 0$ the iterates*

$$x_{i+1} = Tx_i + M^{-1}\beta^{\mu}, \tag{5.1.10}$$

$$y_{i+1} = PTy_i + PM^{-1}\beta^{\mu}. \tag{5.1.11}$$

Then $y_i = Px_i$ for all $i \geq 0$, the matrix PT is convergent, $\rho(PT) < 1$, and

$$\lim_{i\to\infty} y_i = P(\lim_{i\to\infty} x_i) = \alpha^\mu, \tag{5.1.12}$$

where α^μ is the unique solution of the constrained singular system (5.1.8). Moreover, for all $i \geq 0$, each partial sum

$$G_i^\dagger = \sum_{j=0}^{i} (PT)^j PM^{-1}Q, \tag{5.1.13}$$

admits nullspace $N(G_i^\dagger) = S$ and range $R(G_i^\dagger) = \mathcal{C}$. Finally, we have

$$G^\dagger = \sum_{j=0}^{\infty} (PT)^j PM^{-1}Q, \tag{5.1.14}$$

where $G^\dagger$ is the generalized inverse of G with prescribed nullspace $N(G^\dagger) = S$ and range $R(G^\dagger) = \mathcal{C}$.

Proof. By assumption, the matrix T is convergent and $N(G)\oplus\mathcal{C} = \mathbb{R}^\omega$, so that Theorem 5.1.4 applies and the matrix PT is convergent with $\rho(PT) < 1$. Furthermore, since $\beta^\mu \in R(G)$, we have $M^{-1}\beta^\mu \in R(I-T)$ and since T is convergent, the iterative scheme (5.1.10) is convergent by Lemma 5.1.2. In addition, $PM^{-1}\beta^\mu \in R(I-PT)$ since $\rho(PT) < 1$ so that, by Lemma 5.1.2, the iterative scheme (5.1.11) is also convergent. We next claim that $PTP = PT$. Indeed, for $z \in \mathcal{C}$, we have $Pz = z$, and for $z \in N(G)$, we obtain $Tz = z$ and $Pz = 0$. Therefore, PTP and PT coincide on $\mathcal{C}$ and $N(G)$ and hence on $\mathbb{R}^\omega = N(G)\oplus\mathcal{C}$. From $PTP = PT$ we then deduce by induction that $y_i = Px_i$ for all $i \geq 0$. Consequently, we have $\lim_{i\to\infty} y_i = P(\lim_{i\to\infty} x_i)$ and denoting by α^μ this limit, we have $\alpha^\mu \in \mathcal{C}$. Moreover, from the relation $GP = G$ and (5.1.10), we easily deduce that $G\alpha^\mu = G(\lim_{i\to\infty} x_i) = \beta^\mu$ and, therefore, α^μ is the unique solution of (5.1.8). On the other hand, the matrix $\sum_{j=0}^{i}(PT)^j$ is nonsingular since $\sum_{j=0}^{i}(PT)^j(I-PT) = I-(PT)^{i+1}$ and $\rho(PT) < 1$, and, therefore, we have $N(G_i^\dagger) = N(PM^{-1}Q)$. Furthermore, since T is convergent, we obtain from Lemma 5.1.1 that $N(I-T)\cap R(I-T) = \{0\}$, and from $R(M^{-1}Q) = R(I-T)$ and $N(P) = N(G) = N(I-T)$, we then deduce that $N(P)\cap R(M^{-1}Q) = \{0\}$. Consequently, we have $N(PM^{-1}Q) = N(M^{-1}Q)$ and hence, $N(G_i^\dagger) = N(Q) = S$ since M is invertible. In addition, we clearly have $R(G_i^\dagger) \subset \mathcal{C}$, and since the dimensions of $N(G_i^\dagger)$ and $R(G_i^\dagger)$ must sum up to ω, we have $R(G_i^\dagger) = \mathcal{C}$. Finally, for any $\beta^\mu \in R(G)$, we obtain by induction that $y_{i+1} = G_i^\dagger\beta^\mu$ for the particular choice $y_0 = 0$. Moreover, we also deduce from Proposition 4.1.3 that $\alpha^\mu = G^\dagger\beta^\mu$, where

$G^\dagger$ is the generalized inverse of G with prescribed nullspace $N(G^\dagger) = S$ and range $R(G^\dagger) = \mathcal{C}$. Since $\lim_{i\to\infty} y_i = \alpha^\mu$, we obtain $\alpha^\mu = G^\dagger \beta^\mu = \sum_{j=0}^{\infty}(PT)^j PM^{-1}Q\beta^\mu$. The matrices $G^\dagger$ and $\sum_{j=0}^{\infty}(PT)^j PM^{-1}Q$ therefore coincide on $R(G)$. Furthermore, they trivially coincide on S, and hence, (5.1.14) is proven since $R(G) \oplus S = \mathbb{R}^\omega$. □

5.1.3 Convergent Iterative Methods for Positive Semi-Definite Matrices

In this section we consider the special case of symmetric positive semi-definite matrices, where convergent splittings can be characterized by the following theorem due to Keller [Ke65].

Theorem 5.1.6. *(Keller) Let $G \in \mathbb{R}^{\omega,\omega}$ be a symmetric matrix, i.e., $G = G^t$, and let $G = M - Z$ be a splitting. Assume also that $M + Z^t$ is positive definite. Then the matrix $T = M^{-1}Z$ is convergent if and only if G is positive semi-definite. Moreover, we have $\rho(T) < 1$ if and only if G is positive definite.*

Proof. Assume first that G is positive semi-definite and let $\lambda \in \sigma(T)$. Then there exists $u \neq 0$ such that $Tu = \lambda u$ which implies that $Zu = \lambda Mu$ and $Gu = (1-\lambda)Mu$. Assume now that $\lambda \neq 1$. Then $Gu \neq 0$ since M is invertible and thus $\langle Gu, u\rangle > 0$ since G is symmetric positive semi-definite and $u \notin N(G)$. This implies that $\langle Mu, u\rangle - \langle Zu, u\rangle > 0$ and since $M + Z^t$ is positive definite we also have $\langle Mu, u\rangle + \langle Zu, u\rangle > 0$ because $u \neq 0$ and $\langle Zu, u\rangle = \langle Z^t u, u\rangle$. We have thus shown that $-\langle Mu, u\rangle < \langle Zu, u\rangle < \langle Mu, u\rangle$ and we conclude that $|\lambda| < 1$ since we have $\langle Zu, u\rangle = \lambda\langle Mu, u\rangle$ from $Zu = \lambda Mu$. Therefore either $\lambda = 1$ or $|\lambda| < 1$. We now show that T has only elementary divisors corresponding to the eigenvalue 1. Arguing by contradiction, we assume that this property does not hold, so that there exists $u \neq 0$ and $v \neq 0$ such that $Tv = v + u$ and $Tu = u$. Since $Tu = u$ we first obtain that $Mu = Zu$ so that $Gu = 0$ and also $\langle (M + Z^t)u, u\rangle = 2\langle Mu, u\rangle > 0$ since $M + Z^t$ is positive definite. On the other hand, from $Tv = v + u$ we deduce that $Gv = -Mu$ so that $\langle Gv, u\rangle = -\langle Mu, u\rangle$, an obvious contradiction since $\langle Mu, u\rangle > 0$ whereas $\langle Gv, u\rangle = \langle v, Gu\rangle = 0$ since G is symmetric and $Gu = 0$. Therefore, we have shown that T is convergent.

Conversely, we assume that T is convergent and we have to prove that G is positive semi-definite. Arguing by contradiction, assume that there exists $u_0 \neq 0$ such that $\langle Gu_0, u_0\rangle < 0$ and consider the sequence defined by $u_i = Tu_{i-1}$, $i \geq 1$. We first claim that the sequence $\langle Gu_i, u_i\rangle$ is decreasing. Indeed, letting $v_i = u_i - u_{i-1}$, we have

$$\langle Gu_i, u_i\rangle - \langle Gu_{i-1}, u_{i-1}\rangle = \big\langle v_i, 2Gu_{i-1} + Gv_i\big\rangle,$$

and $Mv_i = -Gu_{i-1}$ since $v_i = Tu_{i-1} - u_{i-1}$, so that, using $G = M - Z$ and $\langle Z^t v_i, v_i\rangle = \langle Zv_i, v_i\rangle$, we get

$$\langle Gu_i, u_i\rangle - \langle Gu_{i-1}, u_{i-1}\rangle = -\langle (M + Z^t)v_i, v_i\rangle \le 0,$$

since $M + Z^t$ is positive definite by assumption. Therefore, we have shown that

$$\langle Gu_i, u_i\rangle \le \langle Gu_0, u_0\rangle < 0, \qquad i \ge 1. \tag{5.1.15}$$

Consider then the decomposition $N(I - T) \oplus V = \mathbb{R}^\omega$, where V is the joint of all rootspaces associated with the eigenvalues of T different from unity and note also that $N(G) = N(I - T)$ since M is invertible. We may now decompose u_0 into $u_0 = w + v$ where $w \in N(I - T)$ and $v \in V$, and we then have $u_i = T^i u_0 = T^i(w + v) = w + T^i v$ since $Tw = w$. Since T is convergent, we know that $\gamma(T) < 1$ so that $T^i v \to 0$ as $i \to \infty$ and hence $u_i \to w$ as $i \to \infty$. Passing to the limit in (5.1.15), we deduce that $\langle Gw, w\rangle \le \langle Gu_0, u_0\rangle < 0$, an obvious contradiction since $w \in N(I - T) = N(G)$ and thus $Gw = 0$. We have thus shown that G is positive semi-definite.

Finally, since $Tu = u$ if and only if $Gu = 0$, we also conclude that $\rho(T) < 1$ if and only if G is positive definite, which is a classical result of linear algebra [GV83] [Vr62]. □

We now extend the previous convergence results in the following two theorems, that will be applied to all the transport linear systems. We first consider the case where the matrix G is symmetric positive definite, which corresponds to the nonsingular transport linear systems. Referring to Chapter 4, these systems take the form $G\alpha^\mu = \beta^\mu$ and are associated with the evaluation of the quantity $\mu = \langle \alpha^\mu, \beta^\mu\rangle$.

Theorem 5.1.7. *Let $G \in \mathbb{R}^{\omega,\omega}$ be a symmetric positive definite matrix and $G = M - Z$ be a splitting, and assume that M and $M + Z$ are symmetric positive definite. Let $T = M^{-1}Z$, $\beta^\mu \in \mathbb{R}^\omega$, and $x_0 \in \mathbb{R}^\omega$, and consider for $i \ge 0$ the iterates*

$$x_{i+1} = Tx_i + M^{-1}\beta^\mu. \tag{5.1.16}$$

Then the matrix T is convergent, $\rho(T) < 1$, and we have the following limits

$$\begin{cases} \lim\limits_{i\to\infty} x_i = \alpha^\mu, \\ \lim\limits_{i\to\infty} \langle x_i, \beta^\mu\rangle = \mu, \end{cases} \tag{5.1.17}$$

where α^μ is the unique solution of $G\alpha^\mu = \beta^\mu$. Moreover, for all $i \ge 0$, the partial sums

$$G_i^\dagger = \sum_{j=0}^{i} T^j M^{-1}, \tag{5.1.18}$$

are symmetric positive definite, and we have

$$G^{-1} = \sum_{j=0}^{\infty} T^j M^{-1}. \tag{5.1.19}$$

Consequently, the quantities

$$\mu^{[i]} = \langle G_i^{\dagger} \beta^{\mu}, \beta^{\mu} \rangle, \tag{5.1.20}$$

are positive if $\beta^{\mu} \neq 0$, *and we have*

$$\lim_{i \to \infty} \mu^{[i]} = \langle G^{-1} \beta^{\mu}, \beta^{\mu} \rangle = \mu. \tag{5.1.21}$$

Proof. Since the matrix M is positive definite, $G = M - Z$ is a splitting. Furthermore, the matrix $Z = M - G$ is clearly symmetric, and we thus have $M + Z^t = M + Z$, which is symmetric positive definite by assumptions. We then deduce from Theorem 5.1.6 that the matrix T is convergent, and since G is positive definite, we have $\rho(T) < 1$. We also obtain from Lemma 5.1.2 that the sequence (5.1.16) converges towards α^{μ}, the unique solution of the nonsingular system $G\alpha^{\mu} = \beta^{\mu}$, and therefore, $\lim_{i \to \infty} \langle x_i, \beta^{\mu} \rangle = \mu$. On the other hand, since M^{-1} and Z are symmetric, we have for any $j \geq 0$,

$$(T^j M^{-1})^t = (M^{-1}Z \dots M^{-1}ZM^{-1})^t = M^{-1}Z \dots M^{-1}ZM^{-1} = T^j M^{-1},$$

so that $G_i^{\dagger}$ is symmetric for all $i \geq 0$. Furthermore, using again the symmetry of M and Z, one can show that for any $i \geq 0$ and $z \in \mathbb{R}^{\omega}$, we have

$$\langle T^{2i} M^{-1} z, z \rangle = \langle M T^i w, T^i w \rangle,$$

where $w = M^{-1} z$, and

$$\langle T^{2i+1} M^{-1} z, z \rangle + \langle T^{2i} M^{-1} z, z \rangle = \langle (M + Z) T^i w, T^i w \rangle.$$

From the positive definiteness of M and $M + Z$, one easily deduces that for any $i \geq 1$ and for any $z \in \mathbb{R}^{\omega}$ we have

$$\langle G_{2i}^{\dagger} z, z \rangle \geq \langle G_{2i-1}^{\dagger} z, z \rangle,$$

and

$$\langle G_{2i+1}^{\dagger} z, z \rangle \geq \langle G_{2i-1}^{\dagger} z, z \rangle \geq \dots \geq \langle G_1^{\dagger} z, z \rangle.$$

Furthermore, we have $G_0^{\dagger} = M^{-1}$ and $G_1^{\dagger} = M^{-1}(M + Z)M^{-1}$ which are both positive definite so that all the matrices $G_i^{\dagger}$, $i \geq 0$, are positive definite. One may also easily show by induction that $x_{i+1} = G_i^\dagger \beta^\mu$ for the particular choice $x_0 = 0$, and hence, for all $\beta^\mu \in \mathbb{R}^\omega$, we have $\lim_{i\to\infty} G_i^\dagger \beta^\mu = \alpha^\mu = G^{-1}\beta^\mu$, i.e., $G^{-1} = \sum_{j=0}^{\infty} T^j M^{-1}$. Finally, we easily obtain that the quantities $\mu^{[i]} = \langle G_i^\dagger \beta^\mu, \beta^\mu\rangle$ are positive if $\beta^\mu \neq 0$ and that $\lim_{i\to\infty} \mu^{[i]} = \langle G^{-1}\beta^\mu, \beta^\mu\rangle = \mu$. □

The following theorem generalizes the previous one to the case where the matrix G is symmetric positive semi-definite and singular so that a projected iterative scheme must be used. It will be applied to all the singular transport linear systems. Referring to Chapter 4, these systems take the form $G\alpha^\mu = \beta^\mu$ and $\alpha^\mu \in \mathcal{C}$, and are associated with the evaluation of the quantity $\mu = \langle \alpha^\mu, \beta^\mu\rangle$.

Theorem 5.1.8. *Let $G \in \mathbb{R}^{\omega,\omega}$ be a symmetric positive semi-definite matrix and let $\mathcal{C}$ be a subspace complementary to $N(G)$, i.e., $N(G)\oplus\mathcal{C} = \mathbb{R}^\omega$. Consider also a splitting $G = M - Z$ and assume that M and $M + Z$ are symmetric positive definite and let $T = M^{-1}Z$. Let also P be the oblique projector onto the subspace $\mathcal{C}$ along $N(G)$, $\beta^\mu \in R(G)$, $x_0 \in \mathbb{R}^\omega$, $y_0 = Px_0$, and consider for $i \geq 0$ the iterates*

$$x_{i+1} = Tx_i + M^{-1}\beta^\mu, \tag{5.1.22}$$

$$y_{i+1} = PTy_i + PM^{-1}\beta^\mu. \tag{5.1.23}$$

Then $y_i = Px_i$ for all $i \geq 0$, the matrices T and PT are convergent, $\rho(T) = 1$, $\rho(PT) < 1$, and we have the following limits

$$\begin{cases} \lim\limits_{i\to\infty} y_i = P(\lim\limits_{i\to\infty} x_i) = \alpha^\mu, \\ \lim\limits_{i\to\infty} \langle y_i, \beta^\mu\rangle = \lim\limits_{i\to\infty} \langle x_i, \beta^\mu\rangle = \mu, \end{cases} \tag{5.1.24}$$

where α^μ is the unique solution of the constrained singular system $G\alpha^\mu = \beta^\mu$ and $\alpha^\mu \in \mathcal{C}$. Moreover, for all $i \geq 0$, each partial sum

$$G_i^\dagger = \sum_{j=0}^{i} (PT)^j PM^{-1}P^t, \tag{5.1.25}$$

is symmetric, positive semi-definite, positive definite on $R(G)$, and admits nullspace $N(G_i^\dagger) = \mathcal{C}^\perp$ and range $R(G_i^\dagger) = \mathcal{C}$. In addition, we obtain

$$G^\dagger = \sum_{j=0}^{\infty} (PT)^j PM^{-1}P^t, \tag{5.1.26}$$

where $G^\dagger$ is the generalized inverse of G with prescribed nullspace $N(G^\dagger) = \mathcal{C}^\perp$ and range $R(G^\dagger) = \mathcal{C}$. Consequently, the quantities

$$\mu^{[i]} = \langle G_i^\dagger \beta^\mu, \beta^\mu \rangle, \tag{5.1.27}$$

are positive if $\beta^\mu \neq 0$, and we have

$$\lim_{i\to\infty} \mu^{[i]} = \langle G^\dagger \beta^\mu, \beta^\mu \rangle = \mu. \tag{5.1.28}$$

Proof. Since the matrix G is positive semi-definite and the matrix $M + Z^t = M + Z$ is symmetric positive definite, we deduce from Theorem 5.1.6 that the matrix T is convergent. The first part of the theorem is then a direct consequence of Theorem 5.1.5. Moreover, as in the previous proof, we easily establish that $(T^jM^{-1})^t = T^jM^{-1}$ for $j \geq 0$, that

$$\langle PT^{2j}M^{-1}P^t z, z\rangle = \langle MT^j w, T^j w\rangle,$$

for any $z \in \mathbb{R}^\omega$ and $w = M^{-1}P^t z$, and that

$$\langle PT^{2j+1}M^{-1}P^t z, z\rangle + \langle PT^{2j}M^{-1}P^t z, z\rangle = \langle (M+Z)T^j w, T^j w\rangle.$$

Furthermore, from the relation $PTP = PT$ obtained in the proof of Theorem 5.1.5, one can easily show by induction that $(PT)^jP = PT^j$ for all $j \geq 0$ so that we can write $G_i^\dagger = \sum_{j=0}^{i} PT^jM^{-1}P^t$. From the positive definiteness of M and $M + Z$ and the previous relations, we deduce that for any $i \geq 1$ and $z \in \mathbb{R}^\omega$ we have

$$\langle G_{2i}^\dagger z, z\rangle \geq \langle G_{2i-1}^\dagger z, z\rangle,$$

and

$$\langle G_{2i+1}^\dagger z, z\rangle \geq \langle G_{2i-1}^\dagger z, z\rangle \geq \ldots \geq \langle G_1^\dagger z, z\rangle.$$

Since $\langle G_1^\dagger z, z\rangle = \langle (M+Z)w, w\rangle$ and $\langle G_0^\dagger z, z\rangle = \langle Mw, w\rangle$ with $w = M^{-1}P^t z$, all the matrices $G_i^\dagger$, $i \geq 0$, are positive semi-definite. In addition, we have $\langle G_i^\dagger z, z\rangle = 0$ if and only if $w = 0$ which, in turn, is equivalent to $z \in \mathcal{C}^\perp$. This implies that $N(G_i^\dagger) = \mathcal{C}^\perp$ and that $G_i^\dagger$ is positive definite on $R(G)$ since $R(G) \oplus \mathcal{C}^\perp = \mathbb{R}^\omega$. From the symmetry of $G_i^\dagger$, we next deduce that $R(G_i^\dagger) = \mathcal{C}$. In addition, the quantities $\langle G_i^\dagger \beta^\mu, \beta^\mu\rangle$, $i \geq 0$, are then positive if $\beta^\mu \neq 0$ since $\beta^\mu \in R(G)$. Finally, the limit (5.1.26) results from Theorem 5.1.5, and the limit (5.1.28) is a direct consequence of (5.1.26), since by Proposition 4.1.3 we have $\mu = \langle G^\dagger \beta^\mu, \beta^\mu\rangle$. □

Remark. Note that the projector P is needed for the convergence of the series (5.1.26). Although the partial sums $G_i^\dagger$ in (5.1.25) can be rewritten in the form

$$G_i^\dagger = P(\sum_{j=0}^{i} T^j M^{-1})P^t,$$

the series $\sum_{j=0}^{i} T^j M^{-1}$ has no limit since $\sum_{j=0}^{i} T^j M^{-1}(Mu) = (i+1)u$ for $u \in N(G)$. This is at variance with the projected algorithm involving the matrix PT of spectral radius strictly lower than unity. Furthermore, we have $Tz = z$ for $z \in N(G)$ and $M^{-1}\beta^\mu \in R(I-T)$ since $\beta^\mu \in R(G)$, so that $P_{N(G),R(I-T)}x_i = P_{N(G),R(I-T)}x_0$ for all $i \geq 0$. Consequently, the components of x_0 in $N(G)$, according to the decomposition $N(G) \oplus R(I-T) = \mathbb{R}^\omega$, remain undamped in (5.1.22) and may cause successive roundoff errors.

5.1.4 Conjugate Gradient Methods for Positive Semi-Definite Systems

In this section we present the conjugate gradient method for symmetric positive definite systems [HS52] [GV83] and positive semi-definite systems [LR80]. For the latter systems we also introduce the projected conjugate gradient method. In the following two theorems, we present the preconditioned version of both methods, where the preconditioner matrix is denoted by M.

We first consider the linear system $G\alpha^\mu = \beta^\mu$ associated with the evaluation of the quantity $\mu = \langle \alpha^\mu, \beta^\mu \rangle$.

Theorem 5.1.9. *Let G and $M \in \mathbb{R}^{\omega,\omega}$ be two symmetric positive definite matrices and consider $\beta^\mu \in \mathbb{R}^\omega$. Let $x_0 \in \mathbb{R}^\omega$, $r_0 = \beta^\mu - Gx_0$, $p_0 = 0$, $t_0 = 0$, and consider for $i \geq 1$ the iterates*

$$\begin{cases} p_i = M^{-1}r_{i-1} + t_{i-1}p_{i-1}, \\ s_i = \langle r_{i-1}, M^{-1}r_{i-1}\rangle / \langle p_i, Gp_i\rangle, \\ x_i = x_{i-1} + s_i p_i, \\ r_i = r_{i-1} - s_i G p_i, \\ t_i = \langle r_i, M^{-1}r_i\rangle / \langle r_{i-1}, M^{-1}r_{i-1}\rangle. \end{cases} \tag{5.1.29}$$

Then the sequence of iterates x_i converges towards the solution of $G\alpha^\mu = \beta^\mu$ in at most ω steps, and the quantities

$$\mu^{[i]} = \langle x_i, \beta^\mu \rangle, \tag{5.1.30}$$

converge towards μ in at most ω steps. Furthermore, if $x_0 = 0$ and $\beta^\mu \neq 0$, we have $\mu^{[i]} > 0$ for all $i \geq 1$.

Proof. The proof is classical, and we refer to [GV83] for details. From (5.1.29) we deduce by induction that the vectors r_i, $i \geq 0$, are such that $r_i = \beta^\mu - Gx_i$. One can also show by induction that the following orthogonality properties hold

$$\begin{cases} \langle M^{-1}r_i, r_j\rangle = 0, & i \neq j, \\ \langle p_i, Gp_j\rangle = 0, & i \neq j, \\ \langle r_i, p_j\rangle = 0, & i \geq j. \end{cases} \tag{5.1.31}$$

From the first property in (5.1.31), we deduce that the iterates x_i and $\mu^{[i]}$ converge towards α^μ and μ, respectively, in at most ω steps. Note that a breakdown due to a division by zero can only occur in the iterations (5.1.29) if convergence is achieved [GV83]. Finally, we have

$$\mu^{[1]} = \frac{\langle M^{-1}\beta^\mu, \beta^\mu\rangle^2}{\langle M^{-1}\beta^\mu, GM^{-1}\beta^\mu\rangle} > 0,$$

since G and M are positive definite, and for $i \geq 2$, we obtain $\mu^{[i]} = \mu^{[i-1]} + s_i\langle p_i, \beta^\mu\rangle$ and $\langle p_i, \beta^\mu\rangle = \langle p_i, r_0\rangle = t_{i-1}\langle p_{i-1}, \beta^\mu\rangle \geq 0$, using the relation $\langle M^{-1}r_{i-1}, r_0\rangle = 0$ valid for $i \geq 2$. Therefore, we have $\mu^{[i]} \geq \ldots \geq \mu^{[1]} > 0$ for all $i \geq 1$. □

We next consider the constrained linear system $G\alpha^\mu = \beta^\mu$ and $\alpha^\mu \in C$ associated with the evaluation of the quantity $\mu = \langle \alpha^\mu, \beta^\mu\rangle$.

Theorem 5.1.10. *Let $G \in \mathbb{R}^{\omega,\omega}$ be a symmetric positive semi-definite matrix and let $M \in \mathbb{R}^{\omega,\omega}$ be a symmetric positive definite matrix. Let C be a subspace complementary to $N(G)$, i.e., $N(G) \oplus C = \mathbb{R}^\omega$, let P be the oblique projector onto the subspace C along $N(G)$, and let $\beta^\mu \in R(G)$. Let also $x_0 \in \mathbb{R}^\omega$, $y_0 = Px_0$, $r_0 = \beta^\mu - Gx_0$, $p_0 = 0$, $t_0 = 0$, and consider for $i \geq 1$ the iterates*

$$\begin{cases} p_i = M^{-1}r_{i-1} + t_{i-1}p_{i-1}, \\ s_i = \langle r_{i-1}, M^{-1}r_{i-1}\rangle / \langle p_i, Gp_i\rangle, \\ x_i = x_{i-1} + s_i p_i, \\ y_i = y_{i-1} + P(s_i p_i), \\ r_i = r_{i-1} - s_i G p_i, \\ t_i = \langle r_i, M^{-1}r_i\rangle / \langle r_{i-1}, M^{-1}r_{i-1}\rangle. \end{cases} \tag{5.1.32}$$

Then $y_i = Px_i$ for all $i \geq 0$, the sequence of iterates y_i converges towards the solution of $G\alpha^\mu = \beta^\mu$ and $\alpha^\mu \in C$ in at most rank(G) steps, and the quantities

$$\mu^{[i]} = \langle y_i, \beta^\mu \rangle = \langle x_i, \beta^\mu \rangle, \tag{5.1.33}$$

converge towards μ in at most rank(G) steps. Furthermore, if $x_0 = 0$ and $\beta^\mu \neq 0$, we have $\mu^{[i]} > 0$ for all $i \geq 1$.

Proof. First, we can easily show by induction that we have $y_i = Px_i$ and that $r_i = \beta^\mu - Gy_i = \beta^\mu - Gx_i$, $i \geq 0$, since $GP = G$. One can also prove by induction that the orthogonality relations (5.1.31) hold if G is only positive semi-definite so that the iterates y_i and the quantities $\mu^{[i]}$ converge towards α^μ and μ, respectively, in at most rank(G) steps since the vectors r_i are all in $R(G)$. Finally, we note that $\langle y_i, \beta^\mu \rangle = \langle x_i, \beta^\mu \rangle$ since $y_i = Px_i$ and $P^t \beta^\mu = \beta^\mu$, and the positivity of the quantities $\mu^{[i]}$ is obtained as in the proof of Theorem 5.1.9. □

Remark. A breakdown in (5.1.32) can only occur if convergence is already achieved. Indeed, assuming that $\langle p_i, Gp_i \rangle = 0$, we first obtain that $p_i \in N(G)$, but, on the other hand, we deduce from the relations $p_j = M^{-1} r_{j-1} + t_{j-1} p_{j-1}$, valid for $j = 1, \dots, i$, that $Mp_i \in R(G)$. This yields $\langle p_i, Mp_i \rangle = 0$ since $N(G) = R(G)^\perp$, and since M is positive definite, we obtain that $p_i = 0$. We then deduce from the last orthogonality property in (5.1.31) and $p_i = 0$ that $\langle r_{i-1}, M^{-1} r_{i-1} \rangle = 0$, i.e., $r_{i-1} = 0$, as was to be shown.

5.1.5 Schur Complements

In the framework of the practical approximations presented in Section 2.10 [MM62] [MPM65], the lower-right block of various transport linear systems is diagonal. In this case, it may be interesting to consider its Schur complement and to use an iterative algorithm to solve the associated linear system of smaller size. The corresponding computational costs are discussed in Section 6.1.3. In this section we now regroup the main mathematical results and convergence theorems for Schur complements. Appropriate Schur complements are then introduced for various transport linear systems in Sections 5.3 to 5.6.

We consider a symmetric matrix $G \in \mathbb{R}^{\omega,\omega}$, i.e., $G = G^t$, given by the block-decomposition

$$G = \begin{bmatrix} G^{11} & G^{12} \\ G^{21} & G^{22} \end{bmatrix}, \tag{5.1.34}$$

where $G^{11} \in \mathbb{R}^{\omega_1,\omega_1}$, $G^{12} \in \mathbb{R}^{\omega_1,\omega-\omega_1}$, $G^{21} \in \mathbb{R}^{\omega-\omega_1,\omega_1}$, $G^{22} \in \mathbb{R}^{\omega-\omega_1,\omega-\omega_1}$ for some $0 < \omega_1 < \omega$. Correspondingly, each vector $x \in \mathbb{R}^\omega$ is decomposed into its components $x = (x^1, x^2)$, $x^1 \in \mathbb{R}^{\omega_1}$, $x^2 \in \mathbb{R}^{\omega-\omega_1}$. We assume that the submatrix G^{22} is nonsingular so that we can introduce the Schur complement $G_{[s]} \in \mathbb{R}^{\omega_1,\omega_1}$ given by

$$G_{[s]} = G^{11} - G^{12}\left(G^{22}\right)^{-1}G^{21}. \tag{5.1.35}$$

The Nonsingular Case. We consider the linear system

$$\begin{bmatrix} G^{11} & G^{12} \\ G^{21} & G^{22} \end{bmatrix} \begin{bmatrix} \alpha^{1\mu} \\ \alpha^{2\mu} \end{bmatrix} = \begin{bmatrix} \beta^{1\mu} \\ \beta^{2\mu} \end{bmatrix}, \tag{5.1.36}$$

associated with the evaluation of the quantity $\mu = \langle \alpha^\mu, \beta^\mu \rangle = \langle \alpha^{1\mu}, \beta^{1\mu} \rangle + \langle \alpha^{2\mu}, \beta^{2\mu} \rangle$.

Lemma 5.1.11. *Let $G \in \mathbb{R}^{\omega,\omega}$ be a symmetric positive definite matrix. Then the matrix $G_{[s]}$ given by (5.1.35) is symmetric positive definite. Consequently, the linear system*

$$G_{[s]}\alpha^\mu_{[s]} = \beta^\mu_{[s]}, \tag{5.1.37}$$

where

$$\beta^\mu_{[s]} = \beta^{1\mu} - G^{12}\left(G^{22}\right)^{-1}\beta^{2\mu}, \tag{5.1.38}$$

admits a unique solution $\alpha^\mu_{[s]}$. Finally, the quantity μ is given by

$$\mu = \mu_{[s]} + \mu_{[2]} = \langle \alpha^\mu_{[s]}, \beta^\mu_{[s]} \rangle + \langle \left(G^{22}\right)^{-1}\beta^{2\mu}, \beta^{2\mu} \rangle. \tag{5.1.39}$$

Proof. The symmetry of $G_{[s]}$ directly follows from (5.1.35) and the symmetry of G. Furthermore, for $x \in \mathbb{R}^{\omega_1}$, let $y \in \mathbb{R}^\omega$ be given by the block-decomposition $y = (x, -\left(G^{22}\right)^{-1}G^{21}x)$. We then obtain $\langle G_{[s]}x, x\rangle = \langle Gy, y\rangle$ so that $\langle G_{[s]}x, x\rangle = 0$ if and only if $y = 0$, i.e., $x = 0$. Therefore, $G_{[s]}$ is positive definite, and the system (5.1.37) admits a unique solution $\alpha^\mu_{[s]}$. Finally, a straightforward calculation yields $\alpha^\mu_{[s]} = \alpha^{1\mu}$, where $\alpha^\mu = (\alpha^{1\mu}, \alpha^{2\mu})$ is the unique solution of (5.1.36), and (5.1.39) is then easily obtained from (5.1.38). □

The Singular Case. We consider the constrained linear system

$$\begin{cases} \begin{bmatrix} G^{11} & G^{12} \\ G^{21} & G^{22} \end{bmatrix} \begin{bmatrix} \alpha^{1\mu} \\ \alpha^{2\mu} \end{bmatrix} = \begin{bmatrix} \beta^{1\mu} \\ \beta^{2\mu} \end{bmatrix}, \\ \langle \mathcal{G}^1, \alpha^{1\mu} \rangle + \langle \mathcal{G}^2, \alpha^{2\mu} \rangle = 0, \end{cases} \tag{5.1.40}$$

associated with the evaluation of the quantity $\mu = \langle \alpha^\mu, \beta^\mu \rangle = \langle \alpha^{1\mu}, \beta^{1\mu} \rangle + \langle \alpha^{2\mu}, \beta^{2\mu} \rangle$.

Lemma 5.1.12. *Let $G \in \mathbb{R}^{\omega,\omega}$ be a symmetric positive semi-definite matrix with nullspace $N(G) = \mathbb{R}\mathcal{Z}$. Assume that $\beta^\mu \in R(G)$ and $\langle \mathcal{G}, \mathcal{Z} \rangle \neq 0$. Then the matrix $G_{[s]}$ is symmetric positive semi-definite with $N(G_{[s]}) = \mathbb{R}\mathcal{Z}_{[s]}$ where $\mathcal{Z}_{[s]} = \mathcal{Z}^1$. Furthermore, with the notation of Lemma 5.1.11, the constrained linear system*

$$\begin{cases} G_{[s]}\alpha^\mu_{[s]} = \beta^\mu_{[s]}, \\ \langle \mathcal{G}_{[s]}, \alpha^\mu_{[s]} \rangle = 0, \end{cases} \tag{5.1.41}$$

where

$$\mathcal{G}_{[s]} = \mathcal{G}^1 - G^{12}\big(G^{22}\big)^{-1}\mathcal{G}^2, \tag{5.1.42}$$

admits a unique solution $\alpha^\mu_{[s]}$. Finally, the quantity μ is given by (5.1.39).

Proof. As in the previous proof, for $x \in \mathbb{R}^{\omega_1}$, let $y \in \mathbb{R}^\omega$ be given by the block-decomposition $y = (x, -\big(G^{22}\big)^{-1}G^{21}x)$. We then obtain $\langle G_{[s]}x, x \rangle = \langle Gy, y \rangle \geq 0$ so that $G_{[s]}$ is positive semi-definite. Furthermore, $\langle G_{[s]}x, x \rangle = 0$ if and only if $y \in N(G)$, which is equivalent to $x \in \mathbb{R}\mathcal{Z}^1$ since $\mathcal{Z}^2 = -\big(G^{22}\big)^{-1}G^{21}\mathcal{Z}^1$. Therefore, we have $N(G_{[s]}) = \mathbb{R}\mathcal{Z}_{[s]}$ where $\mathcal{Z}_{[s]} = \mathcal{Z}^1$, and consequently $R(G_{[s]}) = \mathcal{Z}^\perp_{[s]}$ since $G_{[s]}$ is symmetric. In addition, a straightforward calculation yields $\langle \beta^\mu_{[s]}, \mathcal{Z}_{[s]} \rangle = \langle \beta^\mu, \mathcal{Z} \rangle = 0$ so that $\beta^\mu_{[s]} \in R(G_{[s]})$, and similarly $\langle \mathcal{Z}_{[s]}, \mathcal{G}_{[s]} \rangle = \langle \mathcal{Z}, \mathcal{G} \rangle \neq 0$ so that $N(G_{[s]}) \oplus \mathcal{G}^\perp_{[s]} = \mathbb{R}^{\omega_1}$. Therefore, Proposition 4.1.3 applies, and the constrained linear system (5.1.41) admits a unique solution $\alpha^\mu_{[s]}$. Finally, one can easily verify that $\alpha^\mu_{[s]} - \alpha^{1\mu} \in \mathbb{R}\mathcal{Z}_{[s]}$ since $G_{[s]}\alpha^{1\mu} = \beta^\mu_{[s]}$, so that $\langle \alpha^\mu_{[s]}, \beta^\mu_{[s]} \rangle = \langle \alpha^{1\mu}, \beta^\mu_{[s]} \rangle$ since $\langle \beta^\mu_{[s]}, \mathcal{Z}_{[s]} \rangle = 0$. We then easily deduce that the quantity μ is given by (5.1.39). □

Remark. Note that we have used the constraint $\langle \mathcal{G}_{[s]}, \alpha^\mu_{[s]} \rangle = 0$, whereas we have $\langle \mathcal{G}_{[s]}, \alpha^{1\mu} \rangle = -\langle \mathcal{G}^2, \big(G^{22}\big)^{-1}\beta^{2\mu} \rangle$.

Standard Iterative Methods. In the following proposition, that applies to both the nonsingular and the singular cases, we obtain a splitting for the Schur complement $G_{[s]}$ yielding a convergent iteration matrix.

Proposition 5.1.13. *Let $G \in \mathbb{R}^{\omega,\omega}$ be a symmetric positive semi-definite matrix, assume that G admits the block-decomposition (5.1.34), and that the matrix G^{22} is nonsingular. Let $M^{11} \in \mathbb{R}^{\omega_1,\omega_1}$ be a symmetric positive definite matrix and consider the splittings $G_{[s]} = M^{11} - Z_{[s]}$ and $G = M - Z$ where*

$$M = \begin{bmatrix} M^{11} & 0 \\ G^{21} & G^{22} \end{bmatrix}, \tag{5.1.43}$$

and the iteration matrices $T_{[s]} = (M^{11})^{-1} Z_{[s]}$ *and* $T = M^{-1}Z$. *Assume also that* $2M^{11} - G^{11}$ *is positive definite. Then the matrices* $T_{[s]}$ *and* T *are convergent. Furthermore, let* $x_{[s]0} = 0$, $x_0 = 0$, $\mathfrak{b}_{[s]} = (M^{11})^{-1}\beta^{\mu}_{[s]}$, *and consider* $\mathfrak{b}$ *given by the block-decomposition* $\mathfrak{b} = (\mathfrak{b}_{[s]}, \Pi\mathfrak{b}_{[s]})$ *where* $\Pi = -(G^{22})^{-1}G^{21}$. *Consider also the iterates* $x_{[s]i} \in \mathbb{R}^{\omega_1}$ *and* $x_i \in \mathbb{R}^{\omega}$, $i \geq 0$, *given by*

$$\begin{cases} x_{[s]i+1} = T_{[s]} x_{[s]i} + \mathfrak{b}_{[s]}, \\ x_{i+1} = T x_i + \mathfrak{b}. \end{cases} \tag{5.1.44}$$

Then for all $i \geq 0$, *we have*

$$x_i^1 = x_{[s]i}, \qquad x_i^2 = \Pi x_{[s]i}. \tag{5.1.45}$$

Proof. For $x^1 \in \mathbb{R}^{\omega_1}$ we have

$$\begin{aligned} \langle (M^{11} + Z_{[s]})x^1, x^1\rangle &= \langle (2M^{11} - G_{[s]})x^1, x^1\rangle \\ &= \langle (2M^{11} - G^{11})x^1, x^1\rangle + \langle G^{22}x^2, x^2\rangle, \end{aligned}$$

where $x^2 = \Pi x^1 \in \mathbb{R}^{\omega-\omega_1}$, and from the assumptions we then deduce that $M^{11} + Z_{[s]}$ is positive definite. Therefore, since by Lemma 5.1.12 the matrix $G_{[s]}$ is symmetric positive semi-definite, Theorem 5.1.6 applies and the matrix $T_{[s]}$ is convergent. Similarly, the matrix $M + Z^t$ admits the block-decomposition

$$M + Z^t = \begin{bmatrix} 2M^{11} - G^{11} & 0 \\ 0 & G^{22} \end{bmatrix},$$

since G is symmetric. We then deduce that $M + Z^t$ is positive definite and since G is positive semi-definite, the matrix T is convergent by Theorem 5.1.6. Finally, the relation (5.1.45) is obtained by induction. It is trivially satisfied for $i = 0$, and assuming that it holds for $i \geq 1$, we deduce from (5.1.43) that

$$M^{-1}Z = \begin{bmatrix} (M^{11})^{-1}(M^{11} - G^{11}) & -(M^{11})^{-1}G^{12} \\ \Pi(M^{11})^{-1}(M^{11} - G^{11}) & -\Pi(M^{11})^{-1}G^{12} \end{bmatrix}$$

and hence, we obtain

$$x_{i+1}^1 = (M^{11})^{-1}(M^{11} - G_{[s]})x_i^1 + \mathfrak{b}_{[s]} = x_{[s]i+1},$$

and

$$x_{i+1}^2 = \Pi\Big((M^{11})^{-1}(M^{11} - G^{11})x_i^1 - (M^{11})^{-1}G^{12}\Pi x_i^1 + \mathfrak{b}_{[s]}\Big) = \Pi x_{i+1}^1,$$

which completes the proof. □

Using Lemmas 5.1.11 and 5.1.12 and Proposition 5.1.13, the convergence theorems for standard iterative methods (Theorems 5.1.7 and 5.1.8) can then be restated for the systems (5.1.37) and (5.1.41). In particular, this yields a sequence of iterates $\mu_{[s]}^{[i]}$ which converge towards $\mu_{[s]} = \mu - \mu_{[2]}$.

Conjugate Gradient Methods. If the matrix G is symmetric positive definite, we deduce from Lemma 5.1.11 that $G_{[s]}$ shares the same properties, and consequently the conjugate gradient method (Theorem 5.1.9) applies. Similarly, if the matrix G is symmetric positive semi-definite, we deduce from Lemma 5.1.12 that $G_{[s]}$ shares the same properties, and consequently the projected conjugate gradient method (Theorem 5.1.10) applies. In particular, this yields a sequence of iterates $\mu_{[s]}^{[i]}$ which converge towards $\mu_{[s]} = \mu - \mu_{[2]}$.

Rescaled Schur Complements. We conclude this section with a brief discussion on Schur complements for the rescaled transport linear systems. We consider a matrix $\widetilde{G}$, not necessarily symmetric, and we assume that the submatrix $\widetilde{G}^{22}$ is symmetric and nonsingular. We can then introduce the Schur complement $\widetilde{G}_{[s]} \in \mathbb{R}^{\omega_1,\omega_1}$ given by

$$\widetilde{G}_{[s]} = \widetilde{G}^{11} - \widetilde{G}^{12}\big(\widetilde{G}^{22}\big)^{-1}\widetilde{G}^{21}. \tag{5.1.46}$$

In the nonsingular case, we consider the linear system

$$\widetilde{G}_{[s]}\alpha_{[s]}^{\mu} = \widetilde{\beta}_{[s]}^{\mu}, \tag{5.1.47}$$

where

$$\widetilde{\beta}_{[s]}^{\mu} = \widetilde{\beta}^{1\mu} - \widetilde{G}^{12}\big(\widetilde{G}^{22}\big)^{-1}\widetilde{\beta}^{2\mu}, \tag{5.1.48}$$

and one can easily verify that $\widetilde{G}_{[s]}$ is nonsingular if $\widetilde{G}$ is nonsingular. In the singular case, we consider the constrained linear system

$$\begin{cases} \widetilde{G}_{[s]}\alpha_{[s]}^{\mu} = \widetilde{\beta}_{[s]}^{\mu}, \\ \langle \mathcal{G}_{[s]}, \alpha_{[s]}^{\mu} \rangle = 0, \end{cases} \tag{5.1.49}$$

where the constraint vector $\mathcal{G}_{[s]}$ is now given by

$$\mathcal{G}_{[s]} = \mathcal{G}^1 - \big(\widetilde{G}^{21}\big)^t\big(\widetilde{G}^{22}\big)^{-1}\mathcal{G}^2. \tag{5.1.50}$$

Assuming that $N(\widetilde{G}) = \mathbb{R}\mathcal{Z}$, $R(\widetilde{G}) = \widetilde{\mathcal{Z}}^{\perp}$, $\widetilde{\beta}^{\mu} \in R(\widetilde{G})$, and $\langle \mathcal{G}, \mathcal{Z} \rangle \neq 0$, we deduce that $N(\widetilde{G}_{[s]}) = \mathbb{R}\mathcal{Z}_{[s]}$ where $\mathcal{Z}_{[s]} = \mathcal{Z}^1$, and $R(\widetilde{G}_{[s]}) = \widetilde{\mathcal{Z}}_{[s]}^{\perp}$ where $\widetilde{\mathcal{Z}}_{[s]} = \widetilde{\mathcal{Z}}^1$. Using the symmetry of $\widetilde{G}^{22}$, we next obtain the identities $\langle\widetilde{\beta}^{\mu}_{[s]}, \widetilde{\mathcal{Z}}_{[s]}\rangle = \langle\widetilde{\beta}^{\mu}, \widetilde{\mathcal{Z}}\rangle = 0$ and $\langle\mathcal{G}_{[s]}, \mathcal{Z}_{[s]}\rangle = \langle\mathcal{G}, \mathcal{Z}\rangle \neq 0$. Therefore, $\widetilde{\beta}^{\mu}_{[s]} \in R(\widetilde{G}_{[s]})$ and $N(\widetilde{G}_{[s]})\oplus\mathcal{G}^{\perp}_{[s]} = \mathbb{R}^{\omega}$, and hence, by Proposition 4.1.3, the system (5.1.49) admits a unique solution $\alpha^{\mu}_{[s]}$.

Finally, in both nonsingular and singular cases, the quantity μ is given by

$$\mu = \mu_{[s]} + \mu_{[2]} = \langle\alpha^{\mu}_{[s]}, \beta^{\mu}_{[s]}\rangle + \langle(\widetilde{G}^{22})^{-1}\widetilde{\beta}^{2\mu}, \beta^{2\mu}\rangle, \tag{5.1.51}$$

where the vector $\beta^{\mu}_{[s]}$ is now given by

$$\beta^{\mu}_{[s]} = \beta^{1\mu} - (\widetilde{G}^{21})^{t}(\widetilde{G}^{22})^{-1}\beta^{2\mu}, \tag{5.1.52}$$

and the vector $\alpha^{\mu}_{[s]}$ is the unique solution of (5.1.47) or (5.1.49).

5.1.6 Stability of Iterative Algorithms for Nonnegative Mass Fractions

The purpose of this section is to show that the iterative algorithms obtained for positive mass fractions can be rewritten in terms of a rescaled system matrix that is still defined for nonnegative mass fractions. We will use left rescaled system matrices for standard iterative methods and symmetric rescaled system matrices for conjugate gradient methods. Left rescaled and symmetric rescaled system matrices are presented in Chapter 4 for all the transport coefficients. We also point out that instead of considering symmetric rescaled system matrices, one can also apply generalized versions of conjugate gradient methods for nonsymmetric matrices [SS86] [EGKS94] to the left rescaled system matrices, but these algorithms are omitted for brevity.

Proposition 5.1.14. *Let G and $\mathcal{N} \in \mathbb{R}^{\omega,\omega}$ be two matrices, assume that $\mathcal{N}$ is invertible, and consider $\widetilde{G} = \mathcal{N}^{-1}G$. Consider also the splittings $G = M - Z$ and $\widetilde{G} = \widetilde{M} - \widetilde{Z}$, the iteration matrices $T = M^{-1}Z$ and $\widetilde{T} = \widetilde{M}^{-1}\widetilde{Z}$, and assume that $\widetilde{M} = \mathcal{N}^{-1}M$ and $\widetilde{Z} = \mathcal{N}^{-1}Z$. Let β^{μ} and $x_0 \in \mathbb{R}^{\omega}$, $\widetilde{\beta}^{\mu} = \mathcal{N}^{-1}\beta^{\mu}$, $\widetilde{x}_0 = x_0$, and consider for $i \geq 0$ the iterates*

$$x_{i+1} = Tx_i + M^{-1}\beta^{\mu}, \tag{5.1.53}$$

$$\widetilde{x}_{i+1} = \widetilde{T}\widetilde{x}_i + \widetilde{M}^{-1}\widetilde{\beta}^{\mu}. \tag{5.1.54}$$

Then for all $i \geq 0$, we have $\widetilde{T} = T$, $\widetilde{x}_i = x_i$, and $\mu^{[i]} = \langle\widetilde{x}_i, \beta^{\mu}\rangle = \langle x_i, \beta^{\mu}\rangle$.

Proof. We have $\widetilde{T} = M^{-1}\mathcal{N}\mathcal{N}^{-1}Z = M^{-1}Z = T$ and $\widetilde{M}^{-1}\widetilde{\beta}^{\mu} = M^{-1}\mathcal{N}\mathcal{N}^{-1}\beta^{\mu} = M^{-1}\beta^{\mu}$ and therefore, we easily deduce by induction that $\widetilde{x}_i = x_i$ for all $i \geq 0$, and consequently, $\mu^{[i]} = \langle\widetilde{x}_i, \beta^{\mu}\rangle = \langle x_i, \beta^{\mu}\rangle$. □

Proposition 5.1.15. *Keeping the assumptions and notation of Proposition 5.1.14, assume that* $N(G) = N(\widetilde{G}) = \mathbb{R}\mathcal{Z}$, *let* $\mathcal{G} \in \mathbb{R}^{\omega}$ *be such that* $\langle \mathcal{Z}, \mathcal{G} \rangle \neq 0$, *and let* P *denote the oblique projector onto the hyperplane* $\mathcal{G}^{\perp}$ *along* $\mathbb{R}\mathcal{Z}$. *Let* $y_0 \in \mathbb{R}^{\omega}$, $\widetilde{y}_0 = y_0$, *and consider for* $i \geq 0$ *the iterates*

$$y_{i+1} = PTy_i + PM^{-1}\beta^{\mu}, \tag{5.1.55}$$

$$\widetilde{y}_{i+1} = P\widetilde{T}\widetilde{y}_i + P\widetilde{M}^{-1}\widetilde{\beta}^{\mu}. \tag{5.1.56}$$

Then for all $i \geq 0$, *we have* $\widetilde{y}_i = y_i$ *and* $\mu^{[i]} = \langle \widetilde{y}_i, \beta^{\mu} \rangle = \langle y_i, \beta^{\mu} \rangle$.

Proof. It is similar to the previous one. □

Proposition 5.1.16. *Let* G, M, *and* $\mathcal{N} \in \mathbb{R}^{\omega,\omega}$ *be three matrices, assume that* G, M, *and* $\mathcal{N}$ *are symmetric positive definite, and consider* $\widehat{G} = \mathcal{N}^{-1/2} G \mathcal{N}^{-1/2}$ *and* $\widehat{M} = \mathcal{N}^{-1/2} M \mathcal{N}^{-1/2}$. *Let* $\beta^{\mu} \in \mathbb{R}^{\omega}$, $x_0 \in \mathbb{R}^{\omega}$, $r_0 = \beta^{\mu} - Gx_0$, $\widehat{\beta}^{\mu} = \mathcal{N}^{-1/2}\beta^{\mu}$, $\widehat{x}_0 = \mathcal{N}^{1/2} x_0$, $\widehat{r}_0 = \mathcal{N}^{-1/2} r_0$, $p_0 = 0$, $\widehat{p}_0 = 0$, $t_0 = \widehat{t}_0 = 0$, *and consider for* $i \geq 1$ *the iterates*

$$\begin{cases} p_i = M^{-1} r_{i-1} + t_{i-1} p_{i-1}, \\ s_i = \langle r_{i-1}, M^{-1} r_{i-1} \rangle / \langle p_i, G p_i \rangle, \\ x_i = x_{i-1} + s_i p_i, \\ r_i = r_{i-1} - s_i G p_i, \\ t_i = \langle r_i, M^{-1} r_i \rangle / \langle r_{i-1}, M^{-1} r_{i-1} \rangle, \end{cases} \qquad \begin{cases} \widehat{p}_i = \widehat{M}^{-1} \widehat{r}_{i-1} + \widehat{t}_{i-1} \widehat{p}_{i-1}, \\ \widehat{s}_i = \langle \widehat{r}_{i-1}, \widehat{M}^{-1} \widehat{r}_{i-1} \rangle / \langle \widehat{p}_i, \widehat{G} \widehat{p}_i \rangle, \\ \widehat{x}_i = \widehat{x}_{i-1} + \widehat{s}_i \widehat{p}_i, \\ \widehat{r}_i = \widehat{r}_{i-1} - \widehat{s}_i \widehat{G} \widehat{p}_i, \\ \widehat{t}_i = \langle \widehat{r}_i, \widehat{M}^{-1} \widehat{r}_i \rangle / \langle \widehat{r}_{i-1}, \widehat{M}^{-1} \widehat{r}_{i-1} \rangle. \end{cases} \tag{5.1.57}$$

Then for all $i \geq 0$, *we have* $\widehat{p}_i = \mathcal{N}^{1/2} p_i$, $\widehat{r}_i = \mathcal{N}^{-1/2} r_i$, $\widehat{s}_i = s_i$, $\widehat{t}_i = t_i$, $\widehat{x}_i = \mathcal{N}^{1/2} x_i$, *and* $\mu^{[i]} = \langle \widehat{x}_i, \widehat{\beta}^{\mu} \rangle = \langle x_i, \beta^{\mu} \rangle$.

Proof. The proof is obtained by induction using the symmetry of $\mathcal{N}$. □

Proposition 5.1.17. *Keeping the assumptions and notation of Proposition 5.1.16, assume that* $N(G) = \mathbb{R}\mathcal{Z}$ *and* $N(\widehat{G}) = \mathbb{R}\widehat{\mathcal{Z}}$ *where* $\widehat{\mathcal{Z}} = \mathcal{N}^{1/2}\mathcal{Z}$. *Let* $\mathcal{G} \in \mathbb{R}^{\omega}$ *be such that* $\langle \mathcal{Z}, \mathcal{G} \rangle \neq 0$, *let* $\widehat{\mathcal{G}} = \mathcal{N}^{-1/2}\mathcal{G}$, *and let* P *denote the oblique projector onto the hyperplane* $\mathcal{G}^{\perp}$ *along* $\mathbb{R}\mathcal{Z}$ *and* $\widehat{P}$ *the oblique projector onto the hyperplane* $\widehat{\mathcal{G}}^{\perp}$ *along* $\mathbb{R}\widehat{\mathcal{Z}}$. *Let* $y_0 \in \mathbb{R}^{\omega}$, $\widehat{y}_0 = \mathcal{N}^{1/2} y_0$, *and consider for* $i \geq 1$ *the iterates*

$$y_i = y_{i-1} + P(s_i p_i), \tag{5.1.58}$$

$$\widehat{y}_i = \widehat{y}_{i-1} + \widehat{P}(\widehat{s}_i \widehat{p}_i). \tag{5.1.59}$$

Then for all $i \geq 0$, we have $\widehat{y}_i = \mathcal{N}^{1/2} y_i$ and $\mu^{[i]} = \langle \widehat{y}_i, \widehat{\beta}^\mu \rangle = \langle y_i, \beta^\mu \rangle$.

Proof. From $\widehat{\mathcal{Z}} = \mathcal{N}^{1/2}\mathcal{Z}$ and $\widehat{\mathcal{G}} = \mathcal{N}^{-1/2}\mathcal{G}$, we deduce that $\widehat{P} = \mathcal{N}^{1/2} P \mathcal{N}^{-1/2}$, and the proof is then similar to the previous one. □

Remark. In the following sections, Propositions 5.1.14–5.1.17 will be applied to the various transport linear systems with the matrix $\mathcal{N} = \mathrm{diag}\big((X_k)_{(r,k)\in\mathcal{B}^\mu}\big)$, where $\mathcal{B}^\mu$ is the indexing set associated with the transport linear system and X_k, $k \in \mathcal{S}$, the mole fractions, and, in this case, we immediately obtain $\mathcal{N}^{1/2} = \mathrm{diag}\big((X_k^{1/2})_{(r,k)\in\mathcal{B}^\mu}\big)$.

5.2 The Shear Viscosity

5.2.1 Iterative Methods for the System $H\alpha^\eta = \beta^\eta$

In this section we want to solve the linear system $H\alpha^\eta = \beta^\eta$ using either standard iterative or conjugate gradient methods. The system matrix $H \in \mathbb{R}^{n,n}$ is assumed to satisfy the properties $(H0)$–$(H2)$ presented in Section 4.2.1, and the corresponding indexing set is $\mathcal{B}^\eta = \{00\}\times\mathcal{S}$. Furthermore, we consider the matrix $db(H) \in \mathbb{R}^{n,n}$ formed by the diagonal of the only block of the matrix H,

$$db(H) = \mathrm{diag}(H^{0000}), \tag{5.2.1}$$

and from the kinetic theory results obtained in Sections 2.3 and 2.4, we can make the following assumptions.

- $(H3)$ $db(H)$ is symmetric positive definite.
- $(H4)$ $2db(H) - H$ is symmetric positive definite.

Theorem 5.2.1. *Let $H \in \mathbb{R}^{n,n}$ be a matrix satisfying the properties $(H0)$–$(H4)$ and let M be the matrix $db(H) + \mathrm{diag}(\mathfrak{d})$, where $\mathfrak{d} = (\mathfrak{d}_k^r)_{(r,k)\in\mathcal{B}^\eta}$ are coefficients such that $\mathfrak{d} \geq 0$. Consider the splitting $H = M - Z$ and the iteration matrix $T = M^{-1}Z$. Let $\beta^\eta \in \mathbb{R}^n$ be given by (4.2.5), $x_0 \in \mathbb{R}^n$, and consider for $i \geq 0$ the iterates*

$$x_{i+1} = Tx_i + M^{-1}\beta^\eta. \tag{5.2.2}$$

Then the matrix T is convergent, $\rho(T) < 1$, and we have the following limits

$$\begin{cases} \lim\limits_{i\to\infty} x_i = \alpha^\eta, \\ \lim\limits_{i\to\infty} \langle x_i, \beta^\eta \rangle = \eta, \end{cases} \tag{5.2.3}$$

where α^η is the unique solution of the linear system $H\alpha^\eta = \beta^\eta$. Moreover, for all $i \geq 0$, each partial sum

$$\eta^{[i]} = \langle \sum_{j=0}^{i} T^j M^{-1} \beta^\eta, \beta^\eta \rangle, \tag{5.2.4}$$

is positive, and we have

$$\lim_{i\to\infty} \eta^{[i]} = \langle \sum_{j=0}^{\infty} T^j M^{-1} \beta^\eta, \beta^\eta \rangle = \eta. \tag{5.2.5}$$

Proof. We deduce from $(H1)$–$(H2)$ that the matrix H is symmetric positive definite, and from $(H3)$–$(H4)$ and $\mathfrak{d} \geq 0$ that M and $M + Z = (2db(H) - H) + 2\mathrm{diag}(\mathfrak{d})$ are also symmetric positive definite. Therefore, Theorem 5.1.7 applies. □

Theorem 5.2.2. *Let $H \in \mathbb{R}^{n,n}$ be a matrix satisfying the properties $(H0)$–$(H3)$ and let M be the matrix $db(H) + \mathrm{diag}(\mathfrak{d})$, where $\mathfrak{d} = (\mathfrak{d}_k^r)_{(r,k)\in\mathcal{B}^\eta}$ are coefficients such that $\mathfrak{d} \geq 0$. Let $\beta^\eta \in \mathbb{R}^n$ be given by (4.2.5), $x_0 \in \mathbb{R}^n$, $r_0 = \beta^\eta - Hx_0$, $p_0 = 0$, $t_0 = 0$, and consider for $i \geq 1$ the iterates*

$$\begin{cases} p_i = M^{-1} r_{i-1} + t_{i-1} p_{i-1}, \\ s_i = \langle r_{i-1}, M^{-1} r_{i-1} \rangle / \langle p_i, H p_i \rangle, \\ x_i = x_{i-1} + s_i p_i, \\ r_i = r_{i-1} - s_i H p_i, \\ t_i = \langle r_i, M^{-1} r_i \rangle / \langle r_{i-1}, M^{-1} r_{i-1} \rangle. \end{cases} \tag{5.2.6}$$

Then the iterates x_i converge towards α^η in a finite number of steps, and the quantities

$$\eta^{[i]} = \langle x_i, \beta^\eta \rangle, \tag{5.2.7}$$

converge towards η in a finite number of steps. Furthermore, if $x_0 = 0$, we have $\eta^{[i]} > 0$ for all $i \geq 1$.

Proof. We deduce from $(H1)$–$(H2)$ that the matrix H is symmetric positive definite, and from $(H3)$ and $\mathfrak{d} \geq 0$ that the matrix M is symmetric positive definite. Therefore, Theorem 5.1.9 applies. □

Finally, we consider the matrix H resulting from the practical approximations presented in Section 2.10.

Proposition 5.2.3. *Let $\bar{A}_{kl}$ and η_{kl}, $k, l \in S$, be symmetric and positive coefficients, let η_k, $k \in S$, be positive coefficients, and assume that $Y > 0$. Then the matrix H given by (2.10.7)(2.10.8) satisfies $(H0)$–$(H4)$.*

Proof. The properties $(H0)$–$(H2)$ have already been established in Proposition 4.2.2, and $(H3)$ directly follows from $(H2)$. Finally, the matrix $2db(H) - H$ is clearly symmetric, and a straightforward computation yields for $x \in \mathbb{R}^n$

$$\langle x, (2db(H) - H)x\rangle = \sum_{k\in S} \frac{X_k^2}{\eta_k} x_k^2$$

$$+ \sum_{\substack{k,l\in S\\ l\neq k}} \frac{X_k X_l}{\eta_{kl}} \Big(\frac{5}{3\bar{A}_{kl}} \frac{m_k m_l}{(m_k + m_l)^2} (x_k + x_l)^2 + \big(\frac{m_l x_k - m_k x_l}{m_k + m_l}\big)^2 \Big),$$

and this quantity clearly vanishes if and only if $x = 0$. □

5.2.2 Standard Iterative Methods for the System $\widetilde{H}\alpha^\eta = \widetilde{\beta}^\eta$

In this section we want to solve the linear system $\widetilde{H}\alpha^\eta = \widetilde{\beta}^\eta$ using standard iterative methods. The system matrix $\widetilde{H} \in \mathbb{R}^{n,n}$ is assumed to satisfy the properties $(\widetilde{H}0)$–$(\widetilde{H}4)$ presented in Section 4.2.2, and we restate that $\mathcal{B}^\eta = \{00\}\times S$, $\mathcal{X}^\eta = \mathrm{diag}\big((X_k)_{(r,k)\in\mathcal{B}^\eta}\big)$, and $\Gamma^\eta \in \mathbb{R}^{n,n}$ is the permutation matrix associated with the reordering of $\mathcal{B}^\eta$ into $(\mathcal{B}^{\eta+}, \mathcal{B}^{\eta-})$. Furthermore, we consider the matrix $db(\widetilde{H}) \in \mathbb{R}^{n,n}$ formed by the diagonal of the only block of the matrix $\widetilde{H}$,

$$db(\widetilde{H}) = \mathrm{diag}(\widetilde{H}^{0000}), \tag{5.2.8}$$

and from the kinetic theory results obtained in Sections 3.1 and 3.2, we can make the following assumptions.

($\widetilde{H}5$) $db(H^{++})$ and $db(\widetilde{H})$ are symmetric positive definite.

($\widetilde{H}6$) $2db(H^{++}) - H^{++}$ is symmetric positive definite.

Theorem 5.2.4. *Let $\widetilde{H} \in \mathbb{R}^{n,n}$ be a matrix satisfying the properties $(\widetilde{H}0)$–$(\widetilde{H}6)$ and let $\widetilde{M}$ be the matrix $db(\widetilde{H}) + \mathrm{diag}(\widetilde{\mathfrak{d}})$, where $\widetilde{\mathfrak{d}} = (\widetilde{\mathfrak{d}}_k^r)_{(r,k)\in\mathcal{B}^\eta}$ are coefficients such that $\widetilde{\mathfrak{d}} \geq 0$. Consider the splitting $\widetilde{H} = \widetilde{M} - \widetilde{Z}$ and the iteration matrix $T = \widetilde{M}^{-1}\widetilde{Z}$. Let β^η and $\widetilde{\beta}^\eta \in \mathbb{R}^n$ be given by (4.2.5) and (4.2.10), respectively, $x_0 \in \mathbb{R}^n$, and consider for $i \geq 0$ the iterates*

$$x_{i+1} = Tx_i + \widetilde{M}^{-1}\widetilde{\beta}^\eta. \tag{5.2.9}$$

Then the matrix T is convergent, $\rho(T) < 1$, and we have the following limits

$$\begin{cases} \lim_{i\to\infty} x_i = \alpha^\eta, \\ \lim_{i\to\infty} \langle x_i, \beta^\eta \rangle = \eta, \end{cases} \tag{5.2.10}$$

where α^η is the unique solution of the linear system $\widetilde{H}\alpha^\eta = \widetilde{\beta}^\eta$. Moreover, for all $i \geq 0$, each partial sum

$$\eta^{[i]} = \langle \sum_{j=0}^{i} T^j \widetilde{M}^{-1} \widetilde{\beta}^\eta, \beta^\eta \rangle, \tag{5.2.11}$$

is positive, and we have

$$\lim_{i\to\infty} \eta^{[i]} = \langle \sum_{j=0}^{\infty} T^j \widetilde{M}^{-1} \widetilde{\beta}^\eta, \beta^\eta \rangle = \eta. \tag{5.2.12}$$

Finally, for positive mass fractions, the iteration matrix T and the iterates x_i and $\eta^{[i]}$ coincide with the ones in Theorem 5.2.1, provided that $\mathfrak{d} = \mathcal{X}^\eta \widetilde{\mathfrak{d}}$.

Proof. From $(\widetilde{H}1)$ we deduce that the matrix T admits the block-decomposition

$$T = \begin{bmatrix} T^{++} & 0 \\ T^{-+} & T^{--} \end{bmatrix},$$

where $T^{--} = \big(db(\widetilde{H}^{--}) + \operatorname{diag}(\widetilde{\mathfrak{d}}^-)\big)^{-1} \operatorname{diag}(\widetilde{\mathfrak{d}}^-)$ with $\widetilde{\mathfrak{d}} = \Gamma^\eta(\widetilde{\mathfrak{d}}^+, \widetilde{\mathfrak{d}}^-)$. We then obtain that $\rho(T^{--}) < 1$ since for $\lambda \in \sigma(T^{--})$ and $x \neq 0$ a corresponding eigenvector, we have

$$\lambda = \frac{\langle \operatorname{diag}(\widetilde{\mathfrak{d}}^-) x, x \rangle}{\langle \big(db(\widetilde{H}^{--}) + \operatorname{diag}(\widetilde{\mathfrak{d}}^-)\big) x, x \rangle},$$

so that λ is real and $0 \leq \lambda < 1$. Furthermore, we have $T^{++} = \big(M^{++}\big)^{-1} Z^{++}$ where $M^{++} = db(H^{++}) + \operatorname{diag}(\mathfrak{d}^+)$, $H^{++} = M^{++} - Z^{++}$, and $\mathfrak{d}^+ = \mathcal{X}^{\eta++} \widetilde{\mathfrak{d}}^+ \geq 0$. As in the proof of Theorem 5.2.1, we then obtain, using $(\widetilde{H}5)$–$(\widetilde{H}6)$, that M^{++} and $M^{++} + Z^{++} = \big(2db(H^{++}) - H^{++}\big) + 2\operatorname{diag}(\mathfrak{d}^+)$ are positive definite so that T^{++} is convergent and $\rho(T^{++}) < 1$ by Theorem 5.1.6. Therefore, by Lemma 5.1.3, the matrix T is convergent, and we have $\rho(T) = \max\big(\rho(T^{++}), \rho(T^{--})\big) < 1$. Consequently, we have $\lim_{i\to\infty} x_i = \alpha^\eta$, the solution of the linear system $\widetilde{H}\alpha^\eta = \widetilde{\beta}^\eta$, and $\lim_{i\to\infty} \langle x_i, \beta^\eta \rangle = \langle \alpha^\eta, \beta^\eta \rangle = \eta$. In addition, since $\beta^{\eta-} = 0$, using the block-decomposition of the matrices $\widetilde{H}$ and T, we obtain $\eta^{[i]} = \langle \big(H_i^\dagger\big)^{++} \beta^{\eta+}, \beta^{\eta+} \rangle$ where $\big(H_i^\dagger\big)^{++} = \sum_{j=0}^{i} \big(T^{++}\big)^j \big(M^{++}\big)^{-1}$, as may be verified after some algebra. Moreover, we can show as in the proof of Theorem 5.1.7 that $\left(H_i^{\dagger}\right)^{++}$ is symmetric positive definite so that $\eta^{[i]} > 0$ for all $i \geq 0$ since $\beta^{\eta+} \neq 0$. Next, we obtain by induction that $\eta^{[i]} = \langle x_{i+1}, \beta^{\eta} \rangle$, $i \geq 0$, for the particular case $x_0 = 0$, so that $\lim_{i\to\infty} \eta^{[i]} = \eta$ by (5.2.10). Finally, the correspondence with Theorem 5.2.1 results from Proposition 5.1.14. □

Remark. In the case of a pure species state of the mixture, i.e., $n^+ = 1$, we obtain a one-step convergence of the iterates $\eta^{[i]}$ for $\widetilde{\mathfrak{d}} = 0$.

Finally, we consider the matrix $\widetilde{H}$ resulting from the practical approximations presented in Section 2.10.

Proposition 5.2.5. *Let $\bar{\lambda}_{kl}$ and η_{kl}, $k, l \in S$, be symmetric and positive coefficients, let η_k, $k \in S$, be positive coefficients, and assume that $Y \geq 0$, $Y \neq 0$. Then the matrix $\widetilde{H}$ given by (3.7.1)(3.7.2) satisfies $(\widetilde{H}0)$–$(\widetilde{H}6)$.*

Proof. It is similar to the one of Propositions 4.2.5 and 5.2.3. □

5.2.3 Conjugate Gradient Methods for the System $\widehat{H}\widehat{\alpha}^{\eta} = \widehat{\beta}^{\eta}$

In this section we assume that the matrix $\widehat{H}$ satisfies the properties $(\widehat{H}0)$–$(\widehat{H}2)$ given in Section 4.2.3, and we want to obtain the only solution $\widehat{\alpha}^{\eta}$ of the linear system $\widehat{H}\widehat{\alpha}^{\eta} = \widehat{\beta}^{\eta}$ using conjugate gradient methods.

Theorem 5.2.6. *Let $\widehat{H} \in \mathbb{R}^{n,n}$ be a matrix satisfying the properties $(\widehat{H}0)$–$(\widehat{H}2)$ and let $\widehat{M}$ be the matrix $db(\widehat{H}) + \mathrm{diag}(\widehat{\mathfrak{d}})$, where $\widehat{\mathfrak{d}} = (\widehat{\mathfrak{d}}_k^r)_{(r,k)\in\mathcal{B}^{\eta}}$ are coefficients such that $\widehat{\mathfrak{d}} \geq 0$. Let $\widehat{\beta}^{\eta} \in \mathbb{R}^n$ be given by (4.2.16), $\widehat{x}_0 \in \mathbb{R}^n$, $\widehat{r}_0 = \widehat{\beta}^{\eta} - \widehat{H}\widehat{x}_0$, $\widehat{p}_0 = 0$, $t_0 = 0$, and consider for $i \geq 1$ the iterates*

$$\left\{\begin{aligned} \widehat{p}_i &= \widehat{M}^{-1}\widehat{r}_{i-1} + t_{i-1}\widehat{p}_{i-1}, \\ s_i &= \langle \widehat{r}_{i-1}, \widehat{M}^{-1}\widehat{r}_{i-1}\rangle / \langle \widehat{p}_i, \widehat{H}\widehat{p}_i \rangle, \\ \widehat{x}_i &= \widehat{x}_{i-1} + s_i\widehat{p}_i, \\ \widehat{r}_i &= \widehat{r}_{i-1} - s_i\widehat{H}\widehat{p}_i, \\ t_i &= \langle \widehat{r}_i, \widehat{M}^{-1}\widehat{r}_i\rangle / \langle \widehat{r}_{i-1}, \widehat{M}^{-1}\widehat{r}_{i-1}\rangle. \end{aligned}\right. \tag{5.2.13}$$

Then the iterates $\widehat{x}_i$ converge towards $\widehat{\alpha}^{\eta}$ in a finite number of steps, and the quantities

$$\eta^{[i]} = \langle \widehat{x}_i, \widehat{\beta}^{\eta} \rangle, \tag{5.2.14}$$

converge towards η in a finite number of steps. Furthermore, if $\widehat{x}_0 = 0$, we have $\eta^{[i]} > 0$ for all $i \geq 1$. Finally, for positive mass fractions, the iterates $\eta^{[i]}$ coincide with the ones in Theorem 5.2.2 and $\widehat{x}_i = (\mathcal{X}^\eta)^{1/2} x_i$, provided that $\mathfrak{d} = \mathcal{X}^\eta \widehat{\mathfrak{d}}$ and $\widehat{x}_0 = (\mathcal{X}^\eta)^{1/2} x_0$.

Finally, we consider the matrix $\widehat{H}$ resulting from the practical approximations presented in Section 2.10.

Proposition 5.2.7. *Under the assumptions of Proposition 5.2.5, the matrix $\widehat{H}$, obtained from the matrix $\widetilde{H}$ by replacing X_l by $\sqrt{X_k X_l}$ in all the terms $\widetilde{H}_{kl}^{rs}$, $k \neq l$, satisfies $(\widehat{H}0)$–$(\widehat{H}2)$.*

5.3 The Volume Viscosity

In this section we assume that we have $p \geq 1$ and $p^+ \geq 1$ since we have seen in Section 4.3.1 that the volume viscosity of monatomic gas mixtures vanishes [FK72]. Furthermore, the simplified transport linear systems $K_{[10]}\alpha^\kappa_{[10]} = \beta^\kappa_{[10]}$ and $K_{[d]}\alpha^\kappa_{[d]} = \beta^\kappa_{[d]}$ will not be considered since we will see in Chapter 6 that the volume viscosities $\kappa_{[10]}$ and $\kappa_{[d]}$ are not accurate.

5.3.1 Iterative Methods for the System $K\alpha^\kappa = \beta^\kappa$

In this section we want to solve the linear system $K\alpha^\kappa = \beta^\kappa$ using either standard iterative or conjugate gradient methods, and obtain the only solution α^κ such that $\langle \mathcal{K}, \alpha^\kappa \rangle = 0$. The system matrix $K \in \mathbb{R}^{n+p,n+p}$ is assumed to satisfy the properties $(K0)$–$(K5)$ presented in Section 4.3.1, and the corresponding indexing set is $\mathcal{B}^\kappa = \{10\}\times\mathcal{S} \cup \{01\}\times\mathcal{P}$. Furthermore, we consider the matrix $db(K) \in \mathbb{R}^{n+p,n+p}$ formed by the diagonal of the four blocks of the matrix K,

$$db(K) = \begin{bmatrix} \mathrm{diag}(K^{1010}) & \mathrm{diag}(K^{1001}) \\ \mathrm{diag}(K^{0110}) & \mathrm{diag}(K^{0101}) \end{bmatrix}, \tag{5.3.1}$$

and from the kinetic theory results obtained in Sections 2.3 and 2.5, we can make the following assumptions.

- $(K6)$ $db(K)$ is symmetric positive semi-definite; for $n \geq 2$, this matrix is also positive definite, and for $n = 1$ its nullspace is spanned by the vector $(1, 1)$.

($K7$) $2db(K) - K$ is symmetric positive semi-definite; for $n \geq 3$, this matrix is also positive definite; for $n = 2$, its nullspace is spanned by the vector $(1, -1, 1, -1)$ if $p = 2$, and $(1, -1, 1)$ or $(1, -1, -1)$ depending on which species is polyatomic if $p = 1$; for $n = 1$, its nullspace is spanned by the vector $(1, 1)$.

Theorem 5.3.1. *Let $K \in \mathbb{R}^{n+p,n+p}$ be a matrix satisfying the properties ($K0$)–($K7$) and let M be the matrix $db(K) + \mathrm{diag}(\mathfrak{d})$, where $\mathfrak{d} = (\mathfrak{d}_k^r)_{(r,k)\in\mathcal{B}^\kappa}$ are coefficients such that $\mathfrak{d} \geq 0$ in general, and $\mathfrak{d} \neq 0$ in the particular cases $n = 1$ or 2. Consider the splitting $K = M - Z$ and the iteration matrix $T = M^{-1}Z$. Let $\mathcal{V}$, $\mathcal{K}$, and $\beta^\kappa \in \mathbb{R}^{n+p}$ be given by (4.3.6), (4.3.7), and (4.3.9), respectively, and let $P = I - \mathcal{V}\otimes\mathcal{K}/\langle\mathcal{V},\mathcal{K}\rangle$ denote the oblique projector onto $\mathcal{K}^\perp$ along $\mathbb{R}\mathcal{V}$. Let $x_0 \in \mathbb{R}^{n+p}$, $y_0 = Px_0$, and consider for $i \geq 0$ the iterates*

$$x_{i+1} = Tx_i + M^{-1}\beta^\kappa, \tag{5.3.2}$$

$$y_{i+1} = PTy_i + PM^{-1}\beta^\kappa. \tag{5.3.3}$$

Then $y_i = Px_i$ for all $i \geq 0$, the matrices T and PT are convergent, $\rho(T) = 1$, $\rho(PT) < 1$, and we have the following limits

$$\begin{cases} \lim\limits_{i\to\infty} y_i = P(\lim\limits_{i\to\infty} x_i) = \alpha^\kappa, \\ \lim\limits_{i\to\infty} \langle y_i, \beta^\kappa\rangle = \lim\limits_{i\to\infty} \langle x_i, \beta^\kappa\rangle = \kappa, \end{cases} \tag{5.3.4}$$

where α^κ is the unique solution of the constrained linear system $K\alpha^\kappa = \beta^\kappa$ and $\langle\mathcal{K}, \alpha^\kappa\rangle = 0$. Moreover, for all $i \geq 0$, each partial sum

$$\kappa^{[i]} = \langle\sum_{j=0}^{i}(PT)^j PM^{-1}\beta^\kappa, \beta^\kappa\rangle = \langle\sum_{j=0}^{i} T^j M^{-1}\beta^\kappa, \beta^\kappa\rangle, \tag{5.3.5}$$

is positive, and we have

$$\lim_{i\to\infty} \kappa^{[i]} = \langle\sum_{j=0}^{\infty}(PT)^j PM^{-1}\beta^\kappa, \beta^\kappa\rangle = \kappa. \tag{5.3.6}$$

Proof. We deduce from ($K6$) and the assumptions on $\mathfrak{d}$ that $M = db(K)+\mathrm{diag}(\mathfrak{d})$ is the sum of two positive semi-definite matrices whose nullspaces do not contain any common vector except zero. Indeed, for $n \geq 2$ the matrix $db(K)$ is positive definite, and for $n = 1$ the vector $(1, 1)$ is not in the nullspace of $\mathrm{diag}(\mathfrak{d})$ since $\mathfrak{d} \neq 0$. Therefore, M is positive definite, and the same argument holds for $M+Z$ since $M+Z = (2db(K)-K)+2\mathrm{diag}(\mathfrak{d})$,

$2db(K) - K$ is positive definite for $n \geq 3$ and for $n = 1$ or $n = 2$ we have $\mathfrak{d} \neq 0$ and $N(2db(K) - K)$ is spanned by a vector that has only nonzero components. Therefore, using $(K1)$–$(K5)$, we deduce that Theorem 5.1.8 applies with $\mathcal{C} = \mathcal{K}^{\perp}$ and $S = \mathbb{R}\mathcal{V}$. □

Theorem 5.3.2. *Let $K \in \mathbb{R}^{n+p,n+p}$ be a matrix satisfying the properties $(K0)$–$(K6)$ and let M be the matrix $db(K) + \mathrm{diag}(\mathfrak{d})$ or $\mathrm{diag}(K) + \mathrm{diag}(\mathfrak{d})$, where $\mathfrak{d} = (\mathfrak{d}_k^r)_{(r,k)\in\mathcal{B}^\kappa}$ are coefficients such that $\mathfrak{d} \geq 0$ in general, and $\mathfrak{d} \neq 0$ in the particular case $n = 1$. Let $\mathcal{V}$, $\mathcal{K}$, and $\beta^\kappa \in \mathbb{R}^{n+p}$ be given by (4.3.6), (4.3.7), and (4.3.9), respectively, and let $P = I - \mathcal{V}\otimes\mathcal{K}/\langle\mathcal{V},\mathcal{K}\rangle$ denote the oblique projector onto $\mathcal{K}^{\perp}$ along $\mathbb{R}\mathcal{V}$. Let $x_0 \in \mathbb{R}^{n+p}$, $y_0 = Px_0$, $r_0 = \beta^\kappa - Kx_0$, $p_0 = 0$, $t_0 = 0$, and consider for $i \geq 1$ the iterates*

$$\begin{cases} p_i = M^{-1}r_{i-1} + t_{i-1}p_{i-1}, \\ s_i = \langle r_{i-1}, M^{-1}r_{i-1}\rangle / \langle p_i, Kp_i\rangle, \\ x_i = x_{i-1} + s_i p_i, \\ y_i = y_{i-1} + P(s_i p_i), \\ r_i = r_{i-1} - s_i K p_i, \\ t_i = \langle r_i, M^{-1}r_i\rangle / \langle r_{i-1}, M^{-1}r_{i-1}\rangle. \end{cases} \tag{5.3.7}$$

Then $y_i = Px_i$ for all $i \geq 0$, the iterates y_i converge towards α^κ in a finite number of steps, and the quantities

$$\kappa^{[i]} = \langle y_i, \beta^\kappa\rangle = \langle x_i, \beta^\kappa\rangle, \tag{5.3.8}$$

converge towards κ in a finite number of steps. Furthermore, if $x_0 = 0$, we have $\kappa^{[i]} > 0$ for all $i \geq 1$.

Proof. We have already shown in the previous proof that the matrix $db(K) + \mathrm{diag}(\mathfrak{d})$ is positive definite and the proof of the positive definiteness of $\mathrm{diag}(K) + \mathrm{diag}(\mathfrak{d})$ is similar. Therefore, using $(K1)$–$(K5)$, we deduce that Theorem 5.1.10 applies. □

Finally, we consider the matrix K resulting from the practical approximations presented in Section 2.10.

Proposition 5.3.3. *Let $\bar{\lambda}_{kl}$ and η_{kl}, $k, l \in \mathcal{S}$, be symmetric and positive coefficients, let η_k, $k \in \mathcal{S}$, c_l^{int} and ξ_l^{int}, $l \in \mathcal{P}$, be positive quantities, and assume that $Y > 0$. Then the matrix K given by (2.10.9)–(2.10.14) satisfies $(K0)$–$(K7)$.*

Proof. By Proposition 4.3.3, only $(K6)$–$(K7)$ need to be proven. First, for $n = 1$, the property $(K6)$ directly results from $db(K) = K$, and for $n \geq 2$, we deduce from the relation

$$\langle x, db(K)x\rangle = \sum_{k\in\mathcal{S}} \Big\langle \sum_{r\in\mathcal{F}_k} x_k^r e^{rk}, K \sum_{r\in\mathcal{F}_k} x_k^r e^{rk} \Big\rangle,$$

and the equivalence $\langle x, Kx\rangle = 0$ if and only if $x \in \mathbb{R}\mathcal{V}$ that $db(K)$ is positive definite since $\mathcal{V}$ has only nonzero components. Next, the matrix $2db(K) - K$ is obviously symmetric, and an explicit calculation yields for $x \in \mathbb{R}^{n+p}$

$$\langle x, (2db(K) - K)x\rangle = \sum_{k\in\mathcal{P}} \frac{4}{k_{\mathrm{B}}\pi} \frac{X_k^2}{\eta_k} \frac{c_k^{\mathrm{int}}}{\xi_k^{\mathrm{int}}} (x_k^{10} - x_k^{01})^2$$

$$+ \sum_{\substack{k,l\in\mathcal{S}\\ l\neq k}} \frac{5}{2} \frac{X_k X_l}{\bar{A}_{kl}\eta_{kl}} \frac{m_k m_l}{(m_k+m_l)^2} (x_k^{10} + x_l^{10})^2$$

$$+ \sum_{\substack{k\in\mathcal{S},l\in\mathcal{P}\\ k\neq l}} \frac{4}{k_{\mathrm{B}}\pi} \frac{X_k X_l}{\eta_{kl}} \frac{c_l^{\mathrm{int}}}{\xi_l^{\mathrm{int}}} \Big(\frac{m_l x_k^{10} - m_k x_l^{10}}{m_k + m_l} + x_l^{01}\Big)^2,$$

so that $2db(K) - K$ is positive semi-definite. Let us consider a vector x such that $\langle x, (2db(K) - K)x\rangle = 0$. From the above expression, we see that $x_k^{10} = x_k^{01} = -x_l^{10} = -x_l^{01}$ for $k, l \in \mathcal{P}$, $k \neq l$. Similarly, we obtain that if $k \notin \mathcal{P}$, $l \in \mathcal{P}$, $k \neq l$, then $x_k^{10} = -x_l^{10} = -x_l^{01}$, and if $k \notin \mathcal{P}$, $l \notin \mathcal{P}$, $k \neq l$, then $x_k^{10} = -x_l^{10}$. We conclude that if $n \geq 3$, the matrix $2db(K) - K$ is positive definite because it is possible to change the signs an odd number of times in the equalities, i.e., $x_k^{10} = -x_l^{10}$, $-x_l^{10} = x_m^{10}$, and $x_m^{10} = -x_k^{10}$, for $k \neq l$, $l \neq m$, and $m \neq k$. In the particular case $n = 2$, the nullspace of $2db(K) - K$ is spanned by the vector $(1, -1, 1, -1)$ if $p = 2$, and the vector $(1, -1, 1)$ or $(1, -1, -1)$ depending on which species is polyatomic if $p = 1$. Finally, if $n = 1$, we have $2db(K) - K = K$, and the nullspace of these matrices is spanned by $(1, 1)$. □

5.3.2 Iterative Methods for the System $K_{[01]}\alpha_{[01]}^{\kappa} = \beta_{[01]}^{\kappa}$

In this section we want to solve the linear system $K_{[01]}\alpha_{[01]}^{\kappa} = \beta_{[01]}^{\kappa}$ using either standard iterative or conjugate gradient methods. The system matrix $K_{[01]} \in \mathbb{R}^{p,p}$ is assumed to satisfy the properties $(K_{[01]}0)$–$(K_{[01]}2)$ presented in Section 4.3.2, and the corresponding indexing set is $\mathcal{B}_{[01]}^{\kappa} = \{01\}\times\mathcal{P}$. Furthermore, we consider the matrix $db(K_{[01]}) \in \mathbb{R}^{p,p}$ formed by the diagonal of the only block of the matrix $K_{[01]}$,

$$db(K_{[01]}) = \mathrm{diag}(K_{[01]}^{0101}), \tag{5.3.9}$$

and from the kinetic theory results obtained in Sections 2.3 and 2.5, we can make the following assumptions.

$(K_{[01]}3)$ $db(K_{[01]})$ is symmetric positive definite.

$(K_{[01]}4)$ $2db(K_{[01]}) - K_{[01]}$ is symmetric positive definite.

Theorem 5.3.4. *Let $K_{[01]} \in \mathbb{R}^{p,p}$ be a matrix satisfying the properties $(K_{[01]}0)$–$(K_{[01]}4)$ and let M be the matrix $db(K_{[01]}) + \operatorname{diag}(\mathfrak{d})$, where $\mathfrak{d} = (\mathfrak{d}_k^r)_{(r,k)\in\mathcal{B}_{[01]}^{\kappa}}$ are coefficients such that $\mathfrak{d} \geq 0$. Consider the splitting $K_{[01]} = M - Z$ and the iteration matrix $T = M^{-1}Z$. Let $\beta_{[01]}^{\kappa} \in \mathbb{R}^p$ be given by (4.3.18), $x_0 \in \mathbb{R}^p$, and consider for $i \geq 0$ the iterates*

$$x_{i+1} = Tx_i + M^{-1}\beta_{[01]}^{\kappa}. \tag{5.3.10}$$

Then the matrix T is convergent, $\rho(T) < 1$, and we have the following limits

$$\begin{cases} \lim\limits_{i\to\infty} x_i = \alpha_{[01]}^{\kappa}, \\ \lim\limits_{i\to\infty} \langle x_i, \beta_{[01]}^{\kappa}\rangle = \kappa_{[01]}, \end{cases} \tag{5.3.11}$$

where $\alpha_{[01]}^{\kappa}$ is the unique solution of the linear system $K_{[01]}\alpha_{[01]}^{\kappa} = \beta_{[01]}^{\kappa}$. Moreover, for all $i \geq 0$, each partial sum

$$\kappa_{[01]}^{[i]} = \langle \sum_{j=0}^{i} T^j M^{-1}\beta_{[01]}^{\kappa}, \beta_{[01]}^{\kappa}\rangle, \tag{5.3.12}$$

is positive, and we have

$$\lim_{i\to\infty} \kappa_{[01]}^{[i]} = \langle \sum_{j=0}^{\infty} T^j M^{-1}\beta_{[01]}^{\kappa}, \beta_{[01]}^{\kappa}\rangle = \kappa_{[01]}. \tag{5.3.13}$$

Theorem 5.3.5. *Let $K_{[01]} \in \mathbb{R}^{p,p}$ be a matrix satisfying the properties $(K_{[01]}0)$–$(K_{[01]}3)$ and let M be the matrix $db(K_{[01]}) + \operatorname{diag}(\mathfrak{d})$, where $\mathfrak{d} = (\mathfrak{d}_k^r)_{(r,k)\in\mathcal{B}_{[01]}^{\kappa}}$ are coefficients such that $\mathfrak{d} \geq 0$. Let $\beta_{[01]}^{\kappa} \in \mathbb{R}^p$ be given by (4.3.18), $x_0 \in \mathbb{R}^p$, $r_0 = \beta_{[01]}^{\kappa} - K_{[01]}x_0$, $p_0 = 0$, $t_0 = 0$, and consider for $i \geq 1$ the iterates*

$$\begin{cases} p_i = M^{-1}r_{i-1} + t_{i-1}p_{i-1}, \\ s_i = \langle r_{i-1}, M^{-1}r_{i-1}\rangle / \langle p_i, K_{[01]}p_i\rangle, \\ x_i = x_{i-1} + s_i p_i, \\ r_i = r_{i-1} - s_i K_{[01]} p_i, \\ t_i = \langle r_i, M^{-1}r_i\rangle / \langle r_{i-1}, M^{-1}r_{i-1}\rangle. \end{cases} \tag{5.3.14}$$

Then the iterates x_i converge towards $\alpha_{[01]}^{\kappa}$ in a finite number of steps, and the quantities

$$\kappa_{[01]}^{[i]} = \langle x_i, \beta_{[01]}^{\kappa}\rangle, \tag{5.3.15}$$

converge towards $\kappa_{[01]}$ in a finite number of steps. Furthermore, if $x_0 = 0$, we have $\kappa_{[01]}^{[i]} > 0$ for all $i \geq 1$.

Finally, we consider the matrix $K_{[01]}$ resulting from the practical approximations presented in Section 2.10.

Proposition 5.3.6. *Let $\bar{A}_{kl}$ and η_{kl}, $k,l \in \mathcal{S}$, be symmetric and positive coefficients, let η_k, $c_k^{\rm int}$ and $\xi_k^{\rm int}$, $k \in \mathcal{P}$, be positive quantities, and assume that $Y > 0$. Then the matrix $K_{[01]}$ given by (2.10.13)(2.10.14) satisfies ($K_{[01]}0$)–($K_{[01]}4$).*

5.3.3 Standard Iterative Methods for the System $\widetilde{K}\alpha^\kappa = \widetilde{\beta}^\kappa$

In this section we want to solve the linear system $\widetilde{K}\alpha^\kappa = \widetilde{\beta}^\kappa$ using standard iterative methods and obtain the only solution α^κ such that $\langle \mathcal{K}, \alpha^\kappa \rangle = 0$. The system matrix $\widetilde{K} \in \mathbb{R}^{n+p,n+p}$ is assumed to satisfy the properties ($\widetilde{K}0$)–($\widetilde{K}5$) presented in Section 4.3.3, and we restate that $\mathcal{B}^\kappa = \{10\}\times\mathcal{S} \,\cup\, \{01\}\times\mathcal{P}$, $\mathcal{X}^\kappa = \mathrm{diag}\big((X_k)_{(r,k)\in\mathcal{B}^\kappa}\big)$, and $\Gamma^\kappa \in \mathbb{R}^{n+p,n+p}$ is the permutation matrix associated with the reordering of $\mathcal{B}^\kappa$ into $(\mathcal{B}^{\kappa+}, \mathcal{B}^{\kappa-})$. Furthermore, we consider the matrix $db(\widetilde{K}) \in \mathbb{R}^{n+p,n+p}$ formed by the diagonal of the four blocks of the matrix $\widetilde{K}$,

$$db(\widetilde{K}) = \begin{bmatrix} \mathrm{diag}(\widetilde{K}^{1010}) & \mathrm{diag}(\widetilde{K}^{1001}) \\ \mathrm{diag}(\widetilde{K}^{0110}) & \mathrm{diag}(\widetilde{K}^{0101}) \end{bmatrix}, \tag{5.3.16}$$

and from the kinetic theory results obtained in Sections 3.1 and 3.3, we can make the following assumptions.

($\widetilde{K}6$) $db(K^{++})$ and $db(\widetilde{K})$ are symmetric positive semi-definite; for $n^+ \geq 2$, these matrices are also positive definite; for $n^+ = 1$, the nullspace of $db(K^{++})$ is spanned by the vector $\mathcal{V}^+ = (1,1)$, and the nullspace of $db(\widetilde{K})$ is spanned by the vector $\Gamma^\kappa(\mathcal{V}^+, 0)$.

($\widetilde{K}7$) $2db(K^{++}) - K^{++}$ is symmetric positive semi-definite; for $n^+ \geq 3$, this matrix is also positive definite; for $n^+ = 2$, its nullspace is spanned by the vector $(1,-1,1,-1)$ if $p^+ = 2$, and $(1,-1,1)$ or $(1,-1,-1)$ depending on which species is polyatomic if $p^+ = 1$; for $n^+ = 1$, its nullspace is spanned by the vector $(1,1)$.

Theorem 5.3.7. *Let $\widetilde{K} \in \mathbb{R}^{n+p,n+p}$ be a matrix satisfying the properties ($\widetilde{K}0$)–($\widetilde{K}7$), and let $\widetilde{M}$ be the matrix $db(\widetilde{K}) + \mathrm{diag}(\widetilde{\mathfrak{d}})$, where $\widetilde{\mathfrak{d}} = (\widetilde{\mathfrak{d}}_k^r)_{(r,k)\in\mathcal{B}^\kappa}$ are coefficients such that $\widetilde{\mathfrak{d}} \geq 0$ in general, and $\widetilde{\mathfrak{d}}^+ \neq 0$ in the particular cases $n^+ = 1$ and 2. Consider the splitting $\widetilde{K} = \widetilde{M} - \widetilde{Z}$ and the iteration matrix $T = \widetilde{M}^{-1}\widetilde{Z}$. Let $\mathcal{V}$, $\mathcal{K}$, β^κ, and $\widetilde{\beta}^\kappa \in \mathbb{R}^{n+p}$ be given by (4.3.6), (4.3.7), (4.3.9), and (4.3.24), respectively, and let $P = I - \mathcal{V}\otimes\mathcal{K}/\langle\mathcal{V},\mathcal{K}\rangle$ be the oblique projector onto $\mathcal{K}^\perp$ along $\mathbb{R}\mathcal{V}$. Let $x_0 \in \mathbb{R}^{n+p}$, $y_0 = Px_0$, and consider for $i \geq 0$ the iterates*

$$x_{i+1} = Tx_i + \widetilde{M}^{-1}\widetilde{\beta}^\kappa, \tag{5.3.17}$$

$$y_{i+1} = PTy_i + P\widetilde{M}^{-1}\widetilde{\beta}^\kappa. \tag{5.3.18}$$

Then $y_i = Px_i$ for all $i \geq 0$, the matrices T and PT are convergent, $\rho(T) = 1$, $\rho(PT) < 1$, and we have the following limits

$$\begin{cases} \lim_{i\to\infty} y_i = P(\lim_{i\to\infty} x_i) = \alpha^\kappa, \\ \lim_{i\to\infty} \langle y_i, \beta^\kappa\rangle = \lim_{i\to\infty} \langle x_i, \beta^\kappa\rangle = \kappa, \end{cases} \tag{5.3.19}$$

where α^κ is the unique solution of the constrained linear system $\widetilde{K}\alpha^\kappa = \widetilde{\beta}^\kappa$ and $\langle \mathcal{K}, \alpha^\kappa\rangle = 0$. Moreover, for all $i \geq 0$, each partial sum

$$\kappa^{[i]} = \langle \sum_{j=0}^{i} (PT)^j P\widetilde{M}^{-1}\widetilde{\beta}^\kappa, \beta^\kappa\rangle = \langle \sum_{j=0}^{i} T^j \widetilde{M}^{-1}\widetilde{\beta}^\kappa, \beta^\kappa\rangle, \tag{5.3.20}$$

is positive, and we have

$$\lim_{i\to\infty} \kappa^{[i]} = \langle \sum_{j=0}^{\infty} (PT)^j P\widetilde{M}^{-1}\widetilde{\beta}^\kappa, \beta^\kappa\rangle = \kappa. \tag{5.3.21}$$

Finally, for positive mass fractions, the iteration matrix T and the iterates x_i, y_i, and $\kappa^{[i]}$ coincide with the ones in Theorem 5.3.1, provided that $\mathfrak{d} = \mathcal{X}^\kappa\widetilde{\mathfrak{d}}$.

Proof. From ($\widetilde{K}1$), we deduce that the matrix T admits the block-decomposition

$$T = \begin{bmatrix} T^{++} & 0 \\ T^{-+} & T^{--} \end{bmatrix},$$

and we can show as in the proof of Theorem 5.2.1 that $\rho(T^{--}) < 1$. Furthermore, we have $T^{++} = (M^{++})^{-1} Z^{++}$ where $M^{++} = db(K^{++}) + \mathrm{diag}(\mathfrak{d}^+)$, $K^{++} = M^{++} - Z^{++}$, and $\mathfrak{d}^+ = \mathcal{X}^{\kappa++}\widetilde{\mathfrak{d}}^+$. As in the proof of Theorem 5.3.1, we deduce from ($\widetilde{K}6$)–($\widetilde{K}7$) and the assumptions on $\widetilde{\mathfrak{d}}$ that the matrix T^{++} is convergent. Therefore, by Lemma 5.1.3, the matrix T is convergent and $\rho(T) = 1$ since $T\mathcal{V} = \mathcal{V}$. We can then apply Theorem 5.1.5 so that $y_i = Px_i$ for all $i \geq 0$, the matrix PT is convergent, $\rho(PT) < 1$, and $\lim_{i\to\infty} y_i = P(\lim_{i\to\infty} x_i) = \alpha^\kappa$, where α^κ is the solution of the constrained linear system $\widetilde{K}\alpha^\kappa = \widetilde{\beta}^\kappa$ and $\langle \mathcal{K}, \alpha^\kappa\rangle = 0$. Consequently, we also have $\lim_{i\to\infty}\langle y_i, \beta^\kappa\rangle = \langle \alpha^\kappa, \beta^\kappa\rangle = \kappa$ and $\lim_{i\to\infty}\langle x_i, \beta^\kappa\rangle = \kappa$ since $y_i = Px_i$ and $P^t\beta^\kappa = \beta^\kappa$. In addition, since $\beta^{\kappa-} = 0$, we obtain $\kappa^{[i]} = \langle (K_i^\dagger)^{++}\beta^{\kappa+}, \beta^{\kappa+}\rangle$, $i \geq 0$, where

$$(K_i^\dagger)^{++} = \sum_{j=0}^{i} (P^{++}T^{++})^j P^{++} (M^{++})^{-1} (P^{++})^t,$$

as may be verified after some algebra. Moreover, we can show as in the proof of Theorem 5.1.8 that the matrix $\big(K_i^{\dagger}\big)^{++}$ is symmetric, positive semi-definite, and positive definite on the hyperplane $\big(\mathcal{V}^+\big)^{\perp}$. We then deduce that $\kappa^{[i]} > 0$ since $\beta^{\kappa +} \in \big(\mathcal{V}^+\big)^{\perp}$ and $\beta^{\kappa +} \neq 0$. Next, we can easily show by induction that $\kappa^{[i]} = \langle y_{i+1}, \beta^{\kappa} \rangle$, $i \geq 0$, for the particular case $y_0 = 0$ so that $\lim_{i\to\infty} \kappa^{[i]} = \kappa$. Finally, the correspondence with Theorem 5.3.1 results from Proposition 5.1.15. □

Remark. In the case of a pure species state of the mixture, i.e., $n^+ = 1$ and $p^+ = 1$, one can easily verify that a one-step convergence of the iterates $\kappa^{[i]}$ is obtained with the choice $\widetilde{\mathfrak{d}} = \Gamma^{\kappa}(\widetilde{\mathfrak{d}}^+, 0)$ and either $\widetilde{\mathfrak{d}}^+ = (\widetilde{\mathfrak{d}}^{10+}, 0)$, $\widetilde{\mathfrak{d}}^{10+} > 0$, or $\widetilde{\mathfrak{d}}^+ = (0, \widetilde{\mathfrak{d}}^{01+})$, $\widetilde{\mathfrak{d}}^{01+} > 0$.

Finally, we consider the matrix $\widetilde{K}$ resulting from the practical approximations presented in Section 2.10.

Proposition 5.3.8. *Let $\bar{A}_{kl}$ and η_{kl}, $k, l \in \mathcal{S}$, be symmetric and positive coefficients, let η_k, $k \in \mathcal{S}$, c_l^{int} and ξ_l^{int}, $l \in \mathcal{P}$, be positive quantities, and assume that $Y \geq 0$, $Y \neq 0$. Then the matrix $\widetilde{K}$ given by (3.7.3)–(3.7.10) satisfies $(\widetilde{K}0)$–$(\widetilde{K}7)$.*

Proof. It is similar to the one of Propositions 4.3.9 and 5.3.3. □

5.3.4 Conjugate Gradient Methods for the System $\widehat{K}\widehat{\alpha}^{\kappa} = \widehat{\beta}^{\kappa}$

In this section we assume that the matrix $\widehat{K}$ satisfies the properties $(\widehat{K}0)$–$(\widehat{K}5)$ given in Section 4.3.4. We then want to solve the linear system $\widehat{K}\widehat{\alpha}^{\kappa} = \widehat{\beta}^{\kappa}$ using conjugate gradient methods and obtain the only solution $\widehat{\alpha}^{\kappa}$ such that $\langle \widehat{\mathcal{K}}, \widehat{\alpha}^{\kappa} \rangle = 0$.

Theorem 5.3.9. *Let $\widehat{K} \in \mathbb{R}^{n+p,n+p}$ be a matrix satisfying the properties $(\widehat{K}0)$–$(\widehat{K}5)$ and let $\widehat{M}$ be the matrix $db(\widehat{K}) + \mathrm{diag}(\widehat{\mathfrak{d}})$ or $\mathrm{diag}(\widehat{K}) + \mathrm{diag}(\widehat{\mathfrak{d}})$, where $\widehat{\mathfrak{d}} = (\widehat{\mathfrak{d}}_k^r)_{(r,k)\in B^{\kappa}}$ are coefficients such that $\widehat{\mathfrak{d}} \geq 0$ in general, and $\widehat{\mathfrak{d}}^+ \neq 0$ in the particular case $n^+ = 1$. Let $\widehat{\mathcal{K}}$, $\widehat{\mathcal{V}}$, and $\widehat{\beta}^{\kappa} \in \mathbb{R}^{n+p}$ be given by (4.3.35)–(4.3.37), and let also $\widehat{P} = I - \widehat{\mathcal{V}} \otimes \widehat{\mathcal{K}} / \langle \widehat{\mathcal{V}}, \widehat{\mathcal{K}} \rangle$ denote the oblique projector onto $\widehat{\mathcal{K}}^{\perp}$ along $\mathbb{R}\widehat{\mathcal{V}}$. Let $\widehat{x}_0 \in \mathbb{R}^{n+p}$, $\widehat{y}_0 = \widehat{P}\widehat{x}_0$, $\widehat{r}_0 = \widehat{\beta}^{\kappa} - \widehat{K}\widehat{x}_0$, $\widehat{p}_0 = 0$, $t_0 = 0$, and consider for $i \geq 1$ the iterates*

$$\begin{cases} \widehat{p}_i = \widehat{M}^{-1}\widehat{r}_{i-1} + t_{i-1}\widehat{p}_{i-1}, \\ s_i = \langle \widehat{r}_{i-1}, \widehat{M}^{-1}\widehat{r}_{i-1} \rangle / \langle \widehat{p}_i, \widehat{K}\widehat{p}_i \rangle, \\ \widehat{x}_i = \widehat{x}_{i-1} + s_i\widehat{p}_i, \\ \widehat{y}_i = \widehat{y}_{i-1} + \widehat{P}(s_i\widehat{p}_i), \\ \widehat{r}_i = \widehat{r}_{i-1} - s_i\widehat{K}\widehat{p}_i, \\ t_i = \langle \widehat{r}_i, \widehat{M}^{-1}\widehat{r}_i \rangle / \langle \widehat{r}_{i-1}, \widehat{M}^{-1}\widehat{r}_{i-1} \rangle. \end{cases} \tag{5.3.22}$$

Then $\widehat{y}_i = \widehat{P}\widehat{x}_i$ for all $i \geq 0$, the iterates $\widehat{y}_i$ converge towards $\widehat{\alpha}^\kappa$ in a finite number of steps, and the quantities

$$\kappa^{[i]} = \langle \widehat{y}_i, \widehat{\beta}^\kappa \rangle = \langle \widehat{x}_i, \widehat{\beta}^\kappa \rangle, \tag{5.3.23}$$

converge towards κ in a finite number of steps. Furthermore, if $\widehat{x}_0 = 0$, we have $\kappa^{[i]} > 0$ for all $i \geq 1$. Finally, for positive mass fractions, the iterates $\kappa^{[i]}$ coincide with the ones in Theorem 5.3.2, $\widehat{x}_i = (\mathcal{X}^\kappa)^{1/2} x_i$, and $\widehat{y}_i = (\mathcal{X}^\kappa)^{1/2} y_i$, provided that $\mathfrak{d} = \mathcal{X}^\kappa \widehat{\mathfrak{d}}$ and $\widehat{x}_0 = (\mathcal{X}^\kappa)^{1/2} x_0$.

Finally, we consider the matrix $\widehat{K}$ resulting from the practical approximations presented in Section 2.10.

Proposition 5.3.10. *Under the assumptions of Proposition 5.3.8, the matrix $\widehat{K}$, obtained from the matrix $\widetilde{K}$ by replacing X_l by $\sqrt{X_k X_l}$ in all the terms $\widetilde{K}_{kl}^{rs}$, $k \neq l$, satisfies $(\widehat{K}0)$–$(\widehat{K}5)$.*

5.3.5 Standard Iterative Methods for the System $\widetilde{K}_{[01]}\alpha^\kappa_{[01]} = \widetilde{\beta}^\kappa_{[01]}$

In this section we want to solve the linear system $\widetilde{K}_{[01]}\alpha^\kappa_{[01]} = \widetilde{\beta}^\kappa_{[01]}$ using standard iterative methods. The system matrix $\widetilde{K}_{[01]} \in \mathbb{R}^{p,p}$ is assumed to satisfy the properties $(\widetilde{K}_{[01]}0)$–$(\widetilde{K}_{[01]}4)$ presented in Section 4.3.5, and we restate that $\mathcal{B}^\kappa_{[01]} = \{01\} \times \mathcal{P}$, $\mathcal{X}^\kappa_{[01]} = \mathrm{diag}\big((X_k)_{(r,k)\in\mathcal{B}^\kappa_{[01]}}\big)$, and $\Gamma^\kappa_{[01]} \in \mathbb{R}^{p,p}$ is the permutation matrix associated with the reordering of $\mathcal{B}^\kappa_{[01]}$ into $(\mathcal{B}^{\kappa+}_{[01]}, \mathcal{B}^{\kappa-}_{[01]})$. Furthermore, we consider the matrix $db(\widetilde{K}_{[01]}) \in \mathbb{R}^{p,p}$ formed by the diagonal of the only block of the matrix $\widetilde{K}_{[01]}$,

$$db(\widetilde{K}_{[01]}) = \mathrm{diag}(\widetilde{K}^{0101}_{[01]}), \tag{5.3.24}$$

and from the kinetic theory results obtained in Sections 3.1 and 3.3, we can make the following assumptions.

$(\widetilde{K}_{[01]}5)$ $db(K^{++}_{[01]})$ and $db(\widetilde{K}_{[01]})$ are symmetric positive definite.

$(\widetilde{K}_{[01]}6)$ $2db(K^{++}_{[01]}) - K^{++}_{[01]}$ is symmetric positive definite.

Theorem 5.3.11. *Let $\widetilde{K}_{[01]} \in \mathbb{R}^{p,p}$ be a matrix satisfying the properties $(\widetilde{K}_{[01]}0)$–$(\widetilde{K}_{[01]}6)$ and let $\widetilde{M}$ be the matrix $db(\widetilde{K}_{[01]}) + \mathrm{diag}(\widetilde{\mathfrak{d}})$, where $\widetilde{\mathfrak{d}} = (\widetilde{\mathfrak{d}}^r_k)_{(r,k)\in\mathcal{B}^\kappa_{[01]}}$ are coefficients such that $\widetilde{\mathfrak{d}} \geq 0$. Consider the splitting $\widetilde{K}_{[01]} = \widetilde{M} - \widetilde{Z}$ and the iteration matrix $T = \widetilde{M}^{-1}\widetilde{Z}$. Let $\beta^\kappa_{[01]}$ and $\widetilde{\beta}^\kappa_{[01]} \in \mathbb{R}^p$ be given by (4.3.18) and (4.3.43), respectively, $x_0 \in \mathbb{R}^p$, and consider for $i \geq 0$ the iterates*

$$x_{i+1} = T x_i + \widetilde{M}^{-1}\widetilde{\beta}^\kappa_{[01]}. \tag{5.3.25}$$

Then the matrix T is convergent, $\rho(T) < 1$, and we have the following limits

$$\begin{cases} \lim_{i\to\infty} x_i = \alpha^{\kappa}_{[01]}, \\ \lim_{i\to\infty} \langle x_i, \beta^{\kappa}_{[01]}\rangle = \kappa_{[01]}, \end{cases} \tag{5.3.26}$$

where $\alpha^{\kappa}_{[01]}$ is the unique solution of the linear system $\widetilde{K}_{[01]}\alpha^{\kappa}_{[01]} = \widetilde{\beta}^{\kappa}_{[01]}$. Moreover, for all $i \geq 0$, each partial sum

$$\kappa^{[i]}_{[01]} = \langle \sum_{j=0}^{i} T^j \widetilde{M}^{-1}\widetilde{\beta}^{\kappa}_{[01]}, \beta^{\kappa}_{[01]}\rangle, \tag{5.3.27}$$

is positive, and we have

$$\lim_{i\to\infty} \kappa^{[i]}_{[01]} = \langle \sum_{j=0}^{\infty} T^j \widetilde{M}^{-1}\widetilde{\beta}^{\kappa}_{[01]}, \beta^{\kappa}_{[01]}\rangle = \kappa_{[01]}. \tag{5.3.28}$$

Finally, for positive mass fractions, the iteration matrix T and the iterates x_i and $\kappa^{[i]}_{[01]}$ coincide with the ones in Theorem 5.3.4, provided that $\mathfrak{d} = \mathcal{X}^{\kappa}_{[01]}\widetilde{\mathfrak{d}}$.

Finally, we consider the matrix $\widetilde{K}_{[01]}$ resulting from the practical approximations presented in Section 2.10.

Proposition 5.3.12. *Let $\bar{A}_{kl}$ and η_{kl}, $k,l \in S$, be symmetric and positive coefficients, let η_k, c_k^{int} and ξ_k^{int}, $k \in \mathcal{P}$, be positive quantities, and assume that $Y \geq 0$, $Y \neq 0$. Then the matrix $\widetilde{K}_{[01]}$ given by (3.7.9)(3.7.10) satisfies $(\widetilde{K}_{[01]}0)$–$(\widetilde{K}_{[01]}6)$.*

5.3.6 Schur Complements

In this section we assume that the matrices K and $\widetilde{K}$ satisfy the structure properties $(K0)$–$(K7)$ and $(\widetilde{K}0)$–$(\widetilde{K}7)$, respectively, as in the previous sections. In particular, the block K^{0101} is then nonsingular, and in the framework of the practical approximations presented in Section 2.10, this block is also diagonal. It is therefore interesting to introduce the constrained linear system

$$\begin{cases} K_{[s]}\alpha^{\kappa}_{[s]} = \beta^{\kappa}_{[s]}, \\ \langle \mathcal{K}_{[s]}, \alpha^{\kappa}_{[s]}\rangle = 0, \end{cases} \tag{5.3.29}$$

where the matrix $K_{[s]} \in \mathbb{R}^{n,n}$ is given by

$$K_{[s]} = K^{1010} - K^{1001}(K^{0101})^{-1}K^{0110}, \tag{5.3.30}$$

and the vectors $\beta^{\kappa}_{[s]}$ and $\mathcal{K}_{[s]} \in \mathbb{R}^n$ by

$$\beta^{\kappa}_{[s]} = \beta^{10\kappa} - K^{1001}\big(K^{0101}\big)^{-1}\beta^{01\kappa}, \tag{5.3.31}$$

and

$$\mathcal{K}_{[s]} = \mathcal{K}^{10} - K^{1001}\big(K^{0101}\big)^{-1}\mathcal{K}^{01}. \tag{5.3.32}$$

We deduce from Lemma 5.1.12 that the system (5.3.29) admits a unique solution $\alpha^{\kappa}_{[s]}$. The volume viscosity is then given by

$$\kappa = \kappa_{[s]} + \kappa_{[01]} = \langle \alpha^{\kappa}_{[s]}, \beta^{\kappa}_{[s]} \rangle + \langle \beta^{01\kappa}, \big(K^{0101}\big)^{-1}\beta^{01\kappa} \rangle, \tag{5.3.33}$$

and we point out that $\kappa_{[01]}$ coincides with the simplified volume viscosity associated with the matrix $K_{[01]}$.

Consider now the matrix $M_{[s]} \in \mathbb{R}^{n,n}$ given by

$$M_{[s]} = db(K^{1010}) + \mathrm{diag}(\mathfrak{d}), \tag{5.3.34}$$

where $\mathfrak{d}^r_k \geq 0$, $(r,k) \in \{10\} \times \mathcal{S}$. From $(K6)$–$(K7)$ we then deduce that the matrices $M_{[s]}$ and $2M_{[s]} - K_{[s]}$ are symmetric positive definite. A convergence theorem similar to Theorem 5.3.1 can then be stated for the constrained linear system (5.3.29) with the splitting matrix (5.3.34). In particular, this yields a sequence of iterates $\kappa^{[i]}_{[s]}$ such that

$$\lim_{i\to\infty} \kappa^{[i]}_{[s]} = \kappa - \kappa_{[01]}. \tag{5.3.35}$$

Moreover, since the matrix $M_{[s]}$ is symmetric positive definite, a convergence theorem similar to Theorem 5.3.2 can be stated for the constrained linear system (5.3.29) with the preconditioner (5.3.34). This yields a sequence of iterates $\kappa^{[i]}_{[s]}$ which converges towards $\kappa - \kappa_{[01]}$ in a finite number of steps.

Stabilized versions, for vanishing mass fractions, of standard iterative methods are obtained by considering the left rescaled constrained linear system

$$\begin{cases} \widetilde{K}_{[s]}\alpha^{\kappa}_{[s]} = \widetilde{\beta}^{\kappa}_{[s]}, \\ \langle \mathcal{K}_{[s]}, \alpha^{\kappa}_{[s]} \rangle = 0, \end{cases} \tag{5.3.36}$$

where

$$\widetilde{K}_{[s]} = \widetilde{K}^{1010} - \widetilde{K}^{1001}\big(\widetilde{K}^{0101}\big)^{-1}\widetilde{K}^{0110}, \tag{5.3.37}$$

$$\widetilde{\beta}^{\kappa}_{[s]} = \widetilde{\beta}^{10\kappa} - \widetilde{K}^{1001}\big(\widetilde{K}^{0101}\big)^{-1}\widetilde{\beta}^{01\kappa}, \tag{5.3.38}$$

and the vector $\mathcal{K}_{[s]}$ is now given by

$$\mathcal{K}_{[s]} = \mathcal{K}^{10} - \big(\widetilde{K}^{0110}\big)^t \big(\widetilde{K}^{0101}\big)^{-1} \mathcal{K}^{01}. \tag{5.3.39}$$

Assuming $\widetilde{K}^{0101}$ to be symmetric, we deduce from Section 5.1.5 that the system (5.3.36) admits a unique solution $\alpha^{\kappa}_{[s]}$. The volume viscosity is then evaluated from

$$\kappa = \kappa_{[s]} + \kappa_{[01]} = \langle \alpha^{\kappa}_{[s]}, \beta^{\kappa}_{[s]} \rangle + \langle \beta^{01\kappa}, \big(\widetilde{K}^{0101}\big)^{-1} \widetilde{\beta}^{01\kappa} \rangle, \tag{5.3.40}$$

where the vector $\beta^{\kappa}_{[s]}$ is now given by

$$\beta^{\kappa}_{[s]} = \beta^{10\kappa} - \big(\widetilde{K}^{0110}\big)^t \big(\widetilde{K}^{0101}\big)^{-1} \beta^{01\kappa}. \tag{5.3.41}$$

For positive mass fractions, one can easily verify that the solutions of the systems (5.3.29) and (5.3.36) coincide, and so do the volume viscosities (5.3.33) and (5.3.40).

Consider now the splitting $\widetilde{K}_{[s]} = \widetilde{M}_{[s]} - \widetilde{Z}_{[s]}$ where

$$\widetilde{M}_{[s]} = db(\widetilde{K}^{1010}) + \mathrm{diag}(\widetilde{\mathfrak{d}}), \tag{5.3.42}$$

and $\widetilde{\mathfrak{d}}^{r}_{k} \geq 0$, $(r,k) \in \{10\} \times \mathcal{S}$. As in the proof of Theorem 5.3.7, one can show, after some algebra, that the iteration matrix $T_{[s]} = \big(\widetilde{M}_{[s]}\big)^{-1} \widetilde{Z}_{[s]}$ is convergent. A convergence theorem similar to Theorem 5.3.7 can then be stated for the constrained linear system (5.3.36) with the splitting matrix (5.3.42). In particular, this yields a sequence of iterates $\kappa^{[i]}_{[s]}$ such that

$$\lim_{i \to \infty} \kappa^{[i]}_{[s]} = \kappa - \kappa_{[01]}. \tag{5.3.43}$$

Finally, stabilized versions of conjugate gradient methods are obtained by considering the symmetric rescaled matrix $\widehat{K}_{[s]}$, and are omitted for brevity.

5.4 The Diffusion Matrix and the Flux Diffusion Matrix

5.4.1 Iterative Methods for the System $L\alpha^{D_l} = \beta^{D_l}$

In this section we want to solve the linear systems $L\alpha^{D_l} = \beta^{D_l}$, $l \in \mathcal{S}$, using either standard iterative or conjugate gradient methods, and obtain the only solution α^{D_l}, $l \in \mathcal{S}$, such that $\langle \mathcal{L}, \alpha^{D_l} \rangle = 0$. The system matrix $L \in \mathbb{R}^{2n+p,2n+p}$ is assumed to satisfy the properties $(L0)$–$(L5)$ presented in Section 4.4.1, and the corresponding indexing set is $\mathcal{B}^D = \{00, 10\}\times\mathcal{S} \cup \{01\}\times\mathcal{P}$. Furthermore, we consider the matrix $db(L) \in \mathbb{R}^{2n+p,2n+p}$ formed by the diagonal of the nine blocks of the matrix L,

$$db(L) = \begin{bmatrix} \mathrm{diag}(L^{0000}) & \mathrm{diag}(L^{0010}) & \mathrm{diag}(L^{0001}) \\ \mathrm{diag}(L^{1000}) & \mathrm{diag}(L^{1010}) & \mathrm{diag}(L^{1001}) \\ \mathrm{diag}(L^{0100}) & \mathrm{diag}(L^{0110}) & \mathrm{diag}(L^{0101}) \end{bmatrix}, \tag{5.4.1}$$

and from the kinetic theory results obtained in Sections 2.3 and 2.6, we can make the following assumptions.

($L6$) $db(L)$ is symmetric positive semi-definite; for $n \geq 2$, this matrix is also positive definite, and for $n = 1$, its nullspace is spanned by the vector $(1, 0, 0)$.

($L7$) $2db(L) - L$ is symmetric positive semi-definite; for $n \geq 3$, this matrix is also positive definite; for $n = 2$, its nullspace is spanned by the vector $(1, -1, 0, 0, 0, 0)$ if $p = 2$, $(1, -1, 0, 0, 0)$ if $p = 1$, and $(1, -1, 0, 0)$ if $p = 0$; for $n = 1$, its nullspace is spanned by the vector $(1, 0, 0)$.

Theorem 5.4.1. *Let $L \in \mathbb{R}^{2n+p,2n+p}$ be a matrix satisfying the properties ($L0$)–($L7$) and let M be the matrix $db(L) + \mathrm{diag}(\mathfrak{d})$, where $\mathfrak{d} = (\mathfrak{d}^r_k)_{(r,k)\in\mathcal{B}^D}$ are coefficients such that $\mathfrak{d} \geq 0$ in general, and $\mathfrak{d}^{00} \neq 0$ in the particular cases $n = 1$ or 2. Consider the splitting $L = M - Z$ and the iteration matrix $T = M^{-1}Z$. Let $\mathcal{L}$, $\mathcal{U}$, and $\beta^{D_l} \in \mathbb{R}^{2n+p}$, $l \in \mathcal{S}$, be given by (4.4.6), (4.4.7), and (4.4.9), respectively, let $P = I - \mathcal{U}\otimes\mathcal{L}/\langle\mathcal{U}, \mathcal{L}\rangle$ denote the oblique projector onto $\mathcal{L}^\perp$ along $\mathbb{R}\mathcal{U}$, and let $\mathfrak{P} \in \mathbb{R}^{n,2n+p}$ be the rectangular matrix formed by the blocks $\mathfrak{P} = [I, 0, 0]$. Let $x^k_0 \in \mathbb{R}^{2n+p}$, $y^k_0 = Px^k_0$, and consider for $i \geq 0$ and $k \in \mathcal{S}$ the iterates*

$$x^k_{i+1} = Tx^k_i + M^{-1}\beta^{D_k}, \tag{5.4.2}$$

$$y^k_{i+1} = PTy^k_i + PM^{-1}\beta^{D_k}. \tag{5.4.3}$$

Then $y^k_i = Px^k_i$ for all $i \geq 0$, the matrices T and PT are convergent, $\rho(T) = 1$, $\rho(PT) < 1$, and we have the following limits

$$\begin{cases} \lim_{i\to\infty} y^l_i = P(\lim_{i\to\infty} x^l_i) = \alpha^{D_l}, & l \in \mathcal{S}, \\ \lim_{i\to\infty} \langle y^k_i, \beta^{D_l}\rangle = \lim_{i\to\infty} \langle x^k_i, \beta^{D_l}\rangle = D_{kl}, & k, l \in \mathcal{S}, \end{cases} \tag{5.4.4}$$

where α^{D_l} is the unique solution of the constrained linear system $L\alpha^{D_l} = \beta^{D_l}$ and $\langle\mathcal{L}, \alpha^{D_l}\rangle = 0$. Moreover, for all $i \geq 0$, each partial sum

$$\begin{aligned} D^{[i]} &= \mathfrak{P}\Big(\sum_{j=0}^{i}(PT)^j PM^{-1}P^t\Big)\mathfrak{P}^t \\ &= P_{Y^\perp,\mathbb{R}U}\mathfrak{P}\Big(\sum_{j=0}^{i} T^j M^{-1}\Big)\mathfrak{P}^t P_{U^\perp,\mathbb{R}Y}, \end{aligned} \tag{5.4.5}$$

is symmetric, positive semi-definite, positive definite on the hyperplane $U^\perp$, *and admits nullspace* $N(D^{[i]}) = \mathbb{R}Y$ *and range* $R(D^{[i]}) = Y^\perp$. *Finally, we have*

$$\lim_{i\to\infty} D^{[i]} = \mathfrak{P}(\sum_{j=0}^{\infty}(PT)^j PM^{-1}P^t)\mathfrak{P}^t = D. \tag{5.4.6}$$

Proof. We deduce from $(L6)$ and the assumptions on $\mathfrak{d}$ that $M = db(L) + \mathrm{diag}(\mathfrak{d})$ is the sum of two positive semi-definite matrices whose nullspaces do not contain any common vector except zero. Indeed, for $n \geq 2$ the matrix $db(L)$ is positive definite and for $n = 1$ the vector $(1, 0, 0)$ is not in the nullspace of $\mathrm{diag}(\mathfrak{d})$ since $\mathfrak{d}^{00} \neq 0$. Therefore, M is positive definite. Similarly, we can prove that $M + Z = (2db(L) - L) + 2\mathrm{diag}(\mathfrak{d})$ is positive definite. Therefore, using $(L1)$–$(L5)$, Theorem 5.1.8 applies with $C = \mathcal{L}^\perp$ and $S = \mathbb{R}\mathcal{U}$. In particular we obtain that the partial sums $L_i^\dagger = \sum_{j=0}^{i}(PT)^j PM^{-1}P^t$, $i \geq 0$, are symmetric positive semi-definite matrices, have nullspace $N(L_i^\dagger) = \mathbb{R}\mathcal{L}$ and range $R(L_i^\dagger) = \mathcal{L}^\perp$, and are positive definite on the hyperplane $\mathcal{U}^\perp$. Since $D^{[i]} = \mathfrak{P}L_i^\dagger\mathfrak{P}^t$, we immediately obtain that the matrices $D^{[i]}$ are symmetric positive semi-definite. Furthermore, for $x \in \mathbb{R}^n$, we have $\langle x, D^{[i]}x\rangle = 0$ if and only if $\mathfrak{P}^t x \in \mathbb{R}\mathcal{L}$, i.e., if and only if $x \in \mathbb{R}Y$. Therefore, $D^{[i]}$ is positive definite on the hyperplane $U^\perp$, $N(D^{[i]}) = \mathbb{R}Y$, and consequently, $R(D^{[i]}) = Y^\perp$ since $D^{[i]}$ is symmetric. Finally, the limit (5.4.6) is a direct consequence of (5.4.4) since we have $D_{kl}^{[i]} = \langle y_i^k, \beta^{D_l}\rangle$, $k, l \in S$, for the particular choice $y_0^k = 0$, $k \in S$. □

Theorem 5.4.2. *Let* $L \in \mathbb{R}^{2n+p,2n+p}$ *be a matrix satisfying the properties* $(L0)$–$(L6)$ *and let* M *be the matrix* $db(L) + \mathrm{diag}(\mathfrak{d})$ *or* $\mathrm{diag}(L) + \mathrm{diag}(\mathfrak{d})$, *where* $\mathfrak{d} = (\mathfrak{d}_k^r)_{(r,k)\in\mathcal{B}^D}$ *are coefficients such that* $\mathfrak{d} \geq 0$ *and* $\mathfrak{d}^{00} \neq 0$ *in the particular case* $n = 1$. *Let* $\mathcal{L}$, $\mathcal{U}$, *and* $\beta^{D_l} \in \mathbb{R}^{2n+p}$, $l \in S$, *be given by (4.4.6), (4.4.7), and (4.4.9), respectively, and let* $P = I - \mathcal{U}\otimes\mathcal{L}/\langle\mathcal{U}, \mathcal{L}\rangle$ *denote the oblique projector onto* $\mathcal{L}^\perp$ *along* $\mathbb{R}\mathcal{U}$. *Let* $x_0^k \in \mathbb{R}^{2n+p}$, $y_0^k = Px_0^k$, $r_0^k = \beta^{D_k} - Lx_0^k$, $p_0^k = 0$, $t_0^k = 0$, *and consider for* $i \geq 1$ *and* $k \in S$ *the iterates*

$$\begin{cases} p_i^k = M^{-1}r_{i-1}^k + t_{i-1}^k p_{i-1}^k, \\ s_i^k = \langle r_{i-1}^k, M^{-1}r_{i-1}^k\rangle / \langle p_i^k, Lp_i^k\rangle, \\ x_i^k = x_{i-1}^k + s_i^k p_i^k, \\ y_i^k = y_{i-1}^k + P(s_i^k p_i^k), \\ r_i^k = r_{i-1}^k - s_i^k L p_i^k, \\ t_i^k = \langle r_i^k, M^{-1}r_i^k\rangle / \langle r_{i-1}^k, M^{-1}r_{i-1}^k\rangle. \end{cases} \tag{5.4.7}$$

Then $y_i^k = Px_i^k$ *for all* $i \geq 0$, *the iterates* y_i^k *converge towards* α^{D_k} *in a finite number of steps, and the matrices*

$$D_{kl}^{[i]} = \langle y_i^l, \beta^{D_k} \rangle = \langle x_i^l, \beta^{D_k} \rangle, \qquad k, l \in S, \tag{5.4.8}$$

converge towards D in a finite number of steps and are such that $Y^t D^{[i]} = 0$.

Proof. The proof of the positive definiteness of $db(L) + \operatorname{diag}(\mathfrak{d})$ and $\operatorname{diag}(L) + \operatorname{diag}(\mathfrak{d})$ is similar to the previous one. Using $(L1)$–$(L5)$, we then deduce that Theorem 5.1.10 applies, and finally, the relation $Y^t D^{[i]} = 0$ directly results from $\sum_{k \in S} Y_k \beta^{D_k} = 0$. □

Remark. The iterates $D^{[i]}$ given by (5.4.8) are generally neither symmetric nor positive definite on the hyperplane $U^\perp$. Therefore, the conjugate gradient method does not yield at each iteration an approximation for the diffusion matrix that satisfies all the mathematical properties that are important from a thermodynamic viewpoint. This is at variance with all the other transport coefficients and also with standard iterative methods for the diffusion matrix.

Finally, we consider the matrix L resulting from the practical approximations presented in Section 2.10.

Proposition 5.4.3. *Let $\mathcal{D}_{k\,\mathrm{int},l}$, c_k^{int}, ξ_k^{int}, and m_l, $k \in \mathcal{P}$, $l \in S$, be positive quantities, let $\mathcal{D}_{kl}$, $\bar{A}_{kl}$, $\bar{B}_{kl}$, and $\bar{C}_{kl}$, $k, l \in S$, be symmetric and positive coefficients such that $25/4 - 3\bar{B}_{kl} > 0$ and $15\bar{C}_{kl} - 3\bar{B}_{kl} - 9\bar{C}_{kl}^2 > 0$, $k, l \in S$, and assume that $Y > 0$. Then the matrix L given by (2.10.15)–(2.10.25) satisfies $(L0)$–$(L7)$.*

Proof. By Proposition 4.4.6, only $(L6)$–$(L7)$ need to be proven. First, for $n = 1$, the property $(L6)$ directly results from $db(L) = L$, and for $n \geq 2$, we deduce from the relation

$$\langle x, db(L)x \rangle = \sum_{k \in S} \Big\langle \sum_{r \in \mathcal{F}_k} x_k^r e^{rk}, L \sum_{r \in \mathcal{F}_k} x_k^r e^{rk} \Big\rangle,$$

and the equivalence $\langle x, Lx \rangle = 0$ if and only if $x \in \mathbb{R}\mathcal{U}$ that $db(L)$ is positive definite since $\mathcal{U}^{00}$ has only nonzero components and $00 \in \mathcal{F}_k$ for all $k \in S$. Next, the matrix $2db(L) - L$ is obviously symmetric, and an explicit calculation yields for $x \in \mathbb{R}^{2n+p}$

$$\begin{aligned}
\langle x, (2db(L) - L)x \rangle = \sum_{\substack{k,l \in S \\ l \neq k}} \frac{X_k X_l}{2\mathcal{D}_{kl}} \Big[(x_k^{00} + x_l^{00})^2 \\
- (6\bar{C}_{kl} - 5)(x_k^{00} + x_l^{00}) \Big(\frac{m_l x_k^{10} + m_k x_l^{10}}{m_k + m_l} \Big) + \Big(\frac{25}{4} - 3\bar{B}_{kl} \Big) \Big(\frac{m_l x_k^{10} + m_k x_l^{10}}{m_k + m_l} \Big)^2 \Big] \\
+ \sum_{\substack{k,l \in S \\ l \neq k}} \frac{X_k X_l}{\mathcal{D}_{kl}} \Big[\frac{15}{4} \Big(\frac{m_k x_k^{10} + m_l x_l^{10}}{m_k + m_l} \Big)^2 + 2\bar{A}_{kl} \frac{m_k m_l}{(m_k + m_l)^2} (x_k^{10} - x_l^{10})^2 \Big]
\end{aligned}$$

$$+\sum_{k\in\mathcal{S}} 2\bar{A}_{kk}\frac{X_k^2}{\mathcal{D}_{kk}}(x_k^{10})^2 + \sum_{k\in\mathcal{P}}\Big(\sum_{\substack{l\in\mathcal{S}\\ l\neq k}} X_k X_l \frac{c_k^{\rm int}}{k_{\rm B}\mathcal{D}_{k\,{\rm int},l}} + X_k^2 \frac{c_k^{\rm int}}{k_{\rm B}\mathcal{D}_{k\,{\rm int},k}}\Big)(x_k^{01})^2$$

$$+\sum_{k\in\mathcal{P}} \frac{20}{3}\frac{\bar{A}_{kk}}{k_{\rm B}\pi}\frac{X_k^2}{\mathcal{D}_{kk}}\frac{c_k^{\rm int}}{\xi_k^{\rm int}}(x_k^{10}-\frac{3}{5}x_k^{01})^2$$

$$+\sum_{\substack{k\in\mathcal{S},l\in\mathcal{P}\\ l\neq k}} \frac{20}{3}\frac{\bar{A}_{kl}}{k_{\rm B}\pi}\frac{X_k X_l}{\mathcal{D}_{kl}}\frac{c_l^{\rm int}}{\xi_l^{\rm int}}\frac{m_l}{m_k}\Big(\frac{m_k}{m_k+m_l}(x_k^{10}-x_l^{10})+\frac{3}{5}x_l^{01}\Big)^2.$$

This shows as in the proof of Proposition 4.4.6 that the matrix $2db(L)-L$ is positive semi-definite. Let us consider a vector x such that $\langle x, (2db(L)-L)x\rangle = 0$. From the above expression, we see that $x_k^{00} = -x_l^{00}$, $k,l\in\mathcal{S}$, $k\neq l$, $x_k^{10}=0$, $k\in\mathcal{S}$, and $x_l^{01}=0$, $l\in\mathcal{P}$. We conclude that if $n\geq 3$, then $2db(L)-L$ is positive definite because it is possible to change the signs an odd number of times in the equalities $x_k^{00} = -x_l^{00}$, $k,l\in\mathcal{S}$, whereas in the particular case $n=2$, the nullspace of $2db(L)-L$ is spanned by the vector $(1,-1,0,0,0,0)$ if $p=2$, the vector $(1,-1,0,0,0)$ if $p=1$, and the vector $(1,-1,0,0)$ if $p=0$, respectively. Finally, if $n=1$, we have $2db(L)-L=L$, and the nullspace of these matrices is spanned by $(1,0,0)$. □

5.4.2 Iterative Methods for the System $L_{[\rm e]}\alpha_{[\rm e]}^{D_l} = \beta_{[\rm e]}^{D_l}$

In this section we want to solve the constrained linear systems $L_{[\rm e]}\alpha_{[\rm e]}^{D_l} = \beta_{[\rm e]}^{D_l}$, $l\in\mathcal{S}$, using either standard iterative or conjugate gradient methods, and obtain the only solution $\alpha_{[\rm e]}^{D_l}$, $l\in\mathcal{S}$, such that $\langle\mathcal{L}_{[\rm e]},\alpha_{[\rm e]}^{D_l}\rangle = 0$. The system matrix $L_{[\rm e]}\in\mathbb{R}^{2n,2n}$ is assumed to satisfy the properties $(L_{[\rm e]}0)$–$(L_{[\rm e]}5)$ presented in Section 4.4.2, and the corresponding indexing set is $\mathcal{B}_{[\rm e]}^{D} = \{00,{\rm e}\}\times\mathcal{S}$. Furthermore, we consider the matrix $db(L_{[\rm e]})\in\mathbb{R}^{2n,2n}$ formed by the diagonal of the four blocks of the matrix $L_{[\rm e]}$,

$$db(L_{[\rm e]}) = \begin{bmatrix} {\rm diag}(L_{[\rm e]}^{0000}) & {\rm diag}(L_{[\rm e]}^{00{\rm e}}) \\ {\rm diag}(L_{[\rm e]}^{{\rm e}00}) & {\rm diag}(L_{[\rm e]}^{\rm ee}) \end{bmatrix}, \tag{5.4.9}$$

and from the kinetic theory results obtained in Sections 2.3 and 2.6, we can make the following assumptions.

$(L_{[\rm e]}6)$ $db(L_{[\rm e]})$ is symmetric positive semi-definite; for $n\geq 2$, this matrix is also positive definite, and for $n=1$, its nullspace is spanned by the vector $(1,0)$.

($L_{[e]}7$) $2db(L_{[e]}) - L_{[e]}$ is symmetric positive semi-definite; for $n \geq 3$, this matrix is positive definite; for $n = 2$, its nullspace is spanned by the vector $(1, -1, 0, 0)$; for $n = 1$, its nullspace is spanned by the vector $(1, 0)$.

Theorem 5.4.4. *Let $L_{[e]} \in \mathbb{R}^{2n,2n}$ be a matrix satisfying the properties ($L_{[e]}0$)–($L_{[e]}7$) and let M be the matrix $db(L_{[e]}) + \mathrm{diag}(\mathfrak{d})$, where $\mathfrak{d} = (\mathfrak{d}_k^r)_{(r,k)\in\mathcal{B}_{[e]}^D}$ are coefficients such that $\mathfrak{d} \geq 0$ in general, and $\mathfrak{d}^{00} \neq 0$ in the particular cases $n = 1$ or 2. Consider the splitting $L_{[e]} = M - Z$ and the iteration matrix $T = M^{-1}Z$. Let $\mathcal{L}_{[e]}$, $\mathcal{U}_{[e]}$, and $\beta_{[e]}^{D_l} \in \mathbb{R}^{2n}$, $l \in S$, be given by (4.4.22), (4.4.23), and (4.4.25), respectively, let $P = I - \mathcal{U}_{[e]} \otimes \mathcal{L}_{[e]} / \langle \mathcal{U}_{[e]}, \mathcal{L}_{[e]} \rangle$ denote the oblique projector onto $\mathcal{L}_{[e]}^{\perp}$ along $\mathbb{R}\mathcal{U}_{[e]}$, and let $\mathfrak{P}_{[e]} \in \mathbb{R}^{n,2n}$ be the rectangular matrix formed by the blocks $\mathfrak{P}_{[e]} = [I, 0]$. Let $x_0^k \in \mathbb{R}^{2n}$, $y_0^k = Px_0^k$, and consider for $i \geq 0$ and $k \in S$ the iterates*

$$x_{i+1}^k = Tx_i^k + M^{-1}\beta_{[e]}^{D_k}, \tag{5.4.10}$$

$$y_{i+1}^k = PTy_i^k + PM^{-1}\beta_{[e]}^{D_k}. \tag{5.4.11}$$

Then $y_i^k = Px_i^k$ for all $i \geq 0$, the matrices T and PT are convergent, $\rho(T) = 1$, $\rho(PT) < 1$, and we have the following limits

$$\begin{cases} \lim\limits_{i\to\infty} y_i^l = P(\lim\limits_{i\to\infty} x_i^l) = \alpha_{[e]}^{D_l}, & l \in S, \\ \lim\limits_{i\to\infty} \langle y_i^k, \beta_{[e]}^{D_l} \rangle = \lim\limits_{i\to\infty} \langle x_i^k, \beta_{[e]}^{D_l} \rangle = D_{[e]kl}, & k, l \in S, \end{cases} \tag{5.4.12}$$

where $\alpha_{[e]}^{D_l}$ is the unique solution of the constrained linear system $L_{[e]}\alpha_{[e]}^{D_l} = \beta_{[e]}^{D_l}$ and $\langle \mathcal{L}_{[e]}, \alpha_{[e]}^{D_l} \rangle = 0$. Moreover, for all $i \geq 0$, each partial sum

$$\begin{aligned} D_{[e]}^{[i]} &= \mathfrak{P}_{[e]} \Big(\sum_{j=0}^{i} (PT)^j PM^{-1}P^t \Big) \mathfrak{P}_{[e]}^t \\ &= P_{Y^\perp, \mathbb{R}U} \mathfrak{P}_{[e]} \Big(\sum_{j=0}^{i} T^j M^{-1} \Big) \mathfrak{P}_{[e]}^t P_{U^\perp, \mathbb{R}Y}, \end{aligned} \tag{5.4.13}$$

is symmetric, positive semi-definite, positive definite on the hyperplane $U^\perp$, and admits nullspace $N(D_{[e]}^{[i]}) = \mathbb{R}Y$ and range $R(D_{[e]}^{[i]}) = Y^\perp$. Finally, we have

$$\lim_{i\to\infty} D_{[e]}^{[i]} = \mathfrak{P}_{[e]} \Big(\sum_{j=0}^{\infty} (PT)^j PM^{-1}P^t \Big) \mathfrak{P}_{[e]}^t = D_{[e]}. \tag{5.4.14}$$

Theorem 5.4.5. *Let $L_{[e]} \in \mathbb{R}^{2n,2n}$ be a matrix satisfying the properties ($L_{[e]}0$)–($L_{[e]}6$) and let M be the matrix $db(L_{[e]}) + \mathrm{diag}(\mathfrak{d})$ or $\mathrm{diag}(L_{[e]}) + \mathrm{diag}(\mathfrak{d})$, where $\mathfrak{d} = (\mathfrak{d}_k^r)_{(r,k)\in\mathcal{B}_{[e]}^D}$*

are coefficients such that $\mathfrak{d} \geq 0$ *in general, and* $\mathfrak{d}^{00} \neq 0$ *in the particular case* $n = 1$. *Let* $\mathcal{L}_{[e]}$, $\mathcal{U}_{[e]}$, *and* $\beta_{[e]}^{D_l} \in \mathbb{R}^{2n}$, $l \in \mathcal{S}$, *be given by (4.4.22), (4.4.23), and (4.4.25), respectively, and let* $P = I - \mathcal{U}_{[e]} \otimes \mathcal{L}_{[e]} / \langle \mathcal{U}_{[e]}, \mathcal{L}_{[e]} \rangle$ *denote the oblique projector onto* $\mathcal{L}_{[e]}^{\perp}$ *along* $\mathbb{R}\mathcal{U}_{[e]}$. *Let* $x_0^k \in \mathbb{R}^{2n}$, $y_0^k = Px_0^k$, $r_0^k = \beta_{[e]}^{D_k} - L_{[e]} x_0^k$, $p_0^k = 0$, $t_0^k = 0$, *and consider for* $i \geq 1$ *and* $k \in \mathcal{S}$ *the iterates*

$$\begin{cases} p_i^k = M^{-1} r_{i-1}^k + t_{i-1}^k p_{i-1}^k, \\ s_i^k = \langle r_{i-1}^k, M^{-1} r_{i-1}^k \rangle / \langle p_i^k, L_{[e]} p_i^k \rangle, \\ x_i^k = x_{i-1}^k + s_i^k p_i^k, \\ y_i^k = y_{i-1}^k + P(s_i^k p_i^k), \\ r_i^k = r_{i-1}^k - s_i^k L_{[e]} p_i^k, \\ t_i^k = \langle r_i^k, M^{-1} r_i^k \rangle / \langle r_{i-1}^k, M^{-1} r_{i-1}^k \rangle. \end{cases} \tag{5.4.15}$$

Then $y_i^k = Px_i^k$ *for all* $i \geq 0$, *the iterates* y_i^k *converge towards* $\alpha_{[e]}^{D_k}$ *in a finite number of steps, and the matrices*

$$D_{[e]kl}^{[i]} = \langle y_i^l, \beta_{[e]}^{D_k} \rangle = \langle x_i^l, \beta_{[e]}^{D_k} \rangle, \qquad k, l \in \mathcal{S}, \tag{5.4.16}$$

converge towards $D_{[e]}$ *in a finite number of steps and are such that* $Y^t D_{[e]}^{[i]} = 0$.

Remark. The iterates $D_{[e]}^{[i]}$ given by (5.4.16) are generally neither symmetric nor positive definite on the hyperplane $U^{\perp}$.

Finally, we consider the matrix $L_{[e]}$ resulting from the practical approximations presented in Section 2.10.

Proposition 5.4.6. *Let* $\mathcal{D}_{k\,\mathrm{int},l}$, c_k^{int}, ξ_k^{int}, *and* m_l, $k \in \mathcal{P}$, $l \in \mathcal{S}$, *be positive quantities, let* $\mathcal{D}_{kl}$, $\bar{A}_{kl}$, $\bar{B}_{kl}$, *and* $\bar{C}_{kl}$, $k, l \in \mathcal{S}$, *be symmetric and positive coefficients such that* $25/4 - 3\bar{B}_{kl} > 0$ *and* $15\bar{C}_{kl} - 3\bar{B}_{kl} - 9\bar{C}_{kl}^2 > 0$, $k, l \in \mathcal{S}$, *and assume that* $Y > 0$. *Then the matrix* $L_{[e]}$ *given by (2.10.26)–(2.10.31) satisfies* $(L_{[e]}0)$*–*$(L_{[e]}7)$.

5.4.3 Iterative Methods for the System $L_{[00]} \alpha_{[00]}^{D_l} = \beta_{[00]}^{D_l}$

In this section we want to solve the constrained linear systems $L_{[00]} \alpha_{[00]}^{D_l} = \beta_{[00]}^{D_l}$, $l \in \mathcal{S}$, using either standard iterative or conjugate gradient methods, and obtain the only solution $\alpha_{[00]}^{D_l}$, $l \in \mathcal{S}$, such that $\langle Y, \alpha_{[00]}^{D_l} \rangle = 0$. The system matrix $L_{[00]} \in \mathbb{R}^{n,n}$ is assumed to satisfy the properties $(L_{[00]}0)$–$(L_{[00]}5)$ presented in Section 4.4.3, and the corresponding indexing set is $\mathcal{B}_{[00]}^{D} = \{00\} \times \mathcal{S}$. Furthermore, we consider the matrix $db(L_{[00]}) \in \mathbb{R}^{n,n}$ formed by the diagonal of the only block of the matrix $L_{[00]}$,

$$db(L_{[00]}) = \mathrm{diag}(L_{[00]}^{0000}), \tag{5.4.17}$$

and from the kinetic theory results obtained in Sections 2.3 and 2.6, we can make the following assumptions.

($L_{[00]}6$) $db(L_{[00]})$ is symmetric positive semi-definite; for $n \geq 2$, this matrix is positive definite, and for $n = 1$, we have $L_{[00]} = 0$.

($L_{[00]}7$) $2db(L_{[00]}) - L_{[00]}$ is symmetric positive semi-definite; for $n \geq 3$, this matrix is positive definite; for $n = 2$, its nullspace is spanned by the vector $(1, -1)$; for $n = 1$, we have $2db(L_{[00]}) - L_{[00]} = 0$.

Theorem 5.4.7. *Let $L_{[00]} \in \mathbb{R}^{n,n}$ be a matrix satisfying the properties ($L_{[00]}0$)–($L_{[00]}7$) and let M be the matrix $db(L_{[00]}) + \mathrm{diag}(\mathfrak{d})$, where $\mathfrak{d} = (\mathfrak{d}^r_k)_{(r,k)\in\mathcal{B}^D_{[00]}}$ are coefficients such that $\mathfrak{d} \geq 0$ in general, and $\mathfrak{d}^{00} \neq 0$ in the particular cases $n = 1$ or 2. Consider the splitting $L_{[00]} = M - Z$ and the iteration matrix $T = M^{-1}Z$. Let $\beta^{D_l}_{[00]} \in \mathbb{R}^n$ be given by (4.4.36), and let $P = I - U \otimes Y/\langle U, Y\rangle$ denote the oblique projector onto $Y^\perp$ along $\mathbb{R}U$. Let $x^k_0 \in \mathbb{R}^n$, $y^k_0 = Px^k_0$, and consider for $i \geq 0$ and $k \in \mathcal{S}$ the iterates*

$$x^k_{i+1} = Tx^k_i + M^{-1}\beta^{D_k}_{[00]}, \tag{5.4.18}$$

$$y^k_{i+1} = PTy^k_i + PM^{-1}\beta^{D_k}_{[00]}. \tag{5.4.19}$$

Then $y^k_i = Px^k_i$ for all $i \geq 0$, the matrices T and PT are convergent, $\rho(T) = 1$, $\rho(PT) < 1$, and we have the following limits

$$\begin{cases} \lim\limits_{i\to\infty} y^l_i = P(\lim\limits_{i\to\infty} x^l_i) = \alpha^{D_l}_{[00]}, & l \in \mathcal{S}, \\ \lim\limits_{i\to\infty} \langle y^k_i, \beta^{D_l}_{[00]}\rangle = \lim\limits_{i\to\infty} \langle x^k_i, \beta^{D_l}_{[00]}\rangle = D_{[00]kl}, & k, l \in \mathcal{S}, \end{cases} \tag{5.4.20}$$

where $\alpha^{D_l}_{[00]}$ is the unique solution of the constrained linear system $L_{[00]}\alpha^{D_l}_{[00]} = \beta^{D_l}_{[00]}$ and $\langle Y, \alpha^{D_l}_{[00]}\rangle = 0$. Moreover, for all $i \geq 0$, each partial sum

$$D^{[i]}_{[00]} = \sum_{j=0}^{i} (PT)^j PM^{-1}P^t = P\Big(\sum_{j=0}^{i} T^j M^{-1}\Big)P^t, \tag{5.4.21}$$

is symmetric, positive semi-definite, positive definite on the hyperplane $U^\perp$, and admits nullspace $N(D^{[i]}_{[00]}) = \mathbb{R}Y$ and range $R(D^{[i]}_{[00]}) = Y^\perp$. Finally, we have

$$\lim_{i\to\infty} D^{[i]}_{[00]} = \sum_{j=0}^{\infty} (PT)^j PM^{-1}P^t = D_{[00]}. \tag{5.4.22}$$

Theorem 5.4.8. *Let $L_{[00]} \in \mathbb{R}^{n,n}$ be a matrix satisfying the properties $(L_{[00]}0)$–$(L_{[00]}6)$ and let M be the matrix $db(L_{[00]}) + \mathrm{diag}(\mathfrak{d})$, where $\mathfrak{d} = (\mathfrak{d}_k^r)_{(r,k)\in\mathcal{B}^D_{[00]}}$ are coefficients such that $\mathfrak{d} \geq 0$ in general, and $\mathfrak{d}^{00} \neq 0$ in the particular case $n = 1$. Let $\beta^{D_l}_{[00]} \in \mathbb{R}^n$ be given by (4.4.36), and let $P = I - U\otimes Y/\langle U, Y\rangle$ denote the oblique projector onto $Y^\perp$ along $\mathbb{R}U$. Let $x_0^k \in \mathbb{R}^n$, $y_0^k = Px_0^k$, $r_0^k = \beta^{D_k}_{[00]} - L_{[00]}x_0^k$, $p_0^k = 0$, $t_0^k = 0$, and consider for $i \geq 1$ and $k \in \mathcal{S}$ the iterates*

$$\begin{cases} p_i^k = M^{-1}r_{i-1}^k + t_{i-1}^k p_{i-1}^k, \\ s_i^k = \langle r_{i-1}^k, M^{-1}r_{i-1}^k\rangle / \langle p_i^k, L_{[00]}p_i^k\rangle, \\ x_i^k = x_{i-1}^k + s_i^k p_i^k, \\ y_i^k = y_{i-1}^k + P(s_i^k p_i^k), \\ r_i^k = r_{i-1}^k - s_i^k L_{[00]}p_i^k, \\ t_i^k = \langle r_i^k, M^{-1}r_i^k\rangle / \langle r_{i-1}^k, M^{-1}r_{i-1}^k\rangle. \end{cases} \tag{5.4.23}$$

Then $y_i^k = Px_i^k$ for all $i \geq 0$, the iterates y_i^k converge towards $\alpha^{D_k}_{[00]}$ in a finite number of steps, and the matrices

$$D^{[i]}_{[00]kl} = \langle y_i^l, \beta^{D_k}_{[00]}\rangle = \langle x_i^l, \beta^{D_k}_{[00]}\rangle, \qquad k, l \in \mathcal{S}, \tag{5.4.24}$$

converge towards $D_{[00]}$ in a finite number of steps and are such that $Y^t D^{[i]}_{[00]} = 0$.

Remark. The iterates $D^{[i]}_{[00]}$ given by (5.4.24) are generally neither symmetric nor positive definite on the hyperplane $U^\perp$.

Finally, we consider the matrix $L_{[00]}$ resulting from the practical approximations presented in Section 2.10.

Proposition 5.4.9. *Let $\mathcal{D}_{kl}$, $k, l \in \mathcal{S}$, be symmetric and positive coefficients and assume that $Y > 0$. Then the matrix $L_{[00]}$ given by (2.10.15)(2.10.16) satisfies $(L_{[00]}0)$–$(L_{[00]}7)$.*

5.4.4 Standard Iterative Methods for the System $\widetilde{L}\widetilde{\alpha}^{D_l} = \widetilde{\beta}^{D_l}$

In this section we want to solve the linear systems $\widetilde{L}\widetilde{\alpha}^{D_l} = \widetilde{\beta}^{D_l}$, $l \in \mathcal{S}$, using standard iterative methods and obtain the only solution $\widetilde{\alpha}^{D_l}$, $l \in \mathcal{S}$, such that $\langle \mathcal{L}, \widetilde{\alpha}^{D_l}\rangle = 0$. The system matrix $\widetilde{L} \in \mathbb{R}^{2n+p,2n+p}$ is assumed to satisfy the properties $(\widetilde{L}0)$–$(\widetilde{L}5)$ presented in Section 4.4.4, and we restate that $\mathcal{B}^D = \{00, 10\}\times\mathcal{S} \cup \{01\}\times\mathcal{P}$, $\mathcal{X}^D = \mathrm{diag}\big((X_k)_{(r,k)\in\mathcal{B}^D}\big)$, and $\Gamma^D \in \mathbb{R}^{2n+p,2n+p}$ is the permutation matrix associated with the reordering of $\mathcal{B}^D$ into $(\mathcal{B}^{D+},\mathcal{B}^{D-})$. We also denote by $\Upsilon \in \mathbb{R}^{n,n}$ the permutation matrix associated with the reordering of $\mathcal{S}$ into $(\mathcal{S}^+,\mathcal{S}^-)$. Furthermore, we consider the matrix $db(\widetilde{L}) \in \mathbb{R}^{2n+p,2n+p}$ formed by the diagonal of the nine blocks of the matrix $\widetilde{L}$,

$$db(\widetilde{L}) = \begin{bmatrix} \mathrm{diag}(\widetilde{L}^{0000}) & \mathrm{diag}(\widetilde{L}^{0010}) & \mathrm{diag}(\widetilde{L}^{0001}) \\ \mathrm{diag}(\widetilde{L}^{1000}) & \mathrm{diag}(\widetilde{L}^{1010}) & \mathrm{diag}(\widetilde{L}^{1001}) \\ \mathrm{diag}(\widetilde{L}^{0100}) & \mathrm{diag}(\widetilde{L}^{0110}) & \mathrm{diag}(\widetilde{L}^{0101}) \end{bmatrix}, \tag{5.4.25}$$

and from the kinetic theory results obtained in Sections 3.1 and 3.4, we can make the following assumptions.

($\widetilde{L}6$) $db(L^{++})$ and $db(\widetilde{L})$ are symmetric positive semi-definite; for $n^+ \geq 2$, these matrices are also positive definite; for $n^+ = 1$, the nullspace of $db(L^{++})$ is spanned by the vector $\mathcal{U}^+ = (1,0,0)$ if $p^+ = 1$ or $\mathcal{U}^+ = (1,0)$ if $p^+ = 0$, and the nullspace of $db(\widetilde{L})$ is spanned by the vector $\Gamma^D(\mathcal{U}^+,0)$.

($\widetilde{L}7$) $2db(L^{++}) - L^{++}$ is symmetric positive semi-definite; for $n^+ \geq 3$, this matrix is also positive definite; for $n^+ = 2$, its nullspace is spanned by the vector $(1,-1,0,0,0,0)$ if $p^+ = 2$, $(1,-1,0,0,0)$ if $p^+ = 1$ and $(1,-1,0,0)$ if $p^+ = 0$; for $n^+ = 1$, its nullspace is spanned by the vector $(1,0,0)$ if $p^+ = 1$ and $(1,0)$ if $p^+ = 0$.

Theorem 5.4.10. *Let $\widetilde{L} \in \mathbb{R}^{2n+p,2n+p}$ be a matrix satisfying the properties ($\widetilde{L}0$)–($\widetilde{L}7$) and let $\widetilde{M}$ be the matrix $db(\widetilde{L}) + \mathrm{diag}(\widetilde{\mathfrak{d}})$, where $\widetilde{\mathfrak{d}} = (\widetilde{\mathfrak{d}}^r_k)_{(r,k)\in\mathcal{B}^D}$ are coefficients such that $\widetilde{\mathfrak{d}} \geq 0$ in general, and $\widetilde{\mathfrak{d}}^{00+} \neq 0$ in the particular cases $n^+ = 1$ or 2. Consider the splitting $\widetilde{L} = \widetilde{M} - \widetilde{Z}$ and the iteration matrix $T = \widetilde{M}^{-1}\widetilde{Z}$. Let $\mathcal{L}$, $\mathcal{U}$, β^{D_l}, $\widetilde{\beta}^{D_l} \in \mathbb{R}^{2n+p}$, $l \in \mathcal{S}$, be given by (4.4.6), (4.4.7), (4.4.9), and (4.4.45), respectively, let $P = I - \mathcal{U}\otimes\mathcal{L}/\langle\mathcal{U},\mathcal{L}\rangle$ be the oblique projector onto $\mathcal{L}^\perp$ along $\mathbb{R}\mathcal{U}$, and let $\mathfrak{P} \in \mathbb{R}^{n,2n+p}$ the rectangular matrix formed by the blocks $\mathfrak{P} = [I,0,0]$. Let $\widetilde{x}^k_0 \in \mathbb{R}^{2n+p}$, $\widetilde{y}^k_0 = P\widetilde{x}^k_0$, $k \in \mathcal{S}$, and consider for $i \geq 0$ and $k \in \mathcal{S}$ the iterates*

$$\widetilde{x}^k_{i+1} = T\widetilde{x}^k_i + \widetilde{M}^{-1}\widetilde{\beta}^{D_k}, \tag{5.4.26}$$

$$\widetilde{y}^k_{i+1} = PT\widetilde{y}^k_i + P\widetilde{M}^{-1}\widetilde{\beta}^{D_k}. \tag{5.4.27}$$

Then $\widetilde{y}^k_i = P\widetilde{x}^k_i$ for all $i \geq 0$, the matrices T and PT are convergent, $\rho(T) = 1$, $\rho(PT) < 1$, and we have the following limits

$$\begin{cases} \lim\limits_{i\to\infty} \widetilde{y}^l_i = P(\lim\limits_{i\to\infty} \widetilde{x}^l_i) = \widetilde{\alpha}^{D_l}, & l \in \mathcal{S}, \\ \lim\limits_{i\to\infty} \langle\widetilde{y}^k_i, \beta^{D_l}\rangle = \lim\limits_{i\to\infty} \langle\widetilde{x}^k_i, \beta^{D_l}\rangle = \widetilde{D}_{kl}, & k,l \in \mathcal{S}, \end{cases} \tag{5.4.28}$$

where $\widetilde{\alpha}^{D_l}$ is the unique solution of the constrained linear system $\widetilde{L}\widetilde{\alpha}^{D_l} = \widetilde{\beta}^{D_l}$ and $\langle \mathcal{L}, \widetilde{\alpha}^{D_l} \rangle = 0$. Moreover, for all $i \geq 0$, each partial sum

$$\widetilde{D}^{[i]} = P_{U^\perp, \mathbb{R}Y} \mathcal{W}\mathfrak{P}\Big[\sum_{j=0}^{i} (PT)^j P\widetilde{M}^{-1}\Big]^t \mathfrak{P}^t$$
$$= P_{U^\perp, \mathbb{R}Y} \mathcal{W}\mathfrak{P}\Big[\sum_{j=0}^{i} T^j \widetilde{M}^{-1}\Big]^t \mathfrak{P}^t P_{U^\perp, \mathbb{R}Y}, \tag{5.4.29}$$

has nullspace $N(\widetilde{D}^{[i]}) = \mathbb{R}Y$ and range $R(\widetilde{D}^{[i]}) = U^\perp$. In addition, the matrix $\widetilde{D}^{[i]}$ admits the block-decomposition

$$\widetilde{D}^{[i]} = \Upsilon \begin{bmatrix} \widetilde{D}^{[i]++} & \widetilde{D}^{[i]+-} \\ 0 & \widetilde{D}^{[i]--} \end{bmatrix} \Upsilon^t, \tag{5.4.30}$$

where $\widetilde{D}^{[i]--}$ is diagonal and positive definite. Furthermore, we have

$$\lim_{i \to \infty} \widetilde{D}^{[i]} = P_{U^\perp, \mathbb{R}Y} \mathcal{W}\mathfrak{P}\Big[\sum_{j=0}^{\infty} (PT)^j P\widetilde{M}^{-1}\Big]^t \mathfrak{P}^t. \tag{5.4.31}$$

Finally, for positive mass fractions, the iteration matrix T coincides with the one in Theorem 5.4.1 and we have $\widetilde{x}_i^k = Y_k x_i^k$, $\widetilde{y}_i^k = Y_k y_i^k$, and $\widetilde{D}_{kl} = Y_k D_{kl}$, $k, l \in S$, provided that $\mathfrak{d} = \mathcal{X}^D \widetilde{\mathfrak{d}}$ and $\widetilde{x}_0^k = Y_k x_0^k$, $k \in S$.

Proof. We deduce from $(\widetilde{L}1)$ that the matrix T admits the block-decomposition

$$T = \begin{bmatrix} T^{++} & 0 \\ T^{-+} & T^{--} \end{bmatrix},$$

and we can prove as in the proof of Theorem 5.2.1 that $\rho(T^{--}) < 1$. Furthermore, as in the proof of Theorem 5.3.7, we have $T^{++} = \big(M^{++}\big)^{-1} Z^{++}$ where $M^{++} = db(L^{++}) + \mathrm{diag}(\mathfrak{d}^+)$, $L^{++} = M^{++} - Z^{++}$, and $\mathfrak{d}^+ = \mathcal{X}^{D++}\widetilde{\mathfrak{d}}^+$, and we deduce from $(\widetilde{L}6)$–$(\widetilde{L}7)$ and the assumptions on $\widetilde{\mathfrak{d}}$ that the matrix T^{++} is convergent. Therefore, by Lemma 5.1.3, the matrix T is convergent and $\rho(T) = 1$ since $T\mathcal{U} = \mathcal{U}$. We can then apply Theorem 5.1.5, and (5.4.28) is easily obtained since $P^t \beta^{D_l} = \beta^{D_l}$, $l \in S$. Consider now the vectors $\widetilde{\mathcal{L}}$ and $\widetilde{\mathcal{U}}$ given by the block-decompositions $\widetilde{\mathcal{L}}^{00} = W$, $\widetilde{\mathcal{L}}^{10} = 0$, $\widetilde{\mathcal{L}}^{01} = 0$, and $\widetilde{\mathcal{U}}^{00} = X$, $\widetilde{\mathcal{U}}^{10} = 0$, $\widetilde{\mathcal{U}}^{01} = 0$, and the matrices $\widetilde{L}_i^\dagger$ given by $\widetilde{L}_i^\dagger = \sum_{j=0}^{i} (PT)^j P\widetilde{M}^{-1} Q$, where Q is the oblique projector onto $\widetilde{\mathcal{U}}^\perp$ along $\mathbb{R}\mathcal{L}$. One can then verify after some algebra that $\widetilde{D}^{[i]} = \mathcal{W}\mathfrak{P}\big[\widetilde{L}_i^\dagger\big]^t \mathfrak{P}^t$. Furthermore, we obtain from Theorem 5.1.5 that $N(\widetilde{L}_i^\dagger) = \mathbb{R}\widetilde{\mathcal{L}}$ and $R(\widetilde{L}_i^\dagger) = \mathcal{L}^\perp$ from which we deduce that $N(\widetilde{D}^{[i]}) = \mathbb{R}Y$ and $R(\widetilde{D}^{[i]}) = U^\perp$. We also deduce from the block-decomposition of the matrices $\widetilde{M}$, P, Q, and T

that (5.4.30) holds and that the submatrices $\widetilde{D}^{[i]++}$ and $\widetilde{D}^{[i]--}$ are formed by the corresponding submatrices of $\widetilde{M}$, P, Q, and T. The matrices $\widetilde{D}^{[i]--}$, $i \geq 0$, are therefore diagonal and positive definite since $\widetilde{L}^{--}$ and $\widetilde{M}^{--}$ are block-diagonal and positive definite. Next, the limit (5.4.31) is a direct consequence of (5.4.28) since we have $\widetilde{D}_{kl}^{[i]} = \langle \widetilde{y}_i^k, \beta^{D_l} \rangle$, $k, l \in \mathcal{S}$, for the particular choice $\widetilde{y}_0^k = 0$, $k \in \mathcal{S}$. Finally, the correspondence with Theorem 5.4.1 is obtained as in the proof of Proposition 5.1.15, keeping in mind that $\mathcal{X}^D \widetilde{\beta}^{D_k} = Y_k \beta^{D_k}$, $k \in \mathcal{S}$. □

Remark. In the case of a pure species state of the mixture, i.e., $n^+ = 1$, it is easy to check that the initializations $\widetilde{y}_0^k = 0$, $k \in \mathcal{S}$, and any splitting of the form $\widetilde{\mathfrak{d}} = \Gamma^D(\widetilde{\mathfrak{d}}^+, 0)$, $\widetilde{\mathfrak{d}}^{00+} > 0$, $\widetilde{\mathfrak{d}}^{10+} = 0$, and $\widetilde{\mathfrak{d}}^{01+} = 0$ leads to a one-step convergence of the partial sums $\widetilde{D}^{[i]}$. An interesting numerical procedure [KWM83] that can be used to obtain such splittings consists in evaluating perturbed coefficients $D_k^*(\epsilon)$, $k \in \mathcal{S}$, defined by

$$D_k^*(\epsilon) = \sum_{\substack{l \in \mathcal{S} \\ l \neq k}} \frac{(X_l + \epsilon)}{\langle Y, U \rangle} \Big/ \sum_{\substack{l \in \mathcal{S} \\ l \neq k}} \frac{X_l + \epsilon}{\mathcal{D}_{kl}},$$

where ϵ is a small positive constant, typically smaller than the machine precision. For a pure species state, this yields $D_k^*(\epsilon) = D_k^* + \mathcal{O}(\epsilon)$, where $D_k^* = 1/\widetilde{L}_{kk}^{0000}$, $k \in \mathcal{S}^-$, and a positive but arbitrary $D_k^*(\epsilon) > 0$ for $k \in \mathcal{S}^+$. Defining now $\widetilde{M}_{kk}^{0000} = 1/D_k^*(\epsilon)$, $k \in \mathcal{S}$, we obtain a one-step convergence of the iterates $\widetilde{D}^{[i]}$.

Finally, we consider the matrix $\widetilde{L}$ resulting from the practical approximations presented in Section 2.10.

Proposition 5.4.11. *Let $\mathcal{D}_{k\,\mathrm{int},l}$, c_k^{int}, ξ_k^{int}, and m_l, $k \in \mathcal{P}$, $l \in \mathcal{S}$, be positive quantities, let $\mathcal{D}_{kl}$, $\bar{\mathcal{A}}_{kl}$, $\bar{\mathcal{B}}_{kl}$, and $\bar{\mathcal{C}}_{kl}$, $k, l \in \mathcal{S}$, be symmetric and positive coefficients such that $25/4 - 3\bar{\mathcal{B}}_{kl} > 0$ and $15\bar{\mathcal{C}}_{kl} - 3\bar{\mathcal{B}}_{kl} - 9\bar{\mathcal{C}}_{kl}^2 > 0$, $k, l \in \mathcal{S}$, and assume that $Y \geq 0$, $Y \neq 0$. Then the matrix $\widetilde{L}$ given by (3.7.11)–(3.7.26) satisfies ($\widetilde{L}0$)–($\widetilde{L}7$).*

Proof. It is similar to the one of Propositions 4.4.21 and 5.4.3. □

5.4.5 Conjugate Gradient Methods for the System $\widehat{L}\widehat{\alpha}^{D_l} = \widehat{\beta}^{D_l}$

In this section we assume that the matrix $\widehat{L}$ satisfies the properties ($\widehat{L}0$)–($\widehat{L}5$) given in Section 4.4.5. We then want to solve the linear systems $\widehat{L}\widehat{\alpha}^{D_l} = \widehat{\beta}^{D_l}$, $l \in \mathcal{S}$, using conjugate gradient methods and obtain the only solution $\widehat{\alpha}^{D_l}$, $l \in \mathcal{S}$, such that $\langle \widehat{\mathcal{L}}, \widehat{\alpha}^{D_l} \rangle = 0$.

Theorem 5.4.12. *Let $\widehat{L} \in \mathbb{R}^{2n+p,2n+p}$ be a matrix satisfying the properties $(\widehat{L}0)$–$(\widehat{L}5)$ and let $\widehat{M}$ be the matrix $db(\widehat{L}) + \mathrm{diag}(\widehat{\mathfrak{d}})$ or $\mathrm{diag}(\widehat{L}) + \mathrm{diag}(\widehat{\mathfrak{d}})$, where $\widehat{\mathfrak{d}} = (\widehat{\mathfrak{d}}_k^r)_{(r,k)\in\mathcal{B}^D}$ are coefficients such that $\widehat{\mathfrak{d}} \geq 0$ in general, and $\widehat{\mathfrak{d}}^{00+} \neq 0$ in the particular case $n^+ = 1$. Let $\widehat{\mathcal{L}}$, $\widehat{\mathcal{U}}$, and $\widehat{\beta}^{D_l} \in \mathbb{R}^{2n+p}$, $l \in \mathcal{S}$, be given by (4.4.59)–(4.4.61) and let also $\widehat{P} = I - \widehat{\mathcal{U}}\otimes\widehat{\mathcal{L}}/\langle\widehat{\mathcal{U}}, \widehat{\mathcal{L}}\rangle$ denote the oblique projector onto $\widehat{\mathcal{L}}^\perp$ along $\mathbb{R}\widehat{\mathcal{U}}$. Let $\widehat{x}_0^k \in \mathbb{R}^{2n+p}$, $\widehat{y}_0^k = \widehat{P}\widehat{x}_0^k$, $\widehat{r}_0^k = \widehat{\beta}^{D_k} - \widehat{L}\widehat{x}_0^k$, $\widehat{p}_0^k = 0$, $t_0^k = 0$, and consider for $i \geq 1$ the iterates*

$$\begin{cases} \widehat{p}_i^k = \widehat{M}^{-1}\widehat{r}_{i-1}^k + t_{i-1}^k\widehat{p}_{i-1}^k, \\ s_i^k = \langle \widehat{r}_{i-1}^k, \widehat{M}^{-1}\widehat{r}_{i-1}^k\rangle / \langle \widehat{p}_i^k, \widehat{L}\widehat{p}_i^k\rangle, \\ \widehat{x}_i^k = \widehat{x}_{i-1}^k + s_i^k\widehat{p}_i^k, \\ \widehat{y}_i^k = \widehat{y}_{i-1}^k + \widehat{P}(s_i^k\widehat{p}_i^k), \\ \widehat{r}_i^k = \widehat{r}_{i-1}^k - s_i^k\widehat{L}\widehat{p}_i^k, \\ t_i^k = \langle \widehat{r}_i^k, \widehat{M}^{-1}\widehat{r}_i^k\rangle / \langle \widehat{r}_{i-1}^k, \widehat{M}^{-1}\widehat{r}_{i-1}^k\rangle. \end{cases} \tag{5.4.32}$$

Then $\widehat{y}_i^k = \widehat{P}\widehat{x}_i^k$ for all $i \geq 0$, $\widehat{y}_i^k$ converges towards $\widehat{\alpha}^{D_k}$ in a finite number of steps, and the quantities

$$\widehat{D}_{kl}^{[i]} = \langle \widehat{y}_i^k, \widehat{\beta}^{D_l}\rangle = \langle \widehat{x}_i^k, \widehat{\beta}^{D_l}\rangle, \qquad k, l \in \mathcal{S}, \tag{5.4.33}$$

converge towards $\widehat{D}$ in a finite number of steps. Finally, for positive mass fractions, the iterates $\widehat{D}^{[i]}$ are related to the iterates $D^{[i]}$ in Theorem 5.4.2 by $\widehat{D}_{kl}^{[i]} = \sqrt{Y_k Y_l} D_{kl}^{[i]}$ and we have $\widehat{x}_i^k = \sqrt{Y_k}\mathcal{X}^{1/2}x_i^k$ and $\widehat{x}_i^k = \sqrt{Y_k}\mathcal{X}^{1/2}x_i^k$, provided that $\mathfrak{d} = \mathcal{X}^D\widehat{\mathfrak{d}}$ and $\widehat{x}_0^k = \sqrt{Y_k}\mathcal{X}^{1/2}x_0^k$.

Finally, we consider the matrix $\widehat{L}$ resulting from the practical approximations presented in Section 2.10.

Proposition 5.4.13. *Under the assumptions of Proposition 5.4.11, the matrix $\widehat{L}$, obtained from the matrix $\widetilde{L}$ by replacing X_l by $\sqrt{X_k X_l}$ in all the terms $\widetilde{L}_{kl}^{rs}$, $k \neq l$, satisfies $(\widehat{L}0)$–$(\widehat{L}5)$.*

5.4.6 Standard Iterative Methods for the System $\widetilde{L}_{[e]}\widetilde{\alpha}_{[e]}^{D_l} = \widetilde{\beta}_{[e]}^{D_l}$

In this section we want to solve the linear systems $\widetilde{L}_{[e]}\widetilde{\alpha}_{[e]}^{D_l} = \widetilde{\beta}_{[e]}^{D_l}$, $l \in \mathcal{S}$, using standard iterative methods and obtain the only solution $\widetilde{\alpha}_{[e]}^{D_l}$, $l \in \mathcal{S}$, such that $\langle \mathcal{L}_{[e]}, \widetilde{\alpha}_{[e]}^{D_l}\rangle = 0$. The system matrix $\widetilde{L}_{[e]} \in \mathbb{R}^{2n,2n}$ is assumed to satisfy the properties $(\widetilde{L}_{[e]}0)$–$(\widetilde{L}_{[e]}5)$ given in Section 4.4.6, and we restate that $\mathcal{B}_{[e]}^D = \{00, e\}\times\mathcal{S}$, $\mathcal{X}_{[e]}^D = \mathrm{diag}\big((X_k)_{(r,k)\in\mathcal{B}_{[e]}^D}\big)$, and $\Gamma^D_{[e]} \in \mathbb{R}^{2n,2n}$ is the permutation matrix associated with the reordering of $\mathcal{B}^D_{[e]}$ into $(\mathcal{B}^{D+}_{[e]}, \mathcal{B}^{D-}_{[e]})$. We also denote by $\Upsilon \in \mathbb{R}^{n,n}$ the permutation matrix associated with the reordering of $\mathcal{S}$ into $(\mathcal{S}^+, \mathcal{S}^-)$. Furthermore, we consider the matrix $db(\widetilde{L}_{[e]}) \in \mathbb{R}^{2n,2n}$ formed by the diagonal of the four blocks of the matrix $\widetilde{L}_{[e]}$,

$$db(\widetilde{L}_{[e]}) = \begin{bmatrix} \operatorname{diag}(\widetilde{L}^{0000}_{[e]}) & \operatorname{diag}(\widetilde{L}^{00e}_{[e]}) \\ \operatorname{diag}(\widetilde{L}^{e00}_{[e]}) & \operatorname{diag}(\widetilde{L}^{ee}_{[e]}) \end{bmatrix}, \tag{5.4.34}$$

and from the kinetic theory results obtained in Sections 3.1 and 3.4, we can make the following assumptions.

($\widetilde{L}_{[e]}6$) $db(L^{++}_{[e]})$ and $db(\widetilde{L}_{[e]})$ are symmetric positive semi-definite; for $n^+ \geq 2$, these matrices are also positive definite; for $n^+ = 1$, the nullspace of $db(L^{++}_{[e]})$ is spanned by the vector $\mathcal{U}^+_{[e]} = (1,0)$, and the nullspace of $db(\widetilde{L}_{[e]})$ is spanned by the vector $\Gamma^D_{[e]}(\mathcal{U}^+_{[e]}, 0)$.

($\widetilde{L}_{[e]}7$) $2db(L^{++}_{[e]}) - L^{++}_{[e]}$ is symmetric positive semi-definite; for $n^+ \geq 3$, this matrix is also positive definite; for $n^+ = 2$, its nullspace is spanned by the vector $(1,-1,0,0)$; for $n^+ = 1$, its nullspace is spanned by the vector $(1,0)$.

Theorem 5.4.14. *Let $\widetilde{L}_{[e]} \in \mathbb{R}^{2n,2n}$ be a matrix satisfying the properties ($\widetilde{L}_{[e]}0$)–($\widetilde{L}_{[e]}7$) and let $\widetilde{M}$ be the matrix $db(\widetilde{L}_{[e]})+\operatorname{diag}(\widetilde{\mathfrak{d}})$, where $\widetilde{\mathfrak{d}} = (\widetilde{\mathfrak{d}}^r_k)_{(r,k)\in\mathcal{B}^D_{[e]}}$ are coefficients such that $\widetilde{\mathfrak{d}} \geq 0$ in general, and $\widetilde{\mathfrak{d}}^{00+} \neq 0$ in the particular cases $n^+ = 1$ or 2. Consider the splitting $\widetilde{L}_{[e]} = \widetilde{M} - \widetilde{Z}$ and the iteration matrix $T = \widetilde{M}^{-1}\widetilde{Z}$. Let $\mathcal{L}_{[e]}$, $\mathcal{U}_{[e]}$, $\beta^{D_l}_{[e]}$, and $\widetilde{\beta}^{D_l}_{[e]} \in \mathbb{R}^{2n}$, $l \in \mathcal{S}$, be given by (4.4.22), (4.4.23), (4.4.25), and (4.4.69), respectively, let $P = I - \mathcal{U}_{[e]}\otimes\mathcal{L}_{[e]}/\langle\mathcal{U}_{[e]}, \mathcal{L}_{[e]}\rangle$ be the oblique projector onto $\mathcal{L}^\perp_{[e]}$ along $\mathbb{R}\mathcal{U}_{[e]}$, and let $\mathfrak{P}_{[e]} \in \mathbb{R}^{n,2n}$ be the rectangular matrix formed by the blocks $\mathfrak{P}_{[e]} = [I, 0]$. Let $\widetilde{x}^k_0 \in \mathbb{R}^{2n}$, $\widetilde{y}^k_0 = P\widetilde{x}^k_0$, $k \in \mathcal{S}$, and consider for $i \geq 0$ and $k \in \mathcal{S}$ the iterates*

$$\widetilde{x}^k_{i+1} = T\widetilde{x}^k_i + \widetilde{M}^{-1}\widetilde{\beta}^{D_k}_{[e]}, \tag{5.4.35}$$

$$\widetilde{y}^k_{i+1} = PT\widetilde{y}^k_i + P\widetilde{M}^{-1}\widetilde{\beta}^{D_k}_{[e]}. \tag{5.4.36}$$

Then $\widetilde{y}^k_i = P\widetilde{x}^k_i$ for all $i \geq 0$, the matrices T and PT are convergent, $\rho(T) = 1$, $\rho(PT) < 1$, and we have the following limits

$$\begin{cases} \lim_{i\to\infty} \widetilde{y}^l_i = P(\lim_{i\to\infty} \widetilde{x}^l_i) = \widetilde{\alpha}^{D_l}_{[e]}, & l \in \mathcal{S}, \\ \lim_{i\to\infty} \langle \widetilde{y}^k_i, \beta^{D_l}_{[e]} \rangle = \lim_{i\to\infty} \langle \widetilde{x}^k_i, \beta^{D_l}_{[e]} \rangle = \widetilde{D}_{[e]kl}, & k, l \in \mathcal{S}, \end{cases} \tag{5.4.37}$$

where $\widetilde{\alpha}_{[e]}^{D_l}$ is the unique solution of the constrained linear system $\widetilde{L}_{[e]}\widetilde{\alpha}_{[e]}^{D_l} = \widetilde{\beta}_{[e]}^{D_l}$ and $\langle \mathcal{L}_{[e]}, \widetilde{\alpha}_{[e]}^{D_l} \rangle = 0$. Moreover, for all $i \geq 0$, each partial sum

$$\widetilde{D}_{[e]}^{[i]} = P_{U^\perp, \mathbb{R}Y} \mathcal{W} \mathfrak{P}_{[e]} \Big[\sum_{j=0}^{i} (PT)^j P \widetilde{M}^{-1} \Big]^t \mathfrak{P}_{[e]}^t$$
$$= P_{U^\perp, \mathbb{R}Y} \mathcal{W} \mathfrak{P}_{[e]} \Big[\sum_{j=0}^{i} T^j \widetilde{M}^{-1} \Big]^t \mathfrak{P}_{[e]}^t P_{U^\perp, \mathbb{R}Y}, \tag{5.4.38}$$

has nullspace $N(\widetilde{D}_{[e]}^{[i]}) = \mathbb{R}Y$ and range $R(\widetilde{D}_{[e]}^{[i]}) = U^\perp$. In addition, the matrix $\widetilde{D}_{[e]}^{[i]}$ admits the block-decomposition

$$\widetilde{D}_{[e]}^{[i]} = \Upsilon \begin{bmatrix} \widetilde{D}_{[e]}^{[i]++} & \widetilde{D}_{[e]}^{[i]+-} \\ 0 & \widetilde{D}_{[e]}^{[i]--} \end{bmatrix} \Upsilon^t, \tag{5.4.39}$$

where $\widetilde{D}_{[e]}^{[i]--}$ is diagonal and positive definite. Furthermore, we have

$$\lim_{i\to\infty} \widetilde{D}_{[e]}^{[i]} = P_{U^\perp, \mathbb{R}Y} \mathcal{W} \mathfrak{P}_{[e]} \Big[\sum_{j=0}^{\infty} (PT)^j P \widetilde{M}^{-1} \Big]^t \mathfrak{P}_{[e]}^t. \tag{5.4.40}$$

Finally, for positive mass fractions, the iteration matrix T coincides with the one in Theorem 5.4.4, and we have $\widetilde{x}_i^k = Y_k x_i^k$ $\widetilde{y}_i^k = Y_k y_i^k$, and $\widetilde{D}_{[e]kl} = y_k D_{[e]kl}$, $k, l \in \mathcal{S}$, provided that $\mathfrak{d} = \mathcal{X}_{[e]}^D \widetilde{\mathfrak{d}}$ and $\widetilde{x}_0^k = Y_k x_0^k$, $k \in \mathcal{S}$.

Finally, we consider the matrix $\widetilde{L}_{[e]}$ resulting from the practical approximations presented in Section 2.10.

Proposition 5.4.15. *Let $\mathcal{D}_{k\,\mathrm{int},l}$, c_k^{int}, ξ_k^{int}, and m_l, $k \in \mathcal{P}$, $l \in \mathcal{S}$, be positive quantities, let $\mathcal{D}_{kl}$, $\bar{A}_{kl}$, $\bar{B}_{kl}$, and $\bar{C}_{kl}$, $k, l \in \mathcal{S}$, be symmetric and positive coefficients such that $25/4 - 3\bar{B}_{kl} > 0$ and $15\bar{C}_{kl} - 3\bar{B}_{kl} - 9\bar{C}_{kl}^2 > 0$, $k, l \in \mathcal{S}$, and assume that $Y \geq 0$, $Y \neq 0$. Then the matrix $\widetilde{L}_{[e]}$ given by (3.7.27)–(3.7.34) satisfies $(\widetilde{L}_{[e]}0)$–$(\widetilde{L}_{[e]}7)$.*

5.4.7 Standard Iterative Methods for the System $\widetilde{L}_{[00]}\widetilde{\alpha}_{[00]}^{D_l} = \widetilde{\beta}_{[00]}^{D_l}$

In this section we want to solve the linear systems $\widetilde{L}_{[00]}\widetilde{\alpha}_{[00]}^{D_l} = \widetilde{\beta}_{[00]}^{D_l}$, $l \in \mathcal{S}$, using standard iterative methods and obtain the only solution $\widetilde{\alpha}_{[00]}^{D_l}$, $l \in \mathcal{S}$, such that $\langle Y, \widetilde{\alpha}_{[00]}^{D_l} \rangle = 0$. The system matrix $\widetilde{L}_{[00]} \in \mathbb{R}^{n,n}$ is assumed to satisfy the properties $(\widetilde{L}_{[00]}0)$–$(\widetilde{L}_{[00]}5)$ presented in Section 4.4.7, and we restate that $\mathcal{B}_{[00]}^D = \{00\} \times \mathcal{S}$, $\mathcal{X}_{[00]}^D = \mathrm{diag}\big((X_k)_{(r,k)\in\mathcal{B}_{[00]}^D}\big)$, and $\Gamma_{[00]}^D \in \mathbb{R}^{n,n}$ is the permutation matrix associated with the reordering of $\mathcal{B}^D_{[00]}$ into $(\mathcal{B}^{D+}_{[00]}, \mathcal{B}^{D-}_{[00]})$. We also denote by $\Upsilon \in \mathbb{R}^{n,n}$ the permutation matrix associated with the reordering of $\mathcal{S}$ into $(\mathcal{S}^+, \mathcal{S}^-)$. Furthermore, we consider the matrix $db(\widetilde{L}_{[00]}) \in \mathbb{R}^{n,n}$ formed by the diagonal of the only block of the matrix $\widetilde{L}_{[00]}$,

$$db(\widetilde{L}_{[00]}) = \mathrm{diag}(\widetilde{L}^{0000}_{[00]}), \tag{5.4.41}$$

and from the kinetic theory results obtained in Sections 3.1 and 3.4, we can make the following assumptions.

($\widetilde{L}_{[00]}6$) $db(L^{++}_{[00]})$ and $db(\widetilde{L}_{[00]})$ are symmetric positive semi-definite; for $n^+ \geq 2$, these matrices are also positive definite; for $n^+ = 1$, the nullspace of $db(L^{++}_{[00]})$ is spanned by the vector $\mathcal{U}^+_{[00]} = (1)$, and the nullspace of $db(\widetilde{L}_{[00]})$ is spanned by the vector $\Gamma^D_{[00]}(\mathcal{U}^+_{[00]}, 0)$.

($\widetilde{L}_{[00]}7$) $2db(L^{++}_{[00]}) - L^{++}_{[00]}$ is symmetric positive semi-definite; for $n^+ \geq 3$, this matrix is also positive definite; for $n^+ = 2$, its nullspace is spanned by the vector $(1, -1)$; for $n^+ = 1$, its nullspace is spanned by the vector (1).

Theorem 5.4.16. *Let $\widetilde{L}_{[00]} \in \mathbb{R}^{n,n}$ be a matrix satisfying the properties ($\widetilde{L}_{[00]}0$)–($\widetilde{L}_{[00]}7$) and let $\widetilde{M}$ be the matrix $db(\widetilde{L}_{[00]}) + \mathrm{diag}(\widetilde{\mathfrak{d}})$, where $\widetilde{\mathfrak{d}} = (\widetilde{\mathfrak{d}}^r_k)_{(r,k)\in\mathcal{B}^D_{[00]}}$ are coefficients such that $\widetilde{\mathfrak{d}} \geq 0$ in general, and $\widetilde{\mathfrak{d}}^{00+} \neq 0$ in the particular cases $n^+ = 1$ or 2. Consider the splitting $\widetilde{L}_{[00]} = \widetilde{M} - \widetilde{Z}$ and the iteration matrix $T = \widetilde{M}^{-1}\widetilde{Z}$. Let $\beta^{D_l}_{[00]}$ and $\widetilde{\beta}^{D_l}_{[00]} \in \mathbb{R}^n$, $l \in \mathcal{S}$, be given by (4.4.36) and (4.4.84), respectively, and let $P = I - U \otimes Y/\langle U, Y\rangle$ be the oblique projector onto $Y^\perp$ along $\mathbb{R}U$. Let $\widetilde{x}^k_0 \in \mathbb{R}^n$, $\widetilde{y}^k_0 = P\widetilde{x}^k_0$, $k \in \mathcal{S}$, and consider for $i \geq 0$ and $k \in \mathcal{S}$ the iterates*

$$\widetilde{x}^k_{i+1} = T\widetilde{x}^k_i + \widetilde{M}^{-1}\widetilde{\beta}^{D_k}_{[00]}, \tag{5.4.42}$$

$$\widetilde{y}^k_{i+1} = PT\widetilde{y}^k_i + P\widetilde{M}^{-1}\widetilde{\beta}^{D_k}_{[00]}. \tag{5.4.43}$$

Then $\widetilde{y}^k_i = P\widetilde{x}^k_i$ for all $i \geq 0$, the matrices T and PT are convergent, $\rho(T) = 1$, $\rho(PT) < 1$, and we have the following limits

$$\begin{cases} \lim_{i\to\infty} \widetilde{y}^l_i = P(\lim_{i\to\infty} \widetilde{x}^l_i) = \widetilde{\alpha}^{D_l}_{[00]}, & l \in \mathcal{S}, \\ \lim_{i\to\infty} \langle \widetilde{y}^k_i, \beta^{D_l}_{[00]}\rangle = \lim_{i\to\infty} \langle \widetilde{x}^k_i, \beta^{D_l}_{[00]}\rangle = \widetilde{D}_{[00]kl}, & k, l \in \mathcal{S}, \end{cases} \tag{5.4.44}$$

where $\widetilde{\alpha}^{D_l}_{[00]}$ is the unique solution of the constrained linear system $\widetilde{L}_{[00]}\widetilde{\alpha}^{D_l}_{[00]} = \widetilde{\beta}^{D_l}_{[00]}$ and $\langle Y, \widetilde{\alpha}^{D_l}_{[00]}\rangle = 0$. Moreover, for all $i \geq 0$, each partial sum

$$\widetilde{D}^{[i]}_{[00]} = P_{U^\perp,\mathbb{R}Y}\mathcal{W}\Big[\sum_{j=0}^{i}(PT)^j P\widetilde{M}^{-1}\Big]^t = P_{U^\perp,\mathbb{R}Y}\mathcal{W}\Big[\sum_{j=0}^{i} T^j\widetilde{M}^{-1}\Big]^t, \tag{5.4.45}$$

has nullspace $N(\widetilde{D}^{[i]}_{[00]}) = \mathbb{R}Y$ *and range* $R(\widetilde{D}^{[i]}_{[00]}) = U^{\perp}$. *In addition, the matrix* $\widetilde{D}^{[i]}_{[00]}$ *admits the block-decomposition*

$$\widetilde{D}^{[i]}_{[00]} = \Upsilon \begin{bmatrix} \widetilde{D}^{[i]++}_{[00]} & \widetilde{D}^{[i]+-}_{[00]} \\ 0 & \widetilde{D}^{[i]--}_{[00]} \end{bmatrix} \Upsilon^t, \tag{5.4.46}$$

where $\widetilde{D}^{[i]--}_{[00]}$ *is diagonal and positive definite. Furthermore, we have*

$$\lim_{i\to\infty} \widetilde{D}^{[i]}_{[00]} = P_{U^{\perp},\mathbb{R}Y}\mathcal{W}\Big[\sum_{j=0}^{\infty}(PT)^j P\widetilde{M}^{-1}\Big]^t = \widetilde{D}_{[00]}. \tag{5.4.47}$$

Finally, for positive mass fractions, the iteration matrix T *coincides with the one in Theorem 5.4.10, and we have* $\widetilde{x}^k_i = Y_k x^k_i$, $\widetilde{y}^k_i = Y_k y^k_i$, *and* $D_{[00]kl} = Y_k D_{[00]kl}$, $k,l \in \mathcal{S}$, *provided that* $\mathfrak{d} = \mathcal{X}^D_{[00]}\widetilde{\mathfrak{d}}$ *and* $\widetilde{x}^k_0 = Y_k x^k_0$, $k \in \mathcal{S}$.

Finally, we consider the matrix $\widetilde{L}_{[00]}$ resulting from the practical approximations presented in Section 2.10.

Proposition 5.4.17. *Let* $\mathcal{D}_{kl}$, $k,l \in \mathcal{S}$, *be symmetric and positive coefficients and assume that* $Y \geq 0$, $Y \neq 0$. *Then the matrix* $\widetilde{L}_{[00]}$ *given by (3.7.11)(3.7.12) satisfies* $(\widetilde{L}_{[00]}0)$–$(\widetilde{L}_{[00]}7)$.

5.4.8 Schur Complements

In this section we assume that the matrices L and $\widetilde{L}$ satisfy the structure properties $(L0)$–$(L7)$ and $(\widetilde{L}0)$–$(\widetilde{L}7)$, respectively, as in the previous sections. In particular, the block L^{0101} is then nonsingular, and in the framework of the practical approximations presented in Section 2.10, this block is also diagonal. It is therefore interesting to introduce the n constrained linear systems indexed by $l \in \mathcal{S}$

$$\begin{cases} L_{[\mathrm{s}]}\alpha^{D_l}_{[\mathrm{s}]} = \beta^{D_l}_{[\mathrm{s}]}, \\ \langle \mathcal{L}_{[\mathrm{s}]}, \alpha^{D_l}_{[\mathrm{s}]} \rangle = 0, \end{cases} \tag{5.4.48}$$

where the matrix $L_{[\mathrm{s}]} \in \mathbb{R}^{2n,2n}$ is given by

$$L_{[\mathrm{s}]} = \begin{bmatrix} L^{0000} & L^{0010} \\ L^{1000} & L^{1010} \end{bmatrix} - \begin{bmatrix} L^{0001} \\ L^{1001} \end{bmatrix} \left(L^{0101}\right)^{-1} [L^{0100}, L^{0110}], \tag{5.4.49}$$

and the vectors $\beta^{D_l}_{[\mathrm{s}]}$, $l \in \mathcal{S}$, and $\mathcal{L}_{[\mathrm{s}]} \in \mathbb{R}^{2n}$ by the block-decompositions

$$\beta^{00D_l}_{[\mathrm{s}]} = \beta^{00D_l}, \qquad \beta^{10D_l}_{[\mathrm{s}]} = \beta^{10D_l}, \tag{5.4.50}$$

and

$$\mathcal{L}^{00}_{[s]} = \mathcal{L}^{00}, \qquad \mathcal{L}^{10}_{[s]} = \mathcal{L}^{10}, \tag{5.4.51}$$

since $\beta^{01D_l} = 0$ and $\mathcal{L}^{01} = 0$. We deduce from Lemma 5.1.12 that the systems (5.4.48) admit a unique solution $\alpha^{D_l}_{[s]}$, $l \in \mathcal{S}$. The diffusion matrix is then given by

$$D_{kl} = \langle \alpha^{D_k}_{[s]}, \beta^{D_l}_{[s]} \rangle = \alpha^{00D_k}_{[s]l} = \alpha^{00D_l}_{[s]k}, \qquad k, l \in \mathcal{S}. \tag{5.4.52}$$

Consider now the matrix $M_{[s]} \in \mathbb{R}^{2n,2n}$ given by

$$M_{[s]} = \begin{bmatrix} \operatorname{diag}(L^{0000}) & \operatorname{diag}(L^{0010}) \\ \operatorname{diag}(L^{1000}) & \operatorname{diag}(L^{1010}) \end{bmatrix} + \operatorname{diag}(\mathfrak{d}), \tag{5.4.53}$$

where $\mathfrak{d}^r_k \geq 0$, $(r,k) \in \{00, 10\} \times \mathcal{S}$, in general, and $\mathfrak{d}^{00}_k > 0$, $k \in \mathcal{S}$, in the particular cases $n = 1$ or 2. From $(L6)$–$(L7)$ we then deduce that the matrices $M_{[s]}$ and $2M_{[s]} - L_{[s]}$ are symmetric positive definite. A convergence theorem similar to Theorem 5.4.1 can then be stated for the constrained linear system (5.4.48) with the splitting matrix (5.4.53). In particular, this yields a sequence of iterates $D^{[i]}_{[s]}$ such that

$$\lim_{i\to\infty} D^{[i]}_{[s]} = D. \tag{5.4.54}$$

Moreover, since the matrix $M_{[s]}$ is symmetric positive definite, a convergence theorem similar to Theorem 5.4.2 can be stated for the constrained linear system (5.4.48) with the preconditioner (5.4.53). This yields a sequence of iterates $D^{[i]}_{[s]}$ which converges towards D in a finite number of steps.

Stabilized versions, for vanishing mass fractions, of standard iterative methods are obtained by considering the left rescaled constrained linear systems indexed by $l \in \mathcal{S}$

$$\begin{cases} \widetilde{L}_{[s]} \widetilde{\alpha}^{D_l}_{[s]} = \widetilde{\beta}^{D_l}_{[s]}, \\ \langle \mathcal{L}_{[s]}, \widetilde{\alpha}^{D_l}_{[s]} \rangle = 0, \end{cases} \tag{5.4.55}$$

where the matrix $\widetilde{L}_{[s]} \in \mathbb{R}^{2n,2n}$ is given by

$$\widetilde{L}_{[s]} = \begin{bmatrix} \widetilde{L}^{0000} & \widetilde{L}^{0010} \\ \widetilde{L}^{1000} & \widetilde{L}^{1010} \end{bmatrix} - \begin{bmatrix} \widetilde{L}^{0001} \\ \widetilde{L}^{1001} \end{bmatrix} (\widetilde{L}^{0101})^{-1} [\widetilde{L}^{0100}, \widetilde{L}^{0110}], \tag{5.4.56}$$

the vectors $\widetilde{\beta}^{D_l}_{[s]}$, $l \in \mathcal{S}$, by the block-decomposition

$$\widetilde{\beta}^{00D_l}_{[s]} = \widetilde{\beta}^{00D_l}, \qquad \widetilde{\beta}^{10D_l}_{[s]} = \widetilde{\beta}^{10D_l}, \tag{5.4.57}$$

and the vector $\mathcal{L}_{[\mathbf{s}]}$ still by (5.4.51). Assuming $\widetilde{L}^{0101}$ to be symmetric, we deduce from Section 5.1.5 that the systems (5.4.55) admit a unique solution $\widetilde{\alpha}_{[\mathbf{s}]}^{D_l}$ $l \in \mathcal{S}$. The relations

$$\widetilde{D}_{kl} = \langle \widetilde{\alpha}_{[\mathbf{s}]}^{D_k}, \beta_{[\mathbf{s}]}^{D_l} \rangle = \widetilde{\alpha}_{[\mathbf{s}]l}^{00D_k}, \qquad k, l \in \mathcal{S}, \tag{5.4.58}$$

then yield the flux diffusion matrix. For positive mass fractions, one can easily verify that the solutions of the systems (5.4.48) and (5.4.55) are such that $\widetilde{\alpha}_{[\mathbf{s}]}^{D_k} = Y_k \alpha_{[\mathbf{s}]}^{D_k}$, $k \in \mathcal{S}$, and that the diffusion and flux diffusion matrices (5.4.52) and (5.4.58) verify $\widetilde{D} = \mathcal{Y}D$.

Consider now the splitting $\widetilde{L}_{[\mathbf{s}]} = \widetilde{M}_{[\mathbf{s}]} - \widetilde{Z}_{[\mathbf{s}]}$ where

$$\widetilde{M}_{[\mathbf{s}]} = \begin{bmatrix} \mathrm{diag}(\widetilde{L}^{0000}) & \mathrm{diag}(\widetilde{L}^{0010}) \\ \mathrm{diag}(\widetilde{L}^{1000}) & \mathrm{diag}(\widetilde{L}^{1010}) \end{bmatrix} + \mathrm{diag}(\widetilde{\mathfrak{d}}), \tag{5.4.59}$$

and $\widetilde{\mathfrak{d}}_k^r \geq 0$, $(r, k) \in \{00, 10\} \times \mathcal{S}$, in general, and $\widetilde{\mathfrak{d}}_k^{00} > 0$, $k \in \mathcal{S}$, in the particular cases $n^+ = 1$ or 2. As in the proof of Theorem 5.4.10, one can show, after some algebra, that the iteration matrix $T_{[\mathbf{s}]} = (\widetilde{M}_{[\mathbf{s}]})^{-1} \widetilde{Z}_{[\mathbf{s}]}$ is convergent. A convergence theorem similar to Theorem 5.4.10 can then be stated for the constrained linear systems (5.4.55) with the splitting matrix (5.4.59). In particular, this yields a sequence of iterates $\widetilde{D}_{[\mathbf{s}]}^{[i]}$ such that

$$\lim_{i \to \infty} \widetilde{D}_{[\mathbf{s}]}^{[i]} = \widetilde{D}. \tag{5.4.60}$$

Finally, stabilized versions of conjugate gradient methods are obtained by considering the symmetric rescaled matrix $\widehat{L}_{[\mathbf{s}]}$, and are omitted for brevity.

5.5 The Partial Thermal Conductivity and the Thermal Diffusion Vector

5.5.1 Iterative Methods for the System $L\alpha^{\lambda'} = \beta^{\lambda'}$

In this section we want to solve the linear system $L\alpha^{\lambda'} = \beta^{\lambda'}$ using either standard iterative or conjugate gradient methods, and obtain the only solution $\alpha^{\lambda'}$ such that $\langle \mathcal{L}, \alpha^{\lambda'} \rangle = 0$. The system matrix $L \in \mathbb{R}^{2n+p, 2n+p}$ is assumed to satisfy the properties $(L0)$–$(L5)$ and $(L6)$–$(L7)$ given in Sections 4.4.1 and 5.4.1, respectively, and as stated in Section 4.5.1, we have $\mathcal{B}^{\lambda'} = \{00, 10\} \times \mathcal{S} \cup \{01\} \times \mathcal{P}$.

Theorem 5.5.1. *Let $L \in \mathbb{R}^{2n+p,2n+p}$ be a matrix satisfying the properties (L0)–(L7) and let M be the matrix $db(L) + \mathrm{diag}(\mathfrak{d})$, where $\mathfrak{d} = (\mathfrak{d}_k^r)_{(r,k)\in\mathcal{B}^{\lambda'}}$ are coefficients such that $\mathfrak{d} \geq 0$ in general, and $\mathfrak{d}^{00} \neq 0$ in the particular cases $n = 1$ or 2. Consider the splitting $L = M - Z$ and the iteration matrix $T = M^{-1}Z$. Let $\mathcal{L}$, $\mathcal{U}$, β^{D_l}, $l \in \mathcal{S}$, and $\beta^{\lambda'} \in \mathbb{R}^{2n+p}$ be given by (4.4.6), (4.4.7), (4.4.9), and (4.5.6), let $P = I - \mathcal{U}\otimes\mathcal{L}/\langle\mathcal{U},\mathcal{L}\rangle$ denote the oblique projector onto $\mathcal{L}^\perp$ along $\mathbb{R}\mathcal{U}$ and let $\mathfrak{P} \in \mathbb{R}^{n,2n+p}$ be the rectangular matrix formed by the blocks $\mathfrak{P} = [I,0,0]$. Let $x_0 \in \mathbb{R}^{2n+p}$, $y_0 = Px_0$, and consider for $i \geq 0$ the iterates*

$$x_{i+1} = Tx_i + M^{-1}\beta^{\lambda'}, \tag{5.5.1}$$

$$y_{i+1} = PTy_i + PM^{-1}\beta^{\lambda'}. \tag{5.5.2}$$

Then $y_i = Px_i$ for all $i \geq 0$, the matrices T and PT are convergent, $\rho(T) = 1$, $\rho(PT) < 1$, and we have the following limits

$$\begin{cases} \lim\limits_{i\to\infty} y_i = P(\lim\limits_{i\to\infty} x_i) = \alpha^{\lambda'}, \\ \lim\limits_{i\to\infty} \dfrac{\bar{p}}{T}\langle y_i, \beta^{\lambda'}\rangle = \lim\limits_{i\to\infty} \dfrac{\bar{p}}{T}\langle x_i, \beta^{\lambda'}\rangle = \lambda', \\ \lim\limits_{i\to\infty} -\langle y_i, \beta^{D_k}\rangle = \lim\limits_{i\to\infty} -\langle x_i, \beta^{D_k}\rangle = \theta_k, \quad k \in \mathcal{S}, \end{cases} \tag{5.5.3}$$

where $\alpha^{\lambda'}$ is the unique solution of the constrained linear system $L\alpha^{\lambda'} = \beta^{\lambda'}$ and $\langle\mathcal{L}, \alpha^{\lambda'}\rangle = 0$. Moreover, for all $i \geq 0$, each partial sum

$$\lambda'^{[i]} = \frac{\bar{p}}{T}\langle\sum_{j=0}^{i}(PT)^j PM^{-1}\beta^{\lambda'}, \beta^{\lambda'}\rangle = \frac{\bar{p}}{T}\langle\sum_{j=0}^{i} T^j M^{-1}\beta^{\lambda'}, \beta^{\lambda'}\rangle, \tag{5.5.4}$$

is positive, and each partial sum

$$\theta^{[i]} = -\mathfrak{P}\sum_{j=0}^{i}(PT)^j PM^{-1}\beta^{\lambda'} = -P_{Y^\perp,\mathbb{R}U}\mathfrak{P}\sum_{j=0}^{i} T^j M^{-1}\beta^{\lambda'}, \tag{5.5.5}$$

is such that $\langle\theta^{[i]}, Y\rangle = 0$. Finally, we have

$$\begin{cases} \lim\limits_{i\to\infty} \lambda'^{[i]} = \dfrac{\bar{p}}{T}\langle\sum\limits_{j=0}^{\infty}(PT)^j PM^{-1}\beta^{\lambda'}, \beta^{\lambda'}\rangle = \lambda', \\ \lim\limits_{i\to\infty} \theta^{[i]} = -\mathfrak{P}\sum\limits_{j=0}^{\infty}(PT)^j PM^{-1}\beta^{\lambda'} = \theta. \end{cases} \tag{5.5.6}$$

Proof. The first part of the proof is similar to the one of Theorem 5.4.1. Keeping in mind that the matrices $L_i^\dagger = \sum_{j=0}^{i}(PT)^j PM^{-1}P^t$, $i \geq 0$, are symmetric, positive semi-definite, and positive definite on the hyperplane $\mathcal{U}^\perp$, we obtain $\lambda'^{[i]} = (\bar{p}/\bar{T})\langle L_i^\dagger \beta^{\lambda'}, \beta^{\lambda'}\rangle > 0$, since $\beta^{\lambda'} \in \mathcal{U}^\perp$ and $\beta^{\lambda'} \neq 0$. Furthermore, since $R(P_{Y^\perp, \mathbb{R}U}) = Y^\perp$, we have $\langle \theta^{[i]}, Y\rangle = 0$, and finally, the limits (5.5.6) are obtained from (5.5.3). □

Remark. Since the thermal diffusion vector is also given by $\theta_k = -\langle \alpha^{D_k}, \beta^{\lambda'}\rangle$, $k \in S$, one can also consider standard iterative methods for the systems $L\alpha^{D_l} = \beta^{D_l}$ and $\langle \mathcal{L}, \alpha^{D_l}\rangle = 0$, $l \in S$, in order to evaluate θ. By symmetry, one can then easily verify that this yields again the partial sums (5.5.5).

Theorem 5.5.2. *Let $L \in \mathbb{R}^{2n+p,2n+p}$ be a matrix satisfying the properties (L0)–(L6) and let M be the matrix $db(L) + \mathrm{diag}(\mathfrak{d})$ or $\mathrm{diag}(L) + \mathrm{diag}(\mathfrak{d})$, where $\mathfrak{d} = (\mathfrak{d}_k^r)_{(r,k)\in\mathcal{B}^{\lambda'}}$ are coefficients such that $\mathfrak{d} \geq 0$ in general, and $\mathfrak{d}^{00} \neq 0$ in the particular case $n = 1$. Let $\mathcal{L}$, $\mathcal{U}$, β^{D_l}, $l \in S$, and $\beta^{\lambda'} \in \mathbb{R}^{2n+p}$ be given by (4.4.6), (4.4.7), (4.4.9), and (4.5.6), and let $P = I - \mathcal{U}\otimes\mathcal{L}/\langle \mathcal{U}, \mathcal{L}\rangle$ denote the oblique projector onto $\mathcal{L}^\perp$ along $\mathbb{R}\mathcal{U}$. Let $x_0 \in \mathbb{R}^{2n+p}$, $y_0 = Px_0$, $r_0 = \beta^{\lambda'} - Lx_0$, $p_0 = 0$, $t_0 = 0$, and consider for $i \geq 1$ the iterates*

$$\begin{cases} p_i = M^{-1} r_{i-1} + t_{i-1} p_{i-1}, \\ s_i = \langle r_{i-1}, M^{-1} r_{i-1}\rangle / \langle p_i, L p_i\rangle, \\ x_i = x_{i-1} + s_i p_i, \\ y_i = y_{i-1} + P(s_i p_i), \\ r_i = r_{i-1} - s_i L p_i, \\ t_i = \langle r_i, M^{-1} r_i\rangle / \langle r_{i-1}, M^{-1} r_{i-1}\rangle. \end{cases} \tag{5.5.7}$$

Then $y_i = Px_i$ for all $i \geq 0$, the iterates y_i converge towards $\alpha^{\lambda'}$ in a finite number of steps, and the quantities

$$\begin{cases} \lambda'^{[i]} = \dfrac{\bar{p}}{\bar{T}}\langle y_i, \beta^{\lambda'}\rangle = \dfrac{\bar{p}}{\bar{T}}\langle x_i, \beta^{\lambda'}\rangle, \\ \theta_k^{[i]} = -\langle y_i, \beta^{D_k}\rangle = -\langle x_i, \beta^{D_k}\rangle, \quad k \in S, \end{cases} \tag{5.5.8}$$

converge towards λ' and θ, respectively, in a finite number of steps. Finally, we have $\langle \theta^{[i]}, Y\rangle = 0$, $i \geq 0$, and if $x_0 = 0$, we obtain $\lambda'^{[i]} > 0$ for $i \geq 1$.

Proof. It is similar to the one of Theorem 5.4.2. □

5.5.2 Iterative Methods for the System $L_{[e]}\alpha_{[e]}^{\lambda'} = \beta_{[e]}^{\lambda'}$

In this section we want to solve the linear system $L_{[e]}\alpha_{[e]}^{\lambda'} = \beta_{[e]}^{\lambda'}$ using either standard iterative or conjugate gradient methods, and obtain the only solution $\alpha_{[e]}^{\lambda'}$ such that

$\langle \mathcal{L}_{[e]}, \alpha_{[e]}^{\lambda'} \rangle = 0$. The system matrix $L_{[e]} \in \mathbb{R}^{2n,2n}$ is assumed to satisfy the properties $(L_{[e]}0)$–$(L_{[e]}5)$ and $(L_{[e]}6)$–$(L_{[e]}7)$ given in Sections 4.4.2 and 5.4.2, respectively, and as stated in Section 4.5.2, we have $\mathcal{B}_{[e]}^{\lambda'} = \{00, e\} \times \mathcal{S}$.

Theorem 5.5.3. *Let $L_{[e]} \in \mathbb{R}^{2n,2n}$ be a matrix satisfying the properties $(L_{[e]}0)$–$(L_{[e]}7)$ and let M be the matrix $db(L_{[e]}) + \mathrm{diag}(\mathfrak{d})$, where $\mathfrak{d} = (\mathfrak{d}_k^r)_{(r,k) \in \mathcal{B}_{[e]}^{\lambda'}}$ are coefficients such that $\mathfrak{d} \geq 0$ in general, and $\mathfrak{d}^{00} \neq 0$ in the particular cases $n = 1$ or 2. Consider the splitting $L = M - Z$ and the iteration matrix $T = M^{-1}Z$. Let $\mathcal{L}_{[e]}$, $\mathcal{U}_{[e]}$, $\beta_{[e]}^{D_l}$, $l \in \mathcal{S}$, and $\beta_{[e]}^{\lambda'} \in \mathbb{R}^{2n}$ be given by (4.4.22), (4.4.23), (4.4.25), and (4.5.17), respectively, let $P = I - \mathcal{U}_{[e]} \otimes \mathcal{L}_{[e]} / \langle \mathcal{U}_{[e]}, \mathcal{L}_{[e]} \rangle$ denote the oblique projector onto $\mathcal{L}_{[e]}^{\perp}$ along $\mathbb{R}\mathcal{U}_{[e]}$, and let $\mathfrak{P}_{[e]} \in \mathbb{R}^{n,2n}$ be the rectangular matrix formed by the blocks $\mathfrak{P}_{[e]} = [I, 0]$. Let $x_0 \in \mathbb{R}^{2n}$, $y_0 = Px_0$, and consider for $i \geq 0$ the iterates*

$$x_{i+1} = Tx_i + M^{-1}\beta_{[e]}^{\lambda'}, \tag{5.5.9}$$

$$y_{i+1} = PTy_i + PM^{-1}\beta_{[e]}^{\lambda'}. \tag{5.5.10}$$

Then $y_i = Px_i$ for all $i \geq 0$, the matrices T and PT are convergent, $\rho(T) = 1$, $\rho(PT) < 1$, and we have the following limits

$$\begin{cases} \lim_{i \to \infty} y_i = P(\lim_{i \to \infty} x_i) = \alpha_{[e]}^{\lambda'}, \\ \lim_{i \to \infty} \frac{\bar{p}}{\overline{T}} \langle y_i, \beta_{[e]}^{\lambda'} \rangle = \lim_{i \to \infty} \frac{\bar{p}}{\overline{T}} \langle x_i, \beta_{[e]}^{\lambda'} \rangle = \lambda'_{[e]}, \\ \lim_{i \to \infty} -\langle y_i, \beta_{[e]}^{D_k} \rangle = \lim_{i \to \infty} -\langle x_i, \beta_{[e]}^{D_k} \rangle = \theta_{[e]k}, \quad k \in \mathcal{S}, \end{cases} \tag{5.5.11}$$

where $\alpha_{[e]}^{\lambda'}$ is the unique solution of the constrained linear system $L_{[e]}\alpha_{[e]}^{\lambda'} = \beta_{[e]}^{\lambda'}$ and $\langle \mathcal{L}_{[e]}, \alpha_{[e]}^{\lambda'} \rangle = 0$. Moreover, for all $i \geq 0$, each partial sum

$$\lambda'^{[i]}_{[e]} = \frac{\bar{p}}{\overline{T}} \langle \sum_{j=0}^{i} (PT)^j PM^{-1}\beta_{[e]}^{\lambda'}, \beta_{[e]}^{\lambda'} \rangle = \frac{\bar{p}}{\overline{T}} \langle \sum_{j=0}^{i} T^j M^{-1}\beta_{[e]}^{\lambda'}, \beta_{[e]}^{\lambda'} \rangle, \tag{5.5.12}$$

is positive, and each partial sum

$$\theta_{[e]}^{[i]} = -\mathfrak{P}_{[e]} \sum_{j=0}^{i} (PT)^j PM^{-1}\beta_{[e]}^{\lambda'} = -P_{Y^{\perp}, \mathbb{R}U} \mathfrak{P}_{[e]} \sum_{j=0}^{i} T^j M^{-1}\beta_{[e]}^{\lambda'}, \tag{5.5.13}$$

is such that $\langle \theta_{[e]}^{[i]}, Y \rangle = 0$. Finally, we have

$$\begin{cases} \lim_{i \to \infty} \lambda'^{[i]}_{[e]} = \frac{\bar{p}}{\overline{T}} \langle \sum_{j=0}^{\infty} (PT)^j PM^{-1}\beta_{[e]}^{\lambda'}, \beta_{[e]}^{\lambda'} \rangle = \lambda'_{[e]}, \\ \lim_{i \to \infty} \theta_{[e]}^{[i]} = -\mathfrak{P}_{[e]} \sum_{j=0}^{\infty} (PT)^j PM^{-1}\beta_{[e]}^{\lambda'} = \theta_{[e]}. \end{cases} \tag{5.5.14}$$

Remark. Since the thermal diffusion vector is also given by $\theta_{[e]k} = -\langle \alpha_{[e]}^{D_k}, \beta_{[e]}^{\lambda'} \rangle$, $k \in \mathcal{S}$, one can also consider standard iterative methods for the systems $L_{[e]}\alpha_{[e]}^{D_l} = \beta_{[e]}^{D_l}$ and $\langle \mathcal{L}_{[e]}, \alpha_{[e]}^{D_l} \rangle = 0$, $l \in \mathcal{S}$, in order to evaluate $\theta_{[e]}$. By symmetry, one can then easily verify that this yields again the partial sums (5.5.13).

Theorem 5.5.4. *Let $L_{[e]} \in \mathbb{R}^{2n,2n}$ be a matrix satisfying the properties $(L_{[e]}0)$–$(L_{[e]}6)$ and let M be the matrix $db(L_{[e]})+\mathrm{diag}(\mathfrak{d})$ or $\mathrm{diag}(L_{[e]})+\mathrm{diag}(\mathfrak{d})$, where $\mathfrak{d} = (\mathfrak{d}_k^r)_{(r,k)\in\mathcal{B}_{[e]}^{\lambda'}}$ are coefficients such that $\mathfrak{d} \geq 0$ in general, and $\mathfrak{d}^{00} \neq 0$ in the particular case $n = 1$. Let $\mathcal{L}_{[e]}$, $\mathcal{U}_{[e]}$, $\beta_{[e]}^{D_l}$, $l \in \mathcal{S}$, and $\beta_{[e]}^{\lambda'} \in \mathbb{R}^{2n}$ be given by (4.4.22), (4.4.23), (4.4.25), and (4.5.17), respectively, and let $P = I - \mathcal{U}_{[e]} \otimes \mathcal{L}_{[e]} / \langle \mathcal{U}_{[e]}, \mathcal{L}_{[e]} \rangle$ denote the oblique projector onto $\mathcal{L}_{[e]}^{\perp}$ along $\mathbb{R}\mathcal{U}_{[e]}$. Let $x_0 \in \mathbb{R}^{2n}$, $y_0 = Px_0$, $r_0 = \beta_{[e]}^{\lambda'} - L_{[e]}x_0$, $p_0 = 0$, $t_0 = 0$, and consider for $i \geq 1$ the iterates*

$$\begin{cases} p_i = M^{-1} r_{i-1} + t_{i-1} p_{i-1}, \\ s_i = \langle r_{i-1}, M^{-1} r_{i-1} \rangle / \langle p_i, L_{[e]} p_i \rangle, \\ x_i = x_{i-1} + s_i p_i, \\ y_i = y_{i-1} + P(s_i p_i), \\ r_i = r_{i-1} - s_i L_{[e]} p_i, \\ t_i = \langle r_i, M^{-1} r_i \rangle / \langle r_{i-1}, M^{-1} r_{i-1} \rangle. \end{cases} \tag{5.5.15}$$

Then for all $i \geq 0$, $y_i = Px_i$, the iterates y_i converge towards $\alpha_{[e]}^{\lambda'}$ in a finite number of steps, and the quantities

$$\begin{cases} \lambda_{[e]}^{\prime[i]} = \dfrac{\overline{p}}{\overline{\overline{T}}} \langle y_i, \beta_{[e]}^{\lambda'} \rangle = \dfrac{\overline{p}}{\overline{\overline{T}}} \langle x_i, \beta_{[e]}^{\lambda'} \rangle, \\ \theta_{[e]k}^{[i]} = -\langle y_i, \beta_{[e]}^{D_k} \rangle = -\langle x_i, \beta_{[e]}^{D_k} \rangle, \quad k \in \mathcal{S}, \end{cases} \tag{5.5.16}$$

converge towards $\lambda'_{[e]}$ and $\theta_{[e]}$, respectively, in a finite number of steps. Finally, we have $\langle \theta_{[e]}^{[i]}, Y \rangle = 0$, $i \geq 0$, and if $x_0^{\lambda'} = 0$, we obtain $\lambda_{[e]}^{\prime[i]} > 0$ for $i \geq 1$.

5.5.3 Standard Iterative Methods for the System $\widetilde{L}\alpha^{\lambda'} = \widetilde{\beta}^{\lambda'}$

In this section we want to solve the constrained linear system $\widetilde{L}\alpha^{\lambda'} = \widetilde{\beta}^{\lambda'}$ using standard iterative methods and obtain the only solution $\alpha^{\lambda'}$ such that $\langle \mathcal{L}, \alpha^{\lambda'} \rangle = 0$. The system matrix $\widetilde{L} \in \mathbb{R}^{2n+p,2n+p}$ is assumed to satisfy the properties $(\widetilde{L}0)$–$(\widetilde{L}5)$ and $(\widetilde{L}6)$–$(\widetilde{L}7)$ given in Sections 4.4.4 and 5.4.4, respectively. Furthermore, we restate that $\mathcal{B}^{\lambda'} = \{00, 10\} \times \mathcal{S} \;\cup\; \{01\} \times \mathcal{P}$, $\mathcal{X}^{\lambda'} = \mathrm{diag}((X_k)_{(r,k)\in\mathcal{B}^{\lambda'}})$, and $\Gamma^{\lambda'} \in \mathbb{R}^{2n+p,2n+p}$ is the permutation matrix associated with the reordering of $\mathcal{B}^{\lambda'}$ into $(\mathcal{B}^{\lambda'+}, \mathcal{B}^{\lambda'-})$.

Theorem 5.5.5. *Let $\widetilde{L} \in \mathbb{R}^{2n+p,2n+p}$ be a matrix satisfying the properties ($\widetilde{L}0$)–($\widetilde{L}7$) and let $\widetilde{M}$ be the matrix $db(\widetilde{L}) + \mathrm{diag}(\widetilde{\mathfrak{d}})$, where $\widetilde{\mathfrak{d}} = (\widetilde{\mathfrak{d}}_k^r)_{(r,k)\in B^{\lambda'}}$ are coefficients such that $\widetilde{\mathfrak{d}} \geq 0$ in general, and $\widetilde{\mathfrak{d}}^{00+} \neq 0$ in the particular cases $n^+ = 1$ and 2. Consider the splitting $\widetilde{L} = \widetilde{M} - \widetilde{Z}$ and the iteration matrix $T = \widetilde{M}^{-1}\widetilde{Z}$. Let $\mathcal{L}, \mathcal{U}, \beta^{D_l}, l \in \mathcal{S}, \beta^{\lambda'}$, and $\widetilde{\beta}^{\lambda'} \in \mathbb{R}^{2n+p}$ be given by (4.4.6), (4.4.7), (4.4.9), (4.5.6), and (4.5.27), respectively, let $P = I - \mathcal{U}\otimes\mathcal{L}/\langle\mathcal{U},\mathcal{L}\rangle$ the oblique projector onto $\mathcal{L}^\perp$ along $\mathbb{R}\mathcal{U}$, and let $\mathfrak{P} \in \mathbb{R}^{n,2n+p}$ be the rectangular matrix formed by the blocks $\mathfrak{P} = [I, 0, 0]$. Let $x_0 \in \mathbb{R}^{2n+p}$, $y_0 = Px_0$, and consider the iterates*

$$x_{i+1} = Tx_i + \widetilde{M}^{-1}\widetilde{\beta}^{\lambda'}, \qquad i \geq 0, \tag{5.5.17}$$

$$y_{i+1} = PTy_i + P\widetilde{M}^{-1}\widetilde{\beta}^{\lambda'}, \qquad i \geq 0. \tag{5.5.18}$$

Then $y_i = Px_i$ for all $i \geq 0$, the matrices T and PT are convergent, $\rho(T) = 1$, $\rho(PT) < 1$, and we have the following limits

$$\begin{cases} \lim\limits_{i\to\infty} y_i = P(\lim\limits_{i\to\infty} x_i) = \alpha^{\lambda'}, \\ \lim\limits_{i\to\infty} \frac{\overline{p}}{\overline{\overline{T}}}\langle y_i, \beta^{\lambda'}\rangle = \lim\limits_{i\to\infty} \frac{\overline{p}}{\overline{\overline{T}}}\langle x_i, \beta^{\lambda'}\rangle = \lambda', \\ \lim\limits_{i\to\infty} -\langle y_i, \beta^{D_k}\rangle = \lim\limits_{i\to\infty} -\langle x_i, \beta^{D_k}\rangle = \theta_k, \quad k \in \mathcal{S}, \end{cases} \tag{5.5.19}$$

where $\alpha^{\lambda'}$ is the unique solution of the constrained linear system $\widetilde{L}\alpha^{\lambda'} = \widetilde{\beta}^{\lambda'}$ and $\langle\mathcal{L}, \alpha^{\lambda'}\rangle = 0$. Moreover, for all $i \geq 0$, each partial sum

$$\lambda'^{[i]} = \frac{\overline{p}}{\overline{\overline{T}}}\langle\sum_{j=0}^{i}(PT)^j P\widetilde{M}^{-1}\widetilde{\beta}^{\lambda'}, \beta^{\lambda'}\rangle = \frac{\overline{p}}{\overline{\overline{T}}}\langle\sum_{j=0}^{i} T^j\widetilde{M}^{-1}\widetilde{\beta}^{\lambda'}, \beta^{\lambda'}\rangle, \tag{5.5.20}$$

is positive, and each partial sum

$$\theta^{[i]} = -\mathfrak{P}\sum_{j=0}^{i}(PT)^j P\widetilde{M}^{-1}\widetilde{\beta}^{\lambda'} = -P_{Y^\perp,\mathbb{R}U}\mathfrak{P}\sum_{j=0}^{i} T^j\widetilde{M}^{-1}\widetilde{\beta}^{\lambda'}, \tag{5.5.21}$$

is such that $\langle\theta^{[i]}, Y\rangle = 0$. Furthermore, we have

$$\begin{cases} \lim\limits_{i\to\infty} \lambda'^{[i]} = \frac{\overline{p}}{\overline{\overline{T}}}\langle\sum_{j=0}^{\infty}(PT)^j P\widetilde{M}^{-1}\widetilde{\beta}^{\lambda'}, \beta^{\lambda'}\rangle = \lambda', \\ \lim\limits_{i\to\infty} \theta^{[i]} = -\mathfrak{P}\sum_{j=0}^{\infty}(PT)^j P\widetilde{M}^{-1}\widetilde{\beta}^{\lambda'} = \theta. \end{cases} \tag{5.5.22}$$

Finally, for positive mass fractions, the iteration matrix T and the iterates x_i, y_i, $\lambda'^{[i]}$, and $\theta^{[i]}$ coincide with the ones in Theorem 5.5.1, provided that $\mathfrak{d} = \mathcal{X}^{\lambda'}\widetilde{\mathfrak{d}}$.

Proof. The first part of the proof is similar to the one of Theorem 5.4.10. Furthermore, the matrix $(L_i^\dagger)^{++} = \sum_{j=0}^{i}(P^{++}T^{++})^j P^{++}(M^{-1})^{++}(P^{++})^t$ is symmetric, positive semi-definite, and positive definite on the hyperplane $(\mathcal{U}^+)^\perp$, as may easily be verified. Therefore, we have $\lambda'^{[i]} = (\bar{p}/\overline{T})\langle (L_i^\dagger)^{++}\beta^{\lambda'+}, \beta^{\lambda'+}\rangle > 0$, since $\beta^{\lambda'+} \neq 0$ and $\beta^{\lambda'+} \in (\mathcal{U}^+)^\perp$. On the other hand, the relation $\langle \theta^{[i]}, Y\rangle = 0$ directly follows from $R(P_{Y^\perp,\mathbb{R}U}) = Y^\perp$. Furthermore, the limits (5.5.22) result from (5.5.19), since we have $\theta_k^{[i]} = -\langle y_{i+1}, \beta^{D_k}\rangle$, $k \in S$, for the particular choice $y_0 = 0$. Finally, the correspondence with Theorem 5.5.1 is obtained using Proposition 5.1.15. □

Remark. In the case of a pure species state of the mixture, i.e., $n^+ = 1$, one may easily verify that a one-step convergence of the iterates $\lambda'^{[i]}$ and $\theta^{[i]}$ is obtained for the initialization $y_0 = 0$ and any splitting of the form $\widetilde{\mathfrak{d}} = \Gamma^{\lambda'}(\widetilde{\mathfrak{d}}^+, 0)$, where $\widetilde{\mathfrak{d}}^+ = (\widetilde{\mathfrak{d}}^{00+}, 0, 0)$ and $\widetilde{\mathfrak{d}}^{00+} > 0$.

5.5.4 Conjugate Gradient Methods for the System $\widehat{L}\widehat{\alpha}^{\lambda'} = \widehat{\beta}^{\lambda'}$

In this section we assume that the matrix $\widehat{L}$ satisfies the properties ($\widehat{L}$0)–($\widehat{L}$5) given in Section 4.4.5. We then want to solve the constrained linear system $\widehat{L}\widehat{\alpha}^{\lambda'} = \widehat{\beta}^{\lambda'}$ using conjugate gradient methods and obtain the only solution $\widehat{\alpha}^{\lambda'}$ such that $\langle \widehat{\mathcal{L}}, \widehat{\alpha}^{\lambda'}\rangle = 0$.

Theorem 5.5.6. *Let $\widehat{L} \in \mathbb{R}^{2n+p,2n+p}$ be a matrix satisfying the properties ($\widehat{L}$0)–($\widehat{L}$5) and let $\widehat{M}$ be the matrix $db(\widehat{L}) + \mathrm{diag}(\widehat{\mathfrak{d}})$ or $\mathrm{diag}(\widehat{L}) + \mathrm{diag}(\widehat{\mathfrak{d}})$, where $\widehat{\mathfrak{d}} = (\widehat{\mathfrak{d}}_k^r)_{(r,k)\in B^{\lambda'}}$ are coefficients such that $\widehat{\mathfrak{d}} \geq 0$ in general, and $\widehat{\mathfrak{d}}^{00+} \neq 0$ in the particular case $n^+ = 1$. Let $\widehat{\beta}^{\lambda'}$, $\widehat{\mathcal{L}}$, and $\widehat{\mathcal{U}} \in \mathbb{R}^{2n+p}$ be given by (4.5.37), (4.4.59), and (4.4.60), respectively, and let also $\widehat{P} = I - \widehat{\mathcal{U}}\otimes\widehat{\mathcal{L}}/\langle\widehat{\mathcal{U}}, \widehat{\mathcal{L}}\rangle$ denote the oblique projector onto $\widehat{\mathcal{L}}^\perp$ along $\mathbb{R}\widehat{\mathcal{U}}$. Let $\widehat{x}_0 \in \mathbb{R}^{2n+p}$, $\widehat{y}_0 = \widehat{P}\widehat{x}_0$, $\widehat{r}_0 = \widehat{\beta}^{\lambda'} - \widehat{L}\widehat{x}_0$, $\widehat{p}_0 = 0$, $t_0 = 0$, and consider for $i \geq 1$ the iterates*

$$\begin{cases} \widehat{p}_i = \widehat{M}^{-1}\widehat{r}_{i-1} + t_{i-1}\widehat{p}_{i-1}, \\ s_i = \langle \widehat{r}_{i-1}, \widehat{M}^{-1}\widehat{r}_{i-1}\rangle / \langle \widehat{p}_i, \widehat{L}\widehat{p}_i\rangle, \\ \widehat{x}_i = \widehat{x}_{i-1} + s_i\widehat{p}_i, \\ \widehat{y}_i = \widehat{y}_{i-1} + \widehat{P}(s_i\widehat{p}_i), \\ \widehat{r}_i = \widehat{r}_{i-1} - s_i\widehat{L}\widehat{p}_i, \\ t_i = \langle \widehat{r}_i, \widehat{M}^{-1}\widehat{r}_i\rangle / \langle \widehat{r}_{i-1}, \widehat{M}^{-1}\widehat{r}_{i-1}\rangle. \end{cases} \tag{5.5.23}$$

Then $\widehat{y}_i = \widehat{P}\widehat{x}_i$ for all $i \geq 0$, the iterates $\widehat{y}_i$ converge towards $\widehat{\alpha}^{\lambda'}$ in a finite number of steps, and the quantities

$$\begin{cases} \lambda'^{[i]} = \frac{\bar{p}}{\overline{T}}\langle \widehat{y}_i, \widehat{\beta}^{\lambda'}\rangle = \frac{\bar{p}}{\overline{T}}\langle \widehat{x}_i, \widehat{\beta}^{\lambda'}\rangle, \\ \widehat{\theta}_k^{[i]} = -\langle \widehat{y}_i, \widehat{\beta}^{D_k}\rangle = -\langle \widehat{x}_i, \widehat{\beta}^{D_k}\rangle, \quad k \in \mathcal{S}, \end{cases} \tag{5.5.24}$$

converge towards λ' and $\widehat{\theta}$, respectively, in a finite number of steps. Furthermore, if $\widehat{x}_0 = 0$, we have $\lambda'^{[i]} > 0$ for $i \geq 1$. Finally, for positive mass fractions, the iterates $\lambda'^{[i]}$ coincide with the ones in Theorem 5.5.2, and we have $\widehat{\theta}_k^{[i]} = \sqrt{Y_k}\theta_k^{[i]}$, $\widehat{x}_i = \left(\mathcal{X}^{\lambda'}\right)^{1/2} x_i$, and $\widehat{y}_i = \left(\mathcal{X}^{\lambda'}\right)^{1/2} y_i$, provided that $\mathfrak{d} = \mathcal{X}^{\lambda'}\widehat{\mathfrak{d}}$ and $\widehat{x}_0 = \left(\mathcal{X}^{\lambda'}\right)^{1/2} x_0$.

5.5.5 Standard Iterative Methods for the System $\widetilde{L}_{[e]}\alpha_{[e]}^{\lambda'} = \widetilde{\beta}_{[e]}^{\lambda'}$

In this section we want to solve the constrained linear system $\widetilde{L}_{[e]}\alpha_{[e]}^{\lambda'} = \widetilde{\beta}_{[e]}^{\lambda'}$ using standard iterative methods and obtain the only solution $\alpha_{[e]}^{\lambda'}$ such that $\langle \mathcal{L}_{[e]}, \alpha_{[e]}^{\lambda'}\rangle = 0$. The system matrix $\widetilde{L}_{[e]} \in \mathbb{R}^{2n,2n}$ satisfies the properties $(\widetilde{L}_{[e]}0)$–$(\widetilde{L}_{[e]}5)$ and $(\widetilde{L}_{[e]}6)$–$(\widetilde{L}_{[e]}7)$ given in Sections 4.4.6 and 5.4.6, respectively. Furthermore, we restate that $\mathcal{B}_{[e]}^{\lambda'} = \{00, \mathrm{e}\}\times\mathcal{S}$, $\mathcal{X}_{[e]}^{\lambda'} = \mathrm{diag}\big((X_k)_{(r,k)\in\mathcal{B}_{[e]}^{\lambda'}}\big)$, and $\Gamma_{[e]}^{\lambda'} \in \mathbb{R}^{2n,2n}$ is the permutation matrix associated with the reordering of $\mathcal{B}_{[e]}^{\lambda'}$ into $(\mathcal{B}_{[e]}^{\lambda'+}, \mathcal{B}_{[e]}^{\lambda'-})$.

Theorem 5.5.7. *Let $\widetilde{L}_{[e]} \in \mathbb{R}^{2n,2n}$ be a matrix satisfying the properties $(\widetilde{L}_{[e]}0)$–$(\widetilde{L}_{[e]}7)$ and let $\widetilde{M}$ be the matrix $db(\widetilde{L}_{[e]}) + \mathrm{diag}(\widetilde{\mathfrak{d}})$, where $\widetilde{\mathfrak{d}} = (\widetilde{\mathfrak{d}}_k^r)_{(r,k)\in\mathcal{B}_{[e]}^{\lambda'}}$ are coefficients such that $\widetilde{\mathfrak{d}} \geq 0$ in general, and $\widetilde{\mathfrak{d}}^{00+} \neq 0$ in the particular cases $n^+ = 1$ and 2. Consider the splitting $\widetilde{L}_{[e]} = \widetilde{M} - \widetilde{Z}$ and the iteration matrix $T = \widetilde{M}^{-1}\widetilde{Z}$. Let $\mathcal{L}_{[e]}$, $\mathcal{U}_{[e]}$, $\beta_{[e]}^{D_l}$, $l \in \mathcal{S}$, $\beta_{[e]}^{\lambda'}$, and $\widetilde{\beta}_{[e]}^{\lambda'} \in \mathbb{R}^{2n}$ be given by (4.4.22), (4.4.23), (4.4.25), (4.5.17), and (4.5.44), respectively, let $P = I - \mathcal{U}_{[e]}\otimes\mathcal{L}_{[e]}/\langle\mathcal{U}_{[e]}, \mathcal{L}_{[e]}\rangle$ the oblique projector onto $\mathcal{L}_{[e]}^{\perp}$ along $\mathbb{R}\mathcal{U}_{[e]}$, and let $\mathfrak{P}_{[e]} \in \mathbb{R}^{n,2n}$ be the rectangular matrix formed by the blocks $\mathfrak{P}_{[e]} = [I, 0]$. Let $x_0 \in \mathbb{R}^{2n}$, $y_0 = Px_0$, and consider for $i \geq 0$ the iterates*

$$x_{i+1} = Tx_i + \widetilde{M}^{-1}\widetilde{\beta}_{[e]}^{\lambda'}, \tag{5.5.25}$$

$$y_{i+1} = PTy_i + P\widetilde{M}^{-1}\widetilde{\beta}_{[e]}^{\lambda'}. \tag{5.5.26}$$

Then $y_i = Px_i$ for all $i \geq 0$, the matrices T and PT are convergent, $\rho(T) = 1$, $\rho(PT) < 1$, and we have the following limits

$$\begin{cases} \lim_{i\to\infty} y_i = P\big(\lim_{i\to\infty} x_i\big) = \alpha_{[e]}^{\lambda'}, \\ \lim_{i\to\infty} \frac{\bar{p}}{\overline{T}}\langle y_i, \beta_{[e]}^{\lambda'}\rangle = \lim_{i\to\infty} \frac{\bar{p}}{\overline{T}}\langle x_i, \beta_{[e]}^{\lambda'}\rangle = \lambda'_{[e]}, \\ \lim_{i\to\infty} -\langle y_i, \beta_{[e]}^{D_k}\rangle = \lim_{i\to\infty} -\langle x_i, \beta_{[e]}^{D_k}\rangle = \theta_{[e]k}, \quad k \in \mathcal{S}, \end{cases} \tag{5.5.27}$$

where $\alpha_{[e]}^{\lambda'}$ is the unique solution of the constrained linear system $\widetilde{L}_{[e]}\alpha_{[e]}^{\lambda'} = \widetilde{\beta}_{[e]}^{\lambda'}$ and $\langle \mathcal{L}_{[e]}, \alpha_{[e]}^{\lambda'} \rangle = 0$. Moreover, for all $i \geq 0$, each partial sum

$$\lambda_{[e]}'^{[i]} = \frac{\overline{p}}{\overline{\overline{T}}} \langle \sum_{j=0}^{i} (PT)^j P \widetilde{M}^{-1} \widetilde{\beta}_{[e]}^{\lambda'}, \beta_{[e]}^{\lambda'} \rangle = \frac{\overline{p}}{\overline{\overline{T}}} \langle \sum_{j=0}^{i} T^j \widetilde{M}^{-1} \widetilde{\beta}_{[e]}^{\lambda'}, \beta_{[e]}^{\lambda'} \rangle, \tag{5.5.28}$$

is positive, and each partial sum

$$\theta_{[e]}^{[i]} = -\mathfrak{P}_{[e]} \sum_{j=0}^{i} (PT)^j P \widetilde{M}^{-1} \widetilde{\beta}_{[e]}^{\lambda'} = -P_{Y^\perp, \mathbb{R}U} \mathfrak{P}_{[e]} \sum_{j=0}^{i} T^j \widetilde{M}^{-1} \widetilde{\beta}_{[e]}^{\lambda'}, \tag{5.5.29}$$

is such that $\langle \theta_{[e]}^{[i]}, Y \rangle = 0$. Furthermore, we have

$$\begin{cases} \lim\limits_{i\to\infty} \lambda_{[e]}'^{[i]} = \dfrac{\overline{p}}{\overline{\overline{T}}} \langle \sum\limits_{j=0}^{\infty} (PT)^j P \widetilde{M}^{-1} \widetilde{\beta}_{[e]}^{\lambda'}, \beta_{[e]}^{\lambda'} \rangle = \lambda_{[e]}', \\ \lim\limits_{i\to\infty} \theta_{[e]}^{[i]} = -\mathfrak{P}_{[e]} \sum\limits_{j=0}^{\infty} (PT)^j P \widetilde{M}^{-1} \widetilde{\beta}_{[e]}^{\lambda'} = \theta_{[e]}. \end{cases} \tag{5.5.30}$$

Finally, for positive mass fractions, the iteration matrix T and the iterates x_i, y_i, $\lambda_{[e]}'^{[i]}$, and $\theta_{[e]}^{[i]}$ coincide with the ones in Theorem 5.5.3, provided that $\mathfrak{d} = \mathcal{X}_{[e]}^{\lambda'} \widetilde{\mathfrak{d}}$.

5.5.6 Schur Complements

In this section we assume that the matrices L and $\widetilde{L}$ satisfy the structure properties $(L0)$–$(L7)$ and $(\widetilde{L}0)$–$(\widetilde{L}7)$, respectively, as in the previous sections. In particular, the block L^{0101} is then nonsingular, and in the framework of the practical approximations presented in Section 2.10, this block is also diagonal. It is therefore interesting to introduce the constrained linear system

$$\begin{cases} L_{[s]}\alpha_{[s]}^{\lambda'} = \beta_{[s]}^{\lambda'}, \\ \langle \mathcal{L}_{[s]}, \alpha_{[s]}^{\lambda'} \rangle = 0, \end{cases} \tag{5.5.31}$$

where the matrix $L_{[s]} \in \mathbb{R}^{2n,2n}$ is given by (5.4.49), the vector $\beta_{[s]}^{\lambda'} \in \mathbb{R}^{2n}$ by

$$\beta_{[s]}^{\lambda'} = \begin{bmatrix} \beta^{00\lambda'} \\ \beta^{10\lambda'} \end{bmatrix} - \begin{bmatrix} L^{0001} \\ L^{1001} \end{bmatrix} \left(L^{0101}\right)^{-1} \beta^{01\lambda'}, \tag{5.5.32}$$

and the vector $\mathcal{L}_{[s]} \in \mathbb{R}^{2n}$ by (5.4.51). We deduce from Lemma 5.1.12 that the system (5.5.31) admits a unique solution $\alpha_{[s]}^{\lambda'}$. The partial thermal conductivity is then given by

$$\lambda' = \lambda_{[s]}' + \lambda_{[01]}' = \frac{\overline{p}}{\overline{\overline{T}}} \langle \alpha_{[s]}^{\lambda'}, \beta_{[s]}^{\lambda'} \rangle + \frac{\overline{p}}{\overline{\overline{T}}} \langle \beta^{01\lambda'}, \left(L^{0101}\right)^{-1} \beta^{01\lambda'} \rangle, \tag{5.5.33}$$

and the thermal diffusion vector by

$$\theta = -\alpha_{[s]}^{00\lambda'}. \tag{5.5.34}$$

Consider now the matrix $M_{[s]} \in \mathbb{R}^{2n,2n}$ given by (5.4.53). We deduce from Section 5.4.8 that a convergence theorem similar to Theorem 5.5.1 can be stated for the constrained linear system (5.5.31) with the splitting matrix (5.4.53). In particular, this yields sequences of iterates $\lambda_{[s]}^{\prime [i]}$ and $\theta_{[s]}^{[i]}$ such that

$$\begin{cases} \lim_{i\to\infty} \lambda_{[s]}^{\prime [i]} = \lambda' - \lambda'_{[01]}, \\ \lim_{i\to\infty} \theta_{[s]}^{[i]} = \theta. \end{cases} \tag{5.5.35}$$

Moreover, we deduce from Section 5.4.8 that a convergence theorem similar to Theorem 5.5.2 can be stated for the constrained linear system (5.5.31) with the preconditioner (5.4.53). This yields sequences of iterates $\lambda_{[s]}^{\prime [i]}$ and $\theta_{[s]}^{[i]}$ which converge towards $\lambda' - \lambda'_{[01]}$ and θ, respectively, in a finite number of steps.

Stabilized versions, for vanishing mass fractions, of standard iterative methods are obtained by considering the left rescaled constrained linear system

$$\begin{cases} \widetilde{L}_{[s]} \alpha_{[s]}^{\lambda'} = \widetilde{\beta}_{[s]}^{\lambda'}, \\ \langle \mathcal{L}_{[s]}, \alpha_{[s]}^{\lambda'} \rangle = 0, \end{cases} \tag{5.5.36}$$

where the matrix $\widetilde{L}_{[s]} \in \mathbb{R}^{2n,2n}$ is given by (5.4.56), and the vector $\widetilde{\beta}_{[s]}^{\lambda'} \in \mathbb{R}^{2n}$ by

$$\widetilde{\beta}_{[s]}^{\lambda'} = \begin{bmatrix} \widetilde{\beta}^{00\lambda'} \\ \widetilde{\beta}^{10\lambda'} \end{bmatrix} - \begin{bmatrix} \widetilde{L}^{0001} \\ \widetilde{L}^{1001} \end{bmatrix} \left(\widetilde{L}^{0101}\right)^{-1} \widetilde{\beta}^{01\lambda'}. \tag{5.5.37}$$

Assuming $\widetilde{L}^{0101}$ to be symmetric, we deduce from Section 5.1.5 that the system (5.5.36) admits a unique solution $\alpha_{[s]}^{\lambda'}$. The partial thermal conductivity is then evaluated from

$$\lambda' = \lambda'_{[s]} + \lambda'_{[01]} = \frac{\overline{p}}{\overline{T}} \langle \alpha_{[s]}^{\lambda'}, \beta_{[s]}^{\lambda'} \rangle + \frac{\overline{p}}{\overline{T}} \langle \beta^{01\lambda'}, \left(\widetilde{L}^{0101}\right)^{-1} \widetilde{\beta}^{01\lambda'} \rangle, \tag{5.5.38}$$

where the vector $\beta_{[s]}^{\lambda'}$ is now given by

$$\beta_{[s]}^{\lambda'} = \begin{bmatrix} \beta^{00\lambda'} \\ \beta^{10\lambda'} \end{bmatrix} - \begin{bmatrix} \left(\widetilde{L}^{0100}\right)^t \\ \left(\widetilde{L}^{0110}\right)^t \end{bmatrix} \left(\widetilde{L}^{0101}\right)^{-1} \beta^{01\lambda'}, \tag{5.5.39}$$

and the thermal diffusion vector from

$$\theta = -\alpha_{[s]}^{00\lambda'}. \tag{5.5.40}$$

For positive mass fractions, one can easily verify that the solutions of the systems (5.5.31) and (5.5.36), the partial thermal conductivities (5.5.33) and (5.5.38), and the thermal diffusion vectors (5.5.34) and (5.5.40) coincide.

We now deduce from Section 5.4.8 that the splitting matrix (5.4.59) yields a convergent iteration matrix. A convergence theorem similar to Theorem 5.5.5 can then be stated for the constrained linear system (5.5.36). In particular, this yields sequences of iterates $\lambda'^{[i]}_{[s]}$ and $\theta^{[i]}_{[s]}$ such that

$$\begin{cases} \lim_{i\to\infty} \lambda'^{[i]}_{[s]} = \lambda' - \lambda'_{[01]}, \\ \lim_{i\to\infty} \theta^{[i]}_{[s]} = \theta. \end{cases} \tag{5.5.41}$$

Finally, stabilized versions of conjugate gradient methods are obtained by considering the symmetric rescaled matrix $\widehat{L}_{[s]}$, and are omitted for brevity.

5.6 The Thermal Conductivity and the Thermal Diffusion Ratios

5.6.1 Iterative Methods for the System $\Lambda\alpha^\lambda = \beta^\lambda$

In this section we want to solve the linear system $\Lambda\alpha^\lambda = \beta^\lambda$ using either standard iterative methods or conjugate gradient algorithms. The system matrix $\Lambda \in \mathbb{R}^{n+p,n+p}$ is assumed to satisfy the properties $(\Lambda0)$–$(\Lambda2)$ presented in Section 4.6.1, and the corresponding indexing set is $\mathcal{B}^\lambda = \{10\}\times\mathcal{S} \cup \{01\}\times\mathcal{P}$. Furthermore, we consider the matrix $db(\Lambda) \in \mathbb{R}^{n+p,n+p}$ formed by the diagonal of the four blocks of the matrix Λ,

$$db(\Lambda) = \begin{bmatrix} \mathrm{diag}(\Lambda^{1010}) & \mathrm{diag}(\Lambda^{1001}) \\ \mathrm{diag}(\Lambda^{0110}) & \mathrm{diag}(\Lambda^{0101}) \end{bmatrix}, \tag{5.6.1}$$

and from the kinetic theory results obtained in Sections 2.3 and 2.8, we can make the following assumptions.

$(\Lambda3)$ $db(\Lambda)$ is symmetric positive definite.

$(\Lambda4)$ $2db(\Lambda) - \Lambda$ is symmetric positive definite.

Theorem 5.6.1. *Let $\Lambda \in \mathbb{R}^{n+p,n+p}$ be a matrix satisfying the properties $(\Lambda0)$–$(\Lambda4)$, assume that the matrix $[L^{0010}, L^{0001}] \in \mathbb{R}^{n,n+p}$ satisfies (4.6.6), and let M be the matrix $db(\Lambda) + \mathrm{diag}(\mathfrak{d})$, where $\mathfrak{d} = (\mathfrak{d}^r_k)_{(r,k)\in\mathcal{B}^\lambda}$ are coefficients such that $\mathfrak{d} \geq 0$. Consider the* *splitting $\Lambda = M - Z$ and the iteration matrix $T = M^{-1}Z$. Let $\beta^\lambda \in \mathbb{R}^{n+p}$ be given by (4.6.8), $x_0 \in \mathbb{R}^{n+p}$, and consider for $i \geq 0$ the iterates*

$$x_{i+1} = Tx_i + M^{-1}\beta^\lambda. \tag{5.6.2}$$

Then the matrix T is convergent, $\rho(T) < 1$, and we have the following limits

$$\begin{cases} \lim_{i\to\infty} x_i = \alpha^\lambda, \\ \lim_{i\to\infty} \frac{\bar{p}}{\overline{\overline{T}}}\langle x_i, \beta^\lambda\rangle = \lambda, \\ \lim_{i\to\infty} [L^{0010}, L^{0001}]x_i = \chi, \end{cases} \tag{5.6.3}$$

where α^λ is the unique solution of the linear system $\Lambda\alpha^\lambda = \beta^\lambda$. Moreover, for all $i \geq 0$, each partial sum

$$\lambda^{[i]} = \frac{\bar{p}}{\overline{\overline{T}}}\langle \sum_{j=0}^{i} T^j M^{-1}\beta^\lambda, \beta^\lambda\rangle, \tag{5.6.4}$$

is positive, and each partial sum

$$\chi^{[i]} = [L^{0010}, L^{0001}] \sum_{j=0}^{i} T^j M^{-1}\beta^\lambda, \tag{5.6.5}$$

satisfies the relation $\langle \chi^{[i]}, U\rangle = 0$. Finally, we have

$$\begin{cases} \lim_{i\to\infty} \lambda^{[i]} = \frac{\bar{p}}{\overline{\overline{T}}}\langle \sum_{j=0}^{\infty} T^j M^{-1}\beta^\lambda, \beta^\lambda\rangle = \lambda, \\ \lim_{i\to\infty} \chi^{[i]} = [L^{0010}, L^{0001}] \sum_{j=0}^{\infty} T^j M^{-1}\beta^\lambda = \chi. \end{cases} \tag{5.6.6}$$

Proof. It is similar to the one of Theorem 5.2.1, and the relation $\langle \chi^{[i]}, U\rangle = 0$ directly follows from (4.6.6). □

Theorem 5.6.2. *Let $\Lambda \in \mathbb{R}^{n+p,n+p}$ be a matrix satisfying the properties (Λ0)–(Λ3), assume that the matrix $[L^{0010}, L^{0001}] \in \mathbb{R}^{n,n+p}$ satisfies (4.6.6), and let M be the matrix $db(\Lambda) + \mathrm{diag}(\mathfrak{d})$ or $\mathrm{diag}(\Lambda) + \mathrm{diag}(\mathfrak{d})$, where $\mathfrak{d} = (\mathfrak{d}^r_k)_{(r,k)\in B^\lambda}$ are coefficients such that $\mathfrak{d} \geq 0$. Let $\beta^\lambda \in \mathbb{R}^{n+p}$ be given by (4.6.8), $x_0 \in \mathbb{R}^{n+p}$, $r_0 = \beta^\lambda - \Lambda x_0$, $p_0 = 0$, $t_0 = 0$, and consider for $i \geq 1$ the iterates*

$$\begin{cases} p_i = M^{-1}r_{i-1} + t_{i-1}p_{i-1}, \\ s_i = \langle r_{i-1}, M^{-1}r_{i-1}\rangle / \langle p_i, \Lambda p_i\rangle, \\ x_i = x_{i-1} + s_i p_i, \\ r_i = r_{i-1} - s_i\Lambda p_i, \\ t_i = \langle r_i, M^{-1}r_i\rangle / \langle r_{i-1}, M^{-1}r_{i-1}\rangle. \end{cases} \tag{5.6.7}$$

Then the iterates x_i converge towards α^λ in a finite number of steps, and the quantities

$$\begin{cases} \lambda^{[i]} = \frac{\bar{p}}{\overline{T}}\langle x_i, \beta^\lambda\rangle, \\ \chi^{[i]} = [L^{0010}, L^{0001}]x_i, \end{cases} \tag{5.6.8}$$

converge towards λ and χ, respectively, in a finite number of steps. Finally, we have $\langle\chi^{[i]}, U\rangle = 0$, $i \geq 0$, and if $x_0 = 0$, we obtain $\lambda^{[i]} > 0$ for $i \geq 1$.

Proof. It is similar to the one of Theorem 5.2.2, and the relation $\langle\chi^{[i]}, U\rangle = 0$ directly follows from (4.6.6). □

Finally, we consider the matrix Λ resulting from the practical approximations presented in Section 2.10.

Proposition 5.6.3. *Let $\mathcal{D}_{k\,\mathrm{int},l}$, c_k^{int}, ξ_k^{int}, and m_l, $k \in \mathcal{P}$, $l \in \mathcal{S}$, be positive quantities, let $\mathcal{D}_{kl}$, $\bar{A}_{kl}$, $\bar{B}_{kl}$, and $\bar{C}_{kl}$, $k,l \in \mathcal{S}$, be symmetric and positive coefficients such that $25/4 - 3\bar{B}_{kl} > 0$, $k,l \in \mathcal{S}$, and assume that $Y > 0$. Then the matrix Λ given by (2.10.20)–(2.10.25) satisfies (Λ0)–(Λ4).*

Proof. By Proposition 4.6.3, only (Λ3)–(Λ4) need to be proven. First, the property (Λ3) directly results from (Λ2), and an explicit calculation yields for $x \in \mathbb{R}^{n+p}$

$$\begin{aligned}
\langle x, (2db(\Lambda) - \Lambda)x\rangle = &\sum_{\substack{k,l\in\mathcal{S}\\ l\neq k}} \Big(\frac{25}{4} - 3\bar{B}_{kl}\Big)\Big(\frac{m_l x_k^{10} + m_k x_l^{10}}{m_k + m_l}\Big)^2 \\
&+ \sum_{\substack{k,l\in\mathcal{S}\\ l\neq k}} \frac{X_k X_l}{\mathcal{D}_{kl}}\Big[\frac{15}{4}\Big(\frac{m_k x_k^{10} + m_l x_l^{10}}{m_k + m_l}\Big)^2 + 2\bar{A}_{kl}\frac{m_k m_l}{(m_k + m_l)^2}(x_k^{10} - x_l^{10})^2\Big] \\
&+ \sum_{k\in\mathcal{S}} 2\bar{A}_{kk}\frac{X_k^2}{\mathcal{D}_{kk}}\big(x_k^{10}\big)^2 + \sum_{k\in\mathcal{P}}\Big(\sum_{\substack{l\in\mathcal{S}\\ l\neq k}} X_k X_l \frac{c_k^{\mathrm{int}}}{k_{\mathrm{B}}\mathcal{D}_{k\,\mathrm{int},l}} + X_k^2\frac{c_k^{\mathrm{int}}}{k_{\mathrm{B}}\mathcal{D}_{k\,\mathrm{int},k}}\Big)\big(x_k^{01}\big)^2 \\
&+ \sum_{k\in\mathcal{P}} \frac{20}{3}\frac{\bar{A}_{kk}}{k_{\mathrm{B}}\pi}\frac{X_k^2}{\mathcal{D}_{kk}}\frac{c_k^{\mathrm{int}}}{\xi_k^{\mathrm{int}}}\Big(x_k^{10} - \frac{3}{5}x_k^{01}\Big)^2 \\
&+ \sum_{\substack{k\in\mathcal{S},l\in\mathcal{P}\\ l\neq k}} \frac{20}{3}\frac{\bar{A}_{kl}}{k_{\mathrm{B}}\pi}\frac{X_k X_l}{\mathcal{D}_{kl}}\frac{c_l^{\mathrm{int}}}{\xi_l^{\mathrm{int}}}\frac{m_l}{m_k}\Big(\frac{m_k}{m_k + m_l}(x_k^{10} - x_l^{10}) + \frac{3}{5}x_l^{01}\Big)^2,
\end{aligned}$$

so that $\langle x, (2db(\Lambda) - \Lambda)x\rangle = 0$ if and only if $x = 0$. □

5.6.2 Iterative Methods for the System $\Lambda_{[e]}\alpha^{\lambda}_{[e]} = \beta^{\lambda}_{[e]}$

In this section we want to solve the linear system $\Lambda_{[e]}\alpha^{\lambda}_{[e]} = \beta^{\lambda}_{[e]}$ using either standard iterative or conjugate gradient methods. The system matrix $\Lambda_{[e]} \in \mathbb{R}^{n,n}$ is assumed to satisfy the properties $(\Lambda_{[e]}0)$–$(\Lambda_{[e]}2)$ presented in Section 4.6.2, and the corresponding indexing set is $\mathcal{B}^{\lambda}_{[e]} = \{\mathrm{e}\}\times\mathcal{S}$. Furthermore, we consider the matrix $db(\Lambda_{[e]}) \in \mathbb{R}^{n,n}$ formed by the diagonal of the only block of the matrix $\Lambda_{[e]}$,

$$db(\Lambda_{[e]}) = \mathrm{diag}(\Lambda^{\mathrm{ee}}_{[e]}), \tag{5.6.9}$$

and from the kinetic theory results obtained in Sections 2.3 and 2.8, we can make the following assumptions.

$(\Lambda_{[e]}3)$ $db(\Lambda_{[e]})$ is symmetric positive definite.

$(\Lambda_{[e]}4)$ $2db(\Lambda_{[e]}) - \Lambda_{[e]}$ is symmetric positive definite.

Theorem 5.6.4. *Let $\Lambda_{[e]} \in \mathbb{R}^{n,n}$ be a matrix satisfying the properties $(\Lambda_{[e]}0)$–$(\Lambda_{[e]}4)$, assume that the matrix $[L^{00\mathrm{e}}_{[e]}] \in \mathbb{R}^{n,n}$ satisfies (4.6.18), and let M be the matrix $db(\Lambda_{[e]}) + \mathrm{diag}(\mathfrak{d})$, where $\mathfrak{d} = (\mathfrak{d}^r_k)_{(r,k)\in\mathcal{B}^{\lambda}_{[e]}}$ are coefficients such that $\mathfrak{d} \geq 0$. Consider the splitting $\Lambda_{[e]} = M - Z$ and the iteration matrix $T = M^{-1}Z$. Let $\beta^{\lambda}_{[e]} \in \mathbb{R}^n$ be given by (4.6.20), $x_0 \in \mathbb{R}^{n+p}$, and consider for $i \geq 0$ the iterates*

$$x_{i+1} = Tx_i + M^{-1}\beta^{\lambda}_{[e]}. \tag{5.6.10}$$

Then the matrix T is convergent, $\rho(T) < 1$, and we have the following limits

$$\begin{cases} \lim_{i\to\infty} x_i = \alpha^{\lambda}_{[e]}, \\ \lim_{i\to\infty} \frac{\bar{p}}{\overline{T}}\langle x_i, \beta^{\lambda}_{[e]}\rangle = \lambda_{[e]}, \\ \lim_{i\to\infty} [L^{00\mathrm{e}}_{[e]}]x_i = \chi_{[e]}, \end{cases} \tag{5.6.11}$$

where $\alpha^{\lambda}_{[e]}$ is the unique solution of the linear system $\Lambda_{[e]}\alpha^{\lambda}_{[e]} = \beta^{\lambda}_{[e]}$. Moreover, for all $i \geq 0$, each partial sum

$$\lambda^{[i]}_{[e]} = \frac{\bar{p}}{\overline{T}}\langle \sum_{j=0}^{i} T^j M^{-1}\beta^{\lambda}_{[e]}, \beta^{\lambda}_{[e]}\rangle, \tag{5.6.12}$$

is positive, and each partial sum

$$\chi^{[i]}_{[e]} = [L^{00\mathrm{e}}_{[e]}]\sum_{j=0}^{i} T^j M^{-1}\beta^{\lambda}_{[e]}, \tag{5.6.13}$$

satisfies the relation $\langle \chi_{[e]}^{[i]}, U\rangle = 0$. *Finally, we have*

$$\begin{cases} \lim\limits_{i\to\infty} \lambda_{[e]}^{[i]} = \frac{\bar{p}}{\overline{T}}\langle \sum\limits_{j=0}^{\infty} T^j M^{-1}\beta_{[e]}^{\lambda}, \beta_{[e]}^{\lambda}\rangle = \lambda_{[e]}, \\ \lim\limits_{i\to\infty} \chi_{[e]}^{[i]} = [L_{[e]}^{00e}]\sum\limits_{j=0}^{\infty} T^j M^{-1}\beta_{[e]}^{\lambda} = \chi_{[e]}. \end{cases} \tag{5.6.14}$$

Theorem 5.6.5. *Let* $\Lambda_{[e]} \in \mathbb{R}^{n,n}$ *be a matrix satisfying the properties* $(\Lambda_{[e]}0)$–$(\Lambda_{[e]}3)$, *assume that the matrix* $[L_{[e]}^{00e}] \in \mathbb{R}^{n,n}$ *satisfies (4.6.18), and let* M *be the matrix* $db(\Lambda_{[e]}) + \mathrm{diag}(\mathfrak{d})$ *or* $\mathrm{diag}(\Lambda_{[e]}) + \mathrm{diag}(\mathfrak{d})$, *where* $\mathfrak{d} = (\mathfrak{d}_k^r)_{(r,k)\in B_{[e]}^{\lambda}}$ *are coefficients such that* $\mathfrak{d} \geq 0$. *Let* $\beta_{[e]}^{\lambda} \in \mathbb{R}^n$ *be given by (4.6.20),* $x_0 \in \mathbb{R}^n$, $r_0 = \beta_{[e]}^{\lambda} - \Lambda_{[e]}x_0$, $p_0 = 0$, $t_0 = 0$, *and consider for* $i \geq 1$ *the iterates*

$$\begin{cases} p_i = M^{-1}r_{i-1} + t_{i-1}p_{i-1}, \\ s_i = \langle r_{i-1}, M^{-1}r_{i-1}\rangle / \langle p_i, \Lambda_{[e]}p_i\rangle, \\ x_i = x_{i-1} + s_i p_i, \\ r_i = r_{i-1} - s_i \Lambda_{[e]} p_i, \\ t_i = \langle r_i, M^{-1}r_i\rangle / \langle r_{i-1}, M^{-1}r_{i-1}\rangle. \end{cases} \tag{5.6.15}$$

Then the iterates x_i *converge towards* $\alpha_{[e]}^{\lambda}$ *in a finite number of steps, and the quantities*

$$\begin{cases} \lambda_{[e]}^{[i]} = \frac{\bar{p}}{\overline{T}}\langle x_i, \beta_{[e]}^{\lambda}\rangle, \\ \chi_{[e]}^{[i]} = [L_{[e]}^{00e}]x_i, \end{cases} \tag{5.6.16}$$

converge towards $\lambda_{[e]}$ *and* $\chi_{[e]}$, *respectively, in a finite number of steps. Finally, we have* $\langle \chi_{[e]}^{[i]}, U\rangle = 0$, $i \geq 0$, *and if* $x_0 = 0$, *we obtain* $\lambda_{[e]}^{[i]} > 0$ *for* $i \geq 1$.

Finally, we consider the matrix $\Lambda_{[e]}$ resulting from the practical approximations presented in Section 2.10.

Proposition 5.6.6. *Let* $\mathcal{D}_{k\,\mathrm{int},l}$, c_k^{int}, ξ_k^{int}, *and* m_l, $k \in \mathcal{P}$, $l \in \mathcal{S}$, *be positive quantities, let* $\mathcal{D}_{kl}$, $\bar{A}_{kl}$, $\bar{B}_{kl}$, *and* $\bar{C}_{kl}$, $k, l \in \mathcal{S}$, *be symmetric and positive coefficients such that* $25/4 - 3\bar{B}_{kl} > 0$, $k, l \in \mathcal{S}$, *and assume that* $Y > 0$. *Then the matrix* $\Lambda_{[e]}$ *given by (2.10.30)–(2.10.31) satisfies* $(\Lambda_{[e]}0)$–$(\Lambda_{[e]}4)$.

5.6.3 Standard Iterative Methods for the System $\widetilde{\Lambda}\alpha^\lambda = \widetilde{\beta}^\lambda$

In this section we want to solve the linear system $\widetilde{\Lambda}\alpha^\lambda = \widetilde{\beta}^\lambda$ using standard iterative methods. The system matrix $\widetilde{\Lambda} \in \mathbb{R}^{n+p,n+p}$ satisfies the properties ($\widetilde{\Lambda}0$)–($\widetilde{\Lambda}4$) presented in Section 4.6.3 and we restate that $\mathcal{B}^\lambda = \{10\}\times\mathcal{S} \cup \{01\}\times\mathcal{P}$, $\mathcal{X}^\lambda = \operatorname{diag}\big((X_k)_{(r,k)\in\mathcal{B}^\lambda}\big)$, and $\Gamma^\lambda \in \mathbb{R}^{n+p,n+p}$ is the permutation matrix associated with the reordering of $\mathcal{B}^\lambda$ into $(\mathcal{B}^{\lambda+}, \mathcal{B}^{\lambda-})$. Furthermore, we consider the matrix $db(\widetilde{\Lambda}) \in \mathbb{R}^{n+p,n+p}$ formed by the diagonal of the four blocks of the matrix $\widetilde{\Lambda}$,

$$db(\widetilde{\Lambda}) = \begin{bmatrix} \operatorname{diag}(\widetilde{\Lambda}^{1010}) & \operatorname{diag}(\widetilde{\Lambda}^{1001}) \\ \operatorname{diag}(\widetilde{\Lambda}^{0110}) & \operatorname{diag}(\widetilde{\Lambda}^{0101}) \end{bmatrix}, \tag{5.6.17}$$

and from the kinetic theory results obtained in Sections 3.1 and 3.6, we can make the following assumptions.

($\widetilde{\Lambda}5$) $db(\Lambda^{++})$ and $db(\widetilde{\Lambda})$ are symmetric positive definite.

($\widetilde{\Lambda}6$) $2db(\Lambda^{++}) - \Lambda^{++}$ is symmetric positive definite.

Theorem 5.6.7. *Let $\widetilde{\Lambda} \in \mathbb{R}^{n+p,n+p}$ be a matrix satisfying the properties ($\widetilde{\Lambda}0$)–($\widetilde{\Lambda}6$), assume that the matrix $[L^{0010}, L^{0001}] \in \mathbb{R}^{n,n+p}$ satisfies (4.6.6)(4.6.33), and let $\widetilde{M}$ be the matrix $db(\widetilde{\Lambda}) + \operatorname{diag}(\widetilde{\mathfrak{d}})$, where $\widetilde{\mathfrak{d}} = (\widetilde{\mathfrak{d}}_k^r)_{(r,k)\in\mathcal{B}^\lambda}$ are coefficients such that $\widetilde{\mathfrak{d}} \geq 0$. Consider the splitting $\widetilde{\Lambda} = \widetilde{M} - \widetilde{Z}$ and the iteration matrix $T = \widetilde{M}^{-1}\widetilde{Z}$. Let β^λ and $\widetilde{\beta}^\lambda \in \mathbb{R}^{n+p}$ be given by (4.6.8) and (4.6.31), respectively, let $x_0 \in \mathbb{R}^{n+p}$, and consider for $i \geq 0$ the iterates*

$$x_{i+1} = Tx_i + \widetilde{M}^{-1}\widetilde{\beta}^\lambda. \tag{5.6.18}$$

Then the matrix T is convergent, $\rho(T) < 1$, and we have the following limits

$$\begin{cases} \lim\limits_{i\to\infty} x_i = \alpha^\lambda, \\ \lim\limits_{i\to\infty} \dfrac{\bar{p}}{\overline{T}}\langle x_i, \beta^\lambda\rangle = \lambda, \\ \lim\limits_{i\to\infty} [L^{0010}, L^{0001}]x_i = \chi, \end{cases} \tag{5.6.19}$$

where α^λ is the unique solution of the linear system $\widetilde{\Lambda}\alpha^\lambda = \widetilde{\beta}^\lambda$. Moreover, for all $i \geq 0$, each partial sum

$$\lambda^{[i]} = \frac{\bar{p}}{\overline{T}}\langle \sum_{j=0}^{i} T^j \widetilde{M}^{-1}\widetilde{\beta}^\lambda, \beta^\lambda\rangle, \tag{5.6.20}$$

is positive, and each partial sum

$$\chi^{[i]} = [L^{0010}, L^{0001}] \sum_{j=0}^{i} T^j \widetilde{M}^{-1}\widetilde{\beta}^\lambda, \tag{5.6.21}$$

satisfies the relations $\langle \chi^{[i]}, U\rangle = 0$ *and* $\chi^{[i]-} = 0$. *Furthermore, we have*

$$\begin{cases} \lim_{i\to\infty} \lambda^{[i]} = \frac{\bar{p}}{\overline{T}}\langle \sum_{j=0}^{\infty} T^j \widetilde{M}^{-1}\widetilde{\beta}^\lambda, \beta^\lambda\rangle = \lambda, \\ \lim_{i\to\infty} \chi^{[i]} = [L^{0010}, L^{0001}] \sum_{j=0}^{\infty} T^j \widetilde{M}^{-1}\widetilde{\beta}^\lambda = \chi. \end{cases} \tag{5.6.22}$$

Finally, for positive mass fractions, the iteration matrix T *and the iterates* x_i, $\lambda^{[i]}$, *and* $\chi^{[i]}$ *coincide with the ones in Theorem 5.6.1, provided that* $\mathfrak{d} = \mathcal{X}^\lambda \widetilde{\mathfrak{d}}$.

Proof. The relations $\langle \chi^{[i]}, U\rangle = 0$ and $\chi^{[i]-} = 0$ directly follow from the assumptions on the matrices L^{0010} and L^{0001}. Furthermore, the correspondence with Theorem 5.6.1 directly results from Proposition 5.1.14, and the rest of the proof is similar to the previous ones. □

Remark. In the case of a pure species state of the mixture, i.e., $n^+ = 1$, we obtain a one-step convergence of the iterates $\lambda^{[i]}$ and $\chi^{[i]}$ for $\widetilde{\mathfrak{d}} = 0$.

Finally, we consider the matrix $\widetilde{\Lambda}$ resulting from the practical approximations presented in Section 2.10.

Proposition 5.6.8. *Let* $\mathcal{D}_{k\,\mathrm{int},l}$, c_k^{int}, ξ_k^{int}, *and* m_l, $k \in \mathcal{P}$, $l \in \mathcal{S}$, *be positive quantities, let* $\mathcal{D}_{kl}$, $\bar{A}_{kl}$, $\bar{B}_{kl}$, *and* $\bar{C}_{kl}$, $k,l \in \mathcal{S}$, *be symmetric and positive coefficients such that* $25/4 - 3\bar{B}_{kl} > 0$, $k,l \in \mathcal{S}$, *and assume that* $Y \geq 0$, $Y \neq 0$. *Then the matrix* $\widetilde{\Lambda}$ *given by (3.7.18)–(3.7.26) satisfies* $(\widetilde{\Lambda}0)$–$(\widetilde{\Lambda}6)$.

Proof. It is similar to the one of Propositions 4.6.10 and 5.6.3. □

5.6.4 Conjugate Gradient Methods for the System $\widehat{\Lambda}\widehat{\alpha}^\lambda = \widehat{\beta}^\lambda$

In this section we assume that the matrix $\widehat{\Lambda}$ satisfies the properties $(\widehat{\Lambda}0)$–$(\widehat{\Lambda}2)$ given in Section 4.6.4 and we want to obtain the only solution $\widehat{\alpha}^\lambda$ of the linear system $\widehat{\Lambda}\widehat{\alpha}^\lambda = \widehat{\beta}^\lambda$ using conjugate gradient methods.

Theorem 5.6.9. *Let* $\widehat{\Lambda} \in \mathbb{R}^{n+p,n+p}$ *be a matrix satisfying the properties* $(\widehat{\Lambda}0)$–$(\widehat{\Lambda}2)$, *assume that the matrix* $[\widehat{L}^{0010}, \widehat{L}^{0001}] \in \mathbb{R}^{n,n+p}$ *satisfies (4.6.45), and let* $\widehat{M}$ *be the matrix* $db(\widehat{\Lambda}) + \mathrm{diag}(\widehat{\mathfrak{d}})$ *or* $\mathrm{diag}(\widehat{\Lambda}) + \mathrm{diag}(\widehat{\mathfrak{d}})$, *where* $\widehat{\mathfrak{d}} = (\widehat{\mathfrak{d}}_k^r)_{(r,k)\in\mathcal{B}^\lambda}$ *are coefficients such that* $\widehat{\mathfrak{d}} \geq 0$. *Let* $\widehat{\beta}^\lambda \in \mathbb{R}^n$ *be given by (4.6.43),* $\widehat{x}_0 \in \mathbb{R}^{n+p}$, $\widehat{r}_0 = \widehat{\beta}^\lambda - \widehat{\Lambda}\widehat{x}_0$, $\widehat{p}_0 = 0$,

$t_0 = 0$, and consider for $i \geq 1$ the iterates

$$\begin{cases} \widehat{p}_i = \widehat{M}^{-1}\widehat{r}_{i-1} + t_{i-1}\widehat{p}_{i-1}, \\ s_i = \langle \widehat{r}_{i-1}, \widehat{M}^{-1}\widehat{r}_{i-1} \rangle / \langle \widehat{p}_i, \widehat{\Lambda}\widehat{p}_i \rangle, \\ \widehat{x}_i = \widehat{x}_{i-1} + s_i\widehat{p}_i, \\ \widehat{r}_i = \widehat{r}_{i-1} - s_i\widehat{\Lambda}\widehat{p}_i, \\ t_i = \langle \widehat{r}_i, \widehat{M}^{-1}\widehat{r}_i \rangle / \langle \widehat{r}_{i-1}, \widehat{M}^{-1}\widehat{r}_{i-1} \rangle. \end{cases} \tag{5.6.23}$$

Then the iterates $\widehat{x}_i$ converge towards $\widehat{\alpha}^\lambda$ in a finite number of steps and the quantities

$$\begin{cases} \lambda^{[i]} = \dfrac{\overline{p}}{\overline{T}} \langle \widehat{x}_i, \widehat{\beta}^\lambda \rangle, \\ \chi^{[i]} = [\widehat{L}^{0010}, \widehat{L}^{0001}]\widehat{x}_i, \end{cases} \tag{5.6.24}$$

converge towards λ and χ, respectively, in a finite number of steps. Furthermore, we have $\langle \chi^{[i]}, U \rangle = 0$, $i \geq 0$, and if $\widehat{x}_0 = 0$, we obtain $\lambda^{[i]} > 0$ for $i \geq 1$. Finally, for positive mass fractions, the iterates $\lambda^{[i]}$ and $\chi^{[i]}$ coincide with the ones in Theorem 5.6.2 and $\widehat{x}_i = (\mathcal{X}^\lambda)^{1/2} x_i$, provided that $\mathfrak{d} = \mathcal{X}^\lambda \widehat{\mathfrak{d}}$ and $\widehat{x}_0 = (\mathcal{X}^\lambda)^{1/2} x_0$.

Finally, we consider the matrix $\widehat{\Lambda}$ resulting from the practical approximations presented in Section 2.10.

Proposition 5.6.10. *Under the assumptions of Proposition 5.6.8, the matrix $\widehat{\Lambda}$, obtained from the matrix $\widetilde{\Lambda}$ by replacing X_l by $\sqrt{X_k X_l}$ in all the terms $\widetilde{\Lambda}^{rs}_{kl}$, $k \neq l$, satisfies $(\widehat{\Lambda}0)$–$(\widehat{\Lambda}2)$.*

5.6.5 Standard Iterative Methods for the System $\widetilde{\Lambda}_{[e]}\alpha^\lambda_{[e]} = \widetilde{\beta}^\lambda_{[e]}$

In this section we want to solve the linear system $\widetilde{\Lambda}_{[e]}\alpha^\lambda_{[e]} = \widetilde{\beta}^\lambda_{[e]}$ using standard iterative methods. The system matrix $\widetilde{\Lambda}_{[e]} \in \mathbb{R}^{n,n}$ is assumed to satisfy the properties $(\widetilde{\Lambda}_{[e]}0)$–$(\widetilde{\Lambda}_{[e]}4)$ presented in Section 4.6.5, and we restate that $\mathcal{B}^\lambda_{[e]} = \{e\} \times \mathcal{S}$, $\mathcal{X}^\lambda_{[e]} = \mathrm{diag}\big((X_k)_{(r,k) \in \mathcal{B}^\lambda_{[e]}}\big)$, and $\Gamma^\lambda_{[e]} \in \mathbb{R}^{n,n}$ is the permutation matrix associated with the reordering of $\mathcal{B}^\lambda_{[e]}$ into $(\mathcal{B}^{\lambda+}_{[e]}, \mathcal{B}^{\lambda-}_{[e]})$. Furthermore, we consider the matrix $db(\widetilde{\Lambda}_{[e]}) \in \mathbb{R}^{n,n}$ formed by the diagonal of the only block of the matrix $\widetilde{\Lambda}_{[e]}$,

$$db(\widetilde{\Lambda}_{[e]}) = \mathrm{diag}(\widetilde{\Lambda}^{ee}_{[e]}), \tag{5.6.25}$$

and from the kinetic theory results obtained in Sections 3.1 and 3.6, we can make the following assumptions.

$(\widetilde{\Lambda}_{[e]}5)$ $db(\Lambda_{[e]}^{++})$ and $db(\widetilde{\Lambda}_{[e]})$ are symmetric positive definite.

$(\widetilde{\Lambda}_{[e]}6)$ $2db(\Lambda_{[e]}^{++}) - \Lambda_{[e]}^{++}$ is symmetric positive definite.

Theorem 5.6.11. *Let $\widetilde{\Lambda}_{[e]} \in \mathbb{R}^{n,n}$ be a matrix satisfying the properties $(\widetilde{\Lambda}_{[e]}0)$–$(\widetilde{\Lambda}_{[e]}6)$, assume that the matrix $[L_{[e]}^{00e}] \in \mathbb{R}^{n,n}$ satisfies (4.6.18)(4.6.53), and let $\widetilde{M}$ be the matrix $db(\widetilde{\Lambda}_{[e]}) + \mathrm{diag}(\widetilde{\mathfrak{d}})$, where $\widetilde{\mathfrak{d}} = (\widetilde{\mathfrak{d}})_{(r,k)\in\mathcal{B}_{[e]}^{\lambda}}$ are coefficients such that $\widetilde{\mathfrak{d}} \geq 0$. Consider the splitting $\widetilde{\Lambda}_{[e]} = \widetilde{M} - \widetilde{Z}$ and the iteration matrix $T = \widetilde{M}^{-1}\widetilde{Z}$. Let $\beta_{[e]}^{\lambda}$ and $\widetilde{\beta}_{[e]}^{\lambda} \in \mathbb{R}^n$ be given by (4.6.20) and (4.6.51), respectively, let $x_0 \in \mathbb{R}^n$, and consider for $i \geq 0$ the iterates*

$$x_{i+1} = Tx_i + \widetilde{M}^{-1}\widetilde{\beta}_{[e]}^{\lambda}. \tag{5.6.26}$$

Then the matrix T is convergent, $\rho(T) < 1$, and we have the following limits

$$\begin{cases} \lim\limits_{i\to\infty} x_i = \alpha_{[e]}^{\lambda}, \\ \lim\limits_{i\to\infty} \dfrac{\bar{p}}{\overline{T}}\langle x_i, \beta_{[e]}^{\lambda}\rangle = \lambda_{[e]}, \\ \lim\limits_{i\to\infty} [L_{[e]}^{00e}]x_i = \chi_{[e]}, \end{cases} \tag{5.6.27}$$

where $\alpha_{[e]}^{\lambda}$ is the unique solution of the linear system $\widetilde{\Lambda}_{[e]}\alpha_{[e]}^{\lambda} = \widetilde{\beta}_{[e]}^{\lambda}$. Moreover, for all $i \geq 0$, each partial sum

$$\lambda_{[e]}^{[i]} = \frac{\bar{p}}{\overline{T}}\langle \sum_{j=0}^{i} T^j \widetilde{M}^{-1}\widetilde{\beta}_{[e]}^{\lambda}, \beta_{[e]}^{\lambda}\rangle, \tag{5.6.28}$$

is positive, and each partial sum

$$\chi_{[e]}^{[i]} = [L_{[e]}^{00e}] \sum_{j=0}^{i} T^j \widetilde{M}^{-1}\widetilde{\beta}_{[e]}^{\lambda}, \tag{5.6.29}$$

satisfies the relations $\langle \chi_{[e]}^{[i]}, U\rangle = 0$ and $\chi_{[e]}^{[i]-} = 0$. Furthermore, we have

$$\begin{cases} \lim\limits_{i\to\infty} \lambda_{[e]}^{[i]} = \dfrac{\bar{p}}{\overline{T}}\langle \sum_{j=0}^{\infty} T^j \widetilde{M}^{-1}\widetilde{\beta}_{[e]}^{\lambda}, \beta_{[e]}^{\lambda}\rangle = \lambda_{[e]}, \\ \lim\limits_{i\to\infty} \chi_{[e]}^{[i]} = [L_{[e]}^{00e}] \sum_{j=0}^{\infty} T^j \widetilde{M}^{-1}\widetilde{\beta}_{[e]}^{\lambda} = \chi_{[e]}. \end{cases} \tag{5.6.30}$$

Finally, for positive mass fractions, the iteration matrix T and the iterates x_i, $\lambda_{[e]}^{[i]}$, and $\chi_{[e]}^{[i]}$ coincide with the ones in Theorem 5.6.4, provided that $\mathfrak{d} = \mathcal{X}_{[e]}^{\lambda}\widetilde{\mathfrak{d}}$.

Finally, we consider the matrix $\widetilde{\Lambda}_{[e]}$ resulting from the practical approximations presented in Section 2.10.

Proposition 5.6.12. *Let $\mathcal{D}_{k\,\mathrm{int},l}$, c_k^{int}, ξ_k^{int}, and m_l, $k \in \mathcal{P}$, $l \in \mathcal{S}$, be positive quantities, let $\mathcal{D}_{kl}$, $\bar{A}_{kl}$, $\bar{B}_{kl}$, and $\bar{C}_{kl}$, $k, l \in \mathcal{S}$, be symmetric and positive coefficients such that $25/4 - 3\bar{B}_{kl} > 0$, $k, l \in \mathcal{S}$, and assume that $Y \geq 0$, $Y \neq 0$. Then the matrix $\widetilde{\Lambda}_{[\mathrm{e}]}$ given by (3.7.33)–(3.7.34) satisfies ($\widetilde{\Lambda}_{[\mathrm{e}]}0$)–($\widetilde{\Lambda}_{[\mathrm{e}]}6$).*

5.6.6 Schur Complements

In this section we assume that the matrices Λ and $\widetilde{\Lambda}$ satisfy the structure properties (Λ0)–(Λ4) and ($\widetilde{\Lambda}$0)–($\widetilde{\Lambda}$6), respectively, as in the previous sections. In particular, the block Λ^{0101} is then nonsingular, and in the framework of the practical approximations presented in Section 2.10, this block is also diagonal. It may therefore be interesting to introduce the linear system

$$\Lambda_{[\mathrm{s}]}\alpha_{[\mathrm{s}]}^{\lambda} = \beta_{[\mathrm{s}]}^{\lambda}, \tag{5.6.31}$$

where the matrix $\Lambda_{[\mathrm{s}]} \in \mathbb{R}^{n,n}$ is given by

$$\Lambda_{[\mathrm{s}]} = \Lambda^{1010} - \Lambda^{1001}\left(\Lambda^{0101}\right)^{-1}\Lambda^{0110}, \tag{5.6.32}$$

and the vector $\beta_{[\mathrm{s}]}^{\lambda} \in \mathbb{R}^n$ by

$$\beta_{[\mathrm{s}]}^{\lambda} = \beta^{10\lambda} - \Lambda^{1001}\left(\Lambda^{0101}\right)^{-1}\beta^{01\lambda}. \tag{5.6.33}$$

We deduce from Lemma 5.1.11 that the system (5.6.31) admits a unique solution $\alpha_{[\mathrm{s}]}^{\lambda}$. The thermal conductivity is then given by

$$\lambda = \lambda_{[\mathrm{s}]} + \lambda_{[01]} = \frac{\bar{p}}{\bar{T}}\langle\alpha_{[\mathrm{s}]}^{\lambda}, \beta_{[\mathrm{s}]}^{\lambda}\rangle + \frac{\bar{p}}{\bar{T}}\langle\beta^{01\lambda}, \left(\Lambda^{0101}\right)^{-1}\beta^{01\lambda}\rangle, \tag{5.6.34}$$

and the thermal diffusion ratios by

$$\chi = \chi_{[\mathrm{s}]} + \chi_{[01]} = [L_{[\mathrm{s}]}^{0010}]\alpha_{[\mathrm{s}]}^{\lambda} + L^{0001}\left(\Lambda^{0101}\right)^{-1}\beta^{01\lambda}, \tag{5.6.35}$$

where the matrix $[L_{[\mathrm{s}]}^{0010}] \in \mathbb{R}^{n,n}$ is given by

$$[L_{[\mathrm{s}]}^{0010}] = L^{0010} - L^{0001}\left(L^{0101}\right)^{-1}L^{0110}. \tag{5.6.36}$$

Consider now the matrix $M_{[\mathrm{s}]} \in \mathbb{R}^{n,n}$ given by

$$M_{[\mathrm{s}]} = db(\Lambda^{1010}) + \mathrm{diag}(\mathfrak{d}), \tag{5.6.37}$$

where $\mathfrak{d}_k^r \geq 0$, $(r,k) \in \{10\} \times \mathcal{S}$. From ($\Lambda$6)–($\Lambda$7) we then deduce that the matrices $M_{[\mathrm{s}]}$ and $2M_{[\mathrm{s}]} - \Lambda_{[\mathrm{s}]}$ are symmetric positive definite. A convergence theorem similar to Theorem 5.6.1 can then be stated for the linear system (5.6.31) with the splitting matrix (5.6.37). In particular, this yields sequences of iterates $\lambda_{[s]}^{[i]}$ and $\chi_{[s]}^{[i]}$ such that

$$\begin{cases} \lim_{i\to\infty} \lambda_{[s]}^{[i]} = \lambda - \lambda_{[01]}, \\ \lim_{i\to\infty} \chi_{[s]}^{[i]} = \chi - \chi_{[01]}. \end{cases} \tag{5.6.38}$$

Moreover, since the matrix $M_{[s]}$ is symmetric positive definite, a convergence theorem similar to Theorem 5.6.2 can then be stated for the linear system (5.6.31) with the preconditioner (5.6.37). This yields sequences of iterates $\lambda_{[s]}^{[i]}$ and $\chi_{[s]}^{[i]}$ that converge towards $\lambda - \lambda_{[01]}$ and $\chi - \chi_{[01]}$, respectively, in a finite number of steps.

Stabilized versions, for vanishing mass fractions, of standard iterative methods are obtained by considering the left rescaled linear system

$$\widetilde{\Lambda}_{[s]}\alpha_{[s]}^{\lambda} = \widetilde{\beta}_{[s]}^{\lambda}, \tag{5.6.39}$$

where the matrix $\widetilde{\Lambda}_{[s]} \in \mathbb{R}^{n,n}$ is given by

$$\widetilde{\Lambda}_{[s]} = \widetilde{\Lambda}^{1010} - \widetilde{\Lambda}^{1001}\big(\widetilde{\Lambda}^{0101}\big)^{-1}\widetilde{\Lambda}^{0110}, \tag{5.6.40}$$

and the vector $\widetilde{\beta}_{[s]}^{\lambda} \in \mathbb{R}^{n}$ by

$$\widetilde{\beta}_{[s]}^{\lambda} = \widetilde{\beta}^{10\lambda} - \widetilde{\Lambda}^{1001}\big(\widetilde{\Lambda}^{0101}\big)^{-1}\widetilde{\beta}^{01\lambda}. \tag{5.6.41}$$

Assuming $\widetilde{\Lambda}^{0101}$ to be symmetric, we deduce from Section 5.1.5 that the system (5.6.39) admits a unique solution $\alpha_{[s]}^{\lambda}$. The thermal conductivity is then evaluated from

$$\lambda = \lambda_{[s]} + \lambda_{[01]} = \frac{\overline{p}}{T}\langle \alpha_{[s]}^{\lambda}, \beta_{[s]}^{\lambda}\rangle + \frac{\overline{p}}{T}\langle \beta^{01\lambda}, \big(\widetilde{\Lambda}^{0101}\big)^{-1}\widetilde{\beta}^{01\lambda}\rangle, \tag{5.6.42}$$

where the vector $\beta_{[s]}^{\lambda}$ is now given by

$$\beta_{[s]}^{\lambda} = \beta^{10\lambda} - \big(\widetilde{\Lambda}^{0110}\big)^{t}\big(\widetilde{\Lambda}^{0101}\big)^{-1}\beta^{01\lambda}, \tag{5.6.43}$$

and the thermal diffusion ratios from

$$\chi = \chi_{[s]} + \chi_{[01]} = [L_{[s]}^{0010}]\alpha_{[s]}^{\lambda} + L^{0001}\big(\widetilde{\Lambda}^{0101}\big)^{-1}\widetilde{\beta}^{01\lambda}, \tag{5.6.44}$$

where the matrix $[L_{[s]}^{0010}]$ is now given by

$$[L_{[s]}^{0010}] = L^{0010} - L^{0001}\big(\widetilde{L}^{0101}\big)^{-1}\widetilde{L}^{0110}. \tag{5.6.45}$$

For positive mass fractions, one can easily verify that the solutions of the systems (5.6.31) and (5.6.39), the thermal conductivities (5.6.34) and (5.6.42), and the thermal diffusion ratios (5.6.35) and (5.6.44) coincide.

Consider now the splitting $\widetilde{\Lambda}_{[\mathrm{s}]} = \widetilde{M}_{[\mathrm{s}]} - \widetilde{Z}_{[\mathrm{s}]}$ where

$$\widetilde{M}_{[\mathrm{s}]} = db(\widetilde{\Lambda}^{1010}) + \mathrm{diag}(\widetilde{\mathfrak{d}}), \tag{5.6.46}$$

and $\widetilde{\mathfrak{d}}_k^r \geq 0$, $(r,k) \in \{10\} \times \mathcal{S}$. One can show, after some algebra, that the iteration matrix $T_{[\mathrm{s}]} = \big(\widetilde{M}_{[\mathrm{s}]}\big)^{-1} \widetilde{Z}_{[\mathrm{s}]}$ is convergent. A convergence theorem similar to Theorem 5.6.7 can then be stated for the linear system (5.6.39) with the splitting matrix (5.6.46). In particular, this yields sequences of iterates $\lambda_{[\mathrm{s}]}^{[i]}$ and $\chi_{[\mathrm{s}]}^{[i]}$ such that

$$\begin{cases} \lim\limits_{i \to \infty} \lambda_{[\mathrm{s}]}^{[i]} = \lambda - \lambda_{[01]}, \\ \lim\limits_{i \to \infty} \chi_{[\mathrm{s}]}^{[i]} = \chi - \chi_{[01]}. \end{cases} \tag{5.6.47}$$

Finally, stabilized versions of conjugate gradient methods are obtained by considering the symmetric rescaled matrix $\widehat{\Lambda}_{[\mathrm{s}]}$, and are omitted for brevity.

6 Numerical Experiments

In this chapter we perform numerical experiments illustrating the convergence results established in Chapter 5. In addition, we verify the accuracy of the simplified formulations introduced in the previous chapters. We also consider several empirical mixture-averaged formulas for the shear viscosity, the volume viscosity, the partial thermal conductivity, and the thermal conductivity. Some of these formulas correspond to approximations widely used in numerical simulations, but new mixture-averaged formulas are also introduced.

In Section 6.1 we present the mixtures used in our numerical experiments and discuss some issues related to the computational cost of multicomponent transport property evaluation. In Section 6.1.1 we describe the test mixtures which are typical gas mixtures associated with hydrogen and methane combustion applications. Estimates of the magnitude of physical parameters for the present theory to be valid for such applications are given in Appendix E. In Section 6.1.2 we discuss the evaluation of the transport linear system coefficients. In Section 6.1.3 we compare iterative methods versus direct inversions for the transport linear systems associated with a given gas mixture. In Section 6.1.4 we discuss the choice of empirical mixture-averaged formulas versus analytic expressions rigorously derived from the kinetic theory. Finally, in Section 6.1.5 we examine optimization of transport property evaluation in multicomponent flow computations.

The numerical results obtained with standard iterative and conjugate gradient methods are presented in Sections 6.2 to 6.6 for all the transport coefficients. We consider the shear viscosity in Section 6.2, the volume viscosity in Section 6.3, the diffusion matrix and the flux diffusion matrix in Section 6.4, the partial thermal conductivity and the thermal diffusion vector in Section 6.5, and the thermal conductivity and the thermal diffusion ratios in Section 6.6. For each transport coefficient, we discuss various rigorously derived approximate expressions in terms of computational cost and accuracy. When considering other mixtures than the ones described in Section 6.1.1, the accuracies of these expressions may, of course, vary. However, the high convergence rates observed in this chapter indicate that iterative methods constitute a very general, stable, and efficient technique for multicomponent transport evaluation.

6.1 Computational Considerations

6.1.1 Preliminaries

The numerical experiments presented in Sections 6.2 to 6.6 are performed for a nine species mixture used in hydrogen-air flame computations and a twenty-six species mixture used in methane-air flame computations [Er94] [GS89] [SG92], at temperature $\overline{T} = 1000$ K and pressure $\bar{p} = 1$ atm. The chemical system used in the hydrogen-air flame is composed of the $n = 9$ species H_2, O_2, N_2, H_2O, H, O, OH, HO_2, and H_2O_2, and is referred to as "the hydrogen mixture". The chemical system used for the methane-air flame is composed of the $n = 26$ species CH_4, CH_3, CH_2, CH, N_2, H_2, O_2, H_2O, H, O, OH, HO_2, H_2O_2, C_2H_6, C_2H_5, C_2H_4, C_2H_3, C_2H_2, C_2H, CHO, CH_2O, CH_3O, CH_2CO, CO_2, CO, and C_2HO and is referred to as "the methane mixture".

We consider hydrogen and methane mixtures in both positive and nonnegative mass fractions states. For positive mass fractions, we consider three mixtures referred to as mixtures 1, 2, and 3. Mixture 1 is an equimolar hydrogen mixture, i.e., all the mole fractions are set equal to $1/n = 1/9$. In mixture 2, the mole fractions of H_2, O_2, and N_2 are set equal to $1/3 - 2\epsilon$ and the remaining ones set to $\epsilon = 1.0\text{E-}4$. Finally, mixture 3 is an equimolar methane mixture, i.e., all the mole fractions are set equal to $1/n = 1/26$. For nonnegative mass fractions, we consider three additional mixtures referred to as mixtures 4, 5, and 6. Mixture 4 consists of the hydrogen mixture in the state $X_{H_2} = X_{O_2} = X_{N_2} = 1/3$ for which $n = 9$ and $n^+ = 3$. Mixture 5 is the hydrogen mixture in the state $X_{H_2} = X_{O_2} = 1/2$ for which $n = 9$ and $n^+ = 2$. Finally, mixture 6 corresponds to the methane mixture in the state where all the mole fractions are equal to $1/(n-1) = 1/25$, except for the mole fraction of the last species, C_2HO, which is set equal to zero, so that in this case $n = 26$ and $n^+ = 25$.

For mixtures in a positive mass fractions state, i.e., mixtures 1, 2, and 3, we have also compared iterative algorithms for the general system matrix G with the corresponding stabilized version for a rescaled matrix, that is, the left rescaled matrix $\widetilde{G}$ for standard iterative methods and the symmetric rescaled matrix $\widehat{G}$ for conjugate gradient methods. As proven in Section 5.1.6, we have observed the coincidence of the resulting iterates in all cases. Consequently, only the results corresponding to the matrix G will be presented for positive mass fractions.

In addition, for singular system matrices G associated with a constraint vector $\mathcal{G}$, we have also considered iterative algorithms for the symmetric positive definite matrix $G+\mathcal{G}\otimes\mathcal{G}$. We have observed, however, slower convergence rates than the ones obtained with projected iterative algorithms for the singular matrix G. We will therefore omit numerical experiments for the symmetric positive definite forms of the singular system matrices. These matrices are more suited to direct inversions, as discussed in the next section.

6.1.2 Evaluation of the System Coefficients

For the numerical implementation of the iterative algorithms, the transport linear system coefficients have been evaluated using the practical approximations presented in Section 2.10 [MM62] [MPM65]. As proven in Chapters 4 and 5, all the mathematical results and all the convergence theorems are valid in the framework of these approximations. We also restate that the general theory presented in this book does not rely upon these particular approximations, but, on the contrary, it applies to the exact systems and to many other systems using, for instance, different approximations.

With the approximations presented in Section 2.10, the evaluation of the system coefficients is reduced to the computation of the binary diffusion coefficients $\mathcal{D}_{kl}$, $k,l \in \mathcal{S}$; the pure species shear viscosities η_k, $k \in \mathcal{S}$; the diffusion coefficients for internal energy $\mathcal{D}_{k\,\mathrm{int},l}$, $k,l \in \mathcal{S}$; the three ratios of collision integrals $\bar{A}_{kl}$, $\bar{B}_{kl}$, and $\bar{C}_{kl}$, $k,l \in \mathcal{S}$; the collision numbers ξ_k^{int}, $k \in \mathcal{S}$; and the internal heat capacities c_k^{int}, $k \in \mathcal{S}$.

The above quantities depend on the forces of interaction between the colliding molecules. For pairs of nonpolar molecules (k,l), the interaction potential is represented by the Lennard-Jones (12–6) potential

$$\varphi_{kl} = 4\epsilon_{kl}\left[\left(\frac{\sigma_{kl}}{r}\right)^{12} - \left(\frac{\sigma_{kl}}{r}\right)^{6}\right], \tag{6.1.1}$$

where r is the distance between centers, ϵ_{kl} the potential well depth, and σ_{kl} the collision diameter. For pairs of polar molecules (k,l), the interaction potential is represented by the Stockmayer potential

$$\begin{aligned}\varphi_{kl} = 4\epsilon_{kl}&\left[\left(\frac{\sigma_{kl}}{r}\right)^{12} - \left(\frac{\sigma_{kl}}{r}\right)^{6}\right]\\ &- \frac{\mu_k\mu_l}{r^3}\big(2\cos(\theta_1)\cos(\theta_2) - \sin(\theta_1)\sin(\theta_2)\cos(\phi)\big),\end{aligned} \tag{6.1.2}$$

where μ_k and μ_l are the dipole moments of the two molecules, θ_1 and θ_2 the angles of the two dipoles to the line joining the centers of the molecules, and ϕ is the azimuthal angle between them. Note that in the limit as μ_k or $\mu_l \to 0$, (6.1.2) reduces to the Lennard-Jones (12–6) potential. Orientation effects arising from the angle-dependent nature of Stockmayer potentials have been discussed in [MM61], where special sets of orientation-averaged collision integrals have been calculated assuming all orientations to be equally probable.

The parameters ϵ_{kl} and σ_{kl} are characteristic of the colliding species pair (k,l) and are evaluated using combining rules [HCB54] [Di68] [FK72] [Di84]. These rules take the form of empiric relations for the parameters for unlike-molecule potentials in terms of the parameters for like-molecule potentials. In the case where the collision partners are either both polar or nonpolar, we use the following rules

$$\begin{cases} \epsilon_{kl} = \sqrt{\epsilon_k \epsilon_l}, \\ \sigma_{kl} = \dfrac{1}{2}(\sigma_k + \sigma_l), \end{cases} \tag{6.1.3}$$

where ϵ_k and σ_k are the pure species Lennard-Jones parameters. In the case of a polar molecule interacting with a nonpolar, but polarizable, molecule, (6.1.3) is replaced by

$$\begin{cases} \epsilon_{kl} = \xi^2 \sqrt{\epsilon_k \epsilon_l}, \\ \sigma_{kl} = \dfrac{\xi^{-1/6}}{2}(\sigma_k + \sigma_l), \end{cases} \tag{6.1.4}$$

where

$$\xi = 1 + \frac{1}{4}\alpha_k^* (\mu_l^*)^2 \sqrt{\frac{\epsilon_l}{\epsilon_k}}, \tag{6.1.5}$$

and where the subscript k refers to the nonpolar molecule and the subscript l to the polar molecule. Furthermore, α_k^* denotes the reduced polarizability of the nonpolar molecule

$$\alpha_k^* = \frac{\alpha_k}{\sigma_k^3}, \tag{6.1.6}$$

and μ_l^* the reduced dipole moment of the polar molecule

$$\mu_l^* = \frac{\mu_l}{\sqrt{\epsilon_l \sigma_l^3}}. \tag{6.1.7}$$

A list of values of ϵ_k, σ_k, α_k, and μ_k for species of interest in our applications is given in Appendix D [Di68] [Wr83] [Di84].

The binary diffusion coefficients $\mathcal{D}_{kl}$, $k,l \in S$, are evaluated as [FK72]

$$\mathcal{D}_{kl} = \frac{3}{16} \frac{\sqrt{2\pi (k_B T)^3 / m_{kl}}}{\bar{p} \pi \sigma_{kl}^2 \Omega_{kl}^{(1,1)*}}, \tag{6.1.8}$$

where $\Omega_{kl}^{(1,1)*}$ is given by the ratio of the collision integral $\Omega_{kl}^{(1,1)}$ divided by the corresponding rigid sphere value. The pure species shear viscosities η_k, $k \in \mathcal{S}$, are evaluated as [FK72]

$$\eta_k = \frac{5}{16} \frac{\sqrt{\pi m_k k_B \overline{T}}}{\pi \sigma_k^2 \Omega_{kk}^{(2,2)*}}, \tag{6.1.9}$$

with obvious notation. The collision integrals $\Omega_{kl}^{(i,j)*}$ are treated as if collisions were elastic [Di68] [Di84]. The elastic-collision integrals are determined from tables based on Stockmayer potentials averaged over all orientations assumed equally probable [MM61]. The table look-up depends on the reduced temperature

$$\overline{T}_{kl}^* = \frac{k_B \overline{T}}{\epsilon_{kl}}, \qquad k,l \in \mathcal{S}, \tag{6.1.10}$$

and the reduced dipole moment

$$\delta_{kl}^* = \frac{1}{2} \mu_k^* \mu_l^*, \qquad k,l \in \mathcal{S}. \tag{6.1.11}$$

The diffusion coefficients for internal energy are approximated by the ordinary diffusion coefficients [MPM65] [Di84]

$$\begin{cases} \mathcal{D}_{k\,\mathrm{int},l} = \mathcal{D}_{kl}, & k,l \in \mathcal{S}, \quad k \neq l, \\ \mathcal{D}_{k\,\mathrm{int},k} = \dfrac{6}{5} \dfrac{k_B \overline{T}}{\overline{p} m_k} \bar{A}_{kk} \eta_k, & k \in \mathcal{S}. \end{cases} \tag{6.1.12}$$

The three ratios of collision integrals $\bar{A}_{kl}$, $\bar{B}_{kl}$, and $\bar{C}_{kl}$, $k,l \in \mathcal{S}$, are determined from tables based on Stockmayer potentials [MM61] in function of the reduced temperature $\overline{T}_{kl}^*$ and the reduced dipole moment δ_{kl}^*. In a first approximation, the quantities $\bar{A}_{kl}$, $\bar{B}_{kl}$, and $\bar{C}_{kl}$, $k,l \in \mathcal{S}$, are evaluated using sixth-order polynomial fits as a function of the logarithm of the reduced temperature $\overline{T}_{kl}^*$ only [KDWCM86]. The polynomial coefficients are restated for completeness in Appendix D.

The collision numbers ξ_k^{int}, $k \in \mathcal{S}$, are expressed in function of the temperature [Pa59] [BJ70] as

$$\xi_k^{\mathrm{int}}(\overline{T}) = \xi_k^{\mathrm{int}}(298\mathrm{K}) \frac{F_k(298\mathrm{K})}{F_k(\overline{T})}, \qquad k \in \mathcal{S}, \tag{6.1.13}$$

where

$$F_k(\overline{T}) = 1 + \frac{\pi^{3/2}}{2} \left(\overline{T}_{kk}^*\right)^{-1/2} + \left(\frac{\pi^2}{4} + 2\right) \left(\overline{T}_{kk}^*\right)^{-1} + \pi^{3/2} \left(\overline{T}_{kk}^*\right)^{-3/2}, \tag{6.1.14}$$

The collision numbers at the reference temperature, $\xi_k^{\mathrm{int}}(298\mathrm{K})$, are restated for completeness in Appendix D.

Finally, the internal heat capacity of the molecules of the kth species is expressed as

$$c_k^{\text{int}} = c_{\text{v},k} - \frac{3}{2} k_{\text{B}}, \tag{6.1.15}$$

and the constant volume heat capacity of the molecules of the kth species, $c_{\text{v},k}$, is evaluated using the JANAF thermodynamic data base [SP71] [Cal85] and the Chemkin thermodynamic data base [KRM87].

6.1.3 Iterative Methods Versus Direct Inversions

In this section we compare iterative methods versus direct inversions for solving the transport linear systems associated with a given gas mixture. The discussion is based on the computational cost of these approaches which is estimated by an operation count. As usual, we define an operation to be one multiplication plus one addition.

Recall that n and p denote the number of species and polyatomic species present in the mixture, respectively. We assume that n is large, that is, $n \geq 10$, keeping in mind that this condition is met in typical multicomponent flow calculations. Therefore, in this chapter, we will only consider the leading order terms in the number of species n for the operation counts. It is then convenient to introduce the notation $\phi(n) = \mathcal{O}(f(n))$ in order to indicate that the functions $\phi(n)$ and $f(n)$ are such that $\phi(n)/f(n)$ is bounded when n is large.

In order to simplify the operation counts, we assume in this section that we have $p = n$. As a consequence, all the transport linear systems are of size $\mathfrak{s}n$ where $\mathfrak{s} = 1, 2$, or 3. More specifically, the matrices H, $K_{[01]}$, $L_{[00]}$, and $\Lambda_{[\text{e}]}$ are of size n, the matrices K, $L_{[\text{e}]}$, and Λ of size $2n$, and the matrix L of size $3n$. In the general case where $p < n$, the operation counts presented below will still provide a very reasonable estimate for the computational costs. Indeed, only a few monatomic species are, in general, present in mixtures considered in multicomponent flow calculations. Furthermore, in practical implementations, the transport linear systems are often trivially modified by adding "dummy" rows and columns in such a way that all the subblocks of the system matrix are square matrices of size n, thereby yielding a much simpler matrix structure.

Cost of System Matrix Evaluation. Before evaluating the cost of inverting the transport linear systems, we first discuss the cost of evaluating the corresponding system matrix. For this purpose, we restate the fundamental relations (2.3.26)

$$\begin{cases} G_{kk}^{rs} = \sum_{l \in S} n_k n_l G_{kl}^{\prime rs} + n_k^2 G_{kk}^{\prime\prime rs}, & (r,k),(s,k) \in \mathcal{B}^\mu, \\ G_{kl}^{rs} = n_k n_l G_{kl}^{\prime\prime rs}, \quad (r,k),(s,l) \in \mathcal{B}^\mu, & k \neq l, \end{cases} \tag{6.1.16}$$

where the matrices G' and G'' are functions of the state variables, and $\mathcal{B}^\mu$ is the indexing set—which has $\mathfrak{s}n$ elements. The matrices G' and G'' are evaluated first in $\mathcal{O}(\mathfrak{s}^2n^2)$ operations. The block-diagonal coefficients G_{kk}^{rs}, $(r,k),(s,k) \in \mathcal{B}^\mu$, then require $\mathfrak{s}^2n^2 + \mathcal{O}(n)$ operations, whereas the remaining coefficients G_{kl}^{rs}, $(r,k),(s,l) \in \mathcal{B}^\mu$, $k \neq l$, require $\mathfrak{s}^2n(n-1)$ operations. Therefore, the cost of the system matrix evaluation is $\mathcal{O}(\mathfrak{s}^2n^2)$ operations.

Direct Inversions. It is well-known that Gaussian elimination yields the LU decomposition of a system matrix of size $\mathfrak{s}n$ in

$$C_{\mathrm{LU}} = \frac{\mathfrak{s}^3n^3}{3} + \mathcal{O}(n^2) \tag{6.1.17}$$

operations [GV83]. In the particular case of symmetric positive definite matrices, however, the Cholesky algorithm requires half as many operations and constructs the LL^t decomposition of the system matrix in

$$C_{\mathrm{LL}^t} = \frac{\mathfrak{s}^3n^3}{6} + \mathcal{O}(n^2) \tag{6.1.18}$$

operations [GV83]. Therefore, the symmetric positive definite forms of the transport linear systems described in this book can be inverted at half the computational cost of the nonsymmetric forms obtained by Monchick, Yun, and Mason [MYM63] and considered, for instance, in [KDWCM86] for the evaluation of the thermal conductivity, the thermal diffusion coefficients, and the multicomponent diffusion coefficients.

Standard Iterative Methods. Referring to Theorem 5.1.7, the initialization requires the evaluation of the vector $M^{-1}\beta^\mu$. As described in Sections 5.2 to 5.6, the matrix M has a block-diagonal structure since it is obtained from the diagonal of the blocks of the original system matrix G. Consequently, the vector $M^{-1}\beta^\mu$ is evaluated by solving n dense linear systems of size $\mathfrak{s}$, associated with the symmetric subblocks $G_{kk} = (G_{kk}^{rs})_{r,s\in\mathcal{F}_k}$. Recall that the set of function type indices $\mathcal{F}_k = \{\, r \in \mathcal{F},\ (r,k) \in \mathcal{B}^\mu \,\}$ has $\mathfrak{s} = 1$, 2, or 3 elements for the transport linear systems considered in this book. The evaluation of $M^{-1}\beta^\mu$ therefore requires $\mathcal{O}(n)$ operations. Next, at each iteration, we form $x_{i+1} = Tx_i + M^{-1}\beta^\mu$, where $Tx_i = x_i - M^{-1}Gx_i$. The vector Gx_i is computed in $\mathfrak{s}^2n^2$ operations, and we then form $M^{-1}Gx_i$, $x_i - M^{-1}Gx_i$, and x_{i+1} in $\mathcal{O}(n)$ operations. Hence, for large n, we deduce that m steps of the standard iterative method require

$$C_{m,\mathrm{SIM}} = m\,\mathfrak{s}^2n^2 + \mathcal{O}(n) \tag{6.1.19}$$

operations. In addition, we note that this estimate is still valid for the projected version of the algorithm described in Theorem 5.1.8, since only $\mathcal{O}(\mathfrak{s}n)$ operations are needed to form the product Px, where P is the projector matrix and x a given vector.

Conjugate Gradient Methods. Referring to Theorem 5.1.9 and assuming that we take $x_0 = 0$, we immediately obtain $r_0 = \beta^\mu$. At each iteration, we then need $\mathfrak{s}^2n^2$ operations to compute the vector Gp_i, whereas the vector $M^{-1}r_i$, the scalar products and the vector updates are evaluated in $\mathcal{O}(n)$ operations. Consequently, for large n, the cost of m steps of the conjugate gradient method is given by

$$C_{m,\mathrm{CGA}} = m\ \mathfrak{s}^2n^2 + \mathcal{O}(n) \tag{6.1.20}$$

operations, and this estimate is also valid for the projected version of the algorithm described in Theorem 5.1.10.

Comparison. It is fundamental to first observe that the cost of the system matrix evaluation scales as $\mathcal{O}(\mathfrak{s}^2n^2)$, so that it does not dominate asymptotically the cost of performing m steps of an iterative method, i.e., the ratio of these costs is bounded when n is large. Furthermore, since the former cost is the same for direct inversions and iterative methods, it can be omitted for their comparison. We then deduce from (6.1.19)(6.1.20) that m steps of an iterative method, projected or not, are more cost-effective than a direct inversion with the Cholesky algorithm if

$$m \le \frac{\mathfrak{s}n}{6}, \tag{6.1.21}$$

yielding an upper bound on the number of steps in the iterative method. Notice that this upper bound increases linearly with the number of species present in the mixture. As an illustration, consider the cases $n = 10$, i.e., typical hydrogen-air flame applications and $n = 30$, i.e., typical methane-air flame applications. If $n = 10$, we obtain that $m \le 5$ for $\mathfrak{s} = 3$, $m \le 3$ for $\mathfrak{s} = 2$, and $m \le 1$ for $\mathfrak{s} = 1$, whereas if $n = 30$, we obtain that $m \le 15$ for $\mathfrak{s} = 3$, $m \le 10$ for $\mathfrak{s} = 2$, and $m \le 5$ for $\mathfrak{s} = 1$. Finally, note that the upper bound on m is twice as large when comparing iterative methods to the Gaussian elimination required for the nonsymmetric systems obtained in [MYM63].

Schur Complements. We now consider iterative schemes for Schur complements of diagonal matrices, as described in Section 5.1.5. The Schur complements are of the form $G_{[\mathfrak{s}]} = G^{11} - G^{12}\left(G^{22}\right)^{-1}G^{21}$, where the diagonal block G^{22} is of size n and the matrix $G_{[\mathfrak{s}]}$ of size $(\mathfrak{s}-1)n$ for $\mathfrak{s} \ge 2$. It is important to notice that the explicit evaluation of the matrix $G_{[\mathfrak{s}]}$, which would require $2(\mathfrak{s}-1)^2n^3$ operations, is not needed. Indeed, we only have to evaluate the product of the Schur complement with a given vector, which requires $3(\mathfrak{s}-1)^2n^2 + \mathcal{O}(n)$ operations. As a consequence, when n is large, one can easily verify that m steps of an iterative method for the Schur complement require

$$C_{m,\mathrm{Schur}} = 3m(\mathfrak{s}-1)^2n^2 + \mathcal{O}(n) \tag{6.1.22}$$

operations. We then deduce that m steps of an iterative method are more cost-effective than a direct inversion of the original system matrix G with the Cholesky algorithm if

$$m \leq \frac{n}{18}\frac{\mathfrak{s}^3}{(\mathfrak{s}-1)^2}, \tag{6.1.23}$$

yielding again an upper bound for the number of steps. As an illustration, consider again the cases $n = 10$, i.e., typical hydrogen-air flame applications and $n = 30$, i.e., typical methane-air flame applications. If $n = 10$, we obtain that $m \leq 3$ for $\mathfrak{s} = 3$ and $m \leq 4$ for $\mathfrak{s} = 2$, whereas if $n = 30$, we obtain that $m \leq 11$ for $\mathfrak{s} = 3$ and $m \leq 13$ for $\mathfrak{s} = 2$.

Remark. The cost of evaluating the transport linear systems will be omitted in the operation counts presented in Sections 6.2 to 6.6, since it depends on the approximations used for various collision integrals. This cost often scales as a rather large multiple of n^2 operations, so that, in practice, the cost of evaluating the system matrix is more significant than the cost of performing a few steps of an iterative algorithm.

6.1.4 Iterative Methods Versus Mixture-Averaged Formulas

In this section we compare empirical mixture-averaged formulas versus analytic expressions rigorously derived from the kinetic theory, for evaluating the transport properties of a given gas mixture. It is fundamental to first note that the latter expressions necessarily require at least n^2 operations, since each gas interacts with all the other gases present in the mixture. It is therefore natural that the cost of the system matrix evaluation also scales as n^2, as described in the previous section. Consequently, it is impossible to obtain analytic expressions rigorously derived from the kinetic theory at a cost of $\mathcal{O}(n)$ operations. On the other hand, it is possible to obtain *empirical* expressions at a cost of $\mathcal{O}(n)$ operations, and these expressions will be referred to as mixture-averaged formulas. Since they are often considered as an economical alternative for transport property evaluation in multicomponent flow calculations, mixture-averaged formulas are also included in the numerical tests performed in the following sections.

In our numerical tests, we consider approximations widely used in numerical simulations, but new mixture-averaged formulas are also introduced. More specifically, for a given real number t, we consider the average formulas of order t given by

$$\mathcal{M}_t(\mu) = \Big(\sum_{k\in\mathcal{S}} X_k(\mu_k)^t\Big)^{1/t}, \tag{6.1.24}$$

for $t \neq 0$, and for $t = 0$, by

$$\mathcal{M}_0(\mu) = \exp\Big(\sum_{k\in\mathcal{S}} X_k \log(\mu_k)\Big), \tag{6.1.25}$$

where the quantities μ_k, $k \in \mathcal{S}$, denote the transport coefficient obtained for each pure species. The mathematical properties of average formulas of order t are described, for instance, in [RV73], and we only note here that they are continuous with respect to t and have the monotonicity property

$$\mathcal{M}_{t_1}(\mu) \leq \mathcal{M}_{t_2}(\mu), \qquad \text{for} \quad t_1 \leq t_2. \tag{6.1.26}$$

Mixture-averaged formulas are considered in Sections 6.2, 6.3, 6.5, and 6.6 for the shear viscosity, the volume viscosity, the partial thermal conductivity, and the thermal conductivity, respectively, since the pure species properties are defined for these transport coefficients only. Note that the partial thermal conductivity coincides with the thermal conductivity in this case and also that (6.1.24)(6.1.25) are valid for mixtures in both positive and nonnegative mass fractions states.

6.1.5 Optimization of Transport Property Evaluation

In this section we discuss optimization of transport property evaluation based on the trade-off between computational cost and accuracy. We first restate that the transport linear systems are derived using polynomial expansions of the species perturbed distribution functions. Consequently, when only the first terms are kept in these expansions, it is reasonable to expect that the resulting transport linear systems may provide transport coefficients accurate to one part per thousand. We refer to this as 1E-3 accuracy. Therefore, in the analytic expressions of the transport coefficients obtained by truncating convergent series, only the number of terms yielding a similar accuracy should be kept at most. Similarly, only the corresponding number of steps in the iterative algorithms are interesting to perform.

Whenever possible, we distinguish the following three strategies for multicomponent transport property evaluation. First, we consider expressions, referred to as economical, which require the lowest possible computational cost. For scalar coefficients, i.e., $\mu = \eta$, κ, λ', or λ, economical expressions are given by empirical mixture-averaged formulas and require $\mathcal{O}(n)$ operations, as described in Section 6.1.4. On the other hand, for $\mu = D$, θ, and χ, which require evaluation of more coefficients, economical expressions are obtained after 1 or 2 steps of an iterative algorithm in $\mathcal{O}(n^2)$ operations. Second, we consider expressions with moderate computational cost and reasonable accuracy, i.e., between 1E-2 and 5E-2. As illustrated in Sections 6.2 to 6.6, these expressions are, in general, obtained by truncating an iterative algorithm after 1 or 2 steps. Finally, we consider expressions with high computational cost and excellent accuracy, i.e., 1E-3. As illustrated in Sections 6.2 to 6.6, these expressions are, in general, obtained by truncating an iterative algorithm after 3 or 4 steps. Expressions corresponding to all the aforementioned strategies will be summarized for each transport coefficient at the end of the corresponding section. In addition, the high convergence rates observed in our numerical experiments also indicate, for the mixtures considered in this book at least, that iterative algorithms are a cost-effective and robust alternative to direct inversions for transport property evaluation.

Finally, optimizing transport property evaluation in multicomponent flow computations also depends on the problem granularity. For coarse-granularity parallel computations in which each processor is responsible for a different portion of the computational domain, multiple input data subroutines that compute simultaneously transport properties over the whole subdomain are preferable [GD88]. This approach has been implemented in [Er94] [GS89] [SG92] for various combustion problems. On the other hand, for fine-grained distributed parallel architectures, simple input data subroutines that only consider one state of the mixture must be used.

6.2 The Shear Viscosity

6.2.1 Numerical Experiments for the System $H\alpha^\eta = \beta^\eta$

Standard Iterative Methods. We consider the initialization $x_0 = 0$ and the splitting

$$H = M - Z, \qquad M = db(H), \tag{6.2.1}$$

Table 1. Shear viscosity. Standard iterative methods with $M = db(H)$. Reduced errors and spectral radius for various mixtures.

	Mixture 1	Mixture 2	Mixture 3
1	1.61E-1	1.05E-1	1.84E-1
2	2.63E-2	1.10E-2	3.52E-2
3	4.29E-3	1.16E-3	6.70E-3
4	7.00E-4	1.22E-4	1.28E-3
ρ	1.63E-1	1.05E-1	1.90E-1

where $db(H)$ is formed by the diagonal of the matrix H. One can then easily verify that Theorem 5.2.1 applies. The reduced errors

$$e_\eta^{[i]} = \frac{|\eta - \eta^{[i]}|}{\eta}, \qquad i = 1, 2, 3, 4, \tag{6.2.2}$$

and the spectral radius of the iteration matrix $\rho = \rho(M^{-1}Z)$ are presented in Table 1 for mixtures 1, 2, and 3. The first two approximations for η can be written

$$\begin{cases} \eta^{[1]} = \langle \beta^\eta, db(H)^{-1}\beta^\eta \rangle, \\ \eta^{[2]} = \langle db(H)^{-1}\beta^\eta, \big(2db(H) - H\big)db(H)^{-1}\beta^\eta \rangle, \end{cases} \tag{6.2.3}$$

and we see from Table 1 that $\eta^{[2]}$ yields an approximation within 4E-2 accuracy. This type of approximation has been considered implicitly by Hirschfelder, Curtiss, and Bird [HCB54] and Brokaw [Br58] [Br64] when deriving approximate formulas for the shear viscosity.

Conjugate Gradient Methods. We consider the initialization $x_0 = 0$ and the preconditioner $M = db(H)$, for which Theorem 5.2.2 applies. The corresponding results are presented in Table 2. After one iteration only, the resulting shear viscosity given by

$$\eta^{[1]} = \frac{\langle \beta^\eta, db(H)^{-1}\beta^\eta \rangle^2}{\langle db(H)^{-1}\beta^\eta, H db(H)^{-1}\beta^\eta \rangle}, \tag{6.2.4}$$

is within 2E-3 accuracy, which is an order of magnitude lower than the results obtained with standard iterative methods.

Table 2. Shear viscosity. Conjugate gradient methods with $M = db(H)$. Reduced errors for various mixtures.

	Mixture 1	Mixture 2	Mixture 3
1	4.00E-4	6.50E-5	1.71E-3
2	1.18E-7	8.63E-8	3.97E-8
3	3.66E-12	1.23E-12	8.46E-13
4	1.27E-16	3.17E-17	—

Table 3. Shear viscosity. Standard iterative methods with $\widetilde{M} = db(\widetilde{H})$. Reduced errors and spectral radius for various mixtures.

	Mixture 4	Mixture 5	Mixture 6
1	1.05E-1	3.05E-2	1.86E-1
2	1.10E-2	1.58E-3	3.53E-2
3	1.15E-3	4.81E-5	6.72E-3
4	1.21E-4	2.49E-6	1.28E-3
ρ	1.05E-1	3.97E-2	1.90E-1

6.2.2 Numerical Experiments for the System $\widetilde{H}\alpha^\eta = \widetilde{\beta}^\eta$

We consider the initialization $x_0 = 0$ and the splitting

$$\widetilde{H} = \widetilde{M} - \widetilde{Z}, \qquad \widetilde{M} = db(\widetilde{H}), \tag{6.2.5}$$

where the matrix $db(\widetilde{H})$ is formed by the diagonal of the matrix $\widetilde{H}$. One can then easily verify that Theorem 5.2.4 applies. The reduced errors

$$e_\eta^{[i]} = \frac{|\eta - \eta^{[i]}|}{\eta}, \qquad i = 1, 2, 3, 4, \tag{6.2.6}$$

and the spectral radius of the iteration matrix $\rho = \rho(\widetilde{M}^{-1}\widetilde{Z})$ are presented in Table 3 for mixtures 4, 5, and 6. The first two approximations for η can be written

$$\begin{cases} \eta^{[1]} = \langle \beta^\eta, db(\widetilde{H})^{-1}\widetilde{\beta}^\eta \rangle, \\ \eta^{[2]} = \langle \beta^\eta, db(\widetilde{H})^{-1}\big(2db(\widetilde{H}) - \widetilde{H}\big)db(\widetilde{H})^{-1}\widetilde{\beta}^\eta \rangle, \end{cases} \tag{6.2.7}$$

and we see from Table 3 that $\eta^{[2]}$ is within 4E-2 accuracy.

Table 4. Shear viscosity. Conjugate gradient methods with $\widehat{M} = db(\widehat{H})$. Reduced errors for various mixtures.

	Mixture 4	Mixture 5	Mixture 6
1	6.35E-5	6.86E-4	1.28E-3
2	7.93E-8	3.15E-17	4.88E-8
3	1.59E-17	—	4.98E-13

6.2.3 Numerical Experiments for the System $\widehat{H}\widehat{\alpha}^\eta = \widehat{\beta}^\eta$

We consider the initialization $\widehat{x}_0 = 0$ and the preconditioner $\widehat{M} = db(\widehat{H})$, for which Theorem 5.2.6 applies. The reduced errors

$$e_\eta^{[i]} = \frac{|\eta - \eta^{[i]}|}{\eta}, \qquad i = 1, 2, 3, 4, \tag{6.2.8}$$

are reported in Table 4 and indicate high convergence rates. In particular, the first iterate is within 1E-3 accuracy and coincides with (6.2.4) for positive mass fractions.

6.2.4 Mixture-Averaged Formulas

As described in Section 6.1.4, we now consider several mixture-averaged formulas for the shear viscosity based on the pure species shear viscosities η_k, $k \in \mathcal{S}$. In Table 5 we present the reduced errors

$$e_t = \frac{|\eta - \mathcal{M}_t(\eta)|}{\eta}, \tag{6.2.9}$$

for various values of the parameter t and for the six mixtures considered in our numerical experiments. The reduced error for the average formula [KWM83]

$$\mathcal{M}_{-1,1}(\eta) = \frac{1}{2}\big(\mathcal{M}_{-1}(\eta) + \mathcal{M}_1(\eta)\big) = \frac{1}{2}\Big(\big(\sum_{k\in\mathcal{S}} X_k(\eta_k)^{-1}\big)^{-1} + \sum_{k\in\mathcal{S}} X_k\eta_k\Big), \tag{6.2.10}$$

is also included in Table 5, as well as the reduced error for the Wilke approximation

$$e_{[\mathrm{W}]} = \frac{|\eta - \eta_{[\mathrm{W}]}|}{\eta}, \tag{6.2.11}$$

which is often considered in computer simulations and given by [Wi50] [KDWCM86]

$$\eta_{[\mathrm{W}]} = \sqrt{8}\sum_{k\in\mathcal{S}} \frac{X_k\eta_k}{\sum_{l\in\mathcal{S}} X_l(1+m_k/m_l)^{-1/2}\big(1+(\eta_k/\eta_l)^{1/2}(m_l/m_k)^{1/4}\big)^2}. \tag{6.2.12}$$

Table 5. Shear viscosity. Reduced errors for various mixture-averaged formulas.

	Mixture 1	Mixture 2	Mixture 3	Mixture 4	Mixture 5	Mixture 6
0	1.70E-1	2.06E-1	6.45E-2	2.06E-2	2.86E-2	5.91E-2
1	1.20E-1	1.49E-1	4.98E-3	1.49E-1	2.14E-1	1.50E-2
2	8.10E-2	1.03E-1	9.53E-2	1.03E-1	1.48E-1	3.17E-2
6	7.75E-3	9.06E-3	5.74E-1	9.09E-3	6.42E-3	1.98E-1
7	2.09E-2	2.53E-3	6.78E-1	2.50E-3	9.60E-3	2.29E-1
-1,1	1.74E-1	2.08E-1	5.60E-2	2.08E-1	2.83E-1	5.67E-2
$e_{[W]}$	1.62E-2	1.86E-2	1.50E-2	1.86E-2	3.96E-2	1.93E-2

From Table 5, we can see that the mixture-averaged formula of order 6 works well for hydrogen mixtures (mixtures 1, 2, 4, and 5) at variance with methane mixtures (mixtures 3 and 6) for which the formula of order 1 is the most accurate. On the other hand, the expression (6.2.10) yields rather poor approximations, while the Wilke formula yields an approximation for the shear viscosity within 4E-2 accuracy. Note, however, that the computational cost of the Wilke formula is $\mathcal{O}(n^2)$ operations, as opposed to $\mathcal{O}(n)$ for the average formulas of order t.

6.2.5 Discussion

From the preceding numerical experiments, we can draw the following conclusions. For the shear viscosity, there is no empirical average formula of order t that works well. It is, however, possible to use an average formula where the parameter t depends on the mixture, but not on its state, that is, $t = 6$ for hydrogen mixtures and $t = 1$ for methane mixtures. These expressions are evaluated in $\mathcal{O}(n)$ operations and are within 9E-3 and 2E-2 accuracy of η, respectively, for the mixtures considered in this book.

On the other hand, a rigorously derived analytic expression is given by the first iterate of the conjugate gradient method

$$\eta^{[1]} = \frac{\left(\sum_{k\in\mathcal{S}} X_k^2/H_{kk}\right)^2}{\sum_{k,l\in\mathcal{S}} X_k X_l H_{kl}/(H_{kk}H_{ll})}. \qquad (6.2.13)$$

This new expression requires approximately the same computational cost as the traditional Wilke formula often used in computer calculations [Wi50] [KDWCM86], i.e., $\mathcal{O}(n^2)$

Table 6. Volume viscosity. Standard iterative methods for K with M given by (6.3.1). Reduced errors and spectral radius for various mixtures.

	Mixture 1	Mixture 2	Mixture 3
1	1.39E-2	4.48E-3	4.82E-2
2	2.35E-2	7.54E-3	3.06E-2
3	8.02E-4	7.70E-4	4.04E-3
4	2.31E-3	6.99E-4	1.82E-3
ρ	3.22E-1	3.69E-1	2.58E-1

operations, but the former expression is an order of magnitude more accurate than the latter.

6.3 The Volume Viscosity

6.3.1 Numerical Experiments for the System $K\alpha^\kappa = \beta^\kappa$

Standard Iterative Methods. We consider the initialization $x_0 = 0$ and the splitting $K = M - Z$, where the matrix M is given by

$$\begin{cases} M_{kl}^{rs} = \dfrac{K_{kk}^{rs}}{1 - (c_v^{\mathrm{tr}}/c_v)X_k}\delta_{kl}, & rs = 1010, \qquad k, l \in \mathcal{S}, \\ M^{rs} = \mathrm{diag}(K^{rs}), & rs \neq 1010. \end{cases} \tag{6.3.1}$$

Splittings of this form can be derived by identifying some of the block-diagonal elements in the relation $K \simeq MP_{\mathcal{K}^\perp,\mathbb{R}\mathcal{V}}$, where $P_{\mathcal{K}^\perp,\mathbb{R}\mathcal{V}}$ is the oblique projector onto $\mathcal{K}^\perp$ along $\mathbb{R}\mathcal{V}$. One can easily verify that the matrix M is of the form

$$M = db(K) + \mathrm{diag}(\mathfrak{d}), \tag{6.3.2}$$

where the matrix $db(K)$ is formed by the diagonal of the four blocks of the matrix K and $\mathfrak{d}^{10} > 0$, so that Theorem 5.3.1 applies. The reduced errors

$$e_\kappa^{[i]} = \frac{|\kappa - \kappa^{[i]}|}{\kappa}, \qquad i = 1, 2, 3, 4, \tag{6.3.3}$$

and the spectral radius of the projected iteration matrix $\rho = \rho(P_{\mathcal{K}^\perp,\mathbb{R}\mathcal{V}}M^{-1}Z)$ are presented in Table 6. In particular, the first iterate given by

$$\kappa = \langle \beta^\kappa, M^{-1}\beta^\kappa \rangle, \tag{6.3.4}$$

already yields an approximation within 5E-2 accuracy.

Table 7. Volume viscosity. Standard iterative methods for $K_{[s]}$ with $M_{[s]} = db(K^{1010})$. Reduced errors and spectral radius for various mixtures.

	Mixture 1	Mixture 2	Mixture 3
1	2.96E-3	2.30E-3	1.72E-3
2	7.08E-4	4.22E-4	1.88E-4
3	1.20E-4	7.92E-5	2.38E-5
4	2.66E-5	1.57E-5	3.58E-6
ρ	2.23E-1	5.19E-1	1.75E-1

Remark. The diagonal of the matrix M can also be taken in the form

$$M_{kk}^{1010} = \frac{K_{kk}^{1010} + (c_k^{\mathrm{int}}/c_{\mathrm{v}})X_k K_{kk}^{1001}}{1-(c_{\mathrm{v}}^{\mathrm{tr}}/c_{\mathrm{v}})X_k}, \qquad M_{ll}^{0101} = \frac{K_{ll}^{0101} + (c_{\mathrm{v}}^{\mathrm{tr}}/c_{\mathrm{v}})X_l K_{ll}^{0110}}{1-(c_l^{\mathrm{int}}/c_{\mathrm{v}})X_l}, \tag{6.3.5}$$

where $k \in \mathcal{S}$ and $l \in \mathcal{P}$. This yields a first iteration within 3E-2 accuracy. On the other hand, the simpler splitting $K = db(K) - Z$ yields a convergent iteration matrix for $n \geq 3$ only. Our numerical experiments indicate that this splitting yields a higher spectral radius and slightly less accuracy for the iterates.

We now investigate standard iterative methods for the Schur complement

$$K_{[s]} = K^{1010} - K^{1001}\big(K^{0101}\big)^{-1}K^{0110}, \tag{6.3.6}$$

introduced in Section 5.3.6. We consider the initialization $x_0 = 0$ and the splitting $K_{[s]} = M_{[s]} - Z_{[s]}$, where the matrix $M_{[s]}$ is diagonal and given by

$$M_{[s]} = db(K^{1010}). \tag{6.3.7}$$

As stated in Section 5.3.6, the corresponding iteration matrix is convergent, and the iterates $\kappa_{[s]}^{[i]}$ converge towards $\kappa - \kappa_{[01]}$. The reduced errors

$$e_\kappa^{[i]} = \frac{|\kappa - (\kappa_{[s]}^{[i]} + \kappa_{[01]})|}{\kappa}, \qquad i = 1, 2, 3, 4, \tag{6.3.8}$$

and the spectral radius of the projected iteration matrix are presented in Table 7 for mixtures 1, 2, and 3. Thanks to the accuracy of $\kappa_{[01]}$, we obtain, after one iteration only, an expression within 3E-3 accuracy. We will see in Section 6.3.2 that $\kappa_{[01]}$ actually corresponds to the simplified volume viscosity associated with the matrix $K_{[01]}$.

Table 8. Volume viscosity. Conjugate gradient methods for K with $M = \mathrm{diag}(K)$. Reduced errors for various mixtures.

	Mixture 1	Mixture 2	Mixture 3
1	3.05E-2	2.27E-2	4.06E-2
2	1.17E-3	4.26E-5	1.59E-3
3	3.53E-5	8.20E-7	8.29E-5
4	6.80E-7	3.67E-9	4.65E-6

Table 9. Volume viscosity. Conjugate gradient methods for $K_{[s]}$ with $M_{[s]} = db(K^{1010})$. Reduced errors for various mixtures.

	Mixture 1	Mixture 2	Mixture 3
1	3.94E-4	2.94E-6	1.03E-4
2	1.78E-6	6.41E-8	6.85E-7
3	1.44E-8	1.15E-12	1.25E-9
4	1.29E-11	—	4.21E-12

Conjugate Gradient Methods. We consider the initialization $x_0 = 0$ and the preconditioner $M = \mathrm{diag}(K)$, for which Theorem 5.3.2 applies when $n \geq 2$. From the results presented in Table 8, we can see that after one iteration, the volume viscosity given by

$$\kappa^{[1]} = \frac{\langle \beta^\kappa, \mathrm{diag}(K)^{-1}\beta^\kappa \rangle^2}{\langle \mathrm{diag}(K)^{-1}\beta^\kappa, K\mathrm{diag}(K)^{-1}\beta^\kappa \rangle}, \tag{6.3.9}$$

is within 4E-2 accuracy. Similarly, we can also consider the preconditioner $M = db(K)$ which yields faster convergence rates, and in this case, the first iterate is given by

$$\kappa^{[1]} = \frac{\langle \beta^\kappa, db(K)^{-1}\beta^\kappa \rangle^2}{\langle db(K)^{-1}\beta^\kappa, K db(K)^{-1}\beta^\kappa \rangle}, \tag{6.3.10}$$

and is of similar accuracy to (6.3.9).

We also investigate conjugate gradient methods for the Schur complement $K_{[s]}$ given by (6.3.6). The numerical results are presented in Table 9 for the initialization $x_0 = 0$ and the preconditioner $M_{[s]} = db(K^{1010})$, for which the results of Section 5.3.6 apply. Because of the accuracy of $\kappa_{[01]}$, the first iterate already yields a reduced error lower than 4E-4.

Table 10. Volume viscosity. Reduced errors for various simplified formulations.

	Mixture 1	Mixture 2	Mixture 3
$e_{[01]}$	2.14E-2	1.26E-2	3.11E-2
$e_{[10]}$	7.27E-1	8.18E-1	4.45E-1
$e_{[d]}$	6.69E-1	8.20E-1	4.22E-1

6.3.2 Numerical Experiments for the System $K_{[01]}\alpha^{\kappa}_{[01]} = \beta^{\kappa}_{[01]}$

In Table 10, we present the reduced errors

$$e_{[01]} = \frac{|\kappa - \kappa_{[01]}|}{\kappa}, \qquad e_{[10]} = \frac{|\kappa - \kappa_{[10]}|}{\kappa}, \qquad e_{[d]} = \frac{|\kappa - \kappa_{[d]}|}{\kappa}, \tag{6.3.11}$$

obtained with the simplified volume viscosities $\kappa_{[01]}$, $\kappa_{[10]}$, and $\kappa_{[d]}$ introduced in Section 2.5. Only the volume viscosity $\kappa_{[01]}$ is found to be accurate, and therefore, we will only consider the system $K_{[01]}\alpha^{\kappa}_{[01]} = \beta^{\kappa}_{[01]}$ in our numerical experiments. Notice that the corresponding simplified volume viscosity is exactly the quantity $\kappa_{[01]} = \langle \beta^{01\kappa}, (K^{0101})^{-1}\beta^{01\kappa}\rangle$ that was used for the Schur complement $K_{[s]}$ in the previous section.

In our numerical applications [MM62] [MPM65], the matrix $K_{[01]}$ is diagonal so that the splitting induced by $M = db(K_{[01]})$ yields a one-step convergence for standard iterative methods and, similarly, the preconditioner $M = db(K_{[01]})$ yields a one-step convergence for conjugate gradient methods.

The simplified formulation $\kappa_{[01]}$ can also be used to obtain an accurate initialization for iterative methods applied to the system $K\alpha^{\kappa} = \beta^{\kappa}$, by taking

$$\begin{cases} x_0^{10\kappa} = 0, \\ x_0^{01\kappa} = (K^{0101})^{-1}\beta^{01\kappa}. \end{cases} \tag{6.3.12}$$

6.3.3 Numerical Experiments for the System $\widetilde{K}\alpha^{\kappa} = \widetilde{\beta}^{\kappa}$

We consider the initialization $x_0 = 0$ and the splitting $\widetilde{K} = \widetilde{M} - \widetilde{Z}$, where the matrix $\widetilde{M}$ is given by

$$\begin{cases} \widetilde{M}^{rs}_{kl} = \dfrac{\widetilde{K}^{rs}_{kk}}{1 - c^{\mathrm{tr}}_{\mathrm{v}}/c_{\mathrm{v}}X_k}\delta_{kl}, \qquad rs = 1010, \qquad k,l \in \mathcal{S}, \\ \widetilde{M}^{rs} = \mathrm{diag}(\widetilde{K}^{rs}), \qquad rs \neq 1010. \end{cases} \tag{6.3.13}$$

Table 11. Volume viscosity. Standard iterative methods for $\widetilde{K}$ with $\widetilde{M}$ given by (6.3.13). Reduced errors and spectral radius for various mixtures.

	Mixture 4	Mixture 5	Mixture 6
1	4.48E-3	1.17E-3	4.64E-2
2	7.54E-3	2.15E-3	2.99E-2
3	7.71E-4	9.37E-4	3.77E-3
4	6.99E-4	5.44E-4	1.76E-3
ρ	3.69E-1	5.56E-1	2.56E-1

One can easily verify that the splitting (6.3.13) is of the form

$$\widetilde{M} = db(\widetilde{K}) + \mathrm{diag}(\widetilde{\mathfrak{d}}), \tag{6.3.14}$$

where the matrix $db(\widetilde{K})$ is formed by the diagonal of the four blocks of the matrix $\widetilde{K}$ and $\widetilde{\mathfrak{d}}^{10+} > 0$, so that Theorem 5.3.7 applies. The reduced errors

$$e_\kappa^{[i]} = \frac{|\kappa - \kappa^{[i]}|}{\kappa}, \qquad i = 1, 2, 3, 4, \tag{6.3.15}$$

and the spectral radius of the projected iteration matrix $\rho = \rho(P_{\mathcal{K}^\perp,\mathbb{R}\mathcal{V}}\widetilde{M}^{-1}\widetilde{Z})$ are presented in Table 11. In particular, the first iterate given by

$$\kappa = \langle \beta^\kappa, \widetilde{M}^{-1}\widetilde{\beta}^\kappa \rangle. \tag{6.3.16}$$

yields an approximation within 5E-2 accuracy.

We now investigate standard iterative methods for the rescaled Schur complement

$$\widetilde{K}_{[s]} = \widetilde{K}^{1010} - \widetilde{K}^{1001}\big(\widetilde{K}^{0101}\big)^{-1}\widetilde{K}^{0110}, \tag{6.3.17}$$

introduced in Section 5.3.6. We consider the initialization $x_0 = 0$ and the splitting $\widetilde{K}_{[s]} = \widetilde{M}_{[s]} - \widetilde{Z}_{[s]}$, where the matrix $\widetilde{M}_{[s]}$ is diagonal and given by

$$\widetilde{M}_{[s]} = db(\widetilde{K}^{1010}). \tag{6.3.18}$$

As stated in Section 5.3.6, the corresponding iteration matrix is convergent, and the iterates $\kappa_{[s]}^{[i]}$ converge towards $\kappa - \kappa_{[01]}$. The reduced errors

$$e_\kappa^{[i]} = \frac{|\kappa - (\kappa_{[s]}^{[i]} + \kappa_{[01]})|}{\kappa}, \qquad i = 1, 2, 3, 4, \tag{6.3.19}$$

Table 12. Volume viscosity. Standard iterative methods for $\widetilde{K}_{[s]}$ with $\widetilde{M}_{[s]} = db(\widetilde{K}^{1010})$. Reduced errors and spectral radius for various mixtures.

	Mixture 4	Mixture 5	Mixture 6
1	2.30E-3	5.99E-3	1.61E-3
2	4.22E-4	3.91E-3	1.79E-4
3	7.91E-5	2.56E-3	2.24E-5
4	1.57E-5	1.67E-3	3.33E-6
ρ	5.19E-1	6.53E-1	1.73E-1

Table 13. Volume viscosity. Conjugate gradient methods for $\widehat{K}$ with $\widehat{M} = \mathrm{diag}(\widehat{K})$. Reduced errors for various mixtures.

	Mixture 4	Mixture 5	Mixture 6
1	2.27E-2	1.64E-2	3.99E-2
2	4.18E-5	3.84E-4	1.58E-3
3	7.96E-7	3.70E-17	8.13E-5
4	2.76E-10	—	4.42E-6

and the spectral radius of the projected iteration matrix are presented in Table 12 for mixtures 4, 5, and 6. Thanks to the accuracy of $\kappa_{[01]}$, we obtain, after one iteration only, an expression within 2E-3 accuracy. We will see in Section 6.3.5 that $\kappa_{[01]}$ actually corresponds to the simplified volume viscosity associated with the matrix $\widetilde{K}_{[01]}$.

6.3.4 Numerical Experiments for the System $\widehat{K}\widehat{\alpha}^{\kappa} = \widehat{\beta}^{\kappa}$

We consider the initialization $\widehat{x}_0 = 0$ and the preconditioner $\widehat{M} = \mathrm{diag}(\widehat{K})$, for which Theorem 5.3.9 applies when $n^+ \geq 2$. The reduced errors

$$e_{\kappa}^{[i]} = \frac{|\kappa - \kappa^{[i]}|}{\kappa}, \qquad i = 1, 2, 3, 4, \tag{6.3.20}$$

are reported in Table 13 and indicate high convergence rates. In particular, the first iterate is within 4E-2 accuracy and coincides with (6.3.9) for positive mass fractions.

Table 14. Volume viscosity. Reduced errors for $\kappa_{[01]}$.

	Mixture 4	Mixture 5	Mixture 6
$e_{[01]}$	1.26E-2	9.17E-3	2.97E-2

6.3.5 Numerical Experiments for the System $\widetilde{K}_{[01]}\alpha^{\kappa}_{[01]} = \widetilde{\beta}^{\kappa}_{[01]}$

In this section we only consider the simplified volume viscosity $\kappa_{[01]}$ since we have seen in Section 6.3.2 that the simplified volume viscosities $\kappa_{[10]}$ and $\kappa_{[d]}$ yield poor approximations for κ. The reduced errors

$$e_{[01]} = \frac{|\kappa - \kappa_{[01]}|}{\kappa}, \tag{6.3.21}$$

are presented in Table 14 for mixtures 4, 5, and 6, and indicate that the simplified volume viscosity $\kappa_{[01]}$ is within 3E-2 accuracy. Notice that $\kappa_{[01]}$ exactly corresponds to the quantity $\kappa_{[01]} = \langle \beta^{01\kappa}, (\widetilde{K}^{0101})^{-1}\widetilde{\beta}^{01\kappa}\rangle$ used for the Schur complement $\widetilde{K}_{[s]}$ in Section 6.3.3.

In our numerical applications [MM62] [MPM65], the matrix $\widetilde{K}_{[01]}$ is diagonal so that the splitting induced by $\widetilde{M} = db(\widetilde{K}_{[01]})$ yields a one-step convergence for standard iterative methods.

6.3.6 Mixture-Averaged Formulas

As described in Section 6.1.4, we now consider several mixture-averaged formulas for the volume viscosity. These expressions only depend on the polyatomic species volume viscosities κ_k, $k \in \mathcal{P}$, since the pure species volume viscosities of all the monatomic species vanish. Assuming that $p \geq 1$ and rescaling the mole fractions by the quantity

$$\sigma_{\mathcal{P}} = \frac{1}{\sum_{k\in\mathcal{P}} X_k}, \tag{6.3.22}$$

the average formulas of order t then become

$$\mathcal{M}_t(\kappa) = \Big(\sum_{k\in\mathcal{P}} \sigma_{\mathcal{P}} X_k (\kappa_k)^t\Big)^{1/t}, \tag{6.3.23}$$

for $t \neq 0$, and for $t = 0$,

$$\mathcal{M}_0(\kappa) = \exp\Big(\sum_{k\in\mathcal{P}} \sigma_{\mathcal{P}} X_k \log(\kappa_k)\Big). \tag{6.3.24}$$

Table 15. Volume viscosity. Reduced errors for various mixture-averaged formulas.

	Mixture 1	Mixture 2	Mixture 3	Mixture 4	Mixture 5	Mixture 6
1/2	2.18E-1	2.35E-1	2.10E-1	2.35E-2	1.32E-1	2.22E-1
2/3	6.13E-2	1.05E-1	1.07E-1	1.05E-1	2.53E-2	1.16E-1
3/4	2.85E-2	3.66E-2	4.41E-2	3.67E-2	2.64E-2	5.13E-2
1	3.36E-1	1.69E-1	1.94E-1	1.69E-1	1.71E-1	1.93E-1

In Table 15 we present the reduced errors

$$e_t = \frac{|\kappa - \mathcal{M}_t(\kappa)|}{\kappa}, \tag{6.3.25}$$

for various values of the parameter t and for the six mixtures considered in our numerical experiments. From this table we can see that the mixture-averaged formula of order 3/4 yields an approximation for the volume viscosity within 5E-2 accuracy.

6.3.7 Discussion

From the preceding numerical experiments, we can draw the following conclusions. For the volume viscosity, a new empirical average formula is given by the average formula of order 3/4, that is,

$$\kappa = \Big(\sum_{k\in\mathcal{P}} \sigma_{\mathcal{P}} X_k (\kappa_k)^{3/4}\Big)^{4/3}, \tag{6.3.26}$$

which only requires $\mathcal{O}(n)$ operations for its evaluation and is accurate to 5E-2 for the mixtures considered in this book. The quantity $\sigma_{\mathcal{P}}$ is given by $\sigma_{\mathcal{P}} = 1/\sum_{k\in\mathcal{P}} X_k$, and the pure species volume viscosities κ_k, $k \in \mathcal{P}$, are given by

$$\kappa_k = \frac{k_{\mathrm{B}}\pi}{4}\Big(\frac{c_k^{\mathrm{int}}}{c_{\mathrm{v},k}}\Big)^2 \frac{\xi_k^{\mathrm{int}}}{c_k^{\mathrm{int}}}\eta_k, \qquad k \in \mathcal{P}, \tag{6.3.27}$$

where $c_{\mathrm{v},k} = c_{\mathrm{v}}^{\mathrm{tr}} + c_k^{\mathrm{int}}$ is the constant volume heat capacity of the molecules of the kth species.

On the other hand, a new rigorously derived analytic expression is given by the simplified volume viscosity $\kappa_{[01]}$

$$\kappa_{[01]} = \sum_{k\in\mathcal{P}} \Big(\frac{c_{\mathrm{v},k}}{c_{\mathrm{v}}}\Big)^2 \frac{\Omega_{kk}^{22} X_k}{\sum_{l\in\mathcal{S}} X_l \Omega_{kl}^{22}} \kappa_k, \tag{6.3.28}$$

which requires $\mathcal{O}(n^2)$ operations for its evaluation and is accurate to 3E-2. On the other hand, the simplified volume viscosities $\kappa_{[10]}$ and $\kappa_{[d]}$ are not accurate for the mixtures considered in this book.

Finally, a detailed analytic expression is obtained after one standard iteration for the system $K_{[s]}\alpha^{\kappa}_{[s]} = \beta^{\kappa}_{[s]}$. This yields the new expression

$$\kappa^{[1]} = \sum_{k\in\mathcal{S}} \frac{\left(\beta^{10\kappa}_{[s]k}\right)^2}{K^{1010}_{kk}} + \kappa_{[01]}, \tag{6.3.29}$$

which requires $\mathcal{O}(n^2)$ operations and is accurate to 3E-3. On the other hand, two conjugate gradient iterations for the system $K\alpha^{\kappa} = \beta^{\kappa}$, preconditioned by diag(K), yield, in $\mathcal{O}(n^2)$ operations, an expression within 2E-3 accuracy.

6.4 The Diffusion Matrix and the Flux Diffusion Matrix

6.4.1 Numerical Experiments for the System $L\alpha^{D_l} = \beta^{D_l}$

Standard Iterative Methods. We consider the initializations $x_0^k = 0$, $k \in \mathcal{S}$, and the splitting $L = M - Z$, where the matrix M is given by

$$\begin{cases} M^{rs}_{kl} = \dfrac{L^{rs}_{kk}}{1-Y_k}\delta_{kl}, & rs = 0000, \quad k,l \in \mathcal{S}, \\ M^{rs} = \mathrm{diag}(L^{rs}), & rs \neq 0000. \end{cases} \tag{6.4.1}$$

This splitting can be derived by identifying some of the diagonal elements in the relation $L \simeq MP_{\mathcal{L}^\perp,\mathbb{R}\mathcal{U}}$, where $P_{\mathcal{L}^\perp,\mathbb{R}\mathcal{U}}$ is the oblique projector onto $\mathcal{L}^\perp$ along $\mathbb{R}\mathcal{U}$. Note also that the matrix (6.4.1) generalizes the one considered in [Gi91] for the first-order diffusion matrix $D_{[00]}$. Furthermore, one can easily verify that the matrix M is of the form

$$M = db(L) + \mathrm{diag}(\mathfrak{d}), \tag{6.4.2}$$

where the matrix $db(L)$ is formed by the diagonal of the nine blocks of the matrix L and $\mathfrak{d}^{00} > 0$, so that Theorem 5.4.1 applies. The reduced errors

$$e_D^{[i]} = \frac{||D - D^{[i]}||_\infty}{||D||_\infty}, \qquad i = 1,2,3,4, \tag{6.4.3}$$

where $||D||_\infty = \max_{k,l\in\mathcal{S}} |D_{kl}|$ and the spectral radius of the projected iteration matrix $\rho = \rho(P_{\mathcal{L}^\perp,\mathbb{R}\mathcal{U}}M^{-1}Z)$ are presented in Table 16. After one iteration only, we obtain an

Table 16. Diffusion matrix. Standard iterative methods for L with M given by (6.4.1). Reduced errors and spectral radius for various mixtures.

	Mixture 1	Mixture 2	Mixture 3
1	3.03E-2	1.03E-5	8.19E-3
2	2.16E-3	3.47E-5	2.22E-4
3	1.50E-4	7.26E-8	1.40E-5
4	1.99E-5	1.06E-8	3.06E-6
ρ	2.53E-1	1.75E-1	2.74E-1

expression within 3E-2 accuracy, and an additional iteration yields 2E-3 accuracy. The corresponding diffusion matrices are given by

$$\begin{cases} D^{[1]} = P_{Y^\perp,\mathbb{R}U}\mathfrak{P}M^{-1}\mathfrak{P}^t P_{U^\perp,\mathbb{R}Y}, \\ D^{[2]} = P_{Y^\perp,\mathbb{R}U}\mathfrak{P}M^{-1}(2M-L)M^{-1}\mathfrak{P}^t P_{U^\perp,\mathbb{R}Y}, \end{cases} \tag{6.4.4}$$

where $\mathfrak{P} \in \mathbb{R}^{n,2n+p}$ denotes the block-rectangular matrix $\mathfrak{P} = [I,0,0]$. The diffusion velocities $V^{[1]} = -D^{[1]}d$, where d is the diffusion driving force vector, take on the same form as the Hirschfelder-Curtiss approximate diffusion velocities [HC49] with a species independent mass correction velocity [CHe81] [KWM83]. However, second-order effects and the internal energy of the molecules are also taken into account in $V^{[1]}$, as opposed to the Hirschfelder-Curtiss expressions.

Remark. The simpler splitting $L = db(L) - Z$ yields a convergent iteration matrix for $n \geq 3$ only. Our numerical experiments indicate that this splitting yields iteration matrices with higher spectral radius and slower convergence rates for mixture 2.

We now investigate standard iterative methods for the Schur complement

$$L_{[s]} = \begin{bmatrix} L^{0000} & L^{0010} \\ L^{1000} & L^{1010} \end{bmatrix} - \begin{bmatrix} L^{0001} \\ L^{1001} \end{bmatrix} \left(L^{0101}\right)^{-1} [L^{0100}, L^{0110}], \tag{6.4.5}$$

introduced in Section 5.4.8. We consider the initializations $x_0^k = 0$, $k \in \mathcal{S}$, and the splitting $L_{[s]} = M_{[s]} - Z_{[s]}$, where the matrix $M_{[s]}$ consists of four diagonal blocks and given by

$$\begin{cases} M^{rs}_{[s]kl} = \dfrac{L^{rs}_{kk}}{1-Y_k}\delta_{kl}, & rs = 0000, \quad k,l \in \mathcal{S}, \\ M^{rs}_{[s]} = \operatorname{diag}(L^{rs}), & rs = 1000, 0010, 1010. \end{cases} \tag{6.4.6}$$

Table 17. Diffusion matrix. Standard iterative methods for $L_{[s]}$ with $M_{[s]}$ given by (6.4.6). Reduced errors and spectral radius for various mixtures.

	Mixture 1	Mixture 2	Mixture 3
1	3.03E-2	3.47E-5	8.19E-3
2	2.15E-3	1.54E-6	2.20E-4
3	1.46E-4	7.29E-8	1.32E-5
4	1.85E-5	1.05E-8	2.80E-6
ρ	2.46E-1	1.75E-1	2.68E-1

Table 18. Diffusion matrix. Conjugate gradient methods for L with $M = db(L)$. Reduced errors for various mixtures.

	Mixture 1	Mixture 2	Mixture 3
1	2.63E-2	7.16E-6	7.89E-3
2	1.19E-3	1.57E-6	4.93E-4
3	4.43E-5	6.16E-7	1.36E-5
4	2.68E-6	1.23E-7	2.92E-7

As stated in Section 5.4.8, the corresponding iteration matrix is convergent, and the matrices $D_{[s]}^{[i]}$ converge towards D. The reduced errors

$$e_D^{[i]} = \frac{||D - D_{[s]}^{[i]}||_\infty}{||D||_\infty}, \qquad i = 1, 2, 3, 4, \tag{6.4.7}$$

and the spectral radius of the projected iteration matrix are presented in Table 17 for mixtures 1, 2, and 3. Very similar results to the ones obtained with the matrix L are observed. In particular, after one and two iterations, the expressions for the diffusion matrix are within 3E-2 and 2E-3 accuracy, respectively, and take on a form similar to (6.4.4).

Conjugate Gradient Methods. We consider the initializations $x_0^k = 0$, $k \in S$, and the preconditioner $M = db(L)$, for which Theorem 5.4.2 applies when $n \geq 2$. The numerical results are given in Table 18 and indicate high convergence rates. We point out, however, that the iterates are generally neither symmetric nor positive definite on the physical hyperplane of zero sum gradients.

We also consider conjugate gradient methods for the Schur complement $L_{[s]}$ given by (6.4.5). The numerical results are presented in Table 19 with the initializations

Table 19. Diffusion matrix. Conjugate gradient methods for $L_{[s]}$ with $M_{[s]}$ given by (6.4.8). Reduced errors for various mixtures.

	Mixture 1	Mixture 2	Mixture 3
1	2.63E-2	7.16E-6	7.89E-3
2	1.10E-3	1.57E-6	4.60E-4
3	4.28E-5	6.14E-7	1.34E-5
4	2.58E-6	1.17E-7	2.51E-7

Table 20. Diffusion matrix. Reduced errors for $D_{[e]}$.

	Mixture 1	Mixture 2	Mixture 3
$e_{D_{[e]}}$	4.75E-3	6.48E-4	5.45E-3

$x_0^k = 0,\ k \in \mathcal{S}$, and the preconditioner

$$M_{[s]} = \begin{bmatrix} \operatorname{diag}(L^{0000}) & \operatorname{diag}(L^{0010}) \\ \operatorname{diag}(L^{1000}) & \operatorname{diag}(L^{1010}) \end{bmatrix}, \tag{6.4.8}$$

for which the results of Section 5.4.8 apply when $n \geq 2$. Similar convergence rates to the ones obtained with the matrix L are observed.

6.4.2 Numerical Experiments for the System $L_{[e]}\alpha_{[e]}^{D_l} = \beta_{[e]}^{D_l}$

In Table 20 we present the reduced errors

$$e_{D_{[e]}} = \frac{||D - D_{[e]}||_\infty}{||D||_\infty}, \tag{6.4.9}$$

which indicate that the simplified diffusion matrix $D_{[e]}$ is within 5E-3 accuracy for mixtures 1, 2, and 3. This new formulation is therefore an attractive alternative to the full system $L\alpha^{D_l} = \beta^{D_l}$.

Standard Iterative Methods. We consider the initializations $x_0^k = 0,\ k \in \mathcal{S}$, and the splitting $L_{[e]} = M_{[e]} - Z_{[e]}$, where the matrix $M_{[e]}$ is given by

$$\begin{cases} M_{[e]kl}^{rs} = \dfrac{L_{[e]kk}^{rs}}{1 - Y_k}\delta_{kl}, & rs = 0000, \quad k,l \in \mathcal{S}, \\ M_{[e]}^{rs} = \operatorname{diag}(L_{[e]}^{rs}), & rs \neq 0000. \end{cases} \tag{6.4.10}$$

Table 21. Diffusion matrix. Standard iterative methods for $L_{[e]}$ with $M_{[e]}$ given by (6.4.10). Reduced errors and spectral radius for various mixtures.

	Mixture 1	Mixture 2	Mixture 3
1	2.98E-2	1.01E-5	8.07E-3
2	2.11E-3	1.50E-6	2.13E-4
3	1.36E-4	7.58E-8	8.51E-6
4	1.55E-5	9.97E-9	1.21E-6
ρ	2.06E-1	1.38E-1	1.78E-1

This splitting can be derived by identifying some of the diagonal elements in the relation $L_{[e]} \simeq M_{[e]} P_{\mathcal{L}_{[e]}^{\perp}, \mathbb{R}\mathcal{U}_{[e]}}$, where $P_{\mathcal{L}_{[e]}^{\perp}, \mathbb{R}\mathcal{U}_{[e]}}$ is the oblique projector onto $\mathcal{L}_{[e]}^{\perp}$ along $\mathbb{R}\mathcal{U}_{[e]}$. One can easily verify that the matrix $M_{[e]}$ is of the form

$$M_{[e]} = db(L_{[e]}) + \mathrm{diag}(\mathfrak{d}), \tag{6.4.11}$$

where the matrix $db(L_{[e]})$ is formed by the diagonal of the four blocks of the matrix $L_{[e]}$ and $\mathfrak{d}^{00} > 0$, so that Theorem 5.4.4 applies. The reduced errors

$$e_{D_{[e]}}^{[i]} = \frac{||D_{[e]} - D_{[e]}^{[i]}||_\infty}{||D_{[e]}||_\infty}, \qquad i = 1, 2, 3, 4, \tag{6.4.12}$$

and the spectral radius of the projected iteration matrix $\rho = \rho(P_{\mathcal{L}_{[e]}^{\perp}, \mathbb{R}\mathcal{U}_{[e]}} M_{[e]}^{-1} Z_{[e]})$ are presented in Table 21. After one iteration only, we obtain an approximation within 3E-2 accuracy of $D_{[e]}$, and an additional iteration yields 2E-3 accuracy. The corresponding diffusion matrices are given by

$$\begin{cases} D_{[e]}^{[1]} = P_{Y^{\perp}, \mathbb{R}U} \mathfrak{P}_{[e]} M_{[e]}^{-1} \mathfrak{P}_{[e]}^t P_{U^{\perp}, \mathbb{R}Y}, \\ D_{[e]}^{[2]} = P_{Y^{\perp}, \mathbb{R}U} \mathfrak{P}_{[e]} M_{[e]}^{-1} (2M_{[e]} - L_{[e]}) M_{[e]}^{-1} \mathfrak{P}_{[e]}^t P_{U^{\perp}, \mathbb{R}Y}, \end{cases} \tag{6.4.13}$$

where $\mathfrak{P}_{[e]} \in \mathbb{R}^{n,2n}$ denotes the block-rectangular matrix $\mathfrak{P}_{[e]} = [I, 0]$. The diffusion velocities $V_{[e]}^{[1]} = -D_{[e]}^{[1]} d$ take on the same form as the Hirschfelder-Curtiss diffusion velocities [HC49] with a species independent mass correction velocity [CH83][KWM83], but also incorporate second-order effects and the internal energy of the molecules.

Conjugate Gradient Methods. We consider the initializations $x_0^k = 0$, $k \in S$, and the preconditioner $M_{[e]} = db(L_{[e]})$, for which Theorem 5.4.5 applies when $n \geq 2$. The

Table 22. Diffusion matrix. Conjugate gradient methods for $L_{[e]}$ with $M_{[e]} = db(L_{[e]})$. Reduced errors for various mixtures.

	Mixture 1	Mixture 2	Mixture 3
1	2.59E-2	7.16E-6	7.77E-3
2	8.13E-4	1.57E-6	2.47E-4
3	3.64E-5	6.12E-7	9.79E-6
4	1.77E-6	9.33E-8	1.21E-7

Table 23. Diffusion matrix. Reduced errors for $D_{[00]}$.

	Mixture 1	Mixture 2	Mixture 3
$e_{D_{[00]}}$	2.95E-2	3.00E-2	2.86E-2

numerical results are given in Table 22 and indicate high convergence rates. We point out, however, that the iterates are generally neither symmetric nor positive definite on the physical hyperplane of zero sum gradients.

6.4.3 Numerical Experiments for the System $L_{[00]}\alpha_{[00]}^{D_l} = \beta_{[00]}^{D_l}$

In Table 23 we present the reduced errors

$$e_{D_{[00]}} = \frac{||D - D_{[00]}||_\infty}{||D||_\infty}, \tag{6.4.14}$$

which indicate that the first-order diffusion matrix $D_{[00]}$ is within 3E-2 accuracy for mixtures 1, 2, and 3.

Standard Iterative Methods. We consider the initializations $x_0^k = 0$, $k \in \mathcal{S}$, and the splitting $L_{[00]} = M_{[00]} - Z_{[00]}$, where the matrix $M_{[00]}$ is given by

$$M_{[00]kl}^{0000} = \frac{L_{[00]kk}^{0000}}{1 - Y_k}\delta_{kl}, \qquad k, l \in \mathcal{S}. \tag{6.4.15}$$

This splitting has already been considered in [Gi91]. In particular, it can be derived by identifying some of the diagonal elements in the relation $L_{[00]} \simeq M_{[00]}P_{Y^\perp,\mathbb{R}U}$, where $P_{Y^\perp,\mathbb{R}U}$ is the oblique projector onto $Y^\perp$ along $\mathbb{R}U$. One can easily verify that the matrix $M_{[00]}$ is of the form

$$M_{[00]} = db(L_{[00]}) + \mathrm{diag}(\mathfrak{d}), \tag{6.4.16}$$

Table 24. Diffusion matrix. Standard iterative methods for $L_{[00]}$ with $M_{[00]}$ given by (6.4.15). Reduced errors and spectral radius for various mixtures.

	Mixture 1	Mixture 2	Mixture 3
1	2.92E-2	9.92E-6	7.87E-3
2	1.88E-3	1.39E-6	1.91E-4
3	1.01E-4	8.52E-8	6.22E-6
4	6.67E-6	9.06E-9	2.04E-7
ρ	6.44E-2	8.17E-2	3.33E-2

where the matrix $db(L_{[00]})$ is formed by the diagonal of the matrix $L_{[00]}$ and $\mathfrak{d} > 0$, so that Theorem 5.4.7 applies. The reduced errors

$$e^{[i]}_{D_{[00]}} = \frac{||D_{[00]} - D^{[i]}_{[00]}||_\infty}{||D_{[00]}||_\infty}, \qquad i = 1, 2, 3, 4, \tag{6.4.17}$$

and the spectral radius of the projected iteration matrix $\rho = \rho(P_{Y^\perp,\mathbb{R}U} M^{-1}_{[00]} Z_{[00]})$ are presented in Table 24. After one iteration only, we obtain an approximation within 3E-2 accuracy of $D_{[00]}$, and an additional iteration yields 2E-3 accuracy. The corresponding diffusion matrices are given by

$$\begin{cases} D^{[1]}_{[00]} = P_{Y^\perp,\mathbb{R}U} M^{-1}_{[00]} P_{U^\perp,\mathbb{R}Y}, \\ D^{[2]}_{[00]} = P_{Y^\perp,\mathbb{R}U} M^{-1}_{[00]} (2M_{[00]} - L_{[00]}) M^{-1}_{[00]} P_{U^\perp,\mathbb{R}Y}. \end{cases} \tag{6.4.18}$$

The diffusion velocities $V^{[1]}_{[00]} = -D^{[1]}_{[00]} d$ exactly correspond to the Hirschfelder-Curtiss approximate diffusion velocities with a species independent mass corrector, as proven in [Gi91].

Conjugate Gradient Methods. We consider the initializations $x^k_0 = 0$, $k \in S$, and the preconditioner $M_{[00]} = db(L_{[00]})$, for which Theorem 5.4.8 applies when $n \geq 2$. The numerical results are given in Table 25 and indicate high convergence rates. We point out, however, that the iterates are generally neither symmetric nor positive definite on the physical hyperplane of zero sum gradients.

Table 25. Diffusion matrix. Conjugate gradient methods for $L_{[00]}$ with $M_{[00]} = db(L_{[00]})$. Reduced errors for various mixtures.

	Mixture 1	Mixture 2	Mixture 3
1	2.51E-2	7.35E-6	7.56E-3
2	1.86E-4	2.55E-6	4.20E-5
3	3.21E-6	6.46E-7	4.65E-7
4	9.21E-8	2.62E-11	4.93E-9

6.4.4 Numerical Experiments for the System $\widetilde{L}\widetilde{\alpha}^{D_l} = \widetilde{\beta}^{D_l}$

We consider the initializations $x_0^k = 0$, $k \in \mathcal{S}$, and the splitting $\widetilde{L} = \widetilde{M} - \widetilde{Z}$, where the matrix $\widetilde{M}$ is given by

$$\begin{cases} \widetilde{M}_{kl}^{rs} = \dfrac{\widetilde{L}_{kk}^{rs}}{1 - Y_k}\delta_{kl}, & rs = 0000, \quad k, l \in \mathcal{S}, \\ \widetilde{M}^{rs} = \operatorname{diag}(\widetilde{L}^{rs}), & rs \neq 0000. \end{cases} \tag{6.4.19}$$

This matrix generalizes the one considered in [Gi91] for the first-order flux diffusion matrix $\widetilde{D}_{[00]}$. Furthermore, one can easily verify that the matrix $\widetilde{M}$ is of the form

$$\widetilde{M} = db(\widetilde{L}) + \operatorname{diag}(\widetilde{\mathfrak{d}}), \tag{6.4.20}$$

where the matrix $db(\widetilde{L})$ is formed by the diagonal of the nine blocks of the matrix $\widetilde{L}$ and $\widetilde{\mathfrak{d}}^{00+} > 0$, so that Theorem 5.4.10 applies. The reduced errors

$$e_{\widetilde{D}}^{[i]} = \frac{||\widetilde{D} - \widetilde{D}^{[i]}||_\infty}{||\widetilde{D}||_\infty}, \qquad i = 1, 2, 3, 4, \tag{6.4.21}$$

and the spectral radius of the projected iteration matrix $\rho = \rho(P_{\mathcal{L}^\perp, \mathbb{R}\mathcal{U}}\widetilde{M}^{-1}\widetilde{Z})$ are presented in Table 26. After one iteration, we obtain an expression for the flux diffusion matrix which is within 9E-2 accuracy, and an additional iteration yields 8E-3 accuracy. More explicitly, we have

$$\begin{cases} \widetilde{D}^{[1]} = P_{U^\perp, \mathbb{R}Y}\mathcal{W}\mathfrak{P}\left[\widetilde{M}^{-1}\right]^t\mathfrak{P}^t P_{U^\perp, \mathbb{R}Y}, \\ \widetilde{D}^{[2]} = P_{U^\perp, \mathbb{R}Y}\mathcal{W}\mathfrak{P}\left[\widetilde{M}^{-1}(2\widetilde{M} - \widetilde{L})\widetilde{M}^{-1}\right]^t\mathfrak{P}^t P_{U^\perp, \mathbb{R}Y}, \end{cases} \tag{6.4.22}$$

where $\mathcal{W} = \operatorname{diag}\big((m_k/m)_{k\in\mathcal{S}}\big)$ and $\mathfrak{P} \in \mathbb{R}^{n, 2n+p}$ denotes the block-rectangular matrix $\mathfrak{P} = [I, 0, 0]$.

Table 26. Flux diffusion matrix. Standard iterative methods for $\widetilde{L}$ with $\widetilde{M}$ given by (6.4.19). Reduced errors and spectral radius for various mixtures.

	Mixture 4	Mixture 5	Mixture 6
1	5.31E-2	9.22E-2	1.37E-2
2	8.10E-3	2.92E-4	3.37E-4
3	4.31E-4	1.08E-5	2.34E-5
4	6.18E-5	3.67E-7	5.35E-6
ρ	1.75E-1	4.81E-2	2.75E-1

We now investigate standard iterative methods for the rescaled Schur complement

$$\widetilde{L}_{[s]} = \begin{bmatrix} \widetilde{L}^{0000} & \widetilde{L}^{0010} \\ \widetilde{L}^{1000} & \widetilde{L}^{1010} \end{bmatrix} - \begin{bmatrix} \widetilde{L}^{0001} \\ \widetilde{L}^{1001} \end{bmatrix} \left(\widetilde{L}^{0101}\right)^{-1} [\widetilde{L}^{0100}, \widetilde{L}^{0110}], \tag{6.4.23}$$

introduced in Section 5.4.8. We consider the initializations $x_0^k = 0$, $k \in \mathcal{S}$, and the splitting $\widetilde{L}_{[s]} = \widetilde{M}_{[s]} - \widetilde{Z}_{[s]}$, where the matrix $\widetilde{M}_{[s]}$ consists of four diagonal blocks and given by

$$\begin{cases} \widetilde{M}^{rs}_{[s]kl} = \dfrac{\widetilde{L}^{rs}_{kk}}{1-Y_k}\delta_{kl}, & rs = 0000, \quad k,l \in \mathcal{S}, \\ \widetilde{M}^{rs}_{[s]} = \operatorname{diag}(\widetilde{L}^{rs}), & rs = 0010, 1000, 1010. \end{cases} \tag{6.4.24}$$

As stated in Section 5.4.8, the corresponding iteration matrix is convergent, and the matrices $\widetilde{D}^{[i]}_{[s]}$ converge towards $\widetilde{D}$. The reduced errors

$$e^{[i]}_{\widetilde{D}} = \frac{\|\widetilde{D} - \widetilde{D}^{[i]}_{[s]}\|_\infty}{\|\widetilde{D}\|_\infty}, \qquad i = 1,2,3,4, \tag{6.4.25}$$

and the spectral radius of the projected iteration matrix are presented in Table 27 for mixtures 4, 5, and 6. Similar convergence rates to the ones obtained with the matrix $\widetilde{L}$ are observed.

6.4.5 Numerical Experiments for the System $\widehat{L}\widehat{\alpha}^{D_l} = \widehat{\beta}^{D_l}$

We consider the initializations $\widehat{x}_0^k = 0$, $k \in \mathcal{S}$, and the preconditioner $\widehat{M} = db(\widehat{L})$, for which Theorem 5.4.12 applies when $n^+ \geq 2$. Note that the corresponding iterates converge towards the symmetric rescaled diffusion matrix $\widehat{D}$. The reduced errors

$$e^{[i]}_{\widehat{D}} = \frac{\|\widehat{D} - \widehat{D}^{[i]}\|_\infty}{\|\widehat{D}\|_\infty}, \qquad i = 1,2,3,4, \tag{6.4.26}$$

Table 27. Flux diffusion matrix. Standard iterative methods for $\widetilde{L}_{[s]}$ with $\widetilde{M}_{[s]}$ given by (6.4.24). Reduced errors and spectral radius for various mixtures.

	Mixture 4	Mixture 5	Mixture 6
1	5.31E-2	9.22E-2	1.37E-2
2	8.10E-3	2.89E-4	3.35E-4
3	4.31E-4	9.08E-6	2.23E-5
4	6.19E-5	1.80E-7	4.96E-6
ρ	1.75E-1	4.26E-2	2.70E-1

Table 28. Symmetric rescaled diffusion matrix. Conjugate gradient methods for $\widehat{L}$ with $\widehat{M} = db(\widehat{L})$. Reduced errors for various mixtures.

	Mixture 4	Mixture 5	Mixture 6
1	1.22E-2	1.56E-3	1.68E-3
2	1.78E-3	3.46E-6	2.19E-4
3	1.64E-4	2.27E-12	1.86E-5
4	3.09E-6	—	3.52E-7

Table 29. Flux diffusion matrix. Reduced errors for $\widetilde{D}_{[e]}$.

	Mixture 4	Mixture 5	Mixture 6
$e_{\widetilde{D}_{[e]}}$	2.53E-3	2.13E-3	2.07E-3

are reported in Table 28 and indicate high convergence rates.

6.4.6 Numerical Experiments for the System $\widetilde{L}_{[e]}\widetilde{\alpha}_{[e]}^{D_l} = \widetilde{\beta}_{[e]}^{D_l}$

In Table 29 we present the reduced errors

$$e_{\widetilde{D}_{[e]}} = \frac{||\widetilde{D} - \widetilde{D}_{[e]}||_\infty}{||\widetilde{D}||_\infty}, \tag{6.4.27}$$

which indicate that the simplified flux diffusion matrix $\widetilde{D}_{[e]}$ is within 3E-3 accuracy for mixtures 4, 5, and 6. This new formulation is therefore an attractive alternative to the full system $\widetilde{L}\widetilde{\alpha}^{D_l} = \widetilde{\beta}^{D_l}$.

Table 30. Flux diffusion matrix. Standard iterative methods for $\widetilde{L}_{[e]}$ with $\widetilde{M}_{[e]}$ given by (6.4.28). Reduced errors and spectral radius for various mixtures.

	Mixture 4	Mixture 5	Mixture 6
1	5.29E-2	9.16E-2	1.36E-2
2	8.08E-3	2.01E-4	3.18E-4
3	4.09E-4	6.25E-6	1.22E-5
4	5.99E-5	1.17E-7	1.54E-6
ρ	1.38E-1	4.04E-2	1.78E-1

We consider the initializations $x_0^k = 0$, $k \in \mathcal{S}$, and the splitting $\widetilde{L}_{[e]} = \widetilde{M}_{[e]} - \widetilde{Z}_{[e]}$, where the matrix $\widetilde{M}_{[e]}$ is given by

$$\begin{cases} \widetilde{M}^{rs}_{[e]kl} = \dfrac{\widetilde{L}^{rs}_{[e]kk}}{1-Y_k}\delta_{kl}, & rs = 0000, \quad k,l \in \mathcal{S}, \\ \widetilde{M}^{rs}_{[e]} = \operatorname{diag}(\widetilde{L}^{rs}_{[e]}), & rs \neq 0000. \end{cases} \tag{6.4.28}$$

One can easily verify that the matrix $\widetilde{M}_{[e]}$ is of the form

$$\widetilde{M}_{[e]} = db(\widetilde{L}_{[e]}) + \operatorname{diag}(\widetilde{\mathfrak{d}}), \tag{6.4.29}$$

where the matrix $db(\widetilde{L}_{[e]})$ is formed by the diagonal of the four blocks of the matrix $\widetilde{L}_{[e]}$ and $\widetilde{\mathfrak{d}}^{00+} > 0$, so that Theorem 5.4.14 applies. The reduced errors

$$e^{[i]}_{\widetilde{D}_{[e]}} = \frac{\|\widetilde{D}_{[e]} - \widetilde{D}^{[i]}_{[e]}\|_\infty}{\|\widetilde{D}_{[e]}\|_\infty}, \qquad i = 1,2,3,4, \tag{6.4.30}$$

and the spectral radius of the projected iteration matrix $\rho = \rho(P_{\mathcal{L}^\perp_{[e]},\mathbb{R}\mathcal{U}_{[e]}}\widetilde{M}_{[e]}^{-1}\widetilde{Z}_{[e]})$ are presented in Table 30. After one iteration, we obtain an expression for the flux diffusion matrix which is within 9E-2 accuracy of $\widetilde{D}_{[e]}$, and an additional iteration yields 8E-3 accuracy. More explicitly, we have

$$\begin{cases} \widetilde{D}^{[1]}_{[e]} = P_{U^\perp,\mathbb{R}Y}\mathcal{W}\mathfrak{P}_{[e]}\left[\widetilde{M}_{[e]}^{-1}\right]^t\mathfrak{P}^t_{[e]}P_{U^\perp,\mathbb{R}Y}, \\ \widetilde{D}^{[2]}_{[e]} = P_{U^\perp,\mathbb{R}Y}\mathcal{W}\mathfrak{P}_{[e]}\left[\widetilde{M}_{[e]}^{-1}(2\widetilde{M}_{[e]} - \widetilde{L}_{[e]})\widetilde{M}_{[e]}^{-1}\right]^t\mathfrak{P}^t_{[e]}P_{U^\perp,\mathbb{R}Y}, \end{cases} \tag{6.4.31}$$

where $\mathcal{W} = \operatorname{diag}\big((m_k/m)_{k\in\mathcal{S}}\big)$ and $\mathfrak{P}_{[e]} \in \mathbb{R}^{n,2n}$ denotes the block-rectangular matrix $\mathfrak{P}_{[e]} = [I, 0]$.

Table 31. Flux diffusion matrix. Reduced errors for $\widetilde{D}_{[00]}$.

	Mixture 4	Mixture 5	Mixture 6
$e_{\widetilde{D}_{[00]}}$	7.69E-3	6.77E-3	5.28E-3

6.4.7 Numerical Experiments for the System $\widetilde{L}_{[00]}\widetilde{\alpha}^{D_l}_{[00]} = \widetilde{\beta}^{D_l}_{[00]}$

In Table 31 we present the reduced errors

$$e_{\widetilde{D}_{[00]}} = \frac{||\widetilde{D} - \widetilde{D}_{[00]}||_\infty}{||\widetilde{D}||_\infty}, \tag{6.4.32}$$

which indicate that the first-order flux diffusion matrix $\widetilde{D}_{[00]}$ is within 8E-3 accuracy for mixtures 4, 5, and 6.

We consider the initializations $x_0^k = 0$, $k \in \mathcal{S}$, and the splitting $\widetilde{L}_{[00]} = \widetilde{M}_{[00]} - \widetilde{Z}_{[00]}$, where the matrix $\widetilde{M}_{[00]}$ is given by

$$\widetilde{M}^{0000}_{[00]kl} = \frac{\widetilde{L}^{0000}_{[00]kk}}{1 - Y_k}\delta_{kl}, \qquad k, l \in \mathcal{S}. \tag{6.4.33}$$

This splitting has already been considered in [Gi91], and one can easily verify that the matrix $\widetilde{M}_{[00]}$ is of the form

$$\widetilde{M}_{[00]} = db(\widetilde{L}_{[00]}) + \mathrm{diag}(\widetilde{\mathfrak{d}}), \tag{6.4.34}$$

where the matrix $db(\widetilde{L}_{[00]})$ is formed by the diagonal of the matrix $\widetilde{L}_{[00]}$ and $\widetilde{\mathfrak{d}}^{00+} > 0$, so that Theorem 5.4.16 applies. The reduced errors

$$e^{[i]}_{\widetilde{D}_{[00]}} = \frac{||\widetilde{D}_{[00]} - \widetilde{D}^{[i]}_{[00]}||_\infty}{||\widetilde{D}_{[00]}||_\infty}, \qquad i = 1, 2, 3, 4, \tag{6.4.35}$$

and the spectral radius of the projected iteration matrix $\rho = \rho(P_{Y^\perp,\mathbb{R}U}\widetilde{M}^{-1}_{[00]}\widetilde{Z}_{[00]})$ are presented in Table 32. After one iteration, we obtain an expression for the flux diffusion matrix which is within 9E-2 accuracy of $\widetilde{D}_{[00]}$, and an additional iteration yields 8E-3 accuracy. More explicitly, we have

$$\begin{cases} \widetilde{D}^{[1]}_{[00]} = P_{U^\perp,\mathbb{R}Y}\mathcal{W}\Big[\widetilde{M}^{-1}_{[00]}\Big]^t P_{U^\perp,\mathbb{R}Y}, \\ \widetilde{D}^{[2]}_{[00]} = P_{U^\perp,\mathbb{R}Y}\mathcal{W}\Big[\widetilde{M}^{-1}_{[00]}(2\widetilde{M}_{[00]} - \widetilde{L}_{[00]})\widetilde{M}^{-1}_{[00]}\Big]^t P_{U^\perp,\mathbb{R}Y}, \end{cases} \tag{6.4.36}$$

Table 32. Flux diffusion matrix. Standard iterative methods for $\widetilde{L}_{[00]}$ with $\widetilde{M}_{[00]}$ given by (6.4.33). Reduced errors and spectral radius for various mixtures.

	Mixture 4	Mixture 5	Mixture 6
1	5.18E-2	8.93E-2	1.34E-2
2	7.86E-3	1.72E-16	3.02E-4
3	3.37E-4	—	6.99E-6
4	5.11E-5	—	1.66E-7
ρ	8.17E-2	9.62E-4	2.51E-2

where $\mathcal{W} = \mathrm{diag}\big((m_k/m)_{k\in\mathcal{S}}\big)$.

6.4.8 Discussion

Only the numerical experiments concerning projected standard iterative methods will be discussed here. Indeed, the resulting diffusion matrices satisfy the mathematical properties that are important from a thermodynamic viewpoint, that is, symmetry, mass conservation, and positive definiteness on the physical hyperplane of zero sum gradients. On the other hand, conjugate gradient methods can also be considered for diffusion matrix evaluation, since the iterates converge in very few iterations, as indicated by our numerical experiments. However, these iterates are generally neither symmetric nor positive definite on the physical hyperplane of zero sum gradients.

It is interesting to consider the simplified diffusion matrix $D_{[00]}$ which is within 3E-2 accuracy of the standard diffusion matrix D. The matrix $D_{[00]}$ is often referred to as the first-order diffusion matrix and is associated with a system matrix of size n instead of $2n+p$. The first iterate for the system $L_{[00]}\alpha_{[00]}^{D_l} = \beta_{[00]}^{D_l}$ provides the matrix $D_{[00]}^{[1]}$ which takes on the form of a projected diagonal matrix and is given by

$$D_{[00]}^{[1]} = P_{Y^\perp,\mathbb{R}U}\mathrm{diag}(D_1^*/X_1,\dots,D_n^*/X_n)P_{U^\perp,\mathbb{R}Y}, \qquad (6.4.37)$$

where

$$D_k^* = \frac{1-Y_k}{\sum_{\substack{l\in\mathcal{S}\\ l\neq k}} X_l/\mathcal{D}_{kl}}, \qquad k\in\mathcal{S}, \qquad (6.4.38)$$

and is within 3E-2 accuracy of $D_{[00]}$. The resulting diffusion velocities $V_{[00]}^{[1]} = -D_{[00]}^{[1]}d$ correspond to the Hirschfelder-Curtiss approximate diffusion velocities [HC49] with a species independent mass correction velocity [CHe81], as proven in [Gi91]. It is also important to observe that no dense matrix multiplications are needed to form the projected expressions in the right-hand side of (6.4.37), since we have

$$D^{[1]}_{[00]kl} = \frac{D^*_k}{X_k}\delta_{kl} - \frac{1}{\langle Y,U\rangle}\frac{D^*_k}{X_k}Y_k - \frac{1}{\langle Y,U\rangle}\frac{D^*_l}{X_l}Y_l + \frac{1}{\langle Y,U\rangle^2}\sum_{i\in\mathcal{S}}\frac{D^*_i}{X_i}Y_i^2, \qquad k,l\in\mathcal{S}, \tag{6.4.39}$$

so that $D^{[1]}_{[00]}$ is evaluated in $2n^2+\mathcal{O}(n)$ operations.

On the other hand, one can also consider the simplified diffusion matrix $D_{[e]}$ which is within 5E-3 accuracy of the standard diffusion matrix D. The matrix $D_{[e]}$ is new and is associated with a system matrix of size $2n$ instead of $2n+p$. The first iterate for the system $L_{[e]}\alpha^{D_l}_{[e]} = \beta^{D_l}_{[e]}$ yields the matrix $D^{[1]}_{[e]}$ given by

$$D^{[1]}_{[e]} = P_{Y^\perp,\mathbb{R}U}\,\mathrm{diag}(D^*_{[e]1}/X_1,\ldots,D^*_{[e]n}/X_n)P_{U^\perp,\mathbb{R}Y}, \tag{6.4.40}$$

where

$$D^*_{[e]k} = \frac{D^*_k}{1-\Big(\big(L^{00e}_{[e]kk}\big)^2/L^{ee}_{[e]kk}\Big)\big(D^*_k/X_k\big)}, \qquad k\in\mathcal{S}, \tag{6.4.41}$$

and is within 3E-2 accuracy of $D_{[e]}$. The resulting diffusion velocities $V^{[1]}_{[e]} = -D^{[1]}_{[e]}d$ generalize the Hirschfelder-Curtiss approximate diffusion velocities by taking into account second-order effects and the internal energy of the molecules. As for the diffusion matrix $D^{[1]}_{[00]}$, no dense matrix multiplications are needed to form the projected expressions in the right-hand side of (6.4.40) which requires therefore $2n^2+\mathcal{O}(n)$ operations for its evaluation. Note, however, that the cost of evaluating the coefficients $D^*_{[e]k}$, $k\in\mathcal{S}$, is larger than the one associated with the evaluation of D^*_k, $k\in\mathcal{S}$.

A detailed analytic expression for the diffusion matrix is obtained after two iterations for the system $L_{[e]}\alpha^{D_l}_{[e]} = \beta^{D_l}_{[e]}$. The corresponding diffusion matrix $D^{[2]}_{[e]}$ is within 2E-3 accuracy of $D_{[e]}$ and may be expressed as

$$D^{[2]}_{[e]} = P_{Y^\perp,\mathbb{R}U}\mathfrak{P}_{[e]}M^{-1}_{[e]}(2M_{[e]}-L_{[e]})M^{-1}_{[e]}\mathfrak{P}^t_{[e]}P_{U^\perp,\mathbb{R}Y}, \tag{6.4.42}$$

where $\mathfrak{P}_{[e]}\in\mathbb{R}^{n,2n}$ denotes the rectangular matrix formed by the blocks $\mathfrak{P}_{[e]} = [I,0]$, and the matrix $M_{[e]}$ coincides with the matrix $db(L_{[e]})$, except for the upper-left block given by $M^{0000}_{[e]kk} = L^{0000}_{[e]kk}/(1-Y_k)$, $k\in\mathcal{S}$. Since the matrix $M_{[e]}$ consists of four diagonal blocks, the product of $M^{-1}_{[e]}$ with a given matrix only requires $\mathcal{O}(n^2)$ operations, so that the computational cost of $D^{[2]}_{[e]}$ is still $\mathcal{O}(n^2)$ operations.

Another detailed analytic expression for the diffusion matrix is obtained after two iterations for the full system $L\alpha^{D_l} = \beta^{D_l}$. The corresponding diffusion matrix $D^{[2]}$ is within 2E-3 accuracy of D and may be expressed as

$$D^{[2]} = P_{Y^\perp,\mathbb{R}U}\mathfrak{P}M^{-1}(2M - L)M^{-1}\mathfrak{P}^t P_{U^\perp,\mathbb{R}Y}, \tag{6.4.43}$$

where $\mathfrak{P} \in \mathbb{R}^{n,2n+p}$ denotes the rectangular matrix formed by the blocks $\mathfrak{P} = [I, 0, 0]$, and the matrix M coincides with the matrix $db(L)$, except for the upper-left block given by $M_{kk}^{0000} = L_{kk}^{0000}/(1 - Y_k)$, $k \in S$. Since the matrix M consists of nine diagonal blocks, the product of M^{-1} with a given matrix only requires $\mathcal{O}(n^2)$ operations, so that the computational cost of $D^{[2]}$ is still $\mathcal{O}(n^2)$ operations.

Finally, an additional iteration for the system $L\alpha^{D_l} = \beta^{D_l}$ yields the matrix $D^{[3]}$ which is accurate to 2E-4, but requires $\mathcal{O}(n^3)$ operations for its evaluation.

6.5 The Partial Thermal Conductivity and the Thermal Diffusion Vector

6.5.1 Numerical Experiments for the System $L\alpha^{\lambda'} = \beta^{\lambda'}$

Standard Iterative Methods. We consider the initialization $x_0 = 0$ and the splitting induced by (6.4.1) for which Theorem 5.5.1 applies. In Tables 33 and 34 we present the reduced errors

$$e_{\lambda'}^{[i]} = \frac{|\lambda' - \lambda'^{[i]}|}{\lambda'}, \qquad e_{\theta}^{[i]} = \frac{||\theta - \theta^{[i]}||_\infty}{||\theta||_\infty}, \qquad i = 1, 2, 3, 4, \tag{6.5.1}$$

where $||\theta||_\infty = \max_{k\in S} |\theta_k|$. One can see that after two iterations the resulting approximations given by

$$\begin{cases} \lambda'^{[2]} = \frac{\bar{p}}{T}\langle M^{-1}\beta^{\lambda'}, (2M - L)M^{-1}\beta^{\lambda'}\rangle, \\ \theta^{[2]} = -P_{Y^\perp,\mathbb{R}U}\mathfrak{P}M^{-1}(2M - L)M^{-1}\beta^{\lambda'}, \end{cases} \tag{6.5.2}$$

where $\mathfrak{P} = [I, 0, 0] \in \mathbb{R}^{n,2n+p}$, are within 4E-2 and 3E-2 accuracy, respectively.

We now investigate standard iterative methods for the Schur complement $L_{[s]}$ given by (6.4.5). We consider the initialization $x_0 = 0$ and the splitting induced by (6.4.6). As stated in Section 6.4.1, this splitting yields a convergent iteration matrix, and we deduce from Section 5.5.6 that the iterates $\lambda'^{[i]}_{[s]}$ and $\theta^{[i]}_{[s]}$ converge towards $\lambda' - \lambda'_{[01]}$

Table 33. Partial thermal conductivity. Standard iterative methods for L with M given by (6.4.1). Reduced errors for various mixtures.

	Mixture 1	Mixture 2	Mixture 3
1	1.45E-1	7.95E-2	1.34E-1
2	3.40E-2	1.32E-2	3.58E-2
3	8.47E-3	2.29E-3	9.76E-3
4	2.13E-3	4.00E-4	2.67E-3

Table 34. Thermal diffusion vector. Standard iterative methods for L with M given by (6.4.1). Reduced errors for various mixtures.

	Mixture 1	Mixture 2	Mixture 3
1	1.52E-1	1.94E-1	1.08E-1
2	3.28E-2	1.96E-2	2.60E-2
3	7.83E-3	2.79E-3	7.00E-3
4	1.96E-3	5.29E-4	1.92E-3

Table 35. Partial thermal conductivity. Standard iterative methods for $L_{[s]}$ with $M_{[s]}$ given by (6.4.6). Reduced errors for various mixtures.

	Mixture 1	Mixture 2	Mixture 3
1	1.13E-1	6.47E-2	9.78E-2
2	2.61E-2	1.16E-2	2.59E-2
3	6.37E-3	2.03E-3	6.95E-3
4	1.56E-3	3.56E-4	1.87E-3

and θ, respectively. At variance with the simplified volume viscosity $\kappa_{[01]}$, however, the simplified partial thermal conductivity $\lambda'_{[01]}$ yields poor approximations for λ'. The reduced errors

$$e_{\lambda'}^{[i]} = \frac{|\lambda' - (\lambda'^{[i]}_{[s]} + \lambda'_{[01]})|}{\lambda'}, \qquad e_{\theta}^{[i]} = \frac{||\theta - \theta^{[i]}_{[s]}||_\infty}{||\theta||_\infty}, \qquad i = 1,2,3,4, \tag{6.5.3}$$

are presented in Tables 35 and 36, respectively, for mixtures 1, 2, and 3. After two iterations, the expressions for the partial thermal conductivity and the thermal diffusion vector are within 3E-2 accuracy.

Table 36. Thermal diffusion vector. Standard iterative methods for $L_{[s]}$ with $M_{[s]}$ given by (6.4.6). Reduced errors for various mixtures.

	Mixture 1	Mixture 2	Mixture 3
1	1.42E-1	1.89E-1	9.29E-2
2	2.81E-2	1.73E-2	2.14E-2
3	6.49E-3	2.49E-3	5.70E-3
4	1.59E-3	4.92E-4	1.53E-3

Table 37. Partial thermal conductivity. Conjugate gradient methods for L with $M = db(L)$. Reduced errors for various mixtures.

	Mixture 1	Mixture 2	Mixture 3
1	1.75E-2	8.05E-3	2.33E-2
2	2.70E-5	7.20E-6	1.53E-5
3	5.82E-8	1.03E-7	3.36E-9
4	6.68E-11	5.85E-9	8.46E-14

Conjugate Gradient Methods. We consider the initialization $x_0 = 0$ and the preconditioner $M = db(L)$, for which Theorem 5.5.2 applies when $n \geq 2$. The numerical results are given in Tables 37 and 38. After one iteration, the resulting partial thermal conductivity given by

$$\lambda'^{[1]} = \frac{\overline{p}}{\overline{T}} \frac{\langle \beta^{\lambda'}, db(L)^{-1}\beta^{\lambda'}\rangle^2}{\langle db(L)^{-1}\beta^{\lambda'}, L db(L)^{-1}\beta^{\lambda'}\rangle}, \tag{6.5.4}$$

is within 2E-2 accuracy. For the thermal diffusion vector, we obtain an approximation within 2E-2 accuracy after two iterations. The preconditioner $M = \text{diag}(L)$ also yields accurate expressions for the partial thermal conductivity since the first iterate is within 4E-2 accuracy. Slower convergence rates, however, are obtained for the thermal diffusion vector for the first iterations.

We next consider conjugate gradient methods for the Schur complement $L_{[s]}$ with the initialization $x_0 = 0$ and the preconditioner $M_{[s]}$ given by (6.4.8), for which the results of Section 5.5.6 apply when $n \geq 2$. The numerical results are presented in Tables 39 and 40 and indicate slightly faster convergence rates than the ones obtained with the matrix L. The first iterate for the partial thermal conductivity is within 1E-2 accuracy, whereas the second iterate for the thermal diffusion vector is within 2E-2

Table 38. Thermal diffusion vector. Conjugate gradient methods for L with $M = db(L)$. Reduced errors for various mixtures.

	Mixture 1	Mixture 2	Mixture 3
1	7.31E-2	1.64E-1	4.08E-2
2	8.45E-3	2.03E-2	4.77E-3
3	1.01E-3	2.93E-3	1.50E-4
4	1.68E-5	6.65E-4	1.26E-6

Table 39. Partial thermal conductivity. Conjugate gradient methods for $L_{[s]}$ with $M_{[s]}$ given by (6.4.8). Reduced errors for various mixtures.

	Mixture 1	Mixture 2	Mixture 3
1	1.14E-2	6.54E-3	9.28E-3
2	8.01E-6	3.80E-6	7.92E-7
3	2.57E-8	3.28E-8	2.42E-10
4	1.33E-11	1.24E-10	9.12E-14

Table 40. Thermal diffusion vector. Conjugate gradient methods for $L_{[s]}$ with $M_{[s]}$ given by (6.4.8). Reduced errors for various mixtures.

	Mixture 1	Mixture 2	Mixture 3
1	4.89E-2	1.44E-1	1.16E-1
2	9.07E-3	2.19E-2	2.89E-3
3	6.60E-4	4.58E-3	5.02E-5
4	3.21E-5	1.80E-3	8.67E-7

accuracy.

6.5.2 Numerical Experiments for the System $L_{[e]}\alpha^{\lambda'}_{[e]} = \beta^{\lambda'}_{[e]}$

In Table 41 we present the reduced errors

$$e_{\lambda'_{[e]}} = \frac{|\lambda' - \lambda'_{[e]}|}{\lambda'}, \qquad e_{\theta_{[e]}} = \frac{||\theta - \theta_{[e]}||_\infty}{||\theta||_\infty}, \tag{6.5.5}$$

which indicate that the simplified transport coefficients $\lambda'_{[e]}$ and $\theta_{[e]}$ are accurate to

Table 41. Partial thermal conductivity and thermal diffusion vector. Reduced errors for $\lambda'_{[e]}$ and $\theta_{[e]}$.

	Mixture 1	Mixture 2	Mixture 3
$e_{\lambda'_{[e]}}$	7.41E-4	4.07E-4	1.84E-3
$e_{\theta_{[e]}}$	3.33E-3	4.41E-3	8.13E-3

Table 42. Partial thermal conductivity. Standard iterative methods for $L_{[e]}$ with $M_{[e]}$ given by (6.4.10). Reduced errors for various mixtures.

	Mixture 1	Mixture 2	Mixture 3
1	1.55E-1	8.40E-2	1.54E-1
2	3.12E-2	1.20E-2	2.73E-2
3	6.40E-3	1.64E-3	4.86E-3
4	1.32E-3	2.27E-4	8.65E-4

2E-3 and 8E-3, respectively, for mixtures 1, 2, and 3. This new formulation is therefore an attractive alternative to the full system $L\alpha^{\lambda'} = \beta^{\lambda'}$.

Standard Iterative Methods. We consider the initialization $x_0 = 0$ and the splitting induced by (6.4.10), for which Theorem 5.5.3 applies. The reduced errors

$$e^{[i]}_{\lambda'_{[e]}} = \frac{|\lambda'_{[e]} - \lambda'^{[i]}_{[e]}|}{\lambda'_{[e]}}, \qquad e^{[i]}_{\theta_{[e]}} = \frac{||\theta_{[e]} - \theta^{[i]}_{[e]}||_\infty}{||\theta_{[e]}||_\infty}, \qquad i = 1,2,3,4, \tag{6.5.6}$$

are presented in Tables 42 and 43. In particular, we can see that after two iterations, the expressions

$$\begin{cases} \lambda'^{[2]}_{[e]} = \dfrac{\bar{p}}{T}\langle M^{-1}_{[e]}\beta^{\lambda'}_{[e]}, (2M_{[e]} - L_{[e]})M^{-1}_{[e]}\beta^{\lambda'}_{[e]}\rangle, \\ \theta^{[2]}_{[e]} = -P_{Y^\perp,\mathbb{R}U}\mathfrak{P}_{[e]}M^{-1}_{[e]}(2M_{[e]} - L_{[e]})M^{-1}_{[e]}\beta^{\lambda'}_{[e]}, \end{cases} \tag{6.5.7}$$

where $\mathfrak{P}_{[e]} = [I,0] \in \mathbb{R}^{n,2n}$, are both within 3E-2 accuracy of $\lambda'_{[e]}$ and $\theta_{[e]}$, respectively.

Conjugate Gradient Methods. We consider the initialization $x_0 = 0$ and the preconditioner $M_{[e]} = db(L_{[e]})$, for which Theorem 5.5.4 applies when $n \geq 2$. From the numerical results given in Tables 44 and 45, we can see that after only one iteration, we obtain the expression

$$\lambda'^{[1]}_{[e]} = \frac{\bar{p}}{T}\frac{\langle\beta^{\lambda'}_{[e]}, db(L_{[e]})^{-1}\beta^{\lambda'}_{[e]}\rangle^2}{\langle db(L_{[e]})^{-1}\beta^{\lambda'}_{[e]}, L_{[e]}db(L_{[e]})^{-1}\beta^{\lambda'}_{[e]}\rangle}, \tag{6.5.8}$$

Table 43. Thermal diffusion vector. Standard iterative methods for $L_{[e]}$ with $M_{[e]}$ given by (6.4.10). Reduced errors for various mixtures.

	Mixture 1	Mixture 2	Mixture 3
1	1.62E-1	1.99E-1	1.21E-1
2	3.04E-2	1.77E-2	2.12E-2
3	6.17E-3	2.00E-3	3.78E-3
4	1.27E-3	3.26E-4	6.72E-4

Table 44. Partial thermal conductivity. Conjugate gradient methods for $L_{[e]}$ with $M_{[e]} = db(L_{[e]})$. Reduced errors for various mixtures.

	Mixture 1	Mixture 2	Mixture 3
1	9.90E-3	5.89E-3	5.13E-3
2	7.83E-6	4.25E-6	4.72E-7
3	2.10E-8	2.63E-8	1.12E-10
4	6.19E-12	3.45E-10	2.14E-14

Table 45. Thermal diffusion vector. Conjugate gradient methods for $L_{[e]}$ with $M_{[e]} = db(L_{[e]})$. Reduced errors for various mixtures.

	Mixture 1	Mixture 2	Mixture 3
1	4.82E-2	1.65E-1	3.62E-2
2	6.14E-3	1.31E-2	1.07E-3
3	5.67E-4	2.76E-3	3.82E-5
4	1.99E-5	1.09E-3	6.66E-7

which is within 1E-2 accuracy of $\lambda'_{[e]}$. After a second iteration, we obtain an expression within 6E-3 accuracy of $\theta_{[e]}$. The preconditioner $M_{[e]} = \mathrm{diag}(L_{[e]})$ also yields accurate expressions for the partial thermal conductivity since the first iterate is within 4E-2 accuracy. However, slower convergence rates are obtained for the thermal diffusion vector for the first iterates.

Finally, this simplified formulation can be used to obtain an accurate initialization for iterative algorithms applied to the system $L\alpha^{\lambda'} = \beta^{\lambda'}$. Indeed, we deduce from the accuracy of $\lambda'_{[e]}$ and $\theta_{[e]}$ that the solution $\alpha^{\lambda'}$ of the constrained linear system

Table 46. Partial thermal conductivity. Standard iterative methods for $\widetilde{L}$ with $\widetilde{M}$ given by (6.4.19). Reduced errors for various mixtures.

	Mixture 4	Mixture 5	Mixture 6
1	7.95E-2	3.42E-2	1.33E-1
2	1.32E-2	1.75E-3	3.58E-2
3	2.29E-3	7.87E-5	9.81E-3
4	3.99E-4	4.04E-6	2.70E-3

$L\alpha^{\lambda'} = \beta^{\lambda'}$ and $\langle \mathcal{L}, \alpha^{\lambda'} \rangle = 0$ is such that $\alpha_k^{10\lambda'} \simeq \alpha_k^{01\lambda'}$, $k \in \mathcal{P}$. Assuming that $\alpha^{01\lambda'} \simeq (L^{0101})^{-1} \beta^{01\lambda'}$ and considering the limit of this expression when $c_k^{\rm int} \to 0$, we then obtain an estimate for $\alpha_k^{10\lambda'}$ when $k \in \mathcal{S} \setminus \mathcal{P}$. This procedure ultimately yields

$$\begin{cases} x_0^{00\lambda'} = -P_{Y^\perp,\mathbb{R}U} M_{[00]}^{-1} P_{U^\perp,\mathbb{R}Y} L^{0010} x_0^{10\lambda'}, \\ x_{0k}^{10\lambda'} = x_{0k}^{01\lambda'} = \dfrac{c_k^{\rm int}}{k_{\rm B}} \dfrac{X_k}{L_{kk}^{0101}}, \quad k \in \mathcal{P}, \\ x_{0k}^{10\lambda'} = \dfrac{1}{\sum_{l \in \mathcal{S}} X_l / \mathcal{D}_{kl}}, \quad k \in \mathcal{S} \setminus \mathcal{P}. \end{cases} \tag{6.5.9}$$

where the matrix $M_{[00]}$ is given by (6.4.15). The corresponding approximations for λ' and θ are within 1E-1 accuracy.

6.5.3 Numerical Experiments for the System $\widetilde{L}\alpha^{\lambda'} = \widetilde{\beta}^{\lambda'}$

We consider the initialization $x_0 = 0$ and the splitting induced by (6.4.19), for which Theorem 5.5.5 applies. In Tables 46 and 47 we present the reduced errors

$$e_{\lambda'}^{[i]} = \frac{|\lambda' - \lambda'^{[i]}|}{\lambda'}, \qquad e_{\theta}^{[i]} = \frac{||\theta - \theta^{[i]}||_\infty}{||\theta||_\infty}, \qquad i = 1, 2, 3, 4. \tag{6.5.10}$$

One can see that after two iterations the expressions for the partial thermal conductivity and the thermal diffusion vector are within 4E-2 and 3E-2 accuracy, respectively. Furthermore, these expressions coincide with (6.5.2) for positive mass fractions.

We now consider standard iterative methods for the Schur complement $\widetilde{L}_{[s]}$ given by (6.4.23). We consider the initialization $x_0 = 0$ and the splitting induced by (6.4.24). As stated in Section 6.4.4, this splitting yields a convergent iteration matrix, and we deduce from Section 5.5.6 that the iterates $\lambda_{[s]}'^{[i]}$ and $\theta_{[s]}^{[i]}$ converge towards $\lambda' - \lambda'_{[01]}$ and θ, respectively. The reduced errors

$$e_{\lambda'}^{[i]} = \frac{|\lambda' - (\lambda_{[s]}'^{[i]} + \lambda'_{[01]})|}{\lambda'}, \qquad e_{\theta}^{[i]} = \frac{||\theta - \theta_{[s]}^{[i]}||_\infty}{||\theta||_\infty}, \qquad i = 1, 2, 3, 4. \tag{6.5.11}$$

Table 47. Thermal diffusion vector. Standard iterative methods for $\widetilde{L}$ with $\widetilde{M}$ given by (6.4.19). Reduced errors for various mixtures.

	Mixture 4	Mixture 5	Mixture 6
1	1.94E-1	2.55E-1	1.06E-1
2	1.95E-2	1.76E-2	2.54E-2
3	2.79E-3	7.48E-4	6.86E-3
4	5.29E-4	4.02E-5	1.89E-3

Table 48. Partial thermal conductivity. Standard iterative methods for $\widetilde{L}_{[s]}$ with $\widetilde{M}_{[s]}$ given by (6.4.24). Reduced errors for various mixtures.

	Mixture 4	Mixture 5	Mixture 6
1	6.47E-2	2.37E-2	9.91E-2
2	1.16E-2	1.20E-3	2.65E-2
3	2.03E-3	4.38E-5	7.15E-3
4	3.56E-4	2.16E-6	1.93E-3

Table 49. Thermal diffusion vector. Standard iterative methods for $\widetilde{L}_{[s]}$ with $\widetilde{M}_{[s]}$ given by (6.4.24). Reduced errors for various mixtures.

	Mixture 4	Mixture 5	Mixture 6
1	1.89E-1	2.50E-1	9.20E-2
2	1.73E-2	1.47E-2	2.11E-2
3	2.49E-3	4.58E-4	5.65E-3
4	4.91E-4	2.61E-5	1.53E-3

are presented in Tables 48 and 49, respectively, for mixtures 4, 5, and 6. After two iterations, the expressions for the partial thermal conductivity and the thermal diffusion vector are within 3E-2 and 2E-2 accuracy, respectively.

6.5.4 Numerical Experiments for the System $\widehat{L}\widehat{\alpha}^{\lambda'} = \widehat{\beta}^{\lambda'}$

We consider the initialization $\widehat{x}_0 = 0$ and the preconditioner $\widehat{M} = db(\widehat{L})$, for which Theorem 5.5.6 applies when $n^+ \geq 2$. Note that the corresponding iterates converge

Table 50. Partial thermal conductivity. Conjugate gradient methods for $\widehat{L}$ with $\widehat{M} = db(\widehat{L})$. Reduced errors for various mixtures.

	Mixture 4	Mixture 5	Mixture 6
1	8.04E-3	6.03E-4	2.34E-2
2	7.19E-6	3.59E-7	1.48E-5
3	1.02E-7	4.24E-16	3.10E-9
4	5.84E-9	—	6.73E-14

Table 51. Rescaled thermal diffusion vector. Conjugate gradient methods for $\widehat{L}$ with $\widehat{M} = db(\widehat{L})$. Reduced errors for various mixtures.

	Mixture 4	Mixture 5	Mixture 6
1	5.56E-2	6.62E-4	1.49E-1
2	1.64E-2	5.78E-4	7.16E-3
3	2.69E-3	4.66E-10	1.52E-4
4	5.56E-4	—	4.71E-6

towards the partial thermal conductivity λ' and the rescaled thermal diffusion vector $\widehat{\theta}$. The reduced errors

$$e_{\lambda'}^{[i]} = \frac{|\lambda' - \lambda'^{[i]}|}{\lambda'}, \qquad e_{\widehat{\theta}}^{[i]} = \frac{\|\widehat{\theta} - \widehat{\theta}^{[i]}\|_\infty}{\|\widehat{\theta}\|_\infty}, \qquad i = 1, 2, 3, 4, \tag{6.5.12}$$

are reported in Tables 50 and 51, respectively, and indicate high convergence rates. In particular, the first iterate for λ' is within 2E-2 accuracy and coincides with (6.5.4) for positive mass fractions. The second iterate for $\widehat{\theta}$ is within 2E-2 accuracy.

6.5.5 Numerical Experiments for the System $\widetilde{L}_{[e]}\alpha_{[e]}^{\lambda'} = \widetilde{\beta}_{[e]}^{\lambda'}$

In Table 52 we present the reduced errors

$$e_{\lambda'_{[e]}} = \frac{|\lambda' - \lambda'_{[e]}|}{\lambda'}, \qquad e_{\theta_{[e]}} = \frac{\|\theta - \theta_{[e]}\|_\infty}{\|\theta\|_\infty}, \tag{6.5.13}$$

which indicate that the simplified transport coefficients $\lambda'_{[e]}$ and $\theta_{[e]}$ are accurate to 1E-3 and 8E-3, respectively, for mixtures 4, 5, and 6. This new formulation is therefore an attractive alternative to the full system $\widetilde{L}\alpha^{\lambda'} = \widetilde{\beta}^{\lambda'}$.

Table 52. Partial thermal conductivity and thermal diffusion vector. Reduced errors for $\lambda'_{[e]}$ and $\theta_{[e]}$.

	Mixture 4	Mixture 5	Mixture 6
$e_{\lambda'_{[e]}}$	4.07E-4	6.05E-4	1.33E-3
$e_{\theta_{[e]}}$	4.71E-3	3.50E-3	8.36E-3

Table 53. Partial thermal conductivity. Standard iterative methods for $\widetilde{L}_{[e]}$ with $\widetilde{M}_{[e]}$ given by (6.4.28). Reduced errors for various mixtures.

	Mixture 4	Mixture 5	Mixture 6
1	8.40E-2	3.45E-2	1.54E-1
2	1.20E-2	1.59E-3	2.73E-2
3	1.64E-3	5.49E-5	4.87E-3
4	2.26E-4	2.54E-6	8.67E-4

Table 54. Thermal diffusion vector. Standard iterative methods for $\widetilde{L}_{[e]}$ with $\widetilde{M}_{[e]}$ given by (6.4.28). Reduced errors for various mixtures.

	Mixture 4	Mixture 5	Mixture 6
1	1.99E-1	2.56E-1	1.20E-1
2	1.76E-2	1.50E-2	2.07E-2
3	1.99E-3	4.56E-4	3.68E-3
4	3.26E-4	2.41E-5	6.56E-4

We consider the initialization $x_0 = 0$ and the splitting induced by (6.4.28) for which Theorem 5.5.7 applies. The reduced errors

$$e^{[i]}_{\lambda'_{[e]}} = \frac{|\lambda'_{[e]} - \lambda'^{[i]}_{[e]}|}{\lambda'_{[e]}}, \qquad e^{[i]}_{\theta_{[e]}} = \frac{||\theta_{[e]} - \theta^{[i]}_{[e]}||_\infty}{||\theta_{[e]}||_\infty}, \qquad i = 1, 2, 3, 4, \tag{6.5.14}$$

are presented in Tables 53 and 54. In particular, after two iterations, we obtain the quantity $\lambda'^{[2]}_{[e]}$ which is within 3E-2 accuracy of $\lambda'_{[e]}$ and the vector $\theta^{[2]}_{[e]}$ which is within 2E-2 accuracy of $\theta_{[e]}$.

Table 55. Partial thermal conductivity. Reduced errors for various mixture-averaged formulas.

	Mixture 1	Mixture 2	Mixture 3	Mixture 4	Mixture 5	Mixture 6
-1	2.28E-1	3.09E-1	1.25E-1	3.09E-1	3.49E-1	1.23E-1
0	4.84E-2	9.10E-2	2.03E-2	9.11E-2	7.98E-2	2.19E-2
1/4	2.57E-2	1.88E-3	1.94E-2	1.89E-3	1.30E-2	1.76E-2
1/2	1.15E-1	1.02E-1	6.77E-2	1.02E-1	1.10E-1	6.63E-2
1	3.41E-1	3.36E-1	1.97E-1	3.36E-1	3.00E-1	2.00E-1
-1,1	5.62E-2	1.40E-2	3.61E-2	1.40E-2	2.42E-2	3.86E-2

6.5.6 Mixture-Averaged Formulas

As described in Section 6.3.1, we now consider several mixture-averaged formulas for the partial thermal conductivity. These expressions are based on the pure species thermal conductivities λ_k, $k \in \mathcal{S}$, that coincide with the pure species partial thermal conductivities. In Table 55 we present the reduced errors

$$e_t = \frac{|\lambda' - \mathcal{M}_t(\lambda')|}{\lambda'}, \tag{6.5.15}$$

for various values of the parameter t and for the six mixtures considered in our numerical experiments. The reduced error for the approximation introduced in [BW53] and considered, for instance, in [MTS67] [KDWCM86]

$$\mathcal{M}_{-1,1}(\lambda') = \frac{1}{2}(\mathcal{M}_{-1}(\lambda') + \mathcal{M}_1(\lambda')) = \frac{1}{2}\Big(\big(\sum_{k\in\mathcal{S}} X_k(\lambda_k)^{-1}\big)^{-1} + \sum_{k\in\mathcal{S}} X_k\lambda_k\Big), \tag{6.5.16}$$

is also included in Table 55. From this table we can see that the average formula of order 1/4 yields approximations within 3E-2 accuracy. On the other hand, the expression (6.5.16) only yields 6E-2 accuracy.

6.5.7 Discussion

From the preceding numerical experiments, we can draw the following conclusions.

The Partial Thermal Conductivity. The average formula of order 1/4,

$$\lambda' = \big(\sum_{k\in\mathcal{S}} X_k(\lambda_k)^{1/4}\big)^4, \tag{6.5.17}$$

where λ_k, $k \in \mathcal{S}$, are the pure species thermal conductivities, requires only $\mathcal{O}(n)$ operations for its evaluation and is accurate to 3E-2 for the mixtures considered in this book. The average formula of order zero

$$\lambda' = \exp\big(\sum_{k\in\mathcal{S}} X_k \log(\lambda_k)\big), \tag{6.5.18}$$

is accurate to 9E-2 only, but less computationally expensive when the pure species thermal conductivities are fitted in terms of the logarithm of the temperature [KWM83].

Furthermore, the simplified system matrix $L_{[e]}$ is an interesting alternative to the standard system matrix L since it is of size $2n$ instead of $2n+p$. Moreover, the resulting simplified partial thermal conductivity $\lambda'_{[e]}$ is within 2E-3 accuracy for the mixtures considered in this book. A new rigorously derived analytical expression is obtained after one conjugate gradient iteration for the system $L_{[e]}\alpha^{\lambda'}_{[e]} = \beta^{\lambda'}_{[e]}$, preconditioned by the matrix $db(L_{[e]})$. This yields the partial thermal conductivity

$$\lambda'^{[1]}_{[e]} = \frac{\bar{p}}{\overline{T}} \frac{\langle \beta^{\lambda'}_{[e]}, db(L_{[e]})^{-1}\beta^{\lambda'}_{[e]}\rangle^2}{\langle db(L_{[e]})^{-1}\beta^{\lambda'}_{[e]}, L_{[e]}db(L_{[e]})^{-1}\beta^{\lambda'}_{[e]}\rangle}, \tag{6.5.19}$$

which requires $4n^2 + \mathcal{O}(n)$ operations for its evaluation and is within 1E-2 accuracy of $\lambda'_{[e]}$. Note that the matrix $db(L_{[e]})$ consists of four diagonal blocks so that the product of $db(L_{[e]})^{-1}$ with a given vector only requires $\mathcal{O}(n)$ operations.

Another rigorously derived analytic expression is obtained after one conjugate gradient iteration for the full system $L\alpha^{\lambda'} = \beta^{\lambda'}$, preconditioned by the matrix $db(L)$. This yields the new expression

$$\lambda'^{[1]} = \frac{\bar{p}}{\overline{T}} \frac{\langle \beta^{\lambda'}, db(L)^{-1}\beta^{\lambda'}\rangle^2}{\langle db(L)^{-1}\beta^{\lambda'}, Ldb(L)^{-1}\beta^{\lambda'}\rangle}, \tag{6.5.20}$$

which requires $9n^2 + \mathcal{O}(n)$ operations for its evaluation and is within 2E-2 accuracy of λ'. Recalling that the matrix $db(L)$ is formed by the diagonal of the nine blocks of the matrix L, the vector $db(L)^{-1}\beta^{\lambda'}$ is evaluated in $\mathcal{O}(n)$ operations by directly solving n dense linear symmetric systems of size 3 or 2. Instead of the matrix $db(L)$, one can also consider the simpler preconditioner $\mathrm{diag}(L)$ which can be trivially inverted. The expression for the partial thermal conductivity, after one conjugate gradient iteration, is obtained from (6.5.20) by just replacing the matrix $db(L)^{-1}$ by $\mathrm{diag}(L)^{-1}$. Its computational cost is still $9n^2+\mathcal{O}(n)$ operations, and it is accurate to 4E-2. Note, however, that the preconditioner $\mathrm{diag}(L)$ yields slower convergence rates than $db(L)$ for the thermal diffusion vector. If both transport coefficients are to be evaluated simultaneously, it is therefore preferable to use $db(L)$ as a common preconditioner.

More detailed analytic expressions are obtained after two conjugate gradient iterations for the system $L\alpha^{\lambda'} = \beta^{\lambda'}$, preconditioned by the matrix $db(L)$, with an accuracy below 3E-5.

The Thermal Diffusion Vector. The simplified thermal diffusion vector $\theta_{[e]}$ is within 8E-3 accuracy for the mixtures considered in this book. It may therefore be an interesting alternative to the standard thermal diffusion vector θ since it is associated with a linear system of size $2n$ instead of $2n+p$. Two projected standard iterations for the system $L_{[e]}\alpha^{\lambda'}_{[e]} = \beta^{\lambda'}_{[e]}$ yield a new rigorously derived analytic expression which is within 3E-2 accuracy of $\theta_{[e]}$. The corresponding vector $\theta^{[2]}_{[e]}$ is given by

$$\theta^{[2]}_{[e]} = -P_{Y^\perp,\mathbb{R}U}\mathfrak{P}_{[e]}M^{-1}_{[e]}(2M_{[e]} - L_{[e]})M^{-1}_{[e]}\beta^{\lambda'}_{[e]}, \tag{6.5.21}$$

where $\mathfrak{P}_{[e]} \in \mathbb{R}^{n,2n}$ denotes the rectangular matrix formed by the blocks $\mathfrak{P}_{[e]} = [I, 0]$, and the matrix $M_{[e]}$ coincides with the matrix $db(L_{[e]})$, except for the upper-left block given by $M^{0000}_{[e]kk} = L^{0000}_{[e]kk}/(1-Y_k)$, $k \in S$. Since the matrix $M_{[e]}$ is formed by four diagonal blocks, the product of $M^{-1}_{[e]}$ with a given vector only requires $\mathcal{O}(n)$ operations, so that the computational cost of $\theta^{[2]}_{[e]}$ is $4n^2 + \mathcal{O}(n)$ operations.

Another new, rigorously derived, analytic expression is obtained after two projected standard iterations for the full system $L\alpha^{\lambda'} = \beta^{\lambda'}$. The vector $\theta^{[2]}$ is within 3E-2 accuracy of θ and can be expressed as

$$\theta^{[2]} = -P_{Y^\perp,\mathbb{R}U}\mathfrak{P}M^{-1}(2M - L)M^{-1}\beta^{\lambda'}, \tag{6.5.22}$$

where $\mathfrak{P} \in \mathbb{R}^{n,2n+p}$ denotes the rectangular matrix formed by the blocks $\mathfrak{P} = [I, 0, 0]$, and the matrix M coincides with the matrix $db(L)$, except for the upper-left block given by $M^{0000}_{kk} = L^{0000}_{kk}/(1-Y_k)$, $k \in S$. Since the matrix M is formed by nine diagonal blocks, the product of M^{-1} times a given matrix only requires $\mathcal{O}(n^2)$ operations, so that the computational cost of $\theta^{[2]}$ is $9n^2 + \mathcal{O}(n)$ operations. Note also that the vectors $\theta^{[2]}_{[e]}$ and $\theta^{[2]}$ require, for their evaluation, the same matrices as the ones considered for the diffusion matrices $D^{[2]}_{[e]}$ and $D^{[2]}$, respectively.

Two conjugate gradient iterations for the system $L_{[e]}\alpha^{\lambda'}_{[e]} = \beta^{\lambda'}_{[e]}$, preconditioned by the matrix $db(L_{[e]})$, yield, in $8n^2 + \mathcal{O}(n)$ operations, a new, rigorously derived, analytic expression for the thermal diffusion vector within 1E-2 accuracy of $\theta_{[e]}$.

More detailed analytic expressions are derived after three conjugate gradient iterations for the system $L\alpha^{\lambda'} = \beta^{\lambda'}$, preconditioned by the matrix $db(L)$. The new thermal diffusion vector $\theta^{[3]}$ requires $27n^2 + \mathcal{O}(n)$ operations for its evaluation and is within 3E-3 accuracy.

Table 56. Thermal conductivity. Standard iterative methods for Λ with $M = db(\Lambda)$. Reduced errors and spectral radius for various mixtures.

	Mixture 1	Mixture 2	Mixture 3
1	1.46E-1	7.98E-2	1.35E-1
2	3.44E-2	1.34E-2	3.64E-2
3	8.62E-3	2.33E-3	1.00E-2
4	2.18E-3	4.09E-4	2.76E-3
ρ	2.54E-1	1.75E-1	2.76E-1

6.6 The Thermal Conductivity and the Thermal Diffusion Ratios

6.6.1 Numerical Experiments for the System $\Lambda\alpha^\lambda = \beta^\lambda$

Standard Iterative Methods. We consider the initialization $x_0 = 0$ and the splitting

$$\Lambda = M - Z, \qquad M = db(\Lambda), \tag{6.6.1}$$

where $db(\Lambda)$ is formed by the diagonal of the four blocks of the matrix Λ. One can then easily verify that Theorem 5.6.1 applies. The reduced errors

$$e_\lambda^{[i]} = \frac{|\lambda - \lambda^{[i]}|}{\lambda}, \qquad e_\chi^{[i]} = \frac{||\chi - \chi^{[i]}||_\infty}{||\chi||_\infty}, \qquad i = 1,2,3,4, \tag{6.6.2}$$

where $||\chi||_\infty = \max_{k\in S} |\chi_k|$ are presented in Tables 56 and 57, respectively, together with the spectral radius of the iteration matrix $\rho = \rho(M^{-1}Z)$. After two iterations, we obtain an approximation for the thermal conductivity and the thermal diffusion ratios within 4E-2 and 3E-2 accuracy, respectively. The corresponding expressions may be written

$$\begin{cases} \lambda^{[2]} = \frac{\bar{p}}{T}\langle db(\Lambda)^{-1}\beta^\lambda, (2db(\Lambda) - \Lambda)db(\Lambda)^{-1}\beta^\lambda\rangle, \\ \chi^{[2]} = [L^{0010}, L^{0001}](2I - db(\Lambda)^{-1}\Lambda)db(\Lambda)^{-1}\beta^\lambda, \end{cases} \tag{6.6.3}$$

where $I \in \mathbb{R}^{n+p,n+p}$ denotes the identity matrix.

We now investigate standard iterative methods for the Schur complement

$$\Lambda_{[s]} = \Lambda^{1010} - \Lambda^{1001}\left(\Lambda^{0101}\right)^{-1}\Lambda^{0110}, \tag{6.6.4}$$

introduced in Section 5.6.6. We consider the initialization $x_0 = 0$ and the splitting induced by $M_{[s]} = db(\Lambda^{1010})$. As stated in Section 5.6.6, this splitting yields a convergent iteration matrix, and the quantities $\lambda_{[s]}^{[i]}$ and the vectors $\chi_{[s]}^{[i]}$ converge towards

Table 57. Thermal diffusion ratios. Standard iterative methods for Λ with $M = db(\Lambda)$. Reduced errors for various mixtures.

	Mixture 1	Mixture 2	Mixture 3
1	1.17E-1	3.88E-2	1.23E-1
2	2.69E-2	6.40E-3	3.22E-2
3	6.73E-3	1.08E-3	8.88E-3
4	1.70E-3	1.90E-4	2.45E-3

Table 58. Thermal conductivity. Standard iterative methods for $\Lambda_{[s]}$ with $M_{[s]} = db(\Lambda^{1010})$. Reduced errors and spectral radius for various mixtures.

	Mixture 1	Mixture 2	Mixture 3
1	1.14E-1	6.52E-2	9.90E-2
2	2.65E-2	1.18E-2	2.65E-2
3	6.50E-3	2.07E-3	7.16E-3
4	1.61E-3	3.64E-4	1.94E-3
ρ	2.47E-1	1.76E-1	2.70E-1

$\lambda - \lambda_{[01]}$ and $\chi - \chi_{[01]}$, respectively. At variance with the simplified volume viscosity $\kappa_{[01]}$, the simplified thermal conductivity $\lambda_{[01]}$ and thermal diffusion ratios $\chi_{[01]}$ yield poor approximations for λ and χ, respectively. The reduced errors

$$e_{\lambda}^{[i]} = \frac{|\lambda - (\lambda_{[s]}^{[i]} + \lambda_{[01]})|}{\lambda}, \qquad e_{\chi}^{[i]} = \frac{||\chi - (\chi_{[s]}^{[i]} + \chi_{[01]})||_{\infty}}{||\chi||_{\infty}}, \qquad i = 1, 2, 3, 4, \qquad (6.6.5)$$

and the spectral radius of the iteration matrix are presented in Tables 58 and 59, respectively, for mixtures 4, 5, and 6. In particular, one can see that after two iterations the approximation for the thermal conductivity is within 3E-2 accuracy, and after an additional iteration, the one for the thermal diffusion ratios is within 2E-2 accuracy.

Conjugate Gradient Methods. We consider the initialization $x_0 = 0$ and the preconditioner $M = db(\Lambda)$, for which Theorem 5.6.2 applies. From the numerical results given in Tables 60 and 61, we observe that after one iteration, the approximation for the thermal conductivity given by

$$\lambda^{[1]} = \frac{\bar{p}}{\bar{T}} \frac{\langle \beta^{\lambda}, db(\Lambda)^{-1}\beta^{\lambda}\rangle^2}{\langle db(\Lambda)^{-1}\beta^{\lambda}, \Lambda db(\Lambda)^{-1}\beta^{\lambda}\rangle}. \qquad (6.6.6)$$

Table 59. Thermal diffusion ratios. Standard iterative methods for $\Lambda_{[s]}$ with $M_{[s]} = db(\Lambda^{1010})$. Reduced errors for various mixtures.

	Mixture 1	Mixture 2	Mixture 3
1	1.90E-1	3.52E-2	2.70E-1
2	4.17E-2	7.04E-3	7.01E-2
3	1.02E-2	1.26E-3	1.89E-2
4	2.51E-3	2.17E-4	5.11E-3

Table 60. Thermal conductivity. Conjugate gradient methods for Λ with $M = db(\Lambda)$. Reduced errors for various mixtures.

	Mixture 1	Mixture 2	Mixture 3
1	1.77E-2	8.19E-3	2.38E-2
2	2.43E-5	3.39E-6	1.55E-5
3	7.39E-9	9.22E-9	3.15E-9
4	9.17E-13	9.67E-12	2.18E-14

Table 61. Thermal diffusion ratios. Conjugate gradient methods for Λ with $M = db(\Lambda)$. Reduced errors for various mixtures.

	Mixture 1	Mixture 2	Mixture 3
1	8.48E-2	3.61E-2	1.02E-1
2	8.37E-3	8.66E-4	1.10E-2
3	1.13E-4	9.19E-5	2.44E-5
4	1.32E-6	3.27E-6	2.35E-7

is within 2E-2 accuracy. After an additional iteration, the resulting approximation for the thermal diffusion ratios is within 1E-2 accuracy. For the thermal diffusion ratios, we obtain an approximation within 1E-2 accuracy after two iterations. The preconditioner $M = \mathrm{diag}(\Lambda)$ also yields accurate expressions for the thermal conductivity since the first iterate is within 4E-2 accuracy. Slower convergence rates, however, are obtained for the thermal diffusion ratios for the first iterations.

We now investigate conjugate gradient methods for the Schur complement $\Lambda_{[s]}$ given by (6.6.4). We consider the initialization $x_0 = 0$ and the preconditioner $M_{[s]} = db(\Lambda^{1010})$, for which the results of Section 5.6.6 apply. The numerical results are pre-

Table 62. Thermal conductivity. Conjugate gradient methods for $\Lambda_{[s]}$ with $M_{[s]} = db(\Lambda^{1010})$. Reduced errors for various mixtures.

	Mixture 1	Mixture 2	Mixture 3
1	1.15E-2	6.58E-3	9.32E-3
2	3.98E-6	2.03E-7	6.54E-7
3	2.31E-10	2.24E-10	7.80E-11
4	7.28E-14	2.23E-15	2.00E-15

Table 63. Thermal diffusion ratios. Conjugate gradient methods for $\Lambda_{[s]}$ with $M_{[s]} = db(\Lambda^{1010})$. Reduced errors for various mixtures.

	Mixture 1	Mixture 2	Mixture 3
1	1.70E-1	8.48E-2	3.02E-1
2	5.14E-3	6.14E-4	3.24E-3
3	3.36E-5	4.41E-6	3.58E-5
4	5.94E-7	8.31E-10	1.82E-7

Table 64. Thermal conductivity and thermal diffusion ratios. Reduced errors for $\lambda_{[e]}$ and $\chi_{[e]}$.

	Mixture 1	Mixture 2	Mixture 3
$e_{\lambda_{[e]}}$	8.20E-4	7.09E-4	1.90E-3
$e_{\chi_{[e]}}$	1.44E-2	1.67E-2	3.98E-2

sented in Tables 62 and 63. After one iteration, the approximation for the thermal conductivity is within 1E-2 accuracy, whereas, for the thermal diffusion ratios, the iterates are within 5E-3 accuracy after two iterations.

6.6.2 Numerical Experiments for the System $\Lambda_{[e]}\alpha^{\lambda}_{[e]} = \beta^{\lambda}_{[e]}$

In Table 64 we present the reduced errors

$$e_{\lambda_{[e]}} = \frac{|\lambda - \lambda_{[e]}|}{\lambda}, \qquad e_{\chi_{[e]}} = \frac{||\chi - \chi_{[e]}||_\infty}{||\chi||_\infty}, \tag{6.6.7}$$

which indicate that the simplified transport coefficients $\lambda_{[e]}$ and $\chi_{[e]}$ are accurate to 2E-3 and 2E-2, respectively, for mixtures 1, 2, and 3. This new formulation is therefore an attractive alternative to the full system $\Lambda\alpha^{\lambda} = \beta^{\lambda}$.

Table 65. Thermal conductivity. Standard iterative methods for $\Lambda_{[e]}$ with $M_{[e]} = db(\Lambda_{[e]})$. Reduced errors and spectral radius for various mixtures.

	Mixture 1	Mixture 2	Mixture 3
1	1.56E-1	8.47E-2	1.55E-1
2	3.16E-2	1.21E-2	2.78E-2
3	6.49E-3	1.67E-3	4.97E-3
4	1.34E-3	2.31E-4	8.91E-4
ρ	6.62E-2	1.34E-1	2.35E-2

Table 66. Thermal diffusion ratios. Standard iterative methods for $\Lambda_{[e]}$ with $M_{[e]} = db(\Lambda_{[e]})$. Reduced errors for various mixtures.

	Mixture 1	Mixture 2	Mixture 3
1	2.21E-1	3.62E-2	2.92E-1
2	4.28E-2	5.81E-3	5.10E-2
3	8.81E-3	8.20E-4	9.13E-3
4	1.82E-3	1.10E-4	1.63E-3

Standard Iterative Methods. We consider the initialization $x_0 = 0$ and the splitting

$$\Lambda_{[e]} = M_{[e]} - Z_{[e]}, \qquad M_{[e]} = db(\Lambda_{[e]}). \tag{6.6.8}$$

One can then easily verify that Theorem 5.6.4 applies. The reduced errors

$$e^{[i]}_{\lambda_{[e]}} = \frac{|\lambda_{[e]} - \lambda^{[i]}_{[e]}|}{\lambda_{[e]}}, \qquad e^{[i]}_{\chi_{[e]}} = \frac{||\chi_{[e]} - \chi^{[i]}_{[e]}||_\infty}{||\chi_{[e]}||_\infty}, \qquad i = 1,2,3,4, \tag{6.6.9}$$

and the spectral radius of the iteration matrix $\rho = \rho(M^{-1}_{[e]} Z_{[e]})$ are presented in Tables 65 and 66. In particular, after two iterations, the quantity $\lambda^{[2]}_{[e]}$ is within 3E-2 accuracy of $\lambda_{[e]}$, whereas the vector $\chi^{[2]}_{[e]}$ is within 5E-2 accuracy of $\chi_{[e]}$.

Conjugate Gradient Methods. We consider the initialization $x_0 = 0$ and the preconditioner $M_{[e]} = db(\Lambda_{[e]})$, for which Theorem 5.6.5 applies. The numerical results are given in Tables 67 and 68. In particular, we can see that after one iteration, we obtain the expression

$$\lambda^{[1]}_{[e]} = \frac{\bar{p}}{\overline{T}} \frac{\langle \beta^{\lambda}_{[e]}, db(\Lambda_{[e]})^{-1} \beta^{\lambda}_{[e]} \rangle^2}{\langle db(\Lambda_{[e]})^{-1} \beta^{\lambda}_{[e]}, \Lambda db(\Lambda_{[e]})^{-1} \beta^{\lambda}_{[e]} \rangle}, \tag{6.6.10}$$

Table 67. Thermal conductivity. Conjugate gradient methods for $\Lambda_{[e]}$ with $M_{[e]} = db(\Lambda_{[e]})$. Reduced errors for various mixtures.

	Mixture 1	Mixture 2	Mixture 3
1	9.86E-3	5.90E-3	5.12E-3
2	6.00E-6	5.72E-7	4.03E-7
3	1.49E-10	7.30E-11	7.40E-12
4	4.00E-15	3.24E-17	1.12E-16

Table 68. Thermal diffusion ratios. Conjugate gradient methods for $\Lambda_{[e]}$ with $M_{[e]} = db(\Lambda_{[e]})$. Reduced errors for various mixtures.

	Mixture 1	Mixture 2	Mixture 3
1	1.83E-1	7.08E-2	2.28E-1
2	4.01E-3	8.80E-4	2.00E-3
3	2.01E-5	8.95E-7	8.07E-6
4	1.16E-7	2.98E-12	2.64E-8

which is within 1E-2 accuracy of $\lambda_{[e]}$. After a second iteration, the approximation for the thermal diffusion ratios is within 4E-3 accuracy of $\chi_{[e]}$.

Finally, this simplified formulation can be used to obtain an accurate initialization for iterative methods applied to the system $\Lambda\alpha^\lambda = \beta^\lambda$. Using the same ideas as in Section 6.5.2, we consider the initialization

$$\begin{cases} x_{0k}^{10\lambda} = x_{0k}^{01\lambda} = \dfrac{c_k^{\mathrm{int}}}{k_{\mathrm{B}}} \dfrac{X_k}{\Lambda_{kk}^{0101}}, & k \in \mathcal{P}, \\ x_{0k}^{10\lambda} = \dfrac{1}{\sum_{l\in\mathcal{S}} X_l/\mathcal{D}_{kl}}, & k \in \mathcal{S}\setminus\mathcal{P}, \end{cases} \tag{6.6.11}$$

which yields an approximation for λ and χ within 1E-1 accuracy.

6.6.3 Numerical Experiments for the System $\widetilde{\Lambda}\alpha^\lambda = \widetilde{\beta}^\lambda$

We consider the initialization $x_0 = 0$ and the splitting

$$\widetilde{\Lambda} = \widetilde{M} - \widetilde{Z}, \qquad \widetilde{M} = db(\widetilde{\Lambda}), \tag{6.6.12}$$

where $db(\widetilde{\Lambda})$ is formed by the diagonal of the four blocks of the matrix $\widetilde{\Lambda}$. One can then easily verify that Theorem 5.6.7 applies. The reduced errors

$$e_\lambda^{[i]} = \frac{|\lambda - \lambda^{[i]}|}{\lambda}, \qquad e_\chi^{[i]} = \frac{\|\chi - \chi^{[i]}\|_\infty}{\|\chi\|_\infty}, \qquad i = 1,2,3,4, \tag{6.6.13}$$

Table 69. Thermal conductivity. Standard iterative methods for $\widetilde{\Lambda}$ with $\widetilde{M} = db(\widetilde{\Lambda})$. Reduced errors and spectral radius for various mixtures.

	Mixture 4	Mixture 5	Mixture 6
1	7.98E-2	3.35E-2	1.34E-1
2	1.34E-2	1.68E-2	3.63E-2
3	2.33E-3	7.41E-5	1.01E-2
4	4.08E-4	3.72E-6	2.79E-3
ρ	1.75E-1	4.70E-2	2.78E-1

Table 70. Thermal diffusion ratios. Standard iterative methods for $\widetilde{\Lambda}$ with $\widetilde{M} = db(\widetilde{\Lambda})$. Reduced errors for various mixtures.

	Mixture 4	Mixture 5	Mixture 6
1	3.88E-2	3.26E-2	1.24E-1
2	6.39E-3	2.21E-3	3.21E-2
3	1.08E-3	7.21E-5	8.88E-3
4	1.89E-4	4.90E-6	2.47E-3

are presented in Tables 69 and 70, respectively, together with the spectral radius of the iteration matrix $\rho = \rho(\widetilde{M}^{-1}\widetilde{Z})$. After two iterations, we obtain approximations for the thermal conductivity and for the thermal diffusion ratios which are within 4E-2 and 3E-2 accuracy, respectively. These expressions coincide with (6.6.3) for positive mass fractions.

We now consider standard iterative methods for the rescaled Schur complement

$$\widetilde{\Lambda}_{[s]} = \widetilde{\Lambda}^{1010} - \widetilde{\Lambda}^{1001}\big(\widetilde{\Lambda}^{0101}\big)^{-1}\widetilde{\Lambda}^{0110}, \tag{6.6.14}$$

introduced in Section 5.6.6. We consider the initialization $x_0 = 0$ and the splitting induced by $\widetilde{M}_{[s]} = db(\widetilde{\Lambda}^{1010})$. As stated in Section 5.6.6, this splitting yields a convergent iteration matrix, and the quantities $\lambda_{[s]}^{[i]}$ and the vectors $\chi_{[s]}^{[i]}$ converge towards $\lambda - \lambda_{[01]}$ and $\chi - \chi_{[01]}$, respectively. The reduced errors

$$e_\lambda^{[i]} = \frac{|\lambda - (\lambda_{[s]}^{[i]} + \lambda_{[01]})|}{\lambda}, \qquad e_\chi^{[i]} = \frac{||\chi - (\chi_{[s]}^{[i]} + \chi_{[01]})||_\infty}{||\chi||_\infty}, \qquad i = 1, 2, 3, 4, \tag{6.6.15}$$

and the spectral radius of the iteration matrix are presented in Tables 71 and 72. After

Table 71. Thermal conductivity. Standard iterative methods for $\widetilde{\Lambda}_{[s]}$ with $\widetilde{M}_{[s]} = db(\widetilde{\Lambda}^{1010})$. Reduced errors and spectral radius for various mixtures.

	Mixture 4	Mixture 5	Mixture 6
1	6.51E-2	2.31E-2	1.00E-1
2	1.18E-2	1.16E-3	2.70E-2
3	2.07E-3	4.16E-5	7.36E-3
4	3.64E-4	2.00E-6	2.01E-3
ρ	1.76E-1	4.20E-2	2.73E-1

Table 72. Thermal diffusion ratios. Standard iterative methods for $\widetilde{\Lambda}_{[s]}$ with $\widetilde{M}_{[s]} = db(\widetilde{\Lambda}^{1010})$. Reduced errors for various mixtures.

	Mixture 4	Mixture 5	Mixture 6
1	3.52E-2	3.07E-2	2.63E-1
2	7.03E-3	2.36E-3	6.85E-2
3	1.26E-3	5.93E-5	1.86E-2
4	2.17E-4	3.97E-6	5.08E-3

two iterations, we obtain approximations for the thermal conductivity and the thermal diffusion ratios within 3E-2 and 7E-2 accuracy, respectively.

6.6.4 Numerical Experiments for the System $\widehat{\Lambda}\widehat{\alpha}^{\lambda} = \widehat{\beta}^{\lambda}$

We consider the initialization $\widehat{x}_0 = 0$ and the preconditioner $\widehat{M} = db(\widehat{\Lambda})$, for which Theorem 5.6.9 applies. The reduced errors

$$e_{\lambda}^{[i]} = \frac{|\lambda - \lambda^{[i]}|}{\lambda}, \qquad e_{\chi}^{[i]} = \frac{||\chi - \chi^{[i]}||_{\infty}}{||\chi||_{\infty}}, \qquad i = 1,2,3,4, \tag{6.6.16}$$

are reported in Table 73 and 74, respectively, and indicate high convergence rates. In particular, the first iterate $\lambda^{[1]}$ is within 2E-2 accuracy and coincides with (6.6.6) for positive mass fractions. The second iterate $\chi^{[2]}$ is within 1E-2 accuracy.

6.6.5 Numerical Experiments for the System $\widetilde{\Lambda}_{[e]}\alpha_{[e]}^{\lambda} = \widetilde{\beta}_{[e]}^{\lambda}$

In Table 75 we present the reduced errors

$$e_{\lambda_{[e]}} = \frac{|\lambda - \lambda_{[e]}|}{\lambda}, \qquad e_{\chi_{[e]}} = \frac{||\chi - \chi_{[e]}||_{\infty}}{||\chi||_{\infty}}, \tag{6.6.17}$$

Table 73. Thermal conductivity. Conjugate gradient methods for $\widehat{\Lambda}$ with $\widehat{M} = db(\widehat{\Lambda})$. Reduced errors for various mixtures.

	Mixture 4	Mixture 5	Mixture 6
1	8.18E-3	5.99E-4	2.39E-2
2	3.38E-6	3.48E-7	1.51E-5
3	9.21E-9	3.82E-16	2.93E-9
4	9.61E-12	—	1.99E-14

Table 74. Thermal diffusion ratios. Conjugate gradient methods for $\widehat{\Lambda}$ with $\widehat{M} = db(\widehat{\Lambda})$. Reduced errors for various mixtures.

	Mixture 4	Mixture 5	Mixture 6
1	3.61E-2	2.95E-4	9.87E-2
2	8.67E-4	5.78E-4	1.06E-2
3	9.18E-5	4.65E-10	2.00E-5
4	3.26E-6	—	2.22E-7

Table 75. Thermal conductivity and thermal diffusion ratios. Reduced errors for $\lambda_{[e]}$ and $\chi_{[e]}$.

	Mixture 4	Mixture 5	Mixture 6
$e_{\lambda_{[e]}}$	7.09E-4	7.47E-4	1.41E-3
$e_{\chi_{[e]}}$	1.67E-2	8.99E-3	3.70E-2

which indicate that the simplified transport coefficients $\lambda_{[e]}$ and $\chi_{[e]}$ are accurate to 1E-3 and 2E-2, respectively, for mixtures 4, 5, and 6. This new formulation is therefore an attractive alternative to the full system $\widetilde{\Lambda}\alpha^\lambda = \widetilde{\beta}^\lambda$.

We consider the initialization $x_0 = 0$ and the splitting

$$\widetilde{\Lambda}_{[e]} = \widetilde{M}_{[e]} - \widetilde{Z}_{[e]}, \qquad \widetilde{M}_{[e]} = db(\widetilde{\Lambda}_{[e]}). \tag{6.6.18}$$

One can then easily verify that Theorem 5.6.11 applies. The reduced errors

$$e_{\lambda_{[e]}}^{[i]} = \frac{|\lambda_{[e]} - \lambda_{[e]}^{[i]}|}{\lambda_{[e]}}, \qquad e_{\chi_{[e]}}^{[i]} = \frac{||\chi_{[e]} - \chi_{[e]}^{[i]}||_\infty}{||\chi_{[e]}||_\infty}, \qquad i = 1, 2, 3, 4, \tag{6.6.19}$$

and the spectral radius of the iteration matrix $\rho = \rho(\widetilde{M}_{[e]}^{-1}\widetilde{Z}_{[e]})$ are presented in Tables 76 and 77. In particular, after two iterations, we obtain an approximation for the

Table 76. Thermal conductivity. Standard iterative methods for $\widetilde{\Lambda}_{[e]}$ with $\widetilde{M}_{[e]} = db(\widetilde{\Lambda}_{[e]})$. Reduced errors and spectral radius for various mixtures.

	Mixture 4	Mixture 5	Mixture 6
1	8.47E-2	3.41E-2	1.55E-1
2	1.21E-2	1.55E-3	2.78E-2
3	1.67E-3	5.28E-5	4.98E-3
4	2.31E-4	2.40E-6	8.93E-4
ρ	1.34E-1	3.94E-2	2.40E-2

Table 77. Thermal diffusion ratios. Standard iterative methods for $\widetilde{\Lambda}_{[e]}$ with $\widetilde{M}_{[e]} = db(\widetilde{\Lambda}_{[e]})$. Reduced errors for various mixtures.

	Mixture 4	Mixture 5	Mixture 6
1	3.61E-2	3.17E-2	2.86E-1
2	5.80E-3	2.19E-3	5.00E-2
3	8.19E-4	4.92E-5	8.97E-3
4	1.10E-4	3.40E-6	1.61E-3

thermal conductivity within 3E-2 accuracy of $\lambda_{[e]}$, whereas the one for the thermal diffusion ratios is within 5E-2 accuracy of $\chi_{[e]}$.

6.6.6 Mixture-Averaged Formulas

As described in Section 6.1.4, we now consider several mixture-averaged formulas for the thermal conductivity. These expressions are based on the pure species thermal conductivities λ_k, $k \in \mathcal{S}$. In Table 78 we present the reduced errors

$$e_t = \frac{|\lambda - \mathcal{M}_t(\lambda)|}{\lambda}, \tag{6.6.20}$$

for various values of the parameter t and for the six mixtures considered in our numerical experiments. The reduced error for the approximation introduced in [BW53] and considered, for instance, in [MTS67] [KDWCM86]

$$\mathcal{M}_{-1,1}(\lambda) = \frac{1}{2}(\mathcal{M}_{-1}(\lambda) + \mathcal{M}_1(\lambda)) = \frac{1}{2}\Big(\big(\sum_{k\in\mathcal{S}} X_k(\lambda_k)^{-1}\big)^{-1} + \sum_{k\in\mathcal{S}} X_k\lambda_k\Big), \tag{6.6.21}$$

Table 78. Thermal conductivity. Reduced errors for various mixture-averaged formulas.

	Mixture 1	Mixture 2	Mixture 3	Mixture 4	Mixture 5	Mixture 6
-1	2.20E-1	3.00E-1	1.21E-1	3.00E-1	3.41E-1	1.19E-1
0	3.75E-2	8.01E-2	1.60E-2	8.02E-2	6.82E-2	1.74E-2
1/4	3.75E-2	1.01E-2	2.39E-2	1.01E-2	2.58E-2	2.23E-2
1/2	1.28E-1	1.15E-1	7.24E-2	1.15E-1	1.24E-1	7.12E-2
1	3.56E-1	3.52E-1	2.02E-1	3.52E-1	3.17E-1	2.06E-1
-1,1	6.83E-2	2.61E-2	2.82E-2	2.61E-2	1.19E-2	4.34E-2

is also included in Table 78. From this table we can see that the average formula of order 1/4 yields approximations within 4E-2 accuracy, whereas the expression (6.6.21) is accurate to 7E-2 only.

6.6.7 Discussion

From the preceding numerical experiments, we can draw the following conclusions.

The Thermal Conductivity. The average formula of order 1/4,

$$\lambda = \Big(\sum_{k\in\mathcal{S}} X_k(\lambda_k)^{1/4}\Big)^4, \tag{6.6.22}$$

where λ_k, $k \in \mathcal{S}$, are the pure species thermal conductivities, only requires $\mathcal{O}(n)$ operations for its evaluation and is accurate to 4E-2 for the mixtures considered in this book. The average formula of order zero

$$\lambda = \exp\Big(\sum_{k\in\mathcal{S}} X_k \log(\lambda_k)\Big), \tag{6.6.23}$$

is accurate to 8E-2 only, but less computationally expensive when the pure species thermal conductivities are fitted in terms of the logarithm of the temperature [KWM83].

Furthermore, the simplified system matrix $\Lambda_{[e]}$ is an interesting alternative to the standard system matrix Λ, since it is of size n instead of $n+p$. Moreover, the resulting simplified thermal conductivity $\lambda_{[e]}$ is within 2E-3 accuracy for the mixtures considered in this book. A new, rigorously derived, analytic expression is obtained after one conjugate gradient iteration for the system $\Lambda_{[e]}\alpha^{\lambda}_{[e]} = \beta^{\lambda}_{[e]}$, preconditioned by the matrix $db(\Lambda_{[e]})$. This yields the thermal conductivity

$$\lambda^{[1]}_{[e]} = \frac{\bar{p}}{\bar{\bar{T}}} \frac{\langle \beta^{\lambda}_{[e]}, db(\Lambda_{[e]})^{-1}\beta^{\lambda}_{[e]}\rangle^2}{\langle db(\Lambda_{[e]})^{-1}\beta^{\lambda}_{[e]}, \Lambda_{[e]} db(\Lambda_{[e]})^{-1}\beta^{\lambda}_{[e]}\rangle}, \tag{6.6.24}$$

which requires $n^2 + \mathcal{O}(n)$ operations for its evaluation and is within 1E-2 accuracy of $\lambda_{[e]}$. Note that the matrix $db(\Lambda_{[e]})$ consists of one diagonal block so that the product of $db(\Lambda_{[e]})^{-1}$ with a given vector only requires $\mathcal{O}(n)$ operations.

Another rigorously derived analytic expression is obtained after one conjugate gradient iteration for the system $\Lambda\alpha^\lambda = \beta^\lambda$, preconditioned by the matrix $db(\Lambda)$. This yields the new expression

$$\lambda^{[1]} = \frac{\bar{p}}{\overline{T}} \frac{\langle \beta^\lambda, db(\Lambda)^{-1}\beta^\lambda\rangle^2}{\langle db(\Lambda)^{-1}\beta^\lambda, \Lambda db(\Lambda)^{-1}\beta^\lambda\rangle}, \tag{6.6.25}$$

which requires $4n^2 + \mathcal{O}(n)$ operations for its evaluation and is within 2E-2 accuracy of λ. Recalling that the matrix $db(\Lambda)$ is formed by the diagonal of the four blocks of the matrix Λ, the vector $db(\Lambda)^{-1}\beta^\lambda$ is evaluated in $\mathcal{O}(n)$ operations by directly solving n dense linear symmetric systems of size 2 or 1. Instead of the matrix $db(\Lambda)$, one can also consider the simpler preconditioner $\mathrm{diag}(\Lambda)$ which can be trivially inverted. The expression for the thermal conductivity, after one conjugate gradient iteration, is obtained from (6.6.25) by just replacing the matrix $db(\Lambda)^{-1}$ by $\mathrm{diag}(\Lambda)^{-1}$. Its computational cost is still $4n^2 + \mathcal{O}(n)$ operations, and it is accurate to 3E-2. Note, however, that the preconditioner $\mathrm{diag}(\Lambda)$ yields slower convergence rates than $db(\Lambda)$ for the thermal diffusion ratios. If both transport coefficients are to be evaluated simultaneously, it is therefore preferable to use $db(\Lambda)$ as a common preconditioner.

More detailed analytic expressions are obtained after two conjugate gradient iterations for the system $\Lambda\alpha^\lambda = \beta^\lambda$, preconditioned by the matrix $db(\Lambda)$, with an accuracy below 2E-5.

The Thermal Diffusion Ratios. The simplified thermal diffusion ratios $\chi_{[e]}$ are within 4E-2 accuracy for the mixtures considered in this book. They may therefore be an interesting alternative to the standard thermal diffusion ratios χ since they are associated with a linear system of size n instead of $n+p$. Two standard iterations for the system $\Lambda_{[e]}\alpha^\lambda_{[e]} = \beta^\lambda_{[e]}$ yield a new, rigorously derived, analytic expression which is within 5E-2 accuracy of $\chi_{[e]}$. The corresponding vector $\chi^{[2]}_{[e]}$ is given by

$$\chi^{[2]}_{[e]} = [L^{00e}_{[e]}](2I - db(\Lambda_{[e]})^{-1}\Lambda_{[e]})db(\Lambda_{[e]})^{-1}\beta^\lambda_{[e]}, \tag{6.6.26}$$

where the matrix $db(\Lambda_{[e]})$ is formed by the diagonal of the matrix $\Lambda_{[e]}$ and $\chi^{[2]}_{[e]}$ requires, therefore, $2n^2 + \mathcal{O}(n)$ operations for its evaluation.

Another new, rigorously derived, analytic expression is obtained after two standard iterations for the system $\Lambda\alpha^\lambda = \beta^\lambda$. The vector $\chi^{[2]}$ is within 3E-2 accuracy of χ and can be expressed as

$$\chi^{[2]} = [L^{0010}, L^{0001}](2I - db(\Lambda)^{-1}\Lambda)db(\Lambda)^{-1}\beta^{\lambda}. \qquad (6.6.27)$$

The matrix $db(\Lambda)$ is formed by the diagonal of the four blocks of the matrix Λ, so that the product of $db(\Lambda)^{-1}$ times a given vector requires only $\mathcal{O}(n)$ operations, and hence, the computational cost of $\chi^{[2]}$ is $6n^2 + \mathcal{O}(n)$ operations.

More detailed analytic expressions are derived after three conjugate gradient iterations for the system $\Lambda\alpha^{\lambda} = \beta^{\lambda}$, preconditioned by the matrix $db(\Lambda)$. The new thermal diffusion ratios $\chi^{[3]}$ require $14n^2 + \mathcal{O}(n)$ operations for their evaluation and are within 1E-4 accuracy.

7 Concluding Remarks

In this book we have shown that the transport linear systems are symmetric and take either the nonsingular form $G\alpha^\mu = \beta^\mu$ or the constrained singular form $G\alpha^\mu = \beta^\mu$ and $\langle \mathcal{G}, \alpha^\mu \rangle = 0$. The transport coefficient μ is then expressed typically as the scalar product $\mu = \langle \alpha^\mu, \beta^\mu \rangle$. These systems are well posed, i.e., they admit a unique solution, and the singular systems can be cast into a nonsingular form by considering the symmetric positive definite matrix $G + a\mathcal{G}\otimes\mathcal{G}$, where a is a positive real number. Furthermore, the transport linear systems can be appropriately rescaled in order to establish the smoothness, in the limit of vanishing mass fractions, of all the transport coefficients, provided that the diffusion matrix is replaced by the flux diffusion matrix. In particular, we have obtained the expression of all the transport coefficients in the practically important dilution limit.

Iterative algorithms are an interesting and appealing alternative to direct inversions for transport property evaluation. Indeed, they are generally less computationally expensive than direct inversions, especially for mixtures with a large number of species. Moreover, they provide a rigorous and general way to define analytic expressions for the transport coefficients by truncation. The resulting expressions are more accurate than empirical mixture-averaged formulas and can be evaluated at a moderately higher computational cost.

A fundamental matrix for iterative algorithms is the matrix $db(G)$ formed by the diagonal of all the blocks of the system matrix G. The mathematical properties of the matrices $db(G)$ and $2db(G) - G$ have been derived directly from the kinetic theory, and using these properties, convergence theorems for standard iterative and conjugate gradient methods have been proven for all the transport linear systems. A stabilized version, for vanishing mass fractions, of these algorithms has been introduced also, as well as a projected version for the singular systems. On the other hand, all the mathematical results and all the convergence theorems are still valid when the system coefficients are estimated using practical approximations. Finally, various strategies, designed to optimize transport property evaluation in multicomponent flow computations, have been described.

Extensions of several aspects of the present theory could be considered. It is first possible to use approximation polynomials in the different internal energy modes, e.g., rotation and vibration, which leads to larger variational spaces and therefore to larger transport linear systems. Another interesting extension would be to consider the full species vibrational desequilibrium already at the zeroth order governing equations. Furthermore, instead of using the semi-classical kinetic theory of gases as a starting point, one could also consider a fully quantum mechanical, a fully classical, or a discrete framework.

Appendix A

In this appendix we present some notation used for tensors of rank zero, one, and two over $\mathbb{R}^3$. Tensors of rank zero are simply scalars, tensors of rank one are three-dimensional vectors and tensors of rank two are three by three matrices. We denote by $\mathfrak{e}^1 = (1,0,0)$, $\mathfrak{e}^2 = (0,1,0)$, and $\mathfrak{e}^3 = (0,0,1)$ the canonical basis of $\mathbb{R}^3$, and for any $\mathfrak{e}, \mathfrak{f} \in \mathbb{R}^3$, we denote by $\mathfrak{e}\otimes\mathfrak{f}$ the three by three matrix with coefficients $\mathfrak{e}_i\mathfrak{f}_j$, $i, j \in [1,3]$.

For a given tensorial rank a, the space of tensors of rank a over the three-dimensional space is of dimension $\tau = 3^a$. We denote by $\mathcal{T}_\nu$, $\nu \in [1,\tau]$, the canonical basis of the space of tensors of order a over $\mathbb{R}^3$. Any tensor of rank a has then components with respect to the basis $\mathcal{T}_\nu$, $\nu \in [1,\tau]$, which are denoted by x_ν, $\nu \in [1,\tau]$, so that

$$x = \sum_{\nu\in[1,\tau]} x_\nu \, \mathcal{T}_\nu.$$

The canonical basis $\mathcal{T}_\nu$, $\nu \in [1,\tau]$, is described as follows.

Tensors of Rank Zero. In the trivial scalar case $a = 0$, we have $\tau = 1$ and

$$\mathcal{T}_1 = (1).$$

Tensors of Rank One. In the vector case $a = 1$, we have $\tau = 3$ and $\mathcal{T}_i = \mathfrak{e}^i$ so that

$$\mathcal{T}_1 = (1,0,0), \qquad \mathcal{T}_2 = (0,1,0), \qquad \mathcal{T}_3 = (0,0,1).$$

Tensors of Rank Two. In the matrix case $a = 2$, we have $\tau = 9$ and $\mathcal{T}_{i+3(j-1)} = \mathfrak{e}^i\otimes\mathfrak{e}^j$, so that

$$\mathcal{T}_1 = \begin{pmatrix} 1 & 0 & 0 \\ 0 & 0 & 0 \\ 0 & 0 & 0 \end{pmatrix}, \quad \mathcal{T}_2 = \begin{pmatrix} 0 & 0 & 0 \\ 1 & 0 & 0 \\ 0 & 0 & 0 \end{pmatrix}, \quad \mathcal{T}_3 = \begin{pmatrix} 0 & 0 & 0 \\ 0 & 0 & 0 \\ 1 & 0 & 0 \end{pmatrix},$$

$$\mathcal{T}_4 = \begin{pmatrix} 0 & 1 & 0 \\ 0 & 0 & 0 \\ 0 & 0 & 0 \end{pmatrix}, \quad \mathcal{T}_5 = \begin{pmatrix} 0 & 0 & 0 \\ 0 & 1 & 0 \\ 0 & 0 & 0 \end{pmatrix}, \quad \mathcal{T}_6 = \begin{pmatrix} 0 & 0 & 0 \\ 0 & 0 & 0 \\ 0 & 1 & 0 \end{pmatrix},$$

$$\mathcal{T}_7 = \begin{pmatrix} 0 & 0 & 1 \\ 0 & 0 & 0 \\ 0 & 0 & 0 \end{pmatrix}, \qquad \mathcal{T}_8 = \begin{pmatrix} 0 & 0 & 0 \\ 0 & 0 & 1 \\ 0 & 0 & 0 \end{pmatrix}, \qquad \mathcal{T}_9 = \begin{pmatrix} 0 & 0 & 0 \\ 0 & 0 & 0 \\ 0 & 0 & 1 \end{pmatrix},$$

Tensor Contraction. Let x and y be any tensor of rank zero, one or two with respect to the physical three-dimensional space $\mathbb{R}^3$. We then denote by $x \odot y$ the maximum contracted product between the tensors x and y. More specifically, we have $x \odot y = xy$ if either x or y is a scalar, $x \odot y = x \cdot y$ if both x and y are three-dimensional vectors, $x \odot y = x : y$ if both x and y are three by three symmetric matrices, $x \odot y = xy$ if x is a matrix and y a vector, and $x \odot y = y^t x$ if x is a vector and y a matrix, where y^t is the transpose of y. In particular, for $\mathfrak{e}, \mathfrak{f} \in \mathbb{R}^3$, we have $\mathfrak{e} \odot (\mathfrak{f} \otimes \mathfrak{f}) = (\mathfrak{f} \otimes \mathfrak{f}) \odot \mathfrak{e} = (\mathfrak{e} \cdot \mathfrak{f})\mathfrak{f}$ and $(\mathfrak{e} \otimes \mathfrak{e}) \odot (\mathfrak{f} \otimes \mathfrak{f}) = (\mathfrak{f} \otimes \mathfrak{f}) \odot (\mathfrak{e} \otimes \mathfrak{e}) = (\mathfrak{e} \cdot \mathfrak{f})^2$.

Finally, we point out that the notation "$\cdot$", "$:$", and "$\odot$" is restricted, in this book, to tensors of rank zero, one or two with respect to the three-dimensional physical space $\mathbb{R}^3$.

Appendix B

In this appendix we give the expression and properties of the Laguerre and Sonine polynomials, and the Wang Chang and Uhlenbeck polynomials.

The Laguerre and Sonine Polynomials. Let α be a positive number and i be an integer. The Laguerre and Sonine polynomial of order i with parameter α is then defined by

$$S_\alpha^i(x) = \frac{1}{i!} x^{-\alpha} e^x \frac{d^i}{dx^i}(x^{i+\alpha} e^{-x}) = \sum_{j\in[0,i]} \frac{(-1)^j}{j!} \binom{i+\alpha}{i-j} x^j,$$

and S_α^i is of degree i. These polynomials verify the orthogonality relation

$$\int_0^\infty x^\alpha e^{-x} S_\alpha^i(x) S_\alpha^j(x)\, dx = \delta_{ij}\, \frac{\Gamma(i+1+\alpha)}{i!}, \qquad i,j \geq 0,$$

where Γ is the Euler function, and the recurrence relation

$$(i+1)S_\alpha^{i+1}(x) = (2i+1+\alpha-x)S_\alpha^i(x) - (i+\alpha)S_\alpha^{i-1},$$

and can be evaluated from the first iterates given by

$$\begin{cases} S_\alpha^0(x) = 1, \\ S_\alpha^1(x) = -x+1+\alpha. \end{cases}$$

Only these first iterates have been used in this book for $\alpha = 1/2$, $\alpha = 3/2$, and $\alpha = 5/2$.

The Wang Chang and Uhlenbeck Polynomials. The Wang Chang and Uhlenbeck polynomials associated with the kth species are defined as follows. We first consider the finite set of internal reduced energies $\{\, \epsilon_{kK},\ K \in \mathcal{E}_k \,\}$ and the associated positive degeneracies $\{\, a_{kK},\ K \in \mathcal{E}_k \,\}$. For any polynomials P and Q, we define the product

$$((P,Q)) = \frac{1}{Q_k} \sum_{K\in\mathcal{E}_k} a_{kK} P(\epsilon_{kK}) Q(\epsilon_{kK}) \exp(-\epsilon_{kK}),$$

where $Q_k = \sum_{K \in \mathcal{E}_k} a_{kK} \exp(-\epsilon_{kK})$, and $(\!(\ , \ )\!)$ is then a quadratic form. This form is positive definite over the polynomials of degree $d < \mathrm{card}(\mathcal{E}_k)$, where $\mathrm{card}(\mathcal{E}_k)$ is the number of elements of $\mathcal{E}_k$. Indeed, we have

$$(\!(P, P)\!) = 0 \quad \Longleftrightarrow \quad P(\epsilon_{kK}) = 0, \qquad K \in \mathcal{E}_k,$$

and no nonzero polynomial of degree $d < \mathrm{card}(\mathcal{E}_k)$ satisfies this property, since the reduced internal energies are different. We may thus form an orthogonal basis, with respect to this quadratic form, by using the classical orthogonalization procedure. More specifically, the Wang Chang and Uhlenbeck polynomials W_i^k, $0 \le i < \mathrm{card}(\mathcal{E}_k)$, are defined by

$$W_0^k = 1,$$

and the recurrence relation [WT62]

$$W_{i+1}^k = -\Pi W_i^k + \sum_{j \in [0,i]} \frac{(\!(\Pi W_i^k, W_j^k)\!)}{(\!(W_j^k, W_j^k)\!)} W_j^k, \qquad i+1 < \mathrm{card}(\mathcal{E}_k),$$

where $\Pi(x) = x$. The degree of W_i^k is i, and these polynomials verify the orthogonality relations

$$(\!(W_i^k, W_j^k)\!) = \delta_{ij} \, (\!(W_i^k, W_i^k)\!), \qquad 0 \le i, j < \mathrm{card}(\mathcal{E}_k).$$

The first iterates are also given by

$$\begin{cases} W_0^k(x) = 1, \\ W_1^k(x) = \bar{\epsilon}_k - x, \end{cases}$$

where

$$\bar{\epsilon}_k = \sum_{K \in \mathcal{E}_k} a_{kK} \epsilon_{kK} \exp(-\epsilon_{kK}) / Q_k,$$

is the averaged reduced internal energy of the kth molecule. Only these first iterates have been used in this book. Note that the orthogonalization process can still be used for $i = \mathrm{card}(\mathcal{E}_k)$, but then provides the polynomial $W_i^k = \prod_{K \in \mathcal{E}_k} (x - \epsilon_{kK})$ which vanishes over the reduced energies. It is therefore convenient to set $W_i^k = 0$ whenever $i \ge \mathrm{card}(\mathcal{E}_k)$.

The Functions ϕ^{a0cdk}. The basis functions ϕ^{a0cdk} are defined in Section 2.2.2 by

$$\phi^{a0cdk}(c_k, K) = \Big(S_{a+\frac{1}{2}}^{c}(w_k \cdot w_k) \, W_k^d(\epsilon_{kK}) \, \overline{\otimes^a w_k} \, \delta_{ki} \Big)_{i \in S},$$

where a, c, and d are integers, $S^c_{a+1/2}$ is the Laguerre and Sonine polynomial of order c with parameter $a+1/2$, W^d_k the Wang Chang and Uhlenbeck polynomial of order d for the kth species, and $\overline{\otimes^a w_k}$ a tensor of rank a with respect to the three-dimensional space given by $\overline{\otimes^0 w_k} = 1$, $\overline{\otimes^1 w_k} = w_k$, and $\overline{\otimes^2 w_k} = w_k \otimes w_k - \frac{1}{3} w_k \cdot w_k I$ [WT62]. The functions ϕ^{a0cdk} are nonzero provided that $d < \text{card}(\mathcal{E}_k)$. Keeping in mind that for any $a, a', c, c', d, d' \geq 0$ and $k, l \in \mathcal{S}$, the scalar product $\langle\langle \phi^{a0cdk}, \phi^{a'0c'd'l} \rangle\rangle$ is given by

$$\langle\langle \phi^{a0cdk}, \phi^{a'0c'd'l} \rangle\rangle = \sum_{\substack{i \in \mathcal{S} \\ I \in \mathcal{E}_i}} \int \phi_i^{a0cdk} \odot \phi_i^{a'0c'd'l} f_i^0 \, dc_i,$$

it is easy to obtain from the orthogonality relations of the Laguerre and Sonine polynomials, and Wang Chang and Uhlenbeck polynomials, that we have

$$\langle\langle \phi^{a0cdk}, \phi^{a'0c'd'l} \rangle\rangle = \langle\langle \phi^{a0cdk}, \phi^{a0cdk} \rangle\rangle \delta_{aa'} \delta_{cc'} \delta_{dd'} \delta_{kl},$$

for $a, a', c, c', d, d' \geq 0$ and $k, l \in \mathcal{S}$. In addition, a straightforward calculation yields that for $k \in \mathcal{S}$ we have

$$\langle\langle \phi^{0000k}, \phi^{0000k} \rangle\rangle = n_k,$$

$$\langle\langle \phi^{0010k}, \phi^{0010k} \rangle\rangle = \frac{c_{\mathrm{V}}^{\mathrm{tr}}}{k_{\mathrm{B}}} n_k,$$

$$\langle\langle \phi^{0001k}, \phi^{0001k} \rangle\rangle = \frac{c_k^{\mathrm{int}}}{k_{\mathrm{B}}} n_k,$$

$$\langle\langle \phi^{1000k}, \phi^{1000k} \rangle\rangle = \tfrac{3}{2} n_k,$$

$$\langle\langle \phi^{1010k}, \phi^{1010k} \rangle\rangle = \frac{3}{2} \frac{c_{\mathrm{P}}^{\mathrm{tr}}}{k_{\mathrm{B}}} n_k,$$

$$\langle\langle \phi^{1001k}, \phi^{1001k} \rangle\rangle = \frac{3}{2} \frac{c_k^{\mathrm{int}}}{k_{\mathrm{B}}} n_k,$$

$$\langle\langle \phi^{2000k}, \phi^{2000k} \rangle\rangle = \tfrac{5}{2} n_k.$$

Appendix C

In this appendix we give the expression of the partial brackets products for the basis functions introduced in Section 2.2 in terms of various collision integrals.

Calculation of the Partial Brackets. Keeping the notation of Sections 2.1 and 2.2 for the species pair (k,l), the partial bracket are defined as [WT62]

$$n_k n_l[\xi,\zeta]'_{kl} = \sum_{\substack{K,K'\in\mathcal{E}_k\\ L,L'\in\mathcal{E}_l}} \iiiint f_k^0 f_l^0 \xi_k \odot (\zeta_k - \zeta'_k)\bar{\sigma}_{kl}^{KLK'L'} g \sin\chi d\chi d\varphi dc_k d\tilde{c}_l,$$

and

$$n_k n_l[\xi,\zeta]''_{kl} = \sum_{\substack{K,K'\in\mathcal{E}_k\\ L,L'\in\mathcal{E}_l}} \iiiint f_k^0 f_l^0 \xi_k \odot (\tilde{\zeta}_l - \tilde{\zeta}'_l)\bar{\sigma}_{kl}^{KLK'L'} g \sin\chi d\chi d\varphi dc_k d\tilde{c}_l,$$

using the classical notation $\zeta'_k = \zeta_k(c'_k, K')$, $\tilde{\zeta}_l = \zeta_l(\tilde{c}_l, L)$, and $\tilde{\zeta}'_l = \zeta_l(\tilde{c}'_l, L')$. The calculation of the partial bracket products has been performed independently and the resulting expressions have been compared formally with the results of Köhler and 't Hooft [KT79] [MBKK91] obtained for linear molecules in a fully quantum mechanical framework. The agreement has been found to be complete with [MBKK91], and complete with [KT79] after the elimination of two sign misprints.

Notation for Collision Integrals. The partial brackets are expressed in terms of the collision averaging operator $[\![\ ]\!]$ given by

$$[\![\alpha]\!]_{kl} = \Big(\frac{k_B\overline{T}}{2\pi m_{kl}}\Big)^{1/2} \sum_{\substack{K,K'\in\mathcal{E}_k\\ L,L'\in\mathcal{E}_l}} \frac{a_{kK}a_{lL}}{Q_k Q_l} \iiint \alpha\gamma^3 \exp(-\gamma^2 - \epsilon_{kK} - \epsilon_{lL})\bar{\sigma}_{kl}^{KLK'L'} \sin\chi d\chi d\varphi d\gamma,$$

where $\gamma = g(m_{kl}/2k_B\overline{T})^{1/2}$ is integrated over $(0,+\infty)$, χ over $(0,\pi)$, φ over $(0,2\pi)$, and where α stands for any function of γ, $\gamma' = g'(m_{kl}/2k_B\overline{T})^{1/2}$, χ, φ, ϵ_{kK}, $\epsilon_{kK'}$, ϵ_{lL}, and $\epsilon_{lL'}$, keeping the notation of Section 2.1.2. As defined previously, we have

$$\begin{cases} \Delta\epsilon_{kl} = \Delta\epsilon_k + \widetilde{\Delta}\epsilon_l, \\ \Delta\epsilon_k = \epsilon_{kK'} - \epsilon_{kK}, \\ \widetilde{\Delta}\epsilon_l = \epsilon_{lL'} - \epsilon_{lL}, \end{cases}$$

where the extra superscript $\widetilde{}$ is used to distinguish one of the collision partner from the other in the case where k and l are the same, so that for $k = l$ we have

$$\begin{cases} \widetilde{\Delta}\epsilon_k = \epsilon_{kL'} - \epsilon_{kL}, \\ \Delta\epsilon_{kk} = \Delta\epsilon_k + \widetilde{\Delta}\epsilon_k. \end{cases}$$

We also restate that

$$\epsilon_{kK}^0 = \epsilon_{kK} - \bar{\epsilon}_k,$$

is a shifted reduced internal energy, and for a given species pair (k, l) we denote by μ_k and μ_l the mass ratios

$$\begin{cases} \mu_k = \dfrac{m_k}{m_k + m_l}, \\ \mu_l = \dfrac{m_l}{m_k + m_l}. \end{cases}$$

Expressions of the Partial Brackets. The partial brackets involving the basis functions ϕ^{0010k}, $k \in \mathcal{S}$, and ϕ^{0001k}, $k \in \mathcal{P}$, are first given by

$$[\phi^{0010k}, \phi^{0010k}]'_{kl} = 4\mu_l^2 \Big[4\frac{\mu_k}{\mu_l} [\![\gamma^2 - \gamma\gamma' \cos\chi]\!]_{kl} + [\![(\Delta\epsilon_{kl})^2]\!]_{kl} \Big],$$

$$[\phi^{0010k}, \phi^{0010l}]''_{kl} = 4\mu_k\mu_l \Big[-4[\![\gamma^2 - \gamma\gamma' \cos\chi]\!]_{kl} + [\![(\Delta\epsilon_{kl})^2]\!]_{kl} \Big],$$

$$[\phi^{0010k}, \phi^{0001k}]'_{kl} = -4\mu_l [\![\Delta\epsilon_k \Delta\epsilon_{kl}]\!]_{kl},$$

$$[\phi^{0010k}, \phi^{0001l}]''_{kl} = -4\mu_l [\![\widetilde{\Delta}\epsilon_l \Delta\epsilon_{kl}]\!]_{kl},$$

$$[\phi^{0001k}, \phi^{0001k}]'_{kl} = 4[\![(\Delta\epsilon_k)^2]\!]_{kl},$$

$$[\phi^{0001k}, \phi^{0001l}]''_{kl} = 4[\![\Delta\epsilon_k \widetilde{\Delta}\epsilon_l]\!]_{kl}.$$

The partial brackets involving the basis functions ϕ^{1000k}, $k \in \mathcal{S}$, ϕ^{1010k}, $k \in \mathcal{S}$, and ϕ^{1001k}, $k \in \mathcal{P}$, are given by

$$[\phi^{1000k}, \phi^{1000k}]'_{kl} = 8\mu_l [\![\gamma^2 - \gamma\gamma' \cos\chi]\!]_{kl},$$

$$[\phi^{1000k}, \phi^{1000l}]''_{kl} = -8\mu_k^{\frac{1}{2}} \mu_l^{\frac{1}{2}} [\![\gamma^2 - \gamma\gamma' \cos\chi]\!]_{kl},$$

$$[\phi^{1000k}, \phi^{1010k}]'_{kl} = 4\mu_l^2 \Big[5[\![\gamma^2 - \gamma\gamma' \cos\chi]\!]_{kl} - 2[\![\gamma^4 - \gamma^3\gamma' \cos\chi]\!]_{kl} \Big],$$

$$[\phi^{1000k}, \phi^{1010l}]''_{kl} = -4\mu_k^{\frac{3}{2}} \mu_l^{\frac{1}{2}} \Big[5[\![\gamma^2 - \gamma\gamma' \cos\chi]\!]_{kl} - 2[\![\gamma^4 - \gamma^3\gamma' \cos\chi]\!]_{kl} \Big],$$

$$[\phi^{1000k}, \phi^{1001k}]'_{kl} = -8\mu_l \left[\epsilon^0_{kK}(\gamma^2 - \gamma\gamma' \cos\chi)\right]_{kl},$$

$$[\phi^{1000k}, \phi^{1001l}]''_{kl} = 8\mu_k^{\frac{1}{2}}\mu_l^{\frac{1}{2}} \left[\epsilon^0_{lL}(\gamma^2 - \gamma\gamma' \cos\chi)\right]_{kl},$$

$$\begin{aligned}[\phi^{1010k}, \phi^{1010k}]'_{kl} = 8\mu_k\mu_l^2 \Big[& \left(\tfrac{15}{2}\frac{\mu_k}{\mu_l} + \tfrac{25}{4}\frac{\mu_l}{\mu_k}\right)[\![\gamma^2 - \gamma\gamma'\cos\chi]\!]_{kl} \\ & - 5\frac{\mu_l}{\mu_k}[\![\gamma^4 - \gamma^3\gamma'\cos\chi]\!]_{kl} + \frac{\mu_l}{\mu_k}[\![\gamma^6 - \gamma^3\gamma'^3\cos\chi]\!]_{kl} \\ & + 2[\![\gamma^4 - \gamma^2\gamma'^2\cos^2\chi]\!]_{kl} + \tfrac{7}{4}[\![(\Delta\epsilon_{kl})^2]\!]_{kl}\Big],\end{aligned}$$

$$\begin{aligned}[\phi^{1010k}, \phi^{1010l}]''_{kl} = 8\mu_k^{\frac{3}{2}}\mu_l^{\frac{3}{2}} \Big[& -\tfrac{55}{4}[\![\gamma^2 - \gamma\gamma'\cos\chi]\!]_{kl} \\ & + 5[\![\gamma^4 - \gamma^3\gamma'\cos\chi]\!]_{kl} - [\![\gamma^6 - \gamma^3\gamma'^3\cos\chi]\!]_{kl} \\ & + 2[\![\gamma^4 - \gamma^2\gamma'^2\cos^2\chi]\!]_{kl} + \tfrac{7}{4}[\![(\Delta\epsilon_{kl})^2]\!]_{kl}\Big],\end{aligned}$$

$$\begin{aligned}[\phi^{1010k}, \phi^{1001k}]'_{kl} = -4\mu_k\mu_l \Big[& \tfrac{5}{2}[\![\Delta\epsilon_k\Delta\epsilon_{kl}]\!]_{kl} + 5\frac{\mu_l}{\mu_k}\left[\epsilon^0_{kK}(\gamma^2 - \gamma\gamma'\cos\chi)\right]_{kl} \\ & - 2\frac{\mu_l}{\mu_k}\left[\epsilon^0_{kK}(\gamma^4 - \gamma\gamma'^3\cos\chi)\right]_{kl}\Big],\end{aligned}$$

$$\begin{aligned}[\phi^{1010k}, \phi^{1001l}]''_{kl} = -4\mu_k^{\frac{1}{2}}\mu_l^{\frac{3}{2}} \Big[& \tfrac{5}{2}[\![\widetilde{\Delta}\epsilon_l\Delta\epsilon_{kl}]\!]_{kl} - 5\left[\epsilon^0_{lL}(\gamma^2 - \gamma\gamma'\cos\chi)\right]_{kl} \\ & + 2\left[\epsilon^0_{lL}(\gamma^4 - \gamma\gamma'^3\cos\chi)\right]_{kl}\Big],\end{aligned}$$

$$[\phi^{1001k}, \phi^{1001k}]'_{kl} = 8\mu_l\Big[\left[\epsilon^0_{kK}(\epsilon^0_{kK}\gamma^2 - \epsilon^0_{kK'}\gamma\gamma'\cos\chi)\right]_{kl} + \tfrac{3}{4}\frac{\mu_k}{\mu_l}\left[(\Delta\epsilon_k)^2\right]_{kl}\Big],$$

$$[\phi^{1001k}, \phi^{1001l}]''_{kl} = 8\mu_k^{\frac{1}{2}}\mu_l^{\frac{1}{2}}\Big[-\left[\epsilon^0_{kK}(\epsilon^0_{lL}\gamma^2 - \epsilon^0_{lL'}\gamma\gamma'\cos\chi)\right]_{kl} + \tfrac{3}{4}\left[\Delta\epsilon_k\widetilde{\Delta}\epsilon_l\right]_{kl}\Big].$$

Finally, the partial brackets involving the basis functions ϕ^{2000k}, $k \in \mathcal{S}$, are given by

$$\begin{aligned}[\phi^{2000k}, \phi^{2000k}]'_{kl} = 8\mu_l^2\Big[& \tfrac{10}{3}\frac{\mu_k}{\mu_l}[\![\gamma^2 - \gamma\gamma'\cos\chi]\!]_{kl} \\ & + \left[(\gamma^4 - \gamma^2\gamma'^2\cos^2\chi)\right]_{kl} - \tfrac{1}{6}[\![(\Delta\epsilon_{kl})^2]\!]_{kl}\Big],\end{aligned}$$

$$\begin{aligned}[\phi^{2000k}, \phi^{2000l}]''_{kl} = 8\mu_k\mu_l\Big[& -\tfrac{10}{3}[\![\gamma^2 - \gamma\gamma'\cos\chi]\!]_{kl} \\ & + \left[(\gamma^4 - \gamma^2\gamma'^2\cos^2\chi)\right]_{kl} - \tfrac{1}{6}[\![(\Delta\epsilon_{kl})^2]\!]_{kl}\Big].\end{aligned}$$

Remark. Equivalent formulations of the partial brackets can also be obtained by using the properties of the averaging operator. In particular, if α is any function of γ, γ', χ,

φ, ϵ_{kK}, $\epsilon_{kK'}$, ϵ_{lL}, and $\epsilon_{lL'}$, we first have the symmetry property

$$[\![\alpha(\gamma,\gamma',\epsilon_{kK},\epsilon_{kK'},\epsilon_{lL},\epsilon_{lL'},\chi,\varphi)]\!]_{kl} = [\![\alpha(\gamma,\gamma',\epsilon_{kK},\epsilon_{kK'},\epsilon_{lL},\epsilon_{lL'},\chi,\varphi)]\!]_{lk}.$$

By using inverse collisions, we also have

$$[\![\alpha(\gamma,\gamma',\epsilon_{kK},\epsilon_{kK'},\epsilon_{lL},\epsilon_{lL'},\chi,\varphi)]\!]_{kl} = [\![\alpha(\gamma',\gamma,\epsilon_{kK'},\epsilon_{kK},\epsilon_{lL'},\epsilon_{lL},\chi,\varphi)]\!]_{kl},$$

and this implies, for instance, that $[\![\gamma^2 - \gamma\gamma'\cos\chi]\!]_{kl} = [\![\gamma^2 + \gamma'^2 - 2\gamma\gamma'\cos\chi]\!]_{kl}$ and that $\left[(\Delta\epsilon_k)^2\right]_{kl} = -2\left[\epsilon^0_{kK}\Delta\epsilon_k\right]_{kl}$. Finally, in the special case where $k = l$, we have the symmetry property

$$[\![\alpha(\gamma,\gamma',\epsilon_{kK},\epsilon_{kK'},\epsilon_{kL},\epsilon_{kL'},\chi,\varphi)]\!]_{kk} = [\![\alpha(\gamma,\gamma',\epsilon_{kL},\epsilon_{kL'},\epsilon_{kK},\epsilon_{kK'},\chi,\varphi)]\!]_{kk},$$

so that for instance $\left[(\Delta\epsilon_{kk})^2\right]_{kk} = 2\left[\Delta\epsilon_k\Delta\epsilon_{kk}\right]_{kk}$. Notice that it is also possible to express the partial brackets in terms of more classical quantities as for instance the collision integrals

$$\begin{cases} \Omega_{kl}^{(1,1)} = \left[\gamma^2 - \gamma\gamma'\cos\chi\right]_{kl}, \\ \Omega_{kl}^{(2,2)} = \left[\gamma^4 - \gamma^2\gamma'^2\cos^2\chi - \frac{1}{6}(\Delta\epsilon_{kl})^2\right]_{kl}, \\ \Omega_{kl}^{(1,2)} = \left[\gamma^4 - \gamma^3\gamma'\cos\chi\right]_{kl}, \\ \Omega_{kl}^{(1,3)} = \left[\gamma^6 - \gamma^3\gamma'^3\cos\chi\right]_{kl}, \end{cases}$$

and the quantities introduced in Section 2.2.3.

Appendix D

In this appendix we present the molecular parameters used in Chapter 6 in order to evaluate numerically the transport linear system coefficients. The species Lennard-Jones potential well depths ϵ_k and collision diameters σ_k, the dipole moments μ_k of polar molecules, the polarizabilities α_k of nonpolar molecules, and the collision numbers $\xi_k^{\mathrm{int}}(298\mathrm{K})$ are presented in Table D.1 [Di68] [Wr83] [Di84] [KDWCM86].

Table D.1. Species molecular parameters.

Species	ϵ_k/k_B [K]	σ_k [nm]	μ_k [Debye]	$10^{24}\alpha_k$ [cm^3]	$\xi_k^{\mathrm{int}}(298\mathrm{K})$
C_2H	209.0	0.4100	—	—	2.5
C_2H_2	209.0	0.4100	—	—	2.5
C_2H_3	209.0	0.4100	—	—	1.0
C_2H_4	280.8	0.3971	—	—	1.5
C_2H_5	252.3	0.4302	—	—	1.5
C_2H_6	252.3	0.4302	—	—	1.5
CH	80.0	0.2750	—	—	1.0
CH_2	144.0	0.3800	—	—	1.0
CH_2CO	436.0	0.3970	—	—	2.0
CH_2O	498.0	0.3590	—	—	2.0
CH_3	144.0	0.3800	—	—	1.0
CH_3O	417.0	0.3690	1.7	—	2.0
CH_4	141.4	0.3746	—	2.6	13.0
CO	98.1	0.3650	—	1.95	1.8
CO_2	244.0	0.3763	—	2.65	2.1
H	145.0	0.2050	—	—	—
H_2	38.0	0.2920	—	0.79	280.0
H_2O	572.4	0.2605	1.844	—	4.0
H_2O_2	107.4	0.3458	—	—	3.8
$HCCO$	150.0	0.2500	—	—	1.0
HCO	498.0	0.3590	—	—	1.0
HO_2	107.4	0.3458	—	—	1.0
N_2	97.53	0.3621	—	1.76	4.0
O	80.0	0.2750	—	—	—
O_2	107.4	0.3458	—	1.6	3.8
OH	80.0	0.2750	—	—	1.0

The three ratios of collision integrals $\bar{A}_{kl}$, $\bar{B}_{kl}$, and $\bar{C}_{kl}$, $k,l \in \mathcal{S}$, are evaluated using sixth-order polynomial fits in the form

$$\begin{cases} \bar{A}_{kl} = \sum_{i=0}^{6} a_i \left(\log \overline{T}^*_{kl}\right)^i, \\ \bar{B}_{kl} = \sum_{i=0}^{6} b_i \left(\log \overline{T}^*_{kl}\right)^i, \\ \bar{C}_{kl} = \sum_{i=0}^{6} c_i \left(\log \overline{T}^*_{kl}\right)^i, \end{cases}$$

where the reduced temperature $\overline{T}^*_{kl}$, $k,l \in \mathcal{S}$, is given by

$$\overline{T}^*_{kl} = \frac{k_B \overline{T}}{\epsilon_{kl}},$$

and ϵ_{kl} is the Lennard-Jones potential well depth for species pair (k,l) which is estimated as discussed in Section 6.1.2. The polynomial coefficients for the sixth-order fits are given in Table D.2 [KDWCM86].

Table D.2. Coefficients of sixth-order polynomial fits.

i	a_i	b_i	c_i
0	.1106910525E+01	.1199673577E+01	.8386993788E+00
1	-.7065517161E–02	-.1140928763E+00	.4748325276E–01
2	-.1671975393E–01	-.2147636665E–02	.3250097527E–01
3	.1188708609E–01	.2512965407E–01	-.1625859588E–01
4	.7569367323E–03	-.3030372973E–02	-.2260153363E–02
5	-.1313998345E–02	-.1445009039E–02	.1844922811E–02
6	.1720853282E–03	.2492954809E–03	-.2115417788E–03

Appendix E

In this appendix we investigate the validity of the tempered reaction regime, discussed in Section 2.1.6, for a typical chemical reaction mechanism often considered in hydrogen and methane combustion applications [GS89] [SG92] [Er94]. More specifically, we verify numerically that the mean free times of the molecules—the times of free flight—are at least an order of magnitude lower than the corresponding chemistry times of the mixture. In this case, a tempered reaction regime is obtained and transport coefficients can be calculated as if there were no chemical reactions. For other chemically reacting flow applications where transport coefficients are calculated as if there were no chemical reactions, similar verifications should be performed.

Estimates for Mean Free Times. We consider a molecule of the ith species and we are interested in the mean free time between successive collisions of this molecule with a molecule of the jth species. This time can be estimated as [FK72]

$$t_{ij}^{\text{coll}} = \frac{1}{\sigma_{ij}^2 n_j} \sqrt{\frac{m_{ij}}{8\pi k_B T}},$$

where σ_{ij} is the collision diameter for species pair (i,j), n_j the number density of the jth species, and m_{ij} the reduced mass of species pair (i,j) given by

$$m_{ij} = \frac{m_i m_j}{m_i + m_j}.$$

The collision diameter σ_{ij} is estimated by the Lennard-Jones collision diameter which is, in turn, evaluated using the combining rules stated in Section 6.1.2 [Di68] [Di84].

Estimates for Chemistry Times. We consider the chemical reaction mechanism presented in Table E.1 which is often used in hydrogen combustion applications. This mechanism contains 8 reactive species H_2, O_2, H_2O, H, O, OH, HO_2, H_2O_2, participating in 16 reversible elementary reactions. For methane combustion applications, this mechanism is completed by the one presented in Table E.2. The resulting mechanism contains 15 reactive species CH_4, CH_3, H_2, O_2, H_2O, H, O, OH, HO_2, H_2O_2, CHO,

Table E.1. Hydrogen-Air Reaction Mechanism.
Rate coefficients: $k_{\mathfrak{g}} = A_{\mathfrak{g}}\overline{T}^{\beta_{\mathfrak{g}}}\exp(-E_{\mathfrak{g}}/R\overline{T})$.
(units: moles, cubic centimeters, seconds, Kelvins and calories).

	Reaction	A	β	E
1.	$H_2+O_2 \rightleftharpoons 2OH$	1.70E+13	0.000	47780.
2.	$OH+H_2 \rightleftharpoons H_2O+H$	1.17E+09	1.300	3626.
3.	$H+O_2 \rightleftharpoons OH+O$	2.00E+14	0.000	16800.
4.	$O+H_2 \rightleftharpoons OH+H$	1.80E+10	1.000	8826.
5.	$H+O_2+M \rightleftharpoons HO_2+M$	2.10E+18	-1.000	0.
6.	$OH+HO_2 \rightleftharpoons H_2O+O_2$	5.00E+13	0.000	1000.
7.	$H+HO_2 \rightleftharpoons 2OH$	2.50E+14	0.000	1900.
8.	$O+HO_2 \rightleftharpoons O_2+OH$	4.80E+13	0.000	1000.
9.	$2OH \rightleftharpoons O+H_2O$	6.00E+08	1.300	0.
10.	$H_2+M \rightleftharpoons H+H+M$	2.23E+12	0.500	92600.
11.	$O_2+M \rightleftharpoons O+O+M$	1.85E+11	0.500	95560.
12.	$H+OH+M \rightleftharpoons H_2O+M$	7.50E+23	-2.600	0.
13.	$H+HO_2 \rightleftharpoons H_2+O_2$	2.50E+13	0.000	700.
14.	$HO_2+HO_2 \rightleftharpoons H_2O_2+O_2$	2.00E+12	0.000	0.
15.	$H_2O_2+M \rightleftharpoons OH+OH+M$	1.30E+17	0.000	45500.
16.	$H_2O_2+OH \rightleftharpoons H_2O+HO_2$	1.00E+13	0.000	1800.

CH_2O, CH_3O, CO_2, and CO, participating in 43 reversible elementary reactions. Most of the reactions are bimolecular but some require the presence of a third body. The third body can be any species present in the mixture and is denoted by the symbol M in Tables E.1 and E.2. The formalism used for describing the chemical reaction mechanism is classical, and we refer to [Di67] [Bal72] [Bal76] [DiW77] [Wr82] [Wr84] [Bal92] and references therein for more details.

We consider first the case of a chemical reaction $\mathfrak{g} \in \mathfrak{G}$ which does not require the presence of a third body. This reaction typically takes the form

$$\mathfrak{S}_i + \mathfrak{S}_j \rightleftharpoons \text{Products},$$

where $\mathfrak{S}_i$ is the chemical symbol for the ith species. Note that the indices i and j may be the same. The characteristic time for a reactive collision of a molecule of the ith species with a molecule of the jth species is estimated as

$$t_{\mathfrak{g}i}^{\text{chem}} = \frac{\mathcal{N} n_i}{k_{\mathfrak{g}} n_i n_j} = \frac{\mathcal{N}}{k_{\mathfrak{g}} n_j},$$

where $\mathcal{N} = 6.022\times10^{23}\text{mol}^{-1}$ is the Avogadro constant, n_j the number density of the jth species, and $k_{\mathfrak{g}}$ the molar rate constant for reaction $\mathfrak{g}$. For bimolecular reactions,

Table E.2. Methane-Air Reaction Mechanism.
Rate coefficients: $k_{\mathfrak{g}} = A_{\mathfrak{g}} \overline{T}^{\beta_{\mathfrak{g}}} \exp(-E_{\mathfrak{g}}/R\overline{T})$.
(units: moles, cubic centimeters, seconds, Kelvins and calories).

	Reaction	A	β	E
1.	$CH_3+H \rightleftharpoons CH_4$	1.90E+36	-7.000	9050.
2.	$CH_4+O_2 \rightleftharpoons CH_3+HO_2$	7.90E+13	0.000	56000.
3.	$CH_4+H \rightleftharpoons CH_3+H_2$	2.20E+04	3.000	8750.
4.	$CH_4+O \rightleftharpoons CH_3+OH$	1.60E+06	2.360	7400.
5.	$CH_4+OH \rightleftharpoons CH_3+H_2O$	1.60E+06	2.100	2460.
6.	$CH_2O+OH \rightleftharpoons HCO+H_2O$	7.53E+12	0.000	167.
7.	$CH_2O+H \rightleftharpoons HCO+H_2$	3.31E+14	0.000	10500.
8.	$CH_2O+M \rightleftharpoons HCO+H+M$	3.31E+16	0.000	81000.
9.	$CH_2O+O \rightleftharpoons HCO+OH$	1.81E+13	0.000	3082.
10.	$HCO+OH \rightleftharpoons CO+H_2O$	5.00E+12	0.000	0.
11.	$HCO+M \rightleftharpoons H+CO+M$	1.60E+14	0.000	14700.
12.	$HCO+H \rightleftharpoons CO+H_2$	4.00E+13	0.000	0.
13.	$HCO+O \rightleftharpoons OH+CO$	1.00E+13	0.000	0.
14.	$HCO+O_2 \rightleftharpoons HO_2+CO$	3.00E+12	0.000	0.
15.	$CO+O+M \rightleftharpoons CO_2+M$	3.20E+13	0.000	-4200.
16.	$CO+OH \rightleftharpoons CO_2+H$	1.51E+07	1.300	-758.
17.	$CO+O_2 \rightleftharpoons CO_2+O$	1.60E+13	0.000	41000.
18.	$CH_3+O_2 \rightleftharpoons CH_3O+O$	7.00E+12	0.000	25652.
19.	$CH_3O+M \rightleftharpoons CH_2O+H+M$	2.40E+13	0.000	28812.
20.	$CH_3O+H \rightleftharpoons CH_2O+H_2$	2.00E+13	0.000	0.
21.	$CH_3O+OH \rightleftharpoons CH_2O+H_2O$	1.00E+13	0.000	0.
22.	$CH_3O+O \rightleftharpoons CH_2O+OH$	1.00E+13	0.000	0.
23.	$CH_3O+O_2 \rightleftharpoons CH_2O+HO_2$	6.30E+10	0.000	2600.
24.	$CH_3+O_2 \rightleftharpoons CH_2O+OH$	5.20E+13	0.000	34574.
25.	$CH_3+O \rightleftharpoons CH_2O+H$	6.80E+13	0.000	0.
26.	$CH_3+OH \rightleftharpoons CH_2O+H_2$	7.50E+12	0.000	0.
27.	$HO_2+CO \rightleftharpoons CO_2+OH$	5.80E+13	0.000	22934.

the molar rate constant $k_{\mathfrak{g}}$ is related to the molecular rate constant $\mathfrak{F}_{\mathfrak{g}}$, introduced in Section 2.1.6, by the relation $k_{\mathfrak{g}} = \mathfrak{F}_{\mathfrak{g}}/\mathcal{N}$. The case of chemical reactions involving a third body is treated as follows. For triple reactive collisions in the form

$$\mathfrak{S}_i + \mathfrak{S}_j + \mathrm{M} \rightleftharpoons \text{Products},$$

the forward rate constant is multiplied by the factor $\bar{p}_{\mathrm{M}}/(R\overline{T})$ where $\bar{p}_{\mathrm{M}}$ is the third body partial pressure and $R = 8.314\times10^7 \text{ergs}/(\text{mol}\cdot\text{K})$ the universal gas constant. For bimolecular collisions in the form

$$\mathfrak{S}_i + \mathrm{M} \rightleftharpoons \text{Products},$$

we estimate the characteristic chemistry time t_{gi}^{chem} as being the smallest of all the characteristic times for all the species $j \in \mathcal{S}$.

Arrhenius Fits. The chemical reactions considered in Tables E.1 and E.2 are reversible and involve, therefore, two characteristic times, one for the forward reaction and one for the reverse reaction. The rate constant for the forward reaction is modeled using an empiric Arrhenius-type expression

$$k_{\mathfrak{g}} = A_{\mathfrak{g}} \overline{T}^{\beta_{\mathfrak{g}}} \exp \frac{-E_{\mathfrak{g}}}{R\overline{T}},$$

where $A_{\mathfrak{g}}$ is the pre-exponential factor for the $\mathfrak{g}^{\text{th}}$ reaction, $\beta_{\mathfrak{g}}$ the temperature exponent, and $E_{\mathfrak{g}}$ the activation energy. The parameters $A_{\mathfrak{g}}$, $\beta_{\mathfrak{g}}$, and $E_{\mathfrak{g}}$ are tabulated in Tables E.1 and E.2. These parameters are usually determined from experiments and are often updated in the literature so that these tables constitute only a typical example. Note also that these parameters are usually valid in a given temperature range where the experimental measurements have been performed. For more details, we refer to [Di67] [Bal72] [Bal76] [DiW77] [Wr82] [Wr84] [Bal92] and references therein. On the other hand, the rate constant for the reverse reaction is given by the ratio of the rate constant for the forward reaction divided by the equilibrium constant. The equilibrium constant is evaluated using the JANAF thermodynamic data base [SP71] [Cal85] and the Chemkin thermodynamic data base [KRM87].

Comparison. We consider first the case of hydrogen mixtures which correspond to the chemical reactions presented in Table E.1. In this case, we have found that the ratio

$$\frac{t_{ij}^{\text{coll}}}{t_{\mathfrak{g}i}^{\text{chem}}} = \frac{1}{\mathcal{N}\sqrt{8\pi k_B \overline{T}}} \frac{\sqrt{m_{ij}}}{\sigma_{ij}^2} k_{\mathfrak{g}}. \tag{E.1}$$

is always lower than 0.16 for temperatures varying between 300 and 2000K. At temperatures lower than 600K, however, this ratio becomes larger for the reverse reaction 15, but the Arrhenius parameters of Table E.1 are no longer valid for such temperatures [Bal92]. This is a typical difficulty arising when using Arrhenius fits for temperatures falling outside their experimental validity range.

For methane mixtures, we have found that the ratio (E.1) is always lower than 0.16 for temperatures varying between 300 and 2000K with the following exceptions. For temperatures lower than 1000K, this is no longer the case for the forward reaction 25 and the reverse reaction 18. At a temperature of 300K, the ratio (E.1) is also larger for the reverse reaction 8. However, the experimental validity range for the Arrhenius parameters of reactions 8 and 18 is 1600–3000 and 1700–2300K, respectively [Bal92]. For reaction 25, the Arrhenius parameters are known up to a factor of two only [Bal92].

As a conclusion, the assumption of a tempered reaction regime appears as quite reasonable for hydrogen mixtures. In the case of methane mixtures, accurate measurements of the Arrhenius parameters for certain chemical reactions at lower temperatures, e.g., 1000K, would be needed to ascertain the applicability of the tempered reaction regime. Finally, we refer to Zellner [Ze84] for a detailed discussion of eventual internal desequilibrium effects.

Bibliography

[Ah72] W. A. Ahtye, Thermal Conductivity in Vibrationally Excited Gases. *J. Chem. Phys.* **57**, 5542–5555, (1972).

[ACG94] B. V. Alexeev, A. Chikhaoui, and I. T. Grushin, Application of the Generalized Chapman-Enskog Method to the Transport-Coefficient Calculation in a Reacting Gas Mixture. *Phys. Review E* **49**, 2809–2825, (1994).

[An89] J. R. Anderson, *Hypersonic and High Temperature Gas Dynamics.* McGraw-Hill Book Company, New York (1989).

[Bal72] D. L. Baulch, D. D. Drysdale, D. G. Horne, and A. C. Llyod, Evaluated Kinetic Data for High Temperature Reactions, Volume 1: Homogeneous Gas Phase Reactions of the H_2-O_2 System, Butterworth, London (1972).

[Bal76] D. L. Baulch, D. D. Drysdale, J. Duxbury, and S. Grant, Evaluated Kinetic Data for High Temperature Reactions, Volume 3: Homogeneous Gas Phase Reactions of the H_2-O_2 System, the CO-O_2-H_2 System, and of the Sulphur-Containing Species, Butterworth, London (1976).

[Bal92] D. L. Baulch, C. J. Cobos, R. A. Cox, C. Esser, P. Frank, T. Just, J. A. Kerr, M. J. Pilling, J. Troe, R. J. Walker, and J. Warnatz, Evaluated Kinetic Data for Combustion Modelling, J. Phys. Chem. Ref. Data **21**, 411–734, (1992).

[BG74] A. Ben-Israel and T. N. E. Greville, *Generalized Inverses, Theory and Applications.* Wiley, New York (1974).

[BP79] A. Bermann and R. J. Plemmons, *Positive Matrices in the Mathematical Science.* Academic Press, New York (1979).

[BJ70] C. A. Brau and R. M. Jonkman, Classical Theory of Rotational Relaxation in Diatomic Gases. *J. Chem. Phys.* **52**, 477–484, (1970).

[Br58] R. S. Brokaw, Approximate Formulas for the Viscosity and Thermal Conductivity of Gas Mixtures. *J. Chem. Phys.* **29**, 391–397, (1958).

[Br64] R. S. Brokaw, Approximate Formulas for the Viscosity and Thermal Conductivity of Gas Mixtures. II. *J. Chem. Phys.* **42**, 1140–1146, (1964).

[Bu88] R. Brun, Transport Properties in Reactive Gas Flows. AIAA paper, AIAA-88-2655, (1988).

[BW53] J. H. Burgoyne and F. Weinberg, A Method of Analysis of a Plane Combustion Wave. *In Fourth Symposium (International) on Combustion*, pp. 294–302. Williams and Wilkins Co., Baltimore (1953).

[Ce88] C. Cercignani, *The Boltzmann Equation and its Applications.* Springer Verlag, New York (1988).

[CC70] S. Chapman and T. G. Cowling, *The Mathematical Theory of Non-Uniform Gases.* Cambridge Univ. Press, Cambridge (1970).

[Cal85] M. W. Chase Jr., C. A. Davies, J. R. Downey Jr., D. J. Frurip, R. A. McDonald and A. N. Syverud, JANAF Thermochemical Tables, Third Edition, J. Phys. Chem. Ref. Data **14**, Suppl. 1, 1–1856, (1985).

[CHe81] T. P. Coffee and J. M. Heimerl, Transport Algorithms for Premixed, Laminar Steady-State Flames. *Combust. and Flame* **43**, 273–289, (1981).

[Cu68] C. F. Curtiss, Symmetric Gaseous Diffusion Coefficients. *J. Chem. Phys.* **49**, 2917–2919, (1968).

[Cu81] C. F. Curtiss, The Classical Boltzmann Equation of a Gas of Diatomic Molecules. *J. Chem. Phys.* **75**, 376–378, (1981).

[CH49] C. F. Curtiss and J. O. Hirschfelder, Transport Properties of Multicomponent Gas Mixtures. *J. Chem. Phys.* **17**, 550–555, (1949).

[Di67] G. Dixon-Lewis, Flame Structure and Flame Reaction Kinetics, I. Solution of Conservation Equations and Application to Rich Hydrogen-Oxygen Flames, Proc. Roy. Soc. **A 298**, 495–513, (1967).

[Di68] G. Dixon-Lewis, Flame Structure and Flame Reaction Kinetics, II. Transport Phenomena in Multicomponent Systems. *Proc. Roy. Soc.* **A 307**, 111–135, (1968).

[DiW77] G. Dixon-Lewis and D. J. Williams, The Oxydation of Hydrogen and Carbon Monoxyde, *in* C. H. Bamford and C. F. H. Tipper (Eds.), *Comprehensive Chemical Kinetics*, Elsevier Scientific Publishing Co., Amsterdam, 1–234, (1977).

[Di84] G. Dixon-Lewis, Computer Modeling of Combustion Reactions in Flowing Systems with Transport. *In* W. C. Gardiner (Ed.), *Combustion Chemistry*. Springer, New York, 21–125, (1984).

[Er94] A. Ern, Vorticity-Velocity Modeling of Chemically Reacting Flows. Ph. D. dissertation, Yale University, Department of Mechanical Engineering, February 1994.

[EGKS94] A. Ern, V. Giovangigli, D. E. Keyes, and M. D. Smooke, Towards Polyalgorithmic Linear Systems Solvers for Nonlinear Elliptic Problems. *SIAM J. Sci. Stat. Comp.* **15**, (in press) (1994).

[FK72] J. H. Ferziger and H. G. Kaper, *Mathematical Theory of Transport Processes in Gases.* North Holland, Amsterdam (1972).

[Fo36] R. H. Fowler, *Statistical Mechanics.* Cambridge University Press, Cambridge, (1936).

[Gi90] V. Giovangigli, Mass Conservation and Singular Multicomponent Diffusion Algorithms. *Impact Comput. Sci. Eng.* **2**, 73–97, (1990).

[Gi91] V. Giovangigli, Convergent Iterative Methods for Multicomponent Diffusion. *Impact Comput. Sci. Eng.* **3**, 244–276, (1991).

[GD88] V. Giovangigli and N. Darabiha, Vector Computers and Complex Chemistry Combustion. *In* C. Brauner and C. Schmidt-Laine (Eds.), *Proc. Conference Mathematical Modeling in Combustion and Related Topics*, NATO Adv. Sci. Inst. Ser. E, Vol. 140, 491–503. Martinus Nijhoff pub., Dordrecht (1988).

[GS89] V. Giovangigli and M. D. Smooke, Adaptive Continuation Algorithms with Application to Combustion Problems. *Appl. Numer. Math.* **5**, 305–331, (1989).

[GV83] G. H. Golub and C. F. Van Loan, *Matrix Computations.* Johns Hopkins Univ. Press, Baltimore (1983).

[HS52] M. R. Hestenes and E. Stiefel, Methods of Conjugate Gradients for Solving Linear Systems. *J. Res. Nat. Bur. Standards* **49**, 409–436, (1952).

[HC49] J. O. Hirschfelder and C. F. Curtiss, Flame Propagation in Explosive Gas Mixtures. *In Third Symposium (International) on Combustion*, pp. 121–127. Reinhold, New York (1949).

[HCB54] J. O. Hirschfelder, C. F. Curtiss, and R. B. Bird, *Molecular Theory of Gases and Liquids.* Wiley, New York (1954).

[JB81] W. W. Jones and J. P. Boris, An Algorithm for Multispecies Diffusion Fluxes. *Comput. Chem.* **5**, 139–146, (1981).

[KA61] Y. Kagan and A. M. Afanas'ev, On the Kinetic Theory of Gases with Rotational Degrees of Freedom. *J. Explt. Theoret. Phys. (URSS)* **41**, 1536–1545, (1961) and *Soviet Phys. JETP* **14**, 1096–1101, (1962).

[KM66] Y. Kagan and L. M. Maksimov, Kinetic Theory of Gases Taking into Account Rotational Degrees of Freedom in an External Field. *J. Explt. Theoret. Phys. (URSS)* **51**, 1893–1908, (1966), and *Soviet Phys. JETP* **24**, 1272–1281, (1967).

[KDWCM86] R. J. Kee, G. Dixon-Lewis, J. Warnatz, M. E. Coltrin, and J. A. Miller, A Fortran Computer Code Package for the Evaluation of Gas-Phase Multicomponent Transport Properties. SANDIA National Laboratories Report, SAND86-8246 (1986).

[KMJ80] R. J. Kee, J. A. Miller, and T. H. Jefferson, CHEMKIN: A General-Purpose, Problem-Independent, Transportable, Fortran Chemical Kinetics Code Package. SANDIA National Laboratories Report, SAND80-8003 (1980).

[KRM87] R. J. Kee, F. M. Rupley, and J. A. Miller, The Chemkin Thermodynamic Data Base. SANDIA National Laboratories Report, SAND87-8215 (1987).

[KWM83] R. J. Kee, J. Warnatz, and J. A. Miller, A Fortran Computer Code Package for the Evaluation of Gas-Phase Viscosities, Conductivities, and Diffusion Coefficients. SANDIA National Laboratories Report, SAND83-8209 (1983).

[Ke65] H. B. Keller, On the Solution of Singular and Semidefinite Linear Systems by Iteration. *SIAM J. Numer. Anal.* **2**, 281–290, (1965).

[KT79] W. E. Köhler and G. W. 't Hooft, Waldmann-Snider Collision Integrals for Mixtures of Polyatomic Gases. *Zeitschr. Naturforsch.* **34a**, 1255–1268, (1979).

[Ku91] I. Kuščer, Dissociation and Recombination in an Inhomogeneous Gas. *Physica A* **176**, 542–556, (1991).

[KHCBV93] I. Kuščer, L. J. F. Hermans, P. L. Chapowsky, J. J. M. Beenakker, and G. J. van der Meer, Description of Light-Induced Drift in Terms of Transport Mean Paths. *J. Phys. B* **26**, 2837–2852, (1993).

[Le89] V. N. Lebedev, The Influence of Molecular Transfer on the Structure and Extinction of Laminar Strained H_2/Air and CH_4/Air Flames. Internal report, The Institute of Chemical Physics, Chernogolovka (1989).

[LR80] J. G. Lewis and R. R. Rehm, The Numerical Solution of a Nonseparable Elliptic Partial Differential Equation by Preconditioned Conjugate Gradients. *J. Res. Nat. Bur. Stand.* **85**, 367–390, (1980).

[LH60] G. Ludwig and M. Heil, Boundary Layer Theory with Dissociation and Ionization. *In Advances in Applied Mechanics*, Volume **VI**, pp. 39–118. Academic Press, New York (1960).

[MMW83] G. C. Maitland, M. Mustafa, and W. A. Wakeham, Second Order Approximations for the Transport Properties of Dilute Polyatomic Gases. *J. Chem. Soc., Faraday Trans. 2* **79**, 1425–1441, (1983).

[MM62] E. A. Mason and L. Monchick, Heat Conductivity of Polyatomic and Polar Gases. *J. Chem. Phys.* **36**, 1622–1639, (1962).

[MTS67] S. Mathur, P. K. Tondon, and S. C. Saxena, Thermal Conductivity of Binary, Ternary and Quaternary Mixtures of Rare Gases. *Molec. Phys.* **12**, 569–579, (1967).

[MM63] E. A. Mason and L. Monchick, Theory of Transport Properties of Gases. *In Ninth Symposium (International) on Combustion*, pp. 713–724. Academic Press, New York (1963).

[MBKK90] F. R. McCourt, J. J. Beenakker, W. E. Köhler, and I. Kuščer, *Non Equilibrium Phenomena in Polyatomic Gases.* Volume I: Dilute Gases, Clarendon Press, Oxford (1990).

[MBKK91] F. R. McCourt, J. J. Beenakker, W. E. Köhler, and I. Kuščer, *Non Equilibrium Phenomena in Polyatomic Gases.* Volume II: Cross Sections, Scattering and Rarefied Gases, Clarendon Press, Oxford (1991).

[MS64] F. R. McCourt and R. F. Snider, Transport Properties of Gases with Rotational States. *J. Chem. Phys.* **41**, 3185–3194, (1964).

[MS65] F. R. McCourt and R. F. Snider, Transport Properties of Gases with Rotational States. II. *J. Chem. Phys.* **43**, 2276–2283, (1965).

[MP77] C. D. Meyer and R. J. Plemmons, Convergent Powers of a Matrix with Applications to Iterative Methods for Singular Systems. *SIAM J. Numer. Anal.* **14**, 699–705, (1977).

[MVW88] J. Millat, V. Vesovic, and W. A. Wakeham, On the Validity of the Simplified Expression for the Thermal Conductivity of Thijsse et al. *Physica A* **166**, 153–164, (1988).

[Mo64] L. Monchick, Small Periodic Disturbances in Polyatomic Gases. *Phys. Fluids* **7**, 882–896, (1964).

[MM61] L. Monchick and E. A. Mason, Transport Properties of Polar Gases. *J. Chem. Phys.* **35**, 1676–1697, (1961).

[MMM66] L. Monchick, R. J. Munn, and E. A. Mason, Thermal Diffusion in Polyatomic Gases: a Generalized Stefan-Maxwell Diffusion Equation. *J. Chem. Phys.* **45**, 3051–3058, (1966).

[MPM65] L. Monchick, A. N. G. Pereira, and E. A. Mason, Heat Conductivity of Polyatomic and Polar Gases and Gas Mixtures. *J. Chem. Phys.* **42**, 3241–3256, (1965).

[MYM63] L. Monchick, K. S. Yun, and E. A. Mason, Formal Kinetic Theory of Transport Phenomena in Polyatomic Gas Mixtures. *J. Chem. Phys.* **39**, 654–669, (1963).

[MS58] E. W. Montroll and K. E. Shuler, The Application of the Theory of Stochastic Processes to Chemical Kinetics. *In* I. Prigogine (Ed.), *Advances in Chemical Physics*, **1**. Interscience Publishers Inc., New York, 361–399, (1958).

[NP78] M. Neumann and R. J. Plemmons, Convergent Nonnegative Matrices and Iterative Methods for Consistent Linear Systems. *Numer. Math.* **31**, 265–279, (1978).

[Ol40] R. Oldenburger, Infinite Powers of Matrices and Characteristic Roots. *Duke Math J.* **6**, 357–361, (1940).

[OB81] E. S. Oran and J. P. Boris, Detailed Modeling of Combustion Systems. *Prog. Energy Combust. Sci.* **7**, 1–72, (1981).

[Pa59] J. G. Parker, Rotational and Vibrational Relaxation in Diatomic Gases. *Phys. Fluids* **2**, 449–462, (1959).

[Pi22] F. B. Pidduck, The Kinetic Theory of a Special Type of Rigid Molecule. *Proc. Roy. Soc. (London)* **A101**, 101–112, (1922).

[Pr59] R. D. Present, On the Velocity Distribution in a Chemically Reacting Gas. *J. Chem. Phys.* **31**, 747–750, (1959).

[PM50] I. Prigogine and M. Mathieu, Sur la Perturbation de la Distribution de Maxwell par des Réactions Chimiques en Phase Gazeuse. *Physica.* **16**, 51–64, (1950).

[PX49] I. Prigogine and E. Xhrouet, On the Perturbation of Maxwell Distribution Function by Chemical Reactions in Gases. *Physica.* **15**, 913–932, (1949).

[RV73] A. W. Roberts and D. E. Varberg, *Convex Functions*. Academic Press, New York (1973).

[RM60] J. Ross and P. Mazur, Some Deductions from a Formal Statistical Mechanical Theory of Chemical Kinetics. *J. Chem. Phys.* **35**, 19–28, (1961).

[SS86] Y. Saad and M. H. Schultz, GMRES: A Minimal Residual Algorithm for Solving Nonsymmetric Linear Systems. *SIAM J. Sci. Stat. Comp.* **7**, 856–869, (1986).

[SK70] B. Shizgal and M. Karplus, Nonequilibrium Contributions to the Rate of Reaction. I. Perturbation of the Velocity Distribution Function. *J. Chem. Phys.* **52**, 4262–4278, (1970).

[SK71a] B. Shizgal and M. Karplus, Nonequilibrium Contributions to the Rate of Reaction. II. Isolated Multicomponent Systems. *J. Chem. Phys.* **54**, 4345–4356, (1971).

[SK71b] B. Shizgal and M. Karplus, Nonequilibrium Contributions to the Rate of Reaction. III. Isothermal Multicomponent Systems. *J. Chem. Phys.* **54**, 4357–4362, (1971).

[SG92] M. D. Smooke and V. Giovangigli, Numerical Modeling of Axisymmetric Laminar Diffusion Flames. *Impact Comput. Sci. Eng.* **4**, 46–79, (1992).

[Sn60] R. F. Snider, Quantum-Mechanical Modified Boltzmann Equation for Degenerate Internal States. *J. Chem. Phys.* **32**, 1051–1060, (1960).

[Sn71] R. F. Snider, Generalized Boltzmann Equations for Molecules with Internal States. *J. Chem. Phys.* **55**, 1555–1566, (1971).

[SP71] D. R. Stull and H. Prophet, JANAF Thermochemical Tables, Second ed., Washington, NBS NSRDS-NBS37 (1971).

[Ta51] K. Takayanagi, On the Theory of Chemically Reacting Gas. *Prog. Theor. Phys.* **VI**, 486–497, (1951).

[TTCKB79] B. J. Thijsse, G. W. 't Hooft, D. A. Coombe, H. F. P. Knaap, and J. J. M. Beenakker, Some Simplified Expressions for the Thermal Conductivity in an External Field. *Physica A* **98**, 307–312, (1979).

[VK88] R. J. Van den Oord and J. Korving, The Thermal Conductivity of Polyatomic Molecules. *J. Chem. Phys.* **89**, 4333–4338, (1988).

[VDBK88] R. J. Van den Oord, M. C. De Lignie, J. J. M. Beenakker, and J. Korving, The Role of Internal Energy in the Distribution Function of a Heat Conducting Gas. *Physica A* **152**, 199–216, (1988).

[Va67] J. Van de Ree, On the Definition of the Diffusion Coefficients in Reacting Gases. *Physica* **36**, 118–126, (1967).

[Vr62] R. S. Varga, *Matrix Iterative Analysis*. Prentice-Hall, Englewood Cliffs, NJ (1962).

[WV92] W. A. Wakeham and V. Vesovic, Traditional Transport Properties. *In* W. A. Wakeman, A. S. Dickinson, F. R. W. McCourt, and V. Vesovic (Eds.), *Status and Future Developments in the Study of Transport Properties*, NATO Adv. Sci. Inst. Ser. C, Vol. 361, 29–55. Kluwer Acad. pub., Dordrecht (1992).

[Wa47] L. Waldmann, Der Diffusionthermoeffekt II. *Zeitschr. Physik.* **124**, 175–195, (1947).

[Wa57] L. Waldmann, Die Boltzmann-Gleichung für Gase mit Rotierenden Molekülen. *Zeitschr. Naturforschg.* **12a**, 660–662, (1957).

[Wa58] L. Waldmann, Transporterscheinungen in Gasen von Mittlerem Druck. *Handbuch der Physik*, S. Flügge ed., **12**, Springer Verlag, Berlin 295–514, (1958).

[Wa73] L. Waldmann, On Kinetic Equations for Gases with Internal Degrees of Freedom. *In* E. G. D. Cohen and W. Thirring (Eds.), *The Boltzmann Equation, Theory and Applications*, 223–246. Springer, Vienna (1973).

[WT62] L. Waldmann und E. Trübenbacher, Formale Kinetische Theorie von Gasgemischen aus Anregbaren Molekülen. *Zeitschr. Naturforschg.* **17a**, 363–376, (1962).

[WU51] C. S. Wang Chang and G. E. Uhlenbeck, Transport Phenomena in Polyatomic Gases. University of Michigan Engineering Research Report CM-681, (1951).

[WU70] C. S. Wang Chang and G. E. Uhlenbeck, The Kinetic Theory of Gases. *In* J. De Boer and G. E. Uhlenbeck (Eds.), *Studies in Statistical Mechanics 5*, 1–75. North Holland, Amsterdam (1970).

[WUD64] C. S. Wang Chang, G. E. Uhlenbeck, and J. De Boer, The Heat Conductivity and Viscosity of Polyatomic Gases. *In* J. De Boer and G. E. Uhlenbeck (Eds.), ***Studies in Statistical Mechanics 2***, 242–268. North Holland, Amsterdam (1964).

[Wr82] J. Warnatz, Influence of Transport Models and Boundary Conditions on Flame Structure. *In* N. Peters and J. Warnatz (Eds.), ***Numerical Methods in Laminar Flame Propagation***, 87–111. Vieweg Verlag, Braunschweig (1982).

[Wr83] J. Warnatz, The Mechanism of High Temperature Combustion of Propane and Butane. *Comb. Sci. Tech.* **34**, 177–200, (1983).

[Wr84] J. Warnatz, Survey of Rate Coefficients in the C/H/O System, *In* W. C. Gardiner (Ed.), *Combustion Chemistry*. Springer, New York, 197–360, (1984).

[Wi50] C. R. Wilke, A Viscosity Equation for Gas Mixtures. *J. Chem. Phys.* **18**, 517–519, (1950).

[Wl58] F. A. Williams, Elementary Derivation of the Multicomponent Diffusion Equation. *Amer. J. Phys.* **26**, 467–469, (1958).

[Ze84] R. Zellner, Bimolecular Reaction Rate Coefficients, *In* W. C. Gardiner (Ed.), *Combustion Chemistry*. Springer, New York, 127–172, (1984).

[ZE47] B. J. Zwolinsky and H. Eyring, The Non-Equilibrium Theory of Absolute Rates of Reaction. *J. Am. Chem. Soc.* **69**, 2702–2707, (1947).

Index

U

V

W

Springer-Verlag and the Environment

We at Springer-Verlag firmly believe that an international science publisher has a special obligation to the environment, and our corporate policies consistently reflect this conviction.

We also expect our business partners – paper mills, printers, packaging manufacturers, etc. – to commit themselves to using environmentally friendly materials and production processes.

The paper in this book is made from low- or no-chlorine pulp and is acid free, in conformance with international standards for paper permanency.

Lecture Notes in Physics

For information about Vols. 1–394
please contact your bookseller or Springer-Verlag

Vol. 395: H. E. Trease, M. J. Fritts, W. P. Crowley (Eds.), Advances in the Free-Lagrange Method. Proceedings, 1990. XI, 327 pages. 1991.

Vol. 396: H. Mitter, H. Gausterer (Eds.), Recent Aspects of Quantum Fields. Proceedings. XIII, 332 pages. 1991.

Vol. 398: T. M. M. Verheggen (Ed.), Numerical Methods for the Simulation of Multi-Phase and Complex Flow. Proceedings, 1990. VI, 153 pages. 1992.

Vol. 399: Z. Švestka, B. V. Jackson, M. E. Machedo (Eds.), Eruptive Solar Flares. Proceedings, 1991. XIV, 409 pages. 1992.

Vol. 400: M. Dienes, M. Month, S. Turner (Eds.), Frontiers of Particle Beams: Intensity Limitations. Proceedings, 1990. IX, 610 pages. 1992.

Vol. 401: U. Heber, C. S. Jeffery (Eds.), The Atmospheres of Early-Type Stars. Proceedings, 1991. XIX, 450 pages. 1992.

Vol. 402: L. Boi, D. Flament, J.-M. Salanskis (Eds.), 1830-1930: A Century of Geometry. VIII, 304 pages. 1992.

Vol. 403: E. Balslev (Ed.), Schrödinger Operators. Proceedings, 1991. VIII, 264 pages. 1992.

Vol. 404: R. Schmidt, H. O. Lutz, R. Dreizler (Eds.), Nuclear Physics Concepts in the Study of Atomic Cluster Physics. Proceedings, 1991. XVIII, 363 pages. 1992.

Vol. 405: W. Hollik, R. Rückl, J. Wess (Eds.), Phenomenological Aspects of Supersymmetry. VII, 329 pages. 1992.

Vol. 406: R. Kayser, T. Schramm, L. Nieser (Eds.), Gravitational Lenses. Proceedings, 1991. XXII, 399 pages. 1992.

Vol. 407: P. L. Smith, W. L. Wiese (Eds.), Atomic and Molecular Data for Space Astronomy. VII, 158 pages. 1992.

Vol. 408: V. J. Martínez, M. Portilla, D. Sàez (Eds.), New Insights into the Universe. Proceedings, 1991. XI, 298 pages. 1992.

Vol. 409: H. Gausterer, C. B. Lang (Eds.), Computational Methods in Field Theory. Proceedings, 1992. XII, 274 pages. 1992.

Vol. 410: J. Ehlers, G. Schäfer (Eds.), Relativistic Gravity Research. Proceedings, VIII, 409 pages. 1992.

Vol. 411: W. Dieter Heiss (Ed.), Chaos and Quantum Chaos. Proceedings, XIV, 330 pages. 1992.

Vol. 412: A. W. Clegg, G. E. Nedoluha (Eds.), Astrophysical Masers. Proceedings, 1992. XX, 480 pages. 1993.

Vol. 413: Aa. Sandqvist, T. P. Ray (Eds.); Central Activity in Galaxies. From Observational Data to Astrophysical Diagnostics. XIII, 235 pages. 1993.

Vol. 414: M. Napolitano, F. Sabetta (Eds.), Thirteenth International Conference on Numerical Methods in Fluid Dynamics. Proceedings, 1992. XIV, 541 pages. 1993.

Vol. 415: L. Garrido (Ed.), Complex Fluids. Proceedings, 1992. XIII, 413 pages. 1993.

Vol. 416: B. Baschek, G. Klare, J. Lequeux (Eds.), New Aspects of Magellanic Cloud Research. Proceedings, 1992. XIII, 494 pages. 1993.

Vol. 417: K. Goeke P. Kroll, H.-R. Petry (Eds.), Quark Cluster Dynamics. Proceedings, 1992. XI, 297 pages. 1993.

Vol. 418: J. van Paradijs, H. M. Maitzen (Eds.), Galactic High-Energy Astrophysics. XIII, 293 pages. 1993.

Vol. 419: K. H. Ploog, L. Tapfer (Eds.), Physics and Technology of Semiconductor Quantum Devices. Proceedings, 1992. VIII, 212 pages. 1993.

Vol. 420: F. Ehlotzky (Ed.), Fundamentals of Quantum Optics III. Proceedings, 1993. XII, 346 pages. 1993.

Vol. 421: H.-J. Röser, K. Meisenheimer (Eds.), Jets in Extragalactic Radio Sources. XX, 301 pages. 1993.

Vol. 422: L. Päivärinta, E. Somersalo (Eds.), Inverse Problems in Mathematical Physics. Proceedings, 1992. XVIII, 256 pages. 1993.

Vol. 423: F. J. Chinea, L. M. González-Romero (Eds.), Rotating Objects and Relativistic Physics. Proceedings, 1992. XII, 304 pages. 1993.

Vol. 424: G. F. Helminck (Ed.), Geometric and Quantum Aspects of Integrable Systems. Proceedings, 1992. IX, 224 pages. 1993.

Vol. 425: M. Dienes, M. Month, B. Strasser, S. Turner (Eds.), Frontiers of Particle Beams: Factories with $e^+ e^-$ Rings. Proceedings, 1992. IX, 414 pages. 1994.

Vol. 426: L. Mathelitsch, W. Plessas (Eds.), Substructures of Matter as Revealed with Electroweak Probes. Proceedings, 1993. XIV, 441 pages. 1994

Vol. 427: H. V. von Geramb (Ed.), Quantum Inversion Theory and Applications. Proceedings, 1993. VIII, 481 pages. 1994.

Vol. 428: U. G. Jørgensen (Ed.), Molecules in the Stellar Environment. Proceedings, 1993. VIII, 440 pages. 1994.

Vol. 429: J. L. Sanz, E. Martínez-González, L. Cayón (Eds.), Present and Future of the Cosmic Microwave Background. Proceedings, 1993. VIII, 233 pages. 1994.

Vol. 430: V. G. Gurzadyan, D. Pfenniger (Eds.), Ergodic Concepts in Stellar Dynamics. Proceedings, 1993. XVI, 302 pages. 1994.

Vol. 431: T. P. Ray, S. Beckwith (Eds.), Star Formation and Techniques in Infrared and mm-Wave Astronomy. XIV, 314 pages. 1994.

Vol. 432: G. Belvedere, M. Rodonò, G. M. Simnett (Eds.), Advances in Solar Physics. Proceedings, 1993. XVII, 335 pages. 1994.

Vol. 433: G. Contopoulos, N. Spyrou, L. Vlahos (Eds.), Galactic Dynamics and N-Body Simulations. XIV, 417 pages. 1994.

New Series m: Monographs

Vol. m 1: H. Hora, Plasmas at High Temperature and Density. VIII, 442 pages. 1991.

Vol. m 2: P. Busch, P. J. Lahti, P. Mittelstaedt, The Quantum Theory of Measurement. XIII, 165 pages. 1991.

Vol. m 3: A. Heck, J. M. Perdang (Eds.), Applying Fractals in Astronomy. IX, 210 pages. 1991.

Vol. m 4: R. K. Zeytounian, Mécanique des fluides fondamentale. XV, 615 pages, 1991.

Vol. m 5: R. K. Zeytounian, Meteorological Fluid Dynamics. XI, 346 pages. 1991.

Vol. m 6: N. M. J. Woodhouse, Special Relativity. VIII, 86 pages. 1992.

Vol. m 7: G. Morandi, The Role of Topology in Classical and Quantum Physics. XIII, 239 pages. 1992.

Vol. m 8: D. Funaro, Polynomial Approximation of Differential Equations. X, 305 pages. 1992.

Vol. m 9: M. Namiki, Stochastic Quantization. X, 217 pages. 1992.

Vol. m 10: J. Hoppe, Lectures on Integrable Systems. VII, 111 pages. 1992.

Vol. m 11: A. D. Yaghjian, Relativistic Dynamics of a Charged Sphere. XII, 115 pages. 1992.

Vol. m 12: G. Esposito, Quantum Gravity, Quantum Cosmology and Lorentzian Geometries. Second Corrected and Enlarged Edition. XVIII, 349 pages. 1994.

Vol. m 13: M. Klein, A. Knauf, Classical Planar Scattering by Coulombic Potentials. V, 142 pages. 1992.

Vol. m 14: A. Lerda, Anyons. XI, 138 pages. 1992.

Vol. m 15: N. Peters, B. Rogg (Eds.), Reduced Kinetic Mechanisms for Applications in Combustion Systems. X, 360 pages. 1993.

Vol. m 16: P. Christe, M. Henkel, Introduction to Conformal Invariance and Its Applications to Critical Phenomena. XV, 260 pages. 1993.

Vol. m 17: M. Schoen, Computer Simulation of Condensed Phases in Complex Geometries. X, 136 pages. 1993.

Vol. m 18: H. Carmichael, An Open Systems Approach to Quantum Optics. X, 179 pages. 1993.

Vol. m 19: S. D. Bogan, M. K. Hinders, Interface Effects in Elastic Wave Scattering. XII, 182 pages. 1994.

Vol. m 20: E. Abdalla, M. C. B. Abdalla, D. Dalmazi, A. Zadra, 2D-Gravity in Non-Critical Strings. IX, 319 pages. 1994.

Vol. m 21: G. P. Berman, E. N. Bulgakov, D. D. Holm, Crossover-Time in Quantum Boson and Spin Systems. XI, 268 pages. 1994.

Vol. m 22: M.-O. Hongler, Chaotic and Stochastic Behaviour in Automatic Production Lines. V, 85 pages. 1994.

Vol. m 23: V. S. Viswanath, G. Müller, The Recursion Method. X, 259 pages. 1994.

Vol. m 24: A. Ern, V. Giovangigli, Multicomponent Transport Algorithms. XIV, 427 pages. 1994.